MIMO-OFDM for LTE, Wi-Fi and WiMAX

MIMO-OFDM for LTE, Wi-Fi and WiMAX

Coherent versus Non-coherent and Cooperative Turbo-transceivers

Prof. Lajos Hanzo, Dr. Yosef (Jos) Akhtman and Dr. Li Wang
All of
University of Southampton, UK

Dr. Ming Jiang
Currently with
New Postcom Equipment Co., Ltd

A John Wiley and Sons, Ltd, Publication

This edition first published 2011
© 2011 John Wiley & Sons Ltd

Registered office
John Wiley & Sons Ltd, The Atrium, Southern Gate, Chichester, West Sussex, PO19 8SQ,
United Kingdom

For details of our global editorial offices, for customer services and for information about how to apply
for permission to reuse the copyright material in this book please see our website at www.wiley.com.

The right of the author to be identified as the author of this work has been asserted in accordance with
the Copyright, Designs and Patents Act 1988.

Library of Congress Cataloging-in-Publication Data

Hanzo, Lajos, 1952–
    MIMO-OFDM for LTE, WiFi, and WiMAX : coherent versus non-coherent and cooperative
turbo-transceivers / by L. Hanzo, J. Akhtman, L. Wang, M. Jiang.
        p. cm.
    Includes bibliographical references and index.
    ISBN 978-0-470-68669-0 (cloth)
1. Orthogonal frequency division multiplexing. 2. MIMO systems. 3. Wireless LANs–Equipment and
supplies. 4. IEEE 802.11 (Standard). 5. IEEE 802.16 (Standard) 6. Radio–Transmitter-receivers. I.
Akhtman, J. (Jos) II. Wang, L. (Li), 1982- III. Title.
    TK5103.484.H36 2010
    621.382'16–dc22

                                                                                    2010015324

A catalogue record for this book is available from the British Library.

Print ISBN: 9780470686690 (H/B)
ePDF ISBN: 9780470711767
oBook ISBN: 9780470711750

Set in 9/11pt Times by Sunrise Setting Ltd, Torquay, UK.
Printed and bound in Singapore by Markono Print Media Pte Ltd.

We dedicate this monograph to the numerous contributors to this field, many of whom are listed in the Author Index.

The MIMO capacity theoretically increases linearly with the number of transmit antennas, provided that the number of receive antennas is equal to the number of transmit antennas. With the further proviso that the total transmit power is increased proportionately to the number of transmit antennas, a linear capacity increase is achieved on increasing the transmit power. However, under realistic conditions the theoretical MIMO-OFDM performance erodes, hence, to circumvent this degradation, our monograph is dedicated to the design of practical coherent, non-coherent and cooperative MIMO-OFDM turbo-transceivers ...

Contents

Part II   Coherent versus Non-coherent and Cooperative OFDM Systems   271

Part III   Coherent SDM-OFDM Systems   491

About the Authors

Lajos Hanzo FREng, FIEEE, FIET, DSc received his degree in electronics in 1976 and his doctorate in 1983. During his career he has held various research and academic posts in Hungary, Germany and the UK. Since 1986 he has been with the School of Electronics and Computer Science, University of Southampton, UK, where he holds the chair in telecommunications. He has co-authored 19 books on mobile radio communications, totalling in excess of 10 000 published 844 research entries at IEEE Xplore, acted as TPC Chair of IEEE conferences, presented keynote lectures and been awarded a number of distinctions. Currently he is directing an academic research team working on a range of research projects in the field of wireless multimedia communications sponsored by industry, the Engineering and Physical Sciences Research Council (EPSRC) UK, the European IST Programme and the Mobile Virtual Centre of Excellence (VCE), UK. He is an enthusiastic supporter of industrial and academic liaison and he offers a range of industrial courses. He is also an IEEE Distinguished Lecturer as well as a Governor of both the IEEE ComSoc and the VTS. He is the acting Editor-in-Chief of the IEEE Press. For further information on research in progress and associated publications refer to http://www-mobile.ecs.soton.ac.uk

Dr Yosef (Jos) Akhtman received a BSc degree in physics and mathematics from the Hebrew University of Jerusalem, Israel, in June 2000 and the PhD degree in electronics engineering from the University of Southampton in July 2007. He was awarded a full PhD studentship in the University of Southampton as well as an Outstanding Contribution Award for his work as part of the Core 3 research programme of the Mobile Virtual Centre of Excellence in Mobile Communications (MobileVCE). He has also received a BAE Prize for Innovation in Autonomy for his contribution to the Southampton Autonomous Underwater Vehicle (SotonAUV) project. Between January 2007 and December 2009 he conducted research as a senior research fellow in the 5* School of Electronics and Computer Science at Southampton University.

Li Wang received his BEng degree with distinction in information engineering from Chengdu University of Technology (CDUT), Chengdu, China, in 2005 and his MSc degree (with distinction) in radio frequency communication systems from the University of Southampton, UK, in 2006. Between October 2006 and January 2010 he was a PhD student in the Communications Group, School of Electronics and Computer Science, University of Southampton, and participated in the Delivery Efficiency Core Research Programme of the Virtual Centre of Excellence in Mobile and Personal Communications (Mobile VCE). His research interests include space–time coding, channel coding, multi-user detection for future wireless networks. Upon the completion of his PhD in January 2010 he joined the Communications Group as a postdoctoral researcher.

Dr Ming Jiang received his BEng and MEng degrees in electronics engineering in 1999 and 2002 from South China University of Technology (SCUT), China, and a PhD degree in Telecommunications in 2006 from the University of Southampton, UK. From 2002 to 2005, he was involved in the Core 3 research project of the Mobile Virtual Centre of Excellence (VCE), UK, on air-interface algorithms for MIMO-OFDM systems. Since April 2006, Dr Jiang has been with Advanced Technology, Standards and Regulation (ATSR) of the Samsung Electronics Research Institute (SERI), UK, working on the European FP6 WIN-NER project as well as internal projects on advanced wireless communication systems. His research interests fall in the general area of wireless communications, including multi-user detection, channel estimation, space–time processing, heuristic and adaptive optimization, frequency hopping, MIMO-OFDM and OFDMA systems, etc. Dr Jiang has co-authored one IEEE Press book chapter, six IEE/IEEE journal papers and eight IEE/IEEE conference papers. Recently he returned to his native country China and had been working for Nortel.

Other Wiley–IEEE Press Books on Related Topics

For detailed contents and sample chapters please refer to www.wiley.com and
www-mobile.ecs.soton.ac.uk

- R. Steele, L. Hanzo (Ed.): *Mobile Radio Communications: Second and Third Generation Cellular and WATM Systems*, 2nd edition, 1999, 1064 pages

- L. Hanzo, T. H. Liew, B. L. Yeap: *Turbo Coding, Turbo Equalisation and Space-Time Coding*, 2002, 751 pages

- L. Hanzo, C. H. Wong, M. S. Yee: *Adaptive Wireless Transceivers: Turbo-Coded, Turbo-Equalised and Space-Time Coded TDMA, CDMA and OFDM Systems*, 2002, 737 pages

- L. Hanzo, L.-L. Yang, E.-L. Kuan, K. Yen: *Single- and Multi-Carrier CDMA: Multi-User Detection, Space-Time Spreading, Synchronisation, Networking and Standards*, 2003, 1060 pages

- L. Hanzo, M. Münster, T. Keller, B.-J. Choi: *OFDM and MC-CDMA for Broadband Multi-User Communications, WLANs and Broadcasting*, 2003, 978 pages

- L. Hanzo, S.-X. Ng, T. Keller and W. T. Webb: *Quadrature Amplitude Modulation: From Basics to Adaptive Trellis-Coded, Turbo-Equalised and Space-Time Coded OFDM, CDMA and MC-CDMA Systems*, 2004, 1105 pages

- L. Hanzo, T. Keller: *An OFDM and MC-CDMA Primer*, 2006, 430 pages

- L. Hanzo, F. C. A. Somerville, J. P. Woodard: *Voice and Audio Compression for Wireless Communications*, 2007, 858 pages

- L. Hanzo, P. J. Cherriman, J. Streit: *Video Compression and Communications: H.261, H.263, H.264, MPEG4 and HSDPA-Style Adaptive Turbo-Transceivers*, 2007, 680 pages

- L. Hanzo, J. S. Blogh, S. Ni: *3G, HSDPA, HSUPA and FDD Versus TDD Networking: Smart Antennas and Adaptive Modulation*, 2008, 564 pages

- L. Hanzo, O. Alamri, M. El-Hajjar, N. Wu: *Near-Capacity Multi-Functional MIMO Systems: Sphere-Packing, Iterative Detection and Cooperation* 2009, 738 pages

- L. Hanzo, R. G. Maunder, J. Wang and L.-L. Yang: *Near-Capacity Variable-Length Coding: Regular and EXIT-Chart Aided Irregular Designs*, 2010, 496 pages

Preface

The rationale and structure of this volume is centred around the following 'story-line'. The conception of *parallel transmission of data* over dispersive channels dates back to the seminal paper of Doelz *et al.* published in 1957, leading to the OFDM philosophy, which has found its way into virtually all recent wireless systems, such as the Wi-Fi, WiMAX, LTE and DVB as well as DAB broadcast standards. Although *MIMO techniques* are significantly 'younger' than OFDM, they *also reached a state of maturity* and hence the family of recent wireless standards includes the optional employment of MIMO techniques, which motivates the joint study of OFDM and MIMO techniques in this volume.

The research of MIMO arrangements was motivated by the observation that the MIMO capacity increases linearly with the number of transmit antennas, provided that the number of receive antennas is equal to the number of transmit antennas. With the further proviso that the total transmit power is increased proportionately to the number of transmit antennas, a linear capacity increase is achieved upon increasing the transmit power. This is beneficial since, according to the classic Shannon–Hartley law, the achievable channel capacity increases only logarithmically with the transmit power. Thus *MIMO-OFDM may be considered a 'green' transceiver solution.*

This volume therefore sets out to explore the recent research advances in MIMO-OFDM techniques as well as their limitations. The basic types of multiple-antenna-aided OFDM systems are classified and their benefits are characterized. Space-Division Multiple Access (SDMA), Space-Division Multiplexing (SDM) and space–time coding MIMOs are addressed. We also argue that *under realistic propagation conditions*, when for example the signals associated with the MIMO elements become correlated owing to shadow fading, *the predicted performance gains may substantially erode*. Furthermore, owing to the limited dimensions of shirt-pocket-sized handsets, the employment of multiple-antenna elements at the mobile station is impractical.

Hence in practical terms only the family of distributed MIMO elements, which relies on the cooperation of potentially single-element mobile stations, is capable of eliminating the correlation of the signals impinging on the MIMO elements, as will be discussed in the book. The topic of *cooperative wireless communications* cast in the context of distributed MIMOs has recently attracted substantial research interests, but, nonetheless, it *has numerous open problems, before all the idealized simplifying assumptions currently invoked in the literature are eliminated.*

On a more technical note, *we aim at achieving a near-capacity MIMO-OFDM performance*, which requires sophisticated designs, as detailed below:

- A high throughput may be achieved with the aid of a high number of MIMO elements, but this is attained at a *potentially high complexity, which increases exponentially as a function of both the number of MIMO elements and the number of bits per symbol*, when using a full-search-based Maximum Likelihood (ML) multi-stream/multi-user detector.

- In order to approach the above-mentioned near-capacity performance, while circumventing the problem of an exponentially increasing complexity, *we design radical multi-stream/multi-user detectors which 'capture' the ML solution with a high probability at a fraction of the ML complexity.*

- This ambitious design goal is achieved with the aid of sophisticated *soft-decision-based Genetic Algorithm (GA) assisted MUDs or new sphere detectors, which are capable of operating in the high-importance rank-deficient scenarios*, when the number of transmit antennas may be as high as twice the number of receiver antennas.

- The achievable gain of space–time codes is further improved with the aid of *sphere-packing modulation, which allows us to design the space–time symbols of multiple transmit antennas jointly*, while previous designs made no effort to do so. Naturally, this joint design no longer facilitates low-complexity single-stream detection, but our sphere decoders allow us to circumvent this increased detection complexity.

- Sophisticated *joint coding and modulation schemes* are used, which accommodate the parity bits of the channel codec without bandwidth extension, simply by extending the modulation alphabet.

- Estimating the MIMO channel for a high number of transmit and receive antennas becomes extremely challenging, since we have to estimate $N_t \cdot N_r$ channels, although in reality we are only interested in the data symbols, not the channel. *This problem becomes even more grave in the context of the above-mentioned rank-deficient scenarios, since we have to estimate more channels than the number of received streams.* Finally, the pilot overhead imposed by estimating $N_t \cdot N_r$ channels might become prohibitive, which erodes the attainable throughput gains.

- In order to tackle the above-mentioned challenging channel estimation problem, we designed *new iterative joint channel estimation and data detection techniques.* More explicitly, provided that a powerful MIMO MUD, such as the above-mentioned GA-aided or sphere-decoding-based MUD, is available for delivering a sufficiently reliable first data estimate, the power of decision-directed channel estimation may be invoked, which exploits the fact that after a first tentative data decision – in the absence of decision errors – the receiver effectively knows the transmitted signal and hence may then exploit the presence of 100% pilot information for generating a more accurate channel estimate. Again, this design philosophy is detailed in the book in great depth in the context of joint iterative channel estimation and data detection.

- Although the number of studies/papers on cooperative communications has increased exponentially over the past few years, most *investigations stipulate the simplifying assumption of having access to perfect channel information* – despite the fact that, as detailed under the previous bullet point, this is an extremely challenging task even for co-located MIMO elements.

- Thus it is necessary to design new non-coherently detected cooperative systems, which can dispense with the requirement of channel estimation, despite the typical 3 dB performance loss of differential detection. It is demonstrated in the book that *the low-complexity non-coherent detector's potential performance penalty can in fact be recovered by jointly detecting a number of consecutive symbols with the aid of the so-called multiple-symbol differential detector*, although this is achieved at the cost of increased complexity.

- *Thus the proposed sphere detector may be invoked again, but now as a reduced-complexity multiple-symbol differential detector.*

- The above-mentioned cooperative systems require *specifically designed resource allocation*, including the choice of the relaying protocols, the selection of the cooperating partners and the power control techniques.

- It is demonstrated that when the available relaying partners are roaming close to the source, the Decode-and-Forward (DF) protocol is the best cooperating protocol, which avoids potential error precipitation. By contrast, in case the cooperating partners roam closer to the destination, then the Amplify-and-Forward (AF) protocol is preferred for the same reasons. *These complementary features suggest the emergence of a hybrid DF/AF protocol*, which is controlled with the aid of our novel resource allocation techniques.

- The book concludes by outlining a variety of promising *future research directions*.

Our intention in the book is:

1. First, to pay tribute to all researchers, colleagues and valued friends who contributed to the field. Hence this book is dedicated to them, since without their quest for better MIMO-OFDM solutions this monograph could not have been conceived. They are too numerous to name here, but they do appear in the Author Index of the book. Our hope is that the conception of this monograph on the topic will provide an adequate portrayal of the community's research and will further fuel the innovation process.

2. We expect to stimulate further research by exposing open research problems and by collating a range of practical problems and design issues for the practitioners. The coherent further efforts of the wireless research community are expected to lead to the solution of a range of outstanding problems, ultimately providing us with flexible coherent and non-coherent detection-aided as well as cooperative MIMO-OFDM wireless transceivers exhibiting a performance close to information theoretical limits.

Acknowledgements

We are indebted to our many colleagues who have enhanced our understanding of the subject. These colleagues and valued friends, too numerous to be mentioned individually, have influenced our views concerning the subject of the book. We thank them for the enlightenment gained from our collaborations on various projects, papers and books. We are particularly grateful to our academic colleagues Professor Sheng Chen, Dr Soon-Xin Ng, Dr Rob Maunder and Dr Lie-Liang Yang. We would also like to express our appreciation to Sohail Ahmed, Andreas Ahrens, Jos Akhtman, Osamah Alamri, Jon Blogh, Nicholas Bonello, Jan Brecht, Marco Breiling, Marco del Buono, Fasih Muhammad Butt, Sheng Chen, Peter Cherriman, Stanley Chia, Joseph Cheung, Byoung Jo Choi, Jin-Yi Chung, Thanh Nguyen Dang, Sheyam Lal Dhomeja, Dirk Didascalou, Lim Dongmin, Mohammed El-Hajjar, Stephan Ernst, Peter Fortune, Eddie Green, David Greenwood, Chen Hong, Hee Thong How, Bin Hu, Ming Jiang, Thomas Keller, Lingkun Kong, Choo Leng Koh, Ee Lin Kuan, W. H. Lam, C. C. Lee, Chee Siong Lee, Kyungchun Lee, Tong-Hooi Liew, Xiao Lin, Wei Liu, Xiang Liu, Matthias Münster, Song Ni, M. A. Nofal, Noor Shamsiah Othman, Raja Ali Raja Riaz, Vincent Roger-Marchart, Redwan Salami, Clare Sommerville, Professor Raymond Steele, Tim Stevens, David Stewart, Shinya Sugiura, Shuang Tan, Ronal Tee, Jeff Torrance, Spyros Vlahoyiannatos, Jin Wang, Li Wang, William Webb, Chun-Yi Wei, Hua Wei, Stefan Weiss, John Williams, Seung-Hwang Won, Jason Woodard, Choong Hin Wong, Henry Wong, James Wong, Andy Wolfgang, Nan Wu, Chong Xu, Lei Xu, Du Yang, Wang Yao, Bee-Leong Yeap, Mong-Suan Yee, Kai Yen, Andy Yuen, Jiayi Zhang, Rong Zhang, and many others with whom we enjoyed an association.

We also acknowledge our valuable associations with the Virtual Centre of Excellence in Mobile Communications, in particular with its chief executive, Dr Walter Tuttlebee, and other members of its Executive Committee, namely Professor Hamid Aghvami, Dr Keith Baughan, Professor Mark Beach, Professor John Dunlop, Professor Barry Evans, Professor Peter Grant, Dr Dean Kitchener, Professor Steve MacLaughlin, Professor Joseph McGeehan, Dr Tim Moulsley, Professor Rahim Tafazolli, Professor Mike Walker and many other valued colleagues. Our sincere thanks are also due to John Hand and Andrew Lawrence of EPSRC, UK, for supporting our research. We would also like to thank Dr Joao Da Silva, Dr Jorge Pereira, Bartholome Arroyo, Bernard Barani, Demosthenes Ikonomou, and other valued colleagues from the Commission of the European Communities, Brussels, Belgium.

Similarly, our sincere thanks are due to Mark Hammond, Sarah Tilley and their colleagues at Wiley in Chichester, UK. Finally, our sincere gratitude is due to the numerous authors listed in the Author Index – as well as to those whose work was not cited owing to space limitations – for their contributions to the state of the art, without whom this book would not have materialized.

Lajos Hanzo, Jos Akhtman, Li Wang and Ming Jiang
School of Electronics and Computer Science
University of Southampton, UK

List of Symbols

$(\cdot)[n, k]$	The indices indicating the kth subcarrier of the nth OFDM symbol
$(\cdot)^T$	The transposition operation
$(\cdot)^H$	Hermitian transpose
$(\cdot)^*$	Complex conjugate
$\Im$	The imaginary component of a complex number
$\Re$	The real component of a complex number
$\mathcal{I}\{\cdot\}$	Imaginary part of a complex value
$\mathcal{I}$	Mutual information
π	The ratio of the circumference of a circle to the diameter
$\mathcal{R}\{\cdot\}$	Real part of a complex value
$\exp(\cdot)$	The exponential operation
$\mathbf{A}^{(l)}$	The remaining user set for the lth iteration of the subcarrier-to-user assignment process
$\mathbf{A}^{\mathrm{T}}$	Matrix/vector transpose
$\mathbf{A}^{\mathrm{H}}$	Matrix/vector hermitian adjoint, *i.e.* complex conjugate transpose
$\mathbf{A}^*$	Matrix/vector/scalar complex conjugate
$\mathbf{A}^{-1}$	Matrix inverse
$\mathbf{A}^+$	Moore-Penrose pseudoinverse
$\mathrm{tr}(\mathbf{A})$	Trace of matrix, *i.e.* the sum of its diagonal elements
α_P	The user load of an L-user and P-receiver conventional SDMA system
B_{T}	The overall system throughput in bits per OFDM symbol
$(i_{\mathrm{ce}}, i_{\mathrm{det}}, i_{\mathrm{dec}})$	Number of (channel estimation, detection, decoding) iterations
E_b	Energy per transmitted bit
E_s	Energy per transmitted M-QAM symbol
L_f	Number of data-frames per transmission burst
N_d	Number of data SDM-OFDM symbols per data-frame
N_p	Number of pilot SDM-OFDM symbols in burst preamble
T	OFDM symbol duration
T_s	OFDM FFT frame duration
f_{D}	Maximum Doppler frequency
K	Number of OFDM subcarriers
B	Signal bandwidth
β	RLS CIR tap prediction filter forgetting factor
$\mathcal{C}$	Unconstrained capacity
f_{c}	Carrier frequency
η	PASTD aided CIR tap tracking filter forgetting factor
γ	OHRSA search resolution parameter
m_{t}	Number of receive antennas

n_{r}	Number of transmit antennas
ν_τ	OFDM-symbol-normalized PDP tap drift rate
ρ	OHRSA search radius factor parameter
σ_w^2	Gaussian noise variance
τ_{rms}	RMS delay spread
ε	Pilot overhead
ζ	MIMO-CTF RLS tracking filter forgetting factor
b_{l,m_B}	The (m_B)th bit of the lth user's transmitted symbol
r	Size of the transmitted bit-wise signal vector $\mathbf{t}$
$\hat{b}_s^{(l)}[n,k]$	The lth user's detected soft bit
$\hat{\mathbf{b}}_s^{(l)}$	The detected soft bit block of the lth user
$\mathbf{b}^{(l)}$	The information bit block of the lth user
$\mathbf{b}_s^{(l)}$	The coded bit block of the lth user
$\mathbb{C}$	The complex space
$\mathbb{C}^{(x \times y)}$	The $(x \times y)$-dimensional complex space
$\mathrm{CC}(n,k,K)$	Convolutional codes with the number of input bits k, the number of coded bits n and the constraint length K
$\boldsymbol{I}$	Identity matrix
$\mathcal{H}$	Hadamard matrix
$\mathcal{L}$	Log Likelihood Ratio value
$\mathcal{M}$	Set of M-PSK/M-QAM constellation phasors
$c_{g_l}(t)$	The DSS signature sequence assigned to the lth user and associated with the gth DSS group
$\bar{\mathbf{c}}_{G_q}$	The $(1 \times L_q)$-dimensional DSS code vector
$\check{\mathbf{c}}_{G_q}$	The $(G_q \times 1)$-dimensional DSS code vector
$\mathbf{c}_g$	The spreading code sequence associated with the gth DSS group
$\mathbf{c}$	The user signature vector
$\mathbf{c}^{(l)}$	the lth user's code sequence
$\mathbf{c}_{g_l}$	The DSS code vector for the lth user in the gth DSS group
$\check{\mathbf{s}}$	A priori signal vector estimate
$\hat{\mathbf{s}}$	A posteriori signal vector estimate
$\hat{\mathbf{x}}$	Unconstrained a posteriori signal vector estimate
$\mathbf{H}$	Subcarrier-related MIMO CTF matrix
$\mathbf{d}$	Transmitted bit-wise signal
$\mathbf{s}$	Transmitted subcarrier-related SDM signal
$\mathbf{t}$	Transmitted subcarrier-related bit-wise SDM signal
$\mathbf{y}$	Received subcarrier-related SDM signal
$\mathbf{w}$	Gaussian noise sample vector
$\tilde{\mathbf{s}}$	Soft-information aided signal vector estimate
$\Delta_{p,(y,x)}^{(l)}[n,k]$	The random step size for the (p,l)th channel gene during step mutation associated with the xth individual of the yth generation
ϵ	The pilot overhead
F_D	The OFDM-symbol-normalized Doppler frequency
$\mathrm{Cov}\{\cdot,\cdot\}$	Covariance of two random variables
$\mathrm{Var}\{\cdot\}$	Variance of a random variable

$\mathrm{E}\{\cdot\}$	Expectation of a random variable
$\mathrm{Ei}\{\cdot\}$	Exponential integral
$\mathrm{JacLog}(\cdot)$	Jacobian logarithm
κ	Channel estimation efficiency criteria
$\|\cdot\|_2$	Second order norm
$\mathrm{P}\{\cdot\}$	Probability density function
$\mathrm{rms}\{\cdot\}$	Root mean square value
f_d'	Normalized Doppler frequency
f_c	Carrier frequency
f_d	Maximum Doppler frequency
f_q	Carrier frequency associated with the qth subband
$f_{(y,x)}$	The fitness value associated with the xth individual of the yth generation
G	The number of DSS user groups in a DSS/SSCH system
G_q	The total number of different DSS codes used by the users activating the qth subcarrier
$\Gamma_\tau(t)$	The rectangular pulse within the duration of $[0, \tau)$
$H_p^{(l)}$	The FD-CHTF associated with the lth user and the pth receiver antenna element
$H_{p,q}^{(l)}$	The FD-CHTF associated with the specific link between the lth user and the pth receiver at the qth subcarrier
$H_p^{(l)}[n,k]$	The true FD-CHTF associated with the channel link between the lth user and the pth receiver
$\hat{H}_p^{(l)}[n,k]$	The improved a posteriori FD-CHTF estimate associated with the channel link between the lth user and the pth receiver
$\mathbf{H}$	The FD-CHTF matrix
$\mathbf{H}^{(l)}$	The FD-CHTF vector associated with the lth user
$\mathbf{H}_{g,q}^{(l)}$	The $(P \times 1)$-dimensional FD-CHTF vector associated with the transmission paths between the lth user's transmitter antenna and each element of the P-element receiver antenna array, corresponding to the gth DSS group at the qth subcarrier
$\mathbf{H}_p$	The pth row of the FD-CHTF matrix $\mathbf{H}$
$\mathbf{H}_{g,q}$	The $(P \times l_g)$-dimensional FD-CHTF matrix associated with the gth DSS group at the qth subcarrier
$\mathbf{H}_{p,g,q}$	The pth row of the FD-CHTF matrix $\mathbf{H}_{g,q}$ associated with the gth DSS group at the qth subcarrier
$\mathbf{H}_p[n,k]$	The initial FD-CHTF estimate matrix associated with all the channel links between each user and the pth receiver
$\bar{\mathbf{H}}_{p,q}$	The L_q users' $(L_q \times L_q)$-dimensional diagonal FD-CHTF matrix associated with the qth subcarrier at the pth receiver
$\bar{\mathbf{H}}_p[n,k]$	The diagonal FD-CHTF matrix associated with all the channel links between each user and the pth receiver
$\tilde{\mathbf{H}}[n,k]$	The trial FD-CHTF matrix of the GA-JCEMUD
$\tilde{\mathbf{H}}_{(y,x)}[n,k]$	The FD-CHTF chromosome of the GA-JCEMUD individual associated with the xth individual of the yth generation
$\tilde{H}_{p,(y,x)}^{(l)}[n,k]$	The (p,l)th channel gene of the GA-JCEMUD FD-CHTF chromosome associated with the xth individual of the yth generation
$\tilde{H}_p^{(l)}[0,k]$	The initial FD-CHTF estimate associated with the channel link between the lth user and the pth receiver at the kth subcarrier in the first OFDM symbol duration

$\tilde{h}_p^{(l)}[n,k]$	The initial estimate of the CIR-related taps associated with the channel link between the lth user and the pth receiver
$\mathbf{I}$	Identity matrix
K_0	The range of CIR-related taps to be retained
L	Number of simultaneous mobile users supported in a SDMA system
L_q	The number of users that activate the qth subcarrier
$\mathcal{L}_{l,m_B}$	The LLR associated with the (m_B)th bit position of the lth user's transmitted symbol
$\Lambda_q^{(l)}(t)$	The subcarrier activation function
l_g	The number of users in the gth DSS group
λ_{max}	The maximum mutation step size of the step mutation
M_{WHT}	The WHT block size
$\mathcal{M}^L$	The set consisting of 2^{mL} number of $(L \times 1)$-dimensional trial vectors
$\mathcal{M}_{l,m_B,b}^L$	The specific subset associated with the lth user, which is constituted by those specific trial vectors, whose lth element's (m_B)th bit has a value of b
$\mathcal{M}_c$	The set containing the 2^m number of legitimate complex constellation points associated with the specific modulation scheme employed
m_B	The bit position of a constellation symbol
$\overline{\mathrm{MSE}}$	The average FD-CHTF estimation MSE
$\overline{\mathrm{MSE}}[n]$	The average FD-CHTF estimation MSE associated with the nth OFDM symbol
N_T	The total number of OFDM symbols transmitted
$n_p(t)$	The AWGN at the pth receiver
$n_{p,q}$	The noise signal associated with the qth subcarrier at the pth receiver
$\bar{\mathbf{n}}_{p,q}$	The $(G_q \times 1)$-dimensional effective noise vector associated with the qth subcarrier at the pth receiver
$\mathbf{n}$	Noise signal vector
ω_{ij}	The cross-correlation coefficient of the ith DSS group's and the jth DSS group's signature sequence
$\Omega(\cdot)$	The GA's joint objective function for all antennas
$\Omega_{g,q}(\cdot)$	The GA's joint objective function for all antennas associated with the gth DSS group at the qth subcarrier
$\Omega_{p,g,q}(\cdot)$	The GA's objective function associated with the gth DSS group of the pth antenna at the qth subcarrier
$\Omega_p(\cdot)$	The GA's objective function associated with the pth antenna
$\Omega_{y,T}$	The maximum GA objective score generated by evaluating the T individuals in the mating pool
P	Number of receiver antenna elements employed by the BS in SDMA systems
P_T	Transmitted signal power
$\tilde{p}_{mt}^{(ij)}$	The normalized mutation-induced transition probability
$p_{mt}^{(ij)}$	The 1D transition probability of mutating from a 1D symbol s_{Ri} to another 1D symbol s_{Rj}
$p_{mt}^{(ii)}$	The original legitimate constellation symbol's probability of remaining unchanged

$p_{mt}^{(ij)}$	The mutation-induced transition probability, which quantifies the probability of the ith legitimate symbol becoming the jth
p_m	The mutation probability, which denotes the probability of how likely it is that a gene will mutate
$\Phi(\cdot)$	The cost function of the OHRSA MUD
$\Phi_i(\cdot)$	The cumulative sub-cost function of the OHRSA MUD at the ith recursive step
$\varphi^{(l)}$	The lth user's phase angle introduced by carrier modulation
$\phi(\cdot)$	The sub-cost function of the OHRSA MUD
$Q(x)$	The Q-function
$\mathbf{Q}_L$	The L-order full permutation set
Q_c	The number of available subcarriers in conventional or SSCH systems
Q_f	The number of available subbands in SFH systems
Q_g	The number of subcarriers in a USSCH subcarrier group
$\mathbf{q}_k$	The subcarrier vector generated for the kth subcarrier group
$\mathfrak{q}^{(l)}$	The USSCH pattern set of the lth user
R	Code rate
$\mathbf{R_n}$	The $(P \times P)$-dimensional covariance matrix
$\bar{\mathbf{R}}_{G_q}$	The $(G_q \times L_q)$-dimensional cross-correlation matrix of the L_q users' DSS code sequences
$r_p(t)$	The received signal at the pth receiver
$r_{p,q}$	The discrete signal received at the qth subcarrier of the pth receiver during an OFDM symbol duration
$x_{p,g}(t)$	The despread signal of the gth DSS group at the pth receiver
$\hat{s}_i^{(l)}$	The ith constellation point of $\mathcal{M}_c$ as well as a possible gene symbol for the lth user
$s_{g_l,q}^{\prime(l)}(t)$	The transmitted signal at the qth subcarrier associated with the lth user in the gth DSS group
$s^{(l)}$	The transmitted signal of the lth user at a subcarrier
$s_{g_l,q}^{(l)}(t)$	The information signal at the qth subcarrier associated with the lth user in the gth DSS group
s_{Ri}	The ith 1D constellation symbol in the context of real axis
$\bar{\mathbf{s}}_q$	The L_q users' $(L_q \times 1)$-dimensional information signal vector
$\check{\mathbf{s}}$	The candidate trial vector
$\check{\mathbf{s}}_i$	The sub-vector of $\check{\mathbf{s}}$ at the ith OHRSA recursive step
$\hat{\mathbf{s}}^{(l)}$	The lth user's estimated information symbol block of the FFT length
$\hat{\mathbf{s}}_{\mathrm{w}}^{(l)}$	The estimated lth user's WHT-despreading signal block
$\hat{\mathbf{s}}_{\mathrm{w},0}^{(l)}$	The estimated lth user's WHT-despread signal block
$\hat{\mathbf{s}}_{\mathrm{GA}}$	The estimated transmitted symbol vector detected by the GA MUD
$\hat{\mathbf{s}}_{\mathrm{GA}g,q}$	The GA-based estimated $(l_g \times 1)$-dimensional signal vector associated with the gth DSS group at the qth subcarrier
$\hat{\mathbf{s}}_{\mathrm{MMSE}g,q}$	The MMSE-based estimated $(l_g \times 1)$-dimensional signal vector associated with the gth DSS group at the qth subcarrier
$\tilde{\mathbf{s}}[n,k]$	The trial data vector of the GA-JCEMUD
$\tilde{\mathbf{s}}_{(y,x)}$	The xth individual of the yth generation
$\tilde{\mathbf{s}}_{(y,x)}[n,k]$	The symbol chromosome of the GA-JCEMUD individual associated with the xth individual of the yth generation

$\mathbf{s}$	Transmitted signal vector
$\mathbf{s}^{(l)}$	The lth user's information symbol block of the FFT length
$\mathbf{s}_{\mathrm{w}}^{(l)}$	The lth user's WHT-spread signal block
$\mathbf{s}_{\mathrm{w},0}^{(l)}$	The lth user's WHT-spreading signal block
$\mathbf{s}_g$	The $(l_g \times 1)$-dimensional trial symbol vector for the GA's objective function associated with the gth DSS group
$\tilde{s}_{(y,x)}^{(l)}[n,k]$	The lth symbol gene of the GA-JCEMUD symbol chromosome associated with the xth individual of the yth generation
σ_l^2	Signal variance associated with the lth user
σ_n^2	Noise variance
T_h	The FH dwell time
$\mathrm{TC}(n,k,K)$	Turbo convolutional codes with the number of input bits k, the number of coded bits n and the constraint length K
T_r	The reuse time interval of hopping patterns
T_c	The DSS chip duration
$\mathbf{U}_{\mathrm{WHT}_K}$	The K-order WHT matrix
$u_{g_l}[c]$	The cth element of the gth row in the $(G \times G)$-dimensional WHT matrix, which is associated with the lth user
$\mathbf{V}$	The upper-triangular matrix having positive real-valued elements on the main diagonal
ν	CM code memory
W	System bandwidth
W_{sc}	Subcarrier bandwidth
$\mathbf{W}_{\mathrm{MMSE}}$	The MMSE-based weight matrix
$\mathbf{W}_{\mathrm{MMSE}_{g,q}}$	The MMSE-based $(P \times l_g)$-dimensional weight matrix associated with the gth DSS group at the qth subcarrier
X	GA population size
x_p	The received signal at the pth receiver at a subcarrier
$\bar{x}_{p,q}$	The despread signal associated with the qth subcarrier at the pth receiver
$\mathbf{x}$	Received signal vector
$\mathbf{x}_p$	The received symbol block of the FFT length at the pth receiver
$\mathbf{x}_{g,q}$	The $(P \times 1)$-dimensional despread signal vector associated with the gth DSS group at the qth subcarrier
Y	Number of GA generations

Chapter 1

Introduction to OFDM and MIMO-OFDM

1.1 OFDM History

In recent years Orthogonal Frequency-Division Multiplexing (OFDM) [1–4] has emerged as a successful air-interface technique. In the context of wired environments, OFDM techniques are also known as Discrete Multi-Tone (DMT) [5] transmissions and are employed in the American National Standards Institute's (ANSI's) Asymmetric Digital Subscriber Line (ADSL) [6], High-bit-rate Digital Subscriber Line (HDSL) [7], and Very-high-speed Digital Subscriber Line (VDSL) [8] standards as well as in the European Telecommunication Standard Institute's (ETSI's) [9] VDSL applications. In wireless scenarios, OFDM has been advocated by many European standards, such as Digital Audio Broadcasting (DAB) [10], Digital Video Broadcasting for Terrestrial television (DVB-T) [11], Digital Video Broadcasting for Handheld terminals (DVB-H) [12], Wireless Local Area Networks (WLANs) [13] and Broadband Radio Access Networks (BRANs) [14]. Furthermore, OFDM has been ratified as a standard or has been considered as a candidate standard by a number of standardization groups of the Institute of Electrical and Electronics Engineers (IEEE), such as the IEEE 802.11 [15] and the IEEE 802.16 [16] standard families.

The concept of parallel transmission of data over dispersive channels was first mentioned as early as 1957 in the pioneering contribution of Doelz *et al.* [17], while the first OFDM schemes date back to the 1960s, which were proposed by Chang [18] and Saltzberg [19]. In the classic parallel data transmission systems [18, 19], the Frequency-Domain (FD) bandwidth is divided into a number of non-overlapping subchannels, each of which hosts a specific carrier widely referred to as a subcarrier. While each subcarrier is separately modulated by a data symbol, the overall modulation operation across all the subchannels results in a frequency-multiplexed signal. All of the sinc-shaped subchannel spectra exhibit zero crossings at all of the remaining subcarrier frequencies and the individual subchannel spectra are orthogonal to each other. This ensures that the subcarrier signals do not interfere with each other, when communicating over perfectly distortionless channels, as a consequence of their orthogonality [3].

The early OFDM schemes [18–21] required banks of sinusoidal subcarrier generators and demodulators, which imposed a high implementation complexity. This drawback limited the application of OFDM to military systems until 1971, when Weinstein and Ebert [22] suggested that the Discrete Fourier Transform (DFT) can be used for the OFDM modulation and demodulation processes, which significantly reduces the implementation complexity of OFDM. Since then, more practical OFDM research has been carried out. For example, in the early 1980s Peled and Ruiz [23] proposed a simplified FD data transmission method using a cyclic prefix-aided technique and exploited reduced-complexity algorithms for achieving a significantly lower computational complexity than that of classic single-carrier time-domain Quadrature Amplitude Modulation (QAM) [24] modems. Around the same era, Keasler *et al.* [25] invented a high-speed OFDM modem for employment in switched networks, such as the telephone network. Hirosaki designed a subchannel-based equalizer for an orthogonally multiplexed QAM system in 1980 [26] and later introduced the DFT-based implementation of OFDM systems [27], on the basis of which a so-called groupband data modem was developed [28]. Cimini [29] and Kalet [30] investigated the performance of OFDM modems in mobile communication channels. Furthermore, Alard and Lassalle [31] applied OFDM in digital broadcasting systems, which was the pioneering work of the European DAB standard [10] established in the mid-1990s. More recent advances in OFDM transmission were summarized in the state-of-the-art collection of works edited by Fazel and Fettweis [32]. Other important recent OFDM references include the books by Hanzo *et al.* [3] and Van Nee *et al.* [4] as well as a number of overview papers [33–35].

OFDM has some key advantages over other widely used wireless access techniques, such as Time-Division Multiple Access (TDMA) [36], Frequency-Division Multiple Access (FDMA) [36] and Code-Division Multiple Access (CDMA) [37, 38, 40–42]. The main merit of OFDM is the fact that the radio channel is divided into many narrowband, low-rate, frequency-non-selective subchannels or subcarriers, so that multiple symbols can be transmitted in parallel, while maintaining a high spectral efficiency. Each subcarrier may deliver information for a different user, resulting in a simple multiple-access scheme known as Orthogonal Frequency-Division Multiple Access (OFDMA) [43–46]. This enables different media such as video, graphics, speech, text or other data to be transmitted within the same radio link, depending on the specific types of services and their Quality-of-Service (QoS) requirements. Furthermore, in OFDM systems different modulation schemes can be employed for different subcarriers or even for different users. For example, the users close to the Base Station (BS) may have a relatively good channel quality, thus they can use high-order modulation schemes to increase their data rates. By contrast, for those users that are far from the BS or are serviced in highly loaded urban areas, where the subcarriers' quality is expected to be poor, low-order modulation schemes can be invoked [47].

Besides its implementational flexibility, the low complexity required in transmission and reception as well as the attainable high performance render OFDM a highly attractive candidate for high-data-rate communications over time-varying frequency-selective radio channels. For example, in classic single-carrier systems, complex equalizers have to be employed at the receiver for the sake of mitigating the Inter-Symbol Interference (ISI) introduced by multi-path propagation. By contrast, when using a cyclic prefix [23], OFDM exhibits a high resilience against the ISI. Incorporating channel coding techniques into OFDM systems, which results in Coded OFDM (COFDM) [48, 49], allows us to maintain robustness against frequency-selective fading channels, where busty errors are encountered at specific subcarriers in the FD.

However, besides its significant advantages, OFDM also has a few disadvantages. One problem is the associated increased Peak-to-Average Power Ratio (PAPR) in comparison with single-carrier systems [3], requiring a large linear range for the OFDM transmitter's output amplifier. In addition, OFDM is sensitive to carrier frequency offset, resulting in Inter-Carrier Interference (ICI) [50].

As a summary of this section, we outline the milestones and the main contributions found in the OFDM literature in Tables 1.1 and 1.2.

1.1.1   MIMO-Assisted OFDM

1.1.1.1   The Benefits of MIMOs

High-data-rate wireless communications have attracted significant interest and constitute a substantial research challenge in the context of the emerging WLANs and other indoor multimedia networks. Specifically, the employment of multiple antennas at both the transmitter and the receiver, which

Table 1.1: Milestones in the history of OFDM.

Year	Milestone
1957	The concept of parallel data transmission by Doelz *et al.* [17]
1966	First OFDM scheme proposed by Chang [18] for dispersive fading channels
1967	Saltzberg [19] studied a multi-carrier system employing Orthogonal QAM (O-QAM) of the carriers
1970	US patent on OFDM issued [21]
1971	Weinstein and Ebert [22] applied DFT to OFDM modems
1980	Hirosaki designed a subchannel-based equalizer for an orthogonally multiplexed QAM system [26]
	Keasler *et al.* [25] described an OFDM modem for telephone networks
1985	Cimini [29] investigated the feasibility of OFDM in mobile communications
1987	Alard and Lasalle [31] employed OFDM for digital broadcasting
1991	ANSI ADSL standard [6]
1994	ANSI HDSL standard [7]
1995	ETSI DAB standard [10]: the first OFDM-based standard for digital broadcasting systems
1996	ETSI WLAN standard [13]
1997	ETSI DVB-T standard [11]
1998	ANSI VDSL and ETSI VDSL standards [8, 9]
	ETSI BRAN standard [14]
1999	IEEE 802.11a WLAN standard [51]
2002	IEEE 802.11g WLAN standard [52]
2003	Commercial deployment of FLASH-OFDM [53, 54] commenced
2004	ETSI DVB-H standard [12]
	IEEE 802.16-2004 WMAN standard [55]
	IEEE 802.11n draft standard for next generation WLAN [56]
2005	Mobile cellular standard 3GPP Long-Term Evolution (LTE) [57] downlink
2007	Multi-user MIMO-OFDM for next-generation wireless [58]
	Adaptive HSDPA-style OFDM and MC-CDMA transceivers [59]

is widely referred to as the Multiple-Input, Multiple-Output (MIMO) technique, constitutes a cost-effective approach to high-throughput wireless communications.

The concepts of MIMOs have been under development for many years for both wired and wireless systems. One of the earliest MIMO applications for wireless communications dates back to 1984, when Winters [92] published a breakthrough contribution, where he introduced a technique of transmitting data from multiple users over the same frequency/time channel using multiple antennas at both the transmitter and receiver ends. Based on this work, a patent was filed and approved [93]. Sparked off by Winters' pioneering work [92], Salz [94] investigated joint transmitter/receiver optimization using the MMSE criterion. Since then, Winters and others [95–103] have made further significant advances in the field of MIMOs. In 1996, Raleigh [104] and Foschini [105] proposed new approaches for improving the efficiency of MIMO systems, which inspired numerous further contributions [106–114].

As a key building block of next-generation wireless communication systems, MIMOs are capable of supporting significantly higher data rates than the Universal Mobile Telecommunications System (UMTS) and the High-Speed Downlink Packet Access (HSDPA) based 3G networks [115]. As indicated by the terminology, a MIMO system employs multiple transmitter and receiver antennas for delivering parallel data streams, as illustrated in Figure 1.1. Since the information is transmitted through

Table 1.2: Main contributions on OFDM.

Year	Author(s)	Contribution
1966	Chang [18]	Proposed the first OFDM scheme
1967	Saltzberg [19]	Studied a multi-carrier system employing O-QAM
1968	Chang and Gibby [20]	Presented a theoretical analysis of the performance of an orthogonal multiplexing data transmission scheme
1970	Chang [21]	US patent on OFDM issued
1971	Weinstein and Ebert [22]	Applied DFT to OFDM modems
1980	Hirosaki [26]	Designed a subchannel-based equalizer for an orthogonally multiplexed QAM system
	Peled and Ruiz [23]	Described a reduced-complexity FD data transmission method together with a cyclic prefix technique
	Keasler *et al.* [25]	Invented an OFDM modem for telephone networks
1981	Hirosaki [27]	Suggested a DFT-based implementation of OFDM systems
1985	Cimini [29]	Investigated the feasibility of OFDM in mobile communications
1986	Hirosaki *et al.* [28]	Developed a groupband data modem using an orthogonally multiplexed QAM technique
1987	Alard and Lasalle [31]	Employed OFDM for digital broadcasting
1989	Kalet [30]	Analysed multi-tone QAM modems in linear channels
1990	Bingham [1]	Discussed various aspects of early OFDM techniques in depth
1991	Cioffi [6]	Introduced the ANSI ADSL standard
1993–1995	Warner [60], Moose [50] and Pollet [61]	Conducted studies on time and frequency synchronization in OFDM systems
1994–1996	Jones [62], Shepherd [63] and Wulich [64,65]	Explored various coding and post-processing techniques designed for minimizing the peak power of the OFDM signal
1997	Li and Cimini [66,67]	Revealed how clipping and filtering affect OFDM systems
	Hara and Prasad [68]	Compared various methods of combining CDMA and OFDM
1998	Li *et al.* [69]	Designed a robust Minimum Mean Square Error (MMSE) based channel estimator for OFDM systems
	May *et al.* [48]	Carried out a performance analysis of Viterbi decoding in the context of 64-Differential Amplitude and Phase-Shift Keying (64-DAPSK) and 64QAM OFDM signals
1999	Li and Sollenberger [70]	Focused on parameter estimation invoked by an MMSE diversity combiner designed for adaptive antenna array-aided OFDM
	Armour *et al.* [71–73]	Illustrated the combined OFDM equalization-aided receiver and the design of pre-Fast Fourier Transform (FFT) equalizers
	Prasetyo and Aghvami [74,75]	Simplified the transmission frame structure for achieving fast burst synchronization in OFDM systems
	Wong *et al.* [76]	Advocated a subcarrier, bit and power allocation algorithm to minimize the total transmit power of multi-user OFDM
2000	Fazel and Fettweis [32]	A collection of state-of-the-art works on OFDM
	Van Nee and Prasad [4]	OFDM for wireless multimedia communications
	Lin *et al.* [49]	Invoked turbo coding in an OFDM system using diversity
2001–2002	Lu and Wang [77–80]	Considered channel-coded STC-assisted OFDM systems
2003	Hanzo *et al.* [3]	OFDM for broadband multi-user communications, WLANs and broadcasting
2004	Simeone *et al.* [81]	Demonstrated a subspace tracking algorithm used for channel estimation in OFDM systems
	Zhang *et al.* [82]	Adopted an Inter-Carrier Interference (ICI) cancellation scheme to combat the ICI in OFDM systems
	Necker and Stüber [83]	Exploited a blind channel estimation scheme based on the ML principle in OFDM systems
	Doufexi *et al.* [84]	Reflected the benefits of using sectorized antennas in WLANs
	Alsusa *et al.* [85]	Proposed packet-based multi-user OFDM systems using adaptive subcarrier–user allocation
2005	Williams *et al.* [86]	Evaluated a pre-FFT synchronization method for OFDM
2007	Jiang and Hanzo [58]	Multi-user MIMO-OFDM for next-generation wireless
	Hanzo and Choi [59]	Adaptive HSDPA-style OFDM and MC-CDMA transceivers
2009	Fischer and Siegl [87]	Peak-to-average power ratio reduction in single- and multi-antenna OFDM
	Mileounis *et al.* [88]	Blind identification of Hammerstein channels using QAM, PSK and OFDM inputs
	Huang and Hwang [89]	Improvement of active interference cancellation: avoidance technique for OFDM cognitive radio
	Chen *et al.* [90]	Spectrum sensing for OFDM systems employing pilot tones
	Talbot and Farhang-Boroujeny [91]	Time-varying carrier offsets in mobile OFDM

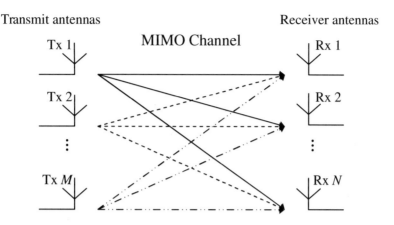

Figure 1.1: Schematic of the generic MIMO system employing M transmitter antennas and N receiver antennas. ©IEEE Jiang & Hanzo 2007 [58]

different paths, a MIMO system is capable of exploiting both transmitter and receiver diversity, hence maintaining reliable communications. Furthermore, with the advent of multiple antennas, it becomes possible to process/combine jointly the multi-antenna signals and thus improve the system's integrity and/or throughput. Briefly, compared with Single-Input, Single-Output (SISO) systems, the two most significant advantages of MIMO systems are:

1. A significant increase of both the system's capacity and spectral efficiency. The capacity of a wireless link increases linearly with the minimum of the number of transmitter or receiver antennas [104, 106]. The data rate can be increased by spatial multiplexing without consuming more frequency resources and without increasing the total transmit power.

2. Dramatic reduction of the effects of fading due to the increased diversity. This is particularly beneficial when the different channels fade independently.

An overview of MIMO techniques covering channel models, performance limits, coding and transceiver designs can be found in [116].

1.1.1.2 MIMO-OFDM

The quality of a wireless link can be described by three basic parameters, namely the transmission rate, the transmission range and the transmission reliability. Conventionally, the transmission rate may be increased by reducing the transmission range and reliability. By contrast, the transmission range may be extended at the cost of a lower transmission rate and reliability, while the transmission reliability may be improved by reducing the transmission rate and range [117]. However, with the advent of MIMO-assisted OFDM systems, the above-mentioned three parameters may be simultaneously improved [117]. Initial field tests of broadband wireless MIMO-OFDM communication systems have shown that an increased capacity, coverage and reliability is achievable with the aid of MIMO techniques [118]. Furthermore, although MIMOs can potentially be combined with any modulation or multiple-access technique, recent research suggests that the implementation of MIMO-aided OFDM is more efficient, as a benefit of the straightforward matrix algebra invoked for processing the MIMO-OFDM signals [117].

MIMO-OFDM, claimed to be invented by Airgo Networks [119], has formed the foundation of all candidate standards proposed for IEEE 802.11n [120]. In recent years, this topic has attracted substantial research efforts, addressing numerous aspects, such as system capacity [121,122], space/time/frequency coding [123–127], Peak-to-Average Power Ratio (PAPR) control [128–130], channel estimation [131–133], receiver design [134–137], etc. Recently, Paulraj *et al.* [116] and Stüber *et al.* [138] have provided

Table 1.3: Main contributions on MIMO-OFDM (Part 1).

Year	Author(s)	Contribution
2001	Piechocki et al. [140]	Reported on the performance benefits of spatial multiplexing using ML decoding for a Vertical Bell Labs Layered Space–Time (V-BLAST) OFDM system
	Blum et al. [123]	Studied improved space–time coding techniques for MIMO-OFDM systems
2002	Li [131]	Exploited optimum training sequence design and simplified channel estimation for improving the performance and reducing the complexity of channel parameter estimation in MIMO-OFDM systems
	Bolckei et al. [121]	Analysed the influence of physical parameters such as the amount of delay spread, cluster angle spread and total angle spread, as well as system parameters such as the number of antennas and the antenna spacing on both the ergodic capacity and outage capacity
	Catreux et al. [141]	Offered an overview of the challenges and promises of link adaptation in future broadband wireless networks
	Piechocki et al. [142]	Presented a performance evaluation of spatial multiplexing and space-frequency-coded modulation schemes designed for WLANs
	Molisch et al. [143]	Proposed a reduced-complexity method for grouping multiple antennas and space–time codes
	Li et al. [134]	Invoked space–time coding and Successive Interference Cancellation (SIC) in MIMO-OFDM systems
	Stamoulis et al. [144]	Revealed the effects of the ICI on MIMO-OFDM
	Doufexi et al. [145]	Characterized the outdoor physical layer performance of a coded MIMO-OFDM system using space–time processing
	Giangaspero et al. [135]	Compared two Co-Channel Interference (CCI) cancellation schemes in the context of MIMO-OFDM
2003	Li et al. [136]	Advocated a CCI cancellation method using angle diversity based on null-steering or minimum variance distortion response beamforming
	Bölcskei et al. [146]	Measured the impact of the propagation environment on the performance of space-frequency-coded MIMO-OFDM
	Barhumi et al. [132]	Described a Least-Squares (LS) channel estimation scheme designed for MIMO-OFDM systems based on pilot tones
	Ganesan and Sayeed [122]	Derived a virtual MIMO framework for single-transmitter, single-receiver multi-path fading channels that enables maximal exploitation of channel diversity at both the transmitter and the receiver
	Gamal et al. [124]	Utilized an OFDM technique to transform the MIMO multi-path channel into a MIMO flat block fading channel, where the associated diversity is exploited by employing space-frequency codes
	Moon et al. [128]	Evaluated the PAPR performance in a MIMO-OFDM-based WLAN system using a Space–Time Block Code (STBC)
	Cai et al. [147]	Developed a technique based on the autocorrelation function for estimating the Doppler spread in Rayleigh fading channels for mobile OFDM systems using multiple antennas
	Leus and Moonen [148]	Employed tone-by-tone-based equalization techniques in MIMO-OFDM systems
	Lee et al. [129]	Investigated the PAPR characteristics in a MIMO-OFDM system using the selective mapping approach
	Piechocki et al. [149]	Devised a blind method for joint detection of space–time-coded MIMO-OFDM.

compelling overviews of MIMO-OFDM communications. Furthermore, Nortel Networks developed a MIMO-OFDM prototype [139] during late 2004, which demonstrates the superiority of MIMO-OFDM over today's networks in terms of the achievable data rate. For the reader's convenience, we have summarized the major contributions on MIMO-OFDM in Tables 1.3, 1.4 and 1.5.

1.1.1.3 SDMA-based MIMO-OFDM Systems

As a subclass of MIMO arrangements, recently the Space-Division Multiple Access (SDMA) [3, 193–195] based techniques have attracted substantial interest. As one of the most promising techniques aimed at solving the capacity problem of wireless communication systems, SDMA enables multiple users to share simultaneously the same bandwidth in different geographical locations. More specifically, the exploitation of the spatial dimension, namely the so-called spatial signature, makes it possible

Table 1.4: Main contributions on MIMO-OFDM (Part 2).

Year	Author(s)	Contribution
2004	Shin *et al.* [133]	Suggested a cyclic comb-type training structure for reducing the Mean Square Errors (MSEs) at the edge subcarriers of MIMO-OFDM signals
	Xia *et al.* [150]	Created an adaptive MIMO-OFDM transmitter by applying an adaptive two-dimensional coder–beamformer with the aid of partial channel knowledge
	Huang and Letaief [151]	Portrayed an OFDM symbol-based space diversity technique
	Butler and Collings [152]	Employed an approximate log-likelihood decoding approach based on a Zero-Forcing (ZF) receiver for bit-interleaved coded-modulation-assisted MIMO-OFDM systems
	Stüber *et al.* [138]	Summarized various physical layer research challenges in MIMO-OFDM system design
	Paulraj *et al.* [116]	Provided an overview of MIMO and/or MIMO-OFDM systems
	Lu *et al.* [153]	Identified the performance of an optimized MIMO-OFDM scheme using Low-Density Parity Check (LDPC) codes
	Van Zelst and Schenk [154]	Implemented MIMO-OFDM processing and evaluated its performance by both simulations and experimental test results
	Pascual-Iserte *et al.* [155]	Conducted studies on maximizing the Signal to Noise and Interference Ratio (SNIR) over the subcarriers subject to a total transmit power constraint.
	Zeng and Ng [156]	Contrived a subspace-based semi-blind method for estimating the channel responses of a multi-user and multi-antenna OFDM uplink system
	Alien *et al.* [157]	Assessed the performance of spatial diversity in an OFDM WLAN for various antenna topologies
	Dayal *et al.* [125]	Introduced space–time channel-sounding training codes designed for multiple-antenna, non-coherent, multiple-block Rayleigh fading channel
	Park and Kang [137]	Adopted a reduced-complexity iterative algorithm for joint Maximum-A-Posteriori (MAP) detection and CCI suppression in MIMO-OFDM systems
	Tan and Stüber [158]	Combined cyclic delay diversity and MIMO-OFDM for achieving full spatial diversity in flat-fading channels
	Wang *et al.* [159]	Illustrated the diversity and coding advantages in terms of the minimum Hamming distance and the minimum squared product distance of the code as well as the relative frequencies
	Pan *et al.* [160]	Discussed dynamic spatial subchannel allocation in conjunction with adaptive beamforming in broadband OFDM wireless systems
	Tepedelenlioğlu and Challagulla [161]	Demonstrated how to achieve high diversity gains in MIMO-OFDM systems with the aid of fractional sampling
	Baek *et al.* [162]	Addressed a time-domain semi-blind channel estimation approach and a PAPR reduction scheme for MIMO-OFDM
	Dubuc *et al.* [139]	Outlined Nortel Networks' MIMO-OFDM concept prototype and provided measured performance results
	Barriac and Madhow [163]	Offered guidelines for optimizing the antenna spacing in MIMO-OFDM systems using feedback of the covariance matrix of the downlink channel

to identify the individual users, even when they are in the same time/frequency/code domains, thus increasing the system's capacity.

In Figure 1.2 we illustrate the concept of SDMA systems. As shown in Figure 1.2, each user exploiting a single-transmitter-antenna-aided Mobile Station (MS) simultaneously communicates with the BS equipped with an array of receiver antennas. Explicitly, SDMA can be considered as a specific branch of the family of MIMO systems, where the transmissions of the multiple-transmitter antennas cannot be coordinated, simply because they belong to different users. Briefly speaking, the major advantages of SDMA techniques are [196]:

- *Range extension*: With the aid of an antenna array, the coverage area of high-integrity reception can be significantly larger than that of any single-antenna-aided systems. In an SDMA system, the number of cells required to cover a given geographic area can be substantially reduced. For example, a 10-element array offers a gain of 10, which typically doubles the radius of the cell and hence quadruples the coverage area.

- *Multi-path mitigation*: Benefitting from the MIMO architecture, in SDMA systems the detrimental effects of multi-path propagations are effectively mitigated. Furthermore, in specific

Table 1.5: Main contributions on MIMO-OFDM (Part 3).

Year	Author(s)	Contribution
2005	Su *et al.* [126, 127]	Designed a general space-frequency block code structure capable of providing full-rate, full-diversity MIMO-OFDM transmission
	Zhang *et al.* [164]	Researched an optimal QR decomposition technique designed for a precoded MIMO-OFDM system using successive cancellation detection
	Yao and Giannakis [165]	Proposed a low-complexity blind Carrier Frequency Offset (CFO) estimator for OFDM systems
	Zheng *et al.* [166]	Extended Time-Division Synchronous CDMA (TD-SCDMA) to Time-Division Code-Division Multiplexing OFDM (TD-CDM-OFDM) for future 4G systems
	Yang [167]	Reviewed the state-of-the-art approaches in MIMO-OFDM air interface
	Zhang and Letaief [168]	Aimed at developing an adaptive resource allocation approach which jointly allocates subcarriers, power and bits for multi-user MIMO-OFDM systems
	Ma [169]	Established a pilot-assisted modulation scheme for CFO and channel estimation in OFDM transmissions over frequency-selective MIMO fading channels
	Fozunbal *et al.* [170]	Calculated a sphere packing lower bound and a pairwise error upper bound of the error probability of space–time frequency-coded OFDM systems using multiple antennas for transmission over block-fading channels
	Nanda *et al.* [171]	Built a MIMO WLAN prototype that provides data rates over 200 Mbps
	Kim *et al.* [172]	Invoked a QR-Decomposition combined with the M-algorithm (QRD-M) for joint data detection and channel estimation in MIMO-OFDM
	Qiao *et al.* [173]	Contrived an iterative LS channel estimation algorithm for MIMO-OFDM.
	Sampath *et al.* [174]	Validated the properties of the transmit correlation matrix through field trial results obtained from a MIMO-OFDM wireless system operated in a macro-cellular environment
	Rey *et al.* [175]	Used a Bayesian approach to design transmit prefiltering matrices for closed-loop schemes, which is robust to channel estimation errors
	Sun *et al.* [176]	Targeted the design of CFO estimator-aided Expectation Maximization (EM) based iterative receivers for MIMO-OFDM systems
	Han and Lee [130]	Provided an overview of PAPR reduction techniques for multi-carrier transmission
	Lodhi *et al.* [177]	Evaluated the complexity and performance of a Multi-Carrier Code-Division Multiple-Access (MC-CDMA) system exploiting STBCs and Cyclic Delay Diversity (CDD)
	Wang *et al.* [178]	Advanced MIMO-OFDM channel estimation using a scheme based on estimating the Time of Arrival (TOA)
	Wen *et al.* [179]	Reported on a low-complexity multi-user angle-frequency coding scheme based on the Fourier basis structure for downlink wireless systems
	Su *et al.* [180]	Performance analysis of MIMO-OFDM systems invoking coding in spatial, temporal and frequency domains
	Tan *et al.* [181]	Advocated a scheme of cross-antenna rotation and inversion utilizing additional degrees of freedom by employing multiple antennas in OFDM systems
	Park and Cho [182]	Characterized a MIMO-OFDM technique based on the weighting factor optimization for reducing the ICI caused by time-varying channels
	Shao and Roy [183]	Maximized the diversity gain achieved over frequency-selective channels by employing a full-rate space-frequency block code for MIMO-OFDM systems
	Schenk *et al.* [184]	Quantified how the transmitter/receiver phase noise affects the performance of a MIMO-OFDM system
	Borgmann and Bölcskei [185]	Contributed to the code designs for non-coherent frequency-selective MIMO-OFDM fading links
	Tarighat and Sayed [186]	Examined the effect of IQ imbalances on MIMO-OFDM systems and developed a digital signal processing framework for combating these distortions
	Jiang *et al.* [187]	Formulated a joint transceiver design combining the Geometric Mean Decomposition (GMD) with ZF-type decoders
	Choi and Heath [188]	Constructed a limited feedback architecture that combines beamforming vector quantization and smart vector interpolation
	Baek *et al.* [189]	Incorporated multiple antennas into high-rate DAB systems
2007	Jiang and Hanzo [58]	Reduced-complexity near-ML SDMA-MUDs and joint iterative channel estimation
	Hanzo and Choi [59]	Near instantaneously adaptive HSPA-style OFDM and MC-CDMA transceivers
2009	Fischer and Siegl [87]	PAPR reduction in single- and multi-antenna OFDM
	Fakhereddin *et al.* [190]	Reduced feedback and random beamforming for MIMI-OFDM
	De *et al.* [191]	Linear prediction-based semi-blind channel estimation for multi-user OFDM
	Haring *et al.* [192]	Fine frequency synchronization in the uplink of multi-user OFDM systems

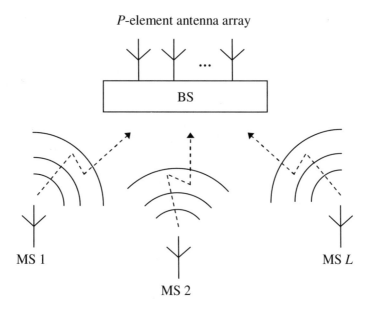

Figure 1.2: Illustration of the generic SDMA system employing a P-element receiver antenna array for supporting L mobile users. ©IEEE Jiang & Hanzo 2007 [58]

scenarios the multi-path phenomenon can even be exploited to enhance the desired users' signals by employing efficient receiver diversity schemes.

- *Capacity increase*: Theoretically, SDMA can be incorporated into any existing multiple-access standard at the cost of a limited increase in system complexity, while attaining a substantial increase in capacity. For instance, by applying SDMA to a conventional TDMA system, two or more users can share the same time slots, resulting in a doubled or higher overall system capacity.

- *Interference suppression*: The interference imposed by other systems and by users in other cells can be significantly reduced by exploiting the desired user's unique, user-specific Channel Impulse Responses (CIRs).

- *Compatibility*: SDMA is compatible with most of the existing modulation schemes, carrier frequencies and other specifications. Furthermore, it can be readily implemented using various array geometries and antenna types.

The combination of SDMA and OFDM results in SDMA-OFDM systems [3, 193, 197, 198], which exploit the merits of both SDMA and OFDM, having attracted more and more interest [198–203]. Tables 1.6 and 1.7 summarize the main contributions on SDMA and SDMA-OFDM found in the open literature.

1.2 OFDM Schematic

In this section we briefly introduce OFDM as a means of dealing with the problems of frequency-selective fading encountered when transmitting over a high-rate wideband radio channel.

In the OFDM scheme of Figure 1.3 the serial data stream of a traffic channel is passed through a serial-to-parallel converter, which splits the data into a number of parallel subchannels. The data in each subchannel are applied to a modulator, such that for M channels there are M modulators whose carrier frequencies are $f_0, f_1, \ldots, f_M$. The difference between adjacent channels is Δf and the overall bandwidth W of the N modulated carriers is $M\Delta f$.

Table 1.6: Main contributions on SDMA (Part 1).

Year	Author(s)	Contribution
1982	Yeh and Reudink [204]	Illustrated that high spectrum efficiencies can be achieved in mobile radio systems using a modest number of space diversity branches
1983	Ko and Davis [205]	Early studies on SDMA in the context of satellite communication networks
1989	Swales *et al.* [206, 207]	Devised a multi-beam adaptive BS antenna in an attempt to mitigate the problem of limited radio resources
1990	Agee *et al.* [208]	Invoked narrowband antenna arrays for blind adaptive signal extraction
1991	Anderson *et al.* [209]	Adopted adaptive antenna techniques to increase the channel capacity
1992	Balaban and Salz [210, 211]	Provided a comprehensive characterization of space diversity reception combined with various equalization techniques
1994	Xu *et al.* [212]	Offered preliminary results of experimental studies on SDMA systems
	Talwar *et al.* [213]	Described an approach for separating and estimating multiple co-channel signals with the aid of an antenna array
1995	Van Der Veen *et al.* [214]	Blindly identified Finite Impulse Response (FIR) channels using oversampling and the finite alphabet property of digital signals
	Khalaj *et al.* [215]	Estimated the spatio-temporal characteristics of the radio channel in coherent direct-sequence spread-spectrum systems
	Anand *et al.* [216]	Established a method of blind separation of co-channel Binary Phase-Shift Keying (BPSK) signals arriving at an antenna array
1997	Liu and Xu [217]	Addressed the SDMA uplink blind channel and sequence estimation problem
	Tsoulos *et al.* [218]	Reported the research of the TSUNAMI project that demonstrated the benefits of SDMA in wireless communications
1998	Deneire and Slock [219]	Derived a subspace fitting and linear prediction method using cyclic statistics of fractionally sampled channels for channel identification in multi-user and multi-antenna systems
	Tsoulos *et al.* [220, 221]	Provided an experimental demonstration of both transmit and receive beamforming supporting SDMA user access
	Barroso *et al.* [222]	Introduced a blind algorithm referred to as Array Channel-Division Multiple Access (AChDMA) for advanced SDMA in mobile communications systems
	Demmerle and Wiesbeck [223]	Designed a biconical multi-beam antenna structure for SDMA communications
	Lindmark [224]	Built a dual-polarized antenna array for an SDMA system working in the 1850–1990 MHz band
	Suard *et al.* [225]	Investigated the channel capacity enhancement of an SDMA system
	Jeng *et al.* [226]	Presented extensive experimental results of spatial signature variation using a smart antenna test bed
	Petrus *et al.* [227]	Proved that capacity improvement can be achieved using adaptive arrays at the BS of an Advanced Mobile Phone Service (AMPS) system
	Xavier *et al.* [228]	Targeted at designing a closed-form estimator for the SDMA-MIMO channel based on second-order statistics
	Farsakh and Nossek [229]	Developed an approach for jointly calculating array weights in such a way that all users receive their signals at a given SINR level
1999	Tsoulos [230]	Provided an overview of smart antennas in the context of current and future personal communication systems
	Piolini and Rolando [231]	Analysed a channel assignment algorithm for SDMA mobile systems
	Vandenameele *et al.* [193, 197, 198]	Advocated a combined SDMA-OFDM approach that couples the capabilities of the two techniques
	Galvan-Tejada and Gardiner [232, 233]	Calculated the theoretical blocking probability resulting from SDMA technology in two different channel allocation schemes
	Tsoulos [234]	Focused on TDMA air-interface techniques combined with SDMA schemes
	Vornefeld *et al.* [235]	Applied SDMA techniques to WATM systems

These M modulated carriers are then combined to give an OFDM signal. We may view the serial-to-parallel converter as applying every Mth symbol to a modulator. This has the effect of interleaving the symbols into each modulator, hence symbols $S_0, S_M, S_{2M}, \ldots$ are applied to the modulator whose carrier frequency is f_1. At the receiver the received OFDM signal is demultiplexed into M frequency bands, and the M modulated signals are demodulated. The baseband signals are then recombined using a parallel-to-serial converter.

Table 1.7: Main contributions on SDMA (Part 2).

Year	Author(s)	Contribution
2000	Djahani and Kahn [236]	Discussed the employment of multi-beam transmitters and imaging receivers in SDMA implementations
2001	Shad *et al.* [237]	Invoked dynamic slot allocation in packet-switched SDMA systems
	Kuehner *et al.* [238]	Considered a BS that communicates with smart-antenna-aided mobiles operating in multi-beam, packet-switched and SDMA modes
2002	Jeon *et al.* [239]	Contrived a smart-antenna-assisted system using adaptive beamforming for broadband wireless communications
	Bellofiore *et al.* [240, 241]	Emphasized the interaction and integration of several critical components of a mobile communication network using smart-antenna techniques
	Fang [242]	Carried out a realistic performance analysis of resource allocation schemes for SDMA systems and obtained analytical results for blocking probability.
	Arredondo *et al.* [243]	Employed a novel synthesis and prediction filter at the smart-antenna-aided BS for predicting vector channels in time-division duplex systems
	Walke and Oechtering [244]	Conducted investigations on the Cumulative Distribution Function (CDF) of the uplink carrier-to-interference ratio in a cellular radio network
	Zwick *et al.* [245]	Proposed a stochastic channel model for indoor propagations in future communication systems equipped with multiple antennas
	Zekavat *et al.* [246]	Combined smart-antenna arrays and MC-CDMA systems
	Pan and Djurić [247]	Suggested sectorized multi-beam cellular mobile communications combined with dynamic channel assignment to beams
	Cavalcante *et al.* [248]	Exploited a blind adaptive optimization criterion for SDMA detection
	Yin and Liu [249]	Developed a Medium Access Control (MAC) protocol for multimedia SDMA/TDMA packet networks
	Thoen *et al.* [199]	Showed that the performance of OFDM/SDMA processors can be significantly enhanced by adapting the constellation size applied on the individual subcarriers to the channel conditions
	Rim [250]	Examined the performance of a high-throughput downlink MIMO-SDMA technique
2003	Thoen *et al.* [200]	Utilized a Constrained Least-Squares (CLS) receiver in multi-user SDMA systems
	Alastalo and Kahola [201]	Reported link-level results of an adaptive-antenna-array-assisted system compatible with IEEE 802.11a WLANs
	Bradaric *et al.* [251]	Characterized a blind nonlinear method for identifying MIMO FIR CDMA and SDMA systems
	Alias *et al.* [202]	Constructed a Minimum Bit Error Rate (MBER) Multi-User Detector (MUD) for SDMA-OFDM systems
	Hanzo *et al.* [3]	Elaborated on channel estimation and multi-user detection techniques designed for SDMA-OFDM systems
2004	Spencer *et al.* [252]	Delivered two constrained solutions referred to as the block diagonalization and the successive optimization schemes contrived for downlink SDMA systems
	Li *et al.* [253]	Explored a low-complexity ML-based detection scheme using a so-called 'sensitive-bits' algorithm
	Choi and Murch [254]	Formulated a pre-Bell Labs Layered Space–Time (BLAST) decision feedback equalization technique for downlink MIMO channels
2005	Ajib and Haccoun [255]	Overviewed the scheduling algorithms proposed for 4G multi-user wireless networks based on MIMO technology
	Dai [203]	Performed an analysis of CFO estimation in SDMA-OFDM systems
	Nasr *et al.* [256]	Researched the estimation of the local average signal level in an indoor environment based on a 'wall-imperfection' model

The main advantage of the above OFDM concept is that because the symbol period has been increased, the channel delay spread is a significantly shorter fraction of a symbol period than in the serial system, potentially rendering the system less sensitive to ISI than the conventional serial system. In other words, in the low-rate subchannels the signal is no longer subject to frequency-selective fading, hence no channel equalization is necessary.

A disadvantage of the OFDM approach shown in Figure 1.3 is the increased complexity over the conventional system caused by employing M modulators and filters at the transmitter and M demodulators and filters at the receiver. It can be shown that this complexity can be reduced by the use of the Discrete Fourier Transform (DFT), typically implemented as a Fast Fourier Transform (FFT) [3].

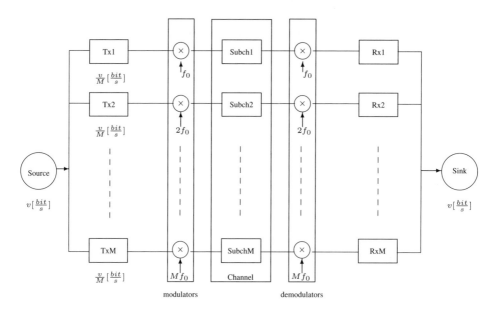

Figure 1.3: Simplified block diagram of the orthogonal parallel modem.

The subchannel modems can use almost any modulation scheme, and 4- or 16-level QAM is an attractive choice in many situations.

A schematic of the FFT-based QAM/FDM modem schematic is portrayed in Figure 1.4. The bits provided by the source are serial/parallel converted in order to form the n-level Gray-coded symbols, M of which are collected in TX buffer 1, while the contents of TX buffer 2 are being transformed by the IFFT in order to form the time-domain modulated signal. The digital-to-analogue (D/A) converted, low-pass filtered modulated signal is then transmitted via the channel and its received samples are collected in RX buffer 1, while the contents of RX buffer 2 are being transformed to derive the demodulated signal. The twin buffers are alternately filled with data to allow for the finite FFT demodulation time. Before the data is Gray coded and passed to the data sink, they can be equalized by a low-complexity method, if there is some dispersion within the narrow subbands. For a deeper tutorial exposure the interested reader is referred to [3].

1.3   Channel Estimation for Multi-carrier Systems

The ever-increasing demand for high data rates in wireless networks requires the efficient utilization of the limited bandwidth available, while supporting a high grade of mobility in diverse propagation environments. OFDM and MC-CDMA techniques [265] are capable of satisfying these requirements. This is a benefit of their ability to cope with highly time-variant wireless channel characteristics. However, as pointed out in [266], the capacity and the achievable integrity of communication systems are highly dependent on the system's knowledge concerning the channel conditions encountered. Thus, the provision of an accurate and robust channel estimation strategy is a crucial factor in achieving a high performance.

Well-documented approaches to the problem of channel estimation are constituted by *pilot-assisted*, *decision-directed* and *blind* channel estimation methods [265, 267], which are briefly summarized in Table 1.8.

Table 1.8: Major contributions addressing channel estimation in multi-carrier systems.

Year	Author(s)	Contribution
1997	Höher *et al.* [257, 258]	Cascaded 1D-FIR Wiener-filter-based channel interpolation
1998	Edfors *et al.* [259]	Detailed analysis of SVD-aided CIR-related domain noise reduction for DDCE
2000	Li [260]	DDCE using DFT-based 2D interpolation and robust prediction
	Li [261]	2D pilot pattern-aided channel estimation using 2D robust frequency-domain Wiener filtering
2001	Yang *et al.* [262]	Detailed discussion of parametric, ESPRIT-assisted channel estimation
2003	Münster and Hanzo [263]	RLS-adaptive PIC-assisted DDCE for OFDM
2004	Otnes and Tüchler [264]	Iterative channel estimation for turbo-equalization

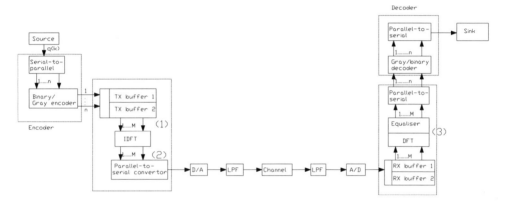

Figure 1.4: FFT-based OFDM modem schematic © Hanzo *et al.* (2003) [3].

The family of *pilot-assisted* channel estimation methods was investigated for example by Li [261], Morelli and Mengali [268], Yang *et al.* [262] as well as Chang and Su [269], where the channel parameters are typically estimated by exploiting the channel-sounding signal. For example, in OFDM and MC-CDMA often a set of frequency-domain pilots are transmitted for estimating the Frequency-Domain Channel Transfer Function (FD-CTF), which are known at the receiver [265]. The main drawback of this method is that the pilot symbols do not carry any useful information and thus they reduce the system's effective throughput.

By contrast, in Decision-Directed Channel Estimation (DDCE) methods both the pilot symbols and all the information symbols are utilized for channel estimation [265]. The simple philosophy of this method is that in the absence of transmission errors we can benefit from the availability of 100% pilot information by using the detected subcarrier symbols as an a posteriori reference signal. The employment of this method allows us to reduce the number of pilot symbols required. This technique is particularly efficient under benign channel conditions, where the probability of a decision error is low, but, naturally, this approach is also prone to error propagation effects. The family of DDCE techniques was investigated for example by van de Beek *et al.* [270], Mignone and Morello [271], Edfors *et al.* [259], Li *et al.* [260], Li and Sollenberg [272] as well as Münster and Hanzo [263, 267, 273, 274].

The class of iterative DDCE schemes, where the channel estimation is carried out through a series of iterations utilizing increasingly refined soft-decision-based feedback, was explored by

Sandell *et al.* [275], Valenti [276], Yeap *et al.* [277], Song *et al.* [278, 279], as well as by Otnes and Tüchler [264, 280].

The closely related class of joint receivers, where the channel parameters and the transmitted information-carrying symbols are estimated jointly, was explored for example by Seshadri [281], developed further by Knickenberg *et al.* [282], recently revisited by Cozzo and Hughes [283] as well as Cui and Tellambura [284, 285].

Finally, the class of *blind* estimation methods eliminates all redundant pilot symbols. Most of these methods rely on the employment of decision feedback and on the exploitation of the redundancy often found in the structure of the modulated signal, as exemplified by the techniques described for example by Antón-Haro *et al.* [286], Boss *et al.* [287], Endres *et al.* [288], Giannakis and Halford [289], Zhou and Giannakis [290] as well as by Necker and Stüber [291].

An additional major subject, closely related to channel estimation, namely the prediction of fast fading channels, was extensively studied by Haykin [292]. A so-called robust predictor was proposed by Li [260] and revised by Münster and Hanzo [274]. An adaptive RLS channel predictor was proposed by Schafhuber and Matz [293].

Subsequently, in this treatise we propose a DDCE scheme, which is suitable for employment in both OFDM and MC-CDMA systems. We analyse the achievable performance of the estimation scheme considered in conjunction with a realistic dispersive Rayleigh fading channel model having a Fractionally Spaced (FS) rather than Symbol-Spaced (SS) Power Delay Profile (PDP).

A basic component of the DDCE schemes proposed in the literature is an a posteriori Least Squares (LS) temporal estimator of the OFDM-subcarrier-related Frequency-Domain Channel Transfer Function (FD-CTF) coefficients [260, 265]. The accuracy of the resultant temporal FD-CTF estimates is typically enhanced using 1D or 2D interpolation exploiting both the time- and the frequency-domain correlation between the desired FD-CTF coefficients. The LS-based temporal FD-CTF estimator was shown to be suitable for QPSK-modulated OFDM systems [260, 265], where the energy of the transmitted subcarrier-related information symbols is constant. However, as will be pointed out in Section 7.3.1 of this treatise, the LS method cannot be readily employed in MC-CDMA systems, where – in contrast to OFDM systems – the energy of the transmitted subcarrier-related information symbols fluctuates as a function of both the modulated sequence and that of the choice of the potentially non-constant-modulus modulation scheme itself. Thus we propose an Minimum Mean Square Error (MMSE) estimation-based DDCE method, which is an appropriate solution for employment in both OFDM and MC-CDMA systems.

The system model and the channel model considered are described in Section 1.7 of this treatise. The difficulty of employing the LS approach to the problem of estimating the OFDM-subcarrier-related FD-CTF coefficients is described in Section 7.3.1. The alternative MMSE FD-CTF estimator circumventing the problem outlined in Section 7.3.1 is analysed in Section 7.3.2. Our discourse evolves further by proposing an MMSE CIR estimator exploiting the frequency-domain correlation of the FD-CTF coefficients in Section 7.4.1 and a reduced-complexity version of the CTF MMSE estimator considered is proposed in Section 7.4.2. The computational complexity of both methods is compared in Section 7.4.3.

In Section 7.4 we continue our discourse with the derivation of both the sample-spaced as well as the fractionally spaced Channel Impulse Response (CIR) estimator. In Section 7.4.5 we then perform a comparison between the two methods considered and demonstrate the advantages of the latter, i.e. fractionally spaced, scheme. Subsequently, in Section 7.5 we develop a method of parametric tracking of the fractionally spaced CIR taps, which facilitates low-complexity channel estimation in realistic channel conditions characterized by a time-variant, fractionally spaced, power delay profile. More specifically, we employ the Projection Approximation Subspace Tracking (PAST) method for the sake of recursive tracking of the covariance matrix of the Channel Transfer Function (CTF) and subsequent tracking of the corresponding CIR taps. We demonstrate that the PAST-aided decision-directed channel estimation scheme proposed exhibits good performance over the entire range of practical conditions.

In Section 7.6 we discuss two major CIR tap prediction strategies. Specifically, in Section 7.6.2 the so-called *robust* implementation of the stationary MMSE CIR predictor is considered. The *robust* CIR predictor [260] assumes a constant-valued, limited-support channel scattering function [265] during the design of the CIR tap prediction filter, and hence relies on the assumption of encountering the worst possible channel conditions. On the other hand, in Section 7.6.4 we discuss the adaptive Recursive Least-Squares (RLS) method of CIR prediction [293]. As opposed to the robust CIR predictor of [260], the RLS CIR predictor does not require any explicit information concerning the channel conditions encountered. Consequently, in Section 7.6.5 we characterize and compare the achievable performance of both methods considered and draw conclusions concerning their relative merits. Specifically, we demonstrate that the RLS prediction technique outperforms its robust counterpart over the entire range of the relevant channel conditions.

In Section 7.7 we characterize the achievable performance of the resultant PAST-aided DDCE scheme. We report an estimation efficiency of $\kappa = -18\,\text{dB}$ exhibited by a system employing 10% of pilots and communicating over a dispersive Rayleigh fading channel having a Doppler frequency of $f_D = 0.003$. Furthermore, we report a BER performance, which is only $3\,\text{dB}$ from the corresponding BER performance exhibited by a similar system assuming perfect channel knowledge.

1.4   Channel Estimation for MIMO-OFDM

In spite of immense interest from both the academic and industrial communities, a practical MIMO transceiver architecture, capable of approaching channel capacity boundaries in realistic channel conditions, remains largely an open problem. In particular, a robust and accurate channel estimation in MIMO systems constitutes a major issue, preventing us from achieving the high capacities predicted by the relevant theoretical analysis.

Some of the major contributions addressing the problem of channel estimation in MIMO systems are summarized in Table 1.9. More specifically, a combined OFDM-SDMA approach was discussed by Vandenameele *et al.* [299]. A pilot-based approach to the problem of MIMO channel estimation has been explored by Jungnickel *et al.* in [300], by Bolcskei *et al.* [301], as well as by Zhu *et al.* [302]. On the other hand, decision-directed iterative channel estimation for MIMO systems was addressed by Li *et al.* [294, 303, 304] as well as Deng *et al.* [296]. Furthermore, s parallel interference cancellation-assisted decision-directed channel estimation scheme for MIMO-OFDM systems was proposed by Münster and Hanzo [297, 305]. Joint decoding and channel estimation for MIMO channels was considered by Grant [306] and further investigated by Cozzo and Hughes [283]. Iterative channel estimation for space–time block-coded systems was addressed by Mai *et al.* [307], while joint iterative DDCE for turbo-coded MIMO-OFDM systems was investigated by Qiao [308]. Blind channel estimation in MIMO-OFDM systems with multi-user interference was explored by Yatawatta and Petropulu [298].

Other closely related issues, namely the iterative tracking of the channel-related parameters using soft decision feedback, was studied by Sandell *et al.* [275], while iterative channel estimation in the context of turbo-equalization was considered by Song *et al.* [279], Mai *et al.* [309], as well as Otnes and Tüchler [264].

Finally, an important overview publication encompassing most major aspects of broadband MIMO-OFDM wireless communications including channel estimation and signal detection, as well as time and frequency synchronization, was contributed by Stüber *et al.* [295].

Against this background, in this treatise we propose a DDCE scheme, which is suitable for employment in a wide range of multi-antenna, multi-carrier systems as well as over the entire range of practical channel conditions. In particular, we consider mobile wireless multi-path channels, which exhibit fast Rayleigh frequency-selective fading and are typically characterized by a time-variant PDP.

We consider a generic MIMO-OFDM system employing K orthogonal frequency-domain subcarriers and having m_t and n_r transmit and receive antennas, respectively. Consequently, our MIMO channel

Table 1.9: Major contributions addressing the problem of channel estimation in MIMO systems.

Year	Author(s)	Contribution
2002	Li *et al.* [294]	MIMO-OFDM for wireless communications: signal detection with enhanced channel estimation
2004	Stüber *et al.* [295]	An important overview encompassing most of the major aspects of the broadband MIMO-OFDM wireless communications, including channel estimation, signal detection as well as time and frequency synchronization
2003	Deng *et al.* [296]	Decision-directed iterative channel estimation for MIMO systems
2003	Cozzo and Hughes [283]	Joint channel estimation and data detection in space–time communications
2005	Münster and Hanzo [297]	Parallel interference cancellation-assisted decision-directed channel estimation for OFDM systems using multiple-transmit antennas
2006	Yatawatta and Petropulu [298]	Blind channel estimation in MIMO-OFDM systems with multi-user interference

estimation scheme comprises an array of K per-subcarrier MIMO-CTF estimators, followed by an $(n_r \times m_t)$-dimensional array of parametric CIR estimators and a corresponding array of $(n_r \times m_t \times L)$ CIR tap predictors, where L is the number of tracked CIR taps per link for the MIMO channel.

In Section 7.8.1 we explore a family of recursive MIMO-CTF tracking methods, which in conjunction with the aforementioned PAST-aided CIR-tracking method of Section 7.5 as well as the RLS CIR tap prediction method of Section 7.6.4, facilitate an effective channel estimation scheme in the context of an MIMO-OFDM system. More specifically, in Section 7.8.1 we consider both hard- and soft-feedback-assisted least mean squares (LMS) and recursive least-squares (RLS) tracking algorithms as well as the modified RLS algorithm, which is capable of improved utilization of the soft information associated with the decision-based estimates.

Finally, in Section 7.8.1.5 we document the achievable performance of the resultant MIMO-DDCE scheme employing recursive CTF tracking followed by the parametric CIR tap tracking and CIR tap prediction. We demonstrate that the MIMO-DDCE scheme proposed exhibits good performance over the entire range of practical conditions.

Both the Bit Error Rate (BER) as well as the corresponding MSE performance of the channel estimation scheme considered are characterized in the context of a turbo-coded MIMO-OFDM system. We demonstrate that the MIMO-DDCE scheme proposed remains effective in channel conditions associated with high terminal speeds of up to 130 km/h, which corresponds to the OFDM-symbol normalized Doppler frequency of 0.006. Additionally, we report a virtually error-free performance of a rate 1/2 turbo-coded 8×8-QPSK-OFDM system, exhibiting a total bit rate of 8 bps/Hz and having a pilot overhead of only 10%, at an SNR of 10 dB and normalized Doppler frequency of 0.003, which corresponds to the mobile terminal speed of roughly 65 km/h.[1]

1.5   Signal Detection in MIMO-OFDM Systems

The demand for both high data rates and improved transmission integrity requires efficient utilization of the limited system resources, while supporting a high grade of mobility in diverse propagation environments. Consequently, the employment of an appropriate modulation format, as well as efficient exploitation of the available bandwidth, constitute crucial factors in achieving a high performance.

---

[1] Additional system parameters are characterized in Table 1.11.

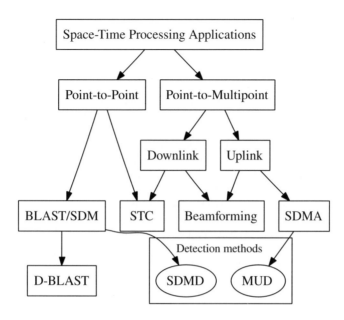

Figure 1.5: Classification of space–time processing techniques.

The OFDM modulation scheme employed in conjunction with an MIMO architecture [265], where multiple antennas are employed at both the transmitter and the receiver of the communication system, constitutes an attractive solution in terms of satisfying these requirements. Firstly, the OFDM modulation technique is capable of coping with the highly frequency-selective, time-variant channel characteristics associated with mobile wireless communication channels, while possessing a high grade of structural flexibility for exploiting the beneficial properties of MIMO architectures.

It is highly beneficial that OFDM and MIMO may be conveniently combined, since the information theoretical analysis predicts [310] that substantial capacity gains are achievable in communication systems employing MIMO architectures. Specifically, if the fading processes corresponding to different transmit–receive antenna pairs may be assumed to be independently Rayleigh distributed,[2] the attainable capacity was shown to increase linearly with the smaller of the numbers of the transmit and receive antennas [310]. Additionally, the employment of MIMO architectures allows for efficient exploitation of the spatial diversity available in wireless MIMO environments, thus improving the system's BER, as well as further increasing the system's capacity.

The family of space–time signal processing methods, which allow for the efficient implementation of communication systems employing MIMO architectures, is commonly referred to as *smart antennas*. In recent years, the concept of smart antennas has attracted intensive research interest in both the academic and industrial communities. As a result, a multiplicity of smart-antenna-related methods has been proposed. These include methods implemented at the transmitter, the receiver or both.

The classification of smart-antenna techniques is illustrated in Figure 1.5. It should be noted, however, that the classification presented here is somewhat informal and its sole purpose is to position appropriately the content of this treatise in the context of the extensive material available on the subject.

Two distinctive system scenarios employing smart antennas can be identified. The first is the so-called Space-Division Multiplexing (SDM) scenario [311], where two *peer* terminals each employ

---

[2]This assumption is typically regarded as valid if the appropriate antenna spacing is larger than of $\lambda/2$, where λ is the corresponding wavelength.

multiple antennas and communicate with each other over a MIMO channel, and the multiple antennas are primarily used for achieving a multiplexing gain, i.e. a higher throughput [312]. The second scenario corresponds to the Space-Division Multiple Access (SDMA) configuration [265], where a single *base station*, employing multiple antennas, communicates simultaneously using a single carrier frequency with multiple *user* terminals, each employing one or several antennas.

The various *point-to-multipoint* smart-antenna applications can be further subdivided into *uplink*- and *downlink*-related applications. The *uplink*-related methods constitute a set of techniques which can be employed in the *base station* in order to detect the signals simultaneously transmitted by multiple *user* terminals. More specifically, provided that the CIR of all users is accurately estimated, it may be used as their unique, user-specific spatial signature for differentiating them, despite communicating within the same frequency band [265]. Hence, the corresponding space–time signal processing problem is commonly referred to as Multi-User Detection (MUD) [265], while the multi-antenna, multi-user systems employing *uplink* space–time MUDs are commonly referred to as SDMA systems [265]. In contrast to the SDM-type systems designed to achieve the highest possible multiplexing gain, the design objective of the SDMA techniques is maximization of the number of users supported. By contrast, the class of beamformers [313] creates angularly selective beams for both the uplink and downlink in the direction of the desired user, while forming nulls towards the interfering users. Finally, the family of Space–Time Codes (STCs) [314] was optimized for achieving the highest possible transmit diversity gain, rather than for multiplexing gain or for increasing the number of users supported. At the time of writing, new research is aiming at achieving both the maximum attainable diversity and multiplexing gain with the aid of eigenvalue decomposition [315].

As stated above, two benefits of employing smart antennas are the system's improved integrity and the increased aggregate throughput. Hence an adequate performance criterion of the particular smart-antenna implementation is a combination of the system's attainable aggregate data throughput and the corresponding data integrity, which can be quantified in terms of the average BER. Consequently, in the context of point-to-multipoint-related smart-antenna applications, the achievable capacity associated with the particular space–time processing method considered may be assessed as a product of the simultaneously supported number of individual users and the attainable data rate associated with each supported user. The measure of data integrity may be the average BER of all the users supported. Thus, the typical objective of the multi-user-related smart-antenna implementations, such as that of an SDMA scheme, is that of increasing the number of simultaneously supported users, while sustaining the highest possible integrity of all the data communicated.

In this treatise, however, we would like to focus our attention on the family of space–time processing methods associated with the *point-to-point* system scenario. The main objective of point-to-point space–time processing is to increase the overall throughput of the system considered, as opposed to increasing the number of individual users simultaneously supported by the system, which was the case in the multi-user SDMA scenario described above. As illustrated in Figure 1.5, the family of time–space processing methods associated with the point-to-point-related smart-antenna applications entail two different approaches, namely that of STCs [314] as well as various layered space–time architectures, best known from the BLAST scheme [312].

The STC methods may be classified in two major categories, namely the Space–Time Block Code (STBC) and the Space–Time Trellis Code (STTC) categories. A simple method of STBC was first presented by Alamouti in [316]. Various STBC techniques were then extensively studied in a series of major publications by Tarokh *et al.* in [317–323] as well as by Ariyavistakul *et al.* in [324, 325]. On the other hand, the original variant of BLAST, known as the Diagonal BLAST (D-BLAST) scheme, was first introduced by Foschini in [312]. A more generic version of the BLAST architecture, the so-called Vertical BLAST (V-BLAST) arrangement, was proposed by Golden *et al.* in [326]. Furthermore, the comparative study of the D-BLAST, as well as the V-BLAST systems employing various detection techniques such as LS- and MMSE-aided Parallel Interference Cancellation (PIC), and LS- and MMSE-aided Successive Interference Cancellation (SIC), was carried out by Sweatman

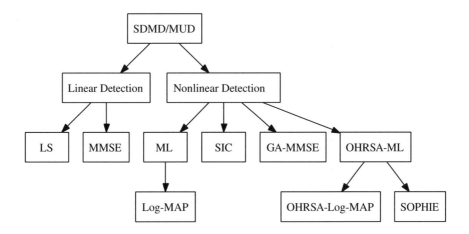

Figure 1.6: Classification of SDM detection methods.

et al. in [327]. Typically, however, the term BLAST refers to the point-to-point single-carrier MIMO architecture employing the SIC detection method, as originally proposed in [312].

For the sake of accuracy, in this work we employ the alternative terminology of SDM in order to refer to a generic MIMO architecture. The corresponding detection methods are referred to as SDM Detection (SDMD) techniques, as opposed to the MUD techniques employed in the context of SDMA systems [265]. Naturally, however, the SDMD and MUD schemes share the same signal detection methods, regardless of whether the signal arrived from multiple antennas of the same or different users. The classification of the most popular SDMD/MUD schemes is depicted in Figure 1.6. The methods considered include the linear LS and MMSE techniques, as well as nonlinear techniques, such as ML, SIC and Genetic-Algorithm-aided MMSE (GA-MMSE) [328, 329], and the novel Optimized Hierarchy Reduced Search Algorithm (OHRSA) methods proposed in this treatise.

In the course of this treatise both the MIMO channel model and the SDM-OFDM system model are described in Section 1.8. The various SDM detection methods considered are outlined in Chapter 15. Specifically, in Section 15.2.1 we demonstrate that the linear increase in capacity, predicted by information theoretical analysis [266], may indeed be achieved by employing a relatively low-complexity linear SDM detection method, such as the MMSE SDM detection technique [330]. Secondly, in Section 15.3.1 we show that a substantially better performance can be achieved by employing a nonlinear ML SDM detector [311, 331, 332], which constitutes the optimal detection method from the point of view of probabilistic sequence estimation. To elaborate a little further, the ML SDM detector is capable of attaining transmit diversity in *fully loaded* systems, where the number of transmit and receive antennas is equal. Moreover, as opposed to the linear detection schemes considered, the ML SDM detector is capable of operating in the *rank-deficient* system configuration, when the number of transmit antennas exceeds that of the receive antennas. Unfortunately, however, the excessive computational complexity associated with the exhaustive search employed by the ML detection method renders it inapplicable to practical implementation in systems having a large number of transmit antennas. Subsequently, in Sections 15.3.2 and 15.3.3 we explore a range of advanced nonlinear SDM detection methods, namely the SIC and GA-aided MMSE detection, respectively, where the latter may potentially constitute an attractive compromise between the low complexity of the linear SDM detection and the high performance of the ML SDM detection schemes. Indeed, we will demonstrate in Section 15.3.3 that the SDM detection method based on the SIC as well as on the GA-MMSE detector [329] are both capable of satisfying these requirements.

Table 1.10: Major contributions addressing sphere-decoder-aided Space–Time processing.

Year	Author(s)	Contribution
1985	Fincke *et al.* [333]	Sphere decoder technique introduced
2000	Damen *et al.* [334]	Sphere decoder was first proposed for employment in the context of space–time processing, where it was utilized for computing the ML estimates of the modulated symbols transmitted simultaneously from multiple transmit antennas.
2003	Hochwald and Brink [335]	The *complex* version of the sphere decoder
2003	Damen *et al.* [336]	Further results on the sphere decoder
2004	Pham *et al.* [337]	Improved version of the complex sphere decoder
2005	Tellambura *et al.* [338]	Multi-stage sphere decoding introduced

In Section 15.4 our discourse evolves further by proposing an enhancement of the SDMD schemes considered by employing both Space-Frequency Interleaving (SFI) and Space-Frequency Walsh–Hadamard Transform (SFWHT) spreading. The performance benefits of employing SFI and SFWHT are quantified in Section 15.4. Finally, our conclusions are summarized in Section 15.6.

PIC-assisted decision-directed channel estimation was also designed or OFDM systems using multiple transmit antennas by Münster and Hanzo [297].

Recently, a family of potent Reduced Search Algorithm (RSA) aided Space–Time processing methods has been explored (see Table 1.10). These new methods utilize the Sphere Decoder (SD) technique introduced by Fincke *et al.* [333]. The SD was first proposed for employment in the context of space–time processing by Damen *et al.* in [334], where it was utilized for computing the ML estimates of the modulated symbols transmitted simultaneously from multiple transmit antennas. The *complex* version of the SD was proposed by Hochwald and Brink in [335]. The subject was further investigated by Damen *et al.* in [336]. Subsequently, an improved version of the Complex Sphere Decoder (CSD) was advocated by Pham *et al.* in [337]. Furthermore, CSD-aided detection was considered by Cui and Tellambura in a joint channel estimation and data detection scheme explored in [284], while a revised version of the CSD method, namely the so-called Multistage Sphere Decoding (MSD), was introduced in [338]. The generalized version of the SD, which is suitable for employment in rank-deficient MIMO systems supporting more transmitters than the number of receive antennas, was introduced by Damen *et al.* in [339] and further refined by Cui and Tellambura in [340]. The so-called *fast* generalized sphere decoding was introduced by Yang *et al.* [341]. Yet another variant of SD algorithms with improved radius search was introduced by Zhao and Giannakis [342]. The subject of approaching MIMO channel capacity using soft detection on hard sphere decoding was explored by Wang and Giannakis [343]. Iterative detection and decoding in MIMO systems using sphere decoding was considered by Vikalo *et al.* [344].

Consequently, a set of novel OHRSA-aided SDM detection methods is outlined in Section 16.1. Specifically, in Section 16.1.1 we derive the OHRSA-aided ML SDM detector, which benefits from the optimal performance of the ML SDM detector [265], while exhibiting a relatively low computational complexity which is only slightly higher than that required by the low-complexity MMSE SDM detector [265]. To elaborate a little further, in Section 16.1.2 we derive a bit-wise OHRSA-aided ML SDM detector, which allows us to apply the OHRSA method of Section 16.1 in high-throughput systems which employ multi-level modulation schemes, such as M-QAM [265].

In Section 16.1.3 our discourse evolves further by deducing the OHRSA-aided Max-Log-MAP SDM detector, which allows for efficient evaluation of the soft-bit information and therefore results in highly efficient turbo decoding. Unfortunately however, in comparison with the OHRSA-aided ML SDM detector of Section 16.1.2, the OHRSA-aided Max-Log-MAP SDM detector of Section 16.1.3

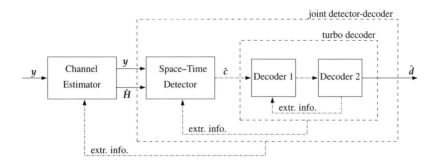

Figure 1.7: Schematic of a joint iterative receiver comprising channel estimator, SDM detector, as well as turbo decoder employing two RCS (Recursive Systematic Convolutional) serially concatenated component codes.

exhibits a substantially higher complexity. Consequently, in Section 16.1.5 we derive an approximate Max-Log-MAP method, which we refer to as Soft-output OPtimized HIErarchy (SOPHIE). The SOPHIE SDM detector combines the advantages of both the OHRSA-aided ML and OHRSA-aided Max-Log-MAP SDM detectors of Sections 16.1.2 and 16.1.3, respectively. Specifically, it exhibits a similar performance to that of the optimal Max-Log-MAP detector, while imposing a modest complexity which is only slightly higher than that required by the low-complexity MMSE SDM detector [265]. The computational complexity as well as the achievable performance of the SOPHIE SDM detector of Section 16.1.5 are analysed and quantified in Sections 16.1.5.1 and 16.1.5.2, respectively.

We will report the achievement of a BER of 10^{-4} at SNRs of $\gamma = 4.2, 9.2$ and 14.5 in high-throughput 8×8 rate $\frac{1}{2}$ turbo-coded $M = 4, 16$ and 64-QAM systems communicating over dispersive Rayleigh fading channels. Additionally, we report the achievement of a BER of 10^{-4} at SNRs of $\gamma = 9.5, 16.3$ and 22.8 in high-throughput rank-deficient 4×4, 6×4 and 8×4 rate $\frac{1}{2}$ turbo-coded 16-QAM systems, respectively.

1.6 Iterative Signal Processing for SDM-OFDM

In spite of immense interest from both the academic and industrial communities, a practical MIMO transceiver architecture, capable of approaching channel capacity boundaries in realistic channel conditions, remains largely an open problem. An important overview publication encompassing most major aspects of broadband MIMO-OFDM wireless communications including channel estimation and signal detection, as well as time and frequency synchronization, was contributed by Stüber *et al.* [295]. Other important publications considering MIMO systems in realistic conditions include those by Münster and Hanzo [297], Li *et al.* [294], Mai *et al.* [307] as well as Qiao *et al.* [308]. Nevertheless, substantial contributions addressing all the major issues inherent to MIMO transceivers, namely error correction, space–time detection and channel estimation in realistic channel conditions, remain scarce.

Against this background, in Section 17.1 we derive an iterative, so-called *turbo* Multi-Antenna, Multi-Carrier (MAMC) receiver architecture. Our turbo-receiver is illustrated in Figure 1.7. Following the philosophy of turbo processing [314], our turbo SDM-OFDM receiver comprises a succession of detection modules, which iteratively exchange soft-bit-related information and thus facilitate a substantial improvement in the overall system performance.

More specifically, our turbo SDM-OFDM receiver comprises three major components, namely the soft-feedback decision-directed channel estimator, discussed in detail in Section 7.8, followed by the soft-input, soft-output OHRSA Log-MAP SDM detector derived in Section 16.1.3 as well as a soft-input, soft-output serially concatenated turbo code [345]. Consequently, in this chapter we would like

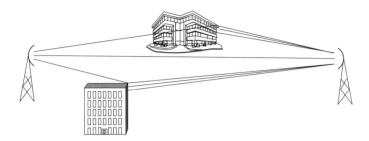

Figure 1.8: Illustration of a wireless multi-path communication link. Note that the non-line-of-sight paths randomly fade as a result of the diffraction induced by scattering surfaces.

to analyse the achievable performance of each individual constituent of our turbo receiver, as well as the achievable performance of the entire iterative system. Our aim is to identify the optimum system configuration, while considering various design trade-offs, such as achievable error-rate performance, achievable data rate and associated computational complexity.

In Section 17.4.2.4 we demonstrate that our turbo SDM-OFDM system employing the MIMO-DDCE scheme of Section 7.8 as well as the OHRSA Log-MAP SDM detector of Section 16.1.3 remains effective in channel conditions associated with high terminal speeds of up to 130 km/h, which corresponds to the OFDM symbol normalized Doppler frequency of 0.006. Additionally, we report a virtually error-free performance for a rate $1/2$ turbo-coded 8×8 QPSK-OFDM system, exhibiting an effective throughput of $8\,\mathrm{MHz} \cdot 8\,\mathrm{bps/Hz} = 64\,\mathrm{Mbps}$ and having a pilot overhead of only 10% at an SNR of 7.5 dB and a normalized Doppler frequency of 0.003, which corresponds to a mobile terminal speed of about 65 km/h.

1.7   System Model

1.7.1   Channel Statistics

A SISO wireless communication link as seen in Figures 1.8 and 1.9 is constituted by a multiplicity of statistically independent components, termed *paths*. Thus, such a channel is referred to as a multi-path channel. A multi-path channel is typically characterized by its Power Delay Profile (PDP), which is a set of parameters constituted by the paths' average powers σ_l^2 and the corresponding relative delays τ_l. Some examples of commonly used PDPs are illustrated in Figure 1.10. The physical interpretation of each individual path is a single distortionless ray between the transmitter and the receiver antennas. While the term PDP corresponds to the average power values associated with the different multi-path channel components, the term CIR refers to the instantaneous state of the dispersive channel encountered and corresponds to the vector of the instantaneous amplitudes $\alpha_l[n]$ associated with different multi-path components. Thus, the statistical distribution of the CIR is determined by the channel's PDP. In the case of independently Rayleigh fading multiple paths we have $\alpha_l[n] \in \mathcal{CN}(0, \sigma_l^2)$, $l = 1, 2, \ldots, L$, where $\mathcal{CN}(0, \sigma^2)$ is a complex Gaussian distribution having mean 0 and variance σ^2.

The individual scattered and delayed signal components usually arise as a result of refraction or diffraction from scattering surfaces, as illustrated in Figure 1.8, and are termed Non-Line-Of-Sight (NLOS) paths. In most recently proposed wireless mobile channel models, each such CIR component α_l associated with an individual channel path is modelled by a Wide-Sense Stationary (WSS) narrowband complex Gaussian process [348] having correlation properties characterized by the cross-correlation function

$$r_\alpha[m, j] = \mathrm{E}\{\alpha_i[n]\alpha_j^*[n - m]\} = r_{t;i}[m]\delta[i - j], \qquad (1.1)$$

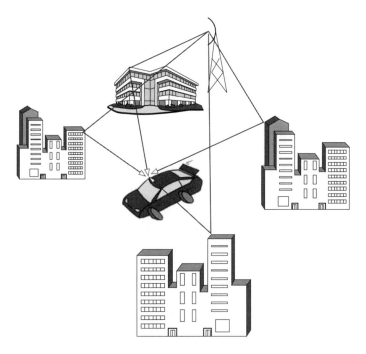

Figure 1.9: Illustration of a wireless multi-path communication link. Note that the non-line-of-sight paths randomly fade as a result of the diffraction induced by scattering surfaces.

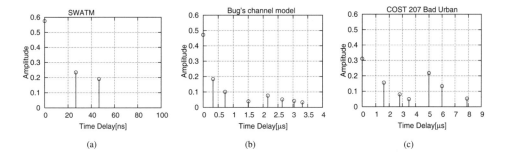

Figure 1.10: PDPs corresponding to three different channel models: namely (a) the Short Wireless Asynchronous Transfer Mode (SWATM) channel model of [265]; (b) Bug's channel model [346]; and (c) the COST-207 Bad Urban (BU) channel model defined for UMTS-type systems, as characterized in [347].

where n is a discrete OFDM-block-related time-domain index and $\delta[\cdot]$ is the Kronecker delta function. The above equation suggests that the different CIR components are assumed to be mutually uncorrelated and each exhibits time-domain autocorrelation properties defined by the time-domain correlation function $r_{t;i}[m]$. The Fourier transform pair of the correlation function $r_t[n]$ associated with each CIR tap corresponds to a band-limited Power Spectral Density (PSD) $p_t(f)$, such that $p_t(f) = 0$ if $|f| > f_D$, where f_D is termed the *maximum Doppler frequency*. The time period $1/f_D$ is the so-called *coherence time* of the channel [348] and usually $1/f_D \gg T$, where T is the duration of the OFDM block.

A particularly popular model of the time-domain correlation function $r_t[n]$ was proposed by Jakes in [349] and is described by

$$r_t[n] = r_J[n] = J_0(nw_d), \tag{1.2}$$

where $J_0(x)$ is a zero-order Bessel function of the first kind and $w_d = 2\pi T f_D$ is the normalized Doppler frequency. The corresponding U-shaped PSD function, termed the Jakes spectrum, is given by [349]

$$p_J(w) = \begin{cases} \dfrac{2}{w_d} \dfrac{1}{\sqrt{1 - (w/w_d)^2}}, & \text{if } |w| < w_d \\ 0, & \text{otherwise.} \end{cases}$$

Generally speaking, the Doppler frequencies f_D can assume different values for different signal paths. However, as was advocated in [260], for the sake of exploiting the time-domain correlation in the context of channel parameter estimation and prediction, it is sufficient to make a worst-case assumption about the nature of time-domain correlation of the channel parameters encountered. The associated worst-case channel time-domain correlation properties can be characterized by an ideally band-limited Doppler PSD function given by [260, 265]

$$p_t(f) = p_{B,unif}(f) = \begin{cases} \dfrac{1}{2f_D}, & \text{if } |f| < f_D \\ 0, & \text{otherwise,} \end{cases} \tag{1.3}$$

where f_D is the assumed value of the maximum Doppler frequency over all channel paths. The corresponding time-domain correlation function can be described as

$$r_t[m] = r_B[m] = \frac{\sin 2\pi f_D m}{2\pi f_D m}. \tag{1.4}$$

We adopt the complex baseband representation of the continuous-time CIR, as given by [348]

$$h(t,\tau) = \sum_l \alpha_l(t) c(\tau - \tau_l), \tag{1.5}$$

where $\alpha_l(t)$ is the time-variant complex amplitude of the lth path and the τ_l is the corresponding path delay, while $c(\tau)$ is the aggregate impulse response of the transmitter–receiver pair, which usually corresponds to the raised-cosine Nyquist filter. From (1.5) the continuous CTF can be described as in [304]

$$\begin{aligned} H(t,f) &= \int_{-\infty}^{\infty} h(t,\tau) e^{-j2\pi f \tau} \, d\tau \\ &= C(f) \sum_l \alpha_l(t) e^{-j2\pi f \tau_l}, \end{aligned} \tag{1.6}$$

where $C(f)$ is the Fourier transform pair of the transceiver impulse response $c(\tau)$ characterized in Figure 1.11.

As was pointed out in [260], in OFDM/MC-CDMA systems using a sufficiently long cyclic prefix and adequate synchronization, the discrete subcarrier-related CTF can be expressed as

$$H[n,k] = H(nT, k\Delta f) = C(k\Delta f) \sum_{l=1}^{L} \alpha_l[n] W_K^{k\tau_l/T_s} \tag{1.7}$$

$$= \sum_{m=0}^{K_0-1} h[n,m] W_K^{km}, \tag{1.8}$$

where $T_s = T/K$ is the baseband sample duration, while K_0 is the length of the cyclic prefix, which normally corresponds to the maximum delay spread encountered, such that $K_0 > \tau_{max}/T_s$.

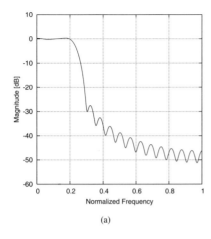

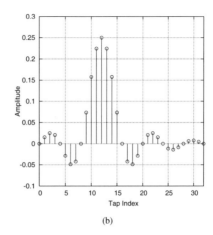

(a) (b)

Figure 1.11: (a) Frequency response and (b) impulse response of an order 8 raised-cosine shaping filter with an oversampling rate of 4, a roll-off factor of 0.2 and a delay of three samples.

Subsequently

$$h[n, m] = h(nT, mT_s) = \sum_{l=1}^{L} \alpha_l[n]c(mT_s - \tau_l) \quad (1.9)$$

is the Sample-Spaced CIR (SS-CIR) and $W_K = \exp(-j2\pi/K)$. Note that in realistic channel conditions associated with non-sample-spaced time-variant path delays $\tau_l(n)$, the receiver will encounter dispersed received signal components in several neighbouring samples owing to the convolution of the transmitted signal with the system's impulse response, which we refer to as leakage. This phenomenon is usually unavoidable and therefore the resultant SS-CIR $h[n, m]$ will be constituted by numerous correlated non-zero taps described by Equation (1.5) and illustrated in Figure 1.12. By contrast, the Fractionally Spaced CIR (FS-CIR) $\alpha_l[n] = \alpha_l(nT)$ will be constituted by a lower number of $L \ll K_0 \ll K$ non-zero, statistically independent taps associated with distinctive propagation paths, as depicted in Figure 1.12.

As shown in [260], the cross-correlation function $r_H[m, l]$, which characterized both the time- and frequency-domain correlation properties of the discrete CTF coefficients $H[n, k]$ associated with different OFDM blocks and subcarriers, can be described as

$$r_H[m, l] = \mathrm{E}\{H[n + m, k + l]H^*[n, k]\}$$
$$= \sigma_H^2 r_t[m]r_f[l], \quad (1.10)$$

where $r_t[m]$ is the time-domain correlation function described by Equation (1.4), while $r_f[i]$ is the frequency-domain correlation function, which can be expressed as follows [261]:

$$r_f[l] = |C(l\Delta f)|^2 \sum_{i=1}^{L} \frac{\sigma_i^2}{\sigma_H^2} e^{-j2\pi l\Delta f \tau_i}, \quad (1.11)$$

where $\sigma_H^2 = \sum_{i=1}^{L} \sigma_i^2$.

1.7.2 Realistic Channel Properties

The majority of existing advanced channel estimation methods rely on a priori knowledge of the channel statistics commonly characterized by the channel's PDP for the sake of estimating the instantaneous CIR

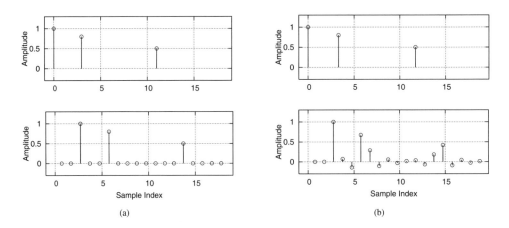

Figure 1.12: The **FS-CIR** (top) and the effective **SS-CIR** (bottom) resulting from the convolution of the original FS-CIR with the raised-cosine FIR of Figure 1.11 for the cases of (a) sample-spaced and (b) fractionally spaced power delay profiles.

and the corresponding CTF. It is evident, however, that in realistic wireless mobile channels, where at least one of the communicating terminals is in motion, the channel's PDP will also become time variant and thus may not be a priori known at the receiver.

For the sake of designing as well as characterizing the performance of an efficient and robust channel estimation scheme, which will be suitable for realistic channel conditions, we propose a channel model which sustains the important characteristics of the realistic wireless mobile channels. More specifically, as opposed to the conventional constant PDP, our channel model is characterized by a time-variant PDP, where both the relative delays τ_l and the corresponding average powers σ_l^2 of different PDP taps vary with time.

Our channel model is dynamically generated using the geometric scattering model illustrated in Figure 1.13. More specifically, the individual scatterers associated with different propagation paths are randomly generated using a Markov statistical model. The corresponding relative delays τ_l and powers σ_l^2 associated with each propagation path are calculated based on the geometrical location of each of the scatterers. Correspondingly, the rate of change in the values of the PDP tap delays τ_l is determined by the speed of the mobile wireless terminal and is characterized by the PDP *tap drift rate* parameter ν_τ. The specific assumptions regarding the practical range of values of the parameter ν_τ are discussed in the next chapter. Furthermore, each propagation path experiences independent fast Rayleigh fading. Finally, the set of parameters characterizing the Markov model employed is chosen such that the average channel statistics correspond to the desired static-PDP channel model.

1.7.3 Baseline Scenario Characteristics

As a baseline scenario we consider a mobile wireless communication system utilizing a frequency bandwidth of $B = 10\,$MHz at a carrier frequency of $f_c = 2.5\,$GHz. Furthermore, we assume an OFDM system having $K = 128$ orthogonal subcarriers. The corresponding FFT frame duration is $T_s = K/B = 16\,\mu$s. We assume a cyclic prefix of $1/4T_s = 4\,\mu$s and thus a total OFDM symbol duration of $T = 20\,\mu$s.

Some other important system-related assumptions include the relative speed of the communicating terminals, which we assume not to exceed $v = 130\,$km/h $= 36\,$m/s. Furthermore, the OFDM-symbol-normalized Doppler frequency f_D relates to the relative speed of the communicating terminals

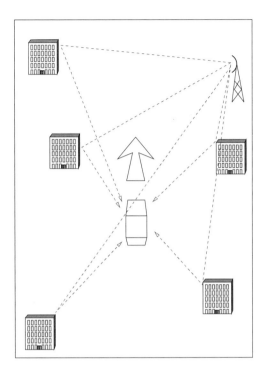

Figure 1.13: Corresponding PDP examples.

as follows:

$$f_D = T\frac{v f_c}{c}, \qquad (1.12)$$

where $c = 3 \times 10^8$ m/s denotes the speed of light. The actual Doppler frequency f_D/T encountered in the mobile wireless environment is assumed to be in the range of 3 to 300 Hz, where the maximum value of 300 Hz corresponds to the relative terminal speed of $v = 130$ km/h and the carrier frequency of $f_c = 2.5$ GHz. Finally, the OFDM-symbol-normalized PDP tap drift speed ν_τ may be calculated as follows:

$$\nu_\tau = T\frac{v}{c}, \qquad (1.13)$$

which suggests that the value of the PDP tap drift speed parameter does not exceed the maximum value of $\nu_\tau = 2.4 \times 10^{-6}\,\mu\text{s} = T \cdot 0.12\,\mu\text{s/s}$.

The resultant baseline scenario system characteristics are summarized in Table 1.11.

1.7.4   MC Transceiver

The transmitter part of the system typically consists of an OFDM/MC-CDMA encoder and modulator, the output of which is a complex-valued baseband time-domain signal. The resultant baseband signal is oversampled and pulse shaped using a Nyquist filter, such as, for example, the root-raised-cosine filter characterized in Figure 1.11. The resultant oversampled signal is then converted into an analogue passband signal using a D/A converter and upconverted to the RF band. At the receiver side a reciprocal process is taking place, where the received RF signal is amplified by the RF frontend and downconverted to an intermediate-frequency passband, then sampled by the A/D converter, downconverted to the

Table 1.11: Baseline scenario system characteristics.

Parameter	Value
Carrier frequency f_c	2.5 GHz
Channel bandwidth B	8 MHz
Number of carriers K	128
FFT frame duration T_s	16 μs
OFDM symbol duration T	20 μs (4 μs of cyclic prefix)
Max. delay spread τ_{max}	4 μs
Max. terminal speed v	130 km/h
Norm. Max. Doppler spread f_D	$0.006 = T \cdot 300$ Hz
Norm. Max. PDP tap drift ν_τ	$2.4 \times 10^{-6} \mu$s $= T \cdot 0.12 \mu$s/s

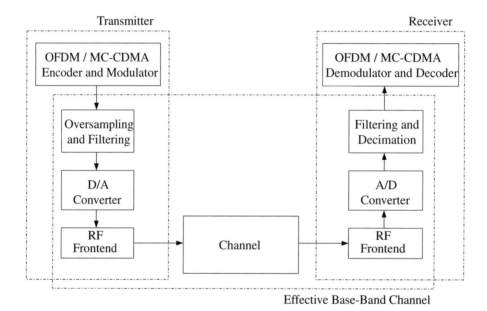

Figure 1.14: Schematic illustration of a typical OFDM/MC-CDMA system's PHY layer.

baseband, filtered by a matched Nyquist filter and finally decimated. The resultant complex-valued baseband signal is processed by the corresponding OFDM/MC-CDMA demodulator and decoder block, where the transmitted information symbols are detected.

In this treatise we consider the link between the output of the MC modulator and the input of the MC demodulator of Figure 1.14 as an *effective baseband channel*. The proof of feasibility for this assumption is beyond the scope of this contribution; however, it can be found for example in [348,350].

The discrete frequency-domain model of the OFDM/MC-CDMA system illustrated in Figure 1.14 can be described as in [304]

$$y[n, k] = H[n, k]x[n, k] + w[n, k], \tag{1.14}$$

for $k = 0, \ldots, K - 1$ and all n, where $y[n, k]$, $x[n, k]$ and $w[n, k]$ are the received symbol, the transmitted symbol and the Gaussian noise sample respectively, corresponding to the kth subcarrier of the nth OFDM block. Furthermore, $H[n, k]$ represents the complex-valued CTF coefficient associated

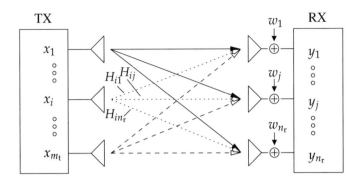

Figure 1.15: Illustration of a MIMO channel constituted by m_t transmit and n_r receive antennas. The corresponding MIMO channel is characterized by the $(n_r \times m_t)$-dimensional matrix $\mathbf{H}$ of CTF coefficients.

with the kth subcarrier and time instance n. Note that in the case of an M-QAM modulated OFDM system, $x[n, k]$ corresponds to the M-QAM symbol accommodated by the kth subcarrier, while in an MC-CDMA system, such as a Walsh–Hadamard Transform (WHT) assisted OFDM scheme using a G-chip WH spreading code and hence capable of supporting G users [265], we have

$$x[n, k] = \sum_{p=0}^{G-1} c[k, p]s[n, p], \tag{1.15}$$

where $c[k, p]$ is the kth chip of the pth spreading code, while $s[n, p]$ is the M-QAM symbol spread by the pth code. Each of the G spreading codes is constituted by G chips.

1.8   SDM-OFDM System Model

1.8.1   MIMO Channel Model

We consider a MIMO wireless communication system employing m_t transmit and n_r receive antennas; hence, the corresponding MIMO wireless communication channel is constituted by $(n_r \times m_t)$ propagation links, as illustrated in Figure 1.15. Furthermore, each of the corresponding $(n_r \times m_t)$ SISO propagation links comprises a multiplicity of statistically independent components, termed *paths*. Thus, each of these SISO propagation links can be characterized as a *multi-path* SISO channel, discussed in detail in Section 1.7.1. Similarly to the SISO case, the multi-carrier structure of our SDM-OFDM transceiver allows us to characterize the broadband frequency-selective channel considered as an OFDM subcarrier-related vector of flat-fading CTF coefficients. However, as opposed to the SISO case, for each OFDM symbol n and subcarrier k the MIMO channel is characterized by an $(n_r \times m_t)$-dimensional matrix $\mathbf{H}[n, k]$ of the CTF coefficients associated with the different propagation links, such that the element $H_{ij}[n, k]$ of the CTF matrix $\mathbf{H}[n, k]$ corresponds to the propagation link connecting the jth transmit and ith receive antennas.

Furthermore, the correlation properties of the MIMO-OFDM channel can be readily derived as a generalization of the SISO-OFDM channel scenario discussed in detail in Section 1.7.1. As shown in [260], the cross-correlation function $r_H[m, l]$, which characterizes both the time- and frequency-domain correlation properties of the discrete CTF coefficients $H_{ij}[n, k]$ associated with the particular (i, j)th propagation link of the MIMO channel, as well as with the different OFDM symbol and subcarrier indices n and k, can be described as

$$r_{H;ij}[m, l] = \mathrm{E}\{H_{ij}^*[n + m, k + l], H_{ij}[n, k]\}$$
$$= \sigma_H^2 r_t[m] r_f[l], \tag{1.16}$$

where $r_t[m]$ is the time-domain correlation function, which may be characterized by the time-domain correlation model proposed by Jakes in [349], where we have

$$r_t[m] = r_J[m] = J_0(nw_d), \tag{1.17}$$

and $J_0(x)$ is a zero-order Bessel function of the first kind, while $w_d = 2\pi T f_D$ is the normalized Doppler frequency. On the other hand, the frequency-domain correlation function $r_f[l]$ can be expressed as follows [261]:

$$r_f[l] = |C(l\Delta f)|^2 \sum_{i=1}^{L} \frac{\sigma_i^2}{\sigma_H^2} e^{-j2\pi l \Delta f \tau_i}, \tag{1.18}$$

where $C(f)$ is the frequency response of the pulse-shaping filter employed by the particular system, σ_i^2 and τ_i, $i = 1, \ldots, L$, are the average power and the corresponding delay of the L-tap PDP encountered, while σ_H^2 is the average power per MIMO channel link, such that $\sigma_H^2 = \sum_{i=1}^{L} \sigma_i^2$.

In this discussion we assume the different MIMO channel links to be mutually uncorrelated. This common assumption is usually valid if the spacing between the adjacent antenna elements exceeds $\lambda/2$, where λ is the wavelength corresponding to the RF signal employed. Thus, the overall cross-correlation function between the (i, j)th and (i', j')th propagation links may be described as

$$r_{H;ij;i'j'}[m, l] = \mathrm{E}\{H_{i'j'}^*[n + m, k + l], H_{ij}[n, k]\}$$
$$= \sigma_H^2 r_t[m] r_f[l] \delta[i - i'] \delta[j - j'], \tag{1.19}$$

where $\delta[i]$ is the discrete Kronecker delta function.

1.8.2 Channel Capacity

While most of the multi-path NLOS channel models can be collectively categorized as Rayleigh fading, different channel models characterized by different PDPs exhibit substantial differences in terms of their *information-carrying capacity* and *potential diversity gain*. The channel's capacity determines the upper bound for the overall system's throughput. On the other hand, the available diversity gain allows the communication system to increase its transmission integrity. Various modulation and coding schemes can be employed by the communication system in order to increase its spectral efficiency and also to take advantage of diversity. Some of these methods are widely discussed in the literature, e.g. in [351], and include the employment of antenna arrays, space–time coding, time- and frequency-domain spreading, channel coding, time- and frequency-domain repetition, etc. The theoretical performance boundaries of such methods are discussed in [266, 352]. Furthermore, the trade-offs between the attainable system capacity gain and the corresponding diversity gain are addressed in [353].

Consequently, the unrestricted capacity of a generic single-carrier ergodic-flat-fading MIMO channel can be expressed as in [335], where

$$\mathcal{C} = \mathrm{E}\left\{\log \det\left[\sigma_w^2 \mathbf{I} + \frac{1}{m_t} \mathbf{H} \mathbf{H}^{\mathrm{H}}\right]\right\}, \tag{1.20}$$

where $\mathbf{H}$ is an $(n_r \times m_t)$-dimensional matrix with independent complex Gaussian-distributed entries.

In realistic communication systems, however, the achievable throughput is limited by the modulation scheme employed. Some examples of such modulation schemes are M-ary PSK or M-ary QAM constellation schemes, where M is the number of complex symbols constituting the constellation map corresponding to the particular modulation scheme employed. The upper bound defining the maximum

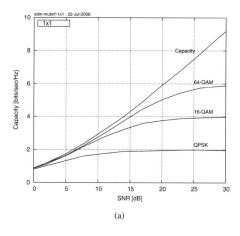 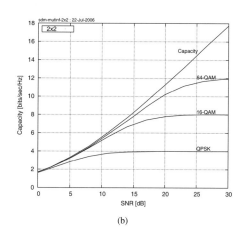

(a)   (b)

Figure 1.16: Capacity $\mathcal{C}$ of Equation (1.20) as well as mutual information $I(\mathbf{s}; \mathbf{y})$ of Equation (1.21) versus SNR for (a) 1×1 and (b) 2×2 systems in Rayleigh uncorrelated flat fading.

throughput achievable by a particular discrete modulation scheme was first discussed by Shannon in [354] and was shown to be determined by the mutual information $I(\mathbf{s}; \mathbf{y})$ exhibited by the modulation scheme employed. The mutual information can be calculated using the following expression:

$$I(\mathbf{s}; \mathbf{y}) = H(\mathbf{y}) - H(\mathbf{y}|\mathbf{s}), \qquad (1.21)$$

where $H(\cdot) = -\mathrm{E}\log p(\cdot)$ denotes the entropy function [354]. In the case of having a Gaussian i.i.d. noise sample vector $\mathbf{w}$ with the corresponding covariance matrix given by $C_w = \sigma_w^2 \mathbf{I}$, the constrained entropy constituent $H(\mathbf{y}|\mathbf{s})$ of Equation (1.21) is expressed as follows [335]:

$$H(\mathbf{y}|\mathbf{x}) = n_r \log 2\pi \sigma_w^2 e, \qquad (1.22)$$

whereas the unconstrained entropy constituent $H(\mathbf{y})$ can be approximated numerically using a Monte Carlo simulation as in [335], where

$$H(\mathbf{y}) = -\mathrm{E}\log\left(\frac{1}{M^{m_t}(2\pi\sigma_w^2)^{n_r}}\sum_{\mathbf{s}}\exp\left[-\frac{1}{2\sigma_w^2}\|\mathbf{y} - \mathbf{H}\mathbf{s}\|^2\right]\right), \qquad (1.23)$$

where the expectation is taken over the three sources of randomness in the choice of $\mathbf{s}, \mathbf{H}$ and $\mathbf{w}$. Moreover, the summation in Equation (1.23) is carried out over all M^{m_t} possible values of $\mathbf{s}$.

Figures 1.16(a) and 1.16(b) characterize both the capacity $\mathcal{C}$ of Equation (1.20) as well as the mutual information $I(\mathbf{s}; \mathbf{y})$ of Equation (1.21) for SISO and 2×2 MIMO systems, respectively. The mutual information plots depicted in both figures correspond to systems employing QPSK as well as 16- and 64-QAM modulations.

1.8.3 SDM-OFDM Transceiver Structure

The schematic of a typical SDM-OFDM system's physical layer is depicted in Figure 1.17. The transmitter of the SDM-OFDM system considered is typically constituted by the encoder and modulator seen in Figure 1.17, generating a set of m_t complex-valued baseband time-domain signals [265]. The modulated baseband signals are then processed in parallel. Specifically, they are oversampled and shaped using a Nyquist filter, such as for example a root-raised-cosine filter. The resultant oversampled signals are then converted into an analogue passband signal using a bank of D/A converters and

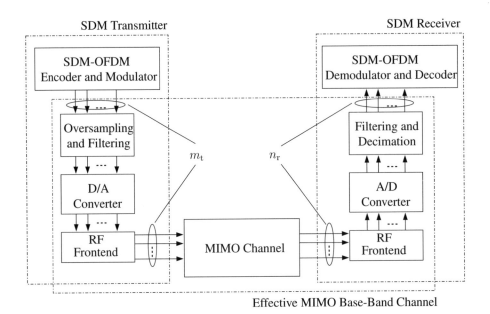

Figure 1.17: Schematic of a typical SDM-OFDM system's physical layer.

upconverted to the RF band. At the receiver side of the SDM-OFDM transceiver, the inverse process takes place, where the set of received RF signals associated with the n_r receive antenna elements is amplified by the RF amplifier and downconverted to an intermediate-frequency passband. The resultant passband signals are then sampled by a bank of A/D converters, downconverted to the baseband, filtered by a matched Nyquist filter and finally decimated, in order to produce a set of discrete complex-valued baseband signals. The resultant set of discrete signals is processed by the corresponding demodulator and decoder module seen in Figure 1.17, where the transmitted information-carrying symbols are detected.

In this treatise we consider the link between the output of the SDM-OFDM modulator and the input of the corresponding SDM-OFDM demodulator of Figure 1.17 as an effective baseband MIMO channel. The proof of feasibility for this assumption is beyond our scope here, but it can be found for example in [348, 350]. The structure of the resultant baseband SDM-OFDM system is depicted in Figure 1.18, where the bold grey arrows illustrate subcarrier-related signals represented by the vectors $\mathbf{x}_i$ and $\mathbf{y}_i$, while the thin black arrows accommodate scalar time-domain signals.

The discrete frequency-domain model of the SDM-OFDM system, illustrated in Figure 1.18, may be characterized as a generalization of the SISO case described in Section 1.7.1. That is, we have

$$y_i[n, k] = \sum_{j=1}^{m_t} H_{ij}[n, k] x_j[n, k] + w_i[n, k], \tag{1.24}$$

where $n = 0, 1, \ldots$ and $k = 0, \ldots, K - 1$ are the OFDM symbol and subcarrier indices, respectively, while $y_i[n, k]$, $x_i[n, k]$ and $w_i[n, k]$ denote the symbol received at the ith receive antenna, the symbol transmitted from the jth transmit antenna and the Gaussian noise sample encountered at the ith receive antenna, respectively. Furthermore, $H_{ij}[n, k]$ represents the complex-valued CTF coefficient associated with the propagation link connecting the jth transmit and ith receive antennas at the kth OFDM subcarrier and time instance n. Note that in the case of an M-QAM modulated OFDM system, $x_j[n, k]$ corresponds to the M-QAM symbol accommodated by the kth subcarrier of the nth OFDM symbol transmitted from the jth transmit antenna element.

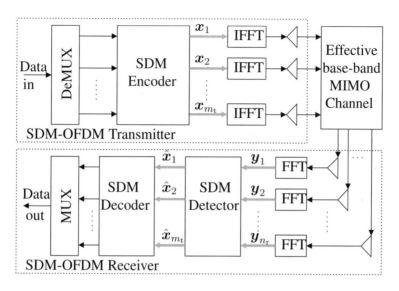

Figure 1.18: Schematic of a generic SDM-OFDM BLAST-type transceiver.

The SDM-OFDM system model described by Equation (1.24) can be interpreted as the per-OFDM-subcarrier vector expression of

$$\mathbf{y}[n, k] = \mathbf{H}[n, k]\mathbf{x}[n, k] + \mathbf{w}[n, k], \tag{1.25}$$

where we introduce the space-division-related vectors $\mathbf{y}[n, k]$, $\mathbf{x}[n, k]$ and $\mathbf{w}[n, k]$, as well as a space-division-related $(n_\mathrm{r} \times m_\mathrm{t})$-dimensional matrix of CTF coefficients $\mathbf{H}[n, k]$. Note that, similarly to the SISO case, the multi-carrier structure of the SDM-OFDM transceiver allows us to represent the broadband frequency-selective MIMO channel as a subcarrier-related vector of flat-fading MIMO-CTF matrices $\mathbf{H}[n, k]$.

1.9 Novel Aspects and Outline of the Book

Having briefly reviewed the OFDM, MIMO-OFDM and SDMA-OFDM literature, let us now outline the organization of this monograph:

- **Chapter 3: Channel Coding Assisted STBC-OFDM Systems**
  As an introductory study, in this chapter we discuss various channel-coded Space–Time Block Codes (STBCs) in the context of single-user and single-carrier OFDM systems. This work constitutes the background for the multi-user systems to be investigated in the following chapters. More specifically, various Turbo Convolutional (TC) codes, Low-Density Parity Check (LDPC) codes and Coded Modulation (CM) schemes are combined with STBCs to improve the performance of the single-user system considered.

- **Chapter 4: Coded Modulation Assisted Multi-user SDMA-OFDM Using Frequency-Domain Spreading**
  In this chapter, we invoke a multi-user MIMO SDMA-OFDM system for uplink communications, where the classic Minimum Mean Square Error (MMSE) Multi-User Detector (MUD) is employed at the BS for separating the different users' signals. The CM schemes discussed in Chapter 3, namely Trellis-Coded Modulation (TCM), Turbo TCM (TTCM), Bit-Interleaved Coded Modulation (BICM) and Iteratively Decoded BICM (BICM-ID), are evaluated and

compared in the context of the SDMA-OFDM system. Furthermore, the performance gain arising from invoking Walsh–Hadamard Transform Spreading (WHTS) across a block of OFDM subcarriers in the Frequency Domain (FD) is studied in both the uncoded SDMA-OFDM and the CM-assisted SDMA-OFDM systems.

- **Chapter 5: Hybrid Multi-user Detection for SDMA-OFDM Systems**
  This chapter focuses on the design of MUDs invoked by the SDMA receiver. Specifically, the Maximum Likelihood Detection (MLD) scheme is found to attain the best performance at the cost of a computational complexity that increases exponentially both with the number of users and with the number of Bits Per Symbol (BPS) transmitted by higher-order modulation schemes. By contrast, the MMSE MUD exhibits a lower complexity at the expense of a performance loss. In order to achieve a good performance–complexity trade-off, Genetic Algorithm (GA) based MUD techniques are proposed for employment in channel-coded SDMA-OFDM systems, where TTCM is used. Moreover, a novel Biased Q-function Based Mutation (BQM) assisted Iterative GA (IGA) MUD is designed. The performance of the proposed BQM-IGA is compared with both that of the optimum MLD and the linear MMSE MUD in the so-called fully loaded and overloaded scenarios, respectively, where the number of users is equal to or higher than the number of receiver antenna elements. Additionally, the computational complexity associated with the various MUD schemes is discussed.

- **Chapter 6: Direct-Sequence Spreading and Slow Subcarrier-Hopping Aided Multi-user SDMA-OFDM Systems**
  This chapter commences with a short review of conventional SDMA-OFDM systems, followed by an introduction to hybrid SDMA-OFDM arrangements, which incorporate Direct-Sequence Spreading (DSS) and/or Frequency-Hopping (FH) techniques into conventional SDMA-OFDM. A novel FH technique referred to as Slow SubCarrier Hopping (SSCH) is designed for hybrid DSS/FH SDMA-OFDM systems using a TTCM scheme. Furthermore, two types of SSCH pattern are discussed, namely the Random SSCH (RSSCH) and the Uniform SSCH (USSCH) patterns. The performance of the proposed TTCM-assisted DSS/SSCH SDMA-OFDM system is evaluated and compared with the conventional SDMA-OFDM and various hybrid SDMA-OFDM configurations.

- **Chapter 7: Channel Estimation for OFDM and MC-CDMA**
  We derive an advanced Decision Directed Channel Estimation (DDCE) scheme, which is capable of recursive tracking and prediction of rapidly fluctuating channel parameters, characterized by time-variant statistics. More specifically, we employ a Projection Approximation Subspace Tracking (PAST) [355] technique for the sake of tracking the channel transfer function's low-rank signal subspace and thus facilitating a high-accuracy tracking of the channel's transfer function, while imposing a relatively low computational complexity.

- **Chapter 8: Iterative Joint Channel Estimation and MUD for SDMA-OFDM Systems**
  The objective of this chapter is to develop an efficient solution to the channel estimation problem of multi-user MIMO-OFDM systems. It is well known that, compared with SISOsystems, channel estimation in the MIMO scenario becomes more challenging, owing to the increased number of independent transmitter–receiver links to be estimated. Against this background, an iterative, joint channel estimation and symbol detection approach is proposed for LDPC-coded MIMO SDMA-OFDM systems. More specifically, the method modifies the GA MUD advocated in Chapter 5 so that it becomes capable of jointly optimizing the Frequency-Domain CHannel Transfer Functions (FD-CHTFs) and the multi-user data symbols. Moreover, an efficient algorithm is derived, which enables the GA to output soft bits for the sake of improving the performance of the LDPC channel decoder.

- **Chapter 9: Reduced-Complexity Sphere Detection for Uncoded SDMA-OFDM Systems**
  The main objective of this chapter is systematically to review the fundamentals of the SD, which is considered to be one of the most promising low-complexity near-optimum detection

techniques at the time of writing. Furthermore, we address the SD-related complexity reduction issues. Specifically, the principle of the Hard-Input, Hard-Output (HIHO) SD is reviewed first in the context of both the depth-first and breadth-first tree search based scenarios, along with that of the GSD, which is applicable to challenging rank-deficient MIMO scenarios. A comprehensive comparative study of the complexity reduction schemes devised for different types of SDs, namely the conventional depth-first SD, the K-best SD and the novel OHRSA detector, is carried out by analysing their conceptual similarities and differences. Finally, their achievable performance and the complexity imposed by the various types of SDs are investigated in comparison with each other.

- **Chapter 10: Reduced-Complexity Iterative Sphere Detection for Channel-Coded SDMA-OFDM Systems**
  The fundamentals of the LSD scheme are studied at the beginning of this chapter in the context of an iterative detection aided channel-coded MIMO-OFDM system. Potentially excessive complexity may be imposed by the conventional LSD, since it has to generate soft information for every transmitted bit, which requires the observation of a high number of hypotheses about the transmitted MIMO symbol. Based on the above-mentioned complexity issue, we contrive a generic centre-shifting SD scheme and the so-called a priori-LLR-threshold assisted SD scheme with the aid of EXIT chart analysis, both of which are capable of effectively reducing the potentially high complexity imposed by the SD-aided iterative receiver. Moreover, we combine the above-mentioned schemes in the interest of further reducing the complexity imposed. In addition, for the sake of enhancing the achievable iterative detention gains and hence improving the bandwidth efficiency, a Unity-Rate Code (URC) assisted three-stage, serially concatenated transceiver employing the so-called Irregular Convolutional Codes (IrCCs) is devised. Finally, the benefits of the proposed centre-shifting SD scheme are also investigated in the context of the above-mentioned three-stage iterative receiver.

- **Chapter 11: Sphere-Packing Modulated STBC-OFDM and its Sphere Detection**
  In this chapter we extend the employment of the turbo-detected Sphere Packing (SP) aided Space–Time Block Coding (STBC) scheme to Multi-User MIMO (MU-MIMO) scenarios, because SP was demonstrated to be capable of providing useful performance improvements over conventionally modulated orthogonal design-based STBC schemes in the context of Single-User MIMO (SU-MIMO) systems. For the sake of achieving a near-MAP performance, while imposing a moderate complexity, we specifically design the K-best SD scheme for supporting the operation of the SP-modulated system, since the conventional SD cannot be directly applied to such a system. Consequently, when relying on our SD, a significant performance gain can be achieved by the SP-modulated system over its conventionally modulated counterpart in the context of MU-MIMO systems.

- **Chapter 12: Multiple-Symbol Differential Sphere Detection for Cooperative OFDM**
  The principle of the MSDSD is first reviewed, as recently proposed for mitigating the time-selective channel-induced performance loss suffered by classic direct transmission schemes employing the Conventional Differential Detection (CDD) scheme. Then, we specifically design the MSDSD for both the Differential Amplify-and-Forward (DAF) and Differential Decode-and-Forward (DDF) assisted cooperative systems based on the multi-dimensional tree search proposed in Chapter 4, which is capable of achieving a significant performance gain for transmission over time-selective channels induced by the relative mobility among the cooperating transceivers.

- **Chapter 13: Resource Allocation for the Differentially Modulated Cooperative UL**
  In this chapter the theoretical BER performance of both the DAF- and DDF-aided cooperative cellular uplinks is investigated. Then, based on the minimum BER criterion, we design efficient Cooperating-User Selection (CUS) and Adaptive-Power Allocation (APA) schemes for the above-mentioned two types of differentially modulated cooperative systems, while requiring

no Channel State Information (CSI) at the receiver. Moreover, we investigate the Cooperative-Protocol Selection (CPS) of the uplink system in conjunction with a beneficial CUS as well as the APA scheme in order to improve further the achievable end-to-end performance, leading to a resource-optimized hybrid cooperative system. Hence, a number of cooperating MSs may be adaptively selected from the available MS candidate pool and the cooperative protocol employed by a specific cooperating MS may also be adaptively selected in the interest of achieving the best possible BER performance.

- **Chapter 14: The Near-Capacity Differentially Modulated Cooperative Cellular Uplink**
  The DDF-aided cooperative system's DCMC capacity is investigated in comparison with that of its classic direct-transmission-based counterpart in order to answer the grave fundamental question of whether it is worth introducing cooperative mechanisms into the development of wireless networks, such as the cellular voice and data networks. Then, we propose a practical framework for designing a cooperative system which is capable of performing close to the network's corresponding non-coherent DCMC capacity. Based on our low-complexity, near-capacity design criterion, a novel Irregular Distributed Hybrid Concatenated Differential (Ir-DHCD) coding scheme is contrived for the DDF cooperative system employing our proposed capacity-achieving, low-complexity, adaptive-window-aided, SISO iterative MSDSD scheme.

- **Chapter 15: Multi-stream Detection for SDM-OFDM Systems**
  The multi-stream detection problem of SDM-OFDM systems is similar to the MUD techniques of SDMA-OFDM arrangements, which are classified and reviewed in this chapter.

- **Chapter 16: Approximate Log-MAP SDM-OFDM Multi-stream Detection**
  We propose the novel family of Optimized Hierarchy Reduced Search Algorithm (OHRSA) aided space–time processing methods, which may be regarded as an advanced extension of the Complex Sphere Decoder (CSD) method, portrayed in [337]. The algorithm proposed extends the potential application range of the CSD methods of [335] and [337], as well as reducing the associated computational complexity. Moreover, the OHRSA-aided SDM detector proposed exhibits the near-optimum performance of the Log-MAP SDM detector, while imposing a substantially lower computational complexity, which renders it an attractive design alternative for practical systems.

- **Chapter 17: Iterative Channel Estimation and Multi-stream Detection for SDM-OFDM**
  Finally, we propose an iterative turbo-receiver architecture, which utilizes both the soft-decision feedback-aided MIMO channel estimation scheme of Chapter 7 as well as the Log-MAP SDM detection method derived in Chapter 16. Additionally, we carry out an analysis of the associated design trade-offs.

- **Chapter 18: Summary, Conclusions and Future Research**
  The major findings of our work are summarized in this chapter, including our suggestions for future research.

1.10 Chapter Summary

The historic development of various MIMO techniques was briefly summarized in Section 1.1.1.1, followed by a rudimentary introduction to MIMO-OFDM systems in Section 1.1.1.2. In Section 1.1.1.3 a concise review of various SDMA and SDMA-OFDM techniques was given, highlighting the associated signal processing problems.

Chapter 2

OFDM Standards

During the past decades, wireless communication has benefitted from substantial advances and it is considered as the key enabling technique of innovative future consumer products. For the sake of satisfying the requirements of various applications, significant technological achievements are required to ensure that wireless devices have appropriate architectures suitable for supporting a wide range of services delivered to the users.

In the foreseeable future, the large-scale employment of wireless devices and the requirements of high-bandwidth applications are expected to lead to tremendous new challenges in terms of the efficient exploitation of the achievable spectral resources. New wireless techniques, such as Ultra WideBand (UWB) [356], advanced source and channel encoding as well as various smart-antenna techniques, e.g. Space–Time Codes (STCs) [314], Space-Division Multiple Access (SDMA) [3] and beamforming, as well as other Multiple-Input, Multiple-Output (MIMO) [92] wireless architectures, are capable of offering substantial improvements over classic communication systems. Hence researchers have focused their attention on the next generation of wireless broadband communications systems, which aim at delivering multimedia services requiring data rates much higher than existing ones. Undoubtedly, supporting such high data rates while maintaining a high robustness against radio channel impairments, such as multi-path fading and frequency-selective fading, requires further enhanced system architectures.

The organization of this chapter is as follows. In Sections 2.1, 2.2 and 2.3, we review various major international standards that adopt OFDM, namely Wi-Fi, the Third-Generation Partnership Project (3GPP), Long-Term Evolution (LTE) and Worldwide Interoperability for Microwave Access (WiMAX), respectively. Finally, we conclude the chapter in Section 2.4.

2.1 Wi-Fi

In 1999, the Wi-Fi Alliance was founded as a global, non-profit organization, aimed at developing a single globally accepted standard for high-speed WLANs. The mission of the Wi-Fi Alliance was to promote Wi-Fi technology and the corresponding Wi-Fi product certification. The Wi-Fi Alliance has now more than 300 members from more than 20 countries and Wi-Fi has achieved huge worldwide success. A study [357] released in September 2008 by the Wi-Fi Alliance found that an increased number of consumers in the United States, the United Kingdom and Japan value the Wi-Fi Certified brand. Developed in March 2000, the certification programme has approved more than 4800 products from various vendors worldwide.

MIMO-OFDM for LTE, Wi-Fi and WiMAX   Lajos Hanzo, Yosef Akhtman, Li Wang and Ming Jiang
© 2011 John Wiley & Sons, Ltd

2.1.1   IEEE 802.11 Standards

Wi-Fi is based on the IEEE 802.11 standard family. The first version of IEEE 802.11 was released in 1997 [358], which was rectified in 1999 [359]. Then, two further supplements of 802.11-1999 were released, the first one being IEEE 802.11a [360], which provides a bit rate of up to 54 Mbps in the 5 GHz band. In comparison with 802.11-1999, where Frequency-Hopping Spread Spectrum (FHSS) or Direct-Sequence Spread Spectrum (DSSS) are used, 802.11a employs an OFDM scheme, which applies to Wireless Asynchronous Transfer Mode (WATM) networks and access hubs. The second supplement to 802.11-1999 was IEEE 802.11b [361], which has a maximum data rate of 11 Mbps in the 2.4 GHz band and uses the same media access method as that defined in 802.11-1999.

Further enhancements to 802.11 were made later. The IEEE 802.11d-2001 standard [362] introduced support for international roaming services. The IEEE 802.11g-2003 standard [363], which exploits the same OFDM modulation scheme as 802.11a, provides a data rate of 20–54 Mbps in the 2.4 GHz band. Other improvements include IEEE 802.11h [364], IEEE 802.11i [365] and IEEE 802.11j [366], which introduced spectrum and transmit power management in the 5 GHz band in Europe, security enhancements and operation in the 4.9–5 GHz band in Japan, respectively. In November 2005, the release of IEEE 802.11e-2005 [367] provided further QoS enhancements.

As the number of standards in the 802.11 family grew, it was proposed by the working group that a single document combining all up-to-date 802.11 specifications should be provided. This resulted in the IEEE 802.11-2007 standard [368], a new release that includes all previous 802.11 amendments. In 2008, another two amendments were completed. The IEEE 802.11k-2008 [369] standard extends 802.11 by specifying mechanisms for Radio Resource Measurement (RRM), and the IEEE 802.11r-2008 [370] standard provides mechanisms for fast Basic Service Set (BSS) transition.

Some of the 802.11 standards are still in the draft stage, such as the IEEE 802.11n standard [56], which aims at developing next-generation WLANs by incorporating MIMO-OFDM techniques. It is expected to offer high-throughput wireless transmission at 100–200 Mbps. The IEEE 802.11y standard [371] will provide support for operation in the 3650–3700 MHz band in the United States.

As a brief summary, we highlight the major 802.11 standards in Tables 2.1, 2.2 and 2.3.

2.2   3GPP LTE

The 3GPP is an international standardization body working on the specification of the 3G Universal Terrestrial Radio Access Network (UTRAN) and on the Global System for Mobile communications (GSM). The latest specification that is being studied and developed in 3GPP is an evolved 3G radio access, widely known as LTE (Long-Term Evolution) or Evolved UTRAN (E-UTRAN), as well as an evolved packet access core network in the System Architecture Evolution (SAE). The initial requirements for LTE were set out in early 2005, which are briefly summarized in Table 2.4 and in [372].

The initial objective of 3GPP was to produce global specifications for a 3G mobile system evolving from the existing GSM core network. This includes the Wideband CDMA (WCDMA) based UTRA Frequency-Division Duplex (FDD) mode and the Time-Division Code-Division Multiple Access (TD-CDMA) based UTRA Time-Division Duplex (TDD) mode [374]. In December 2005, it was decided that the LTE radio access should be based on OFDMA in the downlink (DL) and Single-Carrier Frequency-Division Multiple Access (SC-FDMA) in the uplink (UL). SC-FDMA is also known as Discrete Fourier Transform Spread OFDMA (DFTS-OFDMA). The main PHY parameters of the LTE DL are summarized in Table 2.5.

Briefly, the objective of the SAE is to migrate circuit-switched networks towards packet-switched networks. This is set out in the recent 3GPP releases, where an evolved packet core was defined. The main targets of SAE can be divided into the following aspects [374]:

- high-level user and operational aspects;

- basic capabilities;

Table 2.1: The family of IEEE 802.11 standards (1997–2001).

Year	Standard	Title
1997	802.11-1997 [358]	IEEE Standard for Information Technology – Telecommunications and Information Exchange Between Systems – Local and Metropolitan Area Networks – Specific Requirements – Part 11: Wireless LAN Medium Access Control (MAC) and Physical Layer (PHY) Specifications
1999	802.11-1999 [359]	IEEE Standard for Information Technology – Telecommunications and Information Exchange Between Systems – Local and Metropolitan Area Networks – Specific Requirements – Part 11: Wireless LAN Medium Access Control (MAC) and Physical Layer (PHY) Specifications
	802.11a-1999 [360]	Supplement to IEEE Standard for Information Technology – Telecommunications and Information Exchange Between Systems – Local and Metropolitan Area Networks – Specific Requirements – Part 11: Wireless LAN Medium Access Control (MAC) and Physical Layer (PHY) Specifications: High-speed Physical Layer in the 5 GHz Band
	802.11b-1999 [361]	Supplement to IEEE Standard for Information Technology – Telecommunications and Information Exchange Between Systems – Local and Metropolitan Area Networks – Specific Requirements – Part 11: Wireless LAN Medium Access Control (MAC) and Physical Layer (PHY) Specifications: Higher-speed Physical Layer Extension in the 2.4 GHz Band
2001	802.11d-2001 [362]	IEEE Standard for Information Technology – Telecommunications and Information Exchange Between Systems – Local and Metropolitan Area Networks – Specific Requirement – Part 11: Wireless LAN Medium Access Control (MAC) and Physical Layer (PHY) Specification – Amendment 3: Specifications for Operation in Additional Regulatory Domains

- multi-access and seamless mobility;
- human–machine interface aspects;
- performance requirements for the evolved 3GPP;
- security as well as privacy; and
- charging aspects of the system.

2.3 WiMAX Evolution

The rapidly growing demand for flexible, high-speed broadband services requires advanced communication technologies. The more conventional family of high-rate broadband access techniques has relied on wired access, such as Digital Subscriber Line (DSL), cable modems, Ethernet and optical fibres. However, the extension of the coverage area results in a significantly increased cost imposed by building and maintaining wired networks. This is particularly true for less densely populated zones, e.g. suburban and rural areas.

Hence, Broadband Wireless Access (BWA) techniques have emerged as potent competitors of their conventional wired counterparts, facilitating the provision of broadband services for subscribers that are far from the coverage area of the wired networks. Being flexible, efficient and cost-effective, BWA provides an excellent solution to overcome the above-mentioned coverage problem. During the past decade or so, a number of proprietary wireless access systems have been developed by the wireless industry. Naturally, these proprietary products were based on diverse specifications, which inevitably limited their applications and markets. As a matter of fact, the potential benefits of BWA services were

Table 2.2: The family of IEEE 802.11 Standards (2003–2005).

Year	Standard	Title
2003	802.11g-2003 [363]	IEEE Standard for Information Technology – Telecommunications and Information Exchange Between Systems – Local and Metropolitan Area Networks – Specific Requirements – Part 11: Wireless LAN Medium Access Control (MAC) and Physical Layer (PHY) Specifications – Amendment 4: Further Higher Data Rate Extension in the 2.4 GHz Band
	802.11h-2003 [364]	IEEE Standard for Information Technology – Telecommunications and Information Exchange Between Systems – Local and Metropolitan Area Networks – Specific Requirements – Part 11: Wireless LAN Medium Access Control (MAC) and Physical Layer (PHY) Specifications – Amendment 5: Spectrum and Transmit Power Management Extensions in the 5 GHz band in Europe
2004	802.11i-2004 [365]	IEEE Standard for Information Technology – Telecommunications and Information Exchange Between Systems – Local and Metropolitan Area Networks – Specific Requirements – Part 11: Wireless LAN Medium Access Control (MAC) and Physical Layer (PHY) Specifications – Amendment 6: Medium Access Control (MAC) Security Enhancements
	802.11j-2004 [366]	IEEE Standard for Information Technology – Telecommunications and Information Exchange Between Systems – Local and Metropolitan Area Networks – Specific Requirements – Part 11: Wireless LAN Medium Access Control (MAC) and Physical Layer (PHY) Specifications – Amendment 7: 4.9 GHz–5 GHz Operation in Japan
2005	802.11e-2005 [367]	IEEE Standard for Information Technology – Telecommunications and Information Exchange Between Systems – Local and Metropolitan Area Networks – Specific Requirements – Part 11: Wireless LAN Medium Access Control (MAC) and Physical Layer (PHY) Specifications – Amendment 8: Medium Access Control (MAC) Quality of Service Enhancements

not expected to be widely achieved owing to the lack of a common international standard, until the emergence of the Worldwide Interoperability for Microwave Access (WiMAX) standard [376].

WiMAX is one of the most popular BWA technologies available at the time of writing, aiming to provide high-speed broadband wireless access for Wireless Metropolitan Area Networks (WMANs) [377]. As a standardized technology, WiMAX ensures the interoperability of equipment certified by the WiMAX Forum, resulting in a significant cost reduction for service providers that would like to use products manufactured by diverse vendors. This distinct advantage has paved the way for global broadband wireless services. Another key benefit of WiMAX is that it has been optimized to offer excellent Non-Line-of-Sight (NLOS) coverage with the aid of advanced wireless transmission techniques, such as MIMO transmit/receive diversity and Automatic Retransmission Request (ARQ), etc. [378], combined with Orthogonal Frequency-Division Multiplexing (OFDM) or Orthogonal Frequency-Division Multiple Access (OFDMA) [378].

This section is organized as follows. In Section 2.3.1, we first briefly outline the historic background of WiMAX. Specifically, the IEEE 802.16 standard family is reviewed in Section 2.3.1.1, which has tight links to WiMAX, followed by the introduction of the WiMAX Forum in Section 2.3.1.3. Then a brief introduction of the Korean WiMAX standard, Wireless Broadband (WiBro), is provided in Section 2.3.1.4. In Section 2.3.2, we proceed with discussing the technical aspects of WiMAX, including WiMAX-I in Section 2.3.2.1 and WiMAX-II in Section 2.3.2.2, respectively. The trends concerning the future of WiMAX are summarized in Section 2.3.3.

Table 2.3: The family of IEEE 802.11 Standards (2007–2008).

Year	Standard	Title
2007	802.11-2007 [368]	IEEE Standard for Information Technology – Telecommunications and Information Exchange Between Systems – Local and Metropolitan Area Networks – Specific Requirements – Part 11: Wireless LAN Medium Access Control (MAC) and Physical Layer (PHY) Specifications
2008	802.11k-2008 [369]	IEEE Standard for Information Technology – Telecommunications and Information Exchange Between Systems – Local and Metropolitan Area Networks – Specific Requirements – Part 11: Wireless LAN Medium Access Control (MAC) and Physical Layer (PHY) Specifications – Amendment 1: Radio Resource Measurement of Wireless LANs (Amendment to IEEE 802.11-2007)
	802.11r-2008 [370]	IEEE Standard for Information Technology – Telecommunications and Information Exchange Between Systems – Local and Metropolitan Area Networks – Specific Requirements – Part 11: Wireless LAN Medium Access Control (MAC) and Physical Layer (PHY) Specifications – Amendment 2: Fast Basic Service Set (BSS) Transition (Amendment to IEEE 802.11-2007 and IEEE 802.11k-2008)
	P802.11n-2008 [56]	Draft IEEE Standard for Information Technology – Telecommunications and Information Exchange Between Systems – Local and Metropolitan Area Networks – Specific Requirements – Part 11: Wireless LAN Medium Access Control (MAC) and Physical Layer (PHY) Specifications – Amendment 4: Enhancements for Higher Throughput
	P802.11y-2008 [371]	Draft IEEE Standard for Information Technology – Telecommunications and Information Exchange Between Systems – Local and Metropolitan Area Networks – Specific Requirements – Part 11: Wireless LAN Medium Access Control (MAC) and Physical Layer (PHY) Specifications – Amendment 3: 3650–3700 MHz Operation in USA (Draft Amendment to IEEE 802.11-2007)

2.3.1   Historic Background

In this section, we will commence by briefly reviewing the IEEE 802.16 standard family in Section 2.3.1.1, and portray the brief history of the WiMAX Forum in Section 2.3.1.3. The connection between WiMAX and WiBro is established in Section 2.3.1.4.

2.3.1.1   IEEE 802.16 Standard Family

WiMAX is closely related to the IEEE 802.16 standard family. Before elaborating on WiMAX, let us first briefly review the history of the IEEE 802.16 standards outlined in Table 2.6.

The IEEE 802.16 Working Group (WG) was chartered to define the air interface for BWA systems in certain licensed frequency bands. The 802.16 specifications conform to the family of IEEE 802 standards, governing the Local Area Networks (LANs) and Metropolitan Area Networks (MANs) endorsed by the IEEE in 1990 [388]. As opposed to the IEEE 802.11 WG, which focuses on Wireless Local Area Network (WLAN) standards and applications, the IEEE 802.16 WG has been focused on developing cost-efficient point-to-multipoint BWA architectures that enable multimedia broadband services in MANs and Wide Area Networks (WANs) [388].

2.3.1.2   Early 802.16 Standards

As early as 1998, the IEEE 802.16 group was formed to develop a radio standard for wireless broadband communications. In December 2001, the first member of the IEEE 802.16 standard family was approved, widely referred to as IEEE 802.16-2001 [379]. It focuses on a Line-Of-Sight (LOS) based point-to-multipoint wireless broadband system operating in the 10–66 GHz band. It utilizes a

Table 2.4: System requirements for 3GPP LTE [372]. MBMS: Multimedia Broadcast Multicast Service.

Requirement	Description	
Peak data rate	DL	5 bps/Hz (100 Mbps within 20 MHz)
	UL	2.5 bps/Hz (50 Mbps within 20 MHz)
Control plane latency	<100 ms (transition time from a camped state)	
	<50 ms (transition time between dormant states)	
Control plane capacity	≥200 users per cell	
User plane latency	<5 ms	
Average user throughput	DL	3 to 4 times over Release 6 [373]
	UL	2 to 3 times over Release 6 [373]
Spectrum efficiency	DL	3 to 4 times over Release 6 [373]
	UL	3 to 4 times over Release 6 [373]
Mobility	Optimized performance (0–15 km/h), high performance (15–120 km/h), service maintained (120–350 km/h)	
Coverage	5 km cells	Performance targets met
	30 km cells	Slight degradation
	100 km cells	Not precluded
Enhanced MBMS	Enhanced MBMS shall be supported	
Spectral flexibility	Allocation of different-width bands (1.25–20 MHz) shall be supported. Content can be delivered over an aggregation of resources	
Coexistence with 3GPP RATs	Coexist with GSM EDGE RAN (GERAN) and UTRAN	
Architecture and migration	Packet-based single E-UTRAN architecture to support end-to-end QoS and to minimize 'single points of failure'	
RRM	Enhanced support for end-to-end QoS, load sharing and policy management	
Complexity	Minimize optional functions and remove redundant mandatory features	

Table 2.5: PHY parameters for 3GPP LTE DL [375].

Parameters		Values					
System bandwidth (MHz)		1.25	2.5	5	10	15	20
Time slot duration (ms)					0.675		
Subcarrier spacing (kHz)					15		
Sampling frequency (MHz)		1.92	3.84	7.68	15.36	23.04	30.72
FFT size		128	256	512	1024	1536	2048
No. of used subcarriers		76	151	301	601	901	1201
No. of OFDM symbols per time slot (Short/Long CP)					9/8		
CP length	Short	7.29/14	7.29/28	7.29/56	7.29/112	7.29/168	7.29/224
(μs/samples)	Long	16.67/32	16.67/64	16.67/128	16.67/256	16.67/384	16.67/512
Time slot interval	Short	18	36	72	144	216	288
(samples)	Long	16	32	64	128	192	256

Single Carrier (SC) based physical layer (PHY) in conjunction with a burst Time-Division Multiplexed (TDM) Medium Access Control (MAC) layer [376].

Following the initial release of 802.16-2001, there had been two amendments, namely IEEE 802.16c-2002 [380] in December 2002, which provides detailed system profiles for the 10–66 GHz band, and IEEE 802.16a-2003 [381] in April 2003, which presents some MAC modifications and additional PHY specifications for the 2–11 GHz band. Note that in the standards rectified after

Table 2.6: The family of IEEE 802.16 Standards.

Year	Standard	Title
2001	802.16-2001 [379]	IEEE Standard for Local and Metropolitan Area Networks – Part 16: Air Interface for Fixed Broadband Wireless Access Systems
2002	802.16c-2002 [380]	IEEE Standard for Local and Metropolitan Area Networks – Part 16: Air Interface for Fixed Broadband Wireless Access Systems – Amendment 1: Detailed System Profiles for 10–66 GHz
2003	802.16a-2003 [381]	IEEE Standard for Local and Metropolitan Area Networks – Part 16: Air Interface for Fixed Broadband Wireless Access Systems – Amendment 2: Medium Access Control Modifications and Additional Physical Layer Specifications for 2–11 GHz
2004	802.16d-2004 [55]	IEEE Standard for Local and Metropolitan Area Networks – Part 16: Air Interface for Fixed Broadband Wireless Access Systems
2005	802.16e-2005 [382]	IEEE Standard for Local and Metropolitan Area Networks – Part 16: Air Interface for Fixed and Mobile Broadband Wireless Access Systems – Amendment 2: Physical and Medium Access Control Layers for Combined Fixed and Mobile Operation in Licensed Bands and Corrigendum 1
	802.16f-2005 [383]	IEEE Standard for Local and Metropolitan Area Networks – Part 16: Air Interface for Fixed Broadband Wireless Access Systems – Amendment 1: Management Information Base
2007	802.16k-2007 [384]	IEEE Standard for Local and Metropolitan Area Networks Media Access Control (MAC) Bridges – Amendment 5: Bridging of IEEE 802.16
	802.16g-2007 [385]	IEEE Standard for Local and Metropolitan Area Networks – Part 16: Air Interface for Fixed and Mobile Broadband Wireless Access Systems – Amendment 3: Management Plane Procedure and Services
2008	P802.16h [386]	Draft IEEE Standard for Local and Metropolitan Area Networks – Part 16: Air Interface for Fixed and Mobile Broadband Wireless Access Systems: Improved Coexistence Mechanisms for License-Exempt Operation
	P802.16j [387]	Draft IEEE Standard for Local and Metropolitan Area Networks – Part 16: Air Interface for Fixed and Mobile Broadband Wireless Access Systems Multihop Relay Specification

802.16a-2003, in addition to the SC-based PHY, both the OFDM- and OFDMA-based PHY specifications are also included.

2.3.1.2.1 IEEE 802.16d-2004 – Fixed WiMAX

The IEEE 802.16 standards developed during the early stages tend to describe different parts of the technology. In order to ease future developments, it was decided to merge the previous individual versions into a single one, resulting in IEEE 802.16d-2004 [55], which is also frequently referred to as IEEE 802.16-2004. For operational frequencies spanning from 10 to 66 GHz, the PHY is based on SC modulation. For NLOS propagation conditions at frequencies below 11 GHz, the design alternatives of SC, OFDM or OFDMA modulation can be used [55].

Early WiMAX solutions were based on the Wireless Metropolitan Area Network (WirelessMAN) OFDM PHY of 802.16-2004. Therefore, 802.16-2004 is also known as 'fixed WiMAX', due to the fact that it does not support mobility.

2.3.1.2.2 IEEE 802.16e-2005 – Mobile WiMAX

In order to provide mobility support, the IEEE 802.16 working group continued its developments. In December 2005, an amendment of IEEE 802.16-2004 was approved, which is known as IEEE 802.16e-2005 [382] or 802.16e in brief. In addition to numerous corrections to 802.16-2004 regarding stationary operations, a key enhancement of this standard over its ancestors is that it supports subscriber

Table 2.7: PHY profiles in IEEE 802.16-2004 and IEEE 802.16e-2005.

PHY profile	Air interface	Description
WirelessMAN-SC	SC	Operation in the 10–66 GHz frequency band
WirelessMAN-SCa	SC	For NLOS operation in frequency bands below 11 GHz. For licensed bands, channel bandwidths are limited to the regulatory provisioned bandwidth divided by any power of 2 no less than 1.25 MHz
WirelessMAN-OFDM	OFDM	For NLOS operation in frequency bands below 11 GHz
WirelessMAN-OFDMA	OFDMA	For NLOS operation in frequency bands below 11 GHz. For licensed bands, channel bandwidths are limited to the regulatory provisioned bandwidth divided by any power of 2 no less than 1.0 MHz
WirelessHUMAN	OFDM	Wireless High-speed Unlicensed MAN (WirelessHUMAN) is similar to WirelessMAN-OFDM, but mandates dynamic frequency selection for mainly the Unlicensed National Information Infrastructure (UNII) band [390]

stations moving at vehicular speeds and thereby specifies a system for combined fixed and mobile BWA. The functions required to support higher-layer handover between base stations or sectors are also specified. Its operation is limited to licensed bands suitable for mobility at carrier frequencies below 6 GHz. Furthermore, the previously developed stationary IEEE 802.16 subscriber capabilities are not compromised [382].

Note that 802.16e itself is not a stand-alone document. More specifically, it only includes the differences with respect to the 802.16-2004 document. Although, for the sake of simplicity, 802.16e has been widely referred to as if it were a stand-alone standard, it is more natural to combine it with 802.16-2004 into a single consolidated version. This work is being conducted by the IEEE 802.16's Maintenance Task Group under the IEEE P802.16Rev2 Project. This will result in the second revision of 802.16 since the releases of 802.16-2001 and 802.16-2004. It will consolidate 802.16-2004, 802.16e as well as other 802.16 standards, such as 802.16f/g [383, 385]. The up-to-date working document for this project is P802.16Rev2/D5 [389], which was released in June 2008.

The 802.16e standard formed the basis of WiMAX for nomadic and mobile applications. It is often referred to as 'mobile WiMAX', as compared with 'fixed WiMAX', which is synonymous with IEEE 802.16-2004. Since it evolved from 802.16-2004, 802.16e naturally embraces the different options specified in its predecessor, which were designed to suit a variety of applications and deployment scenarios. A brief summary of these options is provided in Table 2.7.

Apart from the similarities, there are also numerous differences. Some of the significant changes of 802.16e in comparison with 802.16-2004 are [382, 391] as follows:

- The terminology of Mobile Stations (MSs) is introduced. An MS is also a Subscriber Station (SS) as referred to in the standard.

- MAC layer HandOver (HO) procedures are defined, where an MS migrates from the area serviced by one Base Station (BS) to another. In 802.16e, there are two HO variants:

  1. Break-before-make HO ('hard' HO): An HO where communications with the target BS only commence after relinquishing the link with the previous serving BS.

  2. Make-before-break HO ('soft' HO): An HO where communications with the target BS start before disconnection of the service with the previous serving BS. Two types of soft HO are defined, namely the Fast BS Switching (FBSS), where the MS may rapidly switch from one BS to another, and the Macro Diversity HandOver (MDHO), where an MS establishes links with more than one BS.

In order to facilitate power-efficient MS operations and more efficient HOs, two new power-saving modes, namely the sleep mode and the idle mode, are introduced as complements to the active mode already defined in 802.16-2004.

- The concept of Scalable OFDMA (S-OFDMA) is introduced as part of the significant revision of the original WirelessMAN-OFDMA profile in 802.16-2004. The S-OFDMA architecture supports a wide range of bandwidths, ranging from 1.25 to 20 MHz with a fixed subcarrier spacing of 10.94 kHz for both fixed and mobile operations. This has the great benefit of flexibly addressing the need for various spectrum allocation schemes, potentially supporting global requirements.

- In addition to the four scheduling services supported in 802.16-2004, which are the Unsolicited Grant Service (UGS), real-time Polling Service (rtPS), non-real-time Polling Service (nrtPS) and Best Effort (BE), a new class referred to as the extended real-time Polling Service (ertPS) is included in 802.16e. The ertPS scheduling mechanism exploits the advantages of both UGS and rtPS. It is capable of avoiding the latency of a bandwidth request, while catering for dynamic resource allocations.

- New Multicast and Broadcast Services (MBSs) are introduced. Two types of access to MBSs may be supported, namely single-BS access and multi-BS access. Single-BS access is implemented for transmission over multicast and broadcast transport connections within the coverage area of one BS, while multi-BS access is implemented by transmitting data from service flows with the aid of multiple BSs.

- The security sublayer is redefined in order to remove some security 'holes' identified in 802.16-2004, e.g. the non-existence of BS authentication, and to meet dedicated security requirements for mobile services, which are typically more challenging than those designed for stationary scenarios.

- Enhanced PHY technologies. The MIMO and Adaptive Antenna System (AAS) techniques of 802.16-2004 are substantially enhanced with the aid of more detailed implementation guidelines. Low-Density Parity Check (LDPC) codes are included as a further optional channel coding scheme.

2.3.1.2.3   Other 802.16 Standards

Since the publication of 802.16-2004, there have been a few amendments. Besides 802.16e, recently some further amendments have been completed, while others are still progressing at the time of writing, as summarized in Table 2.6. The objective of these amendments to the original 802.16-2004 is to improve specific system-related aspects, such as adding a more efficient handover functionality, or to include other aspects, such as system management information and procedures.

In December 2005, the IEEE 802.16f-2005 standard [383] was approved, which is the first amendment of 802.16-2004. This specification defines a Management Information Base (MIB) for the MAC and PHY layers, together with relevant management procedures, with the aim of providing high-speed unlicensed MAN access.

Other completed 802.16 amendments include the IEEE 802.16k-2007 [384] and IEEE 802.16g-2007 [385] standards, which were approved in August and December 2007, respectively. The former amended the IEEE 802.1D [392] standard to support bridging of the IEEE 802.16 MAC layer. The latter updates and expands 802.16 by defining further management procedures as enhancements to the air interface specified by 802.16 for fixed and mobile broadband wireless systems. It specifies the related management functions, interfaces and protocol procedures.

The draft amendments which are still pending include the IEEE P802.16h [386] and P802.16j [387] projects. The latest P802.16h working document, Draft 7, was released in June 2008. It specifies improved mechanisms, such as policies and MAC enhancements, to enable the coexistence of license-exempt systems based on 802.16 and to facilitate the coexistence of these systems with primary users.

Furthermore, it aims to improve the coexistence of 802.16 systems in non-exclusively assigned bands. Some of the procedures defined could be applied in other licensing cases, which require improved intersystem coexistence [386].

The P802.16j project, on the other hand, specifies OFDMA PHY and MAC enhancements of 802.16 for licensed bands in order to enable the operation of relay stations. It aims at improving the coverage, throughput and system capacity of 802.16 networks by specifying 802.16 multi-hop relay capabilities and functionalities of interoperable relay stations and BSs [387]. The SS specifications are not changed. The most recent working document of P802.16j is Draft 5, which was released in May 2008.

2.3.1.3   WiMAX Forum

In 802.16-2004 and 802.16e, SC, OFDM and OFDMA techniques are used, as summarized in Table 2.7. Accordingly, there are multiple choices for the MAC layer structure, duplexing combinations, etc. Although the variety of design options is sufficiently flexible for diverse application and deployment scenarios, it is not feasible to have a single system that is compatible with all these specifications. This inevitably results in an interoperability problem. To overcome this problem, the standard should have a limited scope, where the number of design and implementation options should be reduced.

Against this background, the WiMAX Forum was established in June 2001. It is an industry-led, not-for-profit organization of more than 520 companies, including over 200 operators [393]. Similar to the Wi-Fi Alliance, which promotes the Wi-Fi standard as well as its product certification, the WiMAX Forum strives to accelerate the global adoption of WiMAX technology for the provision of broadband wireless services. However, Wi-FI and WiMAX are not direct competitors for wireless broadband subscribers or applications. Although both of them aim at providing wireless connectivity and Internet access, they were designed for different application scenarios and thus become more complementary than competitive. Wi-Fi covers a limited LAN area such as a home or an office, while WiMAX is designed to serve a much larger MAN area with a range of kilometres. Wi-Fi uses unlicensed spectra, in contrast to WiMAX, which typically uses licensed spectra. Furthermore, they have different QoS maintenance mechanisms.

The WiMAX Forum works closely with service providers, regulators, manufacturers, etc., to ensure that WiMAX Forum Certified products are fully interoperable and capable of supporting both fixed and mobile broadband services. Its goal is to certify and promote broadband wireless products based upon the harmonized IEEE 802.16 and European Telecommunications Standards Institute (ETSI) HiperMAN standard [393]. The latter is commonly considered [394] as the European equivalent of IEEE 802.16 (or WiMAX), addressing spectrum access in spectral band ranges under 11 GHz. The WiMAX Forum has been using 802.16-2004 and 802.16e to preselect appropriate options and parameter sets, in order to remove the interoperability barrier and to reduce the associated implementational cost, while remaining compatible with the ETSI HiperMAN standard.

More specifically, this is achieved by defining a few system profiles and certification profiles. A system profile is a collection of specific PHY and MAC layer features, which are selected from the 802.16-2004 or 802.16e standards, respectively. Accordingly, this results in two categories: the fixed WiMAX profiles, which are built upon the WirelessMAN-OFDM PHY of 802.16-2004, and the mobile WiMAX profiles, which are based on the scalable WirelessMAN-OFDMA PHY of 802.16e. It is worth pointing out that the mandatory and optional status of a particular feature within a WiMAX system profile may be different from what it is in the original IEEE standard [376]. On the other hand, a WiMAX certification profile is a particular instantiation of a WiMAX system profile where the operating frequency, channel bandwidth and duplexing mode are also specified. WiMAX equipment is certified based on specific certification profiles for meeting interoperability requirements [376]. If a device is WiMAX Forum Certified, it is both compliant with the 802.16 standard and with devices from other vendors, provided that they are also WiMAX Forum Certified. This will greatly reduce the cost for service providers, since they can flexibly 'plug and play' various certified WiMAX equipment, adapting to new business needs without changing their overall infrastructures.

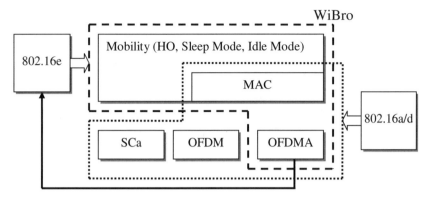

Figure 2.1: WiBro and IEEE 802.16d/e [396].

Table 2.8: Comparison of WiBro, WLAN and cellular systems.

	WiBro	WLAN	Cellular
User data rate	About 1 Mbps	Over 1 Mbps	About 100 kbps
Velocity	120 km/h	Pedestrian	250 km/h
Equipment	Laptop/PDA/cell phone	PC/laptop/PDA	PDA/cell phone
Cell radius	About 1 km	About 100 m	1~3 km

2.3.1.4   WiMAX and WiBro

The Wireless Broadband (WiBro) system constitutes a wireless Internet technology developed by the Korean telecom industry. In February 2002, the Korean government allocated 100 MHz of spectrum in the 2.3 GHz band, and in late 2004, WiBro Phase 1 was standardized by the Telecommunications Technology Association (TTA) of Korea [388]. In June 2006, the first commercial WiBro service was launched in South Korea, by Korean Telecom and SK Telecom Co. Ltd. The service was based on Intel's WiMAX standard and mainly deployed in and around the Seoul area, the capital of South Korea. It is worth pointing out that WiBro is synonymous to mobile WiMAX in Korea. It follows the same standard, namely IEEE 802.16e-2005, the same set of system and certification profiles, as well as the same certification processes as required by mobile WiMAX [395]. A brief comparison of WiBro and the 802.16d/e standards is provided in Figure 2.1.

The features of WiBro, WLAN and cellular systems are given in Table 2.8. It can be seen that WiBro combines the benefits of WLAN and cellular services, providing high data rates, while improving the coverage and mobility. In the future, it will evolve towards 4G networks, where the provision of even higher data rates and mobility are expected, as illustrated in Figure 2.2.

2.3.2   Technical Aspects of WiMAX

WiMAX technology has been based on the IEEE 802.16-2004 and 802.16e-2005 standards, referred to as fixed WiMAX and mobile WiMAX, respectively. Mobile WiMAX is a broadband wireless solution that enables the convergence of mobile and fixed broadband networks with the aid of a common wide area broadband radio access technology and flexible network architecture [398].

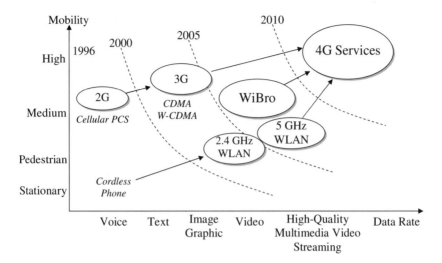

Figure 2.2: WiBro service evolvement [397].

WiMAX technology is based on the S-OFDMA air interface designed for achieving high spectral efficiency and data rates. WiMAX users benefit from broadband connectivity without the need for LOS communications to the BS. A maximum data rate of up to 75 Mbps can be achieved with sufficient bandwidth, simultaneously supporting hundreds of residential and business areas by a single BS [377].

In the following sections, the technical aspects of WiMAX during its previous and ongoing stages of evolution are summarized.

2.3.2.1   WiMAX-I: 802.16-2004 and 802.16e-2005

2.3.2.1.1   OFDMA System Configuration

The concept of scalability was introduced in the IEEE 802.16 WirelessMAN-OFDMA [382] mode by the 802.16 Task Group e (TGe). The S-OFDMA architecture supports a wide range of bandwidth, which spans from 1.25 to 20 MHz, combined with fixed subcarrier spacing for both fixed and portable/mobile uses, in order to address flexibly the need for various spectrum allocation and application requirements.

The scalability is achieved by adjusting the Fast Fourier Transform (FFT) size for various channel bandwidths. In addition to this, 802.16e supports Adaptive Modulation and Coding (AMC) subchannels, Hybrid Automatic Repeat Request (HARQ), efficient uplink subchannelization, MIMO-aided transmit/receive diversity, etc. [399]. Table 2.9 summarizes the key parameters of 802.16e S-OFDMA [382].

2.3.2.1.2   Frame Structure

OFDMA is used for both DL and UL transmissions in 802.16e. The OFDMA PHY mode is based on one of the FFT sizes: 2048 (backward compatible to 802.16-2004 [55]), 1024, 512 and 128. This facilitates the support for the various channel bandwidths. The MS may implement a scanning and search mechanism to detect the DL signal, when performing initial network entry, and this may include the detection of the dynamically configured FFT size and the channel bandwidth employed by the BS.

In licensed bands, the duplexing method shall be either FDD or TDD. FDD MSs may be half-duplex FDD (H-FDD). In license-exempt bands, the duplexing method shall be TDD. Figure 2.3 shows an example of an OFDMA frame (with only mandatory zone) in TDD mode. The OFDMA frame may include multiple zones, such as Partial Usage of SubChannels (PUSC), Full Usage of Subchannels (FUSC),

Table 2.9: WiMAX Release I parameters.

Parameters	Values			
System channel bandwidth (MHz)	5	10	8.75	7
Sampling frequency (F_s in MHz)	5.6	11.2	10	8
FFT size (N_{fft})	512	1024	1024	1024
Number of subchannels	8	16	16	16
Subcarrier frequency spacing (Δf) (kHz)		10.94	9.77	7.81
Useful symbol time ($T_b = 1/\Delta f$) (μs)		91.4	102.4	128
Guard time ($T_g = T_b/8$) (μs)		11.4	12.8	16
OFDMA symbol duration ($T_s = T_b + T_g$) (μs)		102.9	115.2	144
Frame duration (ms)		5	5	5
Number of OFDMA symbols per frame		48	43	34

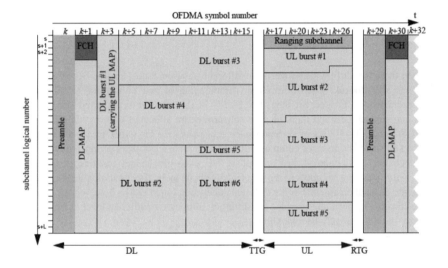

Figure 2.3: Example of a TDD OFDMA frame (with only mandatory zone) [382]. FCH: Feedback Channel; TTG: Transmit Transition Gap; RTG: Receive Transition Gap.

PUSC with all subchannels, optional FUSC, AMC, etc. The transition between zones is indicated in the DL-MAP by the standardized parameter STC_DL_Zone Information Element (IE) or AAS_DL_IE [382]. No DL-MAP or UL-MAP allocations can span multiple zones. Figure 2.4 depicts an OFDMA frame with multiple zones.

2.3.2.1.3 Subcarrier Mapping

For the OFDMA profile of 802.16e, we have $F_s = floor(n \cdot BW/8000) \cdot 8000$ (Hz), where F_s is the sampling frequency and n is the sampling factor, which is dependent on the bandwidth BW. After removing the frequency-domain guard tones or virtual subcarriers from N_{fft}, which is the FFT size, one obtains the set of 'used' subcarriers N_{used}. These used subcarriers are allocated to pilot subcarriers and data subcarriers for both UL and DL.

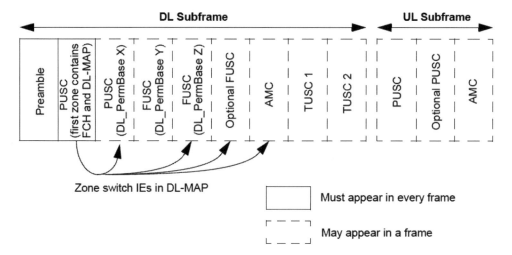

Figure 2.4: Example of a multi-zone OFDMA frame [382]. TUSC: Tile Usage of Subchannels.

However, there is a difference between the different possible zones. For the DL FUSC and PUSC, the pilot tones are allocated first, followed by the mapping of data subcarriers to subchannels (i.e. subbands) exclusively allocated for data. For PUSC in the UL, the set of used subcarriers is first partitioned into subchannels and then the pilot subcarriers are allocated from within each subchannel. Thus, in FUSC, there is one set of common pilot subcarriers for the entire frequency band, while in PUSC of the DL, there is one set of common pilot subcarriers in each major group, which is constituted by a few subchannels. By contrast, in PUSC of the UL, each subchannel contains its own set of pilot subcarriers. After mapping all pilots to the associated subchannels, the remaining used subcarriers will be grouped into data subchannels. For the different zones mentioned above, however, the corresponding subcarrier allocation or permutation rules are different, although a sufficient level of frequency and/or time diversity should generally be maintained.

2.3.2.1.4 Channel Coding

The coding method used as the mandatory scheme in 802.16e is based on tail-biting Convolutional Coding (CC). Optional coding schemes include the Block Turbo Coding (BTC), Convolutional Turbo Codes (CTCs), zero-tailed CC and LDPC codes, which are not included in the earlier 802.16-2004 standard. The encoding block size depends on the number of subchannels allocated and on the modulation scheme specified for the current transmission. For example, the LDPC code specification of 802.16e is summarized in Table 2.10, where n is the codeword length, k is the information block length, and the z factor is the expansion factor which is equal to $n/24$ for a given value of n. Note that, due to the associated subchannelization constraints, the combination of coding parameters and modulation schemes is not arbitrary.

2.3.2.1.5 MIMO Support

The AAS constitutes an integral part of 802.16e, which is included in order to attain a significant system capacity improvement. In 802.16e, the AAS may encompass different MIMO techniques, such as Space–Time Block Coding (STBC), beamforming and Spatial Multiplexing (SM). The STBC adopted is the well-known Alamouti code [400]. For the Open-Loop (OL) AAS, multiple antennas can be used for STBC, SM or their combinations. When the Closed-Loop (CL) AAS is employed, either because we can exploit the channel's reciprocity in the TDD mode, or because the system has explicit receiver

Table 2.10: LDPC block sizes and code rates [55].

n (bits)	n (bytes)	z factor	k (bytes) 1/2	2/3	3/4	5/6	Number of slots QPSK	16-QAM	64-QAM
576	72	24	36	48	54	60	6	3	2
672	84	28	42	56	63	70	7	—	—
768	96	32	48	64	72	80	8	4	—
864	108	36	54	72	81	90	9	—	3
960	120	40	60	80	90	100	10	5	—
1056	132	44	66	88	99	110	11	—	—
1152	144	48	72	96	108	120	12	6	4
1248	156	52	78	104	117	130	13	—	—
1344	168	56	84	112	126	140	14	7	—
1440	180	60	90	120	135	150	15	—	5
1536	192	64	96	128	144	160	16	8	—
1632	204	68	102	136	153	170	17	—	—
1728	216	72	108	144	162	180	18	9	6
1824	228	76	114	152	171	190	19	—	—
1920	240	80	120	160	180	200	20	10	—
2016	252	84	126	168	189	210	21	—	7
2112	264	88	132	176	198	220	22	11	—
2208	276	92	138	184	207	230	23	—	—
2304	288	96	144	192	216	240	24	12	8

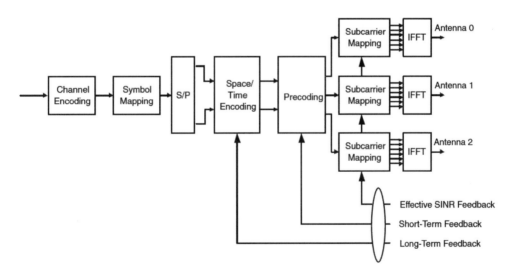

Figure 2.5: The 802.16e-2005 CL MIMO architecture [376].

feedback in the FDD mode, the multiple antennas can be used either for beamforming or for CL MIMO by exploiting transmit antenna precoding techniques.

The general framework of CL MIMO in 802.16e is provided in Figure 2.5. It consists of an OL space–time encoding unit and an MIMO precoding unit. The linear precoding matrix spreads the various parallel streams across the various antennas with the aid of appropriate weighting factors. The precoding matrix is then adapted on a regular basis according to the feedback information gleaned from the receiver.

2.3.2.1.6   Other Aspects

In 802.16e, a preamble is used for initial frame timing and frequency synchronization. Recall that Figure 2.3 shows the OFDMA frame structure in the TDD mode, where the first part of each frame is dedicated to the preamble.

Synchronization is activated during the so-called ranging process, where the BS acquires the signals of new subscribers and adjusts the timing of the existing subscribers through the feedback channel. Synchronization is typically achieved by correlating the received signal against known preamble sequences, taking advantage of the deliberately introduced periodicity of the signal. The results of the correlation evaluation are then passed through a detector to determine whether a legitimate symbol was sent and, if so, to adjust its exact timing.

The development of efficient timing and frequency synchronization algorithms for 802.16e and WiMAX systems is the responsibility of each equipment manufacturer. Some general principles for timing and frequency synchronization can be found in [376].

The 802.16e standard also specifies optional HARQ support, including the following modes:

- Incremental Redundancy (IR) for CTC;

- IR for CC;

- chase combining for all coding schemes.

Chase combining and IR are also referred to as type I and type II HARQ, respectively. These modes can be specified by the normal map and the HARQ map [382].

Other PHY functionalities, such as control mechanisms, channel quality measurements, transmitter/receiver requirements, as well as the specification of the MAC and upper layers, are also detailed in the standard [382]. For the reader's convenience, we summarize the major PHY features of 802.16e in Table 2.11.

2.3.2.2   WiMAX-II: 802.16m

In the International Telecommunications Union – Radio Communications Sector (ITU-R) WP5D meeting held in Dubai during late June 2008, the technical system performance requirements for the IMT-Advanced radio interface [401] were finalized. All IMT-Advanced proposals will be assessed according to this requirement document. The ITU will then assess all technical submissions, review the assessments, select one or more candidate technologies, and develop as well as rectify standards. The evaluation guidelines were finalized in 2008 and it is anticipated that implementation efforts will take place between 2011 and 2012, while a wide deployment may commence by 2015 [402].

In order to meet the anticipated high requirements of IMT-Advanced, in January 2007 the IEEE started the specification of a new version of the 802.16 standard, which aims at increasing the data transmission rates up to 1 Gbps and 100 Mbps for fixed and mobile communications, with improved broadcast, multicast and Voice over Internet Protocol (VoIP) performance, while maintaining backward compatibility with existing WiMAX systems. Under the IEEE 802.16 umbrella, the newly formed Task Group m (TGm) is chartered to develop an amendment of the 802.16 standard, specifying an 'Air Interface for Fixed and Mobile Broadband Wireless Access Systems – Advanced Air Interface', which is referred to as the IEEE 802.16m standard. Also known as WiMAX-II, 802.16m is the only 4G approach evolving from an existing OFDMA technology, namely from 802.16e or WiMAX-I, and will be proposed as a 4G candidate for the ITU's IMT-Advanced systems.

2.3.2.2.1   System Requirements

The ultimate mission of the 802.16m project is to draft an 802.16-compatible standard, which will become a candidate for the IMT-Advanced evaluation process conducted by the ITU-R. To achieve this target, a number of high-level system requirements have been identified by TGm, which are summarized in the 802.16m System Requirement Document (SRD) [403] that was finalized in October 2007.

Table 2.11: Major PHY features of IEEE 802.16e-2005.

Functionality		Configuration/parameters					
Air interface				S-OFDMA			
Modulation				QPSK, 16-QAM, 64-QAM			
System channel bandwidth (MHz)	1.25	5	7	8.75	10		20
Sampling frequency (F_s in MHz)	1.4	5.6	8	10	11.2		22.4
FFT size	128	512	1024	1024	1024		2048
Number of subchannels	2	8	16	16	16		32
Subcarrier spacing (Δf) (kHz)	10.94	10.94	7.81	9.77	10.94		10.94
Useful symbol time ($T_b = 1/\Delta f$) (μs)	91.4	91.4	128	102.4	91.4		91.4
Guard time ($T_g = T_b/8$) (μs)	11.4	11.4	16	12.8	11.4		11.4
OFDMA symbol duration	102.9	102.9	144	115.2	102.9		102.9
($T_s = T_b + T_g$) (μs)							
Number of OFDMA symbols per frame	48	48	34	43	48		48
Guard time ($T_g = T_b/8$) (μs)	11.4	11.4	16	12.8	11.4		11.4
Frame duration (ms)				5			
Duplexing				FDD, TDD			
Frame structure				DL – FUSC/PUSC; UL – PUSC			

Channel coding rates	CC			1/2, 2/3, 3/4		
	BTC			1/2, 3/4		
	CTC	DL		QPSK (1/2, 3/4), 16-QAM (1/2, 3/4), 64-QAM		
				(1/2, 2/3, 3/4, 5/6)		
		UL		QPSK (1/2, 3/4), 16-QAM (1/2, 3/4)		
	LDPC			1/2, 2/3, 3/4, 5/6		

MIMO schemes		AAS with STBC/SM/beamforming
HARQ		IR for CC/CTC
		Chase combining for all channel coding schemes
Channel quality measurements		Mean and standard deviation of RSSI, SINR
Spectral efficiency (bps/Hz)	DL	Raw	2.88
		Useful	1.15
	UL	Raw	2.16
		Useful	0.86
Spectral efficiency per cell (bps/Hz/cell)		DL – 1.2; UL – 0.33
Peak data rates (1×10 MHz 2:1)		DL – 40; UL – 8
(Mbps)		
Average cell throughput (Mbps)		DL – 8; UL – 1
VoIP performance		16 concurrent users/cell/MHz (vehicular speeds up to 120 km/h)
Maximum cell range		$\sim$3.3 km / $\sim$20 km^2
Cell edge performance		Steep drop-off towards cell edge – improved with MIMO
Latency		RTT $<$50 ms

Some of the key 802.16m requirements are [403] as follows:

- *Backward compatibility*: 802.16m shall provide continuing support and interoperability for legacy WirelessMAN-OFDMA [382] equipment, including both MSs and BSs. More specifically, the legacy 802.16e equipment shall be able to coexist with 802.16m equipment without any performance degradation. Additionally, 802.16m shall provide the ability to disable legacy support.

- *Services*: 802.16m should support legacy services more efficiently than the WirelessMAN-OFDMA Reference System [404] as well as facilitate the introduction of new/emerging types of services. Flexible services having different QoS levels should be supported, as required by next-generation mobile networks.

- *Operating frequencies and bandwidths*: 802.16m systems shall operate at RFs less than 6 GHz and be deployable in licensed spectrum allocated to the mobile and fixed broadband services. It shall be able to operate in frequency bands identified for IMT-Advanced. An 802.16m-compliant system shall be capable of coexisting with other IMT-Advanced or IMT-2000 technologies. Scalable bandwidths ranging from 5 to 20 MHz shall be supported.

- *Support of advanced antenna techniques*: 802.16m shall support MIMO, beamforming operation or other advanced antenna techniques for single-user and multi-user scenarios. The minimum number of transmit and receive antennas for the BS and the MS are 2×2 and 1×2, respectively.

- *Minimum peak data rate*: The baseline DL and UL peak data rates are 8 and 2.8 bps/Hz, respectively. When more antennas are used, the rate requirements become 15 and 5.6 bps/Hz for the DL and UL, respectively.

- *Coexistence and co-deployment with other Radio Access Technologies (RATs)*: 802.16m shall support interworking functionality, providing efficient handover and allowing co-deployment with other RATs, including IEEE 802.11 [368], 3GPP GSM/EDGE, UTRA/E-UTRA and 3GPP2 CDMA2000.

- *Mobility and coverage*: 802.16m shall support vehicular speeds of up to 350 km/h and provide a cell coverage of up to even 100 km. For lower mobility and for areas closer to the cell centre, the system's performance should be optimized and degrade gracefully as a function of the vehicular speed and/or the distance from the cell centre.

- *MBS, Location-Based Service (LBS), relaying and self-organization support*: 802.16m shall support Enhanced Multicast and Broadcast Services (E-MBS) for IMT-Advanced multimedia multicast broadcast services in a spectrally efficient manner, provide mechanisms to enable multi-hop relays including those that may involve advanced multiple-antenna techniques, and support self-organizing mechanisms including self-configuration and self-optimization.

For more details on the various aspects of the 802.16m system requirements, we refer to Tables 2.12, 2.13, 2.14 and 2.15.

2.3.2.2.2 System Description

In January 2008, TGm started to develop the 802.16m System Description Document (SDD) [406], which aims to provide a detailed specification of the 802.16m system that meets the requirements set by 802.16m SRD [403]. Since the commencement of the SDD development, significant efforts have been made by the entire project group. A number of Rapporteur Groups (RGs) have been formed to focus on dedicated topics, e.g. the frame structure RG, the multiple access RG, etc. Although technical discussions are still ongoing, some high-level descriptions of the 802.16m system are already in place.

The Network Reference Model (NRM) of the 802.16m system is shown in Figure 2.6. The NRM is a logical representation of the overall network architecture. It identifies functional entities and reference points used for ensuring that interoperability is achieved between functional entities, such as

Table 2.12: The major general requirements for IEEE 802.16m.

Requirement	Description
Backward compatibility	The legacy 802.16e equipment shall be able to coexist with 802.16m equipment without performance degradation; 802.16m shall provide the ability to disable legacy support
Complexity	802.16m should minimize complexity of the architecture and protocols and avoid excessive system complexity. Only the enhancements in those areas where the WirelessMAN-OFDMA Reference System [404] fails to meet the requirements should be provided
Services	802.16m should support legacy services more efficiently than the WirelessMAN-OFDMA Reference System [404] and facilitate new/emerging services. Flexible services with different QoS levels should be supported
Operating frequencies	802.16m systems shall operate at RFs less than 6 GHz in licensed spectrum. It shall be able to operate in frequencies identified for IMT-Advanced. An 802.16m-compliant system shall be capable of coexisting with other IMT-Advanced or IMT-2000 technologies.
Operating bandwidths	Scalable bandwidths from 5 to 20 MHz shall be supported
Duplex schemes	Both TDD and FDD shall be supported, where the FDD mode shall support both full-duplex and half-duplex (H-FDD) MS operation
Support of advanced antenna techniques	802.16m shall support MIMO, beamforming or other advanced antenna techniques for single-user and multi-user scenarios. The minimum number of transmit and receive antennas for BS and MS are 2×2 and 1×2, respectively

the MS, Access Service Network (ASN) and Connectivity Service Network (CSN) [406]. The ASN is defined as a full set of network functions, including 802.16e/m Layer-1 (L1) and Layer-2 (L2) connectivity between an 802.16m BS and an 802.16e/m MS, transfer of Authentication, Authorization and Accounting (AAA) messages to an 802.16e/m subscriber's Home Network Service Provider (H-NSP), network discovery and selection of the subscriber's preferred NSP, relay functionality for establishing Layer-3 (L3) connectivity with an 802.16e/m MS, RRM, ASN/CSN anchored mobility, paging, and so on [406].

An illustration of the 802.16m MS's state transition process is provided in Figure 2.7 [406]. When an MS is switched on, it will enter the initialization state, where the cell selection process is performed by scanning and then synchronizing to a BS's preamble, followed by acquiring the system configuration information through the Broadcast CHannel (BCH). If this is successful, the MS invokes the network entry procedure, requesting for the entry to the selected BS, which results in a number of access state procedures. More specifically, the ranging process is activated first to attain UL synchronization, followed by a 'capability-negotiation' step with the BS. Then the authentication and authorization process will be invoked and the MS will be registered by the BS through the allocation of an 802.16m specific ID. Upon successfully performing the access-state operations of Figure 2.7, the MS will enter the connected state, which consists of three modes, namely the sleep mode, the active mode and the scanning mode. Furthermore, in order to reduce the power consumption, the MS may enter the idle state, which is constituted by two separate modes, namely the 'paging available' mode and the 'paging unavailable' mode. During the idle state, the MS may switch to the access state, if required [406].

IEEE 802.16m supports both TDD and FDD modes, including H-FDD MS operation, in accordance with the 802.16m SRD [403]. OFDMA is adopted as the multiple-access technique in both the DL and UL. Figure 2.8 shows the basic frame structure of 802.16m, where the terminology of superframe and subframe is introduced [406]. Each superframe is 20 ms long and consists of four equal-length radio frames of 5 ms each. Each 5 ms radio frame consists of eight subframes, SF0, ..., SF7, as seen in

Table 2.13: The major functional requirements for IEEE 802.16m.

Requirement	Description	
Peak data rate (bps/Hz)	Baseline	DL (2 × 2): 8.0
		UL (1 × 2): 2.8
	Target	DL (4 × 4): 15.0
		UL (2 × 4): 5.6
Latency (ms)	Data latency	10
	State transition latency	100
	Handover interruption time	30 (intra-frequency mode)
		100 (inter-frequency mode)
QoS	802.16m shall support a range of QoS classes and new applications. When possible, the QoS level should be maintained during handover with other RATs	
RRM	Advanced, efficient RRM shall be supported by 802.16m using appropriate measurement or reporting, interference management and flexible resource allocation mechanisms	
Handover	Handover between 802.16m, legacy systems, and/or other RATs and IEEE 802.21 Media Independent Handover (MIH) Services [405] shall be supported	
Broadcast	E-MBS using a dedicated carrier shall be supported. Switching between broadcast and unicast services shall be supported	
Overhead	802.16m should reduce both user overhead and system overhead when compared with legacy systems without compromising the overall system performance	
Power efficiency	802.16m shall provide support for enhanced power saving functionality for all services and applications	
Coexistence with other RATs	802.16m shall support efficient handover to other RATs, including IEEE 802.11 [368], 3GPP GSM/EDGE, UTRA/E-UTRA and 3GPP2 CDMA2000	

Table 2.14: The major performance requirements for IEEE 802.16m.

Requirement	Description	
User/sector throughputs	These should be two times the WirelessMAN-OFDMA Reference System [404]	
Mobility	0–10 km/h	Optimized system performance
	10–120 km/h	Graceful performance degradation as a function of vehicular speed
	120–350 km/h	Connection should be maintained
Cell coverage	0–5 km	Optimized system performance
	5–30 km	Graceful degradation in system/edge spectral efficiency
	30–100 km	System should be functional (thermal-noise-limited scenario)
MBS/LBS support	Minimum performance requirements for E-MBS (spectral efficiency over 95% coverage areas) are 4 bps/Hz and 2 bps/Hz for an intersite distance of 0.5 km and 1.5 km, respectively. Specific LBS requirements should also be met	

Figure 2.8. Two types of subframes are supported, depending on the length of the cyclic prefix. The so-called type-1 and type-2 subframes are formed by six and seven OFDMA symbols, respectively. The basic frame structure of Figure 2.8 is applied to both TDD and FDD modes, including the H-FDD MS operation. The number of DL/UL switching points in each TDD radio frame is either two or four. The transmission gaps between DL and UL (and vice versa) are required to allow the settling of transients following the switching of the transmitter and receiver circuitry [406].

Table 2.15: The major operational requirements for IEEE 802.16m.

Requirement	Description
Relaying	802.16m should provide mechanisms to enable multi-hop relays with and without advanced antenna techniques
Synchronization	802.16m shall support synchronization of frame timing and frame counters across the entire deployed system, including all BSs and MSs, regardless of carrier frequencies and operators
Co-deployment with other networks	802.16m systems shall be able to be co-deployed in the adjacent licensed frequency bands such as CDMA2000 and 3GPP (GSM, UMTS, HSDPA/HSUPA, LTE), and in unlicensed bands such as 802.11 and 802.15.1 networks, or in the same frequency band on an adjacent carrier such as TD-SCDMA
Self-organization support	802.16m should support self-configuration, enabling real plug-and-play installation of network nodes and cells, and support self-optimization, allowing automated or autonomous optimization of network performance with respect to service availability, QoS, network efficiency and throughput

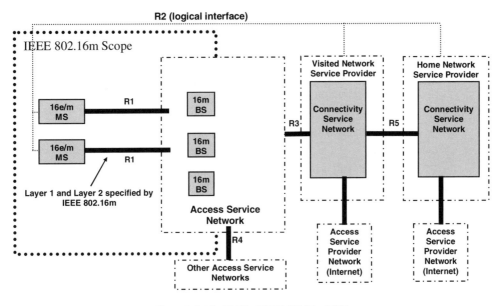

Figure 2.6: The NRM of IEEE 802.16m [406].

In order to ensure backward compatibility with legacy 802.16 equipment, the concept of time zone is introduced in 802.16m, which is applied to both the TDD and FDD modes. The time zone is defined in terms of a non-zero integer number of consecutive subframes [406]. An example of TDD time zones is portrayed in Figure 2.9. More specifically, two zones are multiplexed in the time domain for the DL, one for 802.16m and the other for legacy systems. For UL transmissions, these will be multiplexed in both the time and frequency domains. Note that the legacy MS can only be scheduled in the legacy zones, while the 802.16m MS can be scheduled in both zones. If there is no legacy system in the network, the legacy zones will be replaced by the expanded 802.16m zone for maximizing the system's efficiency [406].

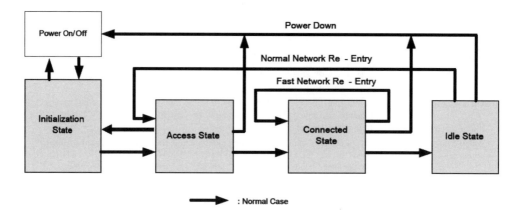

Figure 2.7: The IEEE 802.16m MS state transition diagram [406].

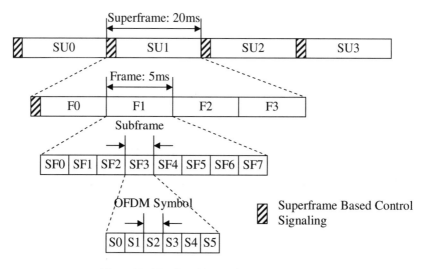

Figure 2.8: The IEEE 802.16m frame structure [406].

2.3.3   The Future of WiMAX

In October 2007, the ITU formally accepted IEEE 802.16e-2005 Mobile WiMAX as the sixth standardized terrestrial radio interface. This specific implementation, known as 'IMT-2000 OFDMA TDD WMAN', is the version of the IEEE 802.16 standard supported in a profile developed for certification purposes by the WiMAX Forum [407]. This will encourage its acceptance by regulatory authorities and operators for allocating cellular spectrum and for future WiMAX deployment.

New products and impressive technical deployments have been stimulating the penetration of WiMAX across the globe. Its rapid speed is driven by new equipment arriving from leading manufacturers and suppliers, as well as by an increasing number of WiMAX trials and deployments supported by telecom service providers. This list includes Alcatel-Lucent, Alvarion, AT&T, BT, Clearwire, Fujitsu, Intel, Korea Telecom, Motorola, Nokia, Nortel, Redline Communications, Samsung, Sequans, SR Telecom, Verizon, etc., who are actively paving the way for WiMAX evolution. It is forecast that by 2010 the worldwide WiMAX market will reach $3.5 billion and account for 4% of all broadband usage [402].

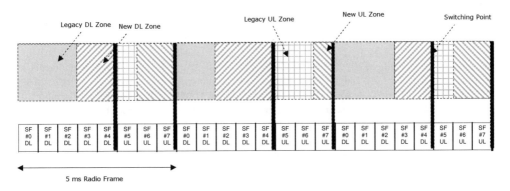

Figure 2.9: The time zones in the IEEE 802.16m TDD mode [406].

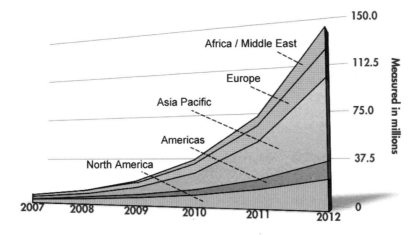

Figure 2.10: Forecast of WiMAX users by region 2007–2012 [408].

At the end of 2007, there were a total of 181 WiMAX operators globally. The WiMAX forum expects this number to rise to 538 by 2012. Among the total of 234 countries in the world, the number of those using WiMAX is anticipated to rise from 94 at the end of 2007 to 201 in 2012 [408]. It is also forecast that by 2012 there will be about 134 million WiMAX users worldwide, with the main growth coming from the Asia Pacific and North America regions, as shown in Figure 2.10 [408].

In June 2008, the WiMAX Forum announced the first WiMAX Forum Certified Mobile WiMAX products, including four BSs and six mobile terminals, provided by eight WiMAX Forum member companies, namely Airspan Networks, Alvarion, Beceem, Intel, Motorola, Samsung, Sequans and ZyXEL [409]. The products are designed for the 2.5 GHz profile.

The latest official schedule for the IEEE 802.16m standardization project was released in July 2008. It aims to be in line with ITU's time line of call for IMT-Advanced proposals – a critical time for the further success of WiMAX in the future.

2.4 Chapter Summary

In this chapter, we have reviewed a range of major international standards that adopt OFDM. We first briefly considered the history, milestones and main contributions in the OFDM literature in

Section 1.1. Then in Section 2.1 we reviewed the Wi-Fi standard family, followed by Section 2.2 where the 3GPP LTE cellular standard was highlighted. A more detailed introduction of WiMAX was presented in Section 2.3. More specifically, we briefly reviewed the history of WiMAX in Section 2.3.1, spanning from the early IEEE 802.16 standards in Section 2.3.1.2 to their recent amendments in Sections 2.3.1.2.1, 2.3.1.2.2 and 2.3.1.2.3. Early WiMAX solutions were based on the WirelessMAN OFDM PHY of 802.16-2004, which is also known as 'fixed WiMAX'. By contrast, as the successor of 802.16-2004, the 802.16e standard incorporates mobility support and thus is referred to as 'mobile WiMAX'. In Section 2.3.1.3, the historic background of the WiMAX Forum was summarized, where its mission and achievements were briefly reviewed. Furthermore, we provided a brief introduction of the Korean WiMAX standard WiBro in Section 2.3.1.4. Being the world's largest commercial mobile WiMAX deployment, the WiBro network is accelerating its evolution towards the 4G era.

In Section 2.3.2, we focused our attention on the technical aspects of WiMAX. More specifically, the major PHY specifications of WiMAX-I, which were based on 802.16-2004 and 802.16e, were provided in Section 2.3.2.1. These included the OFDMA system configuration, frame structure, subcarrier mapping, channel coding, MIMO support, etc. Moreover, in Section 2.3.2.2 the technical requirements and system description of the 802.16m draft standard were presented. It is expected that the 802.16m standard will be finalized during 2010. Finally, a brief discussion of the progress and future of WiMAX was offered in Section 2.3.3.

Part I

Coherently Detected
SDMA-OFDM Systems

Channel Coding Assisted STBC-OFDM Systems

3.1 Introduction

Increasing market expectations for 3G mobile radio systems show a great demand for a wider range of services spanning from voice to high-rate data services required for supporting mobile multimedia communications. This leads to higher technical specifications for existing and future communication systems, which have to support data rates as high as 144 kbps in vehicular, 384 kbps in outdoor-to-indoor and 2 Mbps in indoor and picocelluar environments [410].

The employment of multiple antennas constitutes an effective way of achieving an increased capacity. The classic approach is to use multiple-receiver antennas and exploit Maximum Ratio Combining (MRC) of the received signals for the sake of improving the system's performance [411, 412]. However, the performance improvement of MRC is achieved at the cost of increasing the complexity of the Mobile Stations (MSs). Alternatively, MRC may be employed at the Base Stations (BSs), which support numerous MSs. While this scheme provides diversity gain for the BSs' receivers, the MSs cannot benefit from it.

Employing multiple transmitters rather than receiver antennas at the BSs constitutes a further design option in this context. Since transmitter diversity techniques are proposed for employment at the BSs, it is possible to enhance the system's integrity by upgrading the BSs. Alamouti [400] introduced an attractive scheme, which uses two transmitters in conjunction with an arbitrary number of receivers for communications in non-dispersive Rayleigh fading channels. Tarokh *et al.* [413, 414] generalized Alamouti's scheme to an arbitrary number of transmitters. These schemes introduced Space–Time Block Codes (STBCs), which show remarkable encoding and decoding simplicity, while achieving a good performance.

3.2 Space–Time Block Codes

In this section we will present the basic principles of space–time block codes. Before providing more details, let us first consider a simple STBC system communicating over uncorrelated Rayleigh fading channels, as shown in Figure 3.1.

MIMO-OFDM for LTE, Wi-Fi and WiMAX   Lajos Hanzo, Yosef Akhtman, Li Wang and Ming Jiang
© 2011 John Wiley & Sons, Ltd

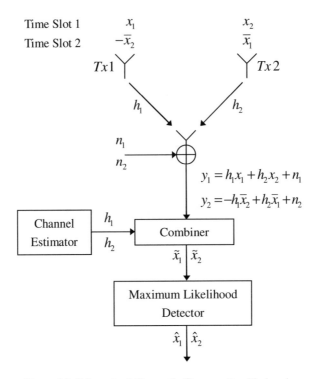

Figure 3.1: Schematic of Alamouti's G_2 space–time block code.

3.2.1   Alamouti's G_2 STBC

The system contains two transmitter antennas and one receiver antenna. The philosophy of Alamouti's G_2 STBC is as follows.

In a conceptually simple approach we could argue that the achievable throughput of the system may be doubled with the aid of the two transmitter antennas, if their signal could be separated by the receiver. This task may be viewed as analogous to multi-user detection, where the two signals would arrive at the BS's receiver from two geographically separated users. From a simple conceptual perspective the Bell Labs Layered Space–Time (BLAST) transmission scheme [105] adopts a similar principle for increasing the achievable throughput of multiple-transmitter antenna-based systems.

By contrast, Alamouti's approach is different, since the aim is to achieve a diversity gain rather than to increase the achievable throughput. This is achieved by extending the duration of time allocated to the transmission of a symbol by a factor of two and transmitting two independently faded and 'appropriately transformed' replicas of the symbol using each of the two antennas. The independence of the two channels may be ascertained by positioning the transmitter antennas $Tx\,1$ and $Tx\,2$ sufficiently far apart, e.g. at a distance of 10λ, where λ is the wavelength. Thus, during the first time slot, the information symbols x_1 and x_2 are transmitted by transmitter antennas $Tx\,1$ and $Tx\,2$, respectively, and again each of the two symbols is transmitted through an independently faded channel. For the sake of avoiding the channel-induced inter-symbol interference between the two time slots, both channels' path gains are assumed to be constituted by a single propagation path given by

$$h_1 = |h_1|e^{j\alpha_1} \tag{3.1}$$

$$h_2 = |h_2|e^{j\alpha_2}, \tag{3.2}$$

where $|h_1|$, $|h_2|$ are the fading magnitudes and α_1, α_2 are the corresponding phase rotations. The complex fading envelopes are assumed to remain constant during the two consecutive time slots [314], which is expressed as

$$h_1 = h_1(t = 1) = h_1(t = 2) \tag{3.3}$$
$$h_2 = h_2(t = 1) = h_2(t = 2). \tag{3.4}$$

Therefore we receive the composite signal y_1 constituted by the superposition of the two transmitted symbols through both channels:

$$y_1 = h_1 x_1 + h_2 x_2 + n_1, \tag{3.5}$$

where n_1 is a complex noise sample. As mentioned above, during the second time slot, a transformed version of the signals x_1 and x_2 is transmitted. Since the consecutive time slots are not faded independently, no diversity gain would be achieved if we mapped these 'transformed' replicas of x_1 and x_2 to the same transmitter antenna as during the first time slot. Hence we now swap the assignment of the time slots to transmitter antennas, since this way their independent fading may be ascertained. More explicitly, during the second time slot, the negative version of the conjugate of x_2 and the conjugate of x_1 are sent by transmitter antennas $Tx\,1$ and $Tx\,2$, respectively. However, as mentioned above, the envelopes of each of the channels associated with the two transmitters are assumed to be the same as during the first time slot, hence we get the second transmitter's signal y_2 as

$$y_2 = -h_1 \bar{x}_2 + h_2 \bar{x}_1 + n_2, \tag{3.6}$$

where n_2 is a complex noise sample. If the channels' characteristics are known to the receiver, i.e. we have a perfect channel estimator, the information symbols x_1 and x_2 can be readily separated in the combiner of Figure 3.1, yielding

$$\begin{aligned}
\tilde{x}_1 &= \bar{h}_1 y_1 + h_2 \bar{y}_2 \\
&= \bar{h}_1 h_1 x_1 + \bar{h}_1 h_2 x_2 + \bar{h}_1 n_1 - h_2 \bar{h}_1 x_2 + h_2 \bar{h}_2 x_1 + h_2 \bar{n}_2 \\
&= (|h_1|^2 + |h_2|^2) x_1 + \bar{h}_1 n_1 + h_2 \bar{n}_2 \tag{3.7} \\
\tilde{x}_2 &= \bar{h}_2 y_1 - h_1 \bar{y}_2 \\
&= \bar{h}_2 h_1 x_1 + \bar{h}_2 h_2 x_2 + \bar{h}_2 n_1 + h_1 \bar{h}_1 x_2 - h_1 \bar{h}_2 x_1 - h_1 \bar{n}_2 \\
&= (|h_1|^2 + |h_2|^2) x_2 + \bar{h}_2 n_1 - h_1 \bar{n}_2, \tag{3.8}
\end{aligned}$$

where $\tilde{x}_1$ and $\tilde{x}_2$ are the extracted noisy signals. Then $\tilde{x}_1$ and $\tilde{x}_2$ will be forwarded to the maximum likelihood detector of Figure 3.1, which determines the most likely transmitted symbols, namely $\hat{x}_1$ and $\hat{x}_2$, by simply outputting the specific legitimate transmitted symbol, which has the lowest Euclidean distance from the received channel-impaired symbol [314].

The equations above, which are based on the one-receiver scheme, can be generalized to multiple-receiver aided schemes, where the received signal y_t^j arriving at receiver j during time slot t is [314]

$$y_t^j = \sum_{i=1}^{p} h_{i,j} x_t^i + n_t^j, \tag{3.9}$$

where p is the number of transmitters, $h_{i,j}$ is the complex-valued path gain between the transmitter i and receiver j, x_t^i is the space–time coded symbol transmitted by transmitter i during time slot t, and n_t^j is the noise sample at receiver j in time slot t. Accordingly, the extracted noisy signals become [314]

$$\tilde{x}_1 = \sum_{j=1}^{q} [(|h_{1,j}|^2 + |h_{2,j}|^2) x_1 + \bar{h}_{1,j} n_1^j - h_{2,j} \bar{n}_2^j] \tag{3.10}$$

$$\tilde{x}_2 = \sum_{j=1}^{q} [(|h_{1,j}|^2 + |h_{2,j}|^2) x_2 + \bar{h}_{2,j} n_1^j - h_{1,j} \bar{n}_2^j]. \tag{3.11}$$

3.2.2   Encoding Algorithm

In the previous section, we provided an example of a simple STBC communication system. Let us now discuss STBCs in more depth.

3.2.2.1   Transmission Matrix

A generic STBC is defined by an $(n \times p)$-dimensional transmission matrix $\mathbf{G}$, where the entries of the matrix $\mathbf{G}$ are linear combinations of the k input symbols $x_1, x_2, \ldots, x_k$ and their conjugates. Each symbol x_i $(i = 1, \ldots, k)$ conveys b original information bits according to the relevant signal constellation that has $M = 2^b$ constellation points, and hence can be regarded as information symbols. Thus, $(k \times b)$ input bits are conveyed by each $(n \times p)$ block. The general form of the transmission matrix of SRBCs is given by Equation (3.12):

$$\mathbf{G} = \begin{bmatrix} g_{11} & g_{12} & \cdots & g_{1p} \\ g_{21} & g_{22} & \cdots & g_{2p} \\ \vdots & \vdots & \vdots & \vdots \\ g_{n1} & g_{n2} & \cdots & g_{np} \end{bmatrix}, \tag{3.12}$$

where the entries g_{ij} $(i = 1, \ldots, n; \ j = 1, \ldots, p)$ represent the linear combinations of the information symbols x_i $(i = 1, \ldots, k)$ and their conjugates. In the transmission matrix $\mathbf{G}$, which can be viewed as a space–time encoding block, the number of rows (namely n) is equal to the number of time slots, while the number of columns (namely p) is equal to the number of transmitter antennas. For example, during time slot $i = 1$, the encoded symbols $g_{11}, g_{12}, \ldots, g_{1p}$ are transmitted simultaneously from transmitter antennas $Tx\,1, Tx\,2, \ldots, Tx\,p$, respectively.

References [400, 413] have defined a range of STBCs. Different designs of the transmission matrix seen in Equation (3.12) will result in different encoding algorithms and code rates. Generally, the code rate is defined as

$$R = k/n, \tag{3.13}$$

where k is the number of possible input information symbols, and n is the number of time slots.

3.2.2.2   Encoding Algorithm of the STBC G_2

From Section 3.2.1 and Equation (3.12) we can readily derive the G_2 transmission matrix in the form of

$$G_2 = \begin{bmatrix} g_{11} & g_{12} \\ g_{21} & g_{22} \end{bmatrix} \tag{3.14}$$

or, more specifically, as

$$G_2 = \begin{bmatrix} x_1 & x_2 \\ -\bar{x}_2 & \bar{x}_1 \end{bmatrix}. \tag{3.15}$$

From Equation (3.15), it can be readily seen that there are $n = 2$ rows in the G_2 matrix associated with two time slots and $p = 2$ columns corresponding to two transmitter antennas, as we have already seen in the example of Figure 3.1. Since there are $k = 2$ input symbols, namely x_1 and x_2, the code rate of G_2 is $R = k/n = 1$.

3.2.2.3   Other STBCs

The G_2 STBC was first proposed by Alamouti [400] in 1998. This code has attracted much attention because of its appealing simplicity compared with the family of Space–Time Trellis Codes (STTCs) proposed in [415–418], although this simplicity was achieved at the cost of a performance loss. Later, Tarokh et al. [413] extended Alamouti's scheme to multiple transmitters, which led to other STBCs, such as the three-transmitter code G_3 and four-transmitter code G_4. The transmission matrix of G_3 [413] is defined as

$$
G_3 = \begin{bmatrix}
x_1 & x_2 & x_3 \\
-x_2 & x_1 & -x_4 \\
-x_3 & x_4 & x_1 \\
-x_4 & -x_3 & x_2 \\
\bar{x}_1 & \bar{x}_2 & \bar{x}_3 \\
-\bar{x}_2 & \bar{x}_1 & -\bar{x}_4 \\
-\bar{x}_3 & \bar{x}_4 & \bar{x}_1 \\
-\bar{x}_4 & -\bar{x}_3 & \bar{x}_2
\end{bmatrix}
\tag{3.16}
$$

and the transmission matrix of G_4 [413] is defined as

$$
G_4 = \begin{bmatrix}
x_1 & x_2 & x_3 & x_4 \\
-x_2 & x_1 & -x_4 & x_3 \\
-x_3 & x_4 & x_1 & -x_2 \\
-x_4 & -x_3 & x_2 & x_1 \\
\bar{x}_1 & \bar{x}_2 & \bar{x}_3 & \bar{x}_4 \\
-\bar{x}_2 & \bar{x}_1 & -\bar{x}_4 & \bar{x}_3 \\
-\bar{x}_3 & \bar{x}_4 & \bar{x}_1 & -\bar{x}_2 \\
-\bar{x}_4 & -\bar{x}_3 & \bar{x}_2 & \bar{x}_1
\end{bmatrix} .
\tag{3.17}
$$

From Equations (3.16) and (3.17), we can see that the code rates of G_3 and G_4 are reduced to a half, which degrades the bandwidth efficiency. However, the STBCs H_3 and H_4 of [413] mitigate this problem, since their code rate is $3/4$. The transmission matrices of H_3 and H_4 [413] are defined as follows:

$$
H_3 = \begin{bmatrix}
x_1 & x_2 & x_3/\sqrt{2} \\
-\bar{x}_2 & \bar{x}_1 & x_3/\sqrt{2} \\
\dfrac{\bar{x}_3}{\sqrt{2}} & \dfrac{\bar{x}_3}{\sqrt{2}} & \dfrac{(-x_1 - \bar{x}_1 + x_2 - \bar{x}_2)}{2} \\
\dfrac{\bar{x}_3}{\sqrt{2}} & -\dfrac{\bar{x}_3}{\sqrt{2}} & \dfrac{(x_2 + \bar{x}_2 + x_1 - \bar{x}_1)}{2}
\end{bmatrix}
\tag{3.18}
$$

$$
H_4 = \begin{bmatrix}
x_1 & x_2 & x_3/\sqrt{2} & x_3/\sqrt{2} \\
-\bar{x}_2 & \bar{x}_1 & x_3/\sqrt{2} & -x_3/\sqrt{2} \\
\dfrac{\bar{x}_3}{\sqrt{2}} & \dfrac{\bar{x}_3}{\sqrt{2}} & \dfrac{(-x_1 - \bar{x}_1 + x_2 - \bar{x}_2)}{2} & \dfrac{(-x_2 - \bar{x}_2 + x_1 - \bar{x}_1)}{2} \\
\dfrac{\bar{x}_3}{\sqrt{2}} & \dfrac{\bar{x}_3}{\sqrt{2}} & \dfrac{(x_2 + \bar{x}_2 + x_1 - \bar{x}_1)}{2} & \dfrac{(-x_1 - \bar{x}_1 - x_2 + \bar{x}_2)}{2}
\end{bmatrix} .
\tag{3.19}
$$

In conclusion, the parameters of the STBCs mentioned above are summarized in Table 3.1.

3.2.3 Decoding Algorithm

In this section, two algorithms are briefly discussed, which are widely used for decoding STBCs. The maximum likelihood (ML) decoding algorithm generates hard-decision outputs, while the Maximum-A-Posteriori (MAP) decoding algorithm is capable of providing soft outputs, which readily lend themselves to channel coding for the sake of achieving further performance improvements.

3.2.3.1 Maximum Likelihood Decoding

ML decoding of STBCs can be achieved using simple linear processing at the receiver, thus maintaining a low decoding complexity. As mentioned in Section 3.2.1, when the space–time coded symbols are transmitted over different channels and arrive at the receiver during time slot t $(t = 1, \ldots, n)$, we will

Table 3.1: Parameters of the STBCs.

Space–time block code	Code rate (R)	Number of transmitters (p)	Number of input symbols (k)	Number of time slots (n)
G_2	1	2	2	2
G_3	1/2	3	4	8
G_4	1/2	4	4	8
H_3	3/4	3	3	4
H_4	3/4	4	3	4

have the received signal y_t^j expressed in Equation (3.9). With the proviso of having a perfect channel estimator, the receiver computes the decision metric

$$\sum_{t=1}^{n}\sum_{j=1}^{q}\left|y_t^j - \sum_{i=1}^{p}h_{i,j}x_t^i\right|^2 \tag{3.20}$$

over all indices $i = 1,\ldots,n;\ j = 1,\ldots,p$, and decides in favour of the specific entry that minimizes the sum.

Alamouti [400] first proposed a simple ML decoding algorithm for the G_2 STBC. Tarokh *et al.* [414] extended it for the STBCs summarized in Table 3.1. The algorithm exploits the orthogonal structure of the STBCs for decoupling the signals transmitted from different antennas rather than requiring their joint detection. According to [414], low-complexity signal processing may be invoked for separating the channel-impaired transmitted signal y_t^j into k decision metrics, each of which corresponds to the channel-impaired version of a specific information symbol of the set x_i, $i = 1,\ldots,k$. For example, for the G_2 STBC, we will minimize the decision metric

$$\left|\left[\sum_{j=1}^{q}(y_1^j\bar{h}_{1,j} + \bar{y}_2^j h_{2,j})\right] - x_1\right|^2 + \left(\sum_{j=1}^{q}\sum_{i=1}^{2}|h_{i,j}|^2 - 1\right)|x_1|^2 \tag{3.21}$$

for decoding x_1 and the decision metric

$$\left|\left[\sum_{j=1}^{q}(y_1^j\bar{h}_{2,j} - \bar{y}_2^j h_{1,j})\right] - x_2\right|^2 + \left(\sum_{j=1}^{q}\sum_{i=1}^{2}|h_{i,j}|^2 - 1\right)|x_2|^2 \tag{3.22}$$

for decoding x_2. The relevant decision metrics for decoding other STBC codes can be found in [414].

3.2.3.2 Maximum-A-Posteriori Decoding

As seen in Equations (3.21) and (3.22), hard decisions would have to be made in order to generate the decoded outputs, which are the most likely transmitted information symbols. In other words, the usual ML detection is a hard-decoding method. However, in most practical systems various channel codes, such as Low-Density Parity Check (LDPC) codes [419] or turbo codes [314, 420, 421], may have to be combined with STBCs for the sake of further improving the system's performance. In this case, the space–time decoder must provide soft outputs, which can be efficiently utilized by the channel decoder.

Bauch [422] presented a simple symbol-by-symbol MAP algorithm for decoding STBCs. According to [422], the a posteriori probability of each information symbol x_i $(i = 1,\ldots,k)$ is

$$\ln P(x_i|\mathbf{y}_1,\mathbf{y}_2,\ldots,\mathbf{y}_q) = \text{const} + \ln P(\mathbf{y}_1,\mathbf{y}_2,\ldots,\mathbf{y}_q|x_i) + \ln P(x_i), \tag{3.23}$$

where $\mathbf{y}_j = [y_1^j, y_2^j,\ldots,y_n^j]$ $(j = 1,\ldots,q)$ represents the received signal vector at receiver j during the period spanning from time slot 1 to time slot n. For example, for the STBC G_2 $(k = 2, n = 2)$ the

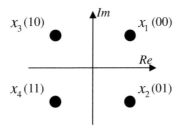

Figure 3.2: Gray-coded QPSK constellation.

a posteriori probabilities are

$$\ln P(x_1|y_1, y_2, \ldots, y_q)$$
$$= \text{const} - \frac{1}{2\sigma^2}\left\{\left|\left[\sum_{j=1}^{q}(y_1^j \bar{h}_{1,j} + \bar{y}_2^j h_{2,j})\right] - x_1\right|^2 + \left(\sum_{j=1}^{q}\sum_{i=1}^{2}|h_{i,j}|^2 - 1\right)|x_1|^2\right\} + \ln P(x_1)$$

(3.24)

$$\ln P(x_2|y_1, y_2, \ldots, y_q)$$
$$= \text{const} - \frac{1}{2\sigma^2}\left\{\left|\left[\sum_{j=1}^{q}(y_1^j \bar{h}_{2,j} - \bar{y}_2^j h_{1,j})\right] - x_2\right|^2 + \left(\sum_{j=1}^{q}\sum_{i=1}^{2}|h_{i,j}|^2 - 1\right)|x_2|^2\right\} + \ln P(x_2).$$

(3.25)

We can see that Equations (3.24) and (3.25) are quite similar to Equations (3.21) and (3.22), respectively. In fact, it can be shown that Bauch's MAP algorithms can be extended for decoding other STBCs, such as G_3, G_4, H_3 and H_4, and the corresponding algorithms also resemble the ML algorithms [314] discussed in Section 3.2.3.1. Given the a posteriori probabilities of the symbols, we can derive the corresponding a posteriori probabilities of the bits (i.e. the corresponding soft outputs) using the symbol-to-bit probability conversion of

$$P(d_i = 0) = \sum_j P(x_j|y_1, y_2, \ldots, y_q), \quad \forall x_j = (d_1 \ldots d_i \ldots d_b), \quad d_i = 0, \quad (3.26)$$

$$P(d_i = 1) = \sum_j P(x_j|y_1, y_2, \ldots, y_q), \quad \forall x_j = (d_1 \ldots d_i \ldots d_b), \quad d_i = 1, \quad (3.27)$$

where $P(d_i = 0)$ or $P(d_i = 1)$ represents the probability of the ith bit, namely d_i, of the b-bit symbol being zero and one, respectively. Let us consider the QPSK modulation scheme for example. The phasor constellation of QPSK is shown in Figure 3.2.

As seen in Figure 3.2, each constellation point consists of 2 bits, hence the constellation points can be represented as

$$x_j = (d_2 d_1), \quad j = \{1, \ldots, 4\}, \quad (3.28)$$

where $d_i = \{0, 1\}$. With the aid of Equations (3.26) and (3.27), we generate the a posteriori probabilities of each bit, taking into account that d_1 assumes a value of zero only in x_1 and x_3, while d_2 in x_1 and x_2, etc., yielding

$$\begin{cases} P(d_1 = 0) = P(x_1|y_1, y_2, \ldots, y_q) + P(x_3|y_1, y_2, \ldots, y_q) \\ P(d_1 = 1) = P(x_2|y_1, y_2, \ldots, y_q) + P(x_4|y_1, y_2, \ldots, y_q) \\ P(d_2 = 0) = P(x_1|y_1, y_2, \ldots, y_q) + P(x_2|y_1, y_2, \ldots, y_q) \\ P(d_2 = 1) = P(x_3|y_1, y_2, \ldots, y_q) + P(x_4|y_1, y_2, \ldots, y_q). \end{cases} \quad (3.29)$$

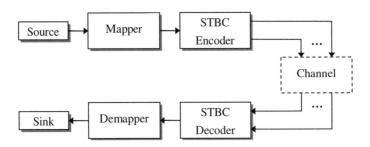

Figure 3.3: Schematic diagram of a simple system employing STBCs.

Then the relevant soft outputs can be forwarded to the channel decoders, which will make a hard decision to finally decode the received signals.

In practical applications, however, the Max-Log-MAP [423, 424] or Log-MAP [425] algorithms are usually preferred [314], since either one can lower the computational complexity to some degree. The Max-Log-MAP algorithm was proposed by both Koch and Baier [423] and Erfanian *et al.* [424] for reducing the complexity of the MAP algorithm. This technique transfers the computation into the logarithmic domain and invokes an approximation for dramatically reducing the complexity imposed. As a consequence of using an approximation, its performance is suboptimal. However, Robertson *et al.* [425] later proposed the Log-MAP algorithm, which partially corrected the approximation invoked in the Max-Log-MAP algorithm. Hence the performance of the Log-MAP algorithm is similar to that of the MAP algorithm, which is achieved at a significantly lower complexity. More details of the Max-Log-MAP and Log-MAP algorithms may be found in [314]. In our STBC soft decoder, the Log-MAP algorithm is employed. In the rest of this section we will characterize the achievable performance of a range of STBC schemes.

3.2.4   System Overview

Figure 3.3 shows the structure of the simulation system used. As seen in the figure, the mapper maps the source information bits to relevant phasor constellation points employed by the specific modulation scheme used. Then the symbols are forwarded to the space–time encoder. The encoded symbols are then transmitted through different antennas and arrive at the receiver(s), where the received signals will be decoded by the space–time decoder. Finally, the demapper converts the STBC-decoded symbols back to the information bits and the Bit Error Rate (BER) is calculated. The parameters of all the STBCs studied are summarized in Table 3.1. The following assumptions were used:

- Signals are transmitted over uncorrelated Rayleigh fading channels.

- The channels are quasi-static so that the path gains are constant across n consecutive time slots, corresponding to the n rows of the STBCs' transmission matrix.

- The average signal power received from each transmitter antenna is the same.

- The receiver has perfect knowledge of the channels' fading amplitudes.

These assumptions simplify the simulations to a degree, therefore the system concerned is not a realistic one. However, since the experimental circumstances are identical for all performance comparisons, the results characterize the relative performance of various STBCs.

3.2.5   Simulation Results

In the previous sections, the basic principles of STBCs as well as the simulation conditions were presented. In this section, our simulation results will be provided for the sake of comparatively studying the performance of the various STBCs of Table 3.1.

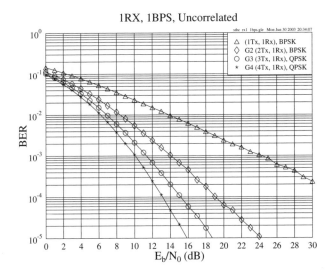

Figure 3.4: The BER versus E_b/N_0 performance of the $\boldsymbol{G_2}$, $\boldsymbol{G_3}$ and $\boldsymbol{G_4}$ STBCs of Table 3.1 at an effective throughput of **1 BPS** using **one receiver** over **uncorrelated** Rayleigh fading channels.

3.2.5.1 Performance over Uncorrelated Rayleigh Fading Channels

Performance at the Throughput of 1 BPS   Figure 3.4 compares the performance of the G_2, G_3 and G_4 STBCs in conjunction with one receiver at the throughput of 1 BPS (Bits Per Symbol) over the uncorrelated Rayleigh fading channel. Binary Phase-Shift Keying (BPSK) modulation is employed in conjunction with the space–time code G_2, while QPSK modulation is considered with the half-rate space–time codes G_3 and G_4 so that the system throughput remains at 1 BPS. From Figure 3.4 we can see that at the BER of 10^{-5} the G_3 and G_4 codes provide an approximately 5.5 and 7.5 dB gain over the G_2 code, respectively. If we add one more receiver in the context of all of these schemes, as seen in Figure 3.5, the relevant E_b/N_0 gain of the G_3 and G_4 schemes over the G_2 arrangement reduces to 2.5 and 3.5 dB, respectively. This may suggest that the G_2 code using two receivers has achieved most of the attainable diversity gain [314], and hence, even if we further increase the number of transmitter antennas, the performance cannot be significantly improved.

Performance at the Throughput of 2 BPS   In Figure 3.6 the performances of the STBCs using one receiver and having a throughput of about 2 BPS over uncorrelated Rayleigh fading channels are compared. In order to meet the 2 BPS throughput criteria, QPSK modulation is used for the G_2 code, while 16-QAM is employed for the G_3 and G_4 arrangements, as the latter ones are half-rate codes. However, for the $\frac{3}{4}$-rate codes H_3 and H_4, an exact throughput of 2 BPS cannot be achieved. Thus we chose 8-PSK and the throughput of the H_3 and H_4 schemes became 2.25 BPS, which is close to 2 BPS.

From Figure 3.6 we can see that when the Signal-to-Noise Ratio (SNR) is low, i.e. E_b/N_0 is below 12.5 dB, the STBC G_2 performs better than other codes, although the performance difference is not significant. When E_b/N_0 increases to a value higher than 12.5 dB, however, the G_4 code outperforms other codes, having an approximately 1 dB gain over the H_4 code at the BER of 10^{-5}. Similarly, the G_3 code achieves an approximately 1 dB gain over the H_3 code at the BER of 10^{-5}. We may note that, although the G_3 and G_4 codes employ a higher-order 16-QAM modulation scheme, which is more vulnerable to channel effects than the lower-order 8-PSK modulation scheme used by the H_3 and H_4 codes, the former performs slightly better than the latter.

In the scenario of using two receivers, however, the G_2 code stands out in comparison with all the candidates, as Figure 3.7 indicates. This result suggests that the attainable diversity gain has

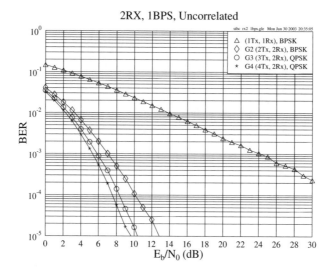

Figure 3.5: The BER versus E_b/N_0 performance of the G_2, G_3 and G_4 STBCs of Table 3.1 at an effective throughput of **1 BPS** using **two receivers** over **uncorrelated** Rayleigh fading channels.

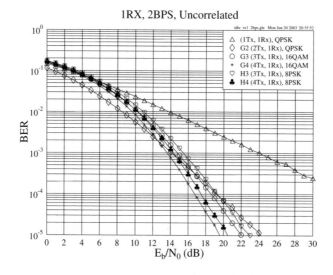

Figure 3.6: The BER versus E_b/N_0 performance of the G_2, G_3, G_4, H_3 and H_4 STBCs of Table 3.1 at an effective throughput of about **2 BPS** using **one receiver** over **uncorrelated** Rayleigh fading channels.

already been achieved by the G_2 code using two receivers. Furthermore, the potential benefit of using more transmitters is eroded by the employment of higher-throughput, but more vulnerable, modulation schemes, which are more prone to transmission errors.

Performance at the Throughput of 3 BPS   The performances of the STBCs using one receiver and having a throughput of 3 BPS over uncorrelated Rayleigh fading channels are compared in Figure 3.8. Again, similar to the scenario of having a throughput of 2 BPS, the G_2 code performs best at a low SNR,

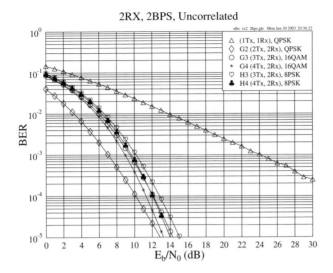

Figure 3.7: The BER versus E_b/N_0 performance of the G_2, G_3, G_4, H_3 and H_4 STBCs of Table 3.1 at an effective throughput of about **2 BPS** using **two receivers** over **uncorrelated** Rayleigh fading channels.

i.e. below about 10 dB, although the performance difference between G_2 and other codes is even smaller than it is in Figure 3.6. As shown in Figure 3.8, at the BER of 10^{-5}, the H_3 and H_4 codes achieve a gain of about 3 dB over the G_3 and G_4 codes, respectively. Since the STBCs themselves do not have an error-correction capability which would allow them to correct the extra errors induced by employing a more vulnerable, higher-order modulation scheme [314], this results in a poorer performance. Furthermore, since the relative increase of the constellation density when changing from 16-QAM to 64-QAM is higher than that from 8-PSK to 16-QAM, the performance degradation imposed by reverting from 16-QAM to 64-QAM is more severe than that imposed by opting for 16-QAM instead of 8-PSK. Therefore it is not surprising that the best code to be used for high-SNR situations becomes the H_4 instead of the G_4 code, which was the best code according to Figure 3.6 at high-SNR scenarios, since the G_4 and H_4 codes are used in conjunction with 64-QAM and 16-QAM, respectively. Moreover, we note that the H_3 code gives approximately 0.4 dB gain over the G_4 code, although the former has a lower diversity order.

If the number of receivers is doubled, as seen in Figure 3.9, the performance degradations of the G_3 and G_4 codes are much more dramatic. In this scenario, even the lower-diversity space–time code G_2 is capable of outperforming the space–time code G_4 having a higher-order diversity by about 1.7 dB at the BER of 10^{-5}.

3.2.5.2 Performance over Correlated Rayleigh Fading Channel

We have compared the performances of the STBCs of Table 3.1 for transmission over uncorrelated Rayleigh fading channels in Section 3.2.5.1. In this section, the performance of the STBCs will be studied based on the same assumptions noted on page 70, except that the channel is assumed to be a correlated Rayleigh fading channel associated with the normalized Doppler frequency of 3.25×10^{-5}.

Specifically, Figure 3.10 compares the performance of the G_2, G_3 and G_4 STBCs in conjunction with one receiver at the throughput of 1 BPS, when communicating over a correlated Rayleigh fading channel. If we compare Figure 3.10 with Figure 3.4, which shows the relevant codes' performance over uncorrelated Rayleigh fading channels, we will note that the performances recorded in these cases are

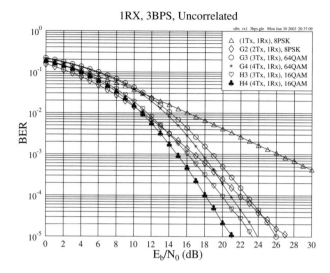

Figure 3.8: The BER versus E_b/N_0 performance of the G_2, G_3, G_4, H_3 and H_4 STBCs of Table 3.1 at an effective throughput of **3 BPS** using **one receiver** over **uncorrelated** Rayleigh fading channels.

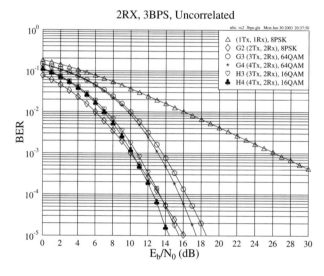

Figure 3.9: The BER versus E_b/N_0 performance of the G_2, G_3, G_4, H_3 and H_4 STBCs of Table 3.1 at an effective throughput of **3 BPS** using **two receivers** over **uncorrelated** Rayleigh fading channels.

almost the same. This is because the receiver is aided by a perfect channel estimator that provides full knowledge of the path gains, and thus the effect imposed on the transmitted signals by the different path gains, regardless of whether they are correlated or uncorrelated, is efficiently counteracted. When the throughput increases to 2 BPS and even further to 3 BPS, it can also be shown that the achievable performances of STBCs communicating over a correlated Rayleigh channel are the same as those of their corresponding counterparts transmitting over uncorrelated Rayleigh channels, respectively.

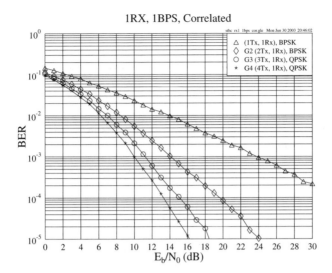

Figure 3.10: The BER versus E_b/N_0 performance of the G_2, G_3 and G_4 STBCs of Table 3.1 at an effective throughput of **1 BPS** using **one receiver** over **correlated** Rayleigh fading channels. The normalized Doppler frequency is 3.25×10^{-5}.

3.2.6 Conclusions

From the discussions and simulation results of Sections 3.2.5.1 and 3.2.5.2, several conclusions can be inferred. Firstly, the encoding and decoding of STBCs have a low complexity. At the receiver end, the ML decoder requires low-complexity linear processing for decoding.

Secondly, from Figures 3.4, 3.6 and 3.8 we note that when the effective throughput is increased, the phasor constellation has to be extended to accommodate the increased number of bits. Hence the performances of the half-rate codes G_3 and G_4 degrade in comparison with that of the unity-rate code G_2.

Thirdly, at the even higher effective throughput of 3 BPS, the H_3 and H_4 codes perform better than the G_3 and G_4 codes, respectively, as shown in Figure 3.8. Moreover, according to Figures 3.4, 3.5, 3.8 and 3.9, when the number of receivers is increased, the performance gain of the G_3, G_4, H_3 and H_4 codes over the G_2 code becomes more modest because much of the attainable diversity gain has already been achieved using the G_2 code employing two receivers.

Last but not least, it was also found that the performances of the space–time codes communicating over uncorrelated and correlated Rayleigh fading channels are similar, provided that the effective throughput is the same.

The achievable coding gains of the STBCs are summarized in Table 3.2. The coding gain is defined as the E_b/N_0 difference, expressed in terms of decibels, at a BER of 10^{-5} between the various space–time block coded and uncoded single-transmitter systems having the same throughput. The best schemes at the effective throughputs of 1, 2 and 3 BPS are printed in bold, respectively.

3.3 Channel-Coded STBCs

In Section 3.2, we presented the basic concepts of the STBCs and provided a range of characteristic performance results. Furthermore, the MAP algorithm invoked for decoding STBCs has also been briefly highlighted. This enables a space–time decoder to provide soft outputs that can be exploited by concatenated channel decoders for further improving the system's performance. In this section, we

Table 3.2: Coding gains of the STBCs using one receiver when communicating over uncorrelated and correlated Rayleigh fading channels. The performance of the best scheme for a specific effective throughput is printed in bold.

				BER			
				E_b/N_0 (dB)		Gain (dB)	
BPS	Code	Code rate	Modem	10^{-3}	10^{-5}	10^{-3}	10^{-5}
1.00	Uncoded	1	BPSK	24.22	44.00	0.00	0.00
	G_2	1	BPSK	14.08	24.22	10.14	19.78
	G_3	1/2	QPSK	11.33	18.71	12.89	25.29
	G_4	1/2	QPSK	**10.10**	**15.85**	**14.12**	**28.15**
2.00	Uncoded	1	QPSK	24.22	44.00	0.00	0.00
	G_2	1	QPSK	14.12	24.22	10.10	19.78
	G_3	1/2	16-QAM	14.78	22.06	9.44	21.94
	G_4	1/2	16-QAM	**13.61**	**19.58**	**10.61**	**24.42**
2.25	H_3	3/4	8-PSK	15.43	23.00	8.79	21.00
≈ 2.00	H_4	3/4	8-PSK	14.31	20.48	9.91	23.52
3.00	Uncoded	1	8-PSK	26.30	46.26	0.00	0.00
	G_2	1	8-PSK	16.80	27.21	9.50	19.05
	G_3	1/2	64-QAM	18.83	26.00	7.47	20.26
	G_4	1/2	64-QAM	17.69	23.92	8.61	22.34
	H_3	3/4	16-QAM	16.06	23.36	10.24	22.90
	H_4	3/4	16-QAM	**14.87**	**21.10**	**11.43**	**25.16**

will concatenate the STBCs with various Low-Density Parity Check (LDPC) channel codes [419] and with a Turbo Convolutional (TC) code [420, 421]. The performances of the different schemes will also be evaluated.

3.3.1 STBCs with LDPC Channel Codes

LDPC codes were devised by Gallager [419] in 1962. During the early evolutionary phase of channel coding, LDPC schemes had a limited impact on the research of the channel coding community, although they showed an unprecedented performance prior to the turbo-coding era. This was because LDPC codes required a relatively high storage space and complexity. However, owing to their capability of approaching Shannon's predicted performance limits [426], research interests in LDPC codes have been rekindled during recent years [426–430].

LDPC codes [419] belong to the family of linear block codes which are defined by a parity check matrix having M rows and N columns. The column weight j and row weight k are typically significantly lower than the dimensions M and N of the parity check matrix. The construction of the parity check matrix is referred to as regular or irregular, depending on whether the Hamming weight per column or row is identical. Reference [428] shows that carefully designed irregular LDPC codes may perform better than their regular counterparts. Furthermore, when the block length is increased, irregular LDPC codes may become capable of outperforming turbo codes [426] at the cost of a higher complexity. Since the details of the decoding of LDPC codes can be found in [431], in the forthcoming sections we are more interested in the performance of LDPC codes than the LDPC decoding algorithm itself.

The number of columns N is given by the number of coded bits hosted by an LDPC codeword, while the number of rows M corresponds to the number of parity check constraints imposed by the design of the LDPC code. The number of information bits encoded by an LDPC codeword is denoted

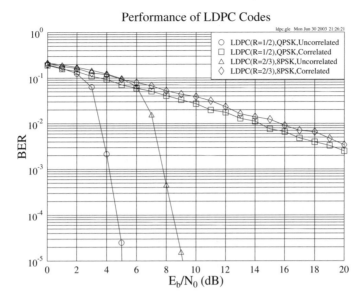

Figure 3.11: The BER versus E_b/N_0 performance of several LDPC codes communicating over both uncorrelated and correlated Rayleigh fading channels. The parameters of the LDPC codes are given in Tables 3.3 and 3.4. The normalized Doppler frequency of the correlated Rayleigh fading channels was 3.25×10^{-5}. The effective throughput of the QPSK and 8-PSK schemes were 1 BPS and 2 BPS, respectively.

by $K = N - M$, yielding a coding rate of K/M [431]. Thus, the LDPC code rate can be adjusted by changing K and/or M.

Figure 3.11 characterizes the performance of several LDPC codes for transmission over both uncorrelated and correlated Rayleigh fading channels. The normalized Doppler frequency of the correlated Rayleigh channel was 3.25×10^{-5}. It was found that the LDPC codes perform far better over uncorrelated than over correlated Rayleigh channels, since the codeword length is short. This characteristic predetermines the expected performance of the LDPC-STBC coded concatenated system to be introduced in the next section and characterized in Section 3.3.1.2.2.

3.3.1.1 System Overview

Figure 3.12 shows a schematic diagram of the system. The source bits are first encoded by the LDPC encoder, whose outputs are modulated and forwarded to the STBC encoder. At the receiver, the noise-contaminated received symbols are decoded by the STBC soft decoder. As discussed in Section 3.2.3.2, the soft outputs constituted by the a posteriori probabilities of the STBC-decoded symbols will be passed on to the soft demapper, where the symbol probabilities are used for generating the resultant bit probabilities. Finally, the LDPC decoder decodes the soft inputs and the channel-decoded information bits are obtained.

In Section 3.3.1.2, we will provide a range of simulation results and compare the achievable performances of the different schemes designed for transmission over both uncorrelated and correlated Rayleigh fading channels. According to [431], when using different column weights j and/or a different number of LDPC decoding iterations, the performance of LDPC codes will change correspondingly. In our system, we fix the column weight and the number of iterations to 3 and 25, respectively. For the scenarios of communicating over correlated Rayleigh fading channels, a fixed-length random channel interleaver was used. Furthermore, for the sake of fair comparisons, we comply with the assumptions

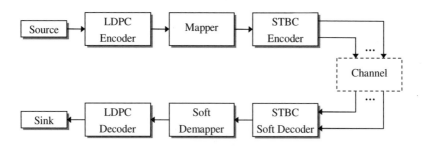

Figure 3.12: System overview of LDPC channel coding aided STBCs.

Table 3.3: The parameters used in the LDPC-STBC coded concatenated schemes for transmissions over **uncorrelated** Rayleigh fading channels.

	STBC		LDPC				
BPS	Code	Code rate	Code rate	Input bits block size	Output bits block size	Column weight	Iterations	Modem
1.00	G_2	1	1/2	1008	2016	3	20	QPSK
	G_3	1/2	1/2	1008	2016	3	20	16-QAM
	G_4	1/2	1/2	1008	2016	3	20	16-QAM
	H_3	3/4	2/3	1344	2016	3	20	QPSK
	H_4	3/4	2/3	1344	2016	3	20	QPSK
2.00	G_2	1	2/3	1344	2016	3	20	8-PSK
	G_2	1	1/2	1008	2016	3	20	16-QAM
	G_2	1	1/3	672	2016	3	20	64-QAM
	G_3	1/2	2/3	1344	2016	3	20	64-QAM
	G_4	1/2	2/3	1344	2016	3	20	64-QAM
2.25	H_3	3/4	1/2	1008	2016	3	20	64-QAM
≈ 2.00	H_4	3/4	1/2	1008	2016	3	20	64-QAM
3.00	G_2	1	3/4	1512	2016	3	20	16-QAM
	G_2	1	1/2	1008	2016	3	20	64-QAM
	H_3	3/4	2/3	1344	2016	3	20	64-QAM
	H_4	3/4	2/3	1344	2016	3	20	64-QAM

outlined on page 70 so that the simulation results of Section 3.2.5 remain comparable in this new context.

The parameters used in our LDPC-STBC coded concatenated system are given in Tables 3.3 and 3.4, while the parameters of the STBCs have been given in Table 3.1. For the sake of maintaining the same effective throughput, the LDPC codec's parameters have to be harmonized with the STBC codec's parameters.

3.3.1.2 Simulation Results

Similar to Section 3.2.5, we will compare the performances of different schemes in the context of the same effective throughput, when communicating over both uncorrelated and correlated Rayleigh fading channels.

Table 3.4: The parameters used in the LDPC-STBC coded concatenated schemes for transmissions over **correlated** Rayleigh fading channels.

	STBC		LDPC						
BPS	Code	Code rate	Code rate	Input bits block size	Output bits block size	Column weight	Iterations	Channel interleaver depth	Modem
1.00	G_2	1	1/2	10 080	20 160	3	20	20 160	QPSK
	G_3	1/2	1/2	10 080	20 160	3	20	20 160	16-QAM
	G_4	1/2	1/2	10 080	20 160	3	20	20 160	16-QAM
	H_3	3/4	2/3	13 440	20 160	3	20	20 160	QPSK
	H_4	3/4	2/3	13 440	20 160	3	20	20 160	QPSK
2.00	G_2	1	2/3	13 440	20 160	3	20	20 160	8-PSK
	G_2	1	1/2	10 080	20 160	3	20	20 160	16-QAM
	G_2	1	1/3	6720	20 160	3	20	20 160	64-QAM
	G_3	1/2	2/3	13 440	20 160	3	20	20 160	64-QAM
	G_4	1/2	2/3	13 440	20 160	3	20	20 160	64-QAM
2.25	H_3	3/4	1/2	10 080	20 160	3	20	20 160	64-QAM
≈ 2.00	H_4	3/4	1/2	10 080	20 160	3	20	20 160	64-QAM
3.00	G_2	1	3/4	15 120	20 160	3	20	20 160	16-QAM
	G_2	1	1/2	10 080	20 160	3	20	20 160	64-QAM
	H_3	3/4	2/3	13 440	20 160	3	20	20 160	64-QAM
	H_4	3/4	2/3	13 440	20 160	3	20	20 160	64-QAM

3.3.1.2.1 Performance over Uncorrelated Rayleigh Fading Channels

Performance at the Throughput of 1 BPS Figure 3.13 compares the achievable performance of the G_2, G_3, G_4, H_3 and H_4 STBCs, which are combined with various LDPC codes, in the context of using one receiver and maintaining a throughput of 1 BPS, while communicating over uncorrelated Rayleigh fading channels. We can see that when E_b/N_0 is lower than about 2.7 dB, the half-rate STBC G_4, using QPSK modulation, which constituted the best system at the throughput of 1 BPS according to Table 3.2, gives the best performance. But when the SNR increases, the situation reverses since the performance of the LDPC-assisted STBCs becomes significantly better than that of the G_4 scheme using no channel coding. Moreover, the scheme constituted by the G_2 code and half-rate LDPC code excels among all the LDPC-coded STBC schemes by a margin of about 2 dB gain over others. Note that in order to maintain the same effective throughput of 1 BPS, the unity-rate STBC G_2 is combined with the half-rate LDPC code employing QPSK modulation, while the half-rate space–time codes G_3 and G_4 assisted by the half-rate LDPC code use 16-QAM modulation. For the $\frac{3}{4}$-rate codes H_3 and H_4, the $\frac{2}{3}$-rate LDPC code is introduced and QPSK modulation is used to meet the criteria of having the same throughput of 1 BPS.

From Figure 3.13, we may arrive at the following conclusions. Provided that a fixed block size of the LDPC codeword is used, a lower LDPC code rate implies that more parity check bits are attached to the original bits sequence, which leads to a better performance. On the other hand, as seen in Figure 3.13, when QPSK modulation is used, the G_2 code employing the half-rate LDPC code outperforms the H_3 and H_4 codes in conjunction with the $\frac{2}{3}$-rate LDPC code, despite the fact that the former has a lower diversity order. Therefore we may surmise for the set of LDPC-coded STBC schemes considered that, when using a specific modulation scheme, the system's performance is predominantly determined by the LDPC code employed instead of the space–time code's diversity order. Another observation from the figure is that the number of bits per symbol used by the specific modulation scheme is a more decisive factor, in terms of determining the achievable performance, than the diversity order, when the same LDPC code is used. In other words, if we want to improve the system's performance, it is more beneficial to reduce the number of bits per symbol instead of increasing the number of transmitter

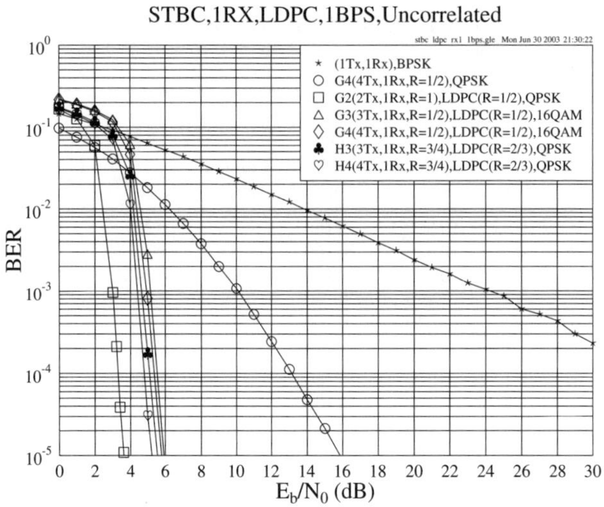

Figure 3.13: The BER versus E_b/N_0 performance of the G_2, G_3, G_4, H_3 and H_4 STBCs of Table 3.1 in conjunction with the different-rate LDPC codes of Table 3.3 at an effective throughput of **1 BPS** using **one receiver** over **uncorrelated** Rayleigh fading channels.

antennas. For example, when employing a half-rate LDPC code, the G_3 and G_4 space–time codes employing 16-QAM perform about 2.2 dB worse than the G_2 code employing QPSK, despite the fact that the former has a higher diversity order.

Performance at the Throughput of 2 BPS   In Figure 3.14 the performance of the LDPC-aided STBCs using one receiver and having a throughput of about 2 BPS, while communicating over uncorrelated Rayleigh fading channels, is studied. Especially, the $\frac{3}{4}$-rate codes H_3 and H_4 are employed in conjunction with the half-rate LDPC code using 64-QAM modulation, and thus achieve an effective throughput of 2.25 BPS, which is close to our target of 2 BPS.

The curves seen in Figure 3.14 can be divided into two groups based on their relative performances. The first group contains the three schemes employing the G_2 code in conjunction with various LDPC codes and gives an average gain of about 3 dB over the members of the second group, in which the $G_3/G_4/H_3/H_4$ schemes are grouped. The performance difference between the two groups tallies with our conclusions derived in the case of aiming for a throughput of 1 BPS. As seen in Figure 3.14, the schemes of the second group suffer performance degradations as a consequence of employing the densely packed 64-QAM constellation, which is more prone to transmission errors than the other modulation schemes. Furthermore, the 64-QAM-based scheme of the first group, i.e. the arrangement employing the G_2 code as well as the $\frac{1}{3}$-rate LDPC code, also performs better than any member scheme of the second group as a benefit of its lower LDPC code rate.

As seen in Figure 3.14, in all the three G_2-based schemes of the superior group, the best design option is the compromise scheme employing the half-rate LDPC-aided G_2 code using 16-QAM modulation. The reason for this phenomenon can be explained from two different aspects. On the one hand, when the number of bits per symbol is moderate, as in relatively lower-order 8-PSK and 16-QAM, for example, the performance trends imposed by the different LDPC codes outweigh those caused by the different modulation schemes. In this case, as expected, if the system uses a lower-rate LDPC code, it will achieve a better performance. This is why the half-rate LDPC-coded scheme using 16-QAM modulation is superior to the $\frac{2}{3}$-rate LDPC-coded scheme using 8-PSK modulation, as seen in Figure 3.14. On the other hand, when the modulation level is increased to 64, for example, as in

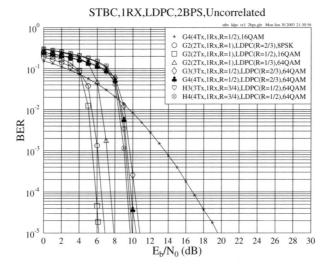

Figure 3.14: The BER versus E_b/N_0 performance of the G_2, G_3, G_4, H_3 and H_4 STBCs of Table 3.1 in conjunction with the different-rate LDPC codes of Table 3.3 at an effective throughput of about **2 BPS** using **one receiver** over **uncorrelated** Rayleigh fading channels.

64-QAM, the situation is reversed and the number of bits per symbol conveyed by the modulation schemes will become the predominant factor. In this case, the combination of lower-rate LDPC codes with high-order modulation arrangements will no longer outperform the scheme that uses higher-rate LDPC codes in conjunction with lower-order modulation constellations. Hence the $\frac{1}{3}$-rate LDPC-coded 64-QAM modulation scheme is outperformed by the half-rate LDPC-coded 16-QAM modulation arrangement, as indicated by Figure 3.14.

Performance at the Throughput of 3 BPS The performance of the LDPC-assisted STBCs using one receiver and having a throughput of 3 BPS for transmissions over uncorrelated Rayleigh fading channels is shown in Figure 3.15. When the half-rate space–time codes G_3 and G_4 are used in conjunction with LDPC codes, even if we employ a high-throughput 64-QAM scheme, the system's effective throughput will be lower than 3 BPS, because the code rate of the LDPC code is below unity. For employment in conjunction with the G_3 and G_4 codes, a high-rate LDPC code has to be used in order to maintain a throughput close to 3 BPS. However, as discussed earlier, when the high-order 64-QAM scheme is used, the achievable performance improvement of LDPC coding remains modest, regardless of the rate of the LDPC code. Hence in this scenario the performance of the LDPC-aided G_3 and G_4 codes is not considered here.

In Figure 3.15, similar to the 2 BPS throughput scenario, it is also found that the four LDPC-aided STBC schemes can be divided into two groups. The G_2 space–time code aided by the $\frac{3}{4}$-rate LDPC code using 16-QAM performs best in high-SNR situations and it only suffers a low performance degradation over the scheme using no channel coding when the SNR is low.

When we increase the number of receiver antennas, the performance gap between the two groups still remains obvious as shown in Figure 3.16. In this scenario, the best scheme is again the one employing the half-rate LDPC-aided G_2 code using 16-QAM. However, the gain achieved by the best LDPC-STBC scheme over the best unprotected STBC scheme using two receivers decreases to about 10 dB compared with the 12.5 dB achieved, while using a single receiver.

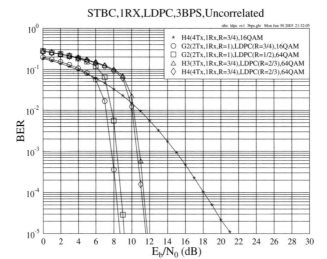

Figure 3.15: The BER versus E_b/N_0 performance of the G_2, H_3 and H_4 STBCs of Table 3.1 in conjunction with the different-rate LDPC codes of Table 3.3 at an effective throughput of **3 BPS** using **one receiver** over **uncorrelated** Rayleigh fading channels.

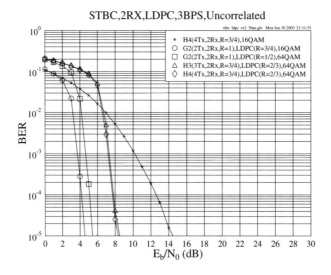

Figure 3.16: The BER versus E_b/N_0 performance of the G_2, H_3 and H_4 STBCs of Table 3.1 in conjunction with the different-rate LDPC codes of Table 3.3 at an effective throughput of **3 BPS** using **two receivers** over **uncorrelated** Rayleigh fading channels.

3.3.1.2.2 Performance over Correlated Rayleigh Fading Channels

We have compared the performance of the LDPC-aided STBCs when communicating over uncorrelated Rayleigh fading channels in Section 3.3.1.2.1. In this section, the performance of the STBCs will be studied based on the same assumptions summarized on page 70, except that the channel is assumed to

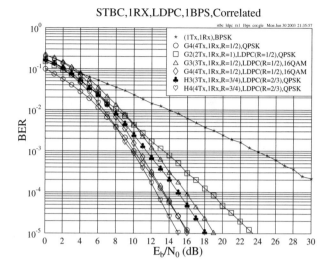

Figure 3.17: The BER versus E_b/N_0 performance of the G_2, G_3, G_4, H_3 and H_4 STBCs of Table 3.1 in conjunction with the different-rate LDPC codes of Table 3.4 at an effective throughput of **1 BPS** using **one receiver** over **correlated** Rayleigh fading channels. The normalized Doppler frequency is 3.25×10^{-5}.

be a correlated Rayleigh fading channel obeying the normalized Doppler frequency of 3.25×10^{-5}. The corresponding parameters are given in Table 3.4.

In Section 3.2.5.2 it was found that if the effective throughput is a fixed constant, the performance of the various space–time codes used for transmission over uncorrelated and correlated Rayleigh fading channels is similar. However, in the context of the LDPC-assisted STBC-coded system, the achievable performances are different over uncorrelated and correlated Rayleigh fading channels. This can be clearly seen by comparing Figures 3.17, 3.18 and 3.19 of this section with Figures 3.13, 3.14 and 3.15 of Section 3.3.1.2.1, respectively. The reason for this phenomenon is that the LDPC codes perform better over uncorrelated rather than correlated Rayleigh fading channels, unless their codeword length is extremely high or long channel interleavers are used. This will be demonstrated during our further discourse.

Performance at the Throughput of 1 BPS   Figure 3.17 shows the achievable performance of the LDPC-assisted G_2, G_3, G_4, H_3 and H_4 codes of Table 3.1 using one receiver at the effective throughput of 1 BPS when communicating over correlated Rayleigh fading channels. We can see that the $\frac{2}{3}$-rate LDPC-coded H_4 and H_3 codes outperform the $\frac{2}{3}$-rate LDPC-coded G_4 and G_3 codes by about 1 dB, respectively. However, when E_b/N_0 is lower than about 8 dB, we notice that the scheme employing the unprotected G_4 code, which is the best design option at the throughput of 1 BPS according to Table 3.2, performs better than the scheme employing the $\frac{2}{3}$-rate LDPC-aided H_4 code. When the SNR is increased to 15 dB, the situation is reversed, since the unprotected G_4 scheme is outperformed by the best LDPC-STBC scheme, namely the $\frac{2}{3}$-rate LDPC-assisted H_4 coded scheme, with about 1 dB E_b/N_0 degradation at the BER of 10^{-5}.

Performance at the Throughput of 2 BPS   At an effective throughput of approximately 2 BPS, the relevant schemes' performances are given in Figure 3.18. It can be seen in the figure that the performance of the schemes employing the H_4 and H_3 codes of Table 3.1 is similar to those of the schemes employing the G_4 and G_3 codes, respectively, although we should bear in mind that the $\frac{3}{4}$-rate H_4 and H_3 codes have an effective throughput of 2.25 BPS, rather than exactly 2 BPS. Furthermore,

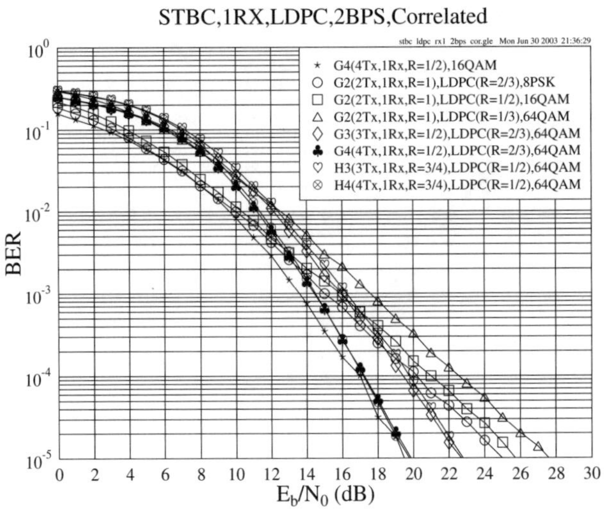

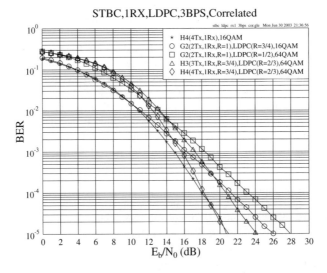

Figure 3.18: The BER versus E_b/N_0 performance of the G_2, G_3, G_4, H_3 and H_4 STBCs of Table 3.1 in conjunction with the different-rate LDPC codes of Table 3.4 at an effective throughput of about **2 BPS** using **one receiver** over **correlated** Rayleigh fading channels. The normalized Doppler frequency is 3.25×10^{-5}.

Figure 3.19: The BER versus E_b/N_0 performance of the G_2, H_3 and H_4 STBCs of Table 3.1 in conjunction with the different-rate LDPC codes of Table 3.4 at an effective throughput of **3 BPS** using **one receiver** over **correlated** Rayleigh fading channels. The normalized Doppler frequency is 3.25×10^{-5}.

an important phenomenon found in Figure 3.18 is that the best unprotected STBC scheme of Figure 3.6 also attains the best performance in this new scenario.

This result may be explained as follows. The performance of the unprotected STBCs remains similar over uncorrelated or correlated Rayleigh fading channels, as indicated in Section 3.2.5.2.

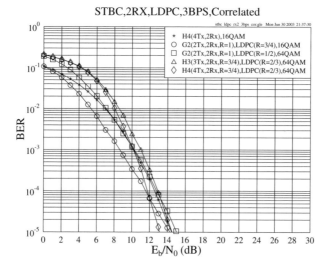

Figure 3.20: The BER versus E_b/N_0 performance of the $\boldsymbol{G_2}$, $\boldsymbol{H_3}$ and $\boldsymbol{H_4}$ STBCs of Table 3.1 in conjunction with the different-rate LDPC codes of Table 3.4 at an effective throughput of **3 BPS** using **two receivers** over **correlated** Rayleigh fading channels. The normalized Doppler frequency is 3.25×10^{-5}.

Furthermore, the performance of the LDPC codes degrades when communicating over correlated rather than uncorrelated Rayleigh fading channels, as seen in Figure 3.11. Hence it is not surprising that the LDPC-STBC coded concatenated system will suffer a performance degradation in the context of correlated Rayleigh channels. In this case, the LDPC codes' relatively poor performance recorded over correlated Rayleigh channels disadvantageously affects the entire system. In other words, the LDPC codes improve the system's performance less dramatically over correlated Rayleigh fading channels than over uncorrelated Rayleigh fading channels, unless the LDPC codeword length is very high or long interleavers are used.

Performance at the Throughput of 3 BPS The performance of the LDPC-assisted STBCs using one receiver and having a throughput of 3 BPS while communicating over correlated Rayleigh fading channel is portrayed in Figure 3.19. Similar to Figure 3.15, the schemes which employ the half-rate space–time codes G_3 and G_4 of Table 3.1 are not considered in this scenario, since they are incapable of achieving an effective throughput of 3 BPS, nor can they achieve a better performance than the candidate schemes characterized in Figure 3.19.

As Figure 3.19 shows, the unprotected STBC H_4 using 16-QAM modulation performs best, giving an approximately 1 dB gain over the best LDPC-aided scheme, namely the STBC H_4 combined with the $\frac{2}{3}$-rate LDPC code using 64-QAM modulation at the BER of 10^{-5}. Similar to the scenario maintaining an effective throughput of 2 BPS, at a relatively lower E_b/N_0 value, i.e. below 14.5 dB, the best STBC-LDPC concatenated scheme is the one that employs the G_2 space–time code combined with a low-order modulation, namely the $\frac{3}{4}$-rate LDPC-coded 16-QAM.

Furthermore, when the number of receivers is increased to two, the $\frac{3}{4}$-rate LDPC-assisted G_2 code outperforms all the other schemes considered, provided that the E_b/N_0 value is below 12 dB, as observed in Figure 3.20. At even higher E_b/N_0 values, i.e. in excess of 12 dB, the $\frac{2}{3}$-rate LDPC-assisted H_4 code exhibits the best performance, although it uses the highest-order 64-QAM modem. It can be also observed from Figure 3.20 that the curves are significantly closer to one another compared with the scenario of using one receiver, which is shown in Figure 3.19.

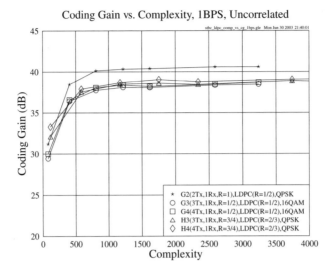

Figure 3.21: Coding gain versus estimated complexity for the LDPC-STBC coded concatenated schemes using one receiver, when communicating over uncorrelated Rayleigh fading channels at the effective throughput of **1 BPS**. The simulation parameters are given in Table 3.3.

3.3.1.3 Complexity Issues

In Section 3.3.1.2, we compared the performance of various LDPC-STBC coded concatenated systems. The best scheme was also identified for each scenario. However, these choices have been made based purely on the achievable performance, and the complexity issue of implementation has not been taken into consideration. In this section, we will briefly address the associated complexity issues.

As discussed in Section 3.2.3.2, the soft decoder of the STBCs employs the Log-MAP algorithm summarized for example in [314]. With the advent of the Log-MAP algorithm, the high-complexity exponential operations are substituted by additions and subtractions carried out in the logarithmic domain. Hence the complexity of the STBC decoder is significantly reduced, while closely matching the performance of the MAP algorithm. In our following discussions, the decoding complexity of the STBCs is considered to be sufficiently low for it to be ignored for the sake of simplifying our comparisons. This will not affect our conclusions, since we will show in the rest of this section that the LDPC-assisted G_2 code, which has the lowest decoding complexity among all the STBCs of Table 3.1, gives the best performance as seen in Figures 3.21, 3.22 and 3.23. Hence, even if the decoding complexity of the STBCs is considered, the G_2 code will still be superior to the other STBCs in the context of the coding gain versus complexity performance.

The decoding complexity per information bit per iteration of the LDPC codes can be calculated as follows [431]:

$$comp\{LDPC\} = \left(\frac{5-R}{1-R}\right) \cdot j^2, \tag{3.30}$$

where R is the code rate of the LDPC code and j is the column weights of the parity check matrix. In our system the value of column weights was fixed to three, thus Equation (3.30) is simplified to

$$comp\{LDPC\} = \left(\frac{45-9R}{1-R}\right). \tag{3.31}$$

According to Equation (3.31), the decoding complexity is essentially based on the code rate of the LDPC code employed. Thus for the rate 1/3, 1/2, 2/3 and 3/4 LDPC codes used in our system, the associated

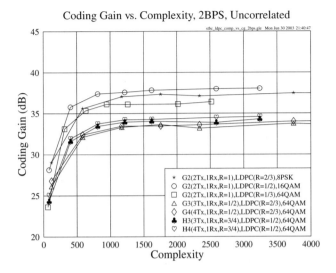

Figure 3.22: Coding gain versus estimated complexity for the LDPC-STBC coded concatenated schemes using one receiver, when communicating over uncorrelated Rayleigh fading channels at the effective throughput of **2 BPS**. The simulation parameters are given in Table 3.3.

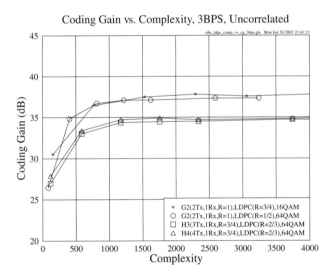

Figure 3.23: Coding gain versus estimated complexity for the LDPC-STBC coded concatenated schemes using one receiver, when communicating over uncorrelated Rayleigh fading channels at the effective throughput of **3 BPS**. The simulation parameters are given in Table 3.3.

decoding complexity per bit per iteration becomes 63, 81, 117 and 153 additions and subtractions, respectively, as summarized in Table 3.5.

Let us now compare the coding gain versus complexity characteristics of the different schemes considered, as seen from Figures 3.21 to 3.23, where the parameters used are given in Table 3.3. The coding gain here is defined as the E_b/N_0 difference, expressed in terms of decibels, at a BER of 10^{-5} between the various channel code assisted STBC systems and the uncoded single-transmitter systems

Table 3.5: Coding gains of the LDPC-STBC coded concatenated schemes using one receiver, when communicating over uncorrelated Rayleigh fading channels. With reference to Figures 3.21, 3.22 and 3.23, the performance of the best scheme is printed in bold for the scenarios of different effective throughputs, respectively.

				BER				
				E_b/N_0 (dB)		Gain (dB)		
BPS	STBC code	LDPC rate	LDPC comp	10^{-3}	10^{-5}	10^{-3}	10^{-5}	Modem
1.00	Uncoded	—	—	24.22	44.00	0.00	0.00	BPSK
	G_4	—	—	10.10	15.85	14.12	28.15	QPSK
	G_2	1/2	81	**2.99**	**3.61**	**21.23**	**40.39**	QPSK
	G_3	1/2	81	5.17	5.92	19.05	38.08	16-QAM
	G_4	1/2	81	4.95	5.79	19.27	38.21	16-QAM
	H_3	2/3	117	4.65	5.56	19.57	38.44	QPSK
	H_4	2/3	117	4.41	5.24	19.81	38.76	QPSK
2.00	Uncoded	—	—	24.22	44.00	0.00	0.00	QPSK
	G_4	—	—	13.61	19.58	10.61	24.42	16-QAM
	G_2	2/3	117	6.05	6.89	18.17	37.11	8-PSK
	G_2	1/2	81	**5.45**	**6.18**	**18.77**	**37.82**	16-QAM
	G_2	1/3	63	7.10	7.90	17.12	36.10	64-QAM
	G_3	2/3	117	9.65	10.81	14.57	33.19	64-QAM
	G_4	2/3	117	9.34	10.40	14.88	33.60	64-QAM
2.25	H_3	1/2	81	9.20	9.96	15.02	34.04	64-QAM
≈ 2.00	H_4	1/2	81	9.02	9.73	15.20	34.27	64-QAM
3.00	Uncoded	—	—	26.30	46.26	0.00	0.00	8-PSK
	H_4	—	—	14.87	21.10	11.43	25.16	16-QAM
	G_2	3/4	153	**7.73**	**8.68**	**18.57**	**37.58**	16-QAM
	G_2	1/2	81	8.33	9.13	17.97	37.13	64-QAM
	H_3	2/3	117	10.85	11.75	15.45	34.51	64-QAM
	H_4	2/3	117	10.58	11.57	15.72	34.69	64-QAM

having the same throughput. All the estimated implementational complexities were calculated based on Equation (3.31), using different numbers of iterations ranging from 1 to 40 with a step of about 5.

At an effective throughput of 1 and 2 BPS, as seen in Figures 3.21 and 3.22, it was found that the best scheme was the half-rate LDPC-coded G_2 space–time code. In the scenario of having an effective throughput of 3 BPS, the performance curves of the half-rate and $\frac{3}{4}$-rate LDPC-coded G_2 space–time code are close to each other, although the former performs slightly better. We may also note that the coding gain increases dramatically in the low-complexity range and tends to saturate in the vicinity of an estimated complexity of about 1200, which corresponds to approximately 19, 15, 10 and 8 iterations for the LDPC codes having a code rate of $\frac{1}{3}$, $\frac{1}{2}$, $\frac{2}{3}$ and $\frac{3}{4}$, respectively. As Figure 3.24 shows, for example, when the number of iterations is increased to about 10 in terms of the half-rate LDPC-aided G_2 code, the performance is already close to the achievable maximum coding gain. This result can be considered as a rule of thumb for setting the number of iterations for the LDPC-STBC coded concatenated schemes, when aiming for a good trade-off in terms of the achievable performance-to-complexity relationships.

In Figure 3.25, we show the E_b/N_0 value required for maintaining a BER of 10^{-5} versus the effective throughput BPS for the unprotected STBCs and the half-rate LDPC-assisted G_2 code, while the associated simulation parameters are summarized in Tables 3.1 and 3.3. The simulation results were obtained using one receiver for communicating over uncorrelated Rayleigh fading channels. It can be observed in Figure 3.25 that the E_b/N_0 value required for maintaining a BER of 10^{-5} increases nearly linearly, as the effective BPS throughput increases. This conclusion was valid for both the unprotected

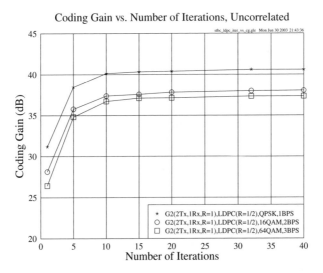

Figure 3.24: Coding gain versus the number of iterations for the half-rate LDPC-assisted G_2 STBC schemes using one receiver, when communicating over uncorrelated Rayleigh fading channels at the effective throughput of 1, 2 and 3 BPS. The parameters used are given in Table 3.3.

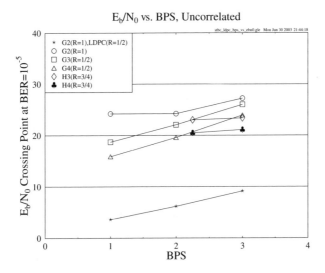

Figure 3.25: The E_b/N_0 value required for maintaining BER $= 10^{-5}$ versus the effective BPS throughput for the STBCs of Table 3.1 and for the half-rate LDPC-assisted G_2 code of Table 3.3, when using one receiver and communicating over uncorrelated Rayleigh fading channels.

STBC-aided schemes and the half-rate LDPC-assisted G_2-coded scheme. Furthermore, the half-rate LDPC-G_2 concatenated scheme achieves a gain of about 15 dB over the best unprotected STBC scheme at the effective throughput values of 1, 2 and 3 BPS, respectively.

3.3.1.4 Conclusions

Having studied Figures 3.13 to 3.19, we may arrive at the following conclusions. First of all, as expected, the LDPC-aided STBC-coded schemes perform significantly better than the unprotected STBC schemes, when transmitting over uncorrelated Rayleigh fading channels. However, over correlated Rayleigh fading channels the performance improvements achieved by the LDPC codes are not as significant as in uncorrelated Rayleigh fading channels, as seen in Figures 3.17 to 3.20. This is because the LDPC codes suffer from their finite codeword length and for a limited tolerable channel interleaver delay, as evidenced by Figure 3.11. This affects the attainable performance of the LDPC-STBC coded concatenated system to some degree. If we use an extremely long codeword or employ a long channel interleaver, however, the achievable performance of the LDPC-STBC concatenated schemes can be improved in the context of correlated Rayleigh fading channels.

Another observation inferred from Figures 3.13 to 3.19 is that when the number of receiver antennas is increased, most of the attainable diversity gain has already been achieved by the LDPC-G_2 coded concatenated schemes. Hence, the employment of an STBC using more transmitter antennas will introduce a higher-throughput modulation mode, which in turn will require an increased E_b/N_0 value and hence degrades the achievable performance.

Furthermore, in the context of uncorrelated Rayleigh fading channels and using a specific modulation scheme, the LDPC-STBC coded system's performance is mainly decided by the code rate and error-correction capability of the LDPC code employed, instead of the space–time codes' diversity order. In this case, using a lower-rate LDPC code will achieve a more substantial performance improvement than increasing the number of transmitters and the associated diversity gain. On the other hand, if the same LDPC code is employed, the throughput of the modulation scheme has more influence on the system's performance than the diversity order. In other words, the benefits brought about by the employment of low-throughput modulation schemes will be more substantial than that offered by a high-order STBC, provided that both schemes are assisted by the same LDPC code. The reason behind this phenomenon is that, when the higher-order STBC codes are used in conjunction with a high number of antennas, more vulnerable high-throughput modulation schemes have to be used, for the sake of maintaining the same effective throughput. Therefore the employment of the latter scenario would result in performance degradations. In summary, the best candidate schemes are the ones using the LDPC-aided G_2 code when communicating over uncorrelated Rayleigh fading channels. When using the same STBC, the complexity of the STBC code can be ignored during the comparisons.

Another useful conclusion can be drawn from Figure 3.24. As seen in the figure, the coding gains of the LDPC-aided schemes tend to remain unimproved, even if the affordable complexity increases to a certain degree, although the validity of this statement depends on the specific choice of the LDPC code used. This result assists us in deciding on the appropriate number of iterations to be used by the LDPC-STBC concatenated schemes, so that the best possible achievable performance-to-complexity trade-off can be achieved.

Before concluding this section, we summarize the achievable performance of the different schemes used in our various candidate systems communicating over uncorrelated Rayleigh fading channels in Table 3.5. For the scenarios having an effective throughput of 1, 2 and 3 BPS, respectively, the corresponding bold numbers denote the best scheme based on the criterion of achieving the best coding gain versus complexity trade-off, as seen in Figures 3.21, 3.22 and 3.23. As a result, the half-rate LDPC-coded STBC G_2 was found to be the best scheme in the scenarios having an effective throughput of 1 and 2 BPS, while the $\frac{3}{4}$-rate LDPC-coded G_2 code performs best in the scenario having an effective throughput of 3 BPS.

3.3.2   LDPC-Aided and TC-Aided STBCs

In Section 3.3.1, we studied the performance of various LDPC-aided STBC systems. It was found that the LDPC codes considerably improve the STBC system's performance over uncorrelated Rayleigh fading channels. However, besides LDPC channel codes, the STBCs can also be concatenated with

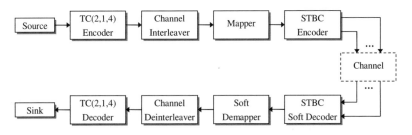

Figure 3.26: Overview of the STBC and TC(2,1,4) channel-coded system.

a range of other channel codes, such as Convolutional Codes (CCs), Turbo Convolutional (TC) codes [420, 421], Turbo Bose–Chaudhuri–Hocquenghem (TBCH) codes (a class of FEC codes) [432], etc. The performance of these various channel-coded G_2 schemes designed for transmission over uncorrelated Rayleigh fading channels has been studied in [314], where the best scheme found was the half-rate TC(2,1,4) code in conjunction with the STBC G_2. Hence, in this section, we will compare the performance of the best LDPC-STBC coded concatenated scheme found in Section 3.3.1.2, namely that of the half-rate LDPC-aided G_2 code, with the half-rate TC(2,1,4)-aided STBC schemes.

3.3.2.1 System Overview

In 1993, Berrou *et al.* [420, 421] proposed a novel channel code, referred to as a turbo code. As detailed in [314], the turbo encoder consists of two component encoders. Generally, convolutional codes are used as the component encoders and the corresponding turbo codes are termed here TC codes. For a TC(n, k, K) code, the three parameters n, k and K have the same meaning as in a convolutional code CC(n, k, K), where k is the number of input bits, n is the number of coded bits and K is the constraint length of the code. More details about TC codes can be found in [314]. A schematic of our experimental system, where the half-rate TC(2,1,4) code is employed, is given in Figure 3.26.

The source bits are first encoded by the half-rate TC(2,1,4) encoder. The Log-MAP decoding algorithm [314] is utilized for iterative turbo decoding, since it operates in the logarithmic domain and thus significantly reduces the computational complexity imposed by the MAP algorithm [425]. The number of turbo iterations is set to eight, since this yields a performance close to the achievable performance associated with an infinite number of iterations. The TC-encoded bits will be interleaved by the channel interleaver, as seen in Figure 3.26. In this case, a random interleaver having a depth of about 20 000 is used. The interleaved bits will then be forwarded to the mapper, followed by the STBC encoder. At the receiver side, the corresponding inverse operations are invoked, as seen in Figure 3.26. The simulation parameters of the TC-STBC coded concatenated system are given in Table 3.6.

3.3.2.2 Complexity Issues

For the sake of fair comparisons, we should calculate and take into account the complexity of the LDPC and TC(2,1,4) codes. The total estimated complexity of the TC codes per information bit per iteration in terms of additions and subtractions to be carried out is [433]

$$comp\{TC(n, 1, K)\} = 40(2^{K-1}) + 12n - 22. \tag{3.32}$$

According to Equation (3.32), the complexity of the TC(2,1,4) code is 322 per information bit per iteration. Since the number of iterations has been set to eight, the total complexity of the TC(2,1,4) code per bit is $322 \times 8 = 2576$ in the context of additions and subtractions, as shown in Table 3.6.

On the other hand, the complexity of the LDPC codes per information bit per iteration can be calculated according to Equation (3.31). Therefore, we can multiply the result of Equation (3.31) by the appropriately selected number of iterations required by the different-rate LDPC codes, so that a

Table 3.6: The parameters used in the TC(2,1,4)-STBC coded concatenated schemes.

| | STBC | | TC(2,1,4) | | | | | | |
BPS	Code	Code rate	Random turbo interleaver depth	Random channel interleaver depth	Code rate	Puncturing pattern	Iterations	Total *comp*	Modem
1.00	G_2	1	10 000	20 000	1/2	10,01	8	2576	QPSK
	G_3	1/2	10 000	20 000	1/2	10,01	8	2576	16-QAM
	G_4	1/2	10 000	20 000	1/2	10,01	8	2576	16-QAM
2.00	G_2	1	10 000	20 000	1/2	10,01	8	2576	16-QAM
3.00	G_2	1	10 002	20 004	1/2	10,01	8	2576	64-QAM

Table 3.7: The parameters used in the LDPC-G_2 coded concatenated schemes invoked for comparison with the TC(2,1,4)-G_2 coded concatenated schemes.

| | STBC | | LDPC | | | | | | | |
BPS	Code	Code rate	Code rate	Input bits block size	Output bits block size	Column weight	Iterations	Total *comp*	Interleaver depth	Modem
1.00	G_2	1	1/2	10 000	20 000	3	32	2592	20 000	QPSK
2.00	G_2	1	1/2	10 000	20 000	3	32	2592	20 000	16-QAM
3.00	G_2	1	1/2	10 002	20 004	3	32	2592	20 004	64-QAM

similar complexity per bit is used for both the LDPC codes and the TC(2,1,4) code. For the half-rate LDPC-coded G_2 scheme, the number of iterations was set to 32 so that the total complexity becomes $81 \times 32 = 2592$, which is close to the estimated complexity of 2576 encountered by the TC(2,1,4) scheme. The parameters of the LDPC-G_2 coded concatenated scheme used in this new scenario are given in Table 3.7.

3.3.2.3 Simulation Results

In this section, the performance of the LDPC- and TC-aided STBC schemes communicating over uncorrelated Rayleigh fading channels will be studied and compared. The parameters of the STBCs, the half-rate TC(2,1,4) code and the LDPC codes are given in Tables 3.1, 3.6 and 3.7, respectively. The simulation results are based on the same assumptions which were outlined in Section 3.2.4 on page 70.

Figure 3.27 compares the performance of the candidate schemes using the parameters summarized in Tables 3.6 and 3.7 operating at an effective throughput of 1 BPS over uncorrelated Rayleigh fading channels. For the half-rate TC(2,1,4)-coded scheme combined with the half-rate G_3 and G_4 codes, the 16-QAM modem is used for maintaining a throughput of 1 BPS. As seen in Figure 3.27, the TC(2,1,4)-aided G_2 code outperforms the others. However, at the BER of 10^{-5}, it only provides an approximately 0.1 dB gain over the LDPC-aided G_2 code, which is the best LDPC-STBC coded concatenated scheme according to Table 3.5. It is also observed that the schemes in which the G_3 and G_4 codes are employed exhibit an inferior performance in comparison with their G_2-code-based counterpart as well as in comparison with the LDPC-aided scheme, because the more densely packed 16-QAM phasor constellation is used.

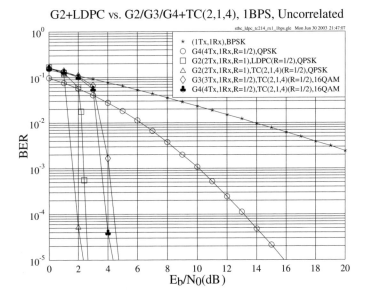

Figure 3.27: The BER versus E_b/N_0 performance of the $\mathbf{G_2}$, $\mathbf{G_3}$ and $\mathbf{G_4}$ STBCs of Table 3.1 in conjunction with the LDPC codes of Table 3.7 or the half-rate TC(2,1,4) code of Table 3.6 at an effective throughput of **1 BPS** using one receiver over uncorrelated Rayleigh fading channels.

Figure 3.28 compares the performance of the candidate schemes using the parameters summarized in Tables 3.6 and 3.7 operating at an effective throughput of 2 and 3 BPS over uncorrelated Rayleigh fading channels. As suggested by Figure 3.28, the TC(2,1,4)-aided schemes perform slightly better than the LDPC-aided schemes, providing an approximately 0.1 dB and 0.4 dB gain at the BER of 10^{-5} in the scenarios of having a throughput of 2 and 3 BPS, respectively.

In Figure 3.29 the achievable coding gain versus complexity is characterized. The coding gains were recorded at the BER of 10^{-5}, as presented in Section 3.3.1.3. From Figure 3.29, we infer that the curves associated with the TC(2,1,4) code are close to those of the LDPC codes, although the former ones perform slightly better than the latter ones in the context of having the same effective throughput of 1, 2 and 3 BPS, respectively. However, in the low-complexity range, i.e. in the complexity range spanning from 0 to 600, the LDPC codes perform better than the TC(2,1,4) code. On the one hand, the achievable lowest complexity of the half-rate TC(2,1,4) code is 322, which is attained when the number of iterations is set to one, while the LDPC codes are capable of providing an even lower complexity down to 81, again, as seen in Figure 3.29. On the other hand, for the LDPC schemes, the achievable coding gain dramatically increases, when the affordable complexity is increased within the range spanning from 0 to about 600. Therefore, the employment of the LDPC-aided schemes may be more attractive in scenarios where the affordable complexity is the most important concern, while the system's performance does not necessarily have to be the best.

3.3.2.4 Conclusions

In Section 3.3.2.3 we presented a range of performance comparisons in the context of the attainable coding gain versus complexity for the various TC(2,1,4)-STBC and LDPC-STBC coded concatenated schemes for transmission over uncorrelated Rayleigh fading channels. As a conclusion, the TC(2,1,4)-assisted STBC G_2 outperforms its LDPC-assisted counterparts for all the three scenarios having different effective throughputs. However, it was found that the associated performance difference is insignificant, i.e. less than 0.3 dB. Furthermore, the LDPC-STBC coded concatenated schemes

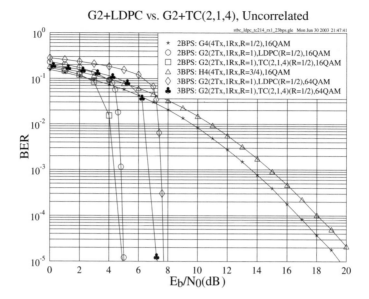

Figure 3.28: The BER versus E_b/N_0 performance of the $\boldsymbol{G_2}$ STBC of Table 3.1 in conjunction with the LDPC codes of Table 3.7 or the half-rate TC(2,1,4) code of Table 3.6 at an effective throughput of **2** and **3 BPS** using one receiver over uncorrelated Rayleigh fading channels.

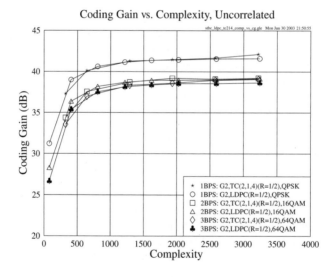

Figure 3.29: Coding gain versus complexity for the LDPC-G_2 concatenated arrangements and for the TC(2,1,4)-G_2 concatenated schemes using one receiver for communicating over uncorrelated Rayleigh fading channels. The associated parameters are given in Tables 3.6 and 3.7.

considered may be preferred for employment in systems where the severity of complexity constraint outweighs the importance of achieving the highest possible performance.

Finally, the performance of the different schemes studied is summarized in Table 3.8. All the results were generated using a single receiver, when communicating over uncorrelated Rayleigh fading

Table 3.8: Coding gains of the LDPC-STBC concatenated schemes and the TC(2,1,4)-STBC concatenated schemes using one receiver for communicating over uncorrelated Rayleigh fading channels. For the scenarios of different effective throughputs, the performance of the best scheme is printed in bold.

				BER				
				E_b/N_0 (dB)		Gain (dB)		
BPS	STBC code	LDPC rate	TC(2,1,4) rate	10^{-3}	10^{-5}	10^{-3}	10^{-5}	Modem
1.00	Uncoded	—	—	24.22	44.00	0.00	0.00	BPSK
	G_4	—	—	10.10	15.85	14.12	28.15	QPSK
	G_2	1/2	—	2.36	2.44	21.86	41.56	QPSK
	G_2	—	1/2	**1.62**	**2.34**	**22.60**	**41.66**	QPSK
	G_3	—	1/2	4.07	4.73	20.15	39.27	16-QAM
	G_4	—	1/2	3.55	4.35	20.67	39.65	16-QAM
2.00	Uncoded	—	—	24.22	44.00	0.00	0.00	QPSK
	G_4	—	—	13.61	19.58	10.61	24.42	16-QAM
	G_2	1/2	—	4.81	5.01	19.41	38.99	16-QAM
	G_2	—	1/2	**4.34**	**4.92**	**19.88**	**39.08**	16-QAM
3.00	Uncoded	—	—	26.30	46.26	0.00	0.00	8-PSK
	H_4	—	—	14.87	21.10	11.43	25.16	16-QAM
	G_2	1/2	—	7.52	7.72	18.78	38.54	64-QAM
	G_2	—	1/2	**6.68**	**7.32**	**19.62**	**38.94**	64-QAM

channels. The performance of the best scheme at the effective throughputs of 1, 2 and 3 BPS is printed in bold, respectively.

3.4 Channel Coding Aided STBC-OFDM

In Section 3.3, we investigated various LDPC channel coding assisted STBC schemes communicating over narrowband fading channels, followed by the performance study of LDPC-aided and TC-aided STBC schemes. Naturally, a range of channel codes can also be combined with the family of STBCs for the sake of improving the system's performance. In this section, various Coded Modulation (CM) [314] assisted STBC schemes will be studied for transmission over multi-path Rayleigh fading channels. Specifically, Trellis-Coded Modulation (TCM) [314, 434], Turbo Trellis-Coded Modulation (TTCM) [314, 435], Bit-Interleaved Coded Modulation (BICM) [314, 436] and iterative joint decoding and demodulation assisted BICM (BICM-ID) [314, 437] will be investigated. Furthermore, the above CM-assisted STBC-aided schemes will be studied in the context of a single-user OFDM [1–4] system. As a well-established technique, OFDM has exhibited a number of advantages over more traditional multiplexing techniques, and has been adopted for both Digital Audio and Video Broadcasting (DAB and DVB) in Europe. It has also been selected as the IEEE 802.11 standards for Wireless Local Area Networks (WLANs). Let us now embark on the investigation of the CM-assisted STBC single-user OFDM system.

3.4.1 CM-Assisted STBCs

Since the signal bandwidth available for wireless communications is limited, one of the most important objectives in the design of digital mobile systems is to make the most of the attainable bandwidth, e.g. with the aid of the CM schemes.

3.4.1.1   CM Principles

The basic principle of CM [314] is that we attach a parity bit to each uncoded information symbol formed by m information bits according to the specific modulation scheme used, hence doubling the number of constellation points to 2^{m+1} compared with that of 2^m in the original modem constellation. This is achieved by extending the modulation constellation, rather than expanding the required bandwidth, while maintaining the same effective throughput of m bits per symbol, as in the case of no channel coding. In other words, the signalling rate remains the same, since the redundant parity bit can be absorbed by the expansion of the constellation. Therefore, when the achievable coding gain of the CM scheme becomes higher than the E_b/N_0 degradation imposed by the more vulnerable higher-order modulation scheme employed, a useful effective coding gain can be achieved.

Among the various CM schemes, TCM [434] was originally designed for transmission over Additive White Gaussian Noise (AWGN) channels. TTCM [435] is a more recent joint coding and modulation scheme which has a structure similar to that of the family of binary turbo codes, but employs TCM schemes as component codes. Both TCM and TTCM employ set-partitioning-based constellation mapping [314], while using symbol-based turbo interleavers and channel interleavers. Another CM scheme, referred to as BICM [436], invokes bit-based channel interleavers in conjunction with grey constellation mapping. Furthermore, iteratively decoded BICM [437] using set partitioning was also proposed. More details about the various CM schemes used can be found in [314]. In this section, we will mostly focus on the performance of the proposed CM-assisted STBC-coded OFDM schemes communicating over wideband Rayleigh fading channels.

3.4.1.2   Inter-symbol Interference and OFDM Basics

If the modulation bandwidth exceeds the coherence bandwidth of the channel, Inter-Symbol Interference (ISI) will be introduced and the consecutive transmitted symbols are distorted, since the past and current symbols of the signals overlap. Hence, at the receiver, channel equalizers have to be employed to remove the effects of ISI [314].

An alternative way of mitigating the effects of ISI is to employ OFDM, which effectively mitigates the detrimental effects of the frequency-selective fading, when transmitting over high-rate wideband channels. The basic principle of OFDM is to split a high-rate data stream into a number of low-rate streams which are transmitted simultaneously over a number of subcarriers. Hence the symbol duration is rendered longer for each of the parallel subcarriers, and thus the relative effects imposed by the multipath channel's delay spread are reduced. In other words, since the system's data throughput is the sum of all the parallel subchannels' throughputs, the data rate per subchannel is only a small fraction of the total data rate of a conventional single-carrier system having the same throughput. This results in the phenomenon where the symbol duration becomes significantly longer than the channel's impulse response, thus it has the potential to disperse with channel equalization. Specifically, if an appropriate-duration cyclic OFDM symbol extension is selected, the ISI between consecutive OFDM symbols can be almost completely eliminated. Furthermore, for a given delay spread, the implementation complexity of an OFDM modem may be significantly lower than that of a single-carrier system employing an equalizer [4].

The schematic of an OFDM modem is shown in Figure 3.30. The source bit stream is first encoded by an STBC or channel encoder and forwarded to the interleaver and the mapper, where the bits are interleaved and may be mapped to non-binary symbols. Some pilot subcarriers may be inserted to assist in the estimation of the channel's frequency-domain transfer function, which is required for the receiver to counteract the effects of the channel's frequency-domain fading. The serial data stream is then converted into a parallel symbol sequence and forwarded to the Inverse Fast Fourier Transform (IFFT) modulator to form the time-domain modulated signal. Again, in order to eliminate the ISI between consecutive OFDM symbols, a cyclic extension has to be added to each OFDM symbol. Then the Digital-to-Analogue Converter (DAC) converts the cyclically extended OFDM signal to the analogue domain, which is finally filtered by a Low-Pass Filter (LPF) and transmitted through the

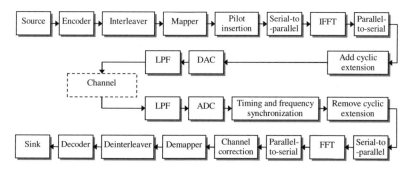

Figure 3.30: Schematic diagram of an OFDM modem.

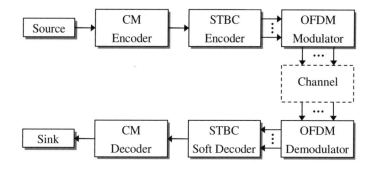

Figure 3.31: Schematic diagram of the proposed CM-assisted STBC-OFDM system.

wideband channel. At the receiver side, the Analogue-to-Digital Converter (ADC) converts the LPF-filtered received signal to the digital domain, where symbol timing and frequency synchronization are the first processing steps [4]. Then the cyclic extension attached to each OFDM symbol is removed and the recovered signal is forwarded to the Fast Fourier Transform (FFT) demodulator, whose output will be processed by the pilot-based frequency-domain channel equalizer in order to compensate the frequency-domain fading imposed by the channel. After symbol demapping and deinterleaving, the received signal is finally passed to the space–time or channel decoder, which outputs the decoded information bits.

3.4.1.3 System Overview

Figure 3.31 shows a schematic of the CM-assisted STBC-OFDM system investigated [3, 314]. As observed in Figure 3.31, the source information bits are first encoded and modulated by the CM encoder followed by the space–time encoder. In our schemes, the STBC employed was the G_2 code of Table 3.1, which invokes two transmitter antennas, and the two space–time-coded samples are mapped to two consecutive OFDM subcarriers and OFDM modulated. Then the frequency-domain symbols are converted to time-domain OFDM symbols by the IFFT-based modulator and the cyclic extension is appended to each individual OFDM symbol. The OFDM symbols are then transmitted via the multi-path fading channel, and the received noise-contaminated symbols are forwarded to the OFDM demodulator, where the FFT operation will be employed for converting the channel-impaired time-domain symbols to their frequency-domain counterparts. The recovered signal is then space–time soft decoded and the soft outputs are fed to the CM decoder for recovering the most likely transmitted information bits.

3.4.1.3.1   Complexity Issues

In order to compare the different candidate schemes under fair conditions, we chose the system parameters so that the decoding complexity of the various CM schemes employed became similar. The complexity imposed by the STBC codec was neglected, since the same G_2 STBC was used for all the CM-STBC concatenated schemes.

The symbol-based Log-MAP decoder [314] is utilized in all the CM schemes considered in our system, i.e. in the TCM, TTCM, BICM and BICM-ID codecs, to reduce the computational complexity imposed by the MAP algorithm [314]. Therefore, the multiplication and addition operations are substituted by additions and by the Jacobian sum operations [425] carried out in the logarithmic domain, respectively. As a result, in terms of the number of additions and subtractions, the total decoding complexity per bit per iteration for the TTCM scheme studied is as follows [431]:

$$comp\{TTCM\} = \frac{10M(2^{\nu+1} - 1)}{m}, \tag{3.33}$$

where m is the number of information bits in a coded information symbol, $M = 2^m$ is the number of legitimate symbols in the mapping constellation set, and ν is the code memory. For example, for the QPSK-based TTCM scheme having a code memory of $\nu = 3$, the associated complexity per bit per iteration is $10 \cdot 2^1 \cdot (2^{3+1} - 1)/1 = 300$, since in this case m is equal to 1. If the number of iterations is 4, the total decoding complexity per bit becomes $300 \cdot 4 = 1200$, as seen in Table 3.9. For the remaining CM schemes used, namely for TCM/BICM/BICM-ID, the corresponding decoding complexity per bit per iteration is

$$comp\{TCM/BICM/BICM\text{-}ID\} = \frac{5M(2^{\nu+1} - 1)}{m}, \tag{3.34}$$

which is half the complexity of that in Equation (3.33). The reason for this is that the TTCM scheme utilizes two Log-MAP decoders, while TCM/BICM/BICM-ID schemes only use one [314], hence the associated complexity of TTCM is doubled. The parameters used by the various CM-STBC concatenated schemes investigated are provided in Table 3.9. From the table, we can see that the total decoding complexity per bit – rather than per bit per iteration - of the four CM schemes is similar.

3.4.1.3.2   Channel Model

As mentioned earlier, we will investigate the proposed system when communicating over dispersive wideband Rayleigh fading channels. Specifically, we consider the Short Wireless Asynchronous Transfer Mode (SWATM) Channel Impulse Response (CIR) given on page 78 of [3], although the Doppler frequency may assume a range of different values. The three-tap SWATM channel is a truncated version of the five-tap Wireless Asynchronous Transfer Mode (WATM) CIR, retaining only the first three impulses [3]. This reduces the total length of the impulse response, where the last path arrives with a delay of 48.9 ns, which corresponds to 11 sample periods. For our simulations each of the three paths experiences independent Rayleigh fading having the normalized Doppler frequency of $f_d' = 1.235 \times 10^{-5}$. Figure 3.32 displays the impulse response of the SWATM channel, while the associated parameters are given in Table 3.10.

For the sake of combating the effects of ISI when communicating over the multi-path Rayleigh fading channel, as discussed in Section 3.4.1, we employ an OFDM modem having 512 subcarriers, while each OFDM symbol is extended by a cyclic prefix of $512/8 = 64$ time-domain samples [3]. Therefore, the length of an OFDM symbol becomes $512 + 64 = 576$ samples. Since the number of subcarriers is sufficiently high, we may assume that each OFDM subcarrier experiences narrowband channel conditions in the frequency domain.

3.4.1.3.3   Assumptions

When the STBCs were employed for transmissions over uncorrelated Rayleigh fading channels, as mentioned in Section 3.2.4, we assumed that the channel was quasi-static so that its path gains remained

Table 3.9: The parameters of the various CM-assisted STBC schemes. The parameters of the STBC G_2 are given in Table 3.1.

| | STBC | | CM | | | | | | | |
| | | Code | CM | Code | Data | | | Symbol-based codeword | Total | |
BPS	Code	rate	scheme	rate	bits	ν	Iterations	length	*comp*	Modem
1.00	G_2	1	—	—	—	—	—	1024	—	QPSK
	G_2	1	TCM	1/2	1	6	—	1024	1270	QPSK
	G_2	1	TTCM	1/2	1	3	4	1024	1200	QPSK
	G_2	1	BICM	1/2	1	6	—	1024	1270	QPSK
	G_2	1	BICM-ID	1/2	1	3	8	1024	1200	QPSK
2.00	G_2	1	—	—	—	—	—	1024	—	8-PSK
	G_2	1	TCM	2/3	2	6	—	1024	1270	8-PSK
	G_2	1	TTCM	2/3	2	3	4	1024	1200	8-PSK
	G_2	1	BICM	2/3	2	6	—	1024	1270	8-PSK
	G_2	1	BICM-ID	2/3	2	3	8	1024	1200	8-PSK
3.00	G_2	1	—	—	—	—	—	1024	—	16-QAM
	G_2	1	TCM	3/4	3	6	—	1024	1693	16-QAM
	G_2	1	TTCM	3/4	3	3	4	1024	1600	16-QAM
	G_2	1	BICM	3/4	3	6	—	1024	1693	16-QAM
	G_2	1	BICM-ID	3/4	3	3	8	1024	1600	16-QAM

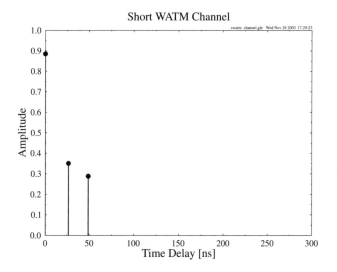

Figure 3.32: The impulse response of the SWATM channel [3]. The corresponding parameters of the channel are summarized in Table 3.10.

constant across for example $n = 2$ consecutive STBC time slots for the G_2 STBC, corresponding to the $n = 2$ rows of the G_2 code's transmission matrix. However, in this new context we can no longer assume that the corresponding frequency-domain subcarrier gains remain identical as a consequence of the wideband channel's frequency-domain fading profile, an issue which will be further discussed in Section 3.4.1.4. This results in a residual error floor for the unprotected G_2 STBC scheme, as seen

Table 3.10: Sampling Rate $1/T_s$, maximum path delay τ_{max}, maximum Doppler frequency f_d, normalized Doppler frequency f'_d, number of paths n, FFT length K and cyclic prefix length cp of the SWATM channel of Figure 3.32.

$1/T_s$	τ_{max}	f_d	f'_d	n	K	cp
225 MHz	48.9 ns	2278 Hz	1.235×10^{-5}	3	512	64

for example in Figure 3.35 below. For the concatenated CM-STBC schemes, however, the error floor experienced may be significantly reduced to a neglectable level.

In Section 3.4.1.4, the performance of the proposed CM-assisted STBC OFDM schemes will be compared. All our simulation results were based on the following assumptions:

- Each path of the multi-path channel employed experiences independent Rayleigh fading.

- The average signal power received from each transmitter antenna is the same.

- The receiver has perfect knowledge of the channels' fading amplitudes.

These assumptions simplify the simulations to a degree, therefore the system concerned is not a realistic one. However, again, since the experimental circumstances are identical for all performance comparisons, the results may be expected to characterize adequately the relative performance of the various schemes used.

3.4.1.4 Simulation Results

In this section, the performance of the CM-assisted STBC-OFDM system considered will be studied. The simulation parameters have been given in Table 3.9. All schemes utilized two transmitter antennas for the G_2 STBC and one receiver antenna. Each OFDM symbol has 512 subcarriers and a cyclic extension of 64 samples.

Performance at an effective throughput of 1 BPS   Figure 3.33 shows the performance of the various CM-assisted G_2 STBC-OFDM schemes communicating over the SWATM channel. In our system, we employ grey-coding-based constellation mapping for the BICM scheme, while using set-partitioning-based constellation mapping for the TCM, TTCM and BICM-ID arrangements [314]. To achieve an effective throughput of 1 BPS, QPSK modulation is used for all the half-rate CM-assisted schemes. As seen in Figure 3.33, the CM-G_2 coded concatenated schemes perform significantly better than the unprotected G_2 scheme, achieving an E_b/N_0 gain of about 14 dB at the BER of 10^{-5}. Among all the CM-assisted schemes, the TTCM-aided arrangement gives the best performance by achieving about 0.5 to 1 dB gain over the other CM-assisted schemes at the BER of 10^{-5}.

Performance at an effective throughput of 2 BPS   The performance comparison of the different CM-STBC concatenated schemes having an effective throughput of 2 BPS for transmissions over the SWATM channel is shown in Figure 3.34. It can be seen in Figure 3.34 that when the E_b/N_0 value encountered is relatively low, i.e. below about 7.5 dB, the unprotected G_2 scheme performs better than the CM-assisted G_2 schemes. However, when the E_b/N_0 value experienced is higher than approximately 7.5 dB, the TTCM-aided G_2 scheme outperforms all the other candidates, achieving a gain of about 1.3 and 12.5 dB over the other CM-aided G_2 schemes and over the unprotected G_2 scheme, respectively, at the BER of 10^{-5}.

Performance at an effective throughput of 3 BPS   If we increase the system's effective throughput to 3 BPS, a residual BER of approximate 6×10^{-5} is observed for the performance curve of the unprotected G_2 scheme, as seen in Figure 3.35. This phenomenon can be explained as follows. In the context of the single-path uncorrelated Rayleigh fading channels mentioned in Section 3.2.4, we assumed that

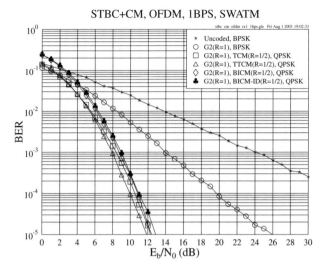

Figure 3.33: The BER versus E_b/N_0 performance of the G_2 STBC of Table 3.1 in conjunction with the various CM schemes of Table 3.9 at an effective throughput of **1 BPS** using one receiver when communicating over the SWATM channel. An OFDM scheme having 512 subcarriers and a cyclic extension of 64 samples is employed.

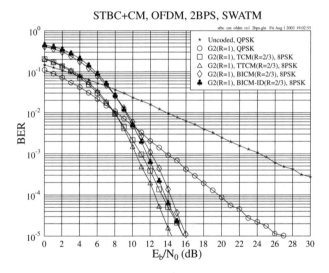

Figure 3.34: The BER versus E_b/N_0 performance of the G_2 STBC of Table 3.1 in conjunction with the various CM schemes of Table 3.9 at an effective throughput of **2 BPS** using one receiver when communicating over the SWATM channel. An OFDM scheme having 512 subcarriers and a cyclic extension of 64 samples is employed.

the channel is quasi-static so that the channel's path gains are constant across n consecutive STBC time slots. For example, we have $n = 2$ for the G_2 STBC, corresponding to the $n = 2$ rows of the STBC's transmission matrix. In the context of wideband channels, for instance the SWATM channel of Figure 3.32, however, the channel's delay spread will have an effect on the associated frequency-domain

STBC+CM, OFDM, 3BPS, SWATM

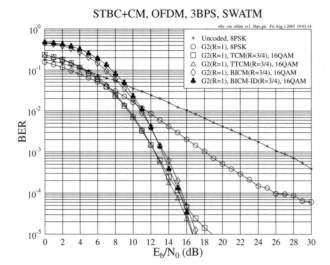

Figure 3.35: The BER versus E_b/N_0 performance of the G_2 STBC of Table 3.1 in conjunction with the various CM schemes of Table 3.9 at an effective throughput of **3 BPS** using one receiver when communicating over the SWATM channel. An OFDM scheme having 512 subcarriers and a cyclic extension of 64 samples is employed.

transfer functions. More specifically, the fading amplitudes vary more rapidly when the delay spread is increased [314]. Since the maximum delay spread of the SWATM channel is as high as $\tau_{max} = 48.9\,ns$, the variation of the frequency-domain fading amplitudes is so dramatic that we can no longer assume that the path gains remain constant during two consecutive STBC time slots. In this case, for the unprotected STBC schemes, the rapid variation of the channel's frequency-domain fading envelope will seriously erode the orthogonality of the G_2 STBC's two components, resulting in a residual error floor, as seen for example in Figure 3.35.

Furthermore, if a higher-order modulation scheme such as 16-QAM is employed, as shown in Figure 3.35, since the signal is mapped to more densely packed constellation phasors which are prone to transmission errors, the error floor imposed by the channel is expected to be higher than that in the scenarios, where a lower-order modulation scheme, such as QPSK or 8-PSK is used, as exhibited by Figures 3.33 and 3.34. More explicitly, comparing Figures 3.33 and 3.34 with Figure 3.35, we can see that the BER error floors observed in Figures 3.33 and 3.34 are below 10^{-5}, while in Figure 3.35 the error floor encountered is about 6×10^{-5}.

With the advent of employing the CM schemes, however, the error floor can be eliminated or reduced to a significantly lower level. As Figure 3.35 shows, the CM schemes significantly improve the STBC-OFDM system's performance and the BER error floor exhibited by the unprotected G_2 scheme has been essentially eliminated. Similar to the scenarios of having an effective throughput of 1 and 2 BPS, the TTCM-G_2 concatenated scheme was found to give the best performance among all the CM-assisted schemes studied, although the E_b/N_0 gain achieved over the other candidate schemes is not significant.

3.4.1.5 Conclusions

In the previous sections we investigated the achievable performance of the various CM-assisted STBC-OFDM schemes for transmissions over the SWATM channel. We first briefly reviewed the basic principles of the CM schemes in Section 3.4.1.1. In Section 3.4.1.2 a rudimentary introduction to OFDM was provided, which was followed by the overview of the simulation arrangement, as detailed

Table 3.11: Performance of the CM-STBC concatenated OFDM schemes using one receiver, when communicating over the SWATM channel. The STBC and CM parameters were given in Table 3.1 and Table 3.9, respectively. An OFDM scheme having 512 subcarriers and a cyclic extension of 64 samples was employed. For the scenarios of having a different effective throughput, the performance of the best scheme is printed in bold.

				BER				
				E_b/N_0 (dB)		Gain (dB)		
	STBC	CM	CM code					
BPS	scheme	scheme	rate	10^{-3}	10^{-5}	10^{-3}	10^{-5}	Modem
1.00	Uncoded	—	—	24.06	44.27	0.00	0.00	BPSK
	G_2	—	—	13.92	25.97	10.14	18.30	BPSK
	G_2	TCM	1/2	8.38	12.44	15.68	31.83	QPSK
	G_2	TTCM	1/2	**7.94**	**11.87**	**16.12**	**32.40**	QPSK
	G_2	BICM	1/2	8.72	12.28	15.34	31.99	QPSK
	G_2	BICM-ID	1/2	8.96	12.89	15.10	31.38	QPSK
2.00	Uncoded	—	—	24.06	44.27	0.00	0.00	QPSK
	G_2	—	—	13.81	27.08	10.25	17.19	QPSK
	G_2	TCM	2/3	10.95	15.73	13.11	28.54	8-PSK
	G_2	TTCM	2/3	**10.36**	**14.43**	**13.70**	**29.84**	8-PSK
	G_2	BICM	2/3	12.10	16.05	11.96	28.22	8-PSK
	G_2	BICM-ID	2/3	11.60	15.73	12.46	28.54	8-PSK
3.00	Uncoded	—	—	26.36	47.17	0.00	0.00	8-PSK
	G_2	—	—	18.09	—	8.27	—	8-PSK
	G_2	TCM	3/4	12.46	18.86	13.90	28.31	16-QAM
	G_2	TTCM	3/4	**12.42**	**16.67**	**13.94**	**30.50**	16-QAM
	G_2	BICM	3/4	13.43	17.11	12.93	30.06	16-QAM
	G_2	BICM-ID	3/4	13.25	16.71	13.11	30.46	16-QAM

in Section 3.4.1.3. Our performance analysis was presented in Section 3.4.1.4, where the CM-assisted STBC schemes were found to improve significantly the system's achievable performance, eliminating the BER floor of the unprotected STBC scheme. Furthermore, the TTCM-STBC concatenated scheme was observed to give the best performance among all the CM-STBC concatenated schemes.

In conclusion, we summarized the performance of the evaluated CM-STBC concatenated schemes in Table 3.11. The coding gains summarized in Table 3.11 were defined as the E_b/N_0 difference, expressed in terms of decibels, at a BER of 10^{-5} between the various channel coding assisted STBC-OFDM systems and the uncoded single-transmitter OFDM system having the same effective throughput. All the results were recorded by using one receiver, while communicating over the SWATM channel of Section 3.4.1.3.2.

3.4.2 CM-Aided and LDPC-Aided STBC-OFDM Schemes

In Section 3.4.1, we studied the performance of different CM-assisted STBC schemes for transmissions over the SWATM channel [3] of Figure 3.32. Instead of the joint coding modulation schemes of Table 3.9, separate channel codes such as LDPC codes [419] can also be incorporated into our STBC-OFDM system to improve the achievable performance. Hence, in this section we will compare the CM-assisted G_2 STBC schemes of Section 3.4.1 with those in which the LDPC codes are combined with the STBC G_2.

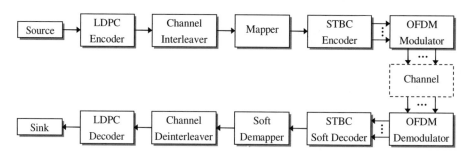

Figure 3.36: Schematic diagram of the proposed LDPC-assisted STBC-OFDM system.

Table 3.12: Parameters of the various LDPC-assisted STBC-OFDM schemes. The parameters of the G_2 STBC are given in Table 3.1.

	STBC		LDPC								
BPS	Code	Code rate	Code rate	Column weight	Iterations	In. Bits block size	Out. Bits block size	Inter-leaver depth	Symbol-based codeword length	Total *comp*	Modem
1.00	G_2	1	1/2	3	15	1024	2048	2048	1024	1215	QPSK
2.00	G_2	1	1/2	3	15	2048	4096	4096	1024	1215	16-QAM
3.00	G_2	1	3/4	3	10	3072	4096	4096	1024	1530	16-QAM

3.4.2.1 System Overview

The LDPC-assisted STBC-OFDM system's schematic is given in Figure 3.36. Compared with Figure 3.31, where the CM-assisted STBC-OFDM system was introduced, we substituted the CM encoder and decoder by an LDPC encoder and decoder, respectively. For the CM schemes, the associated symbol-based channel interleaver and deinterleaver have been integrated in the CM encoder and decoder, respectively. For the LDPC schemes, however, an external bit-based channel interleaver and deinterleaver have to be employed for the sake of further improving the system's performance, as seen in Figure 3.36.

Utilizing Equations (3.31), (3.33), and (3.34), we can calculate the corresponding decoding complexity per bit per iteration for the LDPC and CM schemes, respectively. For the sake of fair comparisons, we have to select the appropriate parameters so that the CM-STBC schemes and the LDPC-STBC schemes exhibit a similar decoding complexity. Specifically, similar to Section 3.4.1.3, the STBC was also chosen to be the G_2 code in all the LDPC-STBC concatenated OFDM schemes investigated, and thus, again, the related decoding complexity of the G_2 code was neglected in order to simplify the comparisons. Furthermore, the symbol-based codeword length of the LDPC-STBC concatenated schemes was fixed to 1024, which is equal to that of the CM-STBC concatenated schemes. Specifically, the same channel model, namely the SWATM channel of Section 3.4.1.3, and the same OFDM modem having 512 subcarriers and a cyclic prefix of 64 samples were employed in this new context.

As mentioned in Section 3.4.1.4, at a specific effective throughput, it was found that the TTCM-assisted G_2-coded scheme gave the best performance. Hence we used the TTCM scheme as the representative of the CM family, while half-rate and $\frac{3}{4}$-rate LDPC codes were chosen for representing the LDPC code family. As a summary, the parameters of the various CM-STBC concatenated OFDM systems are given in Table 3.9, while the parameters of the LDPC-STBC concatenated OFDM systems are provided in Table 3.12.

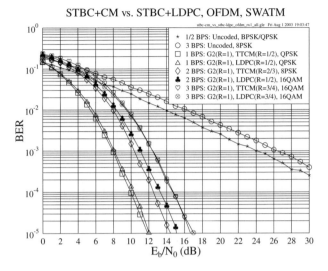

Figure 3.37: The BER versus E_b/N_0 performance of the G_2 STBC of Table 3.1 in conjunction with the TTCM of Table 3.9 or the LDPC codes of Table 3.12 at different effective throughputs using one receiver when communicating over the SWATM channel of Figure 3.32. An OFDM scheme having 512 subcarriers and a cyclic extension of 64 samples is employed.

3.4.2.2 Simulation Results

In this section we compare the TTCM- and LDPC-assisted STBC-OFDM schemes, which are characterized in Figure 3.37. All schemes utilized two transmitter antennas for the G_2 STBC and one receiver antenna. All simulation results were generated based on the assumptions outlined in Section 3.4.1.3.

As seen from Figure 3.37, the TTCM- and LDPC-assisted G_2-coded OFDM schemes have a similar performance. Specifically, when the effective throughput is 1 BPS, the TTCM-assisted scheme performs slightly better than the LDPC-aided candidate system. In the scenario of an effective throughput of 2 BPS, the former outperforms the latter again. In this context, however, we may see that the performance gap between the two competing schemes is larger than that in the scenario of a throughput of 1 BPS. This is because, in order to achieve the same effective throughput of 2 BPS, the TTCM-aided scheme employs 8-PSK modulation in conjunction with set partitioning, while the LDPC-aided candidate has to employ the more vulnerable 16-QAM grey-mapping-based constellation, since the code rate of the TTCM and the LDPC code are $\frac{2}{3}$ and $\frac{1}{2}$, respectively. Nonetheless, it is found in Figure 3.37 that the two corresponding competitors exhibit a similar performance when the throughput is increased to 3 BPS. In this case, however, the LDPC-aided scheme is marginally superior to the TTCM-aided scheme when the E_b/N_0 value is relatively low, i.e. below 11 dB.

In Figure 3.38, the associated coding gain versus complexity results are provided. The coding gain was defined in Section 3.4.1.5, while the complexity of the CM schemes and LDPC codes can be calculated with the aid of Equations (3.33) and (3.31), respectively. Given the same effective throughput, it is found that the coding gain performance of the TTCM-aided G_2 schemes surpasses that of the LDPC-aided G_2 schemes when the affordable complexity is higher than approximately 500, as observed in Figure 3.38. At a low complexity, i.e. below a value of about 500, however, the LDPC-aided schemes tend to achieve a higher coding gain than the TTCM-aided schemes at the specific throughput values considered.

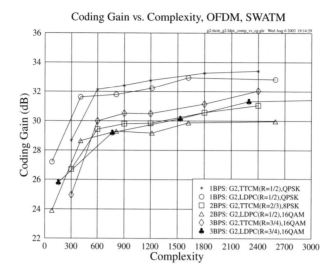

Figure 3.38: Coding gain versus complexity for the TTCM-G_2 concatenated and LDPC-G_2 concatenated schemes using one receiver, when communicating over the SWATM channel of Figure 3.32. An OFDM scheme having 512 subcarriers and a cyclic extension of 64 samples is employed. The simulation parameters are given in Tables 3.9 and 3.12.

3.4.2.3 Conclusions

In Section 3.4.2.2 the performance of the different TTCM- and LDPC-assisted G_2 coded OFDM schemes was studied and compared. As seen from Figure 3.37, the TTCM-assisted G_2 scheme gives a better performance than the LDPC-assisted G_2 scheme. Furthermore, in the context of the achievable coding gain versus complexity performance, it was found that the TTCM-assisted schemes are capable of achieving higher coding gains in the relatively high-complexity range, than the LDPC-assisted candidate schemes.

In conclusion, we summarize the achievable performance of the various schemes discussed in Table 3.13.

3.5 Chapter Summary

The state of the art of various transmission schemes based on multiple transmitters and receivers was briefly reviewed in Section 3.1. A simple communication system invoking the STBC G_2 was introduced in Section 3.2.1, leading to further discussions on various other STBCs. More specifically, Section 3.2.2.1 defined the STBC transmission matrix, while the encoding algorithm of the G_2 and a range of other STBCs was given in Sections 3.2.2.2 and 3.2.2.3, respectively. Section 3.2.3 presented the decoding algorithm of the STBCs considered. More specifically, Section 3.2.3.1 introduced the ML algorithm, while Section 3.2.3.2 discussed the MAP algorithm, which enables the STBC decoder to provide soft outputs. Thus various channel codes can be concatenated with the STBCs to improve the system's performance. In Section 3.2.4, a schematic of the proposed system was presented, and some assumptions used in our simulations were outlined.

The performances of the various STBCs were studied and compared in Section 3.2.5. Specifically, in Sections 3.2.5.1 and 3.2.5.2 the performances of different STBC schemes communicating over both uncorrelated and correlated Rayleigh fading channels were compared, respectively. It was found that the performances of the half-rate codes G_3 and G_4 degraded in comparison with that of the unity-

Table 3.13: Performance of the TTCM- and LDPC-STBC coded concatenated OFDM schemes using one receiver, when communicating over the SWATM channel of Figure 3.32. The STBC, CM and LDPC parameters were given in Tables 3.1, 3.9 and 3.12, respectively. An OFDM scheme having 512 subcarriers and a cyclic extension of 64 samples was employed. For the scenarios of different effective throughput, the performance of the best scheme is printed in bold.

| | | | | BER | | | | |
| | | | | E_b/N_0 (dB) | | Gain (dB) | | |
BPS	STBC scheme	TTCM code rate	LDPC code rate	10^{-3}	10^{-5}	10^{-3}	10^{-5}	Modem
1.00	Uncoded	—	—	24.06	44.27	0.00	0.00	BPSK
	G_2	—	—	13.92	25.97	10.14	18.30	BPSK
	G_2	1/2	—	**7.94**	**11.87**	**16.12**	**32.40**	QPSK
	G_2	—	1/2	8.09	12.04	15.97	32.23	QPSK
2.00	Uncoded	—	—	24.06	44.27	0.00	0.00	QPSK
	G_2	—	—	13.81	27.08	10.25	17.19	QPSK
	G_2	2/3	—	**10.36**	**14.43**	**13.70**	**29.84**	8-PSK
	G_2	—	1/2	11.07	15.10	12.99	29.17	16-QAM
3.00	Uncoded	—	—	26.36	47.17	0.00	0.00	8-PSK
	G_2	—	—	18.09	—	8.27	—	8-PSK
	G_2	3/4	—	**12.42**	**16.67**	**13.94**	**30.50**	16-QAM
	G_2	—	3/4	12.44	17.02	13.92	30.15	16-QAM

rate code G_2, when the effective throughput was increased. The reason is that in order to maintain the same effective throughput, higher-throughput modulation schemes have to be employed in conjunction with the half-rate codes G_3 and G_4, which are more vulnerable to errors. This therefore degrades the performance of the system. The $\frac{3}{4}$-rate codes H_3 and H_4 suffer a lower degradation in this case, as their code rate is higher than that of the G_3 and G_4 codes, therefore a moderate-throughput modulation scheme can be employed. This in turn assists in maintaining the performance advantage achieved by the space–time codes. Additionally, when the number of receivers is increased, the achievable performance gain of the G_3, G_4, H_3 and H_4 codes over the G_2 code becomes lower, as seen in Figures 3.4, 3.5, 3.8 and 3.9. This is because much of the attainable diversity gain has already been achieved using the G_2 code employing two receivers. Another important conclusion is that the performances of the space–time codes communicating over both uncorrelated and correlated Rayleigh fading channels are the same, provided that the effective throughput is the same. The performances of all the STBCs were summarized in Table 3.2 at the end of Section 3.2.6.

The schemes employing STBCs in conjunction with channel codes were studied in Section 3.3, which were divided into two parts: the performance study of LDPC-aided STBCs was presented in Section 3.3.1, while our performance comparisons between LDPC-assisted and TC(2,1,4)-aided STBC schemes were provided in Section 3.3.2.

In Section 3.3.1.1 the LDPC-based system was introduced and the associated simulation parameters were given. The performances of the LDPC-STBC concatenated schemes were provided in Section 3.3.1.2, including the scenarios of both uncorrelated and correlated Rayleigh fading channels in Section 3.3.1.2.1 and Section 3.3.1.2.2, respectively. The implementation complexity issues of the schemes studied were discussed in Section 3.3.1.3, where the coding gain versus complexity at different effective throughputs was shown in Figures 3.21, 3.22 and 3.23. It was found that the LDPC-aided STBC schemes performed significantly better than the STBC-only schemes when communicating over uncorrelated Rayleigh fading channels, while the achievable performance improvement was insignificant when communicating over correlated Rayleigh fading channels. This is because the attainable performances of the LDPC codes were found to be worse when communicating over correlated Rayleigh channels than over uncorrelated Rayleigh channels, unless the LDPC codeword length was sufficiently long or a sufficiently long channel interleaver was used. The phenomenon of achieving different performances over uncorrelated and correlated Rayleigh fading channels was also observed in the context of the LDPC-STBC concatenated system. On the other hand, when the number of receiver antennas was increased, the schemes employing an STBC of a higher-diversity order were found to provide an inferior performance, since most of the attainable diversity gain has already been achieved by the LDPC-aided G_2-coded scheme. It was also found that for transmission over the uncorrelated Rayleigh fading channels, when the same modulation scheme was used, a lower-rate LDPC code benefited the system more than a space–time code of a higher-diversity order did. Furthermore, when the same LDPC code was used, a lower-order modulation scheme tended to offer a higher performance improvement than an STBC of a higher-diversity order did. The performance of different LDPC-aided STBC schemes was summarized in Table 3.5, where the half-rate LDPC-coded STBC G_2 was found to be the best option among all the LDPC-STBC concatenated schemes.

Following Section 3.3.1, where the LDPC-aided space–time-coded system was studied, our comparative study between LDPC-aided and TC(2,1,4)-assisted STBC schemes transmitting over uncorrelated Rayleigh fading channels was presented in Section 3.3.2. The TC(2,1,4)-aided system was introduced in Section 3.3.2.1, while the associated complexity issues were discussed in Section 3.3.2.2, which was followed by the performance analysis in Section 3.3.2.3. From our coding gain versus complexity performance comparisons, it was concluded that the half-rate TC(2,1,4)-assisted STBC G_2 slightly outperforms the LDPC-assisted STBC schemes. However, the LDPC-STBC concatenated schemes may be considered as better design options for complexity-sensitive systems, where the achievable performance does not necessarily have to be the highest possible, since the LDPC-aided schemes are capable of maintaining a lower complexity than the TC(2,1,4)-aided scheme is. In conclusion, the performance of the different schemes studied was summarized in Table 3.8.

Furthermore, channel coding assisted STBC single-user OFDM systems were studied in Section 3.4. This research was divided into two parts. The first part was the investigation of the various CM-assisted STBC-OFDM schemes detailed in Section 3.4.1. The basic OFDM system was introduced in Section 3.4.1.2, followed by the whole system's overview in Section 3.4.1.3. More specifically, a brief complexity analysis was provided in Section 3.4.1.3.1, and the introduction of the SWATM channel model was the subject of Section 3.4.1.3.2. Our simulation results were discussed in Section 3.4.1.4 and summarized in Table 3.11. The latter part of Section 3.4 focused on the performance comparison of the CM- and LDPC-assisted STBC systems considered, which was detailed in Section 3.4.2. The BER versus E_b/N_0 as well as the coding gain versus complexity performances of the two groups of candidate schemes were compared in Section 3.4.2.2, followed by our conclusions in Section 3.4.2.3, where the results were summarized in Table 3.13.

The family of STBCs is readily applicable to employment in downlink systems, where the multiple transmitter antennas are installed at the BS. However, in the context of uplink systems it is impractical to use high-order STBCs at the MSs, since the MSs are expected to have a low implementation complexity and thus cannot afford the added cost of a high number of transmitter antennas. In the next chapter, Space-Division Multiple Access (SDMA) types of uplink multi-user OFDM systems will be investigated, which invoke multiple receiver antenna elements for supporting a multiplicity of MSs, each of which employs a single transmitter antenna only.

Chapter 4

Coded Modulation Assisted Multi-user SDMA-OFDM Using Frequency-Domain Spreading

4.1 Introduction

In recent years Orthogonal Frequency-Division Multiplexing (OFDM) [1, 3, 4, 24] has emerged as a successful air-interface technology for both broadcast and Wireless Local Area Network (WLAN) applications, while Wideband Code-Division Multiple Access (WCDMA) has emerged as the winning candidate for 3G mobile systems. Our research therefore includes exploration of the performance versus complexity trade-offs of a generic class of Multi-Carrier Code-Division Multiple Access (MC-CDMA) [38] systems, which are capable of supporting the interworking of existing as well as future broadcast and personal communication systems.

Space-Division Multiple Access (SDMA) based OFDM [3, 193, 438] communication invoking Multi-User Detection (MUD) [439] techniques has recently attracted intensive research interests. In SDMA Multiple-Input, Multiple-Output (MIMO) systems the transmitted signals of L simultaneous uplink mobile users – each equipped with a single transmitter antenna – are received by the P different receiver antennas of the Base Station (BS). At the BS the individual users' signals are separated with the aid of their unique, user-specific spatial signature constituted by their channel transfer functions or, equivalently, Channel Impulse Responses (CIRs). A variety of MUD schemes, such as the Least-Squares (LS) [3, 439, 440] and Minimum Mean Square Error (MMSE) [3, 193, 197, 439, 440] detectors, or Successive Interference Cancellation (SIC) [3, 193, 197, 439–441], Parallel Interference Cancellation (PIC) [3, 439, 441, 442] and Maximum Likelihood Detection (MLD) [3, 193, 197, 198, 439] schemes may be invoked for the sake of separating the different users at the BS on a per-subcarrier basis. Among these schemes, the MLD arrangement was found to give the best performance, although this was achieved at the cost of a dramatically increased computational complexity, especially in the context of a high number of users and higher-order modulation schemes, such as 16-QAM [441]. By contrast, MMSE detection exhibits the lowest complexity in this set of detectors, while suffering from a performance loss [3, 441].

In order to improve the achievable performance by exploiting the multi-path diversity potential offered by wideband channels, a further technique that is often used in the context of CDMA systems is constituted by the spreading of the subcarrier signals over a number of adjacent subcarriers with the aid of orthogonal spreading codes, such as Walsh–Hadamard Transform (WHT) based codes [32]. This technique may also be employed in multi-user SDMA-OFDM systems in the context of spreading across all or a fraction of the subcarriers [443]. Spreading across all subcarriers using a single Walsh–Hadamard Transform Spreading (WHTS) [3] code is expected to result in a better averaging of the bursty error effects at the cost of a higher WHT complexity.

Furthermore, the achievable performance can be significantly improved, if Forward Error Correction (FEC) schemes, such as for example Turbo Convolutional (TC) codes are incorporated into the SDMA system [3]. Among a number of FEC schemes, Trellis-Coded Modulation (TCM) [314, 434], Turbo TCM (TTCM) [314, 435], Bit-Interleaved Coded Modulation (BICM) [314, 436] and Iteratively Decoded BICM (BICM-ID) [314, 437] have attracted intensive research interests, since they are capable of achieving a substantial coding gain without bandwidth expansion.

In this section, we combine the above-mentioned various Coded Modulation (CM) schemes with a multi-user SDMA-OFDM system, in which WHT-based subcarrier spreading is used. The structure of this section is as follows. The SDMA MIMO channel model is described in Section 4.2.1, while an overview of the CM-assisted multi-user SDMA-WHTS-OFDM system is provided in Section 4.2.2, where the basic principles of CM, MUD and WHT-based spreading (WHTS) are also introduced. Our simulation results are provided in Section 4.3, while our conclusions are summarized in Section 4.4.

4.2 System Model

4.2.1 SDMA MIMO Channel Model

Figure 4.1 shows an SDMA uplink MIMO channel model, where each of the L simultaneous mobile users employs a single transmitter antenna at the Mobile Station (MS), while the BS's receiver exploits P antennas. At the kth subcarrier of the nth OFDM symbol received by the P-element receiver antenna array we have the complex received signal vector $\mathbf{x}[n, k]$, which is constituted by the superposition of the independently faded signals associated with the L mobile users and contaminated by the Additive White Gaussian Noise (AWGN), expressed as

$$\mathbf{x} = \mathbf{H}\mathbf{s} + \mathbf{n}, \tag{4.1}$$

where the $(P \times 1)$-dimensional vector $\mathbf{x}$, the $(L \times 1)$-dimensional vector $\mathbf{s}$ and the $(P \times 1)$-dimensional vector $\mathbf{n}$ are the received, transmitted and noise signals, respectively. Here we have omitted the indices $[n, k]$ for each vector for the sake of notational convenience. Specifically, the vectors $\mathbf{x}$, $\mathbf{s}$ and $\mathbf{n}$ are given by

$$\mathbf{x} = [x_1, x_2, \ldots, x_P]^T, \tag{4.2}$$

$$\mathbf{s} = [s^{(1)}, s^{(2)}, \ldots, s^{(L)}]^T, \tag{4.3}$$

$$\mathbf{n} = [n_1, n_2, \ldots, n_P]^T. \tag{4.4}$$

The $(P \times L)$-dimensional matrix $\mathbf{H}$, which contains the Frequency-Domain CHannel Transfer Functions (FD-CHTFs) of the L users, is given by

$$\mathbf{H} = [\mathbf{H}^{(1)}, \mathbf{H}^{(2)}, \ldots, \mathbf{H}^{(L)}], \tag{4.5}$$

where $\mathbf{H}^{(l)}$ $(l = 1, \ldots, L)$ is the vector of the FD-CHTFs associated with the transmission paths from the lth user's transmitter antenna to each element of the P-element receiver antenna array, which is expressed as

$$\mathbf{H}^{(l)} = [H_1^{(l)}, H_2^{(l)}, \ldots, H_P^{(l)}]^T, \quad l = 1, \ldots, L. \tag{4.6}$$

In Equations (4.1) to (4.6), we assume that the complex signal $s^{(l)}$ transmitted by the lth user has zero mean and a variance of σ_l^2. The AWGN noise signal n_p also exhibits a zero-mean and a variance of σ_n^2.

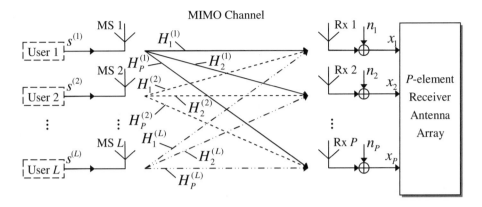

Figure 4.1: Schematic of the SDMA uplink MIMO channel model [3], where each of the L mobile users is equipped with a single transmitter antenna and the BS's receiver is assisted by a P-element antenna frontend.

The FD-CHTFs $H_p^{(l)}$ of the different receivers or users are independent, stationary, complex Gaussian-distributed processes with zero mean and unit variance [443].

4.2.2 CM-Assisted SDMA-OFDM Using Frequency-Domain Spreading

In Section 4.2.1 we briefly reviewed the SDMA MIMO channel model, as shown in Figure 4.1. In Figure 4.2, we present a schematic of the proposed CM-assisted and multi-user detected SDMA-OFDM uplink system employing WHT spreading. At the transmitter end, as seen at the top of Figure 4.2, the information bit sequences of the geographically separated L simultaneous mobile users are forwarded to the CM encoders, where they are encoded into symbols. Each user's encoded signal is divided into a number of WHT signal blocks, denoted by $s_{w,0}^{(l)}$ $(l = 1, \ldots, L)$, which are then forwarded to the subcarrier-based WHT spreader, followed by the OFDM-related Inverse Fast Fourier Transform (IFFT) modulator, which converts the frequency-domain signals to the time-domain modulated OFDM symbols. The OFDM symbols are then transmitted by the MSs to the BS over the SDMA MIMO channel. Then each element of the receiver antenna array shown at the bottom of Figure 4.2 receives the superposition of the AWGN-contaminated transmitted signals and performs Fast Fourier Transform (FFT) OFDM demodulation. The demodulated outputs $\mathbf{x}_{pw}$ $(p = 1, \ldots, P)$ seen in Figure 4.2 are forwarded to the multi-user detector for separating the different users' signals. The separated signals $\hat{\mathbf{s}}_w^{(l)}$ $(l = 1, \ldots, L)$, namely the estimated versions of the transmitted signals, are independently despread based on the inverse WHT (IWHT), resulting in the despread signals of $\hat{\mathbf{s}}_{w,0}^{(l)}$ $(l = 1, \ldots, L)$ which are then decoded by the CM decoders of Figure 4.2.

The further structure of this section is as follows. A brief description of the MMSE MUD employed in our SDMA-OFDM system is given in Section 4.2.2.1. The subcarrier-based WHTS is then introduced in Section 4.2.2.2.

4.2.2.1 MMSE MUD

As mentioned earlier, MUD schemes have to be invoked at the receiver of the SDMA-OFDM system to detect the received signals of different users. From the family of various MUD techniques, represented for example by the Maximum Likelihood Detection (MLD) [3, 193, 197, 198, 439], Parallel Interference Cancellation (PIC) [3, 439, 441, 442], Successive Interference Cancellation (SIC) [3, 193, 197, 439–441], MMSE [3, 193, 197, 439, 440] and LS [3, 439, 440] detectors, ML detection is known to exhibit the optimum performance, although this is achieved at the highest complexity. In order to avoid the

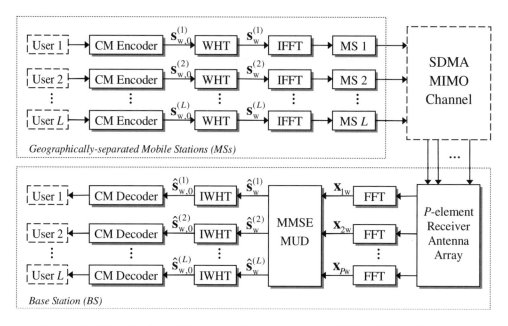

Figure 4.2: Schematic of the CM-assisted and multi-user detected SDMA-OFDM uplink system employing subcarrier-based WHT spreading.

potentially excessive complexity of optimum ML detection, suboptimum detection techniques such as the MMSE-MUD have been devised. Specifically, the MMSE detector exhibits the lowest detection complexity in the set of detectors mentioned above, although this comes at the cost of a Bit Error Ratio (BER) degradation [3, 441].

In the MMSE-based MUD the estimates of the different users' transmitted signals are generated with the aid of the linear MMSE combiner. More specifically, the estimated signal vector $\hat{\mathbf{s}} \in \mathbb{C}^{(L \times 1)}$ generated from the transmitted signal $\mathbf{s}$ of the L simultaneous users, as shown in Figure 4.2, is obtained by linearly combining the signals received by the P different receiver antenna elements with the aid of the array weight matrix, as follows [3]:

$$\hat{\mathbf{s}} = \mathbf{W}_{\text{MMSE}}^{H} \mathbf{x}, \tag{4.7}$$

where the superscript H denotes the Hermitian transpose, and $\mathbf{W}_{\text{MMSE}} \in \mathbb{C}^{(P \times L)}$ is the MMSE-based weight matrix given by [3]

$$\mathbf{W}_{\text{MMSE}} = (\mathbf{H}\mathbf{H}^{H} + \sigma_n^2 \mathbf{I})^{-1} \mathbf{H}, \tag{4.8}$$

while $\mathbf{I}$ is the identity matrix and σ_n^2 is the AWGN noise variance.

4.2.2.2 Subcarrier-Based WHTS

In single- and multi-carrier CDMA systems, the employment of orthogonal codes is vital for the sake of supporting multiple access [32]. In the context of multi-user SDMA-OFDM systems, orthogonal codes may be employed to randomize the wideband channel's frequency-selective fading, rather than for supporting multiple users, since the multiple users are supported with the aid of the SDMA-OFDM system employing a P-element antenna array and appropriate multi-user detection techniques.

A prominent class of orthogonal codes often used in CDMA systems is the family of orthogonal Walsh codes [32], which are particularly attractive, since the operation of spreading with the aid of these codes can be implemented in form of a 'fast' transform which takes advantage of the codes' recursive structure, similar to the FFT [3].

Let us now provide a deeper insight into the operation of the subcarrier-based WHTS [3]. During every OFDM symbol period prior to transmission of the independent user signals, the K data samples associated with the subcarriers, where K is the FFT length, may be spread with the aid of the WHT having a block size of K. This is achieved by left-multiplying the WHT signal block $\mathbf{s}_{w,0}^{(l)}$ ($l = 1, \ldots, L$) of Figure 4.2 by the K-order WHT matrix $\mathbf{U}_{\mathrm{WHT}_K}$ for each user separately:

$$\mathbf{s}_w^{(l)} = \mathbf{U}_{\mathrm{WHT}_K} \mathbf{s}_{w,0}^{(l)}, \quad l = 1, \ldots, L, \tag{4.9}$$

where $\mathbf{s}_w^{(l)}$ is the lth user's spread signal block, and $\mathbf{U}_{\mathrm{WHT}_K}$ is given in a recursive form as

$$\mathbf{U}_{\mathrm{WHT}_K} = \frac{1}{\sqrt{2}} \begin{bmatrix} 1 \cdot \mathbf{U}_{\mathrm{WHT}_{K/2}} & 1 \cdot \mathbf{U}_{\mathrm{WHT}_{K/2}} \\ 1 \cdot \mathbf{U}_{\mathrm{WHT}_{K/2}} & -1 \cdot \mathbf{U}_{\mathrm{WHT}_{K/2}} \end{bmatrix}, \tag{4.10}$$

while the lowest-order WHT unitary matrix is defined by

$$\mathbf{U}_{\mathrm{WHT}_2} = \frac{1}{\sqrt{2}} \begin{bmatrix} 1 & 1 \\ 1 & -1 \end{bmatrix}. \tag{4.11}$$

When the WHT block size is long, e.g. identical to the FFT length of $K = 512$, the computational complexity imposed by the length-K WHTS may be very high. Therefore a more practical solution is to divide the K samples further into K/M_{WHT} interleaved blocks, each of which has a block size of $M_{\mathrm{WHT}} < K$. Specifically, the ith WHT block is constituted by the samples selected from the subcarriers having the indices [3, 443]

$$j = i + r \frac{K}{M_{\mathrm{WHT}}}, \quad 0 \leq i \leq \frac{K}{M_{\mathrm{WHT}}} - 1, \quad 0 \leq r \leq M_{\mathrm{WHT}} - 1, \tag{4.12}$$

where i is the index of the WHT blocks within the same OFDM symbol. In Figure 4.3 we illustrate the operation of the subcarrier-based WHTS, where the number of OFDM subcarriers is 512 and the WHT block size is 32. Therefore in each OFDM symbol we have $512/32 = 16$ frequency-domain interleaved WHT blocks. At the top of Figure 4.3 an OFDM symbol is shown with the subcarriers' indices displayed, while the bottom part of the figure shows the WHT blocks generated, which contain subcarriers of the specified indices, as given by Equation (4.12). As Figure 4.3 shows, for example, the signal sample carried by the second ($r = 1$) slot within the WHT block of index $i = 0$ is selected from the subcarrier of index $j = 16$ within the original OFDM symbol. After the WHT blocks are formed, the WHTS is then invoked with the aid of the length-M_{WHT} WHT matrix given in Equation (4.10).

At the BS receiver seen in Figure 4.2, the despreading operation follows the inverse procedure of that portrayed in Figure 4.3, which is invoked independently for the separated signal of each user. More specifically, we have

$$\hat{\mathbf{s}}_{w,0}^{(l)} = \mathbf{U}_{\mathrm{WHT}_K} \hat{\mathbf{s}}_w^{(l)}, \quad l = 1, \ldots, L, \tag{4.13}$$

where $\hat{\mathbf{s}}_{w,0}^{(l)}$ and $\hat{\mathbf{s}}_w^{(l)}$ are the estimated versions of $\mathbf{s}_{w,0}^{(l)}$ and $\mathbf{s}_w^{(l)}$, respectively, while $\hat{\mathbf{s}}_w^{(l)}$ is achieved by applying Equation (4.7) at each subcarrier of every length-M_{WHT} WHT block. Upon employing the WHTS technique, the detrimental effects imposed on the system's average BER performance by the specific subcarriers corrupted by deep frequency-domain channel fades can be potentially improved, since the effects of the fades are spread over the entire WHT block. Hence the receiver has a high chance of recovering the impaired transmitted signals of the badly affected subcarriers.

4.3 Simulation Results

In this section, we characterize the performance of the proposed CM-assisted MMSE-SDMA-OFDM schemes in conjunction with WHTS. The channel is assumed to be OFDM symbol invariant, implying that the taps of the impulse response are assumed to be constant for the duration of one OFDM symbol,

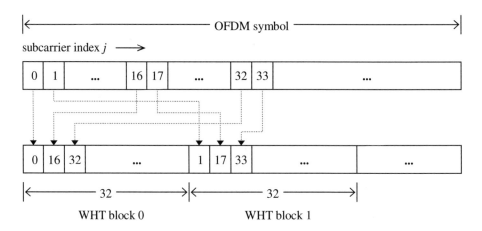

Figure 4.3: Example of the subcarrier-based WHT blocks' generation using a WHT block size of 32, where the total number of OFDM subcarriers is 512.

but they are faded at the beginning of each symbol [2]. Each user's associated transmit power or signal variance is assumed to be the same and normalized to unity, while the complex-valued fading envelope of the different users' signal is assumed to be uncorrelated. To simplify the experimental conditions, the channel's frequency-domain transfer function is assumed to be perfectly known in all simulations. Nonetheless, these performance trends are expected to remain unchanged in case of imperfect channel estimation, in particular, when the turbo-style PIC-aided Recursive Least-Squares (RLS) channel estimators of Chapter 16 in reference [3] are used. This may be made plausible by noting that turbo-style iterative detection techniques have been reported to be capable of achieving a virtually indistinguishable performance from the idealistic system using perfect channel estimation [3, 314].

4.3.1   MMSE-SDMA-OFDM Using WHTS

We commence by considering a multi-user SDMA-OFDM system operating without the assistance of CM communicating over the SWATM channel of [2]. The impulse response of the three-tap SWATM channel was given in Figure 3.32, while the specific channel parameters used were given in Table 3.10. Each of the three paths experiences independent Rayleigh fading having the same normalized Doppler frequency of $f_d' = 1.235 \times 10^{-5}$. A total of 512 subcarriers and a cyclic prefix of 64 samples were used for the OFDM modem.

Figure 4.4 compares the BER versus E_b/N_0 performance of the MMSE-SDMA-OFDM system equipped with two receiver antenna elements, while supporting one or two users both with and without WHTS, respectively. Furthermore, the performance of the unprotected single-user BPSK scheme communicating over an AWGN channel is also provided for reference. As expected, the WHTS-assisted schemes perform better than their non-spread counterparts, both for one and two users. It is also beneficial to view the subcarrier BERs as a function of both the subcarrier index and E_b/N_0, which was portrayed in Figure 4.5 for the SDMA-OFDM system equipped with two BS receiver antennas for supporting two users. The subcarrier BER is defined as the BER averaged over a specific subcarrier of all the consecutive OFDM symbols transmitted by the users. It was found that at a specific E_b/N_0 value, the subcarrier BER curves shown at the top of Figure 4.5 exhibit undulations across the frequency domain owing to deep channel fades at certain subcarriers, which could be potentially eliminated with the aid of WHTS, as observed at the bottom of the figure. This suggests that the system's average BER performance can be potentially improved by using WHTS, since the bursty subcarrier errors can be effectively spread across the subcarriers of the entire WHT block.

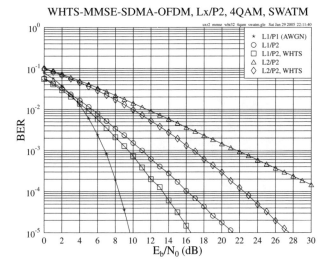

Figure 4.4: **BER** versus E_b/N_0 performance of the **WHTS-assisted MMSE-SDMA-OFDM** system employing a 4-QAM scheme for transmission over the **SWATM** channel, where $L = 1, 2$ users are supported with the aid of $P = 2$ receiver antenna elements. The WHT block size used is **32**.

Table 4.1: The parameters of the various CM schemes used in the multi-user SDMA-OFDM system for communicating over the SWATM channel of Figure 3.32.

CM scheme	Code rate	Data bits	Parity bits	Code memory	Iterations	Codeword length	Modem
TCM	1/2	1	1	6	—	1024	4-QAM
TTCM	1/2	1	1	3	4	1024	4-QAM
BICM	1/2	1	1	6	—	1024	4-QAM
BICMID	1/2	1	1	3	8	1024	4-QAM

4.3.2 CM- and WHTS-assisted MMSE-SDMA-OFDM

Similar to the previous sections, let us first investigate the SDMA-OFDM system's performance while communicating over the SWATM channel [2].

4.3.2.1 Performance over the SWATM Channel

For the various CM schemes used, we select the parameters so that all schemes have the same effective throughput and the same number of decoding states, hence a similar decoding complexity. More specifically, the code memory ν is fixed to 6 for the non-iterative TCM and BICM schemes, so that the number of decoding states becomes $S = 2^\nu = 64$. For the iterative TTCM and BICM-ID schemes, however, ν is fixed to 3, while the number of iterations for these schemes is set to 4 and 8, respectively. Hence the total number of trellis states is $2^3 \cdot 4 \cdot 2 = 64$ for TTCM and $2^3 \cdot 8 \cdot 1 = 64$ for BICM-ID, since there are two eight-state decoders, which are invoked in four iterations in the scenario of TTCM, while only one eight-state decoder is employed in the context of BICM-ID. The generator polynomials expressed in octal format for TCM, TTCM, BICM and BICM-ID are [117 26], [13 6], [133 171] and [15 17], respectively. The parameters of the various CM schemes used are summarized in Table 4.1.

MMSE-SDMA-OFDM, L2/P2, 4QAM, SWATM

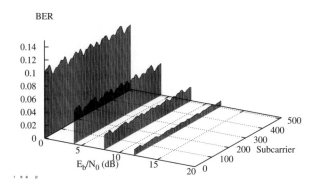

WHTS-MMSE-SDMA-OFDM, L2/P2, 4QAM, SWATM

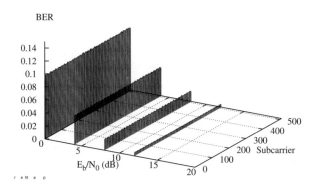

Figure 4.5: BER versus E_b/N_0 performance as a function of the subcarrier index of the **MMSE-SDMA-OFDM (top)** and **WHTS-assisted MMSE-SDMA-OFDM (bottom)** systems employing a 4-QAM scheme for transmission over the **SWATM** channel, where $L = 2$ users are supported with the aid of $P = 2$ receiver antenna elements. The WHT block size used is **32**.

4.3.2.1.1   Two Receiver Antenna Elements

In Section 4.3.1 the beneficial effects of WHTS on the MMSE-SDMA-OFDM system's performance were demonstrated. Let us now combine the various CM schemes considered with the multi-user MMSE-SDMA-OFDM system. The corresponding simulation results are portrayed in Figure 4.6, where the top and bottom of the figure illustrate the BER and CodeWord Error Ratio (CWER) versus E_b/N_0 performance of the proposed CM-MMSE-SDMA-OFDM schemes, respectively.

Here we define the user load of an L-user and P-receiver SDMA-OFDM system as

$$\alpha_P = \frac{L}{P}, \tag{4.14}$$

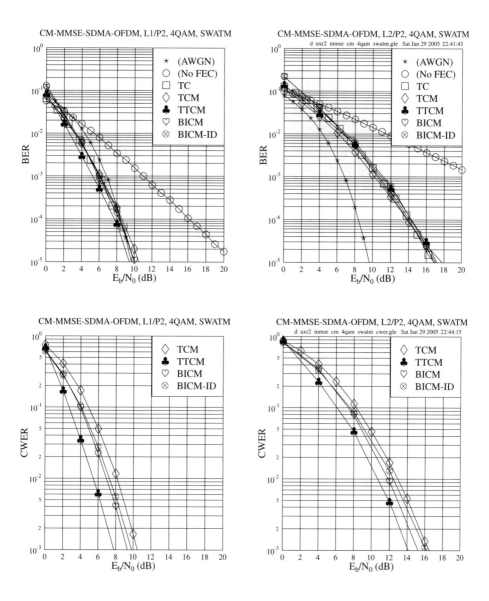

Figure 4.6: BER (top) and **CWER (bottom)** versus E_b/N_0 performance of the **CM-assisted MMSE-SDMA-OFDM** system employing a 4-QAM scheme for transmission over the **SWATM** channel, where $L = 1$ **(left)** or $L = 2$ **(right)** users are supported with the aid of $P = 2$ receiver antenna elements. The CM parameters are given in Table 4.1. The CM codeword length is **1024** symbols. The BER performance of the same 4-QAM MMSE-SDMA-OFDM system assisted by the half-rate TC code [3] (the curves marked with □) is also provided for reference.

which assumes a value of unity in case of full user load, when the number of users is equal to the number of receiver antenna elements. The simulation results generated in the context of $\alpha_2 = 0.5$ and $\alpha_2 = 1$ are plotted at the left and right side of Figure 4.6, respectively. The BER performance of the 32-state TC code assisted MMSE-SDMA-OFDM [3] system is also portrayed as a reference. Furthermore,

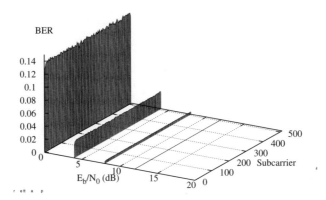

Figure 4.7: BER versus E_b/N_0 performance as a function of the subcarrier index of the **TTCM-assisted MMSE-SDMA-OFDM** system employing a 4-QAM scheme for transmission over the **SWATM** channel, where $L = 2$ users are supported with the aid of $P = 2$ receiver antenna elements. The TTCM codeword length is **1024** symbols.

the performance of the unprotected 4-QAM SDMA-OFDM system and that of the single-user BPSK scheme communicating over an AWGN channel is also provided as a benchmark. As observed in Figure 4.6, the TC-assisted arrangement and the various CM-assisted schemes provide a similar BER performance in both scenarios. As expected, all the FEC-aided schemes perform significantly better than their unprotected counterparts in the BER performance investigations. Furthermore, the TTCM-aided scheme is found to give the best CWER performance from the set of all CM-aided schemes in both the half-loaded and fully loaded scenarios, where α_2 is equal to 0.5 and 1, respectively. This suggests that more transmission errors can be eliminated by TTCM than by the other three CM schemes, although the burst errors inflicted by deep frequency-domain channel fades cannot be recovered completely. However, this effect may be potentially mitigated by employing WHTS, as discussed in Section 4.2.2.2.

As expected, for each of the schemes evaluated, we can see that the performance achieved in the context of $\alpha_2 = 0.5$ is better than that attained when $\alpha_2 = 1$. This phenomenon may be explained as follows. Since P receiver antenna elements are invoked at the BS, there are P uplink paths for each MS user having one transmitter antenna. Hence the achievable spatial diversity order provided by the P paths remains the same for each user, regardless of the total number of simultaneous users supported. However, when the user load is lower, i.e. the number of users supported is lower, the MMSE combiner will benefit from a higher degree of freedom in terms of the choice of the array weights optimized for differentiating the different users' transmitted signal, and thus the system becomes more efficient in terms of suppressing the reduced Multi-User Interference (MUI).

In Figure 4.7, we provide the subcarrier-based BER versus E_b/N_0 performance of the TTCM-assisted MMSE-SDMA-OFDM system in the context of two users and two receiver antenna elements. Comparing Figure 4.5 with Figure 4.7, where the beneficial effects of WHTS have been characterized, we can see that the achievable performance improvement attained by TTCM, or more generally by the CM schemes, is typically higher than that achieved by WHTS.

Having discussed the beneficial effects of WHTS and those of CM on the SDMA-OFDM system, as described in Section 4.3.1 and earlier in this section, respectively, we now combine the MMSE-SDMA-OFDM system with CM and WHTS. The corresponding simulation results are portrayed in Figure 4.8, where the left and right sides of the figure illustrate the scenarios of $\alpha_2 = 0.5$ and $\alpha_2 = 1$, while the top and bottom of the figure show the BER and CWER performance, respectively. Again, the TTCM-aided scheme was found to give the best CWER performance among all the CM-aided schemes considered. Comparing Figure 4.6 and Figure 4.8 shows that in the CM-aided MMSE-SDMA-OFDM system the employment of WHTS having a block size of 32 only insignificantly improves the system's BER and CWER performance, since most of the achievable diversity gain may have already been achieved by using the CM schemes. However, the employment of WHTS has the potential of further enhancing the CM-SDMA-OFDM system's performance in highly dispersive propagation environments, an issue which will be further discussed in Section 4.3.2.2.

Furthermore, if a longer CM codeword length is used, the system's performance can be further improved at the cost of a higher computational complexity. The effects of different CM codeword lengths can be seen in Figure 4.9. As expected, when a higher codeword length is employed, the system's performance becomes better, since a longer CM codeword is capable of better averaging the bursty error effects. However, this performance improvement is achieved at a substantially higher complexity. For examples of the associated complexity issues, the interested reader is referred to Chapter 9 of [314]

4.3.2.1.2 Four Receiver Antenna Elements

In Section 4.3.2.1.1, we compared the various CM- and WHTS-aided schemes in the context of one or two users and two receiver antenna elements. In this section, we investigate a higher-order spatial diversity scenario by increasing the number of receiver antenna elements, and thus supporting a higher number of simultaneous users.

Figures 4.10 and 4.11 show the BER and CWER performance achieved by the CM- and WHTS-aided MMSE-SDMA-OFDM schemes in the scenario where there are four receivers supporting a maximum number of four users. The BER performance of the 32-state TC-assisted MMSE-SDMA-OFDM [3] system is also provided as a reference. As seen in Figure 4.10, similar to the two-receiver scenario of Figure 4.8, the TC-assisted arrangement and the various CM-aided schemes achieve a similar BER performance at a specific user load. However, again, the TTCM-aided scheme stands out among all CM-aided schemes by attaining a better CWER performance.

Furthermore, comparing Figure 4.8 with Figure 4.10, where there are two receivers supporting a maximum of two users, we find that at the same user-load level, e.g. at $\alpha_4 = \alpha_2 = 0.5$ or $\alpha_4 = \alpha_2 = 1.0$, the E_b/N_0 performance achieved by the four-receiver system is approximately 4.5 dB better than that of the two-receiver system, provided that the same CM-assisted scheme is used. This is because, when the number of the BS receiver antenna elements is increased, the SDMA MIMO system becomes capable of providing a higher diversity gain, which may be expected to improve the system's performance for each user.

4.3.2.2 Performance over the COST207 HT Channel

In this section, we will investigate the performance of the CM- and WHTS-assisted MMSE-SDMA-OFDM schemes, while communicating over a more dispersive channel, namely the 12-path COST207 [444] Hilly Terrain (HT) channel channel. The impulse response of the channel model is portrayed in Figure 4.12, while the specific channel parameters are given in Table 4.2. Each of the 12 paths experiences independent Rayleigh fading having the same normalized Doppler frequency of $f'_d = 1.0 \times 10^{-5}$. Compared with the 512-subcarrier OFDM modem used for communication over the SWATM channel investigated in Section 4.3.2.1, we now employ a higher number of 2048 subcarriers and a cyclic prefix of 256, since the maximum path delay of the COST207 HT channel is longer than that of the SWATM channel, and hence it requires a longer cyclic prefix for combating the effects of Inter-Symbol Interference (ISI) [3, 24].

In Sections 4.3.2.2.1 and 4.3.2.2.2, we will compare the corresponding performance of the various CM- and WHTS-assisted MMSE-SDMA-OFDM schemes, when communicating over the COST207 HT channel. The parameters of the CM schemes used are summarized in Table 4.3.

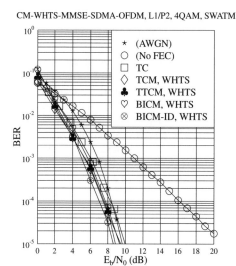

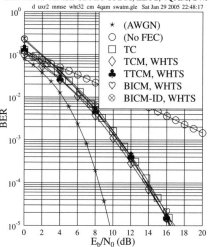

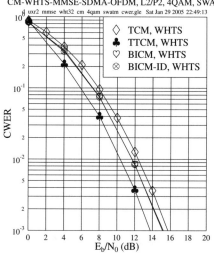

Figure 4.8: BER (top) and **CWER (bottom)** versus E_b/N_0 performance of the **CM- and WHTS-assisted MMSE-SDMA-OFDM** system employing a 4-QAM scheme for transmission over the **SWATM** channel, where $L = 1$ **(left)** or $L = 2$ **(right)** users are supported with the aid of $P = 2$ receiver antenna elements. The CM parameters are given in Table 4.1. The CM codeword length is **1024** symbols and the WHT block size used is **32**. The BER performance of the 4-QAM MMSE-SDMA-OFDM system assisted by the half-rate TC code [3] (the curves indicated by □) is also provided for reference.

4.3.2.2.1 Two Receiver Antenna Elements

We present the BER and CWER performance of the CM-assisted MMSE-SDMA-OFDM system, dispensing with WHTS, for transmission over the COST207 HT channel at the top of Figures 4.13

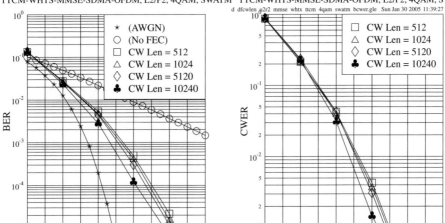

Figure 4.9: BER (left) and **CWER (right)** versus E_b/N_0 performance of the **TTCM- and WHTS-assisted MMSE-SDMA-OFDM** system employing a 4-QAM scheme for transmission over the **SWATM** channel, where $L = 2$ users are supported with the aid of $P = 2$ receiver antenna elements. The TTCM parameters are given in Table 4.1, although a range of **different codeword lengths** is employed. The WHT block size used is **32**.

Table 4.2: Sampling rate $1/T_s$, maximum path delay τ_{max}, maximum Doppler frequency f_d, normalized Doppler frequency f_d', number of paths n, FFT length K and cyclic prefix length cp of the COST207 HT channel of Figure 4.12.

$1/T_s$	τ_{max}	f_d	f_d'	n	K	cp
9.14 MHz	19.9 μs	92.6 Hz	1.0×10^{-5}	12	2048	256

Table 4.3: The parameters of the various CM schemes used in the multi-user SDMA-OFDM system communicating over the COST207 HT channel of Figure 4.12.

CM scheme	Code rate	Data bits	Parity bits	Code memory	Iterations	Codeword length	Modem
TCM	1/2	1	1	6	—	2048	4-QAM
TTCM	1/2	1	1	3	4	2048	4-QAM
BICM	1/2	1	1	6	—	2048	4-QAM
BICMID	1/2	1	1	3	8	2048	4-QAM

and 4.14, respectively. Two receiver antenna elements are employed for supporting a maximum of two users. The simulation results show that the TTCM-aided scheme constitutes the best design option in terms of both the BER and CWER, attaining a coding gain ranging from 2 to 4 dB over the other three CM-aided schemes at the BER of 10^{-5} without the assistance of WHTS. Furthermore, when WHTS

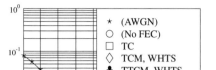

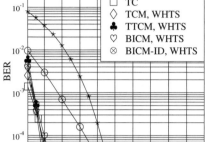

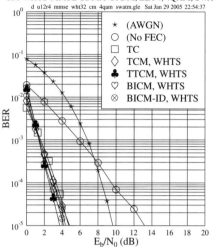

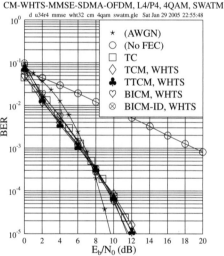

Figure 4.10: **BER** versus E_b/N_0 performance of the **CM- and WHTS-assisted MMSE-SDMA-OFDM** system employing a 4-QAM scheme for transmission over the **SWATM** channel, where $L = 1$ **(top left)**, $L = 2$ **(top right)**, $L = 3$ **(bottom left)** or $L = 4$ **(bottom right)** users are supported with the aid of $P = 4$ receiver antenna elements. The CM parameters are given in Table 4.1. The CM codeword length is **1024** symbols and the WHT block size used is **32**. The BER performance of the 4-QAM MMSE-SDMA-OFDM system assisted by the half-rate TC code [3] (the curves indicated by □) is also provided for reference.

is incorporated into the CM-MMSE-SDMA-OFDM system, as seen in the bottom of Figures 4.13 and 4.14, a further useful E_b/N_0 gain is achieved by most of the four schemes, especially by the TCM-aided arrangement. However, recall that in Section 4.3.2.1 where the SWATM channel was employed,

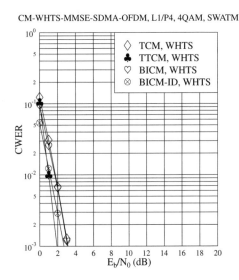

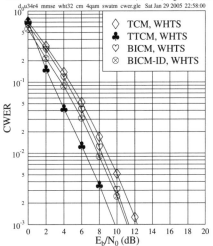

Figure 4.11: CWER versus E_b/N_0 performance of the **CM- and WHTS-assisted MMSE-SDMA-OFDM** system employing a 4-QAM scheme for transmission over the **SWATM** channel, where $L = 1$ **(top left)**, $L = 2$ **(top right)**, $L = 3$ **(bottom left)** or $L = 4$ **(bottom right)** users are supported with the aid of $P = 4$ receiver antenna elements. The CM parameters are given in Table 4.1. The CM codeword length is **1024** symbols and the WHT block size used is **32**.

the additional E_b/N_0 gain achieved by spreading in the context of the various CM- and WHTS-assisted schemes was rather modest. This result may suggest that in highly dispersive environments, such as that characterized by the 12-path COST207 HT channel, the channel-coded SDMA-OFDM system's performance may be further improved by employing WHTS. This spreading-induced E_b/N_0 gain was achieved because the detrimental effects imposed on the system's average BER performance by the

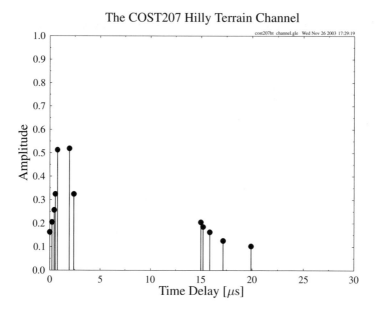

Figure 4.12: COST207 HT channel impulse response. The corresponding parameters of the channel are summarized in Table 4.2.

deeply faded subcarriers has been spread over the entire WHT block, and these randomized or dispersed channel errors may be more readily corrected by the CM decoders.

It transpires from Figures 4.13 and 4.14 that the four CM-aided schemes communicating over the COST207 HT channel attain a different performance. This observation is different from what was noted in Figures 4.6 and 4.8, where the performance of the various CM-aided schemes communicating over the SWATM channel was more similar. The reason for this phenomenon is that the amplitude variation of the FD-CHTFs becomes both more frequent and more dramatic when the channel exhibits a longer path delay [314]. Since the COST207 HT channel's maximum path delay is 19.9 μs, which is significantly longer than the 48.9 ns maximum dispersion of the SWATM channel, the fades occur more frequently in the FD-CHTF of the COST207 HT channel, as indicated by Figure 4.15. Apparently, in the COST207 HT channel displayed on the left side of Figure 4.15, the frequency-domain separation between the neighbouring fades is proportionately lower than that in the SWATM channel shown on the right side of Figure 4.15. This characteristic will result in more uniformly distributed corrupted subcarrier symbols which can be more readily corrected by the channel codes. More specifically, when a deep fade occurs in the FD-CHTF of the SWATM channel, a number of consecutive subcarriers which are located in the corresponding faded block of subcarriers may be seriously affected by the fade.

This implies that the channel codes, e.g. one of the four CM schemes, may have a lower chance of correcting less frequently occurring but prolonged error bursts than more frequently encountered isolated errors. The more prolonged error bursts imposed by the SWATM channel often overload the error-correction capability of the CM schemes, regardless of which of the four CM schemes is used, since, owing to the preponderance of transmission errors, their trellis decoder often opts for choosing the wrong trellis path. This is particularly true for CM schemes using short channel interleavers. Therefore, this phenomenon results in a similar performance for the various CM-aided schemes, when communicating over the SWATM channel, as seen in Figures 4.6 and 4.8.

By contrast, in the context of the COST207 HT channel such prolonged error bursts are unlikely to occur, since the faded subcarriers result in more frequent but less prolonged error bursts, which are

CM-MMSE-SDMA-OFDM, L1/P2, 4QAM, COST207 HT

CM-MMSE-SDMA-OFDM, L2/P2, 4QAM, COST207 HT

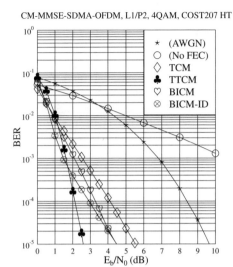

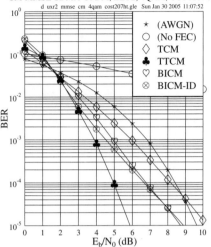

CM-WHTS-MMSE-SDMA-OFDM,L1/P2,4QAM,COST207 HT

CM-WHTS-MMSE-SDMA-OFDM,L2/P2,4QAM,COST207 HT

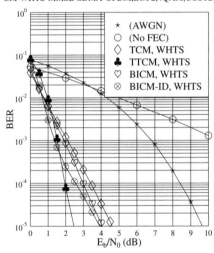

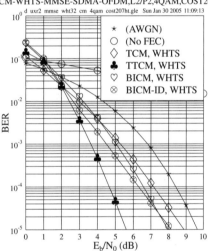

Figure 4.13: BER versus E_b/N_0 performance of the **CM-assisted MMSE-SDMA-OFDM (top)** and **CM- and WHTS-assisted MMSE-SDMA-OFDM (bottom)** systems employing a 4-QAM scheme for transmission over the **COST207 HT** channel, where $L = 1$ **(left)** or $L = 2$ **(right)** users are supported with the aid of $P = 2$ receiver antenna elements. The CM parameters are given in Table 4.3. The CM codeword length is **2048** symbols and the WHT block size used is **32**.

reminiscent of the error distributions experienced in AWGN channels and therefore may have a higher chance of being corrected by the CM decoders used at the receiver. Hence, the different error-correcting capability of the various CM schemes becomes more explicit, as revealed in Figures 4.13 and 4.14.

CM-MMSE-SDMA-OFDM, L1/P2, 4QAM, COST207 HT

CM-MMSE-SDMA-OFDM, L2/P2, 4QAM, COST207 HT

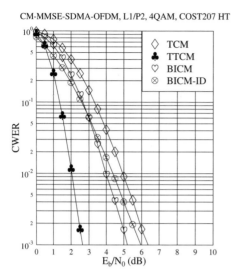

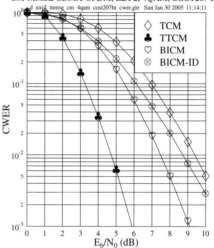

CM-WHTS-MMSE-SDMA-OFDM,L1/P2,4QAM,COST207 HT

CM-WHTS-MMSE-SDMA-OFDM,L2/P2,4QAM,COST207 HT

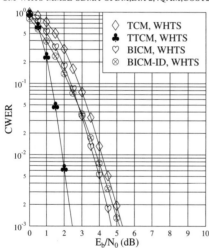

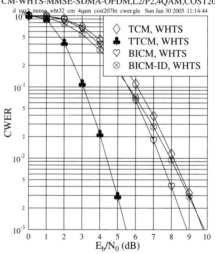

Figure 4.14: CWER versus E_b/N_0 performance of the **CM-assisted MMSE-SDMA-OFDM (top)** and **CM- and WHTS-assisted MMSE-SDMA-OFDM (bottom)** systems employing a 4-QAM scheme for transmission over the **COST207 HT** channel, where $L = 1$ **(left)** or $L = 2$ **(right)** users are supported with the aid of $P = 2$ receiver antenna elements. The CM parameters are given in Table 4.3. The CM codeword length is **2048** symbols and the WHT block size used is **32**.

4.3.2.2.2   Four Receiver Antenna Elements

As mentioned in Section 4.3.2.1.2, when the number of receiver antenna elements is increased to four, the performance of the system becomes better than that experienced in the two-receiver scenario, which was discussed in Section 4.3.2.2.1. In Figures 4.16 and 4.17, we compare both the BER and CWER

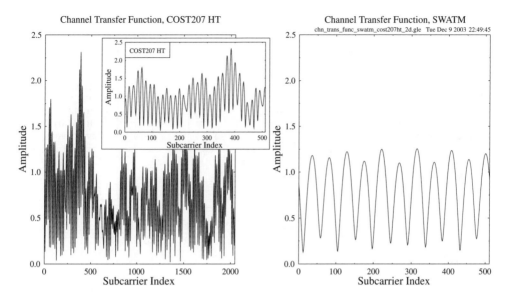

Figure 4.15: The FD-CHTF amplitudes of the **COST207 HT (left)** and **SWATM (right)** channels plotted for the duration of one OFDM symbol.

performance of the different CM- and WHTS-assisted schemes for transmissions over the COST207 HT channel, while employing four receiver antenna elements at user loads of $\alpha_4 = 0.5$ and $\alpha_4 = 1.0$, as shown on the left and right sides of Figures 4.16 and 4.17, respectively. Again, it can be seen in the figures that the performance achieved by the four-receiver SDMA-OFDM system, which has a higher diversity order, is better than that attained by the two-receiver system both at the user loads of $\alpha_{2,4} = 0.5$ and $\alpha_{2,4} = 1.0$, averaging an approximately 3 dB E_b/N_0 improvement for a specific CM-aided scheme. Hence a plausible conclusion is that, at a specific user load α_P, the more receivers the SDMA-MIMO system employs, the higher the attainable grade of diversity becomes and thus an improved performance may be achieved. However, the relative performance improvement achieved by an already high-order system upon doubling the number of receivers is expected to be lower than that in a lower-order system, since most of the attainable gain may have already been achieved, which results in a near-Gaussian performance.

4.3.2.2.3 Performance Comparisons

Table 4.4 summarizes the E_b/N_0 values required by the various CM- and WHTS-assisted MMSE-SDMA-OFDM schemes for achieving a BER of 10^{-5}. The corresponding spreading-induced E_b/N_0 gains achieved by the WHTS-assisted schemes are also provided, which are defined as the E_b/N_0 difference, expressed in terms of decibels, at a BER of 10^{-5} between the WHTS-assisted and the unspreading system having the same effective throughput.

Several useful points can be concluded from Table 4.4. Firstly, when we have a specific user load, the spreading-induced E_b/N_0 gain achieved by a system having a lower diversity order is higher, regardless of whether CM is employed or not. For example, when we have a user load of $\alpha_{2,4} = 0.5$, the spreading-induced E_b/N_0 gains achieved in the No-FEC-, TCM-, TTCM-, BICM- and BICM-ID-assisted two-receiver systems are 6.32, 0.91, 0.13, 0.41 and 0.56 dB, respectively. By contrast, lower spreading-induced E_b/N_0 gains of 3.46, 0.28, 0.03, 0.05 and 0.09 dB are achieved in the corresponding four-receiver systems, respectively. In other words, in relatively lower diversity-order

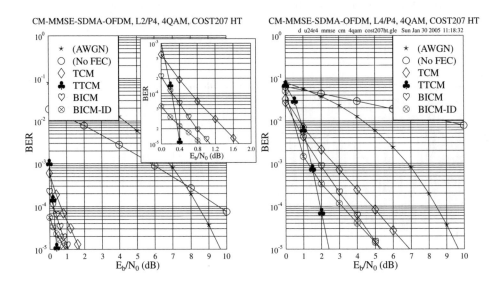

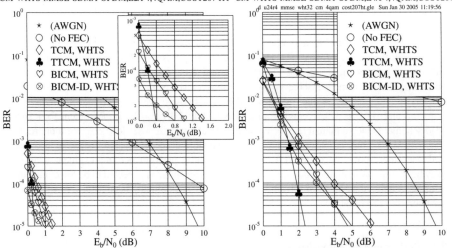

Figure 4.16: BER versus E_b/N_0 performance of the **CM-assisted MMSE-SDMA-OFDM (top)** and **CM- and WHTS-assisted MMSE-SDMA-OFDM (bottom)** systems employing a 4-QAM scheme for transmission over the **COST207 HT** channel, where $L = 2$ **(left)** or $L = 4$ **(right)** users are supported with the aid of $P = 4$ receiver antenna elements. The CM parameters are given in Table 4.3. The CM codeword length is **2048** symbols and the WHT block size used is **32**.

scenarios, the subcarrier-based WHTS technique may be expected to attain a higher system performance improvement.

Furthermore, if the number of users supported is fixed but the number of receiver antenna elements is varied, similar conclusions may be drawn. When there are two simultaneous users, for example, the spreading-induced E_b/N_0 gains achieved in the two-receiver scenario are higher than those achieved

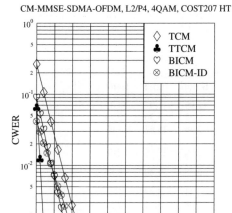

CM-MMSE-SDMA-OFDM, L2/P4, 4QAM, COST207 HT

CM-MMSE-SDMA-OFDM, L4/P4, 4QAM, COST207 HT

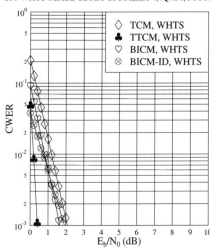

CM-WHTS-MMSE-SDMA-OFDM,L2/P4,4QAM,COST207 HT

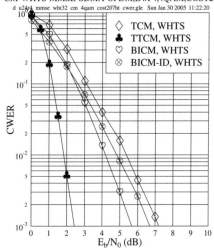

CM-WHTS-MMSE-SDMA-OFDM,L4/P4,4QAM,COST207 HT

Figure 4.17: CWER versus E_b/N_0 performance of the **CM-assisted MMSE-SDMA-OFDM (top)** and **CM- and WHTS-assisted MMSE-SDMA-OFDM (bottom)** systems employing a 4-QAM scheme for transmission over the **COST207 HT** channel, where $L = 2$ **(left)** or $L = 4$ **(right)** users are supported with the aid of $P = 4$ receiver antenna elements. The CM parameters are given in Table 4.3. The CM codeword length is **2048** symbols and the WHT block size used is **32**.

in the four-receiver scenario, as observed in Table 4.4. A plausible explanation for this fact may be as follows. In the SDMA MIMO system, when a higher number of receiver antenna elements is employed, a potentially higher space diversity can be achieved. In this scenario, the benefits of spreading may be less substantial, especially in the CM-aided system, since most of the attainable gain has already been achieved by using the channel codes.

Table 4.4: The E_b/N_0 values required and the spreading-induced gains achieved at the BER of 10^{-5} of the various CM- and WHTS-assisted MMSE-SDMA-OFDM schemes for communicating over the COST207 HT channel. The CM parameters are given in Table 4.3. The CM codeword length is 2048 symbols and the WHT block size used is 32. All data are in dB.

CM schemes	Spreading	$\alpha_{2,4} = 0.5$		$\alpha_{2,4} = 1.0$	
		U1R2	U2R4	U2R2	U4R4
(No FEC)	No WHTS	20.85	12.99	41.97	38.85
	WHTS	14.53	9.53	23.30	20.78
	E_b/N_0 gain	**6.32**	**3.46**	**18.67**	**18.07**
TCM	No WHTS	5.53	1.73	10.28	6.91
	WHTS	4.62	1.45	8.88	6.11
	E_b/N_0 gain	**0.91**	**0.28**	**1.40**	**0.80**
TTCM	No WHTS	2.60	0.41	5.87	2.41
	WHTS	2.47	0.38	5.61	2.35
	E_b/N_0 gain	**0.13**	**0.03**	**0.26**	**0.06**
BICM	No WHTS	4.47	1.08	8.72	5.25
	WHTS	4.06	1.03	7.97	4.73
	E_b/N_0 gain	**0.41**	**0.05**	**0.75**	**0.52**
BICM-ID	No WHTS	4.48	0.88	8.89	5.35
	WHTS	3.92	0.79	8.15	4.96
	E_b/N_0 gain	**0.56**	**0.09**	**0.74**	**0.39**

Secondly, when the BS employs a given number of receiver antenna elements, the spreading-induced E_b/N_0 gains achieved in the context of the fully loaded systems are higher than in the half-loaded systems, regardless of the employment of channel codes. For instance, if we have two receivers installed in the BS, the various schemes having a user load of $\alpha_2 = 1.0$ attain a higher spreading-induced E_b/N_0 gain than their half-loaded counterparts, as seen in Table 4.4. This suggests that more benefits may arise from WHTS, especially in the fully loaded scenarios, where the MUD suffers from a relatively low efficiency in differentiating the different users' signals. Furthermore, if a longer WHT block size is used, the E_b/N_0 gain of WHTS may even be higher, since the detrimental bursty error effects degrading the system's average BER performance will be spread over a higher block length, thus increasing the chances of correcting a higher number of errors, as will be presented in Section 4.3.2.3.

In order to characterize the system's performance at different user loads, as an example we portray the BER performance of the TTCM- and WHTS-aided scheme at a range of different user loads in Figure 4.18. As expected, the system's performance improves when the user load is lower, since the MUD will have a better chance of separating the different users' signals. It is also observed in Figure 4.18 that the relevant E_b/N_0 gain achieved by a lower-load scheme over a higher-load arrangement reduces when the user load decreases. This again shows that most of the achievable E_b/N_0 gain has already been attained at a medium user load.

Figure 4.19 shows the E_b/N_0 crossing points of the various CM-WHTS-MMSE-SDMA-OFDM schemes at the BER of 10^{-5}. It is shown explicitly that the performance gap between the different CM-aided schemes increases as the user load increases. Furthermore, from Figure 4.19 we can see that the TTCM-aided scheme performs best in high user-load scenarios, i.e. for $\alpha_4 \geq 0.5$. In other words, the other three CM schemes, namely the TCM, BICM and BICM-ID arrangements, will suffer a higher performance degradation than TTCM when the MUD's user-separation capability erodes owing to the increased MUI.

In Figure 4.20 we compare the total gain achieved by the four different CM-aided schemes, which includes both the coding gain and the spreading-induced E_b/N_0 gain. As the figure indicates, the

TTCM-WHTS-MMSE-SDMA-OFDM, Lx/P4, 4QAM, COST207 HT

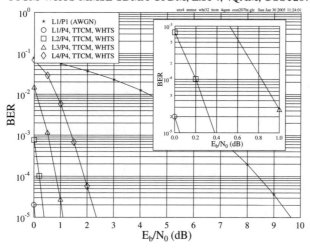

Figure 4.18: BER versus E_b/N_0 performance of the **TTCM- and WHTS-assisted MMSE-SDMA-OFDM** system employing a 4-QAM scheme for transmission over the **COST207 HT** channel, where $L = 1, 2, 3, 4$ users are supported with the aid of $P = 4$ receiver antenna elements. The TTCM parameters are given in Table 4.3. The TTCM codeword length is **2048** symbols and the WHT block size used is **32**.

CM-WHTS-MMSE-SDMA-OFDM, Lx/P4, 4QAM, COST207 HT

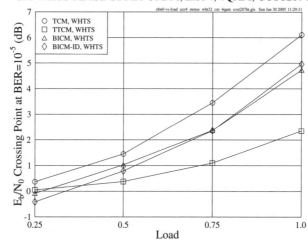

Figure 4.19: The $\mathbb{E}_b/N_0$ **crossing point** at the BER of 10^{-5} versus user-load performance of the **CM- and WHTS-assisted MMSE-SDMA-OFDM** system employing a 4-QAM scheme for transmission over the **COST207 HT** channel, where $L = 1, 2, 3, 4$ users are supported with the aid of $P = 4$ receiver antenna elements. The CM parameters are given in Table 4.3. The CM codeword length is **2048** symbols and the WHT block size used is **32**.

TTCM-aided scheme achieved a further E_b/N_0 gain of 3.76, 2.38 and 2.61 dB over the TCM, BICM and BICM-ID schemes in the fully loaded scenario, respectively. At a relatively low user load, i.e. for

CM-WHTS-MMSE-SDMA-OFDM, Lx/P4, 4QAM, COST207 HT

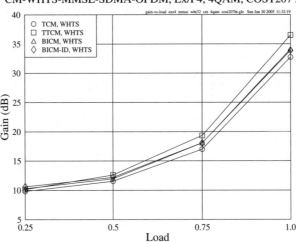

Figure 4.20: Gain at the BER of 10^{-5} versus user-load performance of the **CM- and WHTS-assisted MMSE-SDMA-OFDM** system employing a 4-QAM scheme for transmission over the **COST207 HT** channel, where $L = 1, 2, 3, 4$ users are supported with the aid of $P = 4$ receiver antenna elements. The CM parameters are given in Table 4.3. The CM codeword length is **2048** symbols and the WHT block size used is **32**.

$\alpha_4 \leq 0.5$, the various schemes provide a similar performance, because most of the attainable gain in the four-receiver SDMA-OFDM system has already been achieved, as discussed earlier in this section.

4.3.2.3   Effects of the WHT Block Size

In Sections 4.3.2.1 and 4.3.2.2 we investigated the CM- and WHTS-assisted MMSE-SDMA-OFDM system for transmission over the SWATM and COST207 HT channel, respectively. In this section, we will study how the variation of the WHT block size affects the system's performance, when communicating over the above two channel models.

As mentioned in Section 4.3.2.2.3, when a larger WHT block size is used for the SDMA-OFDM system, the system's performance may potentially be improved, since the signals carried by the subcarriers that are badly affected by deep channel fades could be spread over a larger set of subcarriers, which may mitigate the detrimental channel effects and thus assists the receiver in achieving a better performance. In order to show the effects imposed by different-length WHTS schemes, we provide simulation results generated in the context of different WHT block sizes in both the no-CM and CM-aided scenarios, as shown on the left and right sides of Figure 4.21, respectively. As expected, the system's performance was improved on increasing the WHT block size, regardless of whether CM was employed or not. In the context of the COST207 HT channel of Figure 4.12, similarly, a performance improvement is observed when an increased WHT block size is employed as portrayed in Figure 4.22.

Furthermore, we can see that the spreading-induced E_b/N_0 gains achieved by the no-CM schemes, when using a larger WHT block size, are significantly higher than those attained by the TTCM-aided schemes, as observed in Figures 4.21 and 4.22. This suggests that in the SDMA-OFDM system employing no CM, the performance improvement potential due to the employment of a larger WHT block size is higher than that in the CM-aided system, where most of the achievable diversity gain has been attained by the time diversity of the CM schemes. However, as suggested by Table 4.2, the other three CM-aided schemes, e.g. the TCM-aided arrangement, may be capable of achieving a higher spreading-induced gain than the TTCM-aided scheme with the aid of WHTS. Therefore, owing to their

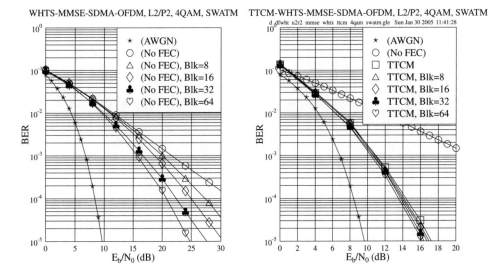

Figure 4.21: BER versus E_b/N_0 performance of the **WHTS-assisted MMSE-SDMA-OFDM (left)** and **TTCM- and WHTS-assisted MMSE-SDMA-OFDM (right)** systems using **different WHT block size** and employing a 4-QAM scheme for transmission over the **SWATM** channel, where $L = 2$ users are supported with the aid of $P = 2$ receiver antenna elements. The TTCM parameters are given in Table 4.1 and the TTCM codeword length is **1024** symbols.

lower time diversity and relatively more modest unspread performance, a potentially higher spreading-induced E_b/N_0 gain may be achieved by combining WHTS with the TCM-, BICM- and BICM-ID-assisted schemes than in conjunction with the TTCM-aided arrangement, when a larger WHT block size is used. Having studied the effects of different WHT block sizes, let us now consider the impact of varying the Doppler frequency.

4.3.2.4 Effects of the Doppler Frequency

In our further investigations we have generated the BER versus E_b/N_0 curves of the TTCM- and WHTS-assisted MMSE-SDMA-OFDM system communicating over the COST207 HT channel, when the maximum Doppler frequency was fixed to a range of different values. For simplicity, here we have assumed perfect channel estimation. As before, the CIR of the 12-path COST207 HT channel of Figure 4.12 was used. To present these results in a compact form, the E_b/N_0 values required to maintain a BER of 10^{-5} were extracted. In Figure 4.23, we show the E_b/N_0 crossing point at the BER of 10^{-5} versus the maximum Doppler frequency for the WHTS-assisted MMSE-SDMA-OFDM system both with and without the aid of TTCM, where two receivers were used for supporting two users. We conclude from the near-horizontal curves shown in Figure 4.23 that the maximum Doppler frequency does not significantly affect the performance of the WHTS-assisted MMSE-SDMA-OFDM system, regardless of the employment of TTCM. This may be a desirable benefit of the error-randomizing effect of WHTS, resulting in a robustness against the variation of the mobile speed. Moreover, as expected, the performance of the TTCM-aided scheme was consistently better than that of the scheme using no channel coding, as evidenced by Figure 4.23.

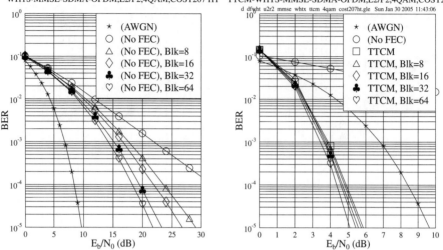

Figure 4.22: BER versus E_b/N_0 performance of the **WHTS-assisted MMSE-SDMA-OFDM (left)** and **TTCM- and WHTS-assisted MMSE-SDMA-OFDM (right)** systems using **different WHT block size** and employing a 4-QAM scheme for transmission over the **COST207 HT** channel, where $L = 2$ users are supported with the aid of $P = 2$ receiver antenna elements. The TTCM parameters are given in Table 4.3 and the TTCM codeword length is **2048** symbols.

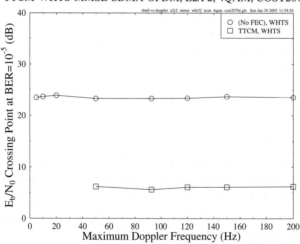

Figure 4.23: The E_b/N_0 **crossing point** at the BER of 10^{-5} versus maximum Doppler frequency performance of the **TTCM- and WHTS-assisted MMSE-SDMA-OFDM** system employing a 4-QAM scheme for transmission over the **COST207 HT** channel, where $L = 2$ users are supported with the aid of $P = 2$ receiver antenna elements. The TTCM parameters are given in Table 4.3. The TTCM codeword length is **2048** symbols and the WHT block size used is **32**.

It is worth noting that when the channel's Doppler frequency is high, the effects of Inter-Carrier Interference (ICI) may become significant,[1] as argued on page 81 of [3]. In this case, for example, the ICI cancellation techniques of Chapter 4 in [3] may be invoked for combating the ICI effects.

4.4 Chapter Summary

The system model of our multi-user uplink SDMA-based OFDM system was presented in Section 4.2, where the SDMA-MIMO channel model was introduced in Section 4.2.1, followed by the detailed description of the CM-assisted multi-user SDMA-OFDM system employing the frequency-domain subcarrier-based WHTS scheme of Section 4.2.2. The theoretical foundations of the MMSE-based MUD were provided in Section 4.2.2.1. Furthermore, a detailed discussion of the WHTS technique was presented in Section 4.2.2.2.

In Section 4.3 the performance of the various CM- and WHTS-assisted MMSE-SDMA-OFDM schemes was studied and compared. The uncoded MMSE-SDMA-OFDM system was first investigated in Section 4.3.1, where it was found that WHTS was capable of improving the system's performance, since the bursty subcarrier errors can be pseudo-randomly spread across the subcarriers of the entire WHT block. In the investigations outlined in Section 4.3.2 our performance comparison among the different CM- and WHTS-assisted MMSE-SDMA-OFDM schemes was detailed. Specifically, the unspread CM-aided SDMA-OFDM system using two receiver antenna elements was studied in Section 4.3.2.1.1, whose performance was compared with that of the WHTS-assisted system, dispensing with the employment of CM. Firstly, it was found that the performance achieved by all half-loaded schemes was better than that attained by the fully loaded arrangements. This is because, when the user load is lower, the MUD will achieve a higher efficiency in differentiating the different users' transmitted signal, since the multi-user interference is lower. Therefore, the system becomes more efficient in terms of suppressing the detrimental fading channel effects. Secondly, it was also noticed that the achievable performance improvement attained by CM was typically higher than that achieved by WHTS, as evidenced by comparing Figure 4.5 with Figure 4.7.

Furthermore, we combined the MMSE-SDMA-OFDM system with both CM and WHTS in Section 4.3.2.1.1. The corresponding simulation results portrayed in Figure 4.8 showed that the TTCM-aided scheme achieved the best CWER performance among all the CM- and WHTS-assisted schemes considered. Moreover, the comparison of Figure 4.6 and Figure 4.8 suggested that in the CM-aided MMSE-SDMA-OFDM system, the employment of WHTS having a block size of 32 subcarriers only insignificantly improved the system's BER and CWER performance, since most of the achievable diversity gain may have already been achieved by the time diversity of the CM schemes. It was also found that when a higher CM codeword length was used, the system's performance was further improved, as observed in Figure 4.9, since a longer CM codeword is capable of more efficiently dispersing and averaging the bursty error effects.

In Section 4.3.2.1.2, the system's performance evaluated in the four-receiver CM scenarios was compared, when communicating over the SWATM channel of Figure 3.32. Comparing Figure 4.8 with Figure 4.10, we found that at the same user-load level, e.g. at $\alpha_4 = \alpha_2 = 0.5$ or $\alpha_4 = \alpha_2 = 1.0$, the E_b/N_0 performance achieved by the four-receiver system was better than that of the two-receiver system, provided that the same CM-assisted scheme was used. The reason for this phenomenon is that when the SDMA-MIMO system's space-diversity order is increased by employing a higher number of receiver antenna elements, it becomes capable of providing a higher diversity gain, which may be expected to improve the system's performance for each user.

---

[1]Since the ICI is caused by the variation of the channel impulse response during the transmission of each OFDM symbol, we introduce the OFDM-symbol-normalized Doppler frequency F_d [3]. In the case of a maximum Doppler frequency of 200 Hz, for example, the corresponding OFDM-symbol-normalized Doppler frequency F_d will become as high as $F_d = f_d \cdot NT_s = f'_d \cdot N = 0.023$, where N is the cyclically extended OFDM symbol's total duration expressed as $N = K + cp$, while the associated values of f_d, f'_d, T_s, K and cp are given in Table 4.2.

The performance of the various CM- and WHTS-assisted MMSE-SDMA-OFDM systems was evaluated in the context of the COST207 HT channel of Figure 4.12 in Section 4.3.2.2, which included the two- and four-receiver scenarios in Section 4.3.2.2.1 and Section 4.3.2.2.2, respectively. It was found that without the assistance of WHTS the TTCM-aided scheme constituted the best design option in terms of both the BER and CWER in comparison with the other three CM-aided schemes. When WHTS was incorporated into the CM-MMSE-SDMA-OFDM system, a further E_b/N_0 gain was achieved by all the schemes, which was different from the scenario of Section 4.3.2.1 recorded for transmission over the SWATM channel, where the additional E_b/N_0 gain achieved by WHTS in the context of the various CM- and WHTS-assisted schemes was rather modest. This may suggest that in highly dispersive environments, such as that characterized by the 12-path COST207 HT channel of Figure 4.12, the channel-coded SDMA-OFDM system's performance may be further improved by employing WHTS. This is because the detrimental fading-induced bursty error effects degrading the system's average BER performance owing to the deeply faded subcarriers can be spread over the entire WHT block, and these dispersed and randomized channel errors may be more readily corrected by the CM decoder.

Furthermore, the four CM-aided schemes communicating over the COST207 HT channel of Figure 4.12 attain a different performance, as shown in Figures 4.13 and 4.14. This conclusion is different from what was noted in Figures 4.6 and 4.8, where the performance of the various CM-aided schemes communicating over the SWATM channel of Figure 3.32 was more similar. This is due to the more frequent and more dramatic amplitude variation of the FD-CHTF in the COST207 HT channel, which exhibits a significantly longer maximum path delay than that of the SWATM channel, as indicated by Figure 4.15. This characteristic resulted in more randomly dispersed rather than bursty corrupted subcarrier symbols, which can be more readily corrected by the channel codes. Hence, in the context of the COST207 HT channel of Figure 4.12, the different error-correcting capability of the various CM schemes becomes more explicit than that exhibited in the scenario of the SWATM channel of Figure 3.32.

In Table 4.4 of Section 4.3.2.2.3 we summarized the E_b/N_0 values required by the various CM- and WHTS-assisted MMSE-SDMA-OFDM schemes for achieving a BER of 10^{-5}, also showing the corresponding gains attained by the WHTS-assisted schemes. We observed that, on the one hand, when we had a specific user load, the spreading-induced E_b/N_0 gain achieved by a system having a lower diversity order was higher, regardless of the employment of CM. In other words, the subcarrier-based WHTS technique may be expected to attain a higher system performance improvement in relatively lower diversity-order scenarios. Moreover, if we supported a fixed number of users but varied the number of receiver antenna elements, similar conclusions might be drawn. A plausible explanation for this fact may be that in the SDMA-MIMO system, a potentially higher space-diversity gain may be achieved when a higher number of receiver antenna elements is employed, and thus the benefits of WHTS may be less substantial. This may be particularly true in the context of the CM-aided systems, since most of the attainable gain has already been achieved by using the channel codes. On the other hand, when a given number of receiver antenna elements was used, the spreading-induced E_b/N_0 gains achieved in the context of the fully loaded systems were higher than in the half-loaded systems, regardless of the employment of channel codes. This suggests that more benefits may arise from WHTS, especially in the fully loaded scenarios, where the MUD suffers from a relatively low efficiency in differentiating the different users' signals.

Additionally, we provided the E_b/N_0 crossing points, and the corresponding total gain achieved by the various CM-WHTS-MMSE-SDMA-OFDM schemes at the BER of 10^{-5}, in Figures 4.19 and 4.20 of Section 4.3.2.2.3, respectively. It was demonstrated that the performance gap between the different CM-aided schemes increased as the user load increased. Furthermore, it was observed in Figures 4.19 and 4.20 that the TCM-, BICM- and BICM-ID-aided schemes suffered a higher performance degradation than the TTCM arrangement, when the MUD's user-separation capability eroded owing to the increased multi-user interference. At a relatively low user load the various schemes provided a similar performance, because most of the attainable gain of the four-receiver SDMA-OFDM system had already been achieved, as discussed earlier in Section 4.3.2.2.3.

The effects of using different WHT block sizes was studied in Section 4.3.2.3, in both the SWATM and COST207 HT channel scenarios, as seen in Figures 4.21 and 4.22, respectively. As expected, it was found that the system's performance was improved, while the WHT block size used was increased. This is because, when a larger WHT block size is used by the SDMA-OFDM system, the data symbols carried by the subcarriers that are badly affected by deep channel fades could be spread over a larger WHT length, which may mitigate the detrimental channel effects and thus assists the receiver in terms of achieving a better error-correction capability. Furthermore, it was suggested by the simulation results of Section 4.3.2.3 that in the SDMA-OFDM system operating without the aid of CM, the performance improvement potential due to the employment of a larger WHT block size was higher than that in the CM-aided system, where most of the achievable diversity gain has been attained by using CM.

Finally, we studied the effects of the Doppler frequency in Section 4.3.2.4. It was concluded that the maximum Doppler frequency does not significantly affect the performance of the WHTS-assisted MMSE-SDMA-OFDM system, regardless of the employment of CM, as for example portrayed in Figure 4.23.

From the investigations conducted, we conclude that the various CM schemes, namely TCM, TTCM, BICM and BICM-ID, are capable of substantially improving the achievable performance of SDMA-OFDM systems. The employment of WHTS has the potential of further enhancing the system's performance in highly dispersive propagation environments. As a result, the TTCM- and WHTS-assisted scheme was found to have the best CWER performance in all the scenarios investigated. Furthermore, it was also the best design option in terms of the achievable E_b/N_0 gain expressed in dB, when communicating in highly dispersive environments, e.g. over the COST207 HT channel of Figure 4.12, while carrying a high user load of $\alpha_P \geq 0.5$.

In the next chapter, our research is targeted at contriving a more sophisticated MUD, namely the Genetic Algorithm (GA) assisted MUD, in order to improve further the SDMA-OFDM system's achievable performance.

Chapter **5**

Hybrid Multi-user Detection for SDMA-OFDM Systems

5.1 Introduction[1]

In the previous chapter, the MMSE MUD was investigated in the context of various CM-assisted SDMA-OFDM systems. Furthermore, the WHT-based frequency-domain spreading technique was incorporated into the CM-assisted MMSE-SDMA-OFDM system to attain performance enhancements. However, the SDMA system's performance is somewhat limited owing to the employment of the low-complexity MMSE MUD, which is devised based on the suboptimal linear MMSE algorithm. On the other hand, the high-complexity optimum Maximum Likelihood (ML) MUD is capable of achieving the best performance owing to the invocation of an exhaustive search. However, the computational complexity of the ML MUD typically increases exponentially with the number of simultaneous users supported by the SDMA-OFDM system, which may render its implementation prohibitive. In the literature, a range of suboptimal nonlinear MUDs have been proposed, such as for example the MUDs based on Successive Interference Cancellation (SIC) [3, 193, 197, 439–441] or Parallel Interference Cancellation (PIC) [3, 439, 441, 442] techniques. Instead of detecting and demodulating the users' signals in a sequential manner, as the MMSE MUD does, the PIC and SIC MUDs invoke an iterative processing technique that combines detection and demodulation. More specifically, the output signal generated during the previous detection iteration is demodulated and fed back to the input of the MUD for the next iterative detection step. Similar techniques invoking decision feedback have been applied also in the context of classic channel equalization. However, since the philosophy of both the PIC and SIC MUDs is based on the principle of removing the effects of the interfering users during each detection stage, they are prone to error propagation occurring during the consecutive detection stages due to the erroneously detected signals of the previous stages [3]. In order to mitigate the effects of error propagation, an attractive design alternative is to *simultaneously* detect all the users' signals, rather than invoke iterative interference cancellation schemes. Recently, another branch of MUD schemes referred to as Sphere Decoder (SD) [445–451] has also been proposed for multi-user systems, which is capable of achieving ML performance at a lower complexity.

As far as we are concerned, most of the above-mentioned techniques were proposed for systems where the number of users is less than or equal to the number of receivers, referred to here as the underloaded or fully loaded scenarios, respectively. However, in practical applications it is possible

---

[1]This chapter is partially based on ©IEEE Jiang & Hanzo 2006 [628].

that the number of users L to be supported exceeds that of the receiver antennas P, which is often referred to as an *overloaded* scenario. In overloaded systems, the $(P \times L)$-dimensional MIMO channel matrix representing the $(P \times L)$ number of channel links becomes singular, thus rendering the degree of freedom of the detector insufficient. This will catastrophically degrade the performance of numerous known detection approaches, such as for example the Vertical Bell Labs Layered Space–Time architecture (V-BLAST) [105, 108, 452] detector of [135], the MMSE algorithm of [3, 439] and the QR Decomposition combined with the M-algorithm (QRD-M) method of [172].

Based on this motivation, in this chapter a sophisticated nonlinear MUD is devised, which exploits the power of genetic algorithms. Genetic Algorithms (GAs) [328, 453–456] have been applied to a number of problems, such as machine learning and modelling adaptive processes. Moreover, GA-based MUD has been proposed by Juntti *et al.* [457] and Wang *et al.* [458], where the analysis was based on the AWGN channel in the absence of diversity techniques. The proposal by Ergün *et al.* [459] utilized GAs as the first stage of a multi-stage, multi-user detector, in order to provide good initial guesses for the subsequent stages. Its employment in Rayleigh fading channels was considered by Yen *et al.* in [38, 460] and [38, 461] in diverse scenarios both with and without the aid of diversity techniques, respectively.

However, most of the GA-aided transceiver research mentioned above was conducted in the context of Code-Division Multiple Access (CDMA) systems [38, 462]. By contrast, in this chapter, we apply GAs in the context of multi-user OFDM schemes, rather than CDMA systems. More specifically, we combine GAs with the MMSE MUD for the sake of contriving a more powerful concatenated MMSE-GA MUD, which is capable of maintaining near-optimum performance in the above-mentioned overloaded systems. Furthermore, TTCM is selected as the FEC scheme for the proposed MMSE-GA MUD in the SDMA-OFDM system, since it generally provides the best performance in the family of CM schemes in the context of MMSE-SDMA-OFDM systems, as demonstrated in Chapter 4. We will show in this chapter that the proposed MMSE-GA assisted TTCM-SDMA-OFDM system is capable of achieving a similar performance to that attained by its optimum ML MUD assisted counterpart at a significantly lower computational complexity, especially at high user loads. Furthermore, the performance of the proposed GA-aided system can be further improved if an enhanced iterative GA MUD is employed. This improvement is achieved at the cost of increased complexity, which is, however, still substantially lower than that imposed by the ML MUD.

The structure of this chapter is as follows. An overview of the GA-assisted TTCM-aided MMSE-SDMA-OFDM system is given in Section 5.2.1, followed by the introduction of the basic principles of the concatenated MMSE-GA MUD of Section 5.2.2. Our simulation results are provided in Section 5.2.3, while the associated complexity issues are discussed in Section 5.2.4. The enhanced GA MUD is introduced in Section 5.3, including the improved mutation scheme of Section 5.3.1 and the enhanced GA MUD framework of Section 5.3.2, while the associated simulation results are provided in Sections 5.3.1.3 and 5.3.2.2, respectively, followed by the associated complexity analysis in Section 5.3.3. Our final conclusions are summarized in Section 5.4.

5.2 GA-Assisted MUD

5.2.1 System Overview

In Figure 5.1, we present a schematic of the proposed concatenated MMSE-GA MUD aided SDMA-OFDM uplink system. At the transmitter end, as seen at the top of Figure 5.1, the information bit sequences of the geographically separated L simultaneous mobile users are forwarded to the TTCM [314] encoders, where they are encoded into symbols. The encoded signals $s^{(l)}$ ($l = 1, \ldots, L$) of the L users are then forwarded to the OFDM-related Inverse Fast Fourier Transform (IFFT) modulator, which converts the frequency-domain signals to the time-domain modulated OFDM symbols. The OFDM symbols are then transmitted by the independent Mobile Stations (MSs) to the Base Station (BS) over the SDMA MIMO channel, which was presented in Section 4.2.1. Then each

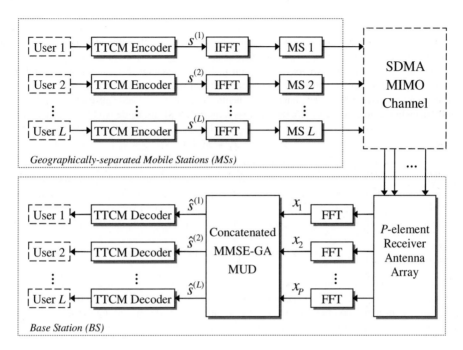

Figure 5.1: Schematic of the MMSE-GA-concatenated MUD SDMA-OFDM uplink system. ©IEEE Jiang & Hanzo 2006 [628]

element of the receiver antenna array shown at the bottom of Figure 5.1 receives the superposition of the transmitted signals faded and contaminated by the channel and performs Fast Fourier Transform (FFT) based OFDM demodulation. The demodulated outputs x_p ($p = 1, \ldots, P$) seen in Figure 5.1 are forwarded to the proposed concatenated MMSE-GA MUD for separating the different users' signals. The separated signals $\hat{s}^{(l)}$ ($l = 1, \ldots, L$), namely the estimated versions of the L users' transmitted signals, are then independently decoded by the TTCM decoders of Figure 5.1.

5.2.2 MMSE-GA-concatenated MUD

5.2.2.1 Optimization Metric for the GA MUD

The optimum ML MUD [3] uses an exhaustive search for finding the most likely transmitted signals. More explicitly, for a ML-detection-assisted SDMA-OFDM system supporting L simultaneous users, a total of 2^{mL} metric evaluations has to be invoked, where m denotes the number of Bits Per Symbol (BPS), in order to detect the L-user symbol vector $\hat{\mathbf{s}}_{\text{ML}}$ that consists of the most likely transmitted symbols of the L users at a specific subcarrier, given by

$$\hat{\mathbf{s}}_{\text{ML}} = \arg\left\{ \min_{\check{\mathbf{s}} \in \mathcal{M}^L} \|\mathbf{x} - \mathbf{H}\check{\mathbf{s}}\|^2 \right\}, \tag{5.1}$$

where the $(P \times 1)$-dimensional received signal vector $\mathbf{x}$ and the $(P \times L)$-dimensional Frequency-Domain CHannel Transfer Function (FD-CHTF) matrix $\mathbf{H}$ are defined by Equations (4.2) and (4.5), respectively. The set $\mathcal{M}^L$ in Equation (5.1), which is constituted by 2^{mL} trial vectors, is formulated as

$$\mathcal{M}^L = \{\check{\mathbf{s}} = [\check{s}^{(1)}, \check{s}^{(2)}, \ldots, \check{s}^{(L)}]^T \mid \check{s}^{(1)}, \check{s}^{(2)}, \ldots, \check{s}^{(L)} \in \mathcal{M}_c\}, \tag{5.2}$$

where $\mathcal{M}_c$ denotes the set containing the 2^m number of legitimate complex constellation points associated with the specific modulation scheme employed. Explicitly, the number of metric evaluations required for detecting the optimum vector increases exponentially with the number of users L.

Furthermore, the optimum ML-based decision metric of Equation (5.1) may also be used in the GA-based MUD to detect the estimated transmitted symbol vector $\hat{\mathbf{s}}_{\text{GA}}$. In the context of the SDMA-OFDM system employing P receiver antenna elements, the decision metric required for the pth receiver antenna, namely the antenna-specific *objective function (OF)* [38] can be derived from Equation (5.1), yielding

$$\Omega_p(\check{\mathbf{s}}) = |x_p - \mathbf{H}_p\check{\mathbf{s}}|^2, \tag{5.3}$$

where x_p is the received symbol at the input of the pth receiver at a specific OFDM subcarrier, while $\mathbf{H}_p$ is the pth row of the channel transfer function matrix $\mathbf{H}$. Therefore the decision rule for the optimum MUD associated with the pth antenna is to choose the specific L-symbol vector $\check{\mathbf{s}}$ which minimizes the metric given in Equation (5.3). Thus, the estimated transmitted symbol vector of the L users based on knowledge of the received signal at the pth receiver antenna and a specific subcarrier is given by

$$\hat{\mathbf{s}}_{\text{GA}_p} = \arg\left\{\min_{\check{\mathbf{s}}}[\Omega_p(\check{\mathbf{s}})]\right\}. \tag{5.4}$$

However, it transpires from the above derivation that we will have P metrics in total for the P receiver antennas. Since the CIRs of each of the P antennas are statistically independent, the L-symbol vector that is considered optimum at antenna 1 may not be considered optimum at antenna 2 etc. In other words, this implies that a decision conflict is encountered, which may be expressed as

$$\arg\left\{\min_{\check{\mathbf{s}}}[\Omega_i(\check{\mathbf{s}})]\right\} = \hat{\mathbf{s}}_{\text{GA}_i} \neq \hat{\mathbf{s}}_{\text{GA}_j} = \arg\left\{\min_{\check{\mathbf{s}}}[\Omega_j(\check{\mathbf{s}})]\right\}, \tag{5.5}$$

where $\forall i, j \in \{1, \ldots, P\}$, $i \neq j$. This decision conflict therefore leads to a so-called multi-objective optimization problem, since the optimization of the P metrics may result in more than one possible L-symbol solution. A similar decision conflict resolution problem was studied in [463] in an attempt to reconcile the decision conflicts of multiple antennas resulting in a decision dilemma. In order to resolve this problem we can adopt a similar approach and amalgamate the P antenna-specific L-symbol metrics into a joint metric as follows:

$$\Omega(\check{\mathbf{s}}) = \sum_{p=1}^{P} \Omega_p(\check{\mathbf{s}}). \tag{5.6}$$

Hence, the decision rule of the GA MUD is to find the specific estimated transmitted L-symbol vector $\hat{\mathbf{s}}_{\text{GA}}$ that minimizes $\Omega(\mathbf{s})$ in Equation (5.6) for every OFDM subcarrier considered.

5.2.2.2   Concatenated MMSE-GA MUD

The BER performance of the MMSE MUD is somewhat limited, since it is the total mean square estimation error imposed by the different simultaneous users that is minimized, rather than directly optimizing the BER performance. Therefore, the MMSE-SDMA-OFDM system's BER performance may be potentially further improved with the aid of a concatenated GA-aided MUD, which is capable of exploiting the output provided by the MMSE MUD of Section 4.2.2.1 in its initial population. For the sake of brevity, we will portray the philosophy of the proposed system in as simple terms as possible. However, readers who are unfamiliar with GAs might like to consult Appendix A.1 for a rudimentary introduction to GA-based optimization in the context of multi-user SDMA-OFDM systems.

The GA invoked in the SDMA-OFDM system commences its search for the optimum L-symbol solution at the initial *generation* with the aid of the MMSE combiner. In other words, using GA parlance, the so-called *individuals* of the $y = 1$st generation having a *population* size of X are created from the estimated length-L transmitted symbol vector provided by the MMSE combiner, where the xth ($x = 1, \ldots, X$) individual is expressed as $\tilde{\mathbf{s}}_{(y,x)} = [\tilde{s}_{(y,x)}^{(1)}, \tilde{s}_{(y,x)}^{(2)}, \ldots, \tilde{s}_{(y,x)}^{(L)}]$, and we have

$\tilde{s}^{(l)}_{(y,x)} \in \mathcal{M}_c$ ($l = 1, \ldots, L$). Note that here a complex symbol representation of the individuals is employed, which is derived from the classic *binary encoding* technique [38], where a binary vector constituted by binary zeros and ones is used to represent an individual. Then the GA-based optimization selects some of the L-symbol candidates from a total of X legitimate individuals in order to create a so-called *mating pool* of a number T of L-symbol *parent* vectors [38]. Two L-symbol parent vectors are then combined using specific GA operations to create two L-symbol *offspring* [38] and this 'genetic-evolution-like' process of generating new L-symbol offspring continues over a number Y of consecutive generations, so that the optimum L-symbol solution may be found.

The selection of the L-symbol individuals for creating the mating pool containing a number T of L-symbol parents is vital in determining the GA's achievable quality of optimization [464]. In our research the individual selection strategy based on the concept of the so-called *Pareto-Optimality* [328] was employed. This strategy favours the so-called *non-dominated* individuals and ignores the so-called *dominated* individuals [38]. More specifically, the uth L-symbol individual is considered to be dominated by the vth individual, if we have [465]

$$\forall i \in \{1, \ldots, P\} : \Omega_i(\tilde{\mathbf{s}}_{(y,v)}) \leq \Omega_i(\tilde{\mathbf{s}}_{(y,u)})$$
$$\wedge \, \exists j \in \{1, \ldots, P\} : \Omega_j(\tilde{\mathbf{s}}_{(y,v)}) < \Omega_j(\tilde{\mathbf{s}}_{(y,u)}). \tag{5.7}$$

If an individual is not dominated in the sense of Equation (5.7) by any other individuals in the population, then it is considered to be non-dominated. All the non-dominated individuals are then selected and placed in the mating pool, which will have a size of $2 < T \leq X$ [38]. Two of the number T of L-symbol individuals in the mating pool are then selected as parents based on their corresponding diversity-based *fitness* values calculated with the aid of Equation (5.6) according to the so-called *fitness-proportionate* selection scheme [38], which is described as follows. Firstly, the so-called *windowing-mapping* [458] technique is invoked in order to get the fitness value $f_{(y,x)}$ associated with the xth individual, which is given by

$$f_{(y,x)} = \Omega_{y,T} - \Omega(\tilde{\mathbf{s}}_{(y,x)}) + c, \tag{5.8}$$

where

$$\Omega_{y,T} = \max_{t \in \{1, \ldots, T\}} \{\Omega(\tilde{\mathbf{s}}_{(y,t)})\} \tag{5.9}$$

is the maximum *Objective Score (OS)*[2] achieved by evaluating the T individuals in the mating pool at the yth generation, and c is a small positive constant, which is used to ensure the positiveness of $f_{(y,x)}$. Then the fitness-proportionate selection probability p_x of the xth individual can be formulated as

$$p_x = \frac{f_{(y,x)}}{\sum_{t=1}^{T} f_{(y,t)}}. \tag{5.10}$$

When two L-symbol parents are selected, the so-called *uniform cross-over, mutation* and *elitism* operations [38] are invoked for offering a chance of evolving the parents' one or more element symbols to other symbols of the set $\mathcal{M}_c$, resulting in two offspring. The above operation is repeated until a new population consisting of X offspring is created. Furthermore, the so-called *incest prevention* [38] technique was invoked during the selection process, which only allows different individuals to be selected for the cross-over operation. Finally, the GA terminates after Y generations and thus the L-symbol individual having the highest diversity-based fitness value will be considered as the detected L-user transmitted symbol vector corresponding to the specific OFDM subcarrier considered.

From the above arguments, we note that in the GA MUD the different users' signals are jointly detected. This mechanism is different from that of the SIC or PIC MUDs, where each user's estimated transmitted signal is inferred by removing the interference imposed by the others. Therefore, there is no error propagation between the different users' signal detections in the GA MUD.

---

[2]Note that the individual having the maximum OS out of the pool of the T candidates is considered as the worst solution in the context of the current mating pool, since the GA searches for the optimum solution which minimizes Equation (5.6).

Table 5.1: The various techniques and parameters used in the simulations of Section 5.2.3.

TTCM	Modem	4-QAM
	Code rate	0.5
	Code memory ν	3
	Octal generator polynomial	[13 6]
	Codeword length	1024 symbols
	Channel interleaver depth	1024 symbols
	Number of turbo iterations	4
GA	Population initialization method	MMSE
	Mating pool creation strategy	Pareto-Optimality
	Selection method	Fitness-proportionate
	Cross-over	Uniform cross-over
	Mutation	M-ary mutation
	Mutation probability p_m	0.1
	Elitism	Enabled
	Incest prevention	Enabled
	Population size X	Varied
	Generations Y	Varied
Channel	CIRs	SWATM [3]
	Paths	3
	Maximum path delay	48.9 ns
	Normalized Doppler frequency f_d	1.235×10^{-5}
	Subcarriers K	512
	Cyclic prefix	64

5.2.3 Simulation Results

In this section, we characterize the performance of the proposed TTCM-assisted SDMA-OFDM system using the concatenated MMSE-GA MUD. The channel is assumed to be 'OFDM symbol invariant', implying that the taps of the impulse response are assumed to be constant for the duration of one OFDM symbol, but are faded at the beginning of each OFDM symbol [3]. The simulation results were obtained using a 4-QAM scheme communicating over the SWATM CIR of Figure 3.32, assuming that the channels' transfer functions are perfectly known. Each of the paths experiences independent Rayleigh fading having the same normalized Doppler frequencies of $f'_d = 1.235 \times 10^{-5}$. The OFDM modem employed $K = 512$ subcarriers and a cyclic prefix of 64 samples, which is longer than the maximum channel delay spread. For the iterative TTCM scheme [314] employed, the code memory ν is fixed to 3, while the number of iterations is set to 4. Hence the total number of trellis states is $2^3 \cdot 4 \cdot 2 = 64$, since there are two eight-state decoders which are invoked in four iterations. The generator polynomial expressed in octal format for the TTCM scheme considered is [13 6], while the codeword length and channel interleaver depth are fixed at 1024 symbols. The various techniques and parameters used in our simulations discussed in this section are summarized in Table 5.1.

The BER performance of the TTCM-assisted MMSE-GA-SDMA-OFDM system employing a 4-QAM scheme for transmission over the SWATM channel, where six users are supported with the aid of six receiver antenna elements, is portrayed in Figure 5.2. The performance of the TTCM-assisted MMSE-detected SDMA-OFDM system, the TTCM-aided optimum ML-detected system, and the uncoded single-user scheme employing either a single receiver or invoking Maximum Ratio Combining (MRC) when communicating over an AWGN channel, are also provided for reference, respectively. The numbers in brackets seen in the legends of Figure 5.2 represent the total GA or ML complexity.[3] It is observed from Figure 5.2 that the BER performance of the TTCM-assisted MMSE-SDMA-OFDM

---

[3]The quantification of the GA or ML complexity will be given in Section 5.2.4.

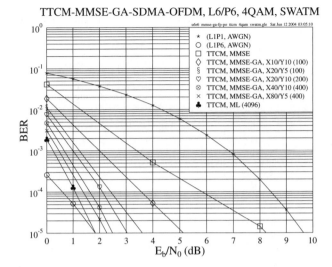

Figure 5.2: BER versus E_b/N_0 performance of the **TTCM-assisted MMSE-GA-SDMA-OFDM** system employing a 4-QAM scheme for transmission over the SWATM channel, where $L = 6$ users are supported with the aid of $P = 6$ receiver antenna elements. The basic simulation parameters are given in Table 5.1.

system was significantly improved with the aid of the GA, having a sufficiently large population size X and/or a larger number of generations Y. This improvement was achieved because a larger population may contain a higher variety of L-symbol individuals, and, similarly, a larger number of generations imply that, again, a more diverse set of individuals may be evaluated, thus extending the GA's search space, which may be expected to increase the chance of finding a lower-BER solution. On the other hand, it can be observed that when we have the same total number of $(X \times Y)$ correlation metric evaluations according to Equation (5.3), the performance improvement achieved by increasing the population size X was more substantial than that achieved by increasing the Y generations. For example, when we have $X \times Y = 100$, the GA-assisted scheme employing a population size $X = 20$ and $Y = 5$ generations achieved about 1.5 dB E_b/N_0 gain over its corresponding counterpart that has $X = 10$ and $Y = 10$, as evidenced by Figure 5.2. This may suggest that, in the TTCM-assisted MMSE-SDMA-OFDM system investigated, the GA's convergence speed tends to be faster when we have a larger population size X instead of a higher number of generations Y. However, when the affordable complexity increases, the improvement achieved by a larger-population GA at a certain value of $(X \times Y)$ becomes modest. For instance, given the maximum affordable complexity of $X \times Y = 400$, the system associated with $X = 80$ and $Y = 5$ brought about a modest E_b/N_0 improvement of 0.25 dB over the system associated with $X = 40$ and $Y = 10$, as shown in Figure 5.2. This is because most of the achievable performance gain of the system is likely to have been attained.

Figure 5.3 shows the performance achieved by the proposed TTCM-MMSE-GA-SDMA-OFDM system in the scenario where the number of supported users and receiver antenna elements was increased to eight. As shown in Figure 5.3, again, the TTCM-aided MMSE-SDMA-OFDM system's performance was significantly improved by employing the GA. We can also note that at a specific computational complexity, the E_b/N_0 gain achieved by the GA MUD was decreased, when compared with that attained in the six-user, six-receiver scenario of Figure 5.2. For example, when we have $X = Y = 10$ and $L = P = 6$, the E_b/N_0 gain achieved by the TTCM-MMSE-GA-SDMA-OFDM system over the TTCM-MMSE-SDMA-OFDM system was about 3 dB, as seen in Figure 5.2. By contrast, Figure 5.3 shows that this gain decreased to about 1.8 dB when we used the same GA

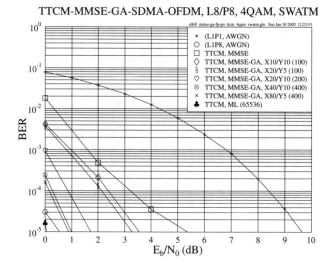

Figure 5.3: BER versus E_b/N_0 performance of the **TTCM-assisted MMSE-GA-SDMA-OFDM** system employing a 4-QAM scheme for transmission over the SWATM channel, where $L = 8$ users are supported with the aid of $P = 8$ receiver antenna elements. The basic simulation parameters are given in Table 5.1.

and $L = P = 8$. This phenomenon may be explained as follows. On the one hand, when the number of users increases, the separation of the different users' signal becomes more challenging, since the interference imposed by the undesired users becomes stronger. Therefore, a higher-complexity GA MUD has to be employed if we aim at maintaining the same E_b/N_0 gain achieved by the system having a lower user load. On the other hand, when we have more receiver antenna elements installed at the BS, i.e. when using a 'higher-order' SDMA system, a higher spatial diversity gain may be achieved. Hence, in a 'higher-order' SDMA system, a higher probability of achieving the total attainable gain may be expected than in a 'lower-order' SDMA system, potentially approaching the best possible AWGN performance. Hence, this trend also results in a less significant GA-induced performance gain.

Furthermore, it can be seen in Figures 5.2 and 5.3 that the MMSE-GA-detected TTCM-SDMA-OFDM system was slightly outperformed by its optimum ML-detected counterpart, since the GAs are unable to guarantee that the optimum ML solution would be found [38]. However, the near-optimum performance of the GA-aided TTCM-SDMA-OFDM system was achieved at a significantly lower computational complexity than that imposed by the ML-aided system, as will be demonstrated in Section 5.2.4.

5.2.4 Complexity Analysis

In this section, an analysis of the associated computational complexity imposed by the optimum ML MUD-aided SDMA-OFDM system and the GA-aided MMSE-SDMA-OFDM system will be presented. For the sake of simplicity, we only compare the optimum ML MUD's complexity with that of the GA MUD, since the simple MMSE MUD is used for providing a single initial solution for the GA's initial population and imposes a significantly lower complexity than that of its concatenated GA-aided counterpart. More specifically, since the proposed GA-aided MUD optimizes the metric of Equation (5.3),[4] we will quantify the complexity imposed in terms of the number of metric computations required by the optimization process.

---

[4]Similarly, the ML-aided MUD optimizes the metric of Equation (5.1), from which Equation (5.3) is derived.

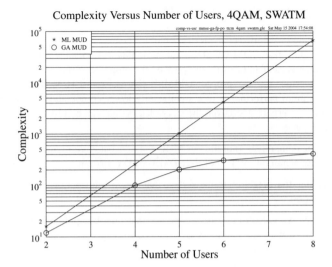

Figure 5.4: Performance comparison of the MUD complexity in terms of the number of metric evaluations versus the number of users for the 4-QAM TTCM-MMSE-GA-SDMA-OFDM and TTCM-ML-SDMA-OFDM systems. The number of receiver antenna elements employed is equivalent to the number of users supported, i.e. $L = P$.

As mentioned in Section 5.2.2.1, for the ML MUD, 2^{mL} metric computations have to be carried out to find the optimum solution [3], namely the most likely transmitted L-user vector, where m denotes the number of bits per symbol. By contrast, our proposed GA MUD requires a maximum of $(X \times Y)$ metric evaluations, since a number X of L-symbol vectors are evaluated during each of the Y generations, as shown in brackets in the legends of Figures 5.2 and 5.3. Furthermore, the number of such metric evaluations may readily be reduced by avoiding repeated evaluations of identical individuals, either within the same generation or across the entire iterative process, provided that the receiver has the necessary memory for storing the corresponding evaluation history. In Figure 5.4, we compare both the ML- and the GA-aided schemes in terms of their complexity, i.e. the number of metric computations. At a specific user load, we always select an appropriate GA-aided scheme for comparison, which suffers from less than 1 dB E_b/N_0 loss at the BER of 10^{-5} compared with the ML-aided system. As shown in Figure 5.4, the ML-aided system imposes an exponentially increasing complexity of the order of $O(2^{mL})$, when the number of users increases, while the complexity of the GA-aided system required for maintaining a near-optimum performance increases only slowly.

5.2.5 Conclusions

From the investigations conducted, we conclude that the GA-assisted TTCM-aided MMSE-SDMA-OFDM system is capable of achieving a similar performance to that of the optimum ML-assisted TTCM-SDMA-OFDM system. Furthermore, this is attained at a significantly lower computational complexity than that imposed by the ML-assisted system, especially when the number of users is high. For example, a complexity reduction in excess of a factor of 100 can be achieved by the proposed system for $L = P = 8$, as evidenced by Figure 5.4.

5.3 Enhanced GA-based MUD

In Section 5.2, we presented a detailed characterization of the concatenated MMSE-GA MUD designed for the TTCM-assisted multi-user SDMA-OFDM system. In this section, an enhanced GA-based MUD will be introduced, which is capable of further improving the proposed system's performance.

As discussed in Section 5.2, the proposed MMSE-GA-assisted TTCM-SDMA-OFDM system achieves a close-to-optimum performance at a significantly lower computational complexity than its optimum ML-assisted counterpart. Moreover, the GA-aided system can be further improved in each or both of the following aspects:

- By optimizing the component(s) of the GA MUD to find a better configuration that may improve the GA's performance in the context of the SDMA-OFDM system.

- By invoking an iterative detection framework so that the system's performance may be improved iteration by iteration.

In the following sections we will discuss the techniques that may be applied in terms of the above-mentioned two aspects for achieving an improved system performance.

5.3.1 Improved Mutation Scheme

In the context of GA-based detection techniques, the efficiency of the mutation scheme employed is important for the success of the entire evolutionary procedure, since it provides a chance for the individuals of the current population to influence the forthcoming ones, so that new areas of the total search space may be explored and thus the chance of finding the optimum solution increases [466]. An efficient mutation scheme is expected to be capable of guiding the search process in the correct direction to find the global optimum, rather than the local ones. In the context of the GA-assisted multi-user SDMA-OFDM system, when the number of users L increases or a high-throughput modulation scheme is used, the total search space consisting of 2^{mL} L-user symbol vectors would become excessive. In such cases, the role of mutation may become vital for the success of the overall system, since the GA may get trapped in local optima without appropriate assistance from the mutation scheme.

In Section 5.3.1.1, we will first discuss a widely used conventional mutation scheme as well as its drawbacks, followed by the introduction of an improved new mutation mechanism in Section 5.3.1.2.

5.3.1.1 Conventional Uniform Mutation

In Section 5.2.3, M-ary mutation was employed by the GA MUD. More specifically, each *gene* $\hat{s}^{(l)}$ ($l = 1, \ldots, L$) of a length-L GA individual $\hat{s}$ in the X-element population is represented by a specific symbol in $\mathcal{M}_c$, where $\mathcal{M}_c$ is the set containing the 2^m legitimate constellation points. In other words, the lth gene denotes the lth user's estimated transmitted symbol – which is a hard-decoded version of the complex signal – at the subcarrier considered. During the genetic evolution, when a gene is subjected to mutation it will be substituted by a different symbol in $\mathcal{M}_c$ based on a uniform *mutation-induced transition probability* $p_{mt}^{(ij)}$,[5] which quantifies the probability of the ith legitimate symbol becoming the jth. For the sake of brevity, from now on we refer to this probability as the *transition probability*. Furthermore, we will refer to the mutation scheme employing uniformly distributed $p_{mt}^{(ij)}$ values as *Uniform Mutation (UM)*, which is a widely used conventional mutation scheme known in the literature and was also employed by the GAs invoked in [38].

More specifically, UM mutates to all the candidate symbols in $\mathcal{M}_c$ with the same probability. For example, let us consider the 4-QAM modem constellation shown in Figure 5.5, where $\hat{s}_i^{(l)} \in \mathcal{M}_c$ ($i = 1, \ldots, 4$) are the constellation points as well as possible gene candidates for the lth user at a specific

---

[5]Note that the mutation probability p_m of Table 5.1 is different from the probability $p_{mt}^{(ij)}$ of mutating to a specific symbol in $\mathcal{M}_c$. The former denotes the probability of how likely it is that a gene will mutate, while the latter specifies how likely it is that a specific symbol in $\mathcal{M}_c$ becomes the mutated gene.

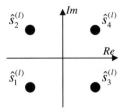

Figure 5.5: A 4-QAM constellation.

subcarrier. When $\hat{s}_i^{(l)}$ is subjected to mutation, we have

$$\hat{s}_j^{(l)} = MUTATION(\hat{s}_i^{(l)} \mid p_{mt}^{(ij)}), \quad i, j \in \{1, \ldots, 4\}, \quad i \neq j. \tag{5.11}$$

For the UM scheme, the transition probability $p_{mt}^{(ij)} = 1/(2^m - 1)$ is equal for all $i, j \in \{1, \ldots, 4\}$, $i \neq j$.

Based on the mechanism of UM, the GA has a chance of successfully identifying the actually transmitted symbol of the lth user at the subcarrier considered. However, UM has a drawback that may prevent the GA from rapid convergence under certain conditions. To explain this further, let us again consider the example of Figure 5.5. Without loss of generality, we can make the following assumptions:

(a) $\hat{s}_1^{(l)}$ is the received symbol, which is the original gene to be mutated; and

(b) $\hat{s}_1^{(l)}$ is *not* the transmitted symbol.

Hence, the task of mutation is to find the actually transmitted symbol from the set of three candidates, namely $\hat{s}_i^{(l)}$ ($i = 2, \ldots, 4$). According to UM, the probability that $\hat{s}_1^{(l)}$ hops to $\hat{s}_2^{(l)}, \hat{s}_3^{(l)}$ or $\hat{s}_4^{(l)}$ is equal. In other words, in this case the chance of finding the true transmitted symbol of the specific user during a single UM operation is $1/3$. However, this fixed uniform transition probability fails to reflect the realistic channel condition that the system is subjected to. More precisely, at different Signal-to-Noise Ratio (SNR) levels, some symbols in $\mathcal{M}_c$ should not constitute high-probability mutation targets. For example, at high SNRs, the chances are that $\hat{s}_2^{(l)}$ or $\hat{s}_3^{(l)}$ is more likely to be the transmitted symbol, rather than $\hat{s}_4^{(l)}$, since the noise effects are insignificant and thus the signal corruption from the most distant symbol $\hat{s}_4^{(l)}$ to the received symbol $\hat{s}_1^{(l)}$ is rare. Hence, it may be more reasonable to consider $\hat{s}_2^{(l)}$ and $\hat{s}_3^{(l)}$ only as the potential mutation candidates, and assign a modified transition probability $p_{mt}^{(1j)} = 0.5$ ($j \in \{2, 3\}$). This fact implies that at different SNR levels we may restrict the *effective* GA search space with the aid of a biased mutation, which ignores the constellation points that are far from the received symbol. This is especially beneficial for the system employing high-throughput modulation schemes such as 16-QAM, where the total search space is exponentially expanded as a function of the number of BPS compared with lower-throughput modems. In such a system, the UM-aided GA which allows mutation to all legitimate symbols may suffer from a slow convergence speed and might result in a high residual error floor, since a considerable portion of the GA's searching power may be wasted on mutating to highly unlikely gene candidates, especially in high-SNR scenarios. By contrast, the above-mentioned biased mutation-guided GA is expected to achieve a better performance, since it searches for the optimum solution in a more efficient way, as will be demonstrated in Section 5.3.1.2.

5.3.1.2 Biased Q-function-Based Mutation

The conventional UM and its drawbacks were discussed in Section 5.3.1.1. In this section, an improved novel mutation scheme will be presented, which we will refer to as *Biased Q-function-based Mutation (BQM)*.

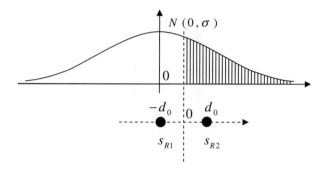

Figure 5.6: Illustration of the 1D transition probability $p_{mt}^{\overline{(ij)}}$ for 4-QAM.

5.3.1.2.1   Theoretical Foundations

As discussed in Section 5.3.1.1, for an original gene to be mutated, an SNR-related *biased* transition probability $p_{mt}^{(ij)}$ has to be assigned to each of the target candidate symbols in $\mathcal{M}_c$. The calculation of $p_{mt}^{(ij)}$ may be carried out with the aid of the widely known Q-function [467]:

$$Q(x) = \frac{1}{\sqrt{2\pi}} \int_x^\infty e^{-t^2/2} \, dt, \quad x \geq 0. \tag{5.12}$$

For the sake of easy explanation, let us first consider a simple One-Dimensional (1D) scenario. In Figure 5.6 we plotted the 1D real component of the constellation symbols $\hat{s}_i^{(l)}$ in the context of the 4-QAM modem constellation seen in Figure 5.5. The horizontal axis is then divided into two zones, each of which represents one specific 1D constellation symbol s_{Ri} $(i = 1, \ldots, 2)$, as separated by the vertical dashed line of Figure 5.6. If s_{R1} is the original gene to be mutated, the Gaussian distribution $N(0, \sigma)$ may be centred at the position of s_{R1}, where σ is the noise variance at a given SNR level. In this specific example, s_{R2} is the only mutation target and the 1D transition probability of mutating from s_{R1} to s_{R2}, i.e. $p_{mt}^{\overline{(12)}}$, is characterized by the shadow area shown in Figure 5.6, which is given by

$$p_{mt}^{\overline{(12)}} = Q\left(\frac{d_0}{\sigma}\right), \tag{5.13}$$

where d_0 is half of the distance between the neighbouring constellation symbols. Similarly, we have

$$p_{mt}^{\overline{(21)}} = Q\left(\frac{d_0}{\sigma}\right). \tag{5.14}$$

Furthermore, we also have a certain probability for the original gene to remain unchanged, which can also be expressed in terms of the Gaussian distribution as

$$p_{mt}^{\overline{(11)}} = p_{mt}^{\overline{(22)}} = 1 - Q\left(\frac{d_0}{\sigma}\right). \tag{5.15}$$

The above-mentioned 1D transition probabilities are summarized in Table 5.2. The corresponding Two-Dimensional (2D) symbol transition probability $p_{mt}^{(ij)}$ can be derived by combining the 1D real and imaginary transition probabilities.[6] Let us again consider the 4-QAM modem of Figure 5.5 as an example, which is replotted in Figure 5.7. In Figure 5.7, for instance, the 2D transition probability of mutating from the constellation symbol $\hat{s}_1^{(l)}$ to $\hat{s}_2^{(l)}$, namely $p_{mt}^{(12)}$, can be calculated by multiplying the

---

[6]Note that the 1D transition probability $p_{mt}^{\overline{(ij)}}$ is different from the transition probability $p_{mt}^{(ij)}$, which is based on the 2D constellation symbols.

Table 5.2: The 1D transition probabilities for 4-QAM.

$\{ij\}$	$p_{mt}^{(ij)}$
$\{12\}, \{21\}$	$Q(d_0/\sigma)$
$\{11\}, \{22\}$	$1 - Q(d_0/\sigma)$

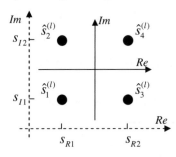

Figure 5.7: Illustration of the 2D transition probability $p_{mt}^{(ij)}$ for 4-QAM, which is the product of the relevant 1D transition probabilities. s_{Ri} and s_{Ii} ($i \in \{1, 2\}$) denote the 1D constellation symbols in the context of the real and imaginary components of the 4-QAM constellation symbols, respectively.

two relevant 1D transition probabilities according to[7]

$$p_{mt}^{(12)} = p_{mt}^{(11)} \cdot p_{mt}^{(12)} = \left[1 - Q\left(\frac{d_0}{\sigma}\right)\right] \cdot Q\left(\frac{d_0}{\sigma}\right), \tag{5.16}$$

while the associated 2D probability of remaining in the current state is

$$p_{mt}^{(11)} = p_{mt}^{(11)} \cdot p_{mt}^{(11)} = \left[1 - Q\left(\frac{d_0}{\sigma}\right)\right]^2. \tag{5.17}$$

For higher-throughput modems, e.g. for 16-QAM and 64-QAM, the same algorithm can be invoked for calculating the corresponding 1D and 2D transition probabilities. In Figure 5.8 an example associated with 16-QAM is provided. According to Figure 5.8, assuming that s_{R1} is the original gene to be mutated, while s_{R2} is the mutation target, we have the following 1D transition probability of

$$p_{mt}^{(12)} = Q\left(\frac{d_0}{\sigma}\right) - Q\left(\frac{3d_0}{\sigma}\right). \tag{5.18}$$

Similarly, we can derive the remaining 1D transition probabilities for 16-QAM, which are summarized in Table 5.3. For the sake of brevity, we omit here the derivation of the corresponding 2D transition probabilities. Note that the proposed BQM scheme can be readily extended to M-Dimensional (MD) constellations, since the MD transition probability associated with a specific MD symbol can be readily derived upon multiplying the M corresponding 1D transition probabilities.

However, when a mutation takes place during the evolution, the mutating gene or constellation symbol should not be allowed to be mutated to itself. Hence, the effect of the probability of mutating a symbol to itself should be removed. This can be achieved by normalizing the 2D transition probability

---

[7]Note that the superscripts i and j of the 2D transition probability $p_{mt}^{(ij)}$ denote the 2D constellation symbols $\hat{s}_i^{(l)}$ and $\hat{s}_j^{(l)}$, while the underlined superscripts i and j of the 1D transition probability $p_{mt}^{(ij)}$ represent the 1D constellation symbols s_{Ri} and s_{Rj}, respectively.

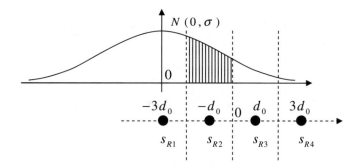

Figure 5.8: Illustration of the 1D transition probability $p_{mt}^{(ij)}$ for 16-QAM.

Table 5.3: The 1D transition probabilities for 16-QAM.

$\{ij\}$	$p_{mt}^{(ij)}$
$\{34\}, \{21\}$	$Q(d_0/\sigma)$
$\{24\}, \{31\}$	$Q(3d_0/\sigma)$
$\{14\}, \{41\}$	$Q(5d_0/\sigma)$
$\{11\}, \{44\}$	$1 - Q(d_0/\sigma)$
$\{22\}, \{33\}$	$1 - 2Q(d_0/\sigma)$
$\{12\}, \{23\}, \{32\}, \{43\}$	$Q(d_0/\sigma) - Q(3d_0/\sigma)$
$\{13\}, \{42\}$	$Q(3d_0/\sigma) - Q(5d_0/\sigma)$

$p_{mt}^{(ij)}$ $(i \neq j)$ with the aid of the original gene's probability of remaining unchanged, namely $p_{mt}^{(ii)}$, following the principles of conditional probability theory [468]. For more details concerning the normalization process, the interested reader is referred to Appendix A.2.

5.3.1.2.2 Simplified BQM

In Section 5.3.1.2.1 we provided a detailed explanation of the mechanism of BQM. Furthermore, the proposed BQM scheme can be effectively simplified when only a subset of all the theoretically possible mutation target symbols is considered. More precisely, for the original gene subjected to mutation, we may only consider its adjacent neighbouring constellation symbols as mutation target candidates, since the original transmitted symbol is less unlikely to be corrupted to a relatively distant constellation symbol.

An example of the simplified BQM designed for 16-QAM is provided in Figure 5.9. As shown in Figure 5.9, for example, we assume that $\hat{s}_1^{(l)}$ is the original gene subjected to mutation, and $\hat{s}_i^{(l)}$ $(i = 2, \ldots, 9)$ represents the adjacent neighbours of $\hat{s}_1^{(l)}$, while the symbols represented by the dashed circles are ignored. Therefore, the GA's entire search space is reduced. Moreover, the search space can be further compressed when we consider only the nearest neighbours of $\hat{s}_1^{(l)}$. In this case, only the symbols $\hat{s}_i^{(l)}$ $(i = 3, 5, 6, 8)$ printed in grey in Figure 5.9 will be regarded as legitimate mutation candidates, each of which is assigned an equal 2D transition probability $p_{mt}^{(1j)} = 1/4$ $(j = 3, 5, 6, 8)$, while all other constellation symbols printed in white are neglected. Since the transition probability for each selected mutation candidate is equally fixed, the BQM scheme is simplified to a scheme similar to UM, which we may refer to as the *Closest-Neighbour Uniform Mutation (CNUM)* scheme. Note that, similar to the scenario of BQM, in CNUM the corresponding transition probability value is also dependent on the

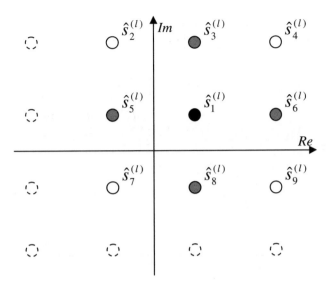

Figure 5.9: An example of the simplified BQM for 16QAM.

Table 5.4: Possible transition probability values for the CNUM scheme.

Modem	Transition probability value set
4-QAM	1/2
16-QAM	1/2, 1/3, 1/4
64-QAM	1/2, 1/3, 1/4

location of the original gene. For instance, if the original gene is located in one of the four corners of the constellation map plotted in Figure 5.9, the relevant transition probability $p_{mt}^{(ij)}$ becomes 1/2, since in this case only two nearest-neighbour symbols exist. The CNUM-related transition probability values of the different modems are summarized in Table 5.4. Hence, by introducing the simplified BQM scheme or the CNUM arrangement, the computational complexity of BQM can be reduced. This issue will be discussed in Section 5.3.3.

5.3.1.3 Simulation Results

In this section, we provide our simulation results characterizing the achievable performance of the TTCM-assisted MMSE-GA-SDMA-OFDM system employing UM or BQM. For the BQM-based schemes, all the parameters used, including the TTCM-, GA- and channel-related ones, are the same as those specified in Table 5.1, except that the UM component was substituted by the BQM in the GA MUD.

5.3.1.3.1 BQM Versus UM

In Figure 5.10, the BER performances of the UM- and BQM-aided GA-assisted TTCM-MMSE-SDMA-OFDM systems employing a 4-QAM scheme for transmission over the SWATM channel of Table 3.10 are compared, where six users are supported with the aid of six receiver antenna elements. Again, the performance of the TTCM-assisted MMSE-SDMA-OFDM system, the TTCM-aided optimum ML-

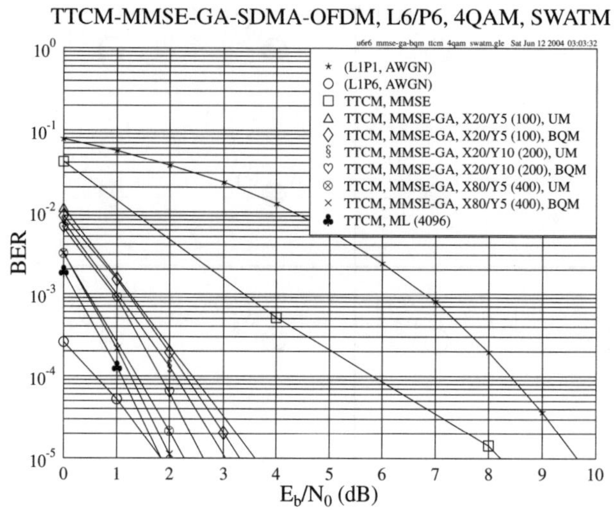

TTCM-MMSE-GA-SDMA-OFDM, L6/P6, 4QAM, SWATM

Figure 5.10: BER versus E_b/N_0 performance comparison of the **TTCM-assisted MMSE-GA-SDMA-OFDM** system using **UM** or **BQM**, while employing a **4-QAM** scheme for transmission over the SWATM channel, where $L = 6$ users are supported with the aid of $P = 6$ receiver antenna elements. The basic simulation parameters are given in Table 5.1.

detected system, and the uncoded single-user scheme employing either a single receiver or invoking MRC when communicating over an AWGN channel, are also provided for reference, respectively. As expected, we can see from Figure 5.10 that the BER performance of the TTCM-assisted MMSE-SDMA-OFDM system was further improved at relatively higher SNRs, when BQM rather than UM was used.

Furthermore, a higher performance improvement can be achieved by the BQM-aided scheme when we have a higher user load or higher throughput, as seen in Figure 5.11. More specifically, the left-hand side of Figure 5.11 shows the simulation results attained in the scenario, where a higher number of $L = 10$ users are supported. In this case, an E_b/N_0 gain of about 1 dB was attained by the employment of BQM for a GA having a size of $X = 20$ and $Y = 5$ at the BER of 10^{-5}. This E_b/N_0 gain was achieved because a higher user load results in a larger search space and hence the conventional UM suffers more from its inefficient mutation mechanism, while BQM may more readily be able to guide the mutation in the desirable direction even at low SNRs, especially in the scenarios having higher user loads. On the other hand, when a high-throughput modem such as for example 16-QAM is employed, BQM may significantly outperform UM, as evidenced on the right-hand side of Figure 5.11. As seen in the figure, the UM-aided scheme yielded a high residual error floor, since the GA apparently settled in local minima during its search owing to the less efficient mutation strategy. By contrast, BQM significantly improved the GA's performance by lowering the error floor by about two orders of magnitude to the BER of 10^{-5}. Note that at SNRs higher than 20 dB, however, a cross-over point appeared on the curves, where a GA having $X = 160$ and $Y = 5$ could no longer improve the MMSE-aided system's performance when the SNR was increased. The reason for encountering this phenomenon at a high SNR level is that it becomes more difficult for a moderate-complexity GA to mitigate the associated symbol errors among the increased number of the 16-QAM constellation symbols, in comparison with the 4-QAM constellation symbols. Hence, to improve the attainable performance of systems employing high-throughput modems, we may either use a more complex GA, which is less attractive for complexity-sensitive systems, or invoke an improved MUD framework, as will be discussed in Section 5.3.2.

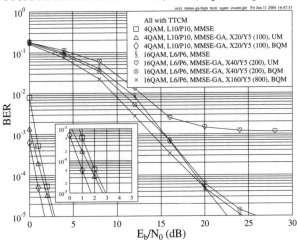

TTCM-MMSE-GA-SDMA-OFDM, Lx/Px, 4QAM/16QAM, SWATM

Figure 5.11: BER versus E_b/N_0 performance comparison of the **TTCM-assisted MMSE-GA-SDMA-OFDM** system using **UM** or **BQM**, while employing a **4-QAM** or **16-QAM** scheme for transmission over the SWATM channel, where $L = 6$ or $L = 10$ users are supported with the aid of $P = 6$ or $P = 10$ receiver antenna elements, respectively. The basic simulation parameters are given in Table 5.1.

5.3.1.3.2 BQM Versus CNUM

As discussed in Section 5.3.1.2.2, the BQM scheme may be simplified to the CNUM arrangement, which mutates to only one of the closest neighbours of the original gene, thus incurring a lower complexity in comparison with BQM. However, CNUM does not necessarily degrade the system's performance dramatically. Figure 5.12 provides a comparison of CNUM and BQM for both low- and high-throughput systems. As observed in Figure 5.12, the BQM-GA-assisted system achieved a slightly better performance than its CNUM-GA-assisted counterpart. This may suggest that in such scenarios the CNUM scheme may become an attractive alternative to the BQM scheme for further decreasing the complexity imposed.

5.3.2 Iterative MUD Framework

In Section 5.3.1, we presented an enhanced mutation scheme, namely BQM, which is capable of improving the GA MUD's performance, especially in systems having high user loads and/or employing high-throughput modems. This may be regarded as GA-related improvement in the context of the GA MUD. In this section, we will focus our attention on an enhanced iterative MUD framework, so that the system's performance may be further improved in all the scenarios considered so far.

5.3.2.1 MMSE-Initialized Iterative GA MUD

In the literature, iterative techniques such as SIC [197, 439–441] and PIC [439, 441, 442] have been designed for multi-user OFDM systems. Following the philosophy of iterative detections, we propose an MMSE-initialized Iterative GA (IGA) MUD for multi-user SDMA-OFDM systems. Figure 5.13 shows the proposed IGA MUD-assisted multi-user SDMA-OFDM uplink system. On comparing Figure 5.1 and Figure 5.13, we can see that the concatenated MMSE-GA MUD used in the BS of Figure 5.1 is replaced by the MMSE-assisted IGA MUD seen in the middle–bottom part of Figure 5.13, while the detailed structure of the IGA MUD is outlined in Figure 5.14. More specifically, the received length-P

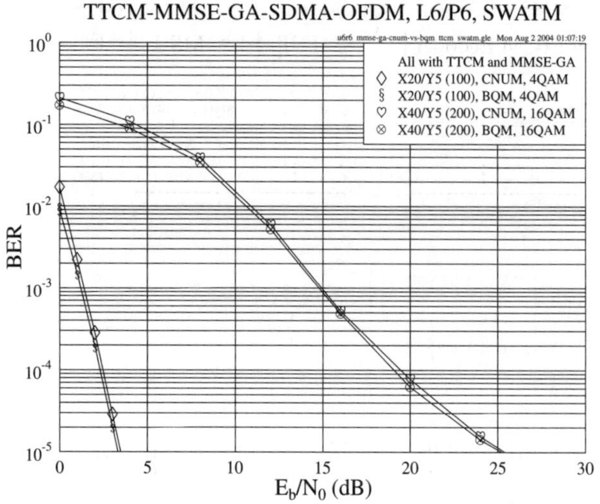

TTCM-MMSE-GA-SDMA-OFDM, L6/P6, SWATM

Figure 5.12: BER versus E_b/N_0 performance comparison of the **TTCM-assisted MMSE-GA-SDMA-OFDM** system using **CNUM** or **BQM**, while employing a **4-QAM** or **16-QAM** scheme for transmission over the SWATM channel, where $L = 6$ users are supported with the aid of $P = 6$ receiver antenna elements, respectively. The basic simulation parameters are given in Table 5.1.

symbol vector $\mathbf{x}$ of Equation (4.2) is first detected by the MMSE MUD, which outputs the L MMSE-detected symbols $\hat{s}_{\text{MMSE}}^{(l)}$ ($l = 1, \ldots, L$) of the L users, and forwards them to L independent TTCM decoders. The TTCM-decoded L-symbol vector, which is more reliable than the MMSE MUD's output, is then fed into the concatenated GA MUD to assist in the creation of the initial population. Then the genetically enhanced output symbol vector $\hat{\mathbf{s}}_{\text{GA}}$, which may be expected to become more reliable, will be fed back to the TTCM decoders in order to improve further the signal's quality, invoking a number of iterations. Following the last iteration, the final GA solution will be decoded by the TTCM decoders, and the hard-decision version of the estimated information bits of the L independent users is forwarded to the output, which is only enabled at the final iteration by the switch seen in Figure 5.14.

Therefore, two improvements have been achieved by the MMSE-IGA MUD. Firstly, more accurate initial knowledge of the transmitted signals, i.e. the output of the TTCM decoders rather than that of the MMSE MUD, is supplied for the GA MUD. This reliable improvement therefore offers a better starting point for the GA's search. Secondly, the iterative processing ensures that the detected L-user symbol vector can be optimized in two dimensions, as demonstrated in Figure 5.15. During every iteration, on the one hand, each L-symbol vector at a specified subcarrier slot is optimized by the GA in the context of the *user domain*. On the other hand, the entire TTCM-coded frame of each user is optimized by the TTCM decoder in the context of the TTCM-related *codeword domain*, or more specifically the *frequency domain*. Hence, as the iterative processing continues, an information exchange takes place between the two domains, resulting in a 2D optimization which may be expected to improve the system's performance.

5.3.2.2 Simulation Results

In this section, we combine the IGA MUD with BQM, which was presented in Section 5.3.1.2, and compare the associated simulation results with our previous results. Note that for the sake of fairness, we have halved the number of TTCM decoding iterations for the IGA-aided scheme, so that the total TTCM-related complexity remains approximately the same as in the non-iterative system. For

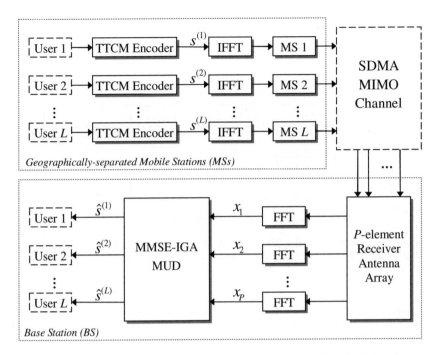

Figure 5.13: Schematic of the IGA MUD-assisted multi-user SDMA-OFDM uplink system.

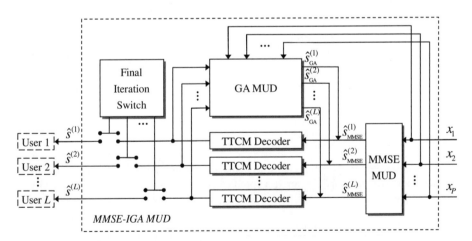

Figure 5.14: Structure of the MMSE-initialized iterative GA MUD used at the BS. ©IEEE Jiang & Hanzo 2006 [628]

the convenience of the reader, we summarize in Table 5.5 the basic simulation parameters used for generating the results provided in this section.

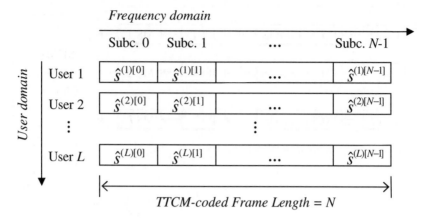

Figure 5.15: The 2D optimization provided by the MMSE-IGA MUD of Figure 5.14. The square brackets [·] denote the subcarrier indices in the TTCM-coded frame of length N.

Table 5.5: The various techniques and parameters used in the simulations of Section 5.3.2.2.

TTCM	Modem	4-QAM
	Code rate	0.5
	Code memory ν	3
	Octal generator polynomial	[13 6]
	Codeword length	1024 symbols
	Channel interleaver depth	1024 symbols
	Number of turbo iterations	2
GA	Population initialization method	MMSE
	Mating pool creation strategy	Pareto-Optimality
	Selection method	Fitness-proportionate
	Cross-over	Uniform cross-over
	Mutation	UM or BQM
	Mutation probability p_m	0.1
	Elitism	Enabled
	Incest prevention	Enabled
	Population size X	Varied
	Generations Y	Varied
	Number of IGA iterations	Varied
Channel	CIRs	SWATM [3]
	Paths	3
	Maximum path delay	48.9 ns
	Normalized Doppler frequency f_d	1.235×10^{-5}
	Subcarriers K	512
	Cyclic prefix	64

5.3.2.2.1 Performance in Underloaded and Fully Loaded Scenarios

In this section, we will investigate the system's achievable performance generated in the so-called underloaded and fully loaded scenarios, respectively, where the number of users L is less than or equal to the number of receiver antennas P.

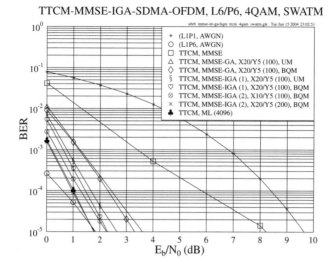

Figure 5.16: **BER** versus E_b/N_0 performance comparison of the **iterative** or **non-iterative TTCM-assisted MMSE-GA-SDMA-OFDM** system using **UM** or **BQM**, while employing a **4-QAM** scheme for transmission over the SWATM channel, where $L = 6$ users are supported with the aid of $P = 6$ receiver antenna elements, respectively. The basic simulation parameters are given in Table 5.5.

5.3.2.2.1.1   BQM-IGA Performance

Figure 5.16 shows the BER performance achieved by the various schemes considered. The numbers in brackets seen in the legends of Figure 5.16 denote the associated number of IGA MUD iterations and the total GA or ML complexity, respectively. From the results of Figure 5.16 two conclusions may be derived. Firstly, an improved performance can be achieved when the GA commences its operation from a better initial population, regardless of the different mutation schemes used. For example, at the same GA complexity, the single-iteration IGA MUD-assisted systems outperformed their non-iterative GA-aided counterparts, since the initial GA populations of the former systems were created based on the first-iteration outputs of the TTCM decoders, rather than on the less reliable MMSE MUD, regardless of whether UM or BQM was employed.

Secondly, the system employing the BQM-aided two-iteration IGA MUD was capable of achieving the same performance as the optimum ML-aided system at an even lower complexity compared with the UM- or BQM-aided non-iterative GA MUD, which was characterized in Figure 5.10. For instance, Figure 5.16 shows that the two-iteration BQM-IGA MUD having a complexity of 200 achieved virtually undistinguishable performance in comparison with its ML-aided counterpart, while the non-iterative UM/BQM-GA MUD having a complexity of 400 attained a slightly inferior performance in comparison with the ML-aided arrangement, as observed in Figure 5.10.

Having investigated the system using the 4-QAM modem, let us now consider various high-throughput scenarios. As mentioned in Section 5.3.1.3, the non-iterative GA MUD may result in a high residual error floor in high-throughput scenarios, even if BQM is employed. However, with the aid of the IGA MUD, the error floor can be effectively reduced, as seen in Figure 5.17. A 16-QAM modem was employed in all the schemes[8] characterized in Figure 5.17, except for the uncoded single-user benchmark scheme communicating over the AWGN channel, which used 8-PSK to maintain the same effective throughput of 3 BPS as the other TTCM-coded schemes. Similar to the 4-QAM scenario

---

[8]Note that in this case the associated complexity of the ML-aided scheme is as high as order $O(2^{mL}) = O(2^{4 \cdot 6}) = O(16\,777\,216)$, which imposes excessive complexity and hence cannot be simulated.

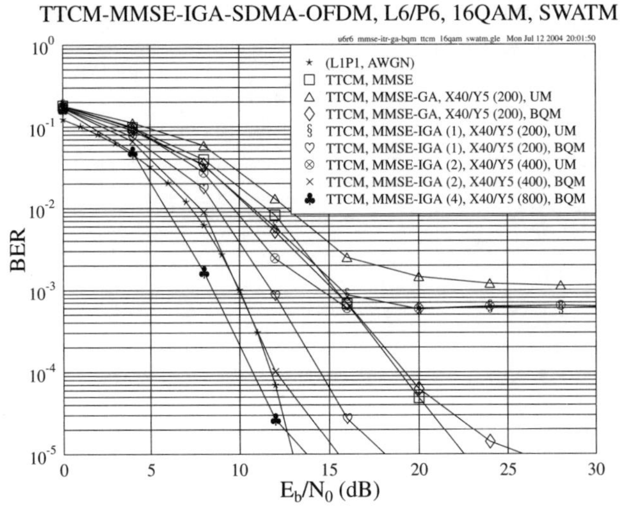

TTCM-MMSE-IGA-SDMA-OFDM, L6/P6, 16QAM, SWATM

Figure 5.17: BER versus E_b/N_0 performance comparison of the **iterative** or **non-iterative TTCM-assisted MMSE-GA-SDMA-OFDM** system using **UM** or **BQM**, while employing a **16-QAM** scheme for transmission over the SWATM channel, where $L = 6$ users are supported with the aid of $P = 6$ receiver antenna elements, respectively. The basic simulation parameters are given in Table 5.5.

of Figure 5.16, it can be seen in Figure 5.17 that a better initial population resulted in an improved performance in both the UM- and BQM-aided systems. However, we ca see that even when assisted by reliable initial knowledge of the transmitted symbols, the UM-aided single-iteration IGA MUD was unable substantially to decrease the error floor. Even when we increased the number of UM-IGA MUD iterations, the situation remained the same, except for the modest improvements achieved at SNRs lower than 16 dB. By contrast, the BQM-aided scheme is capable of substantially exploiting the benefits arising from both a better initial GA population and an increased number of IGA MUD iterations. More specifically, on the one hand, the improved initial population provides a good starting point for the GA, thus assisting the BQM, which in turn benefits the entire detection process, resulting in a substantial performance improvement. On the other hand, the iterative processing invoked by the IGA MUD further enhances the system's performance with the aid of the 2D optimization, as discussed in Section 5.3.2.1, since the beneficial information exchange between the user domain and the frequency domain assists both the GA MUD and the TTCM decoder in eliminating more and more errors found in the received signal, as the iterative procedure continues.

5.3.2.2.1.2 Effects of the Number of IGA MUD Iterations

Figure 5.18 shows the E_b/N_0 gain achieved by the BQM-IGA assisted TTCM-MMSE-SDMA-OFDM systems employing 16-QAM at the BER of 10^{-5}, while using different numbers of IGA MUD iterations. The E_b/N_0 gain is defined here as the E_b/N_0 difference measured at the BER of 10^{-5} between the systems employing the BQM-IGA MUD or the MMSE MUD. As expected, when there are a higher number of IGA MUD iterations, a higher E_b/N_0 gain was attained. It is also found in Figure 5.18 that most of the achievable gain may be attained when the number of IGA MUD iterations reaches 8. Furthermore, when the complexity of the GA MUD increases, because for example a higher population size is employed, a higher gain can be achieved, as seen in Figure 5.18. Moreover, we can also see that when the number of IGA MUD iterations was increased, the difference between the E_b/N_0 gains achieved by the higher-complexity and the lower-complexity IGAs tends to be larger. For example, as observed in Figure 5.18, when we had only one IGA MUD iteration, the E_b/N_0 gain

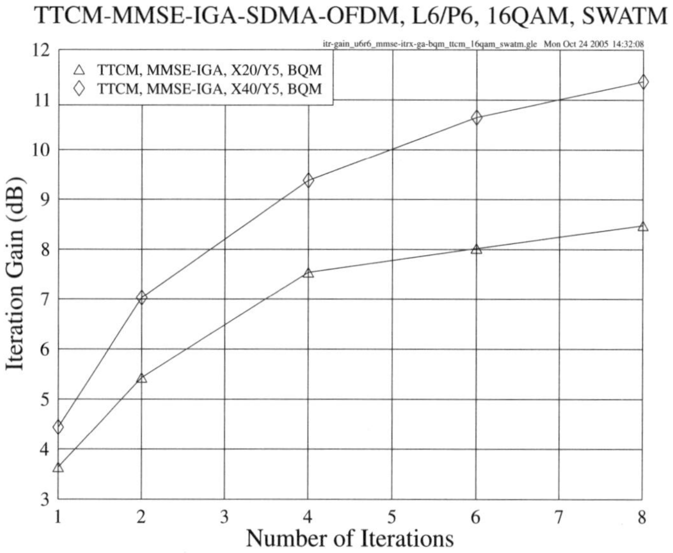

TTCM-MMSE-IGA-SDMA-OFDM, L6/P6, 16QAM, SWATM

Figure 5.18: Performance of **Iteration gain** at the BER of 10^{-5} versus number of IGA MUD iterations for the **TTCM-assisted MMSE-IGA-SDMA-OFDM** system using **BQM**, while employing a **16-QAM** scheme for transmission over the SWATM channel, where $L = 6$ users are supported with the aid of $P = 6$ receiver antenna elements. The basic simulation parameters are given in Table 5.5.

difference between the two curves was about 1 dB, while this value increased to about 3 dB when the number of iterations was increased to 8. This suggests that a high-complexity IGA may benefit more from a higher number of IGA MUD iterations than its lower-complexity counterpart.

5.3.2.2.1.3 Effects of the User Load

Figure 5.19 exhibits the corresponding E_b/N_0 gains achieved by the SDMA-OFDM system exploiting $P = 6$ receiver antenna elements at the BER of 10^{-5} in the scenarios where the user load varies. The user load of SDMA-OFDM systems was defined by Equation (4.14) in Section 4.3.2.1.1. As observed in Figure 5.19, firstly, it is shown that when the user load becomes higher, a higher gain can be attained. For example, for the single-iteration IGA-aided system, a further gain of about 3.5 dB is achieved when the number of users increases from four to six. This is because, when more users were accommodated by the SDMA-OFDM system, the reference MMSE MUD suffered a higher performance degradation than the IGA MUD, and thus a higher E_b/N_0 gain was attained by the IGA MUD. Secondly, a higher number of IGA MUD iterations provide a higher E_b/N_0 gain for the system. For instance, in the full-user-load scenario, i.e. for $L = P = 6$, the two-iteration IGA-aided system achieves a further gain of about 2.7 dB over its single-iteration counterpart, providing an overall E_b/N_0 gain of 7 dB over the baseline TTCM-MMSE-SDMA-OFDM benchmark system dispensing with the GA MUD.

5.3.2.2.2 Performance in Overloaded Scenarios

Recall that in Section 5.1 we pointed out that in practical applications the number of users L may be higher than that of the receiver antennas P, resulting in the overloaded scenario. However, most of the existing MUD techniques, such as for example the MMSE algorithm of [3, 439] and the QRD-M algorithm of [172], suffer from a significant performance degradation in overloaded scenarios, owing to an insufficient degree of detection freedom at the receiver. By contrast, we will show in this section that the proposed IGA MUD is capable of adequately performing in overloaded scenarios.

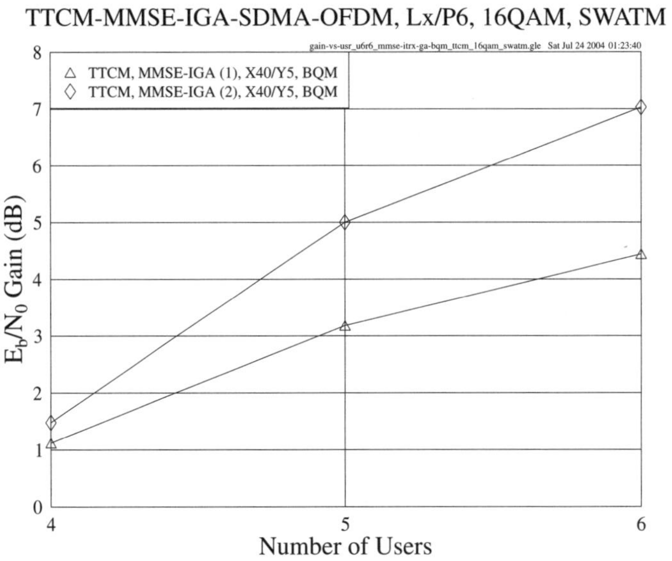

TTCM-MMSE-IGA-SDMA-OFDM, Lx/P6, 16QAM, SWATM

Figure 5.19: Performance of E_b/N_0 **gain** at the BER of 10^{-5} versus number of users for the **TTCM-assisted MMSE-IGA-SDMA-OFDM** system using **BQM**, while employing a **16-QAM** scheme for transmission over the SWATM channel, where $L = 4, 5, 6$ users are supported with the aid of $P = 6$ receiver antenna elements, respectively. The basic simulation parameters are given in Table 5.5.

5.3.2.2.2.1 Overloaded BQM-IGA

Figure 5.20 shows the performance achieved by the BQM-IGA aided TTCM-SDMA-OFDM system using 4-QAM, when six, seven and eight users are supported by six receiver antenna elements, respectively. It can be seen in Figure 5.20 that in the so-called overloaded scenarios, where the number of users exceeds the number of receiver antenna elements, the linear MMSE MUD suffered from an insufficient degree of freedom for separating the different users, since the high number of users incurred excess Multi-User Interference (MUI). This results in a significant performance degradation in the context of the system using the MMSE MUD, when the number of users increased from six to eight, as observed in Figure 5.20. However, in such cases the system employing the proposed BQM-IGA MUD was still capable of maintaining a near-ML performance. For example, when $L = 8$, the two-iteration-based BQM-IGA MUD reduced the BER measured at 3 dB by four orders of magnitude in comparison with the MMSE-aided benchmark system, as evidenced by Figure 5.20. This result characterizes the robustness of the BQM-IGA MUD, which has successfully suppressed the high MUI experienced in overloaded scenarios.

Figure 5.21 shows the iteration gain achieved by the BQM-IGA assisted TTCM-MMSE-SDMA-OFDM system employing 4-QAM at the BER of 10^{-5}, while using different numbers of IGA MUD iterations. The iteration gain is defined here as the E_b/N_0 difference of the systems employing different numbers of IGA MUD iterations measured at the BER of 10^{-5} in comparison with the baseline system employing a single IGA MUD iteration. It is found in Figure 5.21 that when more users are supported, higher iteration gains may be obtained by iterative detection. For example, a gain of about 6 dB was attained by the eight-user system at the second IGA MUD iteration, while that attained by the six-user system was only about 0.5 dB. Furthermore, as the number of iterations was increased from two to six, the former scheme provided a further gain of about 1 dB, while no explicit gain was achieved by the latter arrangement, as shown in Figure 5.21. It is also seen in Figure 5.21 that most of the achievable iteration gain has been attained at the second IGA MUD iteration for all the schemes.

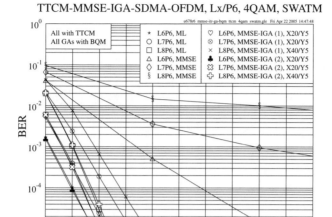

Figure 5.20: BER versus E_b/N_0 performance comparison of the **TTCM-assisted MMSE-IGA-SDMA-OFDM** system using **BQM**, while employing a **4-QAM** scheme for transmission over the SWATM channel, where $L = 6, 7, 8$ users are supported with the aid of $P = 6$ receiver antenna elements, respectively. The basic simulation parameters are given in Table 5.5.

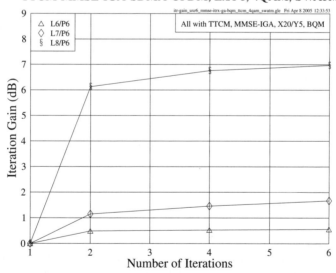

Figure 5.21: Performance of **iteration gain** at the BER of 10^{-5} versus number of IGA MUD iterations for the **TTCM-assisted MMSE-IGA-SDMA-OFDM** system using **BQM**, while employing a **4-QAM** scheme for transmission over the SWATM channel, where $L = 6, 7, 8$ users are supported with the aid of $P = 6$ receiver antenna elements, respectively. The basic simulation parameters are given in Table 5.5. ©IEEE Jiang & Hanzo 2006 [628]

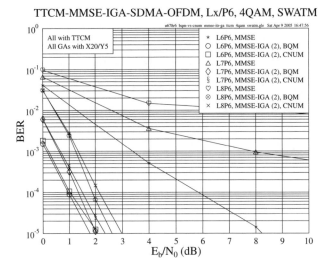

Figure 5.22: **BER** versus E_b/N_0 performance comparison of the **TTCM-assisted MMSE-IGA-SDMA-OFDM** system using **BQM** or **CNUM**, while employing a **4-QAM** scheme for transmission over the SWATM channel, where $L = 6, 7, 8$ users are supported with the aid of $P = 6$ receiver antenna elements, respectively. The basic simulation parameters are given in Table 5.5.

5.3.2.2.2.2 BQM Versus CNUM

In Section 5.3.1.3.2, we described the performance of the CNUM arrangement discussed in Section 5.3.1.2.2 in a fully loaded scenario. In Figure 5.22 we characterize the CNUM-aided system's performance achieved in an overloaded scenario, where six receiver antennas were used. As seen in Figure 5.22, the BQM-IGA-aided system slightly outperformed the CNUM-IGA-aided system. This suggests that, similar to the case of fully loaded scenarios, the CNUM scheme may also be employed in overloaded scenarios for achieving a further complexity reduction over the BQM scheme without suffering from a significant performance loss.

5.3.2.2.3 Performance under Imperfect Channel Estimation

As a further investigation, we provide the simulation results generated in the scenario where the Channel State Information (CSI) was assumed to be imperfect. The estimated CIRs $\hat{h}_i$ were generated by adding random Gaussian noise to the true CIR taps h_i as

$$\hat{h}_i[n] = h_i[n] + \sqrt{\frac{\sigma_n^2}{\varepsilon}} n_i[n], \quad i = 1, \ldots, \mathcal{L}, \tag{5.19}$$

where ε is the effective noise factor, σ_n^2 is the noise variance at the specific SNR level, n_i is an AWGN sample having zero mean and a variance of unity, $\mathcal{L}$ is the number of CIR taps and $[n]$ denotes the nth OFDM symbol. In the scenarios associated with imperfect CIRs, ε was set to 64 and $\mathcal{L}$ was set to 3 for the three-path SWATM channel used. In this case, the effective noise power added to the true CIR taps during each OFDM symbol to simulate imperfect channel estimation was $\sigma_n^2 \cdot \mathcal{L}/\varepsilon = \sigma_n^2 \times 4.69\%$. A snapshot of the SWATM channel is portrayed in Figure 5.23, which shows both the real and imaginary components of the FD-CHTFs associated with both perfect and imperfect CIRs.

Our performance comparison of the proposed BQM-IGA aided TTCM-MMSE-SDMA-OFDM system under the assumptions of both perfect and imperfect CSI is provided in Figure 5.24. As seen in Figure 5.24, the proposed system was capable of attaining an acceptable performance even

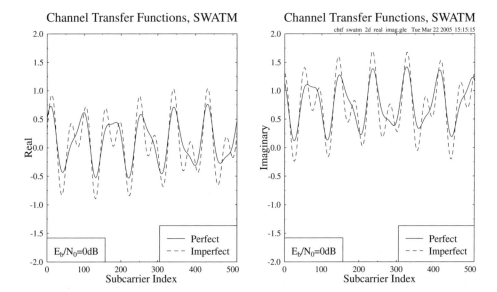

Figure 5.23: The real and imaginary components of the FD-CHTFs of the SWATM channel measured during one OFDM symbol at an E_b/N_0 value of 0 dB in terms of both perfect and imperfect CIRs.

without accurate channel knowledge. Moreover, it was found that when imperfect channel estimation was assumed, the BQM-IGA-aided system outperformed its ML-aided counterpart, especially in the scenarios associated with higher user loads. This phenomenon may be explained as follows. When the CSI is imperfect, the ML-detected signal becomes less reliable than that detected in the scenario benefitting from perfect CSI. The relatively unreliable output of the ML MUD may readily mislead the TTCM decoder owing to error propagation, resulting in a performance degradation. However, the detrimental effects of imperfect CSI may be mitigated by the proposed IGA MUD. More specifically, the IGA MUD optimizes the detected signal in two dimensions, namely in both the user domain and the frequency domain, as discussed in Section 5.3.2.1. The beneficial information exchange offered by the IGA MUD between the two domains may effectively assist the concatenated detection–decoding procedure in counteracting the detrimental effects of imperfect channel estimation. This therefore results in a better system performance in comparison with that achieved by the ML-aided system. Furthermore, when a higher number of users had to be supported, the ML-aided system using imperfect CSI suffered more from the inaccurate MUD, while a more robust behaviour was exhibited by the IGA-aided system, as shown in Figure 5.24.

5.3.3 Complexity Analysis

Compared with the conventional UM scheme, BQM is capable of significantly improving the GA's performance, especially in high-throughput or high-SNR scenarios, as discussed in Section 5.3.1.3.1. Furthermore, this performance improvement was achieved at the cost of a modest complexity increase and a modest memory requirement. More specifically, at different SNR levels, for each of the 2^m constellation symbols, a specific set containing $(2^m - 1)$ normalized 2D transition probabilities has to be created. However, this only imposes a modest 'once-for-all' calculation, since we can derive the associated transition probabilities with the aid of offline experiments for a number of typical SNR levels, where the calculated data can be stored in the BS's memory, hence incurring no further computational complexity. Furthermore, by introducing the simplified BQM scheme of

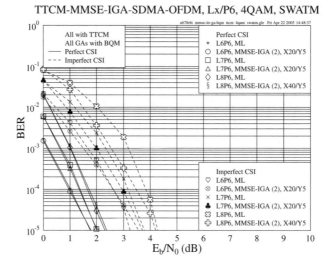

Figure 5.24: BER versus E_b/N_0 performance comparison of the **TTCM-assisted MMSE-IGA-SDMA-OFDM** system using **BQM** with perfect or imperfect CSI, while employing a **4-QAM** scheme for transmission over the SWATM channel, where $L = 6, 7, 8$ users are supported with the aid of $P = 6$ receiver antenna elements, respectively. The basic simulation parameters are given in Table 5.5.

Section 5.3.1.2.2, the associated complexity and memory cost may be dramatically reduced, especially for high-throughput modems such as 16-QAM or 64-QAM, since the number of mutation target candidates decreases and thus fewer transition probability calculations are required. Moreover, if the CNUM scheme is employed, the associated complexity can be further decreased, since in this case there is no need to calculate the transition probabilities, which are already available in Table 5.4. This may significantly reduce the associated complexity and memory requirement, while still maintaining a similar performance to that of the BQM scheme, as seen in Figures 5.12 and 5.22.

As shown in Figures 5.18 and 5.21, the system's performance can be further improved when the number of IGA MUD iterations is increased. When the other parameters remain the same, using a higher number of IGA MUD iterations will result in a further increased complexity. However, this may still be significantly lower than that imposed by the ML-aided scheme. Figure 5.25 provides our comparison of the TTCM-assisted MMSE-SDMA-OFDM, ML-SDMA-OFDM and MMSE-BQM-IGA-SDMA-OFDM systems in the context of their MUD complexity, which was quantified in terms of the number of complex additions and multiplications imposed by the different MUDs on a per-user basis. As illustrated in Figure 5.25, the complexity of the ML MUD is significantly higher than that of the MMSE MUD or the IGA MUD. Furthermore, the IGA MUD's complexity does not significantly vary at different E_b/N_0 values and depends on the number of IGA MUD iterations as well as on the GA's parameters, e.g. the population size. In Figure 5.26 the complexity of the various systems is compared in terms of different user loads at an E_b/N_0 value of 0 dB. At a specific user load, we always select an appropriate GA-aided scheme for comparison, which achieved a similar performance compared with the ML-aided system at the BER of 10^{-5}. As seen in Figure 5.26, the ML-aided system imposes a linearly increasing complexity on a logarithmic scale, which corresponds to an exponential increase when the number of users increases. By contrast, the complexity of the IGA-aided system required for maintaining a near-optimum performance increases only moderately.

In order to characterize the advantage of the BQM-IGA scheme in terms of the performance–complexity trade-off, we summarize in Table 5.6 the computational complexity imposed by the different MUDs assuming an E_b/N_0 value of 3 dB. As observed in Table 5.6, the complexity of the ML MUD

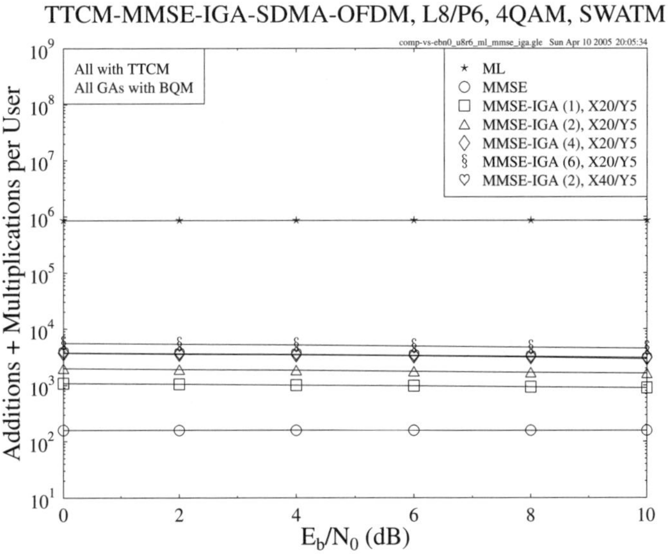

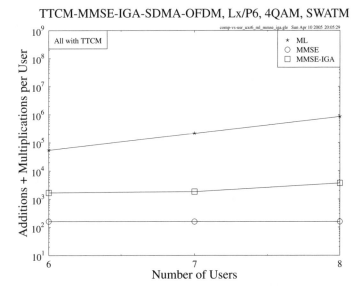

Figure 5.25: Complexity per user versus E_b/N_0 performance comparison of the **TTCM-assisted MMSE-SDMA-OFDM**, **ML-SDMA-OFDM** and **MMSE-BQM-IGA-SDMA-OFDM** systems, while employing a **4-QAM** scheme for transmission over the SWATM channel, where $L = 8$ users are supported with the aid of $P = 6$ receiver antenna elements. The basic simulation parameters are given in Table 5.5.

Figure 5.26: Performance comparison of **complexity** per user versus number of users for the **TTCM-assisted MMSE-SDMA-OFDM**, **ML-SDMA-OFDM** and **MMSE-BQM-IGA-SDMA-OFDM** systems, while employing a **4-QAM** scheme for transmission over the SWATM channel, where $L = 6, 7, 8$ users are supported with the aid of $P = 6$ receiver antenna elements. The basic simulation parameters are given in Table 5.5.

Table 5.6: Comparison of MUD complexity in terms of number of complex additions and multiplications measured at $E_b/N_0 = 3\,\mathrm{dB}$ on a per-user basis in the 4-QAM TTCM-SDMA-OFDM system.

L	MUD	$+$	$\times$	BER
6	ML	2.8×10^4	2.7×10^4	1.8×10^{-7}
	IGA	8.1×10^2	7.9×10^2	2.2×10^{-7}
	MMSE	7.1×10^1	9.0×10^1	1.5×10^{-3}
7	ML	1.1×10^5	1.1×10^5	5.1×10^{-7}
	IGA	8.7×10^2	8.5×10^2	6.2×10^{-7}
	MMSE	7.1×10^1	8.8×10^1	7.5×10^{-3}
8	ML	4.3×10^5	4.2×10^5	8.5×10^{-7}
	IGA	1.8×10^3	1.7×10^3	9.8×10^{-7}
	MMSE	7.1×10^1	8.7×10^1	2.2×10^{-2}

is significantly higher than that of the MMSE MUD or the IGA MUD, especially in highly overloaded scenarios. By contrast, the IGA MUD reduced the BER by up to five orders of magnitude in comparison with the MMSE MUD at a moderate complexity.

5.3.4 Conclusions

In Sections 5.3.1 and 5.3.2 we proposed specific techniques designed for further enhancing the achievable performance of the TTCM-assisted MMSE-GA-SDMA-OFDM system. The novel BQM scheme is capable of improving the GA's search at a modest complexity increase, thus significantly increasing the chances of finding the optimum GA solution in high-SNR and/or high-throughput scenarios. On the other hand, the 2D optimization provided by the proposed IGA MUD has been shown to be beneficial for the SDMA-OFDM system in both the frequency and user domains. Finally, the scheme that combines BQM with the IGA MUD yields the best and near-optimum performance in all scenarios considered, including the so-called overloaded scenario, where the performance of most of the conventional detection techniques such as the classic linear MMSE MUD significantly degrades, owing to the insufficiently high degree of freedom. Furthermore, this superior performance of the proposed scheme is achieved at a significantly lower computational complexity than that imposed by the ML-assisted system, especially when the number of users is high. For example, a complexity reduction of three orders of magnitude can be achieved by the proposed BQM-IGA-aided system in the overloaded scenario associated with $L = 8$, as evidenced by Figure 5.26. Moreover, we demonstrate that the proposed scheme is capable of providing a satisfactory performance even when the channel estimation is imperfect.

5.4 Chapter Summary

In this chapter, we proposed a TTCM-assisted MMSE-GA MUD designed for SDMA-OFDM systems. In Section 5.2.1 we provided a system overview of the proposed GA-assisted TTCM-MMSE-SDMA-OFDM system. The optimization metric designed for the proposed GA MUD was described in Section 5.2.2.1. Section 5.2.2.2 outlined the concatenated MMSE-GA MUD, while its performance was evaluated in Section 5.2.3, where the GA-based schemes were shown to be capable of achieving a near-optimum performance. Furthermore, a complexity comparison between the proposed GA MUD and the optimum ML MUD was provided in Section 5.2.4, where we showed that the complexity of the GA MUD was significantly lower than that of the ML MUD.

To improve further improve the performance of the TTCM-assisted MMSE-GA-SDMA-OFDM system, an enhanced GA MUD was proposed in Section 5.3. This was described in two steps. Firstly, the novel BQM scheme was proposed in Section 5.3.1, including a review of the conventional UM scheme, followed by a detailed explanation of the BQM mechanism, which were the subjects of Sections 5.3.1.1 and 5.3.1.2, respectively. The BQM-aided GA MUD exploits an effective mutation strategy and thus is capable of achieving a better performance in comparison with its UM-aided counterpart, especially at high SNRs or high user loads, as evidenced by the simulation results given in Section 5.3.1.3. Moreover, this was achieved at a modest complexity increase. Secondly, an MMSE-initialized IGA MUD was introduced in Section 5.3.2. The theoretical foundations of the IGA MUD were presented in Section 5.3.2.1, where the IGA framework as well as its optimization capability were characterized. Our related simulation results were provided in Section 5.3.2.2, where the combined BQM-IGA-assisted system was found to give the best performance in all scenarios considered, while maintaining a modest computational complexity. In low-throughput scenarios, e.g. a six-user system employing a 4-QAM modem, a two-iteration BQM-IGA MUD associated with $X = 20$ and $Y = 5$ was capable of achieving the same performance as the optimum ML-aided system at a complexity of 200, which is only about 50% and 5% of the MUD-related complexity imposed by the conventional UM-aided single-iteration IGA MUD and the optimum ML MUD, respectively. On the other hand, in high-throughput six-user systems employing for example a 16-QAM modem, a two-iteration BQM-IGA MUD associated with $X = 40$ and $Y = 5$ achieved an E_b/N_0 gain of about 7 dB over the MMSE MUD benchmark at the BER of 10^{-5}, while the UM-aided GA or IGA MUDs suffered from a high residual error floor even when the iterative framework was employed. Furthermore, the associated E_b/N_0 gain was attained at a modest complexity of 400, which is only 0.002 38% of the excessive complexity imposed by the ML MUD that cannot be simulated in this case.

Moreover, the proposed BQM-IGA MUD is capable of providing a near-optimum performance even in the so-called overloaded scenarios, where the number of users is higher than the number of receiver antenna elements, while many conventional detection techniques suffer from an excessively high error floor. For example, with $L = 8$ users and $P = 6$ receivers, the two-iteration-based BQM-IGA MUD reduced the BER recorded at an E_b/N_0 value of 3 dB by four orders of magnitude in comparison with the classic MMSE-MUD-aided benchmark system, as shown in Figure 5.20. This result characterizes the robustness of the BQM-IGA MUD, which has successfully suppressed the high MUI experienced in overloaded scenarios. As a further investigation, we demonstrated in Section 5.3.2.2.3 that the proposed system is capable of achieving a satisfactory performance even in the case of imperfect channel estimation. Furthermore, the complexity of the proposed detection scheme is only moderately higher than that imposed by the linear MMSE MUD, and is substantially lower than that imposed by the optimum ML MUD, as discussed in Section 5.3.3. We also showed that in both the fully loaded scenario of Section 5.3.1.3.2 and in the overloaded scenario of Section 5.3.2.2.2.2 the complexity of the BQM approach can be further reduced by employing its simplified version, namely the CNUM scheme of Section 5.3.1.2.2, at the cost of a slightly degraded system performance.

Note that the system parameters of the IGA framework, such as the number of TTCM iterations, the number of IGA MUD iterations and the GA-related parameter settings, are all readily configurable, enabling us to strike an attractive trade-off between the achievable performance and the complexity imposed. For specific scenarios, the TTCM scheme used in the system can also be conveniently substituted by other FEC schemes, e.g. the TC codes. Therefore, the facility provided by the proposed IGA MUD may make it possible for applications in multi-mode terminals, where good performance, low complexity and easy flexibility are all important criteria. It is also worth pointing out that the proposed BQM-aided IGA MUD can be readily incorporated into multi-user CDMA systems, e.g. those of [38]. In this case, the initial detected signal supplied to the GA MUD for creating the first GA population is provided by the bank of matched filters installed at the CDMA BS, rather than by the MMSE MUD. However, the BQM scheme may remain unchanged.

In the next chapter, our attention will be focused on a TTCM-assisted MMSE-IGA MUD SDMA-OFDM system employing a new type of Frequency-Hopping (FH) technique for the sake of achieving further performance enhancements.

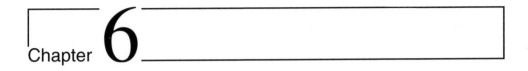

Chapter 6

Direct-Sequence Spreading and Slow Subcarrier-Hopping Aided Multi-user SDMA-OFDM Systems

6.1 Conventional SDMA-OFDM Systems[1]

In Chapters 4 and 5, Coded Modulation (CM) [314] assisted SDMA-OFDM systems invoking both Minimum Mean Square Error (MMSE) and Genetic Algorithm (GA) based Multi-User Detection (MUD) have been investigated, respectively. Specifically, in terms of the bandwidth-sharing strategy, the SDMA-OFDM systems discussed in these chapters are referred to here as the *conventional* SDMA-OFDM systems [3, 193], where all the users exploit the entire system bandwidth for their communications. However, this bandwidth-sharing strategy exhibits a few drawbacks.

On the one hand, the conventional SDMA-OFDM systems can exploit little frequency diversity, since each user activates all available subcarriers. This limitation can be mitigated by combining both Frequency-Hopping (FH) and SDMA-OFDM techniques, resulting in the FH/SDMA-OFDM systems. In these FH/SDMA-OFDM systems the total system bandwidth is divided into several subbands, each of which hosts a number of consecutive subcarriers, and a so-called FH pattern is used for controlling the subband allocation for the different users. Since each user activates different subbands from time to time, the achievable frequency diversity improves as the width of the subbands is reduced.

On the other hand, when the number of users becomes higher in conventional SDMA-OFDM systems, a higher Multi-User Interference (MUI) is expected across the entire bandwidth and hence all users will suffer from a performance degradation. Unfortunately, the same phenomenon is encountered also in FH/SDMA-OFDM systems at those subbands that are shared by excessive numbers of users. Undoubtedly, the best solution to eliminate the MUI is to avoid subband collisions between the different users by assigning each subband exclusively to a single user. This scheme of 'one subband for one user' will inevitably reduce the system's overall throughput. The attainable system throughput can be increased with the aid of higher-order modems, which are more vulnerable to transmission errors as well as impose an increased MUD complexity at the receivers, which is undesirable. Therefore, subcarrier-reuse-based SDMA-OFDM using efficient FH techniques is preferable, since it is capable

---

[1]This chapter is partially based on ©IET Jiang & Hanzo 2006 [469].

MIMO-OFDM for LTE, Wi-Fi and WiMAX   Lajos Hanzo, Yosef Akhtman, Li Wang and Ming Jiang
© 2011 John Wiley & Sons, Ltd

of maintaining a sufficiently high overall system throughput even with the employment of a relatively low-order, low-complexity modem, while effectively suppressing the associated high MUI.

In this chapter, we will introduce a new bandwidth-efficient approach for employment in SDMA-OFDM systems designed for solving the two problems mentioned above.

6.2   Introduction to Hybrid SDMA-OFDM

During the last few decades, a range of Time-Division Multiple Access (TDMA) [36], Frequency-Division Multiple Access (FDMA) [36] and Code-Division Multiple Access (CDMA) [37, 38, 40–42] schemes have found employment in first-, second- and third-generation wireless systems. Spread-Spectrum Multiple Access (SSMA) [470–473] schemes have been widely investigated, since they exhibit a range of attractive properties, including the ability to combat various types of interference. The well-known Direct-Sequence Code-Division Multiple Access (DS-CDMA) [38] scheme is resilient against both narrowband interference and multi-path fading. Another classic SSMA scheme is constituted by Frequency-Hopped SSMA (FH/SSMA) [38, 474–477], where the total available system bandwidth is divided into a number of subbands shared by a number of simultaneous users. An appropriate number of subcarriers can be assigned to each of these subbands, which may experience different channel qualities and hence may deliver different types of services. The so-called FH pattern is used to control a frequency synthesizer involved for the purpose of activating different subcarrier frequencies. The data of each of the simultaneous users modulate the subcarriers independently, and the entire system bandwidth can be allocated on a demand basis. The bandwidth of the subbands may be arbitrarily small or large, depending on the type of services to be delivered or the bit rate to be supported. This flexibility is attractive when aiming at supporting future multimedia services, where variable bit rates associated with different Quality-of-Service (QoS) are required by different applications. Moreover, in FH-aided systems the different subbands of a particular user do not necessarily have to be contiguously allocated. This flexibility is attractive in scenarios where several systems operated by different service providers have to coexist and/or fractional bandwidths have to be exploited.

FH can be effectively amalgamated with a range of well-established techniques, e.g. the family of CDMA systems, resulting in the FH/CDMA systems [38, 478–481]. Furthermore, OFDM [1, 3, 4, 24] also benefits from invoking various FH schemes [482, 483]. The FH-aided OFDM systems may also be combined with time-domain Direct-Sequence Spreading (DSS) techniques for creating hybrid systems, where the multiple users' modulated DSS signals are frequency hopped according to their user-specific FH patterns. Hybrid DSS/FH systems are attractive, because the advantages of both the DSS and FH techniques may be combined, while eliminating or mitigating some of their disadvantages [480, 484]. For example, a hybrid DSS/FH system is capable of combining the interference resilience of DS/SS systems with the attractive partial-band-jamming mitigation features of FH/SS systems [480].

More specifically, hybrid DSS/FH systems can be beneficially amalgamated with the conventional SDMA-OFDM systems [3,193] invoking MUD [439] techniques, which exhibit a number of advantages over more traditional multiple-access techniques, resulting in hybrid DSS/FH-aided SDMA-OFDM systems. Furthermore, the performance of the hybrid system can be significantly improved if the proposed novel type of Slow Frequency-Hopping (SFH) technique, referred to here as the Slow SubCarrier-Hopping (SSCH) scheme, is employed. From a general point of view, the philosophy of the SubCarrier-Hopping (SCH) technique is reminiscent of the concept of Clustered OFDM (ClOFDM) [485–487] and Orthogonal Frequency-Division Multiple Access (OFDMA) [43–46], both of which are conceptually similar. In contrast to the systems which combine OFDM with traditional multiple-access schemes, the intrinsic nature of the orthogonal subcarriers enables OFDM itself to support multiple access, where the subcarriers are allocated to a number of subbands and assigned to different simultaneous users [43, 44, 488], resulting in the concept of OFDMA. OFDMA was initially proposed for Cable TV (CATV) systems [489] and now has been ratified as the IEEE 802.16 standard for Broadband Wireless Multiple Access (BWMA) systems [16].

In the proposed hybrid DSS/SSCH SDMA-OFDM system, each subcarrier is shared by a certain number of users in the context of the OFDM symbol's frequency-domain representation, while each user's time-domain OFDM signal is further spread with the aid of DSS. Moreover, using a simple but efficient Uniform SSCH (USSCH) pattern, the hybrid DSS/SSCH-aided SDMA-OFDM system is capable of achieving a high frequency diversity and hence exhibits a high robustness to the MUI experienced, resulting in a significant performance improvement. In the literature, substantial research efforts have been invested in designing subcarrier allocation algorithms for Single-Input, Single-Output (SISO) OFDM systems, which are subjected to various design constraints, such as requiring the minimal overall transmit power [76, 490–494], achieving the maximum capacity [494–496] or complying with specific QoS criteria [497]. A subcarrier and bit allocation algorithm designed to minimize the overall transmit power for a MIMO OFDM system was proposed in [498]. A number of subband/subcarrier allocation schemes were also proposed for systems based on ClOFDM [485–487] or OFDMA [499–503]. However, many of these algorithms were derived under the assumption that a subband or a subcarrier can be used by one user only, resulting in an MUI-free scenario. By contrast, our proposed USSCH algorithm allows multiple constant-rate users to activate the same subcarrier. More explicitly, each subcarrier can be activated by different users during different hopping intervals, while it is desirable to ensure that the average Frequency-Domain (FD) separation of the subcarriers activated by the same user is sufficiently large, in order to experience uncorrelated fading and hence a high diversity gain. Furthermore, the number of users activating each subcarrier should ideally be similar so as to ensure that the MUI encountered at each subcarrier becomes similar, hence eliminating the MUI 'peaks' across the system bandwidth. Moreover, each subcarrier may be 'overloaded'[2] by being shared by a high number of users for the sake of maintaining a high overall system throughput.

In order to exploit fully the potential benefits of the USSCH pattern, the employment of Forward Error Correction (FEC) schemes is necessary. For the sake of convenient performance comparisons, again, the Turbo Trellis-Coded Modulation (TTCM) [314, 435] scheme used in Chapters 4 and 5 is selected, although we point out that other FEC schemes are also applicable. It will be shown in our forthcoming discourse that with the aid of the advocated USSCH pattern, the MUI experienced by the TTCM-aided conventional SDMA-OFDM system can be effectively suppressed, resulting in a significant performance improvement, especially in high-throughput or overloaded scenarios, while a computational complexity similar to that of the conventional SDMA-OFDM systems can be maintained.

The structure of the rest of this chapter is as follows. A comparison of various frequency resource allocation strategies used in the conventional SDMA-OFDM, SFH/SDMA-OFDM and SSCH/SDMA-OFDM systems is provided in Section 6.3, followed by the introduction of the proposed hybrid DSS/SSCH-aided SDMA-OFDM scheme in Section 6.4. More specifically, a system overview is presented in Section 6.4.1, while the transmitter and receiver structures of the proposed system are elaborated on in Sections 6.4.1.1 and 6.4.1.2, respectively. In Section 6.4.2, two different SSCH pattern assignment strategies are discussed. Section 6.4.3 provides an insight into the DSS despreading and SSCH demapping processes invoked at the SDMA receiver, followed by a brief description of the MUD process in the context of the presence of DSS/SSCH. The numerical results characterizing a range of different SDMA-OFDM systems in different scenarios are provided in Section 6.5, while the complexity issues are discussed in Section 6.6. Finally, Section 6.7 concludes our findings.

6.3 Subband Hopping Versus Subcarrier Hopping

Figure 6.1 shows the frequency resource allocation strategies of the conventional SDMA-OFDM [3, 193], SFH/SDMA-OFDM and SSCH/SDMA-OFDM systems. As illustrated in Figure 6.1, in conventional SDMA-OFDM systems, which do not employ any hopping techniques, the total available system

---

[2]The terminology for an overloaded system in this chapter is slightly different from that discussed in Chapter 5. The former indicates that each of the specific subcarriers used in the DSS/SSCH SDMA-OFDM system is shared by a sufficiently high number of users, while the latter refers to the overloaded conventional SDMA-OFDM system, where the number of users supported exceeds that of the receiver antenna elements.

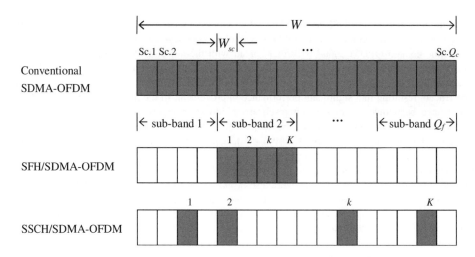

Figure 6.1: Comparison between the conventional SDMA-OFDM, SFH/SDMA-OFDM and SSCH/SDMA-OFDM in the context of frequency resource allocation for a single user. Each block in the figure represents a subcarrier, while those in grey denote the user-activated subcarriers. In this example, random hopping patterns are used for both the SFH- and the SSCH-aided systems. ©IET Jiang & Hanzo 2006 [469]

bandwidth W is partitioned into a total of $Q_c = W/W_{sc}$ subcarriers, where W_{sc} is the subcarrier bandwidth,[3] and each user activates all the Q_c subcarriers for communication.

However, in the SFH-assisted OFDM systems the total system bandwidth W is divided into a number of subbands denoted by $q = 1, \ldots, Q_f$, as seen in Figure 6.1. Each subband is assigned a carrier frequency f_q, which is used for carrying the OFDM signal to be transmitted in this subband, when it is activated. Furthermore, each of the Q_f subbands has K subcarriers, each having a subcarrier bandwidth of W_{sc}, and $W = W_{sc} \times K \times Q_f$. In the context of the SFH/SDMA-OFDM system, during an FH dwell time T_h, each of the L simultaneous users supported is assigned one of the Q_f subbands associated with a specific carrier frequency f_q, which is activated by a frequency synthesizer according to either the pseudo-random or the deterministic user-specific hopping pattern set used. In contrast to Fast Frequency-Hopping (FFH) systems, where the symbol duration T_s obeys $\beta_F = T_s/T_h > 1$, in SFH systems $\beta_S = T_h/T_s > 1$, where both β_F and β_S are integers, while T_h is the hopping dwell period. Using SFH makes it feasible to use coherent demodulation at the receiver side, since the hopping rate is slower than the data rate.

Note that the SFH subbands are not necessarily contiguously allocated in the frequency domain. However, if a subband is activated by a user, all the K subcarriers within this subband are assigned to the specific user. While each user can only activate one subband during each FH dwell time T_h, the same subband can be exploited by more than one user. Therefore, if an excessive number of users happen to activate the same subband, resulting in heavy MUI, severe signal corruption will occur across all the consecutive subcarriers within this specific subband.

In comparison with SFH/SDMA-OFDM systems, where subband-based SFH is employed, in the proposed SSCH/SDMA-OFDM system a subcarrier-based SFH technique is invoked. Similar to conventional SDMA-OFDM systems, in SSCH/SDMA-OFDM systems we also have a total of $Q_c = W/W_{sc}$ subcarriers. However, each user activates only K of the Q_c available subcarriers, where $0 < K < Q_c$. While an SFH user exploits all the K subcarriers of an SFH subband during the hopping dwell time of T_h, an SSCH user employing the same number of K subcarriers can potentially

---

[3]The subcarrier bandwidth W_{sc} represents the FD bandwidth hosting the main spectral lobe of the sinc-function-shaped subcarrier spectrum.

select any K of the Q_c available subcarriers without decreasing the throughput, as seen in Figure 6.1. Furthermore, in SSCH/SDMA-OFDM systems a high MUI is expected to contaminate the dispersed subcarriers activated by a number of users, which is different from the situation experienced in the SFH/SDMA-OFDM scenario, where potentially all the consecutive subcarriers of a subband hosting a number of users may be at a risk of being severely corrupted. Moreover, the SSCH scheme is capable of more efficiently exploiting the benefits of frequency diversity in comparison with the SFH arrangement, since a deep fade which may corrupt many consecutive subcarriers of an SFH user may only affect a small fraction of the subcarriers used by an SSCH user, as an added bonus of the employment of discontinuous subcarriers. Explicitly, when an appropriately designed hopping pattern is used, the 'interleaving-like' FD fading randomization characteristic of the SSCH system disperses the originally bursty FD errors and hence enhances the chances of the channel decoder to correct the residual errors.

The SSCH-aided OFDM may be viewed as being somewhat similar to OFDM/OFDMA with interleaving, since both of these techniques provide a method to exploit the achievable frequency diversity. However, they are different in that in the former system different subcarriers are assigned to different users at different hopping dwell instants, while in the latter system each user employs all the subcarriers (or the same set of subcarriers in the case of OFDMA), although the signal carried by each subcarrier is interleaved. In hostile propagation scenarios some subcarriers may consistently encounter deep channel fades despite using the above-mentioned anti-fading measures, thus the specific user of the interleaved OFDM/OFDMA system who happens to activate these bad subcarriers may suffer from a consistently poor performance. By contrast, in the SSCH-aided OFDM system this rarely happens, since each user will be assigned different subcarriers from time to time.

We point out that by assigning a different number of subcarriers to different users, a flexible multi-rate system can be created, which is capable of satisfying the users' QoS profiles. For the sake of simplicity, in this chapter we assume that each user has the same constant bit rate.

6.4 System Architecture

Having outlined the basic concepts of the SSCH technique, let us now focus our attention on the structure of the proposed TTCM-assisted hybrid DSS/SSCH-aided SDMA-OFDM system. The organization of this section is as follows. In Section 6.4.1 an overview of the basic TTCM-assisted DSS/SSCH-aided SDMA-OFDM system employing the classic MMSE MUD is provided, followed by the introduction of the associated transmitter and receiver structures, which are detailed in Sections 6.4.1.1 and 6.4.1.2, respectively. The SSCH pattern design is discussed in Section 6.4.2. Furthermore, the operations of DSS despreading and SSCH demapping invoked at the receiver are elaborated on in Section 6.4.3. The last part of this section focuses on MUD design in the context of the SSCH/SDMA-OFDM system, as detailed in Section 6.4.4.

6.4.1 System Overview

In Figure 6.2 we show the schematic of a TTCM-assisted hybrid DSS/SSCH-aided SDMA-OFDM system. As an example, the MMSE-based MUD [3] of Chapter 4 is used. As seen in the top of Figure 6.2, for each of the L geographically dispersed mobile users, the information bit sequence is first mapped into Quadrature Amplitude Modulation (QAM) [24] symbols with the aid of the TTCM encoder. The encoded QAM symbols $\mathbf{s}^{(l)} = [s_1^{(l)}, s_2^{(l)}, \ldots, s_K^{(l)}]^T$ $(l = 1, \ldots, L)$, where $K < Q_c$ is the number of subcarriers activated by each user, are then forwarded to the SSCH mapper under the control of a specific SSCH pattern. The mapped symbols are then forwarded to the OFDM modulator, where they are converted to OFDM symbols based on the Q_c-point Inverse Fast Fourier Transform (IFFT) algorithm. The OFDM-modulated signal is then delivered to the DSS spreader for time-domain-based spreading. Afterwards, this subcarrier-hopped and spread signal is transmitted to the BS by the single-antenna-aided MS over the SDMA MIMO channel described by Figure 4.1. At the BS shown in the bottom part of Figure 6.2, P receiver antenna elements are employed, where the despreading

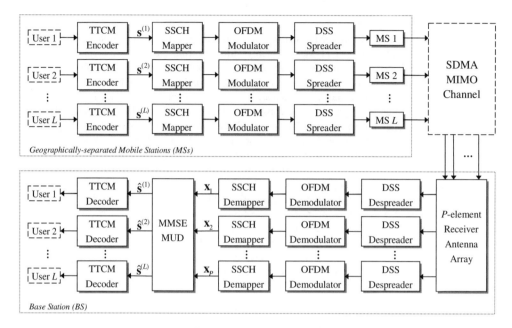

Figure 6.2: Schematic of a TTCM-assisted multi-user DSS/SSCH-aided SDMA-OFDM uplink system using the classic MMSE MUD. ©IET Jiang & Hanzo 2006 [469]

process is first invoked, followed by the OFDM demodulation using the Fast Fourier Transform (FFT) algorithm. Given knowledge of the user-specific SSCH patterns, the despread signal is dehopped by the SSCH demapper. The resultant outputs $\mathbf{x}_p = [x_{p,1}, x_{p,2}, \ldots, x_{p,K}]^T$ $(p = 1, \ldots, P)$ are then forwarded to the MMSE-based MUD for separating the different users' signals. The MUD signals $\hat{\mathbf{s}}^{(l)} = [\hat{s}_1^{(l)}, \hat{s}_2^{(l)}, \ldots, \hat{s}_K^{(l)}]^T$ $(l = 1, \ldots, L)$, namely the estimated versions of the transmitted signals, are then independently channel decoded by the TTCM decoders of Figure 6.2.

In Sections 6.4.1.1 and 6.4.1.2, we will provide further insights into the transmitter and receiver structures of the hybrid DSS/SSCH MMSE-SDMA-OFDM system, respectively. Note that for the sake of simplicity, the procedures of adding and removing the cyclic OFDM prefix [3,24] are omitted in both sections.

6.4.1.1 Transmitter Structure

The transmitter structure of the TTCM-assisted DSS/SSCH SDMA-OFDM system is portrayed in Figure 6.3. As seen in Figure 6.3, the TTCM-coded symbols are first S/P converted and forwarded to the SSCH mapper. More specifically, the K information symbols $s_{g_l,k}^{(l)}$ $(l = 1, \ldots, L; \ k = 1, \ldots, K)$ of a user are mapped to K out of Q_c SSCH subcarriers, where the activation strategy of the set of K subcarriers depends on the specific SSCH pattern used. For example, if pseudo-random hopping is employed, each user can independently select K subcarriers according to the action of a pseudo-random subcarrier selector, as illustrated in Figure 6.3. Since each of the L users activates $K < Q_c$ subcarriers, there will be $Q_c - K$ deactivated subcarriers for each user. Therefore, an ON–OFF-type signalling scheme may be invoked [478], where the activated and deactivated statuses of specific subcarriers represent the ON and OFF states, respectively. Then the total number of Q_c subcarriers can be processed by a Q_c-point IFFT. Since the deactivated subcarriers deliver no information, the transmit power of each SSCH user is the same as that of an SFH user, if the same K activated subcarriers are employed by both of them. In order to exploit coherent demodulation at the receiver, the SSCH dwell period T_h should

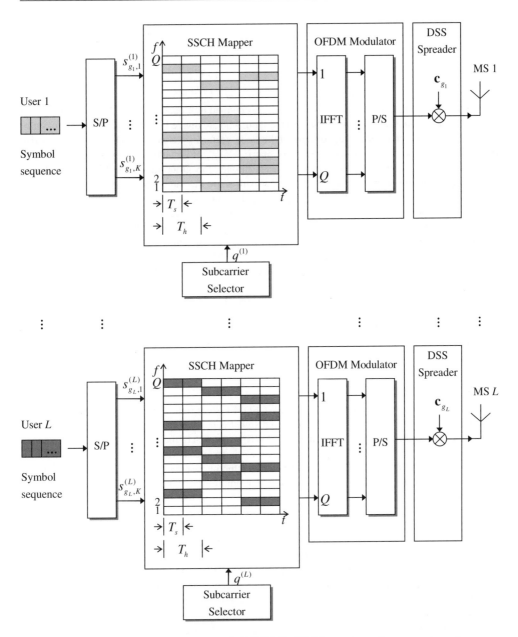

Figure 6.3: Illustration of the L-users' DSS/SSCH SDMA-OFDM transmitters. As an example, random SSCH patterns are used.

be longer than the symbol duration T_s. In the example of Figure 6.3, each user's SSCH pattern remains constant during the period of T_h for two consecutive symbol periods.

Following the IFFT-based OFDM modulation, the user's signal is forwarded to the DSS spreader seen in Figure 6.3, where the time-domain spreading operation is invoked with the aid of orthogonal spreading sequences for achieving time diversity. In this thesis Walsh–Hadamard Transform Spreading

(WHTS) codes are considered. The G-order Walsh–Hadamard Transform (WHT) matrix $\mathbf{U}_{\mathrm{WHT}_G}$ is given in a recursive form as stated in Equation (4.10), which is repeated here for the convenience of the reader:

$$\mathbf{U}_{\mathrm{WHT}_G} = \frac{1}{\sqrt{2}} \begin{pmatrix} 1 \cdot \mathbf{U}_{\mathrm{WHT}_{G/2}} & 1 \cdot \mathbf{U}_{\mathrm{WHT}_{G/2}} \\ 1 \cdot \mathbf{U}_{\mathrm{WHT}_{G/2}} & -1 \cdot \mathbf{U}_{\mathrm{WHT}_{G/2}} \end{pmatrix}, \tag{6.1}$$

while the lowest-order WHT unitary matrix $\mathbf{U}_{\mathrm{WHT}_2}$ is defined by Equation (4.11). More specifically, a total of L users is supported by the DSS/SSCH system, who are divided into the number G of DSS user groups, each of which is assigned a different DSS code vector $\mathbf{c}_{g_l}$ based on WHTS, which is carried out according to one of the G rows in the WHT matrix $\mathbf{U}_{\mathrm{WHT}_G}$ of Equation (6.1), as shown in the DSS spreader block of Figure 6.3. The number of users in one DSS group may be different from that in another; however, the users within the same group will share the same DSS code. From the point of view of the users, the lth user's DSS code may or may not be the same as the other users', depending on whether they belong to the same DSS group or not. Explicitly, the users employing the same DSS code cannot be differentiated in the time domain. However, they are separable in the context of the SDMA architecture with the aid of their unique user-specific spatial signatures, i.e. by their CIRs [3]. More precisely, this is achieved by employment of the SDMA MUD, e.g. the MMSE MUD of Section 4.2.2.1. Following the DSS process, the spread OFDM signal will be transmitted by the MS over the channel using a specific carrier frequency f_c.

Note that the quasi-synchronous uplink operation of MSs may be established with the aid of sufficiently accurate timing advance control by advancing the mobiles' transmission instants according to their estimated time delays [504]. Furthermore, the single-user OFDM synchronization techniques of [3] may be further developed for multi-user systems. Alternatively, the techniques of [505–507] may also be applicable to the SSCH/SDMA-OFDM system.

6.4.1.2   Receiver Structure

The receiver structure of the DSS/SSCH SDMA-OFDM system follows the inverse of the transmitter structure, as illustrated in Figure 6.4. The received signal is the superposition of all users' transmitted signals plus the Additive White Gaussian Noise (AWGN). At each of the P SDMA BS receiver antenna elements, first the received signal r_p $(p = 1, \ldots, P)$ is despread with the aid of the different DSS codes, resulting in the corresponding number of G different despread signal groups, each of which is constituted by the received signals of the users that employ the same specific DSS code. Note that each subcarrier may be activated[4] by different users employing either the same or different DSS codes, depending on the SSCH pattern assignment strategy used. Hence, with knowledge of the SSCH patterns, an active subcarrier selector can be employed at the BS for controlling the despreading operation, so that the appropriate DSS codes are used at the different subcarriers. For example, Figure 6.4 shows the processing invoked for the gth $(g \in \{1, \ldots, G\})$ DSS group, where the gth DSS code is used for separating this DSS group's signal from the others'. Then, FFT-based OFDM demodulation is invoked for each of the different DSS groups at each of the P receiver antenna elements, followed by SSCH demapping. The BS's active subcarrier selector simultaneously selects the active subcarriers at the P receivers, and the resultant composite multi-user output signal is forwarded to the MUD for detection. As an example, the MMSE MUD is pictured in Figure 6.4, but other MUDs such as for example the MMSE-IGA MUD of Chapter 5 are equally applicable. The separated different users' signals are then P/S converted and forwarded to the user-specific TTCM channel decoders.

6.4.2   Subcarrier-Hopping Strategy Design

In the SSCH/SDMA-OFDM system, the appropriate choice of the SSCH strategy is a crucial factor of system design. The SSCH pattern decides upon the choice of the subcarriers to which the different users' signals are mapped and thus has a direct impact on the amount of MUI inflicted. In order to

---

[4] A subcarrier is referred to as being *activated* if it is assigned to at least one user for transmission.

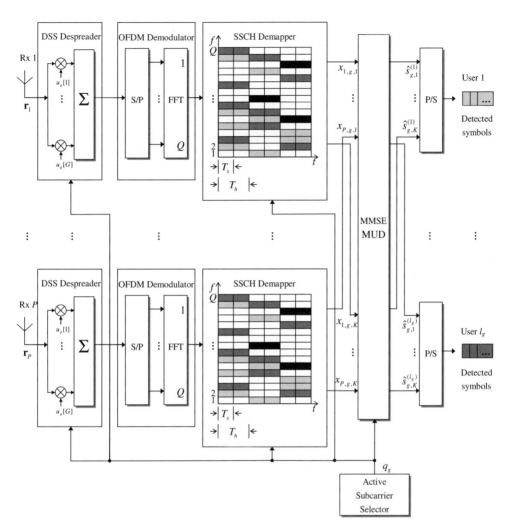

Figure 6.4: Illustration of the L-users' DSS/SSCH SDMA-OFDM receivers. As an example, the MMSE MUD and random SSCH patterns are used. In this figure, the processing invoked in the context of the gth ($g \in \{1, \ldots, G\}$) DSS group containing l_g users is portrayed.

eliminate the MUI, the best solution is to avoid subcarrier collisions between the different users by assigning each subcarrier exclusively to a single user, as proposed for a number of systems based on ClOFDM [485] or OFDMA [499–502]. However, we prefer a subcarrier-reuse system, where the system's overall throughput can be further increased. Furthermore, to design an appropriate SSCH pattern the following two aspects should be taken into account. On the one hand, for the sake of combating the MUI, a meritorious SSCH pattern should ensure that each user's high-MUI subcarriers are dispersed across the FD, rather than being concentrated in the FD. On the other hand, in order to mitigate FD fading, the subcarriers activated by the same user should be uniformly distributed across the entire bandwidth, rather than being consecutively mapped to a small fraction of the entire bandwidth, so that frequency diversity can be efficiently exploited and hence the detrimental effects of deep FD fades can be mitigated. Both of these requirements result in a more random distribution of the originally bursty FD errors incurred by either a high MUI or a deep FD fade, thus enhancing the chances of the channel decoder to correct the residual errors.

Based on the above motivation, two types of SSCH pattern will be considered in this section.

6.4.2.1 Random SSCH

We refer to the first scheme as the Random SSCH (RSSCH) pattern, where each user independently selects *any* K out of the total Q_c available subcarriers during each SSCH dwell time period of T_h. Employing a random hopping strategy is a convenient solution for the simultaneous mobile users, since each user may activate any of the available subcarriers without restrictions imposed by the others. However, the RSSCH strategy also has some drawbacks. Specifically, in RSSCH-based systems some subcarriers may be activated by a smaller number of users, hence the MUI level is low at these subcarriers, which in turn benefits all the active users that use them. Unfortunately, this benefit is achieved at the cost of increasing the MUI imposed on the other subcarriers, which are assigned to an excessive number of users. Therefore, those disadvantaged users will suffer from a high MUI. Furthermore, the benefits gained from the high-quality subcarriers may not be expected to compensate for the detrimental effects arising from the severely corrupted ones, and thus result in an increased average Bit Error Ratio (BER).

6.4.2.2 Uniform SSCH

For the sake of mitigating the above-mentioned deficiencies of the RSSCH scheme, we will introduce another SSCH strategy, referred to as the Uniform SSCH (USSCH) scheme, which is capable of effectively counteracting the above problem. In the proposed USSCH-based systems all users' hopping patterns are jointly designed so that the number of interfering users at each subcarrier – which also depends on the total number of users L supported by the SSCH system – is as similar as possible, and thus the users can be uniformly distributed to the Q_c subcarriers, which in turn satisfies both the requirements mentioned above.

In Section 6.4.2.2.1, we will provide a detailed discussion on the design of the proposed USSCH pattern.

6.4.2.2.1 Design of the USSCH Pattern

During each SSCH dwell interval of T_h, the algorithm used to create a set of USSCH patterns for all the users is as follows. Firstly, we recall that in SSCH systems the total bandwidth W is shared by Q_c subcarriers, each of which has a subcarrier bandwidth of $W_{sc} = W/Q_c$. Then we partition the Q_c subcarriers into K subcarrier groups, where each group has $Q_g = Q_c/K$ number of subcarriers. For example, if we have a total of $Q_c = 512$ available subcarriers, where each user activates $K = 128$ subcarriers for transmission, this will result in $K = 128$ subcarrier groups, each of which has $Q_g = 512/128 = 4$ subcarriers, as illustrated in Figure 6.5. Then an L-iteration USSCH pattern assignment algorithm is invoked for each of the K groups, i.e. from the first to the Kth group, to generate a specific length-K pattern set for each of the L users. Note that this principle is different from that of the RSSCH scheme, where each user is randomly assigned K subcarriers. By contrast, in the USSCH arrangement each of the K subcarriers assigned to a specific user is chosen from a different subcarrier group. In other words, each user's specific set of K activated subcarriers is uniformly distributed across the entire system bandwidth.

Figure 6.6 illustrates the proposed algorithm invoked in the kth ($k = 1, \ldots, K$) subcarrier group for assigning the Q_g subcarriers to the L users. More precisely, initially a *remaining user set* $\mathbf{A}^{(1)}$ containing all the L users is created, as shown in Figure 6.6. During the lth ($l = 1, \ldots, L$) iteration, a subcarrier index $q_k^{(l)}$ is generated by calculating $l \bmod Q_g$, where 'mod' represents the modulo

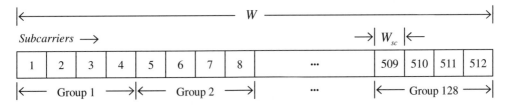

Figure 6.5: An example showing the formation of the subcarrier groups, where the system bandwidth W is shared by a total of $Q_c = 512$ subcarriers, while each user activates $K = 128$ subcarriers for transmission, and each of the 128 subcarrier groups has $Q_g = 4$ subcarriers.

operation. A user $u^{(l)}$ is then randomly selected from $\mathbf{A}^{(l)}$ based on the uniform probability of

$$p^{(l)} = \frac{1}{L - l + 1}, \tag{6.2}$$

and the $(q_k^{(l)})$th subcarrier within this specific subcarrier group will be assigned to user $u^{(l)}$. The remaining user set $\mathbf{A}^{(l)}$ is then updated by removing user $u^{(l)}$ from it, resulting in $\mathbf{A}^{(l+1)}$, which contains the remaining $(L - l)$ users. Then the subcarrier assignment process proceeds to the next, i.e. to the $(l + 1)$th iteration, allocating the next subcarrier to the next randomly selected user, as seen in Figure 6.6. This iterative subcarrier-to-user assignment process continues until the Lth iteration is completed. By this time, each of the L users has been assigned a single subcarrier of the kth subcarrier group. Hence, a vector $\mathbf{q}_k = [q_k^{(1)}, q_k^{(2)}, \ldots, q_k^{(L)}]$ is generated, where $q_k^{(l)}$ $(l = 1, \ldots, L)$ constitutes the kth subcarrier of the lth user's length-K USSCH pattern set.[5] Then the vector-generating routine described above is invoked for the next, i.e. for the $(k+1)$th subcarrier group. When all the K subcarrier groups have been processed, a set containing K length-L vectors $\mathbf{q}_k$ $(k = 1, \ldots, K)$ has been generated, explicitly indicating which specific subcarrier of group k has been assigned to which of the L users. Viewing this subcarrier allocation from the users' perspective, the lth user's specific length-K USSCH pattern set $\mathbf{q}^{(l)}$ is created by choosing that specific subcarrier corresponding to the lth element of each of the K vectors, which has been assigned to user l, resulting in $\mathbf{q}^{(l)} = [q_1^{(l)}, q_2^{(l)}, \ldots, q_K^{(l)}], l \in \{1, \ldots, L\}$.

Note that the above subcarrier-group-based algorithm employs the full permutation theory [508]. More explicitly, each of the K length-L vectors $\mathbf{q}_k$ $(k = 1, \ldots, K)$ associated with the kth subcarrier group is an element of the L-order full permutation set [508], and is generated at a constant probability of

$$
\begin{aligned}
p_L &= \prod_{l=1}^{L} p^{(l)} \\
&= \frac{1}{L} \cdot \frac{1}{L-1} \cdot \cdots \cdot \frac{1}{2} \cdot 1 \\
&= \frac{1}{L!},
\end{aligned} \tag{6.3}
$$

where $(\cdot)!$ represents the factorial operation. More explicitly, given a total number of L users, the associated L-order full permutation set can be expressed as

$$\mathbf{Q}_L = \{\mathbf{q}_1, \mathbf{q}_2, \ldots, \mathbf{q}_{L!}\}, \tag{6.4}$$

where the element vectors are given by $\mathbf{q}_k = [q_k^{(1)}, q_k^{(2)}, \ldots, q_k^{(L)}]$ $(k = 1, \ldots, L!)$. Thus, the employment of the USSCH pattern assignment algorithm of Figure 6.6 is conceptually equivalent to

---

[5]Recall that each user activates K subcarriers for transmission and thus has to be assigned a length-K USSCH pattern set.

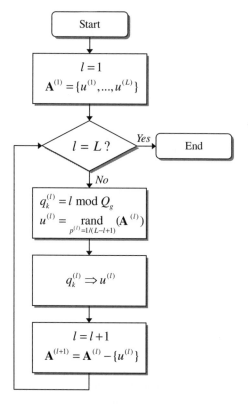

Figure 6.6: Flowchart of the USSCH pattern assignment algorithm invoked in the kth subcarrier group. ©IET Jiang & Hanzo 2006 [469]

invoking the algorithm, which randomly selects one element vector from the set $\mathbf{Q}_L$ described by Equation (6.4) for each of the K subcarrier groups, based on the probability given by Equation (6.3).

In order to offer some insight into the relation between the USSCH pattern assignment algorithm and the full permutation theory, an example is provided in Figure 6.7, which illustrates the USSCH pattern assignment operation invoked for the kth subcarrier group of the example given by Figure 6.5, where $K = 128$ and $Q_g = 4$. Without loss of generality, we assume in this example that $L = 6$ users are supported by the system. According to the algorithm of Figure 6.6, during the lth ($l = 1, \ldots, 6$) iteration a user $u^{(l)}$ is randomly selected from the remaining user set $\mathbf{A}^{(l)}$ based on the selection probability $p^{(l)}$ calculated by Equation (6.2) and is assigned the subcarrier with the index of $q_k^{(l)}$, as observed in Figure 6.7. Note that since $L = 6 > Q_g = 4$, each of the last two users has to activate a subcarrier that has already been assigned to a previous user. The subcarriers activated by more than one user are printed in grey in Figure 6.7. After $L = 6$ iterations, a vector $\mathbf{q}_k$ having six elements is generated. Furthermore, the probability of generating $\mathbf{q}_k$ is the product of the user-selection probability $p^{(l)}$, expressed as

$$
\begin{aligned}
p &= p^{(1)} \cdot p^{(2)} \cdot \ldots \cdot p^{(5)} \cdot p^{(6)} \\
&= \frac{1}{6} \cdot \frac{1}{5} \cdot \ldots \cdot \frac{1}{2} \cdot 1 \\
&= \frac{1}{6!},
\end{aligned}
\tag{6.5}
$$

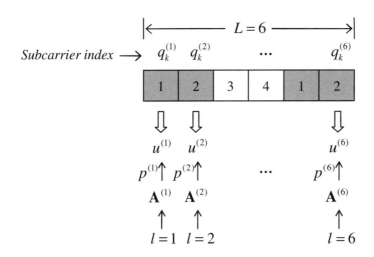

Figure 6.7: An example showing the USSCH pattern assignment operation invoked for the first subcarrier group of Figure 6.5, where $L = 6$ and $Q_g = 4$. For the notations used in this figure we refer to Figure 6.6.

which yields the same result as that calculated by Equation (6.3) in conjunction with $L = 6$. In other words, the generation of the vector $\mathbf{q}_k$ has the same effect as selecting it from the six-order full permutation set $\mathbf{Q}_6$ based on the probability given by Equation (6.3), as pointed out previously.

6.4.2.2.2 Discussions

As informed by Figure 6.7, when we have $L > Q_g$ and $L \bmod Q_g \neq 0$, in each of the K subcarrier groups, there will be $Q_g - (L \bmod Q_g)$ subcarriers that have to be activated by *one* more user than the other subcarriers in the same group. Thus, a higher MUI is expected at these subcarriers. However, since the users that are assigned to the higher-MUI subcarriers are randomly selected, it is unlikely for these more MUI contaminated subcarriers of different subcarrier groups to be always assigned to the same user. In other words, from each user's point of view, among the total K activated subcarriers, the specific subcarriers that encounter a higher MUI can be uniformly dispersed across the entire system bandwidth. Furthermore, since in USSCH/SDMA-OFDM systems each user's activated subcarriers are uniformly distributed over the entire bandwidth, frequency diversity can be efficiently exploited. Thus, both requirements mentioned at the beginning of Section 6.4.2 are satisfied by employing the proposed USSCH pattern assignment algorithm.

On the other hand, if we have $L \bmod Q_g = 0$, each subcarrier is activated by the same number of L/Q_g users. Moreover, when $L < Q_g$, the USSCH system is actually MUI-free, since each subcarrier is assigned to at most one user. In both of these two scenarios, the MUI encountered in the USSCH/SDMA-OFDM system is identical at all the activated subcarriers, since each subcarrier hosts the same number of users. This situation is similar to that encountered in conventional SDMA-OFDM systems [3, 193] in terms of the amount of MUI inflicted at each subcarrier. However, the USSCH/SDMA-OFDM system benefits from the USSCH pattern assignment algorithm of Figure 6.6, which has the ability to exploit efficiently the achievable frequency diversity offered by the system, hence becoming capable of outperforming conventional SDMA-OFDM systems. Furthermore, the advantage of the USSCH pattern design will become even more significant in overloaded scenarios, where an excessively high MUI is expected, as we will demonstrate in Section 6.5.

6.4.2.3 Random and Uniform SFH

Note that the principles of the RSSCH and USSCH strategies can also be applied to SFH/SDMA-OFDM, resulting in Random SFH (RSFH) and Uniform SFH (USFH) aided SDMA-OFDM systems, respectively. More specifically, in USFH/SDMA-OFDM systems subbands of subcarriers rather than individual subcarriers are jointly assigned to the different users, following a philosophy similar to that of Figure 6.6.

It is worth pointing out that the USFH/SDMA-OFDM scheme is capable of outperforming the conventional SDMA-OFDM arrangement, as a benefit of the proposed uniform pattern assignment algorithm. However, given the same system bandwidth and the same total system throughput, the USFH/SDMA-OFDM system is unable to outperform its USSCH-aided counterpart. This is because in USFH-aided systems a subband – which accommodates a number of subcarriers and may be viewed as an 'entire' bandwidth for a 'reduced-size' conventional SDMA-OFDM system – is still more vulnerable both to high MUI and to deep fades than a subcarrier of USSCH-aided systems. The above arguments will be corroborated by our simulation results to be presented in Section 6.5.

6.4.2.4 Offline Pattern Pre-computation

It is also worth pointing out that the USFH/USSCH patterns can be acquired by offline pre-computation, since their choice is not based on any channel knowledge. Furthermore, the patterns can be reused, provided that the reuse time interval is sufficiently long, so that the frequency diversity can be sufficiently exploited. The reuse intervalis defined by

$$T_r = \mu T_h, \tag{6.6}$$

where the reuse factor μ is a positive integer. More explicitly, during the offline pattern pre-computation, μ USFH/USSCH patterns are generated for each of the L users, each of which is associated with one of the μ FH dwell time periods during the time interval of T_r. These USFH/USSCH pattern sets can then be reused during every reuse interval T_r, implying that the real-time signalling of the USFH/USSCH patterns from the transmitters to the receivers is unnecessary. This imposes a significantly lower computational complexity than that required by other adaptive algorithms exploiting real-time channel knowledge.

Having discussed the design of SSCH patterns, in the next section we will detail the DSS despreading and SSCH demapping processes invoked at the DSS/SSCH SDMA-OFDM receiver.

6.4.3 DSS Despreading and SSCH Demapping

In the L-user TTCM-assisted DSS/SSCH SDMA-OFDM system employing P receiver antenna elements, the lth ($l = 1, \ldots, L$) user's transmitted signal at the qth ($q = 1, \ldots, Q$) subcarrier can be expressed as

$$s_{g_l,q}'^{(l)}(t) = \sqrt{2P_T}\Lambda_q^{(l)}(t)s_{g_l,q}^{(l)}(t)c_{g_l}(t)\cos(2\pi f_c t + \varphi^{(l)}),$$
$$l \in \{1, \ldots, L\}, \quad g_l \in \{1, \ldots, G\}, \quad q \in \{1, \ldots, Q\}, \tag{6.7}$$

where P_T is the transmitted power, $s_{g_l,q}^{(l)}(t)$ is the user's information signal, f_c is the carrier frequency and $\varphi^{(l)}$ is the phase angle introduced by carrier modulation, while $\Lambda_q^{(l)}(t)$ is the subcarrier activation function defined as

$$\Lambda_q^{(l)}(t) = \begin{cases} 1, & \text{the } q\text{th subcarrier activated by the } l\text{th user,} \\ 0, & \text{otherwise,} \end{cases} \quad 0 \leq t < T_s. \tag{6.8}$$

Furthermore, $c_{g_l}(t)$ of Equation (6.7) denotes the DSS signature sequence assigned to the lth user and associated with the gth DSS group, given by

$$c_{g_l}(t) = \sum_{c=1}^{G} u_{g_l}[c] \Gamma_{T_c}(t - cT_c), \quad 0 \le t < T_s, \quad g_l \in \{1, \ldots, G\}, \tag{6.9}$$

where T_c is the chip duration, T_s is the OFDM symbol duration and $\Gamma_\tau(t)$ is a rectangular pulse[6] defined as

$$\Gamma_\tau(t) = \begin{cases} 1, & 0 \le t < \tau \\ 0, & \text{otherwise} \end{cases}, \tag{6.10}$$

while $u_{g_l}[c]$ of Equation (6.9) is the cth element of the gth row in the $(G \times G)$-dimensional WHT matrix defined by Equation (6.1). Therefore, the received signal at the pth receiver antenna element is

$$r_p(t) = \sum_{q=1}^{Q} \sum_{l=1}^{L} H_{p,q}^{(l)} s_{g_l,q}^{\prime(l)}(t) + n_p(t), \quad p \in \{1, \ldots, P\}, \tag{6.11}$$

where $H_{p,q}^{(l)}$ represents the FD-CHTFs associated with the specific link between the lth user and the pth receiver antenna element in the context of the qth subcarrier, which is assumed to remain constant during the period of T_s, while $n_p(t)$ denotes the AWGN at the pth receiver. To separate the signals of the G different DSS groups, the despreading operation is then invoked with the aid of the DSS codes at each of the P receiver antennas, yielding

$$x_{p,g}(t) = \int_0^{T_s} c_g(t) r_p(t) \, dt, \quad g \in \{1, \ldots, G\}. \tag{6.12}$$

Note that the above despreading operation can be restricted to the activated subcarriers only, if RSSCH patterns are used, or if USSCH patterns are employed under the condition of $L < Q_g$. In either scenario there exists inactivated subcarriers which do not carry any information signal and thus do not need to be processed. If USSCH patterns are used and we have $L \ge Q_g$, all subcarriers are activated and thus despreading is required across the entire bandwidth. As a generalized case, let us assume that the qth subcarrier is activated by L_q users, where $1 \le L_q \le L$. Furthermore, at this specific subcarrier, a total of G_q $(1 \le G_q \le G)$ different DSS codes are used by the L_q users, where each user's code may or may not be the same as the others'. Without loss of generality, we assume that the gth $(g = 1, \ldots, G_q)$ of the G_q different DSS groups contains l_g users, where we have

$$L_q = \sum_{g=1}^{G_q} l_g. \tag{6.13}$$

Thus, the discrete signal received at the qth subcarrier of the pth receiver during an OFDM symbol duration can be represented as

$$r_{p,q} = \bar{\mathbf{c}}_{G_q} \bar{\mathbf{H}}_{p,q} \bar{\mathbf{s}}_q + n_{p,q}, \quad p \in \{1, \ldots, P\}, \tag{6.14}$$

where the $(1 \times L_q)$-dimensional DSS code vector $\bar{\mathbf{c}}_{G_q}$ and the L_q users' $(L_q \times 1)$-dimensional information signal vector $\bar{\mathbf{s}}_q$ are given by

$$\bar{\mathbf{c}}_{G_q} = [\underbrace{\mathbf{c}_1, \mathbf{c}_1, \ldots, \mathbf{c}_1}_{l_1}, \underbrace{\mathbf{c}_2, \mathbf{c}_2, \ldots, \mathbf{c}_2}_{l_2}, \ldots, \underbrace{\mathbf{c}_{G_q}, \mathbf{c}_{G_q}, \ldots, \mathbf{c}_{G_q}}_{l_{G_q}}], \tag{6.15}$$

$$\bar{\mathbf{s}}_q = [s_{1,q}^{(1)}, s_{1,q}^{(2)}, \ldots, s_{1,q}^{(l_1)}, s_{2,q}^{(1)}, s_{2,q}^{(2)}, \ldots, s_{2,q}^{(l_2)}, s_{G_q,q}^{(1)}, s_{G_q,q}^{(2)}, \ldots, s_{G_q,q}^{(l_{G_q})}]^T, \tag{6.16}$$

---

[6]In practical systems, the chip pulse shape of $\Gamma_\tau(t)$ is a band-limited waveform, such as a raised-cosine Nyquist pulse. However, for the sake of simplicity in our analysis and simulation, we will assume here that $\Gamma_\tau(t)$ is an ideal rectangular pulse.

where $(\cdot)^T$ denotes the transposition operation. The corresponding spreading code sequence $\mathbf{c}_g$ ($g \in \{1, \ldots, G_q\}$) specified in Equation (6.15), which is associated with the gth DSS group in the context of the qth subcarrier, is defined by

$$\mathbf{c}_g = [u_g[1], u_g[2], \ldots, u_g[G]], \quad g = 1, \ldots, G_q. \tag{6.17}$$

Furthermore, the L_q users' $(L_q \times L_q)$-dimensional diagonal FD-CHTF matrix associated with the qth subcarrier at the pth receiver antenna, namely $\bar{\mathbf{H}}_{p,q}$ of Equation (6.14), is expressed as

$$\bar{\mathbf{H}}_{p,q} = \mathrm{diag}[H_{p,1,q}^{(1)}, H_{p,1,q}^{(2)}, \ldots, H_{p,1,q}^{(l_1)},$$
$$H_{p,2,q}^{(1)}, H_{p,2,q}^{(2)}, \ldots, H_{p,2,q}^{(l_2)}, H_{p,G_q,q}^{(1)}, H_{p,G_q,q}^{(2)}, \ldots, H_{p,G_q,q}^{(l_{G_q})}]. \tag{6.18}$$

Therefore, the corresponding despread signal associated with the qth subcarrier of the pth receiver antenna can be given by

$$\bar{\mathbf{x}}_{p,q} = \check{\mathbf{c}}_{G_q} r_{p,q}$$
$$= \bar{\mathbf{R}}_{G_q} \bar{\mathbf{H}}_{p,q} \bar{\mathbf{s}}_q + \bar{\mathbf{n}}_{p,q}, \quad p \in \{1, \ldots, P\}, \tag{6.19}$$

where the $(G_q \times 1)$-dimensional despread received signal vector $\bar{\mathbf{x}}_{p,q}$ and the $(G_q \times 1)$-dimensional effective noise vector $\bar{\mathbf{n}}_{p,q}$ are expressed as

$$\bar{\mathbf{x}}_{p,q} = [x_{p,1,q}, x_{p,2,q}, \ldots, x_{p,G_q,q}]^T, \tag{6.20}$$
$$\bar{\mathbf{n}}_{p,q} = [n_{p,1,q}, n_{p,2,q}, \ldots, n_{p,G_q,q}]^T, \tag{6.21}$$

while the $(G_q \times 1)$-dimensional DSS code vector $\check{\mathbf{c}}_{G_q}$ is given by

$$\check{\mathbf{c}}_{G_q} = [\mathbf{c}_1, \mathbf{c}_2, \ldots, \mathbf{c}_{G_q}]^T. \tag{6.22}$$

Moreover, $\bar{\mathbf{R}}_{G_q}$ in Equation (6.19) is the $(G_q \times L_q)$-dimensional cross-correlation matrix of the L_q users' DSS code sequences, represented as

$$\bar{\mathbf{R}}_{G_q} = \begin{bmatrix} \underbrace{\begin{matrix} \omega_{11} & \omega_{11} & \cdots & \omega_{11} \\ \omega_{21} & \omega_{21} & \cdots & \omega_{21} \\ & \vdots & \\ \omega_{G_q 1} & \omega_{G_q 1} & \cdots & \omega_{G_q 1} \end{matrix}}_{l_1} & \underbrace{\begin{matrix} \omega_{12} & \omega_{12} & \cdots & \omega_{12} \\ \omega_{22} & \omega_{22} & \cdots & \omega_{22} \\ & \vdots & \\ \omega_{G_q 2} & \omega_{G_q 2} & \cdots & \omega_{G_q 2} \end{matrix}}_{l_2} & \begin{matrix} \cdots \\ \cdots \\ \vdots \\ \cdots \end{matrix} & \underbrace{\begin{matrix} \omega_{1G_q} & \omega_{1G_q} & \cdots & \omega_{1G_q} \\ \omega_{2G_q} & \omega_{2G_q} & \cdots & \omega_{2G_q} \\ & \vdots & \\ \omega_{G_q G_q} & \omega_{G_q G_q} & \cdots & \omega_{G_q G_q} \end{matrix}}_{l_{G_q}} \end{bmatrix}, \tag{6.23}$$

where ω_{ij} is the cross-correlation coefficient of the ith DSS group's and the jth DSS group's signature sequence, defined as

$$\omega_{ij} = \int_0^{T_s} c_i(t) c_j(t) \, dt, \quad i, j \in \{1, \ldots, G_q\}. \tag{6.24}$$

Therefore, for each of the G_q DSS groups at the qth subcarrier, the despread signals can be combined from all the P receiver antenna elements and then be forwarded to the SDMA MUD, where the signals of different users in the same DSS group are separated in the spatial domain.

6.4.4 MUD

In Chapters 4 and 5, we investigated the MMSE- and the GA-based MUDs, respectively. The MMSE MUD [3] of Chapter 4 exhibits a rather low complexity, while suffering from a performance loss. By contrast, the GA-based MUD of Chapter 5 is capable of achieving a similar performance to that attained by the optimum ML MUD at a significantly lower complexity, especially at high user loads.

Furthermore, the MMSE-assisted Iterative GA (IGA) MUD of Chapter 5 employing Biased Q-function-based Mutation (BQM) was found to outperform the conventional GA-based MUDs [38] at a similar computational complexity, especially in high-throughput scenarios.

The concatenated MMSE-IGA MUD can also be employed in the hybrid DSS/SSCH SDMA-OFDM system to improve the attainable performance. This is achieved by further processing the output provided by the MMSE MUD as the GA's initial detection knowledge at the active subcarriers, and then invoking the GA-assisted iterative detection technique. More specifically, at the first step of the iterative detection procedure, an initial estimate of the different users' transmitted signals is generated with the aid of the linear MMSE MUD. This operation is similar to that invoked in the GA-aided conventional SDMA-OFDM systems described in Chapter 5. However, in the context of the SDMA-OFDM system employing DSS and SSCH techniques, typically only the signals that arrive at the active subcarriers rather than spread across the entire system bandwidth will be processed.

More specifically, at the qth subcarrier of each of the G_q DSS groups, an initial estimated signal vector $\hat{s}_{\text{MMSE}g,q} \in \mathbb{C}^{(l_g \times 1)}$ $(g = 1, \ldots, G_q)$ associated with the l_g users of this group is generated by linearly combining the $(P \times 1)$-dimensional despread signal vector $x_{g,q}$, which is constituted by the specific despread symbols generated from the gth DSS group at all the P receiver antennas with the aid of the MMSE MUD's array weight matrix, as follows:

$$\hat{s}_{\text{MMSE}g,q} = \mathbf{W}^H_{\text{MMSE}g,q} x_{g,q}, \tag{6.25}$$

where

$$\hat{s}_{\text{MMSE}g,q} = [\hat{s}^{(1)}_{\text{MMSE}g,q}, \hat{s}^{(2)}_{\text{MMSE}g,q}, \ldots, \hat{s}^{(l_g)}_{\text{MMSE}g,q}]^T, \quad g = 1, \ldots, G_q, \tag{6.26}$$

$$x_{g,q} = [x_{1,g,q}, x_{2,g,q}, \ldots, x_{P,g,q}]^T, \quad g = 1, \ldots, G_q. \tag{6.27}$$

The MMSE-based weight matrix $\mathbf{W}_{\text{MMSE}g,q} \in \mathbb{C}^{(P \times l_g)}$ of Equation (6.25) is given by [3]

$$\mathbf{W}_{\text{MMSE}g,q} = (\mathbf{H}_{g,q}\mathbf{H}^H_{g,q} + \sigma_n^2 \mathbf{I})^{-1}\mathbf{H}_{g,q}, \tag{6.28}$$

where $\mathbf{I}$ is the identity matrix and σ_n^2 is the AWGN's variance, while the $(P \times l_g)$-dimensional FD-CHTF matrix associated with the gth DSS group is expressed as

$$\mathbf{H}_{g,q} = (\mathbf{H}^{(1)}_{g,q}, \mathbf{H}^{(2)}_{g,q}, \ldots, \mathbf{H}^{(l_g)}_{g,q}), \quad g = 1, \ldots, G_q, \tag{6.29}$$

where $\mathbf{H}^{(l)}_{g,q}$ $(l = 1, \ldots, l_g)$ is the vector of the FD-CHTFs associated with the transmission paths between the lth user's transmitter antenna and each element of the P-element receiver antenna array, which is expressed as

$$\mathbf{H}^{(l)}_{g,q} = (H^{(l)}_{1,g,q}, H^{(l)}_{2,g,q}, \ldots, H^{(l)}_{P,g,q})^T, \quad l = 1, \ldots, l_g. \tag{6.30}$$

Once the MMSE-based MUD is completed, the resultant output can be forwarded to the concatenated IGA MUD of Chapter 5 for the second-stage iterative detection in the context of the gth DSS group at the qth subcarrier.

More precisely, the IGA MUD evaluates a decision metric associated with the P receivers, which is derived from the optimum ML-based decision metric [3], in order to detect the symbol vector $\hat{s}_{\text{GA}g,q}$ that consists of the estimated transmitted signals of the l_g users in the gth DSS group, which is expressed as

$$\hat{s}_{\text{GA}g,q} = (\hat{s}^{(1)}_{\text{GA}g,q}, \hat{s}^{(2)}_{\text{GA}g,q}, \ldots, \hat{s}^{(l_g)}_{\text{GA}g,q})^T, \quad g = 1, \ldots, G_q. \tag{6.31}$$

The decision metric required for evaluation at the pth receiver antenna is defined by

$$\Omega_{p,g,q}(s_g) = |x_{p,g,q} - \mathbf{H}_{p,g,q}s_g|^2, \tag{6.32}$$

where $x_{p,g,q}$ is the despread symbol of the gth DSS group at the qth subcarrier of the pth receiver, while $\mathbf{H}_{p,g,q}$ is the pth row of the matrix $\mathbf{H}_{g,q}$ related to the same subcarrier, which is defined by Equation (6.29). Therefore, the decision rule for the optimum MUD associated with the pth antenna is to choose that specific length-l_g symbol vector $\mathbf{s}_g$ which minimizes the metric given in Equation (6.32). Since there are P receivers, the following combined metric is invoked:

$$\Omega_{g,q}(\mathbf{s}_g) = \sum_{p=1}^{P} \Omega_{p,g,q}(\mathbf{s}_g). \tag{6.33}$$

Hence, the joint decision rule is to find that specific estimated transmitted l_g-symbol vector $\hat{\mathbf{s}}_{\mathrm{GA}g,q}$ associated with the gth DSS group, which minimizes $\Omega_{g,q}(\mathbf{s}_g)$ in Equation (6.33) for the qth active subcarrier, which is formulated as

$$\hat{\mathbf{s}}_{\mathrm{GA}g,q} = \arg\left\{\min_{\mathbf{s}_g}[\Omega_{g,q}(\mathbf{s}_g)]\right\}. \tag{6.34}$$

Hence, after MUD invoked independently for each of the G_q DSS groups at the qth subcarrier, all the estimated signals of the L_q users that activate this subcarrier will be attained. Here we point out that the IGA-based detection process is similar to that discussed in Chapter 5, except that the operations are separately invoked for each of the G_q DSS groups at the qth activated subcarrier.

6.5 Simulation Results

In this section, we characterize the performance of the TTCM-assisted DSS/SSCH SDMA-OFDM system using the MMSE-IGA MUD of Chapter 5. Its performance is compared with that of its counterparts employing the RSSCH, RSFH and USFH patterns, as well as with that of a conventional SDMA-OFDM system [3, 193] dispensing with the employment of any FH techniques. It was assumed in all SFH- or SSCH-aided schemes that the BS has perfect knowledge of the hopping patterns. The simulation results were obtained using a 4-QAM scheme communicating over the three-path Short Wireless Asynchronous Transfer Mode (SWATM) CIR of Figure 3.32, assuming that the channel transfer functions were perfectly known. Each of the paths experiences independent Rayleigh fading having the same normalized Doppler frequencies of $f'_d = 1.235 \times 10^{-5}$. The channel is assumed to be 'OFDM symbol invariant', implying that the taps of the impulse response are assumed to be constant for the duration of one OFDM symbol, but they are faded at the beginning of each OFDM symbol [3].

The GA-related parameters used were fixed in all the simulations, where the MMSE-IGA MUD was employed. Specifically, the BQM scheme of Section 5.3.1.2 was used, while the IGA MUD invoked a single iteration. For the iterative TTCM scheme [314] employed, the associated parameters were the same as those specified in Table 5.5 of Section 5.3.2.2. For the convenience of the reader, the basic simulation parameters are summarized in Table 6.1.

Recall that in SFH- or SSCH-aided SDMA-OFDM systems the number of activated subcarriers $K < Q_c$ is lower than that of the conventional SDMA-OFDM system [3, 193], where all the users employ all the Q_c available subcarriers for communication. For the sake of fair comparisons, the total system bandwidth was fixed and the number of users L supported by the various SFH/SSCH-aided systems was increased, so that the same total system throughput of B_{T} bits per OFDM symbol was maintained, which is calculated by

$$B_{\mathrm{T}} = K \cdot L \cdot BPS \text{ [bits/OFDM symbol]}, \tag{6.35}$$

where BPS represents the number of Bits Per Symbol per subcarrier. When 4-QAM is employed, $BPS = 2$. For the reader's convenience, the notations used in the figures of this section are summarized in Table 6.2.

Furthermore, according to the classic spread-spectrum philosophy [509], in order to maintain a fixed OFDM symbol duration of T_s while employing time-domain DSS, the bandwidth of the OFDM

Table 6.1: Basic simulation parameters used in Section 6.5.

TTCM	Modem	4-QAM
	Code rate	0.5
	Code memory ν	3
	Octal generator polynomial	[13 6]
	Codeword length	1024 symbols
	Channel interleaver depth	1024 symbols
	Number of turbo iterations	2
GA	Population initialization method	MMSE
	Mating pool creation strategy	Pareto-Optimality
	Selection method	Fitness-proportionate
	Cross-over	Uniform cross-over
	Mutation	BQM
	Mutation probability p_m	0.1
	Elitism	Enabled
	Incest prevention	Enabled
	Population size X	20
	Generations Y	5
	Number of IGA iterations	1
Channel	CIRs	SWATM [3]
	Paths	3
	Maximum path delay	48.9 ns
	Normalized Doppler frequency	1.235×10^{-5}
	Total available subcarriers Q_c	128/256/512
	Cyclic prefix	16/32/64

Table 6.2: Notations used in the figures of simulation results presented in Section 6.5.

Notation	Description
K	Number of subcarriers employed by each user
L	Number of users
P	Number of receiver antenna elements
Q_f	Number of available subbands in SFH systems
Q_c	Number of available subcarriers in SSCH systems
B_T	Total throughput per OFDM symbol of the L-user system
RSFH	Random SFH systems
USFH	Uniform SFH systems
RSSCH	Random SSCH systems
USSCH	Uniform SSCH systems

symbols will have to be proportionately expanded according to the DS spreading factor, since the DSS codes' chip duration T_c is shorter than T_s [38]. Therefore, in the DSS-assisted SDMA-OFDM systems investigated in this section, where a fixed total system bandwidth was assumed, the total number of available subcarriers Q_c was decreased. Equivalently, the subcarrier bandwidth W_{sc} was increased, when a longer DSS code was used to support more DSS groups. For example, compared with an SSCH system having $Q_c = 512$ subcarriers and dispensing with the employment of DSS, a hybrid DSS/SSCH system supporting $G = 2$ DSS groups will have $Q_c = 256$ subcarriers, since in this case we have $T_s = 2T_c$.

TTCM-(DSS/SSCH)-SDMA-OFDM, Lx/P4, 4QAM, SWATM

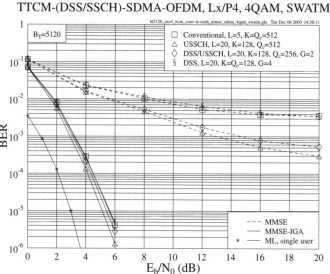

Figure 6.8: BER versus E_b/N_0 performance of the TTCM-assisted **conventional SDMA-OFDM**, **USSCH/SDMA-OFDM**, **DSS/SDMA-OFDM** and **hybrid DSS/USSCH SDMA-OFDM** systems using the **MMSE** or the **MMSE-IGA** MUD, while employing a 4-QAM scheme for transmission over the SWATM channel [3], where L users were supported with the aid of $P = 4$ receiver antenna elements. The associated overall system throughput was $B_T = 5120$ bits. The basic simulation parameters and the notations used in the figure are summarized in Tables 6.1 and 6.2, respectively.

6.5.1   MMSE-Aided Versus MMSE-IGA-Aided DSS/SSCH SDMA-OFDM

In Figure 6.8 we compare the BER performance of the TTCM-assisted USSCH/SDMA-OFDM, DSS/USSCH SDMA-OFDM and conventional SDMA-OFDM systems, where the MMSE and MMSE-IGA MUDs were employed, respectively. The performance of the single-user SDMA-OFDM system employing the optimum ML MUD is also provided for reference. The same total throughput of $B_T = 5120$ bits per OFDM symbol was maintained for all schemes, except for the single-user arrangement, where the throughput was 1024 bits per OFDM symbol, since we assumed that $Q_c = 512$ was the maximum number of subcarriers that can be accommodated in the fixed system bandwidth. Note that in the context of the conventional SDMA-OFDM benchmark system, since $L = 5$ users were supported by $P = 4$ receiver antenna elements, this scenario can be viewed as a moderately overloaded scenario. Explicitly, for the DSS and/or SSCH-aided counterpart systems, the same implication of the overloaded scenario was also applicable, because the same total system throughput was maintained in these systems, which were equivalently overloaded.

From Figure 6.8 we can see that the MMSE-IGA-aided systems significantly outperformed the MMSE-aided systems. It can also be observed that in the scenario where the MMSE MUD was employed, the USSCH/SDMA-OFDM and DSS/USSCH SDMA-OFDM systems achieved a similar performance. The DSS/SDMA-OFDM and the conventional SDMA-OFDM systems also performed similarly to each other. However, both USSCH-aided schemes were found to attain a better performance than those dispensing with USSCH. On the other hand, when using the MMSE-IGA MUD, the performances of the various systems were similar. This suggests that most of the MUI encountered can be effectively suppressed by the MMSE-IGA MUD in the moderately overloaded scenario associated with a throughput of $B_T = 5120$ bits.

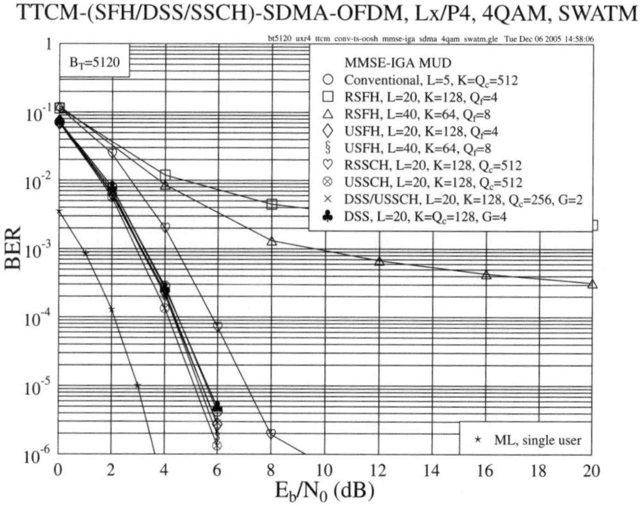

TTCM-(SFH/DSS/SSCH)-SDMA-OFDM, Lx/P4, 4QAM, SWATM

Figure 6.9: BER versus E_b/N_0 performance of the TTCM-assisted **conventional SDMA-OFDM, RSFH/USFH/RSSCH/USSCH SDMA-OFDM, DSS/SDMA-OFDM** and **hybrid DSS/USSCH SDMA-OFDM** systems employing a 4-QAM scheme for transmission over the SWATM channel [3], where L users were supported with the aid of $P = 4$ receiver antenna elements. The associated overall system throughput was $B_T = 5120$ bits. The basic simulation parameters and the notations used in the figure are summarized in Tables 6.1 and 6.2, respectively.

6.5.2 SDMA-OFDM Using SFH and Hybrid DSS/SSCH Techniques

In Section 6.5.1 we have discussed the attractive performance of the MMSE-IGA MUD in the context of the hybrid SDMA-OFDM system. In this section, we will compare the attainable performance of various MMSE-IGA-aided arrangements employing the SFH and the DSS/SSCH schemes.

6.5.2.1 Moderately Overloaded Scenarios

Figure 6.9 compares the BER performance of the various RSFH/RSSCH/USSCH-aided SDMA-OFDM, DSS/USSCH-aided SDMA-OFDM and conventional SDMA-OFDM systems, where TTCM and the MMSE-IGA MUD were employed. Note that, according to the different number of subcarriers K activated by the various schemes, the number of users was proportionally adjusted so that a total system throughput of $B_T = 5120$ bits per OFDM symbol was maintained for all the systems, apart from the ML-aided single-user reference scheme.

It can be seen in Figure 6.9 that the RSFH/SDMA-OFDM scheme employing $K = 64$ subcarriers achieved a better performance than its counterpart, which used $K = 128$ subcarriers, when the E_b/N_0 value became higher than about 4 dB. This is because in high-SNR scenarios the residual errors encountered are more likely to have been engendered by the average MUI, rather than by the noise. Furthermore, the RSFH system employing a lower number of subcarriers, or, more precisely, a larger number of subbands, results in less average MUI than the RSFH system, where fewer subbands are used, owing to supporting a reduced average number of users within each subband. More explicitly, this is because, when a large number of users activate the same subband, all users' subcarriers located within this subband are severely corrupted and the signal carried by them may become unrecoverable, even with the aid of channel codes. Moreover, at these high-MUI subbands, the heavier the MUI, the larger the number of active users affected. Therefore, the RSFH scheme associated with wider subbands, each of which accommodates more subcarriers, will always suffer more from an MUI-induced performance degradation.

Moreover, an increased number of subbands also implies a potentially better exploitation of frequency diversity. In frequency-selective channels, when a deep frequency-selective channel fade occurs across a subband, a large portion of the subcarriers within this specific subband may be affected and the corresponding signal may become obliterated – a situation similar to that of a heavily MUI-infested subband. Although the effects imposed by a deep fade may not be worse than that by severe MUI, especially when the faded frequency band is narrower than the subband's bandwidth, it does inflict a detrimental effect on the system and should be avoided or mitigated. This situation can be improved by employing a large number of subbands, each having a low number of subcarriers. Explicitly, this facilitates a better exploitation of the frequency diversity offered by the system. Furthermore, if the subcarriers of a given user are arranged to be sufficiently far apart from each other in the FD, they will experience independent FD fading, which in turn increases the TTCM channel decoder's chance of removing most – if not all – transmission errors.

This suggests that if the SFH subband of a specific user is further split into narrower bands, each of which is ideally as far from the others as possible, an improved performance may be achieved as a benefit of the associated frequency diversity. This argument is confirmed by the RSSCH-aided SDMA-OFDM system, which significantly outperformed both RSFH-aided arrangements, as a result of employing the subcarrier-based rather than the subband-based hopping approach, as observed in Figure 6.9.

However, note that the performance of the RSFH- or RSSCH-aided SDMA-OFDM system was worse than that of the conventional SDMA-OFDM system, since the use of random patterns inevitably results in an increased average BER, owing to the severe MUI arising from the subcarriers or subbands activated by an excessive number of users, as discussed in Section 6.4.2.1. By contrast, the employment of the uniform patterns calculated using the algorithm of Section 6.4.2.2 has the potential to improve the conventional system's performance, as shown in Figure 6.9. Furthermore, the achievable performance improvement becomes more significant when the total system throughput is further increased, which will be discussed in the next section.

6.5.2.2 Highly Overloaded Scenarios

In this section, we investigate the various systems in the scenario where the total system throughput was further increased to $B_{\mathrm{T}} = 6144$ bits per OFDM symbol, as portrayed in Figure 6.10. From the figure, we can see that the USFH-, the USSCH- and the hybrid DSS/USSCH-aided systems were capable of achieving a significantly better performance in comparison with the other systems, including the conventional SDMA-OFDM arrangement. More precisely, the RSFH/RSSCH SDMA-OFDM, the DSS/SDMA-OFDM and the conventional SDMA-OFDM systems gravely suffered from the excessive MUI and thus exhibited the corresponding error floors. By contrast, the schemes employing USFH/USSCH successfully suppressed the MUI and eliminated the associated error floor, as seen in Figure 6.10, while the USSCH/SDMA-OFDM system was found to be the best performer. This significant BER performance improvement was achieved as a benefit of the USSCH strategy presented in Section 6.4.2.2, which succeeded in exploiting the frequency diversity by dispersing the users' signals across the entire system bandwidth in a uniform random manner. Hence, the TTCM decoder was capable of successfully correcting most of the near-uniformly scattered errors inherent in each user's transmitted signal, resulting in a significant BER performance improvement. For example, at an E_b/N_0 value of 10 dB, the USSCH/SDMA-OFDM system reduced the BER by about two, three and four orders of magnitude in comparison with the conventional SDMA-OFDM, the RSSCH/SDMA-OFDM and the RSFH/SDMA-OFDM systems, respectively, as plotted in Figure 6.10.

Furthermore, as expected, the performance of the RSFH- or USFH-aided SDMA-OFDM system was further improved as the number of subbands was decreased. Moreover, the former system was outperformed by the RSSCH-aided scheme and the latter by the USSCH-aided arrangement. On the other hand, the hybrid DSS/USSCH SDMA-OFDM system attained a performance between those of the USSCH/SDMA-OFDM and the DSS/SDMA-OFDM systems, while the DSS/SDMA-OFDM

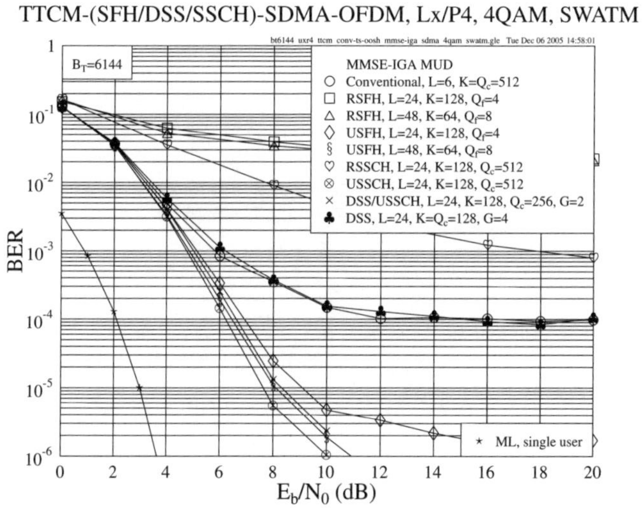

TTCM-(SFH/DSS/SSCH)-SDMA-OFDM, Lx/P4, 4QAM, SWATM

Figure 6.10: **BER** versus E_b/N_0 performance of the TTCM-assisted **conventional SDMA-OFDM**, **RSFH/USFH/RSSCH/USSCH SDMA-OFDM**, **DSS/SDMA-OFDM** and **hybrid DSS/USSCH SDMA-OFDM** systems employing a 4-QAM scheme for transmission over the SWATM channel [3], where L users were supported with the aid of $P = 4$ receiver antenna elements. The associated overall system throughput was $B_T = 6144$ bits. The basic simulation parameters and the notations used in the figure are summarized in Tables 6.1 and 6.2, respectively.

scheme achieved a similar performance to that of the conventional SDMA-OFDM system. Recall that when the total system bandwidth is fixed, the increased length of the DSS code, which implies a better exploitation of time diversity, is achieved at the cost of decreasing the number of available subcarriers, i.e. decreasing the attainable frequency diversity gain. Hence, the results shown in Figure 6.10 suggest that the frequency diversity benefits achieved by USSCH may be more significant than those of the time diversity attained by DSS. This implies that in such scenarios the fixed-bandwidth hybrid system should avoid using long DSS codes that result in a wider subcarrier bandwidth, so that a sufficiently high number of $Q > K$ subcarriers becomes available for the sake of maintaining a sufficiently high frequency diversity.

This characteristic is also confirmed by Figure 6.11, which shows the BER versus total system throughput performances of the various SDMA-OFDM systems, where a different number of users was supported at the fixed E_b/N_0 value of 6 dB. As inferred from Figure 6.11, when the total system throughput was increased, the performance of all the schemes degraded owing to the increased MUI. However, the systems using USSCH were found to outperform their counterparts dispensing with USSCH. Again, the USSCH/SDMA-OFDM system attained the best performance by exhibiting the highest robustness against MUI, which was, again, a direct benefit of the USSCH pattern assignment strategy of Section 6.4.2.2. This merit of the USSCH/SDMA-OFDM scheme was further evidenced by Figure 6.12, which shows the maximum total system throughput that can be supported by the various SFH/SSCH-aided schemes without exceeding the target BER at different E_b/N_0 values. The number of subcarriers activated by each user was set to $K = 128$ for all schemes for the sake of fair comparison. Specifically, the results attained for the target BER of 10^{-3} and 10^{-5} are portrayed on the left- and right-hand sides of Figure 6.12, respectively. It can be seen that, in both scenarios for E_b/N_0 values beyond 8 dB, the RSFH system was unable to tolerate more users, because its performance was limited by the MUI. By contrast, the other schemes were capable of providing a significant user capacity increase, while the USSCH/SDMA-OFDM system distinguished itself by providing the highest capacity.

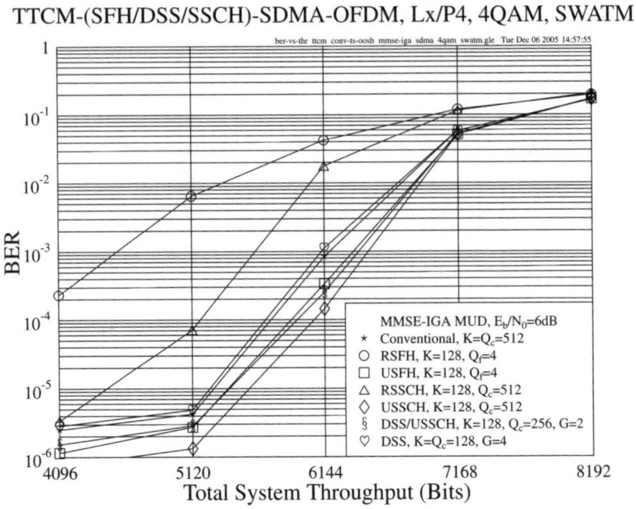

TTCM-(SFH/DSS/SSCH)-SDMA-OFDM, Lx/P4, 4QAM, SWATM

Figure 6.11: BER versus total system throughput performance of the TTCM-assisted **conventional SDMA-OFDM, RSFH/USFH/RSSCH/USSCH SDMA-OFDM, DSS/SDMA-OFDM** and **hybrid DSS/USSCH SDMA-OFDM** systems employing a 4-QAM scheme for transmission over the SWATM channel [3], where $P = 4$ receiver antenna elements were used. The number of users supported increased proportionally to the total system throughput. The basic simulation parameters and the notations used in the figure are summarized in Tables 6.1 and 6.2, respectively.

For example, at the E_b/N_0 value of 12 dB, a capacity increase of about 4%, 13% and 44% was achieved by the USSCH scheme compared with the USFH, the RSSCH and the RSFH arrangement at the target BER of 10^{-3}, and 4%, 14% and 78% at the BER of 10^{-5}, respectively, as shown in Figure 6.12.

6.5.3 Performance Enhancements by Increasing Receiver Diversity

In the previous sections, the performance of the different SFH- and SSCH-aided SDMA-OFDM systems was evaluated. Furthermore, the achievable performance of the different schemes can be potentially enhanced with the aid of a higher-order SDMA receiver diversity. Recall that in the context of SDMA-OFDM systems, the different users are distinguished by their spatial signatures, namely the associated channel CIRs. Hence, given a specific number of simultaneous users, an increased receiver diversity order achieved by increasing the number of BS receiver antennas will directly attain a higher diversity gain for all the users supported by the system, and hence an overloaded system may be turned into one having a lower user load. Therefore, the MUD's efficiency can be enhanced with the advent of a higher degree of freedom in the context of MUD.

In Figure 6.13, the various SFH/SSCH schemes are compared in the four-receiver and six-receiver SDMA-OFDM systems. In both scenarios, the throughput was fixed at $B_T = 5120$ bits per OFDM symbol. Clearly, when the number of receiver antenna elements was increased from four to six, the performance of all systems was improved, especially that of the RSFH/SDMA-OFDM arrangement. More precisely, the corresponding residual error floor observed in the four-antenna scenario was dramatically reduced or completely removed in the six-antenna scenario. Again, in both scenarios, the best solution was the USSCH-assisted system. However, the performance differences between the various schemes were substantially reduced in the higher-diversity scenario, since most of the achievable diversity gain of the SDMA system may be attained by increasing the number of BS receiver antennas, although at the cost of an increased hardware implementation complexity, especially in terms of channel estimation.

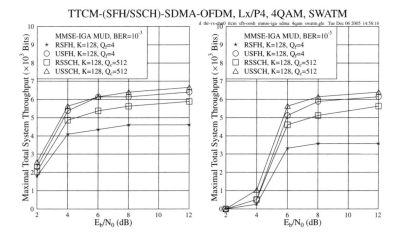

Figure 6.12: Maximal total system throughput versus E_b/N_0 performance of the TTCM-assisted **conventional SDMA-OFDM** and **RSFH/USFH/RSSCH/USSCH SDMA-OFDM** systems employing a 4-QAM scheme for transmission over the SWATM channel [3], where $P = 4$ receiver antenna elements were used. The number of users supported increased proportionally to the maximal total system throughput. The basic simulation parameters and the notations used in the figure are summarized in Tables 6.1 and 6.2, respectively. ©IET Jiang & Hanzo 2006 [469]

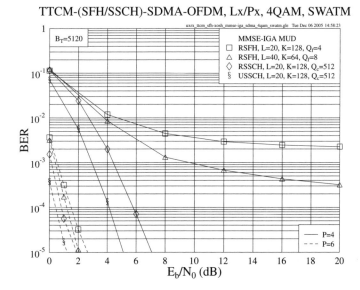

Figure 6.13: BER versus E_b/N_0 performance of the TTCM-assisted **RSFH/RSSCH/USSCH SDMA-OFDM** systems employing a 4-QAM scheme for transmission over the SWATM channel [3], where L users were supported with the aid of $\boldsymbol{P} = \boldsymbol{4}$ or $\boldsymbol{P} = \boldsymbol{6}$ receiver antenna elements. The associated overall system throughput was $\boldsymbol{B_T} = \boldsymbol{5120}$ bits. The basic simulation parameters and the notations used in the figure are summarized in Tables 6.1 and 6.2, respectively.

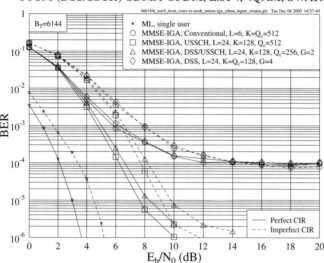

Figure 6.14: BER versus E_b/N_0 performance of the TTCM-assisted **conventional SDMA-OFDM**, **DSS SDMA-OFDM** and **hybrid DSS/USSCH SDMA-OFDM** systems employing a 4-QAM scheme for transmission over the SWATM channel [3], where L users were supported with the aid of $P = 4$ receiver antenna elements. The associated overall system throughput was $B_T = 6144$ bits. The basic simulation parameters and the notations used in the figure are summarized in Tables 6.1 and 6.2, respectively.

6.5.4   Performance under Imperfect Channel Estimation

Similar to Section 5.3.2.2.3, we provide further simulation results for the scenario where channel estimation was assumed to be imperfect. Again, the estimated SWATM CIRs were generated using Equation (5.19), while a snapshot of the SWATM channel has been plotted in Figure 5.23, which shows both the real and imaginary components of the FD-CHTFs associated with both perfect and imperfect CIRs. The performance comparison of the various systems under the assumptions of both perfect and imperfect channel knowledge is given in Figure 6.14, where the total throughput was $B_T = 6144$ bits per OFDM symbol. As seen in Figure 6.14, the various DSS- and/or USSCH-aided SDMA-OFDM systems did not suffer a higher performance degradation than the single-user scheme when the channel estimation was imperfect. This implies that the proposed hybrid system is capable of attaining an acceptable performance even without accurate channel knowledge.

6.6   Complexity Issues

The Q_c-point IFFT employed by the SSCH-aided system will inevitably impose a higher computational complexity than its SFH-aided counterpart, where a K-point IFFT is used, regardless of whether the RSSCH or USSCH scheme is employed, since $K < Q_c$. However, it was shown in Section 6.5 that SSCH systems typically achieve a significant performance improvement over the SFH arrangements considered, hence their additional complexity cost may be deemed justified. Furthermore, in SSCH systems the OFDM symbols modulate a single carrier frequency f_c, hence the frequency synthesizers necessary for both the SFH transmitter and receiver are eliminated, which simplifies the hardware implementation of the SSCH systems. Note that in SFH systems the subcarrier activation/deactivation operations may also be invoked in the baseband, so that the employment of frequency synthesizers becomes unnecessary. However, in this case a Q_c-point IFFT will have to be invoked in the baseband-processing-aided SFH systems, resulting in a similar complexity to that of the SSCH systems.

On the other hand, concerning the RSSCH and the USSCH systems, a similar SSCH transmitter invoking a Q_c-point IFFT is employed, thus the transmitter's complexity is similar in both systems. Nonetheless, the RSSCH arrangement imposes a potentially higher complexity at the receiver side. This is because the number of active users at each subcarrier is a random variable spanning from 1 to L, since a random SSCH pattern is employed. Therefore, an MUD that is capable of detecting a variable number of active users has to be employed, which will impose an increased complexity. Naturally, this problem exists in all the FH/SCH-aided SDMA-OFDM systems using random hopping patterns. By contrast, a USSCH receiver has a complexity similar to that of conventional SDMA-OFDM systems, which employ an MUD designed for a fixed number of active users. This implementation benefit arises from the characteristics of the USSCH system, since the users are uniformly distributed across the subcarriers, so that the number of users activating each subcarrier is similar. Specifically, if the condition of $L \bmod Q_g = L \bmod (Q_c/K) = 0$ is satisfied, i.e. if the number of users supported by the system can be divided by the number of subcarriers within a subcarrier group, the number of users assigned to all subcarriers will be the same, as discussed in Section 6.4.2.2.2.

Note that without channel feedback the proposed USSCH pattern assignment algorithm may not be optimal in terms of combating the deleterious effects of deep channel fades. However, our proposed scheme has the advantage of low complexity and high MUI resistance, since the USSCH patterns can be generated using offline pre-computation and yet achieve an attractive performance, as discussed in Section 6.4.2.4. This imposes a significantly lower computational complexity than that required by complicated adaptive algorithms requiring near-instantaneously channel knowledge.

6.7 Conclusions

From the investigations conducted in Section 6.5, we conclude that the proposed USSCH-aided SDMA-OFDM system was capable of achieving the best performance in comparison with all the other schemes considered, especially in high-throughput scenarios, when TTCM was employed. For example, in the overloaded scenario associated with $B_T = 6144$ bits per OFDM symbol, the USSCH/SDMA-OFDM system reduced the BER recorded at an E_b/N_0 value of 8 dB by one to four orders of magnitude in comparison with the DSS/USSCH SDMA-OFDM, the USFH/SDMA-OFDM, the DSS/SDMA-OFDM, the conventional SDMA-OFDM, the RSSCH/SDMA-OFDM and the RSFH/SDMA-OFDM systems, respectively, as shown in Figure 6.10. More specifically, the USSCH-aided system exhibits a high resilience against the excessive MUI incurred in high-throughput scenarios, owing to the characteristics of the proposed subcarrier assignment algorithm of Section 6.4.2.2. We also show that the attainable system performance may be further improved if a higher-order receiver diversity is provided by the SDMA-OFDM system.

Furthermore, when compared with the conventional SDMA-OFDM system, the only additional computational complexity imposed by the USSCH/SDMA-OFDM system arises from the low-complexity USSCH algorithm. Moreover, it allows offline pre-computation of the patterns, as discussed in Section 6.4.2.4. Thus, the desirable high performance of the proposed USSCH/SDMA-OFDM system is not achieved at the cost of a significantly increased computational complexity, in comparison with its conventional counterpart.

6.8 Chapter Summary

In this chapter, we proposed a TTCM-assisted DSS/USSCH-aided OFDM system operating with the aid of the MMSE-IGA MUD designed for the SDMA MIMO uplink channel introduced in Section 4.2.1, and compared its performance with a range of different SDMA-OFDM systems.

In Section 6.1 the conventional SDMA-OFDM system was briefly reviewed, where we highlighted two of its disadvantages. On the one hand, the conventional SDMA-OFDM cannot efficiently exploit frequency diversity, because all the users share all available subcarriers simultaneously. On the other hand, a high number of users will inevitably result in a high MUI across the entire bandwidth, which degrades all users' performance. In Section 6.2, an introduction to hybrid SDMA-OFDM systems was given, discussing how both of the above-mentioned problems encountered by the conventional SDMA-OFDM system may be overcome. Furthermore, in Section 6.3 we provided a comparison of the conventional SDMA-OFDM [3, 193], the SFH/SDMA-OFDM and the SSCH/SDMA-OFDM systems in terms of their frequency resource allocation strategies, where the SSCH scheme was considered to have more advantages than the others, such as for example its higher efficiency in terms of exploiting frequency diversity.

In Section 6.4, the proposed hybrid SDMA-OFDM system was introduced, which incorporated both DSS and subcarrier-based FH techniques into conventional SDMA-OFDM systems. More specifically, an overview of the system's architecture was presented in Section 6.4.1, including the transmitter and receiver designs in Sections 6.4.1.1 and 6.4.1.2, respectively. Furthermore, two different SSCH pattern assignment strategies were considered in Section 6.4.2, namely the RSSCH scheme of Section 6.4.2.1 and the USSCH scheme of Section 6.4.2.2, where the USSCH strategy was considered to be more meritorious. This is because the USSCH algorithm designed in Section 6.4.2.2.1 assigns uniformly distributed subcarriers to all users, so that the number of users activating each subcarrier becomes similar. Therefore, the average MUI across the whole system bandwidth can be minimized. In addition, with the advent of the USSCH strategy, the system's frequency diversity can be efficiently exploited. The random and uniform pattern assignment strategies can also be applied in SFH/SDMA-OFDM systems, as discussed in Section 6.4.2.3. Moreover, we pointed out in Section 6.4.2.4 that the USFH/USSCH patterns can be generated by offline pre-computation, which imposes a significantly lower computational complexity than that required by other adaptive algorithms benefitting from real-time channel knowledge. In order to offer some insight into the system design, we detailed in Section 6.4.3 the DSS despreading and SSCH demapping operations invoked at the SDMA receiver, followed by a discussion of the MUD process in the context of the hybrid DSS/SSCH SDMA-OFDM system in Section 6.4.4.

Our simulation-based performance results associated with the SSCH/SDMA-OFDM system were provided in Section 6.5. Specifically, in Section 6.5.1 the attainable performance of the DSS- and/or SSCH-aided SDMA-OFDM systems were compared when using the classic MMSE MUD [3] discussed in Chapter 4 and the MMSE-IGA MUD proposed in Chapter 5. It was observed in Figure 6.8 that in the moderately overloaded scenario associated with a total system throughput of $B_T = 5120$ bits, most of the MUI encountered can be effectively suppressed by the MMSE-IGA MUD, while the overloaded MMSE MUD results in a high error floor. In Section 6.5.2, the performance of the hybrid DSS/SSCH SDMA-OFDM system was characterized and compared with that of the conventional SDMA-OFDM and SFH-aided SDMA-OFDM systems, while employing the MMSE-IGA MUD. More specifically, a moderately overloaded scenario was considered in Section 6.5.2.1. The RSFH-, the USFH- and the RSSCH-aided schemes were found to suffer from using random hopping patterns, which resulted in an increased average BER owing to the excessive MUI arising from the subcarriers or subbands activated by a high number of users. By contrast, the systems using the uniform patterns generated by the algorithm of Section 6.4.2.2 were capable of improving the achievable performance, as evidenced in Figure 6.9.

Furthermore, we demonstrated in Section 6.5.2.2 that the proposed uniform pattern assignment algorithm excelled when the total system throughput was further increased. In the highly overloaded scenario, where the total system throughput was increased to $B_T = 6144$ bits per OFDM symbol, the USFH-, the USSCH- and the hybrid DSS/USSCH-aided systems were capable of achieving a significantly better performance in comparison with the conventional SDMA-OFDM, the DSS/SDMA-OFDM as well as the RSFH/RSSCH-aided SDMA-OFDM systems, as portrayed in Figure 6.10. More explicitly, the USSCH/SDMA-OFDM system was found to be the best design option, which successfully suppressed the MUI and eliminated the associated error floor as a benefit of the USSCH strategy presented in Section 6.4.2.2. For example, at an E_b/N_0 value of 10 dB, the USSCH/SDMA-OFDM system reduced the BER by about two, three and four orders of magnitude in comparison with the conventional SDMA-OFDM, the RSSCH/SDMA-OFDM and the RSFH/SDMA-OFDM systems, respectively, as evidenced by Figure 6.10. On the other hand, it was concluded that the frequency diversity benefits achieved by USSCH may be more significant than those of the time diversity attained by DSS, since the hybrid DSS/USSCH SDMA-OFDM system attained a performance between those of the USSCH/SDMA-OFDM and the DSS/SDMA-OFDM systems, while the DSS/SDMA-OFDM scheme achieved a similar performance to that of the conventional SDMA-OFDM system. This suggests that the fixed-bandwidth hybrid system should avoid using long DSS codes that result in a wider subcarrier bandwidth, so that a sufficiently high number of $Q > K$ subcarriers becomes available to maintain a sufficiently high frequency diversity. The proposed USSCH scheme's high robustness against MUI was further confirmed by the results of Figures 6.11 and 6.12. For example, at the E_b/N_0 value of 12 dB, a capacity increase of about 4%, 13% and 44% was achieved by the USSCH scheme compared with the USFH, the RSSCH and the RSFH arrangements, when aiming at the target BER of 10^{-3}, and 4%, 14% and 78% for the target BER of 10^{-5}, respectively, as shown in Figure 6.12. We also demonstrated that the attainable performance of the SSCH-aided SDMA-OFDM system can be further enhanced when the receiver diversity order becomes higher, as discussed in Section 6.5.3. In Section 6.5.4, we demonstrated that the proposed hybrid system is capable of achieving an acceptable performance even without accurate channel knowledge. Finally, we pointed out that the superior performance of the USSCH-aided SDMA-OFDM system is achieved at a similar complexity to that of the conventional SDMA-OFDM arrangement, since the additional computational complexity imposed by the USSCH algorithm manifests itself in terms of the offline pre-computation, as discussed in Section 6.6.

From the perspective of practical applications, the efficient exploitation of frequency diversity and the robust MUI-suppression capability render the channel-coded USSCH technique an attractive design option, resulting in flexible implementations. In fact, the USSCH/SDMA-OFDM scheme can be readily extended to a variable-rate system offering a high grade of flexibility, as required by future wireless multimedia services, where variable bit rates and different QoS requirements have to be satisfied. More specifically, each user may activate a different number of subcarriers depending on the type of service to be delivered or the bit rate to be supported. A modified version of the USSCH algorithm of Section 6.4.2.2 can be invoked to generate the users' uniform patterns, which takes into account their different rates. Furthermore, in SSCH-based systems the different subcarriers do not have to be contiguously allocated, which is a further attractive property, especially in scenarios where several systems operated by different service providers have to coexist and/or fractional bandwidths have to be exploited. In addition, the different types of hybrid SFH/SSCH-assisted SDMA-OFDM systems can be readily implemented by exploiting different number of subbands having different bandwidths, depending on the specific system requirements.

So far, most of our investigations have been conducted under the assumption of perfect channel estimation. Naturally, this assumption is impractical in real applications. Hence, in the next chapter, we will continue our study by developing novel channel estimation techniques for SDMA-OFDM systems.

Chapter 7

Channel Estimation for OFDM and MC-CDMA

7.1 Pilot-Assisted Channel Estimation

In this treatise we concentrate our attention on the derivation and the performance analysis of decision-directed channel estimation methods, although we would also like to give a brief performance comparison between decision-directed and pilot-aided channel estimation methods. Our motivation is that any technique applicable to decision-directed channel estimation can be equally employed in the context of pilot-aided schemes and the difference in attainable performance can be predicted as outlined below.

The attainable performance of both the Decision-Directed (DD) and Pilot-Assisted (PA) channel estimation methods can be compared in the following simple way.

The performance of any PA channel estimation method expressed in terms of the achievable Mean Square Error (MSE) is upper bounded by the expression

$$MSE_{\text{PA}} > \frac{N_0 L}{E_p},\tag{7.1}$$

where E_p is the total power associated with the transmitted pilots, N_0 is the Gaussian noise variance and L is the number of non-zero CIR components.

On the other hand, in the case of DD channel estimation the corresponding performance bound, using the assumption of error-free decisions, can be described by

$$MSE_{\text{DD}} > \frac{N_0}{E_s} \frac{L}{K},\tag{7.2}$$

where E_s is the average signal energy per transmitted complex baseband sample and K is the number of OFDM subcarriers, while N_0 and L are as defined previously.

Thus the resultant performance gain may be quantified as

$$\frac{MSE_{\text{DD}}}{MSE_{\text{PA}}} = \frac{E_p}{E_s K}.\tag{7.3}$$

We would also like to emphasize the trade-off between the PA channel estimator's performance and the system's spectral efficiency loss associated with the allocation of valuable signal power to pilot

---

MIMO-OFDM for LTE, Wi-Fi and WiMAX Lajos Hanzo, Yosef Akhtman, Li Wang and Ming Jiang
© 2011 John Wiley & Sons, Ltd

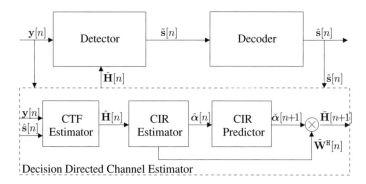

Figure 7.1: Schematics of a generic receiver employing a decision-directed channel estimator constituted by an a posteriori decision-directed CTF estimator, followed by a CIR estimator and an a priori CIR predictor.

symbols. The corresponding data-rate loss can be quantified by a simple expression similar to that of Equation (7.3):

$$r_{\text{loss}} = \frac{E_p}{E_s K}. \tag{7.4}$$

7.2   Decision-Directed Channel Estimation

A schematic of the channel estimation method considered is depicted in Figure 7.1. The symbols $\boldsymbol{y}[n]$ and $\hat{\boldsymbol{s}}[n]$ in the figure represent the received vector of the subcarrier-related samples and the a posteriori decision-based estimated vector of the transmitted information-carrying symbols $\boldsymbol{s}[n]$, respectively. Furthermore, symbols $\boldsymbol{H}[n+1]$, $\alpha[n]$ and $\alpha[n+1]$ represent the CTF and the CIR vectors corresponding to time instants n and $n + 1$, respectively.[1] Finally, the accents $\check{x}$ and $\hat{x}$ represent the a priori predicted and a posteriori estimated values of the variable x, respectively. Figure 7.1 corresponds to the general case of the CIR estimation and both sample-spaced as well as fractionally spaced cases may be considered. We will commence by considering the simpler case of the SS-CIR. Thus, the CIR vector $\alpha[n]$ in Figure 7.1 can be substituted by its sample-spaced projection $\boldsymbol{h}[n]$ described by Equation (1.8).

Our channel estimator is constituted by what we refer to as an a posteriori decision-directed CTF estimator followed by a CIR estimator and an a priori CIR predictor [265]. As seen in Figure 7.1, the task of the CTF estimator is to evaluate the tentative values of the subcarrier-related CTF coefficients of Equation (1.14). Correspondingly, the task of the CIR estimator is to estimate the SS-CIR taps of Equation (1.8). In the case of the SS-CIR-aided channel estimation, discussed in this section, the Inverse Fast Fourier Transform (IFFT) transformation from the subcarrier-related frequency domain to the SS-CIR-related time domain is invoked in order to exploit the frequency-domain correlation of the subcarrier-related CTF coefficients as well as to reduce the computational complexity associated with the CTF prediction process, because the SS-CIR typically has a lower number of $K_0 \ll K$ taps, which have to be predicted, than the number K of FD-CTF coefficients. Hence the overall channel estimation

---

[1]In this study we invoke the analysis of relatively complex multi-dimensional vector structures associated with both time- and spatially-multiplexed composite signals. In some cases, the two aforementioned multiplexing stages have to be addressed concurrently within a single bound of mathematical derivation. In order to emphasize the necessary distinction between the time- and spatial-domain constituents of the composite signals, we will utilize *bold italic* lower-case and upper-case letters $\boldsymbol{v}$ and $\boldsymbol{A}$ to denote the time-domain signals and operators, and **bold** lower-case and upper-case letters $\mathbf{v}$ and $\mathbf{A}$ to describe the spatial-domain signals and operators, respectively.

complexity is reduced, even when the complexity of the FD-CTF to CIR transformation and its inverse are taken into account.[2]

As can be seen in Figure 7.1, the a posteriori CTF estimator inputs are the subcarrier-related signal $y[n]$ and the decision-based estimate $\hat{s}[n]$. The transformation from the frequency to the time domain is performed within the CIR estimator of Figure 7.1 and its output is an a posteriori estimate $\hat{s}[n, k]$ of the CIR taps of Equation (1.8), which is fed into the low-rank time-domain CIR tap predictor of Figure 7.1 to produce an a priori estimate $\check{h}[n + 1, l]$, $l = 0, 1, \ldots, K_0 - 1$, of the next SS-CIR on an SS-CIR tap-by-tap basis [265]. Finally, the predicted SS-CIR is converted to the subcarrier-related CTF estimates with the aid of the FFT. The resultant FD-CTF is employed by the receiver to detect and decode the next OFDM symbol. Note that this principle requires the transmission of a pilot-based channel sounding sequence, such as for example a pilot-assisted OFDM block, during the initialization stage.

7.3 A Posteriori FD-CTF Estimation

In order to emphasize the major difference between the OFDM and MC-CDMA systems in the context of the associated channel estimation scheme, we would first like to analyse the performance of the temporal estimator of the subcarrier-related FD-CTF coefficients $H[n, k]$ based on the a posteriori decision-aided estimates of the transmitted subcarrier-related samples $s[n, k]$ of Equation (1.14). In Section 7.3.1 we will show that the least-squares approach typically employed in DDCE-aided OFDM systems [260, 265] is not applicable in the case of MC-CDMA systems. In Section 7.4.1 we propose an MMSE estimator which renders the DDCE philosophy discussed in [260, 265] suitable for MC-CDMA systems. However, the estimator introduced in Section 7.4.1 exhibits a computational complexity which is significantly higher than the computational complexity of the conventional least-squares-based estimator of [260, 265]. Thus a reduced-complexity approximation of the MMSE estimator of Section 7.4.1 is proposed in Section 7.4.2.

7.3.1 Least-Squares CTF Estimator

Following Equation (1.14), the Least Squares (LS) approach [330] to the problem of estimating the discrete-abscissa FD-CTF coefficients $H[n, k]$, based on knowledge of the decision-aided estimates $\hat{s}[n, k]$ of the transmitted frequency-domain samples $s[n, k]$ of Equation (1.14), can be expressed as

$$\tilde{H}[n, k] = \frac{y[n, k]}{\hat{s}[n, k]} = H[n, k] \cdot \frac{s[n, k]}{\hat{s}[n, k]} + \frac{w[n, k]}{\hat{s}[n, k]}, \qquad (7.5)$$

where $H[n, k]$ represents the Rayleigh-distributed FD-CTF coefficients having a variance of σ_H^2, while $s[n, k]$ denotes the transmitted subcarrier-related samples having zero mean and a variance of σ_s^2. The distribution of the samples $s[n, k]$ is dependent on the particular modulation scheme employed by the system. For instance, in an MC-CDMA system using an arbitrary modulation scheme, the samples $s[n, k]$ are complex-Gaussian distributed, having a Rayleigh-distributed amplitude $|x[n, k]|$ and uniformly distributed phase $\theta[n, k]$. By contrast, in an M-PSK-modulated OFDM system the samples $s[n, k]$ are uniformly distributed within the set of M-PSK symbols having a constant amplitude $|s[n, k]| = \sigma_s$ and a discrete uniform distributed phase $\theta[n, k] = 2\pi(m/M)$, $m = 0, 1, \ldots, M - 1$. Finally, the noise samples $w[n, k]$ are independent and identically distributed (i.i.d.) complex-Gaussian variables having a zero mean and a variance of σ_w^2.

---

[2]The computational complexity associated with the prediction of the K CTF coefficients is of order $O(K^2 N_{prd})$, where N_{prd} is the order of the prediction filter. On the other hand, the CIR prediction combined with the FFT and IFFT operations can be associated with the computational complexity of order $O(K_0^2 N_{prd} + 2K \log_2 K)$. It is evident that, in the typical case of $N_{prd} < K_0 \ll K$, the overall estimation complexity is reduced if the aforementioned method of the CIR prediction is employed.

Under the assumption of carrying out error-free decisions, we have $\hat{s}[n, k] = s[n, k]$ and Equation (7.5) may be simplified to

$$\tilde{H}[n, k] = \frac{y[n, k]}{\hat{s}[n, k]} = H[n, k] + \frac{w[n, k]}{\hat{s}[n, k]}. \tag{7.6}$$

The Mean Square Error (MSE) associated with the LS FD-CTF estimator of (7.6) is given by

$$MSE_{LS} = E\{|H[n, k] - \tilde{H}[n, k]|^2\} = E\left\{\left|\frac{w[n, k]}{s[n, k]}\right|^2\right\}. \tag{7.7}$$

The less ambiguous measure of the estimator's performance is the Normalized Mean Square Error (NMSE), which is defined as the MSE normalized by the variance of the parameter being estimated. The NMSE corresponding to the estimator of Equation (7.6) is given by

$$NMSE_{LS} = \frac{1}{\sigma_H^2}E\left\{\left|\frac{w[n, k]}{s[n, k]}\right|^2\right\}. \tag{7.8}$$

The AWGN samples $w[n, k]$ are known to be i.i.d. complex-Gaussian and hence the MSE of Equation (7.7) is determined by the statistical distribution of the transmitted subcarrier-related samples $s[n, k]$. The NMSE encountered assumes its minimum value when $|s[n, k]|^2 = \sigma_s^2$ is constant, as in the case of an M-PSK-modulated OFDM system. Thus, we have

$$NMSE_{LS,min} = \frac{1}{\sigma_H^2\sigma_s^2}E\{|w[n, k]|^2\} = \frac{\sigma_w^2}{\sigma_H^2\sigma_s^2} = \frac{1}{\gamma}, \tag{7.9}$$

where

$$\gamma = \frac{1}{\sigma_w^2}E\{|H[n, k]s[n, k]|^2\} = \frac{\sigma_H^2\sigma_s^2}{\sigma_w^2} \tag{7.10}$$

is the average SNR level. On the other hand, the NMSE value will increase substantially if the energy of the transmitted samples $s[n, k]$ varies as in the case of M-ary Quadrature Amplitude Modulation (M-QAM) based OFDM or MC-CDMA. In fact, in the case of strictly Gaussian-distributed samples $s[n, k]$, which corresponds to encountering an MC-CDMA system having a sufficiently long spreading code, the NMSE value of Equation (7.8) does not exist, since the variance of the resultant Cauchy-distributed variable associated with the ratio of two Gaussian-distributed variables $s[n, k]$ and $w[n, k]$ of Equation (7.8) cannot be defined [510]. The NMSE of the LS estimator of Equation (7.6) derived for QPSK, 16-, 64- and 256-QAM-modulated OFDM, as well as QPSK-modulated MC-CDMA, is depicted in Figure 7.2(a). The solid line in Figure 7.2(a) corresponds to the lower NMSE bound described by Equation (7.9).

The performance degradation of the LS estimator of Equation (7.6) was imposed by the energy fluctuation of the near-Gaussian-distributed subcarrier-related samples $s[n, k]$, which renders the LS estimator inapplicable for employment in MC-CDMA systems. Therefore, to mitigate this performance degradation we will turn our attention to the MMSE estimation approach.

7.3.2 MMSE CTF Estimator

In order to derive an FD-CTF estimator which is suitable for employment in an MC-CDMA system, where the energy fluctuation of the subcarrier-related samples $s[n, k]$ is near Gaussian, we turn to the MMSE approach. Following the Bayesian linear model theory of [330], the MMSE estimator of the FD-CTF coefficients $H[n, k]$ of the scalar linear model described by Equation (1.14), where the parameters $H[n, k]$ are assumed to be complex-Gaussian distributed with a zero mean and a variance of σ_H^2, is given by [330]

$$\tilde{H}_{MMSE}[n, k] = \left(\frac{x^*[n, k]s[n, k]}{\sigma_w^2} + \frac{1}{\sigma_H^2}\right)^{-1} \cdot \frac{x^*[n, k]y[n, k]}{\sigma_w^2} = \frac{s^*[n, k]y[n, k]}{|s[n, k]|^2 + \sigma_w^2/\sigma_H^2}. \tag{7.11}$$

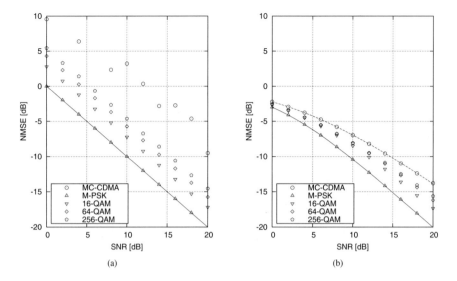

(a)   (b)

Figure 7.2: **NMSE** associated with (a) **Least Squares (LS)** and (b) **Minimum Mean Square Error (MMSE)** estimators of the uncorrelated Rayleigh-distributed subcarrier-related CTF coefficients $H[n,k]$ of Equation (1.14) corresponding to the various statistical distributions of the transmitted subcarrier-related samples $x[n,k]$. The markers on the plot correspond to the simulated cases of M-PSK, 16-, 64- and 256-QAM-modulated OFDM as well as M-QAM-modulated MC-CDMA, while the lines correspond to the analytically calculated performance recorded for the cases of M-PSK OFDM (solid) and MC-CDMA (dashed), which represent the lower and upper NMSE bounds, respectively. Note that the upper bound for the LS estimator in conjunction with MC-CDMA does not exist.

The corresponding NMSE can be expressed as [330]

$$NMSE_{\text{MMSE}} = \frac{1}{\sigma_H^2} \left(\frac{1}{\sigma_H^2} + \frac{|s[n,k]|^2}{\sigma_w^2} \right)^{-1}$$

$$= \frac{\sigma_w^2}{\sigma_H^2 |s[n,k]|^2 + \sigma_w^2} = \frac{1}{\gamma |s[n,k]/\sigma_x|^2 + 1}, \qquad (7.12)$$

where γ is the average SNR level defined by Equation (7.10). As we have seen previously in the context of Equation (7.12), the NMSE is determined by the statistical distribution of the transmitted subcarrier-related samples $s[n,k]$ and assumes its minimum value when the energy of these samples $|s[n,k]|^2 = \sigma_s^2$ is constant. On the other hand, in contrast to the NMSE of the LS estimator of Equation (7.5), the NMSE of the MMSE estimator of Equation (7.11) is upper bounded, which is evidenced by Figure 7.2(b). The NMSE assumes its maximum value when the samples $s[n,k]$ are complex-Gaussian distributed, as in the case of an MC-CDMA system having a sufficiently high spreading factor. Explicitly, the maximum NMSE may be derived as follows:

$$NMSE_{H,\text{max}}(\gamma) = \underset{x \in N(0,\sigma_x^2)}{E} \{NMSE(\gamma,x)\}$$

$$= \underset{r=|x/\sigma_x|^2 \in \chi^2}{E} \{NMSE(\gamma,r)\}$$

$$= \int_0^\infty \frac{1}{\gamma r + 1} e^{-r} \, dr = \frac{1}{\gamma} e^{1/\gamma} \text{Ei}\left(\frac{1}{\gamma} \right), \qquad (7.13)$$

where we integrate, i.e. average, the first multiplicative term upon weighting it by the χ^2-distributed Probability Density Function (PDF) of the NMSE described by $P_{\chi^2}(r) = e^{-r}$ over its entire range spanning from 0 to ∞ and define the *exponential integral* function as

$$\text{Ei}(x) = \int_x^\infty \frac{e^{-t}}{t}\, dt. \tag{7.14}$$

7.3.3 A Priori Predicted-Value-Aided CTF Estimator

In MC-CDMA systems employing spreading codes having a relatively low spreading factor G, there is a finite probability of encountering zero-energy subcarrier-related samples $s[n, k] = 0$ when superimposing the chips of various users corresponding to the subcarrier considered. This probability decreases with increasing spreading factor G as the corresponding power distribution function approaches a Gaussian distribution. As can be seen from Equation (7.11), this will result in a corresponding CTF coefficient MMSE estimate $\tilde{H}[n, k] = 0$, which is unrelated to the actual value of $H[n, k]$ encountered. This problem can be circumvented in the context of the DDCE scheme of Figure 7.1, where the a priori predicted estimate of the subcarrier-related coefficient $H[n, k]$ is readily available, by performing a Maximum Ratio Combining (MRC) of the a posteriori MMSE estimate $\tilde{H}[n, k]$ of Equation (7.11) and the corresponding a priori estimate $\check{H}[n, k]$. It can be shown that the resultant MRC-aided CTF estimator can be expressed as

$$\tilde{H}[n, k] = \frac{\check{H}[n, k] + \hat{s}^*[n, k]y[n, k]}{1 + |\hat{s}[n, k]|^2 + (K_0/K)(\sigma_w^2/\sigma_H^2)}. \tag{7.15}$$

In the following section we employ the following vectorial notation: $\boldsymbol{v}[n] = (v[n, 1], \ldots, v[n, K])^{\text{T}}$.

7.4 A Posteriori CIR Estimation

7.4.1 MMSE SS-CIR Estimator

We will commence our portrayal of the proposed channel estimation philosophy by rendering the DDCE OFDM scheme of [260, 265] as also applicable for employment in MC-CDMA with the derivation of the a posteriori MMSE SS-CIR estimator of Figure 7.1.

By substituting the FD-CTF of Equation (1.7) into (1.14) we arrive at

$$y[n, k] = \sum_{l=0}^{K_0-1} W_K^{kl} h[n, l] x[n, k] + w[n, k], \tag{7.16}$$

which can be expressed in matrix form as

$$\boldsymbol{y}[n] = \text{diag}(x[n, k])\boldsymbol{W}\boldsymbol{h}[n] + \boldsymbol{w}[n], \tag{7.17}$$

where we define the $(K \times K)$-dimensional matrix $\text{diag}(v[k])$ as a diagonal matrix having the corresponding elements of the vector $v[k]$ on the main diagonal, as well as the $(K \times K_0)$-dimensional Fourier transform matrix $\boldsymbol{W}$, which corresponds to the Fourier transform of the zero-padded SS-CIR vector $\boldsymbol{h}[n]$ and is defined by $W_{kl} = W_K^{kl}$ for $k = 0, 1, \ldots, K - 1$ and $l = 0, 1, \ldots, K_0 - 1$.

As before, the SS-CIR taps $h[l]$ are assumed to be uncorrelated complex-Gaussian-distributed variables having a zero mean and a covariance matrix given by

$$\boldsymbol{C}_h = \text{diag}(\sigma_l^2). \tag{7.18}$$

The MMSE estimator of the SS-CIR taps $h[n, l]$ of the linear vector model described by Equation (7.17) is given by [330]

$$\hat{\boldsymbol{h}} = \left(\text{diag}\left(\frac{1}{\sigma_l^2}\right) + \frac{1}{\sigma_w^2} \boldsymbol{W}^{\text{H}} \text{diag}(|\hat{x}[k]|^2)\boldsymbol{W} \right)^{-1}$$

$$\times \frac{1}{\sigma_w^2} \boldsymbol{W}^{\text{H}} \text{diag}(\hat{x}^*[k])\boldsymbol{y}, \tag{7.19}$$

where we omit the time-domain OFDM-block-spaced index n for the sake of notational simplicity. Following the assumptions made in Section 1.7.1 about the nature of the channel model considered, some of the parameters σ_l^2 may assume a zero value. Hence to avoid division by zero, we rewrite Equation (7.19) in a more practical form as follows:

$$\hat{h} = (\sigma_w^2 I + \text{diag}(\sigma_l^2) W^H \text{diag}(|\hat{x}[k]|^2) W)^{-1}$$
$$\times \text{diag}(\sigma_l^2) W^H \text{diag}(\hat{x}^*[k]) y. \tag{7.20}$$

The covariance matrix of the vector $\hat{h}$ of the MMSE SS-CIR estimates can be expressed as [330]

$$C_{\hat{h}|\hat{x}} = \left(I + \text{diag}\left(\frac{\sigma_l^2}{\sigma_w^2}\right) W^H \text{diag}(|\hat{x}[k]|^2) W \right)^{-1}$$
$$\times \text{diag}(\sigma_l^2). \tag{7.21}$$

The corresponding NMSE associated with the lth MMSE SS-CIR tap estimate $\hat{h}[l]$ can be found be approximating the lth diagonal element of the covariance matrix $C_{\hat{h}|\hat{x}}$ of Equation (7.21) and normalizing it by the average channel output power σ_H^2. The above-mentioned approximation is performed by replacing the matrix $\text{diag}(|\hat{x}[k]|^2)$ in Equation (7.21) by its average value $\sigma_x^2 I$. Thus, we arrive at

$$NMSE_{\text{MMSE},l} = \frac{\sigma_l^2}{\sigma_H^2} \frac{\sigma_w^2}{\sigma_w^2 + K\sigma_l^2\sigma_x^2}$$
$$= \frac{\sigma_w^2}{\sigma_H^2\sigma_x^2} \frac{\sigma_l^2}{(\sigma_w^2/\sigma_x^2) + K\sigma_l^2} = \frac{1}{\gamma} \frac{1}{(\sigma_w^2/\sigma_x^2\sigma_l^2) + K}. \tag{7.22}$$

The overall NMSE corresponding to the MMSE SS-CIR estimator of Equation (7.20) may be found by summing all the lth NMSE contributions in Equation (7.22) over the K_0 taps of the CIR encountered, which can be expressed as

$$NMSE_{\text{MMSE}} = \frac{1}{\gamma} \sum_{l=0}^{K_0-1} \frac{1}{(\sigma_w^2/\sigma_x^2\sigma_l^2) + K} \approx \frac{1}{\gamma} \frac{L}{K}, \tag{7.23}$$

where, as before, K is the number of OFDM subcarriers and γ is the average SNR value, while L is the number of non-zero SS-CIR taps encountered. The resultant NMSE described by Equation (7.23) is depicted in Figure 7.3.

7.4.2 Reduced-Complexity SS-CIR Estimator

As can be seen from Equation (7.19), the direct MMSE approach to the problem of estimating the SS-CIR taps $h[n, l]$ involves a time-variant matrix inversion, which introduces a relatively high computational complexity [265]. In order to reduce the associated computational complexity, we introduce a two-step low-complexity SS-CIR estimator invoking an approach which bypasses the computationally intensive matrix inversion operation encountered in Equation (7.19). We will show that the method proposed first employs a scalar MMSE estimator of the subcarrier-related FD-CTF coefficients $H[n, k]$ of Equation (7.11), followed by a simplified MMSE SS-CIR estimator, which exploits the average MSE expression of Equation (7.13) associated with the scalar MMSE FD-CTF estimator of the first processing step.

Following the Bayesian estimation theory of [330], the MMSE CTF estimates $\tilde{H}_{\text{MMSE}}[n, k]$ of Equation (7.11) may be modelled as complex-Gaussian-distributed variables having a mean identical to that of $H[n, k]$, which represents the actual FD-CTF coefficients encountered, and a variance of $\sigma_v^2 = \sigma_H^2 NMSE_{max}$, where σ_H^2 is the average channel output power and $NMSE_{max}$ is the average

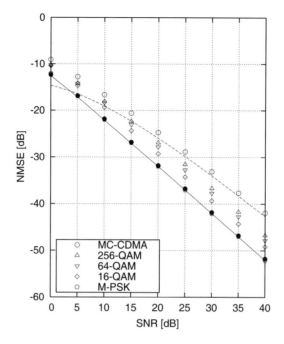

Figure 7.3: NMSE associated with both the **Minimum Mean Square Error (MMSE)** and the **Reduced Complexity (RC) MMSE** SS-CIR estimators described by Equations (7.20) and (7.30), respectively. The markers on the plot correspond to the simulated cases of M-PSK, 16-, 64- and 256-QAM-modulated OFDM, as well as M-QAM-modulated MC-CDMA in conjunction with MMSE (bold) and RC-MMSE (open) SS-CIR estimators, while the lines correspond to the analytically calculated NMSE lower bounds for the cases of MC-CDMA in conjunction with both the MMSE (solid) and the RC-MMSE (dashed) estimators evaluated using Equations (7.23) and (7.32), respectively. Note that the markers associated with the different modulation schemes and RC-MMSE estimator coincide.

NMSE quantified in Equation (7.13). Thus we can write

$$\tilde{H}_{\text{MMSE}}[n, k] = H[n, k] + v[n, k], \tag{7.24}$$

where $v[n, k]$ represents the i.i.d. complex-Gaussian noise samples having a zero mean and a variance of σ_v^2.

By substituting (1.8) into (7.24) we arrive at

$$\tilde{H}_{\text{MMSE}}[n, k] = \sum_{l=0}^{K_0-1} W_K^{kl} h[n, k] + v[n, k], \tag{7.25}$$

where $W_K = e^{-j2\pi(1/K)}$, which can be rewritten in matrix form as

$$\tilde{\boldsymbol{H}}_{\text{MMSE}}[n] = \boldsymbol{W}\boldsymbol{h}[n] + \boldsymbol{v}[n], \tag{7.26}$$

where the $(K \times K_0)$-dimensional matrix $\boldsymbol{W}$ corresponds to the Fourier transform of the zero-padded SS-CIR vector $\boldsymbol{h}[n]$ and is defined by $W_{kl} = W_K^{kl}$ for $k = 0, 1, \ldots, K-1$ and $l = 0, 1, \ldots, K_0 - 1$.

The MMSE estimator of the SS-CIR taps $h[n, k]$ of the linear vector model described by Equation (7.26) is given by [330]

$$\hat{\boldsymbol{h}} = (\boldsymbol{C}_h^{-1} + \boldsymbol{W}^{\text{H}} \boldsymbol{C}_v^{-1} \boldsymbol{W})^{-1} \boldsymbol{W}^{\text{H}} \boldsymbol{C}_v^{-1} \tilde{\boldsymbol{H}}_{\text{MMSE}}, \tag{7.27}$$

where we omit the time-domain OFDM-block-spaced index n for the sake of notational simplicity and define $\boldsymbol{C}_h$ and $\boldsymbol{C}_v$ as the covariance matrices of the SS-CIR vector $\boldsymbol{h}$ and the scalar-MMSE FD-CTF estimator's noise vector $\boldsymbol{v}$, respectively. The elements of the noise vector $\boldsymbol{v}$ are assumed to be complex-Gaussian i.i.d. samples and therefore we have $\boldsymbol{C}_v = \sigma_v^2 \boldsymbol{I}$. On the other hand, as follows from the assumption of having uncorrelated SS-CIR taps, the SS-CIR taps' covariance matrix is a diagonal matrix $\boldsymbol{C}_h = \mathrm{diag}(\sigma_l^2)$, where $\sigma_l^2 = \mathrm{E}\{|h[n,l]|^2\}$. Substituting $\boldsymbol{C}_h$ and $\boldsymbol{C}_v$ into Equation (7.27) yields

$$
\begin{aligned}
\hat{\boldsymbol{h}} &= \left(\mathrm{diag}\left(\frac{1}{\sigma_l^2} \right) + \frac{1}{\sigma_v^2} \boldsymbol{W}^{\mathrm{H}} \boldsymbol{W} \right)^{-1} W^{\mathrm{H}} \frac{1}{\sigma_v^2} \tilde{\boldsymbol{H}}_{\mathrm{MMSE}} \\
&= \left(\mathrm{diag}\left(\frac{\sigma_v^2}{\sigma_l^2} \right) + K\boldsymbol{I} \right)^{-1} \boldsymbol{W}^{\mathrm{H}} \tilde{\boldsymbol{H}}_{\mathrm{MMSE}} \\
&= \mathrm{diag}\left(\frac{\sigma_l^2}{\sigma_v^2 + K\sigma_l^2} \right) \boldsymbol{W}^{\mathrm{H}} \tilde{\boldsymbol{H}}_{\mathrm{MMSE}},
\end{aligned}
\tag{7.28}
$$

where we have exploited the fact that

$$
[\boldsymbol{W}^{\mathrm{H}} \boldsymbol{W}]_{l,l'} = \sum_{k=0}^{K-1} e^{-j2\pi k(l-l')/K} = K\delta[l - l']
\tag{7.29}
$$

and therefore $\boldsymbol{W}^{\mathrm{H}} \boldsymbol{W} = K\boldsymbol{I}$, where $\boldsymbol{I}$ is a $(K_0 \times K_0)$-dimensional identity matrix.

Finally, upon substituting Equation (7.11) into Equation (7.28) we arrive at a scalar expression for the Reduced Complexity (RC) a posteriori MMSE SS-CIR estimator in the form of

$$
\hat{h}[n,l] = \frac{\sigma_l^2}{\sigma_v^2 + K\sigma_l^2} \sum_{k=0}^{K-1} W_K^{kl} \frac{\hat{x}^*[n,k]y[n,k]}{|\hat{x}[n,k]|^2 + (\sigma_w^2/\sigma_H^2)}.
\tag{7.30}
$$

The corresponding NMSE associated with the lth RC-MMSE SS-CIR tap estimate $\hat{h}[l]$ is given by [330]

$$
\begin{aligned}
NMSE_{\mathrm{RCMMSE},l} &= \frac{\sigma_v^2}{\sigma_H^2} \frac{\sigma_l^2}{\sigma_v^2 + K\sigma_l^2} \\
&= \frac{\sigma_v^2}{\sigma_H^2} \frac{1}{(\sigma_v^2/\sigma_l^2) + K},
\end{aligned}
\tag{7.31}
$$

where $\sigma_v^2 = \sigma_H^2 NMSE_{H,\max}$ is the variance of the noise samples $v[k]$ in Equation (7.24), while $NMSE_{H,\max}$ is the maximum NMSE of the scalar MMSE FD-CTF estimator of Equation (7.11). The overall NMSE corresponding to the MMSE SS-CIR estimator of Equation (7.30) can be found similarly to Equation (7.23) by summing all of the lth contributions quantified by Equation (7.31) over the K_0 taps of the CIR encountered, which can be expressed using Equation (7.13) as

$$
\begin{aligned}
NMSE_{\mathrm{RCMMSE}} &= \frac{1}{\gamma} \exp\left(\frac{1}{\gamma} \right) \mathrm{Ei}\frac{1}{\gamma} \sum_{l=0}^{K_0-1} \frac{1}{(\sigma_v^2/\sigma_l^2) + K} \\
&\approx \frac{1}{\gamma} \exp\frac{1}{\gamma} \mathrm{Ei}\left(\frac{1}{\gamma} \right) \frac{L}{K},
\end{aligned}
\tag{7.32}
$$

where, as before, K is the number of OFDM subcarriers and γ is the average SNR value, while L is the number of non-zero SS-CIR taps encountered. The resultant NMSE described by Equation (7.32) represents the lower bound of the NMSE exhibited by the RC-MMSE SS-CIR estimator in conjunction with complex-Gaussian-distributed transmitted samples $x[n,k]$ typically encountered in an MC-CDMA system having a high spreading factor. The resultant NMSE performance is depicted in Figure 7.3 using a dashed line.

7.4.3 Complexity Study

As was shown in Section 7.3, the LS approach to the problem of DDCE-aided OFDM schemes [265] is not suitable in the case of MC-CDMA systems. The MMSE approach of Section 7.4.1 constitutes an appropriate solution; however, it exhibits a relatively high computational complexity imposed by the evaluation and inversion of the $(K_0 \times K_0)$-dimensional matrix $(\boldsymbol{A} + \boldsymbol{W}^{\mathrm{H}}\mathrm{diag}(|x[k]|^2)\boldsymbol{W})$ in Equation (7.20). More explicitly, the MMSE SS-CIR estimator of Equation (7.20) has a computational complexity of the order of $O(K^2K_0 + KK_0^2 + K_0^3)$, where K is the number of OFDM subcarriers and K_0 is the number of SS-CIR taps encountered. By contrast, the RC SS-CIR estimator of Equation (7.30), which avoids the matrix inversion operation, has a complexity of the order of $O(K + K\log_2 K + K_0)$, which is similar to the complexity associated with the conventional LS estimator employed in [265]. It can be seen that the difference between the proposed estimation methods expressed in terms of the associated computational complexity is substantial. In the next section we will derive an alternative RC-MMSE estimator which is capable of estimating the FS-CIR taps of Equation (1.7) using an approach similar to that described above.

7.4.4 MMSE FS-CIR Estimator

The first constituent component of our estimator, namely the scalar MMSE CTF estimator, is identical to that derived in Section 7.4.2 and described by Equation (7.11). Furthermore, our approach used for deriving the MMSE FS-CIR estimator is similar to that utilized in Section 7.4.2, but it exhibits several substantial differences, as detailed below.

By substituting the FD-CTF of Equation (1.7) into (7.24) we arrive at

$$\tilde{H}[n,k] = C(k\Delta f)\sum_{l=1}^{L}\alpha_l[n]W_K^{k\tau_l/T_s} + v[n,k], \qquad (7.33)$$

where, as previously, $C(f)$ is the frequency response of the transceiver's pulse-shaping filter, $W_K \triangleq e^{-\jmath 2\pi(1/K)}$, while $\alpha_l[n]$ and τ_l are the amplitudes and the relative delays of the FS-CIR taps, respectively. Equation (7.33) can be expressed in matrix form as

$$\begin{aligned}\tilde{\boldsymbol{H}}[n] &= \mathrm{diag}(C[k])\boldsymbol{W}\boldsymbol{\alpha}[n] + \boldsymbol{v}[n]\\ &= \boldsymbol{T}\boldsymbol{\alpha}[n] + \boldsymbol{v}[n],\end{aligned} \qquad (7.34)$$

where we define the $(K{\times}L)$-dimensional matrix $\boldsymbol{T} \triangleq \mathrm{diag}(C[k])\boldsymbol{W}$, in which $\mathrm{diag}(C[k])$ is a $(K{\times}K)$-dimensional diagonal matrix with the corresponding elements of the vector $C[k]$ on the main diagonal, while $\boldsymbol{W}$ is the Fourier transform matrix defined by $W_{kl} \triangleq W_K^{k\tau_l/T_s}$ for $k = -K/2, \ldots, K/2 - 1$ and $l = 1, \ldots, L$.

The MMSE estimator of the FS-CIR taps $\alpha_l[n]$ of the linear vector model described by (7.34) is given by [330]

$$\hat{\boldsymbol{\alpha}} = (\boldsymbol{C}_\alpha^{-1} + \boldsymbol{T}^{\mathrm{H}}\boldsymbol{C}_v^{-1}\boldsymbol{T})^{-1}\boldsymbol{T}^{\mathrm{H}}\boldsymbol{C}_v^{-1}\tilde{\boldsymbol{H}}, \qquad (7.35)$$

where we omit the time-domain OFDM-block-spaced index n for the sake of notational simplicity and define $\boldsymbol{C}_\alpha$ and $\boldsymbol{C}_v$ as the covariance matrices of the FS-CIR vector $\boldsymbol{\alpha}$ and CTF-estimator noise vector $\boldsymbol{v}$, respectively. The elements of the noise vector $\boldsymbol{v}$ are assumed to be i.i.d. complex-Gaussian-distributed samples and therefore we have $\boldsymbol{C}_v = \sigma_v^2\boldsymbol{I}$. On the other hand, as follows from Equation (1.1), the FS-CIR taps' covariance matrix is a diagonal matrix $\boldsymbol{C}_h = \mathrm{diag}(\sigma_l^2)$, where $\sigma_l^2 \triangleq \mathrm{E}\{|\alpha_l[n]|^2\}$. Substituting $\boldsymbol{C}_\alpha$ and $\boldsymbol{C}_v$ into (7.35) yields

$$\begin{aligned}\hat{\boldsymbol{\alpha}} &= \left(\mathrm{diag}\left(\frac{1}{\sigma_l^2}\right) + \frac{1}{\sigma_v^2}\boldsymbol{T}^{\mathrm{H}}\boldsymbol{T}\right)^{-1}\boldsymbol{T}^{\mathrm{H}}\frac{1}{\sigma_v^2}\tilde{\boldsymbol{H}}\\ &= (\sigma_v^2\boldsymbol{I} + \mathrm{diag}(\sigma_l^2)\boldsymbol{T}^{\mathrm{H}}\boldsymbol{T})^{-1}\mathrm{diag}(\sigma_l^2)\boldsymbol{T}^{\mathrm{H}}\tilde{\boldsymbol{H}} = \boldsymbol{A}\tilde{\boldsymbol{H}}.\end{aligned} \qquad (7.36)$$

The matrix inversion operation associated with the process of evaluating the estimator matrix $\boldsymbol{A}$ in Equation (7.36) cannot be avoided as opposed to the case of the SS-CIR estimation scheme of Section 7.4.2. However, the estimator matrix $\boldsymbol{A}$ is data independent and may be calculated only once for the case of encountering Wide-Sense Stationary (WSS) channel statistics. In the case of non-WSS channels, where the average FS-CIR taps' magnitudes σ_l^2 and the corresponding relative delays τ_l are time variant, the estimator matrix $\boldsymbol{A}$ can be tracked using the low-complexity Projection Approximation Subspace Tracking (PAST) techniques discussed for example in [355] and [511].

The corresponding covariance matrix associated with the FS-CIR estimate vector $\hat{\boldsymbol{\alpha}}$ can be expressed as in [330]

$$\boldsymbol{C}_\alpha = \sigma_v^2(\sigma_v^2\boldsymbol{I} + \mathrm{diag}(\sigma_l^2)\boldsymbol{T}^H\boldsymbol{T})^{-1}\mathrm{diag}(\sigma_l^2) \tag{7.37}$$

and the resultant NMSE of the RC-MMSE FS-CIR estimator proposed is given by

$$NMSE_\alpha = \frac{\sigma_v^2}{\sigma_H^2}\,\mathbf{tr}((\sigma_v^2\boldsymbol{I} + \mathrm{diag}(\sigma_l^2)\boldsymbol{T}^H\boldsymbol{T})^{-1}\mathrm{diag}(\sigma_l^2)), \tag{7.38}$$

where $\mathbf{tr}(()\boldsymbol{A})$ is the *trace* of the matrix $\boldsymbol{A}$.

7.4.5 Performance Analysis

The performance criteria $NMSE_h$ and $NMSE_\alpha$ of Equations (7.32) and (7.38) respectively cannot be compared directly, since they refer to the estimation processes of different sets of parameters, namely the SS-CIR taps $h[n,k]$, $k = 0,\ldots,K_0 - 1$, and the FS-CIR taps $\alpha_l[n]$, $l = 1,\ldots L$. In order to perform a meaningful comparison of the methods considered we used the NMSE between the two CTFs corresponding to the encountered CIR and the estimated CIR, thus

$$NMSE_H \triangleq \mathrm{E}\{|H[n,k] - \hat{H}[n,k]|^2\}. \tag{7.39}$$

In the case of the SS-CIR estimator we have

$$
\begin{aligned}
NMSE_{H;SS} &= \frac{1}{\sigma_H^2}\mathrm{E}\{\mathbf{tr}((\boldsymbol{H} - \hat{\boldsymbol{H}})(\boldsymbol{H} - \hat{\boldsymbol{H}})^H)\}\\
&= \frac{1}{K\sigma_H^2}\mathbf{tr}(\boldsymbol{W}\mathrm{E}\{(\boldsymbol{h} - \hat{\boldsymbol{h}})(\boldsymbol{h} - \hat{\boldsymbol{h}})^H\}\boldsymbol{W}^H)\\
&= NMSE_h, \tag{7.40}
\end{aligned}
$$

where the Discrete Fourier Transform (DFT) $\boldsymbol{W}$ of Equation (7.25) is a unitary matrix. On the other hand, for the case of the FS-CIR estimator we have

$$
\begin{aligned}
NMSE_{H;FS} &= \frac{1}{\sigma_H^2}\mathrm{E}\{\mathbf{tr}((\boldsymbol{H} - \hat{\boldsymbol{H}})(\boldsymbol{H} - \hat{\boldsymbol{H}})^H)\}\\
&= \frac{1}{K\sigma_H^2}\mathbf{tr}(\boldsymbol{T}\mathrm{E}\{(\boldsymbol{\alpha} - \hat{\boldsymbol{\alpha}})(\boldsymbol{\alpha} - \hat{\boldsymbol{\alpha}})^H\}\boldsymbol{T}^H)\\
&= \frac{\sigma_v^2}{K\sigma_H^2}\,\mathbf{tr}(\boldsymbol{T}\boldsymbol{C}_\alpha\boldsymbol{T}^H), \tag{7.41}
\end{aligned}
$$

where $\boldsymbol{C}_\alpha$ is the covariance matrix of the FS-CIR taps' estimates described by Equation (7.37).

The theoretical $NMSE_H$ performances of both the SS-CIR and the FS-CIR RC-MMSE estimators discussed in Sections 7.4.2 and 7.4.4 and described by Equations (7.40) and (7.41) respectively are depicted in Figure 7.4. As expected, the FS-CIR estimator exhibits a lower NMSE over the whole range of the delay spread RMS values τ_{rms}, which demonstrates its robustness in severe channel conditions exhibiting time-variant delay spread.

In this section, we present our simulation results for both the OFDM and the MC-CDMA systems employing the channel estimation schemes considered.

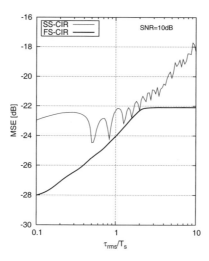

Figure 7.4: Mean Square Error (MSE) exhibited by a posteriori SS- and FS-CIR-based CTF estimators as a function of the channel's sample-rate-normalized RMS delay value τ_{rms}/T_s. The channel encountered corresponds to the eight-path Rayleigh-fading Bug's channel model characterized in [346] having a Gaussian noise variance of 10 dB. The results were evaluated from Equations (7.40) and (7.41).

Our simulations were performed in the baseband frequency domain and the system configuration characterized in Table 7.1 is to a large extent similar to that used in [260]. We assume a total bandwidth of 800 kHz. In the OFDM mode, the system utilizes 128-QPSK-modulated orthogonal subcarriers. In the MC-CDMA mode we employ eight concatenated 16-chip Walsh–Hadamard (WH) codes for frequency-domain interleaved spreading of the QPSK-modulated bits over the $8 \cdot 16 = 128$ orthogonal subcarriers. All the 128 WH spreading codes, each constituted by eight interleaved groups of 16 codes, are assigned to a single user and hence the effective data rate is similar in both the OFDM and MC-CDMA modes. For Forward Error Correction (FEC) we use $\frac{1}{2}$-rate turbo coding [314] employing two constraint-length $K = 3$ Recursive Systematic Convolutional (RSC) component codes and the standard 124-bit WCDMA UMTS code interleaver of [512]. The octally represented RCS generator polynomials of $(7, 5)$ were used.

Firstly, we will demonstrate the achievable performance of the system considered under the assumption of perfect channel knowledge, where knowledge of the frequency-domain subcarrier-related coefficients $H[n, k]$ is available in the receiver. Figure 7.6 characterizes both (a) the uncoded and (b) the turbo-coded Bit Error Rate (BER) exhibited by the QPSK-modulated OFDM and MC-CDMA systems in conjunction with the three different channel models discussed in Section 1.7.1, namely the SWATM channel [265], the COST207 BU channel [347] and Bug's channel characterized in [346]. As expected, in the uncoded OFDM scenario the achievable BER is similar to the BER associated with a flat Rayleigh-fading channel, regardless of the actual channel model encountered. This can be explained by the fact that the uncoded OFDM system effectively experiences flat Rayleigh fading on each frequency-domain subcarrier. In an uncoded OFDM system the adjacent information-carrying symbols are demodulated independently and thus the associated system's BER performance is dominated by the error rates associated with the severely faded subcarriers. In other words, such a system is incapable of exploiting the potential frequency-domain diversity gains available in the dispersive channel, as discussed in Section 1.8.2. By contrast, the uncoded MC-CDMA system avoids this phenomenon with the aid of frequency-domain spreading of the information-carrying symbols. Furthermore, different channel models characterized by different PDPs result in different potential frequency-domain diversity

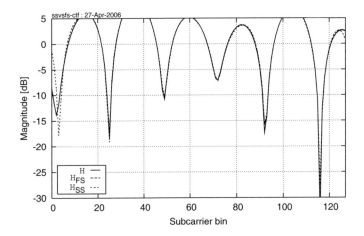

Figure 7.5: Snapshot of both fractionally and sample-spaced estimates of the channel transfer function at $\mathrm{SNR} = 10\,\mathrm{dB}$.

Table 7.1: System parameters.

Parameter	OFDM	MC-CDMA
Channel bandwidth	800 kHz	
Number of carriers K	128	
Symbol duration T	160 μs	
Max. delay spread $\tau_{\max}$	40 μs	
Channel interleaver	WCDMA [512]	—
	248 bit	
Modulation	QPSK	
Spreading scheme	—	WH
FEC	Turbo code [314], rate 1/2	
Component codes	RSC, K $= 3(7, 5)$	
Code interleaver	WCDMA (124 bit)	

gains. As illustrated in Figure 1.10, the SWATM channel model is characterized by a CIR having three taps, where most of the signal power is accommodated by the first tap, hence it behaves similar to a non-dispersive channel and results in a relatively low potential frequency diversity gain, as confirmed by the results depicted in Figure 7.6. By contrast, both the COST207 BU and Bug's channel models have seven- and eight-tap CIRs respectively and thus allow the MC-CDMA system to benefit from a relatively high frequency diversity gain. Similar conclusions can be inferred from Figure 7.6(b), where both the OFDM and MC-CDMA systems benefit from the available frequency diversity gain with the aid of turbo coding. It can be seen in Figure 7.6(b) that the MC-CDMA system slightly outperforms its OFDM counterpart as a result of averaging the error effects with the aid of frequency-domain spreading of the information-carrying symbols.

7.4.5.1 RC-MMSE SS-CIR Estimator Performance

Here we employed the eight-path Rayleigh-fading Bug channel model characterized in [346], using the delay spread of $\tau_{\mathrm{rms}} = 1\,\mu$s and the OFDM-symbol-normalized Doppler frequency of $f_D = 0.01$.

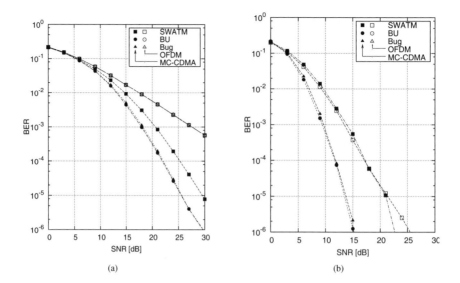

(a)　　　　　　　　　　　　　　　　　(b)

Figure 7.6: Bit Error Rate (BER) exhibited by the (a) **uncoded** and (b) **turbo-coded** QPSK-modulated OFDM and MC-CDMA systems under channel conditions described by SWATM, COST207 Bad–Urban (BU) and Bug channel models.

Figure 7.7(a) characterizes the NMSE exhibited by the DDCE scheme of Figure 7.1 using both the full-complexity MMSE SS-CIR estimator and the RC-MMSE SS-CIR estimator of Sections 7.4.1 and 7.4.2, respectively. Furthermore, the achievable turbo-coded BER of the corresponding QPSK-modulated OFDM and MC-CDMA systems is depicted in Figure 7.7(b). The simulations were carried out over the period of $100\,000$ QPSK-modulated $K = 128$-subcarrier OFDM/MC-CDMA symbols. It can be seen in Figure 7.7(a) that the RC-MMSE method outperforms its MMSE counterpart in the context of both the OFDM and MC-CDMA systems considered. This result can be explained by the fact that in our RC-MMSE CIR estimator we employ the MRC-aided MMSE CTF estimator of Equation (7.15), which takes advantage of the available a priori predicted CTF estimates $\breve{H}[n, k]$ and enhances the performance of the RC-MMSE CIR estimator in comparison with the pure a posteriori full-complexity MMSE CIR estimator of Section 7.4.1. Moreover, as becomes evident from Figure 7.7(b), the MMSE/RC-MMSE SS-CIR operating in the context of the MC-CDMA system outperforms its OFDM counterpart.

7.4.5.2　Fractionally Spaced CIR Estimator Performance

In this section we consider the achievable performance of our DDCE scheme employing both the SS-CIR RC-MMSE estimator of Section 7.4.2 and the Fractionally Spaced RC-MMSE CIR estimator advocated in Section 7.4.4 in the context of both OFDM and MC-CDMA systems communicating over Bug's eight-path dispersive Rayleigh-fading channel characterized in [346]. Here we employ an FS-CIR as opposed to the SS-CIR considered in Section 7.4.5.1. Their difference may be informally assessed with the aid of Figure 7.5.

Figure 7.8(a) portrays the NMSE exhibited by the DDCE scheme of Figure 7.1 employing both the SS-CIR estimator described in Section 7.4.2 and that of the FS-CIR estimator derived in Section 7.4.4 in the context of both the OFDM and MC-CDMA systems considered. The corresponding achievable BER performance is depicted in Figure 7.8(b). The simulations were carried out over the period of $100\,000$ QPSK-modulated $K = 128$-subcarrier OFDM/MC-CDMA symbols. Comparing the results of Figures 7.7 and 7.8, we may conclude that the DDCE employing the a posteriori SS-CIR RC-MMSE

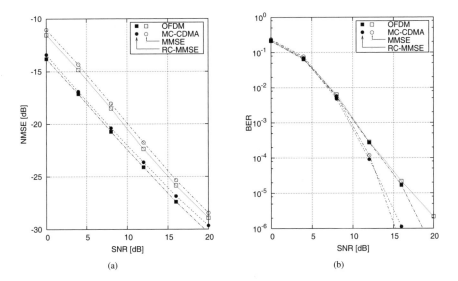

Figure 7.7: (a) **Normalized Mean Square Error (NMSE)** and (b) **Bit Error Rate (BER)** exhibited by the channel estimator which follows the philosophy of Figure 7.1 and employs the Minimum Mean Square Error (MMSE) and the Reduced-Complexity MMSE a posteriori SS-CIR estimators of Equations (7.19) and (7.27), respectively. The a priori prediction is performed using the robust SS-CIR predictor [265] assuming matching propagation conditions described by the COST207 BU channel model having a normalized Doppler frequency of $f_D = 0.01$. The turbo-coded QPSK-modulated OFDM and MC-CDMA modes are identified using the □ and ○ markers, respectively.

method suffers from a substantial performance degradation when assessed in conjunction with the channel characterized by an FS-CIR. Furthermore, the DDCE scheme utilizing the SS-CIR estimator for communicating over a channel characterized by an FS-CIR exhibits an irreducible noise floor at high SNR values. In order to explain this result we will refer to the *leakage* effect discussed in Section 1.7.1 and illustrated in Figure 1.12. Let us recall that a channel characterized by an FS-PDP results in numerous correlated non-zero SS-CIR taps. As a result, the a priori CIR predictor of Section 7.6 designed to track and predict a relatively low number of non-zero uncorrelated CIR taps fails to exploit the leakage-induced correlation observed between the adjacent SS-CIR taps. Furthermore, the correlation of the SS-CIR taps becomes different from the time-domain correlation model assumed during the predictor design and described in Section 1.7.1, which results in a biased channel estimation process. On the other hand, as can be seen in Figure 7.8, the DDCE employing the a posteriori FS-CIR RC-MMSE method of Section 7.4.2 does not experience any performance degradation and outperforms its SS-CIR estimator-based counterpart over the entire range of the SNR values considered. In addition, the achievable NMSE of the DDCE employed in an OFDM system is slightly lower than that exhibited by its MC-CDMA counterpart. This effect is caused by the energy distribution of the subcarrier-related samples $x[n, k]$ used in the channel estimation process. This effect was discussed in Section 7.4 and is illustrated in Figure 7.3.

This conclusion is further substantiated by Figure 7.9, where both the NMSE performance of the channel estimator schemes considered and the corresponding achievable BER performance of the MC-CDMA system are plotted as a function of the channel's Root Mean Square (RMS) delay spread value τ_{rms}. It can be seen in Figure 7.9(a) that the NMSE performance of the SS-CIR estimator-based DDCE scheme exhibits substantial sensitivity to the channel's delay spread, which is also confirmed by the theoretical results depicted in Figure 7.4. This effect can be explained by the fact that the SS-CIR estimator estimates the projections of the actual FS-CIR taps onto the adjacent SS-CIR taps. As can

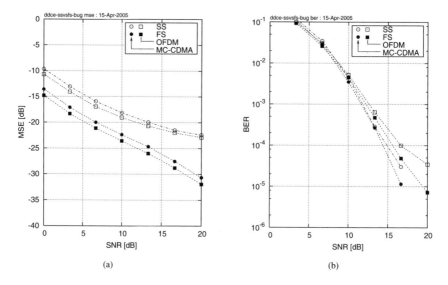

(a)   (b)

Figure 7.8: (a) **MSE** exhibited by the DDCE of Section 7.2 in the context of QPSK-modulated OFDM and MC-CDMA systems and (b) the corresponding achievable **BER** performance. Both performance curves are shown as a function of the average SNR at the receiver antenna. The *frame-variant* fading channel characterized by Bug's channel model [346] was associated with the OFDM symbol-normalized Doppler frequency of $f_D = 0.01$.

be seen in Figure 7.4, the accuracy of this process is highly sensitive to the delays and the amplitudes of the actual FS-CIR taps encountered. Furthermore, as the channel's RMS delay spread increases, the number of effective non-zero SS-CIR taps increases and hence the associated estimation accuracy degrades. On the other hand, the a posteriori FS-CIR estimator exhibits a higher robustness against the channel's delay spread variations, since the channel estimator tends to estimate only the actual FS-CIR taps encountered regardless of the specific values of the RMS delay spread. Additionally, as expected, the corresponding BER of the MC-CDMA system increases upon increasing the RMS delay spread τ_{rms}, because the frequency diversity rank tends to increase when τ_{rms} increases.

7.5   Parametric FS-CIR Estimation

7.5.1   Projection Approximation Subspace Tracking

Let $\boldsymbol{H}[n] \in \mathbb{C}^K$ be the vector of the subcarrier-related CTF coefficients associated with the channel model of Equation (1.14). As described in Section 1.7.1, the CIR associated with the CTF coefficient vector $\boldsymbol{H}[n]$ is constituted by a relatively low number of $L \ll K$ statistically independent Rayleigh-fading paths. The corresponding CIR components are related to the CTF coefficients $H[n, k]$ by means of Equation (1.7). The motivation for employing the so-called subspace technique [513] here is that usually we have $L \ll K$ and thus it is more efficient to estimate a low number of CIR-related taps in the low-dimensional signal subspace than estimating all the K FD-CTF coefficients.

Let λ_l and $\boldsymbol{u}_l$ be the eigenvalues and the corresponding eigenvectors of the CTF's covariance matrix $\boldsymbol{C}_H$, which is defined as follows:

$$\boldsymbol{C}_H = \underset{n}{E}\{\boldsymbol{H}[n]\boldsymbol{H}^{\mathrm{H}}[n]\}. \tag{7.42}$$

Then, $\boldsymbol{C}_H = \boldsymbol{U}\boldsymbol{\Sigma}\boldsymbol{U}^{\mathrm{H}}$, where $\boldsymbol{\Sigma} = \mathrm{diag}(\lambda_l)$ and $\boldsymbol{U} = [\boldsymbol{u}_1 \ldots \boldsymbol{u}_K]$.

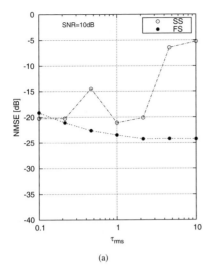
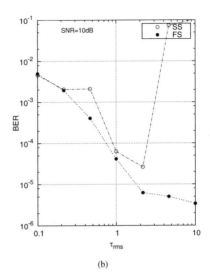

(a)   (b)

Figure 7.9: (a) **NMSE** exhibited by the DDCE of Section 7.2 as a function of the sample-period-normalized RMS delay spread τ_{rms} and (b) the corresponding achievable **BER** performance of the MC-CDMA system employing the aforementioned channel estimation scheme. Both curves correspond to Bug's channel model associated with the OFDM-symbol-normalized Doppler frequency of 0.01 and the average SNR of 10 dB recorded at the receive antenna.

The eigenvalues aligned in a descending order may be expressed as

$$\lambda_1 \geq \cdots \geq \lambda_L > \lambda_{L+1} = \cdots = \lambda_K = \sigma_w^2, \tag{7.43}$$

where the first L dominant eigenvalues $\lambda_1, \ldots, \lambda_L$ in conjunction with the L corresponding eigenvectors $u_1, \ldots, u_L$ may be termed the *signal* eigenvalues and eigenvectors, respectively [355]. The remaining eigenvalues $\lambda_{L+1}, \ldots, \lambda_K$ and eigenvectors $u_{L+1}, \ldots, u_K$ are termed the *noise* eigenvalues and eigenvectors. The resultant sets of *signal* and *noise* eigenvectors, which are column vectors, span the mutually orthogonal *signal* and *noise* subspaces U_S and U_N, such that

$$U_S = [u_1, \ldots, u_L] \quad \text{and} \quad U_N = [u_{L+1}, \ldots, u_K]. \tag{7.44}$$

The corresponding time-domain-related L-tap estimate of the FS-CIR vector $\alpha[n]$ may be obtained as follows:

$$\hat{\alpha} = U_S^{\mathrm{H}}[n]\tilde{H}[n]. \tag{7.45}$$

Furthermore, the reduced-noise estimate of the CTF vector $H[n]$ may reconstructed using

$$\hat{H}[n] = U_S[n]\hat{\alpha}[n]. \tag{7.46}$$

To evaluate and track the potentially time-variant signal subspace $U_S[n]$ we employ the subspace tracking method developed by Yang [355]. More specifically, we consider the following real-valued scalar objective function having the matrix argument of $W \in \mathbb{C}^{K \times L}$:

$$J(W) = \mathrm{E}\{\|H - WW^{\mathrm{H}}H\|^2\}$$
$$= \mathrm{tr}(C_H) - 2\,\mathrm{tr}(W^{\mathrm{H}}C_H W) + \mathrm{tr}(W^{\mathrm{H}}C_H W \cdot W^{\mathrm{H}}W). \tag{7.47}$$

As demonstrated by Yang in [355], the objective function $J(W)$ of Equation (7.47) exhibits the following important properties:

1. W is a stationary point of $J(W)$ if and only if $W = U_L Q$, where $U_L \in \mathbb{C}^{K \times L}$ contains any L distinct eigenvectors of C_H and $Q \in \mathbb{C}^{L \times L}$ is an arbitrary unitary matrix. Furthermore, at each stationary point, $J(W)$ equals the sum of these particular eigenvalues, whose eigenvectors are not involved in U_L [355, Theorem 1].

2. All stationary points of $J(W)$ are local saddle points except when U_L contains the L dominant eigenvectors of C_H. In this case, $J(W)$ attains the global minimum [355, Theorem 2].

3. The global convergence of W is guaranteed by using iterative minimization of $J(W)$ and the columns of the resultant value of W will span the signal subspace of C_H.

4. The use of an iterative algorithm to minimize $J(W)$ will always converge to an orthonormal basis of the signal subspace of C_H without invoking any orthonormalization operations during the iterations.

5. The global minimum of $J(W)$, W, does not necessarily contain the signal eigenvectors, but an arbitrary orthogonal basis of the signal subspace of C_H as indicated by the unitary matrix Q introduced in Property 1. In other words, $W = \arg \min J(W)$ if and only if $W = U_S Q$, where Q is an arbitrary unitary matrix.

6. For the simple scalar case of $L = 1$, the solution minimizing $J(W)$ is given by the most dominant normalized eigenvector of C_H.

Subsequently, Yang [355] proposes an iterative RLS algorithm for tracking the signal subspace of the channel's covariance matrix C_H. Specifically, upon replacing the expectation value in Equation (7.47) by the exponentially weighted sum of the RLS algorithm, we arrive at the following new objective function:

$$
\begin{aligned}
J(W[n]) &= \sum_{i=1}^{n} \eta^{n-i} \| H[i] - W[n]W^H[n]H[i] \|^2 \\
&= \mathrm{tr}(C_H) - 2\,\mathrm{tr}(W^H[n]C_H[n]W[n]) \\
&\quad + \mathrm{tr}(W^H[n]C_H[n]W[n] \cdot W^H[n]W[n]),
\end{aligned}
\tag{7.48}
$$

where $\eta \in (0,1)$ is the so-called *forgetting factor*, which accounts for possible deviations of the actual channel statistics encountered from the WSS assumption. Observe that the sole difference between the objective functions of Equations (7.47) and (7.48) is the introduction of the time-variant exponentially weighted sample covariance matrix [355], which may be expressed as

$$
C_H[n] = \sum_{m=1}^{n} \eta^{n-m} H[m]H^H[m] = \eta C_H[n-1] + H[n]H^H[n]
\tag{7.49}
$$

instead of the time-invariant matrix $C_H = \mathrm{E}\{HH^H\}$ of Equation (7.42).

The PAST algorithm may be derived by approximating the expression $W^H[n]H[m]$ in Equation (7.48), which may be interpreted as a projection of the vector $H[m]$ onto the column space of the matrix $W[n]$, by the readily available a posteriori vector $\alpha[m] = W^H[m]H[m]$. The resultant modified cost function may be formulated as

$$
J'(W[n]) = \sum_{m=1}^{n} \eta^{n-m} \| H[m] - W[n]\alpha[m] \|^2.
\tag{7.50}
$$

As is argued in [355], for stationary or slowly varying signals, the aforementioned projection approximation, hence the name PAST, does not substantially change the error surface associated with the corresponding cost function of Equation (7.50) and therefore does not significantly affect the convergence properties of the derived algorithm.

Similar to other RLS estimation schemes [292, 330], the cost function $J'(W[n])$ is minimized if

$$W = C_{H\alpha}[n]C_{\alpha\alpha}^{-1}[n], \qquad (7.51)$$

where

$$C_{H\alpha}[n] = \sum_{i=1}^{n} \eta^{n-i} H[i]\alpha^H[i] = \eta C_{H\alpha}[n-1] + H[n]\alpha^H[n] \qquad (7.52)$$

and

$$C_{\alpha\alpha}[n] = \sum_{i=1}^{n} \eta^{n-i} \alpha[i]\alpha^H[i] = \eta C_{\alpha\alpha}[n-1] + \alpha[n]\alpha^H[n]. \qquad (7.53)$$

Following the RLS approach [355], a low-complexity solution of the computational problem associated with minimizing the cost function $J'(W[n])$ of Equation (7.50) may be obtained using recursive updates of the matrix $W[n]$. More specifically, we have

$$W[n] = W[n-1] + e[n]k^H[n], \qquad (7.54)$$

where $e[n]$ is the estimation error vector, which may be recursively obtained as

$$e[n] = H[n] - W[n-1]\alpha[n-1], \qquad (7.55)$$

while

$$k[n] = \frac{P[n-1]\alpha[n]}{\eta + \alpha^H[n]P[n-1]\alpha[n]} \qquad (7.56)$$

denotes the RLS gain vector. Furthermore, the matrix $P[n]$ is the inverse of the CIR-related taps' $(L \times L)$-dimensional covariance matrix $C_{\alpha\alpha}$, which can be recursively calculated as follows:

$$P[n] = \frac{1}{\eta}\text{Tri}\{(I - k[n]\alpha^H[n])P[n-1]\}, \qquad (7.57)$$

where the operator $\text{Tri}\{\cdot\}$ indicates that only the upper triangular part of $P[n]$ is calculated and its Hermitian conjugate version is copied to the lower triangular part [355]. The resultant PAST algorithm is summarized in Algorithm 7.1, where we introduced an additional quantity $g[n] = P[n-1]H[n]$ to reduce the associated complexity further.

Algorithm 7.1 Projection approximation subspace tracking

$$\hat{\alpha}[n] = W^H[n-1]\hat{H}[n] \qquad (7.58a)$$

$$g[n] = P[n-1]\hat{\alpha}[n] \qquad (7.58b)$$

$$k[n] = \frac{g[n]}{\eta + \hat{\alpha}^H[n]g[n]} \qquad (7.58c)$$

$$P[n] = \frac{1}{\eta}\text{Tri}\{P[n-1] - k[n]g^H[n]\} \qquad (7.58d)$$

$$e[n] = \hat{H}[n] - W[n-1]\hat{\alpha}[n-1] \qquad (7.58e)$$

$$W[n] = W[n-1] + e[n]k^H[n] \qquad (7.58f)$$

7.5.2 Deflation PAST

In this work, however, we aim at maintaining the lowest possible complexity, hence we are particularly interested in the *deflation*-based version of the PAST algorithm derived in [355], which is referred to as the PASTD algorithm. The simple philosophy of the deflation method is the sequential estimation of the principal components of the CTF covariance matrix C_H [514]. Consequently, we first update the most dominant eigenvector $w_1[n]$ by applying the PAST method of Algorithm 7.1 in conjunction with $L = 1$. Subsequently, the projection of the current sample vector $H[n]$ onto the updated eigenvector $w_1[n]$ is subtracted from itself, resulting in a modified (deflated) version of the CTF vector in the following form: $H_2[n] = H[n] - w_1[n]w_1^H[n]H[n]$. The second most dominant eigenvector $w_2[n]$ has now become the most dominant one and therefore may be updated similarly to $w_1[n]$. By repeatedly applying this procedure, all the desired eigencomponents may be estimated.

The resultant PASTD method is summarized in Algorithm 7.2. Observe that Equations (7.60b–f) of Algorithm 7.2 constitute the PAST estimation procedure of Algorithm 7.1 in conjunction with $L = 1$. Note that the vector expressions of Equations (7.58b–d) in Algorithm 7.1 are substituted by the simple scalar expression of Equation (7.60c), where the new quantity $\lambda_l[n]$ constitutes an exponentially weighted estimate of the corresponding lth eigenvalue and can be identified as a scalar version of the $(L \times L)$-dimensional covariance matrix $C_{\alpha\alpha}[n] = P^{-1}[n]$ of Algorithm 7.1.

A particularly important property of the PASTD method of Algorithm 7.2 is that, as opposed to the PAST method of Algorithm 7.1, it enables the explicit tracking of the time-variant eigencomponents of the channel covariance matrix $C_H[n]$, namely the eigenvectors $w_l[n]$, as well as of the corresponding eigenvalues $\lambda_l[n]$ according to

$$w_l[n] = w_l[n-1] + \frac{\alpha_l^*[n]}{\lambda_l[n]}(H_l[n] - w_l[n-1]\alpha_l[n]), \tag{7.59}$$

where $\alpha_l[n] = w_l^H[n-1]H[n]$ and $\lambda_l[n] = \beta\lambda_l[n-1] + |\alpha_l[n]|^2$.

Algorithm 7.2 Deflation PAST

$$H_1[n] = H[n] \tag{7.60a}$$

$\text{for} \quad l = 1, 2, \ldots, L \quad \text{do}$

$$\alpha_l[n] = w_l^H[n-1]H_l[n] \tag{7.60b}$$

$$\lambda_l[n] = \beta\lambda_l[n-1] + |\alpha_l[n]|^2 \tag{7.60c}$$

$$e_l[n] = H_l[n] - w_l[n-1]\alpha_l[n] \tag{7.60d}$$

$$w_l[n] = w_l[n-1] + e_l[n](\alpha_l^*[n]/\lambda_l[n]) \tag{7.60e}$$

$$H_{l+1}[n] = H_l[n] - w_l[n]\alpha_l[n] \tag{7.60f}$$

end for

7.5.3 PASTD-Aided FS-CIR Estimation

In this section we will utilize the PASTD method detailed in Section 7.5.2 in the context of the channel estimation scheme characterized by Figure 7.1. More specifically, we consider a PASTD-aided a posteriori FS-CIR estimator which corresponds to the CIR estimator module of Figure 7.1. In order to analyse the achievable performance of the CIR estimator derived, we conceive a channel estimation scheme comprising the MMSE CTF estimator of Section 7.3.2 followed by the PASTD-aided CIR estimator of Section 7.5.2.

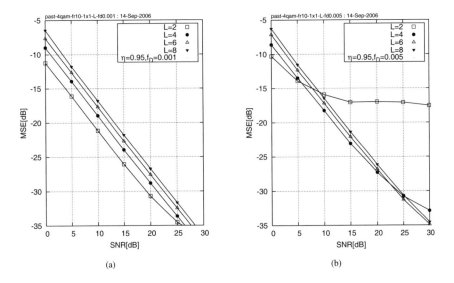

Figure 7.10: The **Mean Square Error (MSE)** exhibited by the **4-QAM-OFDM** system employing the **PASTD** CIR estimator of Algorithm 7.2 and tracking $L = 2, 4, 6$ and 8 CIR taps. The value of the PASTD forgetting factor was $\eta = 0.95$. We considered the scenarios of encountering the Doppler frequencies of (a) $f_D = 0.001$ and (b) $f_D = 0.005$. The abscissa represents the average SNR recorded at the receive antenna elements.

The achievable performance of the subspace tracking method of Section 7.5.2 is characterized in Figures 7.11 and 7.12, where we define the MSE performance criterion as follows:

$$MSE = E\left\{\sum_l |e_l[n]|^2\right\}, \tag{7.61}$$

where e_l is the FD-CTF tracking error defined by Equation 7.55. In our simulations we consider an OFDM system having $K = 128$ orthogonal QPSK-modulated subcarriers. The system characteristics are outlined in Table 1.11. We employ an OFDM-frame-variant channel model having a time-variant eight-tap PDP characterized by the COST207 BU channel model [347], as detailed in Section 1.7.2. Additionally, each individual propagation path undergoes fast Rayleigh fading with a corresponding OFDM-symbol-normalized Doppler frequency of either $f_D = 0.001$ or $f_D = 0.005$. The resultant channel can be characterized as a multi-path Rayleigh-fading channel with slowly varying PDP, where the relative delays τ_l associated with different PDP taps vary with time at a rate determined by the drift rate parameter ν_τ defined in Section 1.7.2.

Firstly, Figure 7.10 characterizes the achievable FD-CTF MSE performance of the PASTD method of Algorithm 7.2 for different ranks L of the estimated subspace, while assuming a constant value of $\eta = 0.95$ for the forgetting factor. Figures 7.10(a) and 7.10(b) correspond to encountering the Doppler frequencies of $f_D = 0.001$ and 0.005, respectively. From Figure 7.10, we may conclude that a high CIR estimator performance may be achieved when assuming that the estimated CTF signal subspace has a rank of $L = 4$, regardless of the actual number of paths constituting the multi-path channel encountered.

Secondly, Figure 7.11 characterizes the achievable MSE performance of the PASTD method of Algorithm 7.2 for different values of the forgetting factor η, while assuming a constant rank of $L = 4$ for the estimated subspace. Figures 7.11(a) and 7.11(b) correspond to encountering the Doppler frequencies

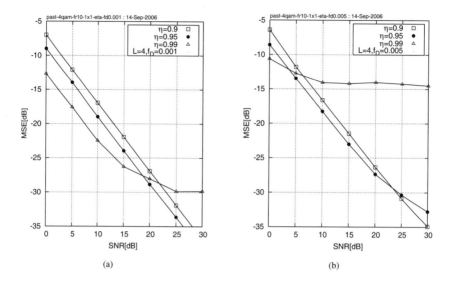

Figure 7.11: The **Mean Square Error (MSE)** exhibited by the **4-QAM-OFDM** system employing the **PASTD** method of Algorithm 7.2. The values of the PASTD forgetting factor were $\eta = 0.9, 0.95$ and 0.9. We considered the scenarios of encountering the Doppler frequencies of (a) $f_D = 0.001$ and (b) $f_D = 0.005$. The abscissa represents the average SNR recorded at the receive antenna elements.

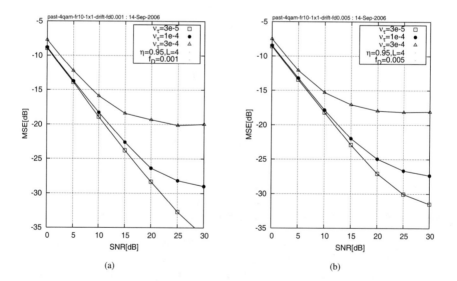

Figure 7.12: The **Mean Square Error (MSE)** exhibited by the **4-QAM-OFDM** system employing the **PASTD** method of Algorithm 7.2, while encountering different values of the PDP tap drift rate $\nu_\tau = 3 \times 10^{-5}, 10^{-4}$ and 3×10^{-4} as well as different values of the Doppler frequencies of (a) $f_D = 0.001$ and (b) $f_D = 0.005$. The abscissa represents the average SNR recorded at the receive antenna elements.

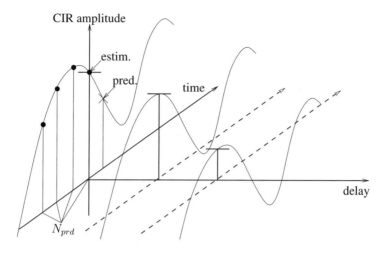

Figure 7.13: Stylized illustration of the estimation and prediction filter, both operating in the CIR-related domain using a number N_{prd} of previous a posteriori CIR-related tap estimates © [265]

of $f_D = 0.001$ and 0.005, respectively. As may be concluded from Figure 7.11, the optimum value of the forgetting factor η is largely dependent on the SNR as well as on the Doppler frequency encountered. Nevertheless, the compromise value of $\eta = 0.95$ appears to constitute a relatively good choice in the practical ranges of both SNR values and Doppler frequencies.

Finally, Figure 7.12 characterizes the achievable MSE performance of the PASTD method of Algorithm 7.2 for different values of the OFDM-symbol-normalized PDP tap drift rate ν_τ. Figures 7.12(a) and 7.12(b) correspond to encountering the Doppler frequencies of $f_D = 0.001$ and 0.005, respectively. Observe that the specific values of the parameter ν_τ assumed in Figure 7.12 substantially exceed the maximum value considered in the baseline scenario outlined in Table 1.11. Consequently, we may conclude that the CIR tracking method of Algorithm 7.2 exhibits an adequate performance over the entire range of practical channel conditions.

In order to complete the design of the DDCE scheme of Figure 7.1 we employ an a priori CIR predictor [265]. The CIR-related tap predictor considered can be employed in conjunction with both the SS-CIR and the FS-CIR estimators of Sections 7.4.2 and 7.4.4, as well as in combination with the parametric PASTD-aided CIR estimator of Section 7.5.2. Observe, however, that the low-rank PASTD-aided CIR estimator of Section 7.5.2 will require the prediction of a substantially lower number of $L \ll K_0$ CIR-related taps. More specifically, in the case of the system characterized by Table 1.11, the SS-CIR estimator of Section 7.4.2 will require the prediction of $K_0 = 32$ SS-CIR taps. This should be contrasted to the PASTD-aided CIR estimator of Section 7.5.2, which will require the prediction of only $L = 4$ FS-CIR-related taps, regardless of the actual number of paths encountered.

In the next section we present an overview of the major CIR tap prediction methods discussed in the literature [260, 265, 292, 293]. We analyse the achievable performance of each method with the aid of extensive simulations and conduct a comparative study aimed at identifying the most promising approaches.

7.6   Time-Domain A Priori CIR Tap Prediction

The philosophy of the a priori CIR predictor considered is illustrated in Figure 7.13. Our aim is to predict the SS/FS-CIR taps $\{\alpha_1[n+1], \ldots, \alpha_L[n+1]\}$ associated with the future channel conditions, given the history of the previous CIRs, namely the a posteriori estimates $\{\{\hat{\alpha}_l[n]\}, \{\hat{\alpha}_l[n-1]\}, \ldots\}$.

7.6.1   MMSE Predictor

As portrayed in Section 1.7.1, the lth CIR component $\alpha_l[n]$ undergoes a narrowband time-domain fading process characterized by the associated cross-correlation properties, which can be described by

$$E\{\alpha_l^*[n]\alpha_{l'}[n-m]\} = r_t[m]\delta[l-l'], \tag{7.62}$$

where $r_t[n]$ is the corresponding time-domain correlation function and $\delta[\cdot]$ is the Kronecker delta function.

This WSS narrowband process can be approximately modelled as a Finite Impulse Response (FIR) auto-regressive process of the order N_{prd} [265], yielding

$$\alpha_l[n+1] = \sum_{m=0}^{N_{\mathrm{prd}}-1} q[m]\alpha_l[n-m] + v_l[n+1], \tag{7.63}$$

where $q[m]$ represents the auto-regressive coefficients and $v_l[n]$ is the model noise.

Let us define the following column vectors:

$$\boldsymbol{\alpha}_l[n] = (\alpha_l[n], \alpha_l[n-1], \ldots, \alpha_l[n-N_{\mathrm{prd}}+1])^{\mathrm{T}}$$
$$\boldsymbol{q} = (q[0], q[1], \ldots, q[N_{\mathrm{prd}}-1])^{\mathrm{T}} \tag{7.64}$$

and rewrite Equation (7.63) in vectorial form as

$$\alpha_l[n+1] = \boldsymbol{\alpha}_l[n]^{\mathrm{T}}\boldsymbol{q} + v[n+1]. \tag{7.65}$$

Left-multiplying both sides of (7.65) by the complex conjugate of the column vector $\boldsymbol{\alpha}_l[n, l]$ and obtaining the expectation value over the time-domain index n yields

$$E\{\boldsymbol{\alpha}_l^*[n]\alpha_l[n+1]\} = E\{\boldsymbol{\alpha}_l^*[n](\boldsymbol{\alpha}_l^{\mathrm{T}}[n]\boldsymbol{q} + v[n+1])\}, \tag{7.66}$$

which can be represented as a set of Yule–Walker equations in the following form [515]:

$$\boldsymbol{r}_{\mathrm{apr}} = \boldsymbol{R}_{l;\mathrm{apt}}\boldsymbol{q}_l, \tag{7.67}$$

where the vector $\boldsymbol{r}_{\mathrm{apr}}$ is the autocorrelation vector of the predicted a priori CIR taps defined by

$$\boldsymbol{r}_{\mathrm{apr}} = \frac{1}{\sigma_l^2}E\{\boldsymbol{\alpha}_l^*[n]\alpha_l[n+1]\}, \tag{7.68}$$

and the matrix $\boldsymbol{R}_{\mathrm{apt}}$ is the autocorrelation matrix of the a posteriori CIR taps described in [265]

$$\boldsymbol{R}_{l;\mathrm{apt}} = \frac{1}{\sigma_l^2}E\{\hat{\boldsymbol{\alpha}}_l[n]\hat{\boldsymbol{\alpha}}_l^{\mathrm{H}}[n]\}$$
$$= \boldsymbol{R}_{\mathrm{apr}} + \rho_l\boldsymbol{I}, \tag{7.69}$$

where

$$\boldsymbol{R}_{\mathrm{apr}} = \frac{1}{\sigma_l^2}E\{\boldsymbol{\alpha}_l[n]\boldsymbol{\alpha}_l^{\mathrm{H}}[n]\} \tag{7.70}$$

and ρ_l is the parameter determined by the variance of the effective estimation noise imposed by the a posteriori CIR estimator employed, σ_{apt}^2, as well as the expectation magnitude of the CIR tap predicted σ_l^2, such that $\rho_l = \sigma_{\mathrm{apt}}^2/\sigma_l^2$.

The optimal solution of Equation (7.67) evaluated in the MSE sense is given by

$$\boldsymbol{q}_{l;\mathrm{prd}} = \boldsymbol{R}_{l;\mathrm{apt}}^{-1}\boldsymbol{r}_{\mathrm{apr}}. \tag{7.71}$$

In the specific scenario when the channel is described by Jakes' model [349], the a priori autocorrelation vector $\boldsymbol{r}_{\mathrm{apr}}$ can be formulated as $r_{\mathrm{apr}}[n] = r_J[n] = J_0(2\pi f_D n)$, $n = 1, 2, \ldots, N_{\mathrm{prd}}$, where $J_0(x)$ is a zero-order Bessel function of the first kind. The corresponding a posteriori autocorrelation matrix $\boldsymbol{R}_{\mathrm{apr}}$ is given by $R_{\mathrm{apr}}[n, m] = r_J[n - m] + \rho\delta[n - m]$, $n, m = 0, 1, \ldots, N_{\mathrm{prd}} - 1$, while the CIR predictor's coefficient vector is described by (7.71) and the prediction is performed according to

$$\check{\alpha}_l[n + 1] = \boldsymbol{q}_{l;\mathrm{prd}}^{\mathrm{T}}\hat{\alpha}_l[n], \quad l = 1, 2, \ldots, L. \tag{7.72}$$

The corresponding performance can be characterized using the frequency-domain NMSE criterion as derived in [265]

$$NMSE_{H;\mathrm{apr}} = \frac{1}{\sigma_H^2}\mathrm{E}\{|H[n, k] - \check{H}[n, k]|^2\}, \tag{7.73}$$

where $H[n, k]$ and $\check{H}[n, k]$ are the CTFs corresponding to the encountered CIR and the a priori predicted CIR $\check{\alpha}_l[n]$, respectively. From [265] we have

$$NMSE_{H;\mathrm{apr}} = \frac{1}{K\sigma_H^2}\sum_{l=1}^{L} MSE_{l;\mathrm{apr}}, \tag{7.74}$$

where

$$MSE_{l;\mathrm{apr}} = \sigma_l^2 - \boldsymbol{q}_{l;\mathrm{prd}}^{\mathrm{T}}\boldsymbol{r}_{l;\mathrm{apt}}^* - \boldsymbol{q}_{l;\mathrm{prd}}\boldsymbol{r}_{l;\mathrm{apt}}^{\mathrm{H}} + \boldsymbol{q}_{l;\mathrm{prd}}^{\mathrm{H}}\boldsymbol{R}_{l,\mathrm{apt}}\boldsymbol{q}_{l;\mathrm{prd}}. \tag{7.75}$$

The attainable NMSE performance of the a priori CIR predictor of Equation (7.72), evaluated for the scenario when the Doppler frequency assumed in the design of the receiver matches the actual Doppler frequency encountered, i.e. when $f_D = f_{D;\mathrm{prd}}$, is depicted in Figures 7.14 and 7.15. More specifically, in Figure 7.14 we demonstrate the NMSE of the CIR prediction method considered using a prediction filter of length $N_{\mathrm{prd}} = 1, 2, 4, 8, 16, 32$ and 64 as a function of the average SNR recorded at the receive antenna. As expected, the performance of the estimator improves when the prediction filter length N_{prd} increases, although the corresponding additional NMSE reduction becomes more modest for high values of the prediction filter length and hence a trade-off between the desired NMSE performance and the associated computational complexity has to be found. A similar system behaviour can be observed in Figure 7.15, where the NMSE is evaluated as a function of the OFDM-symbol-normalized Doppler frequency f_D.

7.6.2 Robust Predictor

The CIR tap prediction process described in the previous section exhibits a high CIR tap estimation performance under the assumption of perfect knowledge of the channel statistics. However, it suffers from a significant performance degradation when the actual channel statistics deviate from the model assumed, such as for example Jakes' model. The issue of statistical mismatch becomes increasingly detrimental in diverse wireless environments, where the channel conditions and the corresponding statistics are time dependent and cannot be assumed to be WSS.

As shown in [260] and [265], the MSE exhibited by the linear CIR predictor of (7.72) is upper bounded by the MSE encountered when communicating over an ideally band-limited channel having a perfect low-pass Doppler PSD function given by

$$p_{B;\mathrm{unif}}(f) = \begin{cases} 1/2f_D, & \mathrm{if}\ |f| < f_D \\ 0, & \mathrm{otherwise.} \end{cases} \tag{7.76}$$

Hence, we arrive at the concept of designing Li's [260] so-called *robust* linear predictor [265], which assumes encountering the worst-possible channel statistics. As pointed out in [267], such a robust channel predictor, optimized for the worst-case PSD of Equation (7.76), can be designed by using

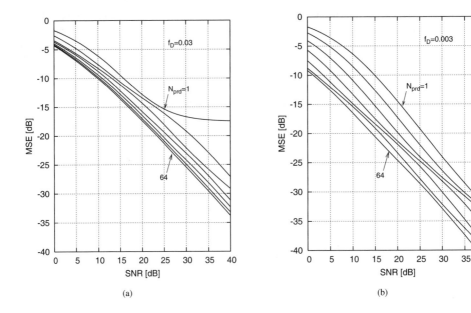

Figure 7.14: Mean Square Error (MSE) exhibited by the **robust a priori CIR predictor** as a function of the average SNR at the receive antenna. The curves on the plot correspond to the prediction filter lengths of $N_{\text{prd}} = 1, 2, 4, 8, 16, 32$ and 64 from top to bottom respectively. The Bug channel model with the OFDM-symbol-normalized Doppler frequency of (a) 0.003 and (b) 0.03 was considered. The results were evaluated from Equation (7.74).

the corresponding sinc-shaped a priori autocorrelation vector $\boldsymbol{r}_{\text{apr;rob}}$, which is given by

$$r_{\text{apr;rob}}[n] = r_B[n] = \frac{\sin 2\pi f_D n}{2\pi f_D n}, \quad n = 1, 2, \ldots, N_{\text{prd}} \qquad (7.77)$$

and by invoking the corresponding a posteriori autocorrelation matrix $\boldsymbol{R}_{\text{apt;rob}}$ defined by

$$R_{\text{apt;rob}}[n, m] = r_B[n - m] + \rho\delta[n - m], \qquad (7.78)$$

where we have $n, m = 0, 1, \ldots N_{\text{prd}} - 1$.

In Figure 7.16 we characterize the attainable NMSE performance of the robust a priori CIR predictor of Equation (7.72) for the scenario when the Doppler frequency $f_{D;\text{prd}}$ assumed in the design of the receiver does not match the actual Doppler frequency f_D encountered. It can be seen that the estimation method considered is robust against a mismatch between the assumed and the encountered Doppler frequency, as long as the encountered Doppler frequency does not exceed the assumed value, i.e. as long as $f_D \leq f_{D;\text{prd}}$.

7.6.3   MMSE Versus Robust Predictor Performance Comparison

The achievable performance of the DDCE scheme of Figure 7.1 employing the **robust** a priori CIR predictor of Section 7.6.2 under **matched** time-domain correlation conditions is quantified in Figure 7.17, when the assumed OFDM-symbol-normalized Doppler frequency $f_{D;\text{prd}}$ matches the actual value encountered. The **NMSE** exhibited by the channel estimation scheme considered is depicted in Figure 7.17(a), while the corresponding BER exhibited by the turbo-coded QPSK-modulated MC-CDMA system is shown in Figure 7.17(b). It can be seen that, while the estimation

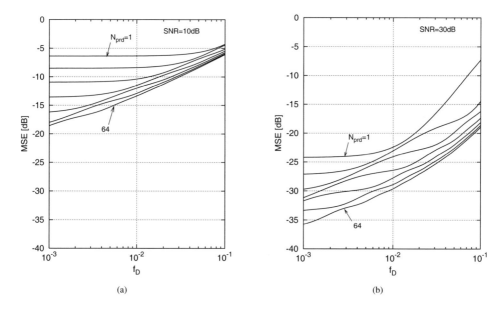

Figure 7.15: Mean Square Error (MSE) exhibited by the **robust a priori CIR predictor** as a function of the OFDM-symbol-normalized Doppler frequency f_D. The curves on the plot correspond to the prediction filter lengths of $N_{\mathrm{prd}} = 1, 2, 4, 8, 16, 32$ and 64 from top to bottom, respectively. Bug's channel model associated with the receive antenna SNRs of (a) 10 and (b) 30 dB was considered. The results were evaluated from Equation (7.74).

accuracy decreases upon increasing the Doppler frequency, the corresponding BER performance remains relatively unaffected.

Finally, Figure 7.18 illustrates the achievable performance of QPSK-modulated MC-CDMA employing the DDCE scheme of Figure 7.1 under **unmatched** time correlation conditions. Our simulations were performed at a constant value of the OFDM-symbol-normalized Doppler frequency assumed at the receiver, i.e. at $f_{D;\mathrm{prd}} = 0.03$. Furthermore, four different values of the actual normalized Doppler frequencies were used, i.e. $f_D = 0.03, 0.01, 0.003$ and 0.001. Figure 7.18(a) characterizes the NMSE performance of the DDCE scheme employed by the MC-CDMA system considered, while the corresponding turbo-coded BER is depicted in Figure 7.18(b). The achievable BER performance in the case of perfect Channel State Information (CSI), i.e. when the CTF is perfectly known at the receiver, is also depicted in Figure 7.18(a). It can be seen that the performance of the CIR predictor advocated is indeed tolerant to the mismatch of the actual Doppler frequency and that assumed during the predictor design, as long as the actual Doppler frequency does not exceed the value assumed in the predictor's design. Furthermore, the results depicted in Figure 7.18(a) substantiate our conclusion that the performance of the MC-CDMA system employing the channel estimation scheme of Figure 7.1 closely approaches the corresponding performance of the MC-CDMA system in the case of perfect CTF knowledge at the receiver. More explicitly, the BER performance corresponding to the different values of the Doppler frequency f_D fall within 1 dB of the BER performance associated with the perfect CSI scenario.

7.6.4 Adaptive RLS Predictor

On the other hand, in the RLS-based adaptive CIR tap prediction approach of [292,293], no assumptions were made concerning the channel's stationarity. The RLS estimation procedure is summarized in

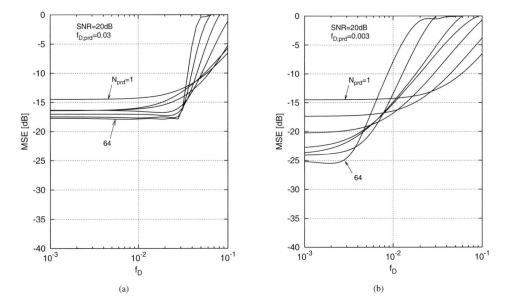

Figure 7.16: Mean Square Error (MSE) exhibited by the **robust a priori CIR predictor** as a function of the encountered OFDM-symbol-normalized Doppler frequency f_D. The results correspond to the case when the Doppler frequency assumed in the receiver does not match the actual value encountered. The assumed Doppler frequencies of (a) $f_D = 0.03$ and (b) 0.003 have been considered and different curves on each plot correspond to the prediction filter lengths of $N_{\rm prd} = 1, 2, 4, 8, 16, 32$ and 64 from top to bottom, respectively. The Bug channel model with the average receive antenna SNR of 20 dB is considered. The results were evaluated from Equation (7.74).

Algorithm 7.3. Consequently, the time-variant lth CIR tap's predictor filter coefficient vector $\boldsymbol{q}_l[n]$ is calculated by minimizing the following scalar cost function:

$$J_{\rm RLS;l}[n] = \sum_{i=1}^{n} \beta^{n-i} |\alpha_l[i+1] - \boldsymbol{q}_l^{\rm H}[n]\boldsymbol{\alpha}_l[i]|^2, \tag{7.79}$$

where $\beta \in (0,1)$ is the so-called *forgetting factor* [292], which accounts for possible deviations of the fading process encountered from the WSS assumption. The resultant recursive update for $\boldsymbol{q}_l[n]$ is given by

$$\boldsymbol{q}_l[n] = \boldsymbol{q}_l[n-1] + \boldsymbol{k}_l[n-1]e_l^*[n], \tag{7.80}$$

where

$$e_l[n] = \hat{\alpha}_l[n] - \boldsymbol{q}_l^{\rm H}[n-1]\hat{\boldsymbol{\alpha}}_l[n-1] \tag{7.81}$$

is the prediction error, while

$$\boldsymbol{k}_l[n] = \frac{\boldsymbol{P}_l[n-1]\hat{\boldsymbol{\alpha}}_l[n]}{\beta + \hat{\boldsymbol{\alpha}}_l^{\rm H}[n]\boldsymbol{P}_l[n-1]_l\hat{\boldsymbol{\alpha}}_l[n]} \tag{7.82}$$

denotes the RLS gain vector. Furthermore, the matrix $\boldsymbol{P}_l[n]$ is the inverse of the lth CIR tap's ($N_{\rm prd} \times N_{\rm prd}$)-dimensional sample covariance matrix, which can be recursively calculated as follows:

$$\boldsymbol{P}_l[n] = \frac{1}{\beta}(\boldsymbol{I} - \boldsymbol{k}_l[n]\hat{\boldsymbol{\alpha}}_l^{\rm H}[n])\boldsymbol{P}_l[n-1]. \tag{7.83}$$

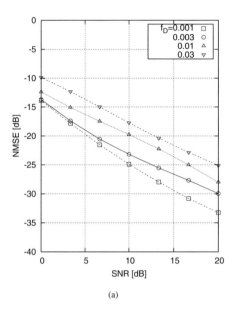 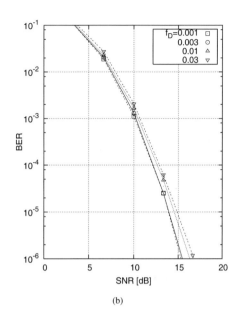

(a) (b)

Figure 7.17: (a) **NMSE** exhibited by the DDCE employing the **RC-MMSE FS-CIR** a posteriori estimator of Section 7.4.4 and the **robust** a priori CIR predictor of Section 7.6.2 as a function of the average SNR recorded at the receiver and (b) the **BER** exhibited by the corresponding QPSK-modulated **turbo-coded** MC-CDMA system. The results correspond to **matched** Doppler conditions, when the Doppler frequency $f_{D;\text{prd}}$ assumed in the receiver matches the actual value encountered. The *frame-variant* Bug channel model was assumed.

As pointed out in [516], the choice of the forgetting factor's value β has only a moderate effect on the performance of the resultant predictor. Specifically, in our simulations we used the value suggested in [516], i.e. $\beta = 0.99$.

Algorithm 7.3 The RLS prediction algorithm

$$e[n] = \hat{\alpha}[n] - \check{\alpha}[n] = \hat{\alpha}[n] - \boldsymbol{q}^{\text{H}}[n-1]\hat{\boldsymbol{\alpha}}[n-1] \tag{7.84a}$$

$$\boldsymbol{q}[n] = \boldsymbol{q}[n-1] + \boldsymbol{k}[n-1]e^*[n] \tag{7.84b}$$

$$\check{\alpha}[n+1] = \boldsymbol{q}^{\text{H}}[n]\hat{\boldsymbol{\alpha}}[n] \tag{7.84c}$$

$$\boldsymbol{g}[n] = \boldsymbol{P}[n-1]\hat{\boldsymbol{\alpha}}[n] \tag{7.84d}$$

$$\boldsymbol{k}[n] = \frac{\boldsymbol{g}[n]}{\beta + \hat{\boldsymbol{\alpha}}^{\text{H}}[n]\boldsymbol{g}[n]} \tag{7.84e}$$

$$\boldsymbol{P}[n] = \frac{1}{\beta}(\boldsymbol{I} - \boldsymbol{k}[n]\hat{\boldsymbol{\alpha}}^{\text{H}}[n])\boldsymbol{P}[n-1] \tag{7.84f}$$

7.6.5 Robust Versus Adaptive Predictor Performance Comparison

Figure 7.19 illustrates the achievable MSE performance of the CIR prediction methods considered as a function of the Doppler frequency f_D encountered. It can be seen that the MMSE CIR predictor, which relies on perfect a priori knowledge of the underlying channel statistics represents the upper bound for

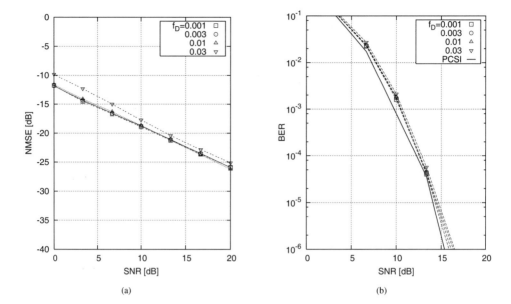

Figure 7.18: (a) **NMSE** exhibited by the DDCE employing the **RC-MMSE FS-CIR** a posteriori estimator of Section 7.4.4 and the **robust** a priori CIR predictor of Section 7.6.2 as a function of the average SNR recorded at the receiver and (b) the corresponding **BER** exhibited by the **turbo-coded** QPSK-modulated MC-CDMA system. The results correspond to **unmatched** Doppler conditions associated with the assumed Doppler frequency of $f_{D;\mathrm{prd}} = 0.03$ and the actual encountered values of $f_D = 0.001, 0.003, 0.01$ and 0.03. The bold line on the BER curve (b) portrays the BER performance of the MC-CDMA system considered in the case of perfect CSI.

the MSE performance achievable by a linear predictor. Furthermore, the robust CIR predictor exhibits a relatively high performance, as long as the actual Doppler frequency encountered does not exceed that assumed. Finally, the RLS CIR predictor, which does not require any explicit knowledge concerning the channel statistics, exhibits a near-optimum performance over the entire range of the values of f_D. Furthermore, Figure 7.20 illustrates the achievable MSE performance of the CIR prediction methods considered as a function of the SNR encountered. Once again, the MMSE CIR predictor exhibits the highest achievable performance. The robust CIR predictor exhibits a relatively high performance, as long as the SNR encountered does not exceed the value $1/\rho$ assumed. On the other hand, the RLS predictor exhibits a near-optimum performance over the whole range of the SNR values. Additionally, the order of the computational complexity associated with both CIR predictors considered in the context of a DDCE-OFDM system and quantified in terms of the total number of complex multiplications and additions per OFDM symbol may be expressed as $O(K \log_2 K + LN_{\mathrm{prd}})$ and $O(K \log_2 K + LN_{\mathrm{prd}} + LN_{\mathrm{prd}}^2)$ for the robust [265] and RLS [293] CIR predictors, respectively.[3] Explicitly, the order of complexity imposed by the RLS CIR predictor is only slightly higher than that associated with the Robust CIR predictor.

7.7   PASTD-Aided DDCE

A detailed schematic of the channel estimation scheme proposed is depicted in Figure 7.21. Our channel estimator consists of a bank of the per-subcarrier a posteriori MMSE CTF estimators outlined in

---

[3] K denotes the number of subcarriers comprising the OFDM symbol, while L is the number of non-zero CIR taps encountered.

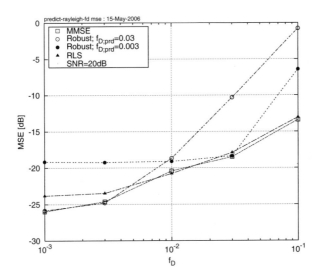

Figure 7.19: **MSE** exhibited by the **MMSE, robust and RLS** a priori CIR predictors as a function of the symbol-normalized Doppler frequency encountered. Two cases of Robust prediction, when $f_{D;\text{prd}} = 0.03$ and $f_{D;\text{prd}} = 0.003$, are considered. The results correspond to the SNR level of 20 dB.

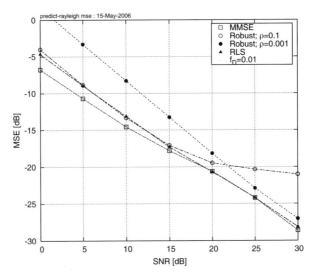

Figure 7.20: **MSE** exhibited by the **MMSE, robust and RLS** a priori CIR predictors as a function of the **SNR** encountered. Two cases of robust prediction, when $\rho = 0.1$ and $\rho = 0.001$, are considered. The results correspond to the symbol-rate-normalized Doppler frequencies of $f_D = 0.01$.

Section 7.3, followed by the PASTᴅ-aided CIR estimator of Section 7.5.2 and by the a priori RLS CIR predictor of Section 7.6.4. The task of the CTF estimator seen in Figure 7.21 is to estimate the subcarrier-related CTF coefficients $H[n, k]$ of Equation (1.7). The resultant estimated subcarrier-related samples $\tilde{H}[n, k]$, which serve as an observation vector of the FD-CTF coefficients $H[n, k]$, are fed to the PASTᴅ subspace-based tracking module, which performs recursive tracking of the channel's covariance matrix C_H signal subspace and the associated CIR-related taps. The output of the PASTᴅ

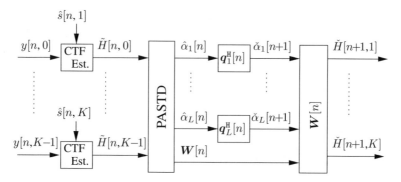

Figure 7.21: Detailed structure of the 2D channel estimator corresponding to the DDCE module of Figure 7.1. The channel estimator comprises a PAST module, which performs recursive tracking of the CIR. The resultant CIR-related taps $\hat{\alpha}_l[n]$ are filtered by the adaptive RLS-based prediction filter resulting in the a priori estimates of the CIR-related taps $\check{\alpha}_l[n+1]$. Finally, the a priori estimates of the subcarrier-related coefficients $H[n+1,k]$ are obtained by applying the transform matrix $\boldsymbol{W}[n]$ provided by the PASTD module.

module consists of the instantaneous CIR-related tap estimates $\hat{\alpha}_l[n]$ and the corresponding estimate of the transformation matrix $\boldsymbol{W}[n]$ of Equation (7.54). The CIR-related estimate vector $\hat{\alpha}_l[n]$ is then fed into the low-rank time-domain CIR-related tap predictor of Figure 7.21 to produce an a priori estimate $\check{\alpha}_l[n+1]$, $l = 1,\ldots,L$, of the next CIR-related tap vector on a tap-by-tap basis [265]. Finally, the predicted CIR is converted to the subcarrier-related CTF with the aid of the transformation matrix $\boldsymbol{W}[n]$ provided by the PASTD module of Figure 7.21. The resultant FD-CTF is employed by the receiver to detect and decode the next OFDM symbol. Note that this principle requires the transmission of a frequency-domain pilot-based channel sounding sequence, such as for example a pilot-assisted OFDM symbol, during the initialization stage. The operation of the resultant DDCE scheme illustrated in Figure 7.21 is summarized in Algorithm 7.4.

In order to characterize the performance of the resultant channel estimation scheme, we will introduce an estimation efficiency criterion κ, which is defined as follows:

$$\kappa = \frac{1}{\sigma_e^2 \gamma} \frac{L}{K}, \tag{7.85}$$

where σ_e^2 and γ are the estimation MSE and SNR, respectively, while K and L are the number of OFDM subcarriers and the number of estimated CIR taps.

The achievable performance of the PASTD-aided DDCE scheme of Algorithm 7.4 is characterized in Figure 7.22. In our simulations we considered an OFDM system having $K = 128$ QPSK-modulated orthogonal subcarriers. The system characteristics are outlined in Table 1.11. We employ an OFDM-frame-variant channel model associated with a time-variant seven-tap PDP characterized by the COST207 BU channel model [347], as detailed in Section 1.7.2. Additionally, each individual propagation path undergoes fast Rayleigh fading having an OFDM-symbol-normalized Doppler frequency of $f_D = 0.003$. We assumed the values $L = 4$ and $\eta = 0.95$ for the PASTD module-related subspace rank and forgetting factor parameters respectively, as well as the value of $\beta = 0.9$ for the RLS CIR tap predictor-related forgetting factor.

Figure 7.22(a) portrays the achievable MSE performance of the PASTD-aided DDCE scheme of Algorithm 7.4 for the pilot overhead ratios $\varepsilon = 0.03, 0.1, 0.3$ and 1.0, where $\varepsilon = 0.03$ and $\varepsilon = 1.0$ correspond to having 3% and 100% pilots, respectively. Specifically, we may identify an estimation efficiency of $\kappa = 5 - 10 = -5\,\text{dB}$.

Furthermore, Figure 7.22(b) portrays the corresponding BER performance of the rate $\frac{1}{2}$ turbo-coded QPSK-modulated OFDM system.

7.8 Channel Estimation for MIMO-OFDM

The main challenge associated with the estimation of the MIMO-CTF coefficients in the context of multi-antenna, multi-carrier systems resides in the fact that, as opposed to the SISO scenario outlined in Section 7.3.2, the estimation of the MIMO-CTFs constitutes a highly rank-deficient problem. More specifically, let us consider the SDM-OFDM system model associated with the kth subcarrier of the nth SDM-OFDM symbol, which may be characterized as follows:

$$\mathbf{y}[n,k] = \mathbf{H}[n,k]\mathbf{s}[n,k] + \mathbf{w}[n,k], \tag{7.86}$$

where $\mathbf{s}[n,k], \mathbf{y}[n,k], \mathbf{w}[n,k]$ and $\mathbf{H}[n,k]$ are the signals associated with the kth subcarrier of the nth SDM-OFDM symbol. Specifically, $\mathbf{s}[n,k]$ is the m_t-dimensional signal vector transmitted from the m_t transmit antennas, $\mathbf{y}[n,k]$ and $\mathbf{w}[n,k]$ are the n_r-dimensional signal and noise vectors recorded at the n_r receive antennas, while $\mathbf{H}[n,k]$ is the $(n_r \times m_t)$-dimensional matrix which characterizes the MIMO-CTFs encountered. Let us assume a relatively simple MIMO scenario of $m_t = n_r = 4$ transmit and receive antennas. The corresponding MIMO-CTF matrix is constituted by $4 \times 4 = 16$ uncorrelated coefficients, which have to be calculated using four recorded samples comprising the received signal $\mathbf{y}[n,k]$, as well as four pilots or decision-based symbols estimating the transmitted signal $\mathbf{s}[n,k]$. Notice that even in the presence of the a priori known pilot-based transmitted signal $\mathbf{s}[n,k]$, the MIMO-CTF matrix $\mathbf{H}[n,k]$ may not be estimated reliably using a linear solution reminiscent of that derived in Section 7.3.2. Consequently, the estimation of the $(n_r \times m_t)$-dimensional MIMO-CTF matrix $\mathbf{H}[n,k]$ requires a sufficiently sophisticated exploitation of both the time- and the frequency-domain correlation properties of the MIMO-CTF coefficients.

In this treatise we propose a MIMO channel estimation scheme which follows the DDCE philosophy of Figure 7.1, as employed in Section 7.2 for SISO multi-carrier systems.

Similar to the SISO case of Section 7.2, our MIMO channel estimation scheme comprises an array of K per-subcarrier MIMO-CTF estimators, followed by an $(n_r \times m_t)$-dimensional array of parametric CIR estimators and a corresponding array of $(n_r \times m_t \times L)$ CIR tap predictors, where L is the number of tracked CIR taps per link for the MIMO channel. The structure of both the parametric PASTD-aided MIMO-CIR tap estimators and that of the RLS MIMO-CIR tap predictors is to a large extent identical to those devised in Sections 7.5.2 and 7.6.4, respectively, in the context of our SISO channel estimation scheme advocated in Section 7.7. On the other hand, our MIMO-CTF estimators exhibit a substantially different structure which reflects the rank-deficient nature of the MIMO channel estimation problem.

In order to exploit the time-domain correlation properties of the MIMO-CTF coefficients matrix $\mathbf{H}[n,k]$, we employ an iterative tracking approach instead of the MMSE estimation method of Section 7.3.2.

7.8.1 Soft Recursive MIMO-CTF Estimation

Analogous to the SISO channel estimator architecture outlined in Section 7.7, at the first stage of our MIMO channel estimation scheme we employ an array of K per-subcarrier MIMO-CTF estimators which function independently of each other. Consequently, for the sake of notational simplicity we omit the subcarrier-related index k in the following section.

7.8.1.1 LMS MIMO-CTF Estimator

The Least Mean Square (LMS) estimation method, which constitutes a simple approximation of the stochastic gradient algorithm [292], was invoked for the iterative tracking of the channel parameters in the context of turbo equalization [264]. More specifically, following the LMS approach, we seek to minimize the MSE-based cost function J_{LMS}, which may be expressed as follows:

$$J_{\text{LMS}} = \sum_{m=1}^{n} \mathbf{e}^{\text{H}}[m]\mathbf{e}[m], \tag{7.87}$$

Algorithm 7.4 PAST$_D$-aided DDCE

Signal detection:

$$\hat{x}[n] = \texttt{Detect}(y[n], \check{H}[n]) \tag{7.88a}$$

CTF estimation:

```
for   k = 1, 2, ..., K   do
```
$$\tilde{H}[n,k] = \frac{y[n,k]\hat{x}^*[n,k]}{|\hat{x}[n,k]|^2 + \sigma_w^2}, \qquad k = 0, \ldots, K-1 \tag{7.88b}$$
```
end for
```

Subspace tracking-aided CIR estimation:

$$H_1[n] = \tilde{H}[n] \tag{7.88c}$$
```
for   l = 1, 2, ..., L   do
```
$$\hat{\alpha}_l[n] = w_l^{\mathrm{H}}[n-1]H_l[n] \tag{7.88d}$$
$$\lambda_l[n] = \eta\lambda_l[n-1] + |\hat{\alpha}_l[n]|^2 \tag{7.88e}$$
$$e_l[n] = H_l[n] - w_l[n-1]\hat{\alpha}_l[n] \tag{7.88f}$$
$$w_l[n] = w_l[n-1] + e_l[n](\alpha_l^*[n]/\lambda_l[n]) \tag{7.88g}$$
$$H_{l+1}[n] = H_l[n] - w_l[n]\hat{\alpha}_l[n] \tag{7.88h}$$
```
end for
```

CIR tap prediction:

```
for   l = 1, 2, ..., L   do
```
$$e[n] = \hat{\alpha}_l[n] - \check{\alpha}_l[n] = \hat{\alpha}_l[n] - q_l^{\mathrm{H}}[n-1]\hat{\alpha}_l[n-1] \tag{7.88i}$$
$$q_l[n] = q_l[n-1] + k_l[n-1]e^*[n] \tag{7.88j}$$
$$\check{\alpha}_l[n+1] = q_l^{\mathrm{H}}[n]\hat{\alpha}_l[n] \tag{7.88k}$$
$$g[n] = P_l[n-1]\hat{\alpha}_l[n] \tag{7.88l}$$
$$k_l[n] = \frac{g[n]}{\beta + \hat{\alpha}_l^{\mathrm{H}}[n]g[n]} \tag{7.88m}$$
$$P_l[n] = \frac{1}{\beta}(I - k_l[n]\hat{\alpha}_l^{\mathrm{H}}[n])P_l[n-1] \tag{7.88n}$$
```
end for
```

CTF reconstruction:

$$\check{H}[n+1] = W[n]\check{\alpha}[n+1] \tag{7.88o}$$

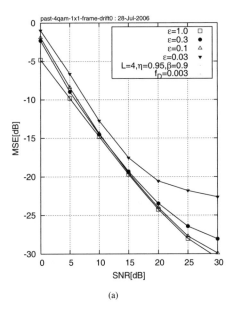 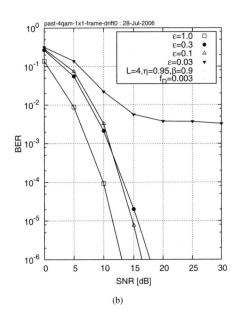

(a)   (b)

Figure 7.22: The (a) **Mean Square Error (MSE)** and (b) **Bit Error Rate (BER)** exhibited by the **4-QAM-OFDM** system employing the **PASTD**-aided DDCE scheme of Algorithm 7.4. The values of the parameters $L = 4, \eta = 0.95$ and $\beta = 0.9$ have been assumed. We considered the scenarios of encountering the Doppler frequency $f_D = 0.003$. The abscissa represents the average SNR recorded at the receive antenna elements.

where $\mathbf{e}[m]$ denotes the error signal, which is given by

$$\mathbf{e}[m] = \mathbf{y}[m] - \tilde{\mathbf{H}}[m]\hat{\mathbf{s}}[m], \tag{7.89}$$

where $\mathbf{y}[m]$ is the signal vector recorded at the n_r transmit antennas, while $\hat{\mathbf{s}}$ is the corresponding estimate of the m_t-dimensional transmitted signal.

Hence, analogous to the solution derived in [264], the LMS estimate of the $(n_r \times m_t)$-dimensional MIMO-CTF coefficient matrix associated with the kth subcarrier of the nth OFDM symbol may be obtained as follows:

$$\tilde{\mathbf{H}}[n] = \tilde{\mathbf{H}}[n-1] + (1 - \zeta)\mathbf{e}[n]\hat{\mathbf{s}}^H[n], \tag{7.90}$$

where we define the forgetting factor ζ. The resultant LMS MIMO-CTF tracking method is summarized in Algorithm 7.5.

Algorithm 7.5 A posteriori LMS MIMO-CTF tracking

$$\hat{\mathbf{s}}[n, k] = \mathtt{Detect}\{\mathbf{y}[n, k], \check{\mathbf{H}}[n, k]\} \tag{7.91a}$$

$$\mathbf{e}[n, k] = \mathbf{y}[n, k] - \tilde{\mathbf{H}}[n, k]\hat{\mathbf{s}}[n, k] \tag{7.91b}$$

$$\tilde{\mathbf{H}}[n, k] = \tilde{\mathbf{H}}[n-1, k] + (1 - \zeta)\mathbf{e}[n, k]\hat{\mathbf{s}}^H[n, k] \tag{7.91c}$$

7.8.1.2 RLS MIMO-CTF Estimator

The RLS algorithm [330] constitutes a rapidly converging LS algorithm. The RLS method was considered in the context of recursive channel parameter estimation and tracking by multiple authors [264, 275, 279, 309]. As opposed to the LMS approach outlined in Section 7.8.1.1, the RLS method attempts to minimize the cost function created from the exponentially weighted and windowed sum of the squared error. That is, we have

$$J_{\mathrm{RLS}}[n] = \sum_{m=1}^{n} \zeta^{n-m} \mathbf{e}^{\mathrm{H}}[m,n]\mathbf{e}[m,n], \tag{7.92}$$

where, analogous to the LMS method of Section 7.8.1.1, the corresponding error signal is given by

$$\mathbf{e}[m,n] = \mathbf{y}[m] - \tilde{\mathbf{H}}[n]\mathbf{s}[m], \tag{7.93}$$

while ζ denotes the forgetting factor. The corresponding RLS estimate of the $(n_{\mathrm{r}} \times m_{\mathrm{t}})$-dimensional MIMO-CTF coefficient matrix associated with the kth subcarrier of the nth OFDM symbol may be calculated as follows [264]:

$$\tilde{\mathbf{H}}[n] = (\mathbf{\Phi}^{-1}[n]\boldsymbol{\theta}[n])^{\mathrm{H}}, \tag{7.94}$$

where we define the MIMO-CTF estimator's input autocorrelation function $\mathbf{\Phi}[n]$, which may be calculated recursively as follows:

$$\mathbf{\Phi}[n] = \sum_{m=1}^{n} \zeta^{n-m} \mathbf{s}[m]\mathbf{s}^{\mathrm{H}}[m] = \zeta\mathbf{\Phi}[n-1] + \mathbf{s}[n]\mathbf{s}^{\mathrm{H}}[n], \tag{7.95}$$

while the MIMO-CTF estimator's input–output cross-correlation matrices $\boldsymbol{\theta}[n]$ as follows:

$$\boldsymbol{\theta}[n] = \sum_{m=1}^{n} \zeta^{n-m} \mathbf{s}[m]\mathbf{y}^{\mathrm{H}}[m] = \zeta\boldsymbol{\theta}[n-1] + \mathbf{s}[n]\mathbf{y}^{\mathrm{H}}[n]. \tag{7.96}$$

The resultant RLS MIMO-CTF tracking method is summarized in Algorithm 7.6.

Algorithm 7.6 A posteriori RLS MIMO-CTF tracking

$$\hat{\mathbf{s}}[n,k] = \mathtt{Detect}\{\mathbf{y}[n,k], \tilde{\mathbf{H}}[n,k]\} \tag{7.97a}$$

$$\mathbf{\Phi}[n,k] = \zeta\mathbf{\Phi}[n-1,k] + \hat{\mathbf{s}}[n,k]\hat{\mathbf{s}}^{\mathrm{H}}[n,k] \tag{7.97b}$$

$$\boldsymbol{\theta}[n,k] = \zeta\boldsymbol{\theta}[n-1,k] + \hat{\mathbf{s}}[n,k]\mathbf{y}^{\mathrm{H}}[n,k] \tag{7.97c}$$

$$\tilde{\mathbf{H}}[n,k] = (\mathbf{\Phi}^{-1}[n,k]\boldsymbol{\theta}[n,k])^{\mathrm{H}} \tag{7.97d}$$

7.8.1.3 Soft-Feedback-Aided RLS MIMO-CTF Estimator

As suggested by the *decision-directed* philosophy of the channel estimation scheme outlined in Section 7.8, the transmitted signal vector $\mathbf{s}[n]$ may not always be readily available at the receiver. More specifically, the transmitted signal vector $\mathbf{s}[n]$ may be assumed to be known a priori if and only if $\mathbf{s}$ constitutes a *pilot symbol*, which occupies a small portion of the transmitted data stream. Alternatively, whenever an information-carrying data symbol is transmitted, the decision-based estimates $\hat{\mathbf{s}}[n]$ become available instead. Unfortunately, however, the decision-based estimates $\hat{\mathbf{s}}$ are prone to decision errors which may potentially result in *error propagation* and thus in a substantial performance degradation.

Consequently, as pointed out in [264, 275, 278], it is highly beneficial to exploit the probability-related soft information available at the output of the MIMO-OFDM system's detector. More specifically, in addition to the *hard*-decision-based values of the transmitted signal estimates $\hat{s}[n]$ we may utilize the associated *soft*-information-related quantities, such as the expectations and the variances of the elements of the estimated transmitted signal vector $\hat{\mathbf{s}} = [\hat{s}_1, \ldots, \hat{s}_{m_t}]^T$. Specifically, the expectation of the ith transmitted symbol may be expressed as follows:

$$\tilde{s}_i = \mathrm{E}\{\hat{s}_i\} = \sum_{c \in \mathcal{M}} c \, p\{s_i = c\}, \tag{7.98}$$

while the corresponding variance is given by

$$v_i = \mathrm{Var}\{\hat{s}_i\} = \left(\sum_{c \in \mathcal{M}} cc^* \, p\{s_i = c\} \right) - \tilde{s}_i \tilde{s}_i^*. \tag{7.99}$$

Subsequently, we may define the following alternative error signals:

$$\hat{\mathbf{e}}[m, n] = \mathbf{y}[m] - \tilde{\mathbf{H}}[n]\hat{\mathbf{s}}[m], \tag{7.100}$$

$$\tilde{\mathbf{e}}[m, n] = \mathbf{y}[m] - \tilde{\mathbf{H}}[n]\tilde{\mathbf{s}}[m]. \tag{7.101}$$

The error signals of Equations (7.100) and (7.101) may be substituted into the LMS and RLS algorithms of Sections 7.8.1.1 and 7.8.1.2 in order to yield the hard- and soft-decision-based LMS and RLS CTF tracking algorithms, respectively.

7.8.1.4 Modified RLS MIMO-CTF Estimator

A further improved version of the soft-decision-based RLS tracking algorithm, namely the so-called modified RLS algorithm, was proposed by Otnes [264]. More specifically, in the modified RLS method the cost function of Equation (7.92) associated with the classic RLS method of Algorithm 7.6 is substituted by a cost function which takes into account the ambiguity inherent in the decision-based estimates $\hat{s}[n]$. Firstly, for the sake of notational convenience the following covariance matrices were defined in [264]:

$$\mathbf{D}[n] = \mathrm{Cov}\{\tilde{\mathbf{s}}[n], \tilde{\mathbf{s}}[n]\} = \mathrm{E}\{\tilde{\mathbf{s}}[n]\tilde{\mathbf{s}}^H[n]\} = \mathrm{diag}(\mathbf{v}[n]) \tag{7.102}$$

and

$$\boldsymbol{U}[n] = \mathrm{E}\{\mathbf{s}[n]\mathbf{s}^H[n]\} = \tilde{\mathbf{s}}[n]\tilde{\mathbf{s}}^H[n] + \mathbf{D}[n], \tag{7.103}$$

where the elements of the variance vector $\mathbf{v}[n]$ are given by Equation (7.99). The corresponding modified RLS cost function may be expressed as follows [264]:

$$J_{\mathrm{modRLS}}[n] = \sum_{m=1}^{n} \zeta^{n-m} \mathrm{E}\{\tilde{\mathbf{e}}^H[m, n]\tilde{\mathbf{e}}[m, n] \mid \mathbf{y}[m], \tilde{\mathbf{s}}[m], \mathbf{D}[m], \tilde{\mathbf{H}}[n]\}, \tag{7.104}$$

where, as previously, ζ denoted the forgetting factor. Observe that, as opposed to the RLS cost function of Equation 7.92, the modified RLS cost function of Equation 7.104 takes into account the ambiguity associated with both the estimated CTF matrix $\hat{\mathbf{H}}[n]$ and the estimated transmitted signal vector $\hat{\mathbf{s}}[n]$.

Finally, following the approach proposed in [264], the modified RLS MIMO-CTF estimate $\hat{\mathbf{H}}[n]$ may be calculated using Equation (7.94), which is repeated here for convenience. Specifically, we have

$$\hat{\mathbf{H}}[n] = (\boldsymbol{\Phi}^{-1}[n]\boldsymbol{\theta}[n])^H, \tag{7.105}$$

where the corresponding covariance matrices $\mathbf{\Phi}[n]$ and $\boldsymbol{\theta}[n]$ may be reformulated using the quantities $\mathbf{D}[n]$ and $\mathbf{U}[n]$ of Equations (7.102) and (7.103), respectively. That is, we have

$$\mathbf{\Phi}[n] = \sum_{m=1}^{n} \zeta^{n-m} \mathbf{U}[m] = \zeta \mathbf{\Phi}[n-1] + \mathbf{U}[n] \tag{7.106}$$

and

$$\boldsymbol{\theta}[n] = \sum_{m=1}^{n} \zeta^{n-m} (\tilde{\mathbf{s}}[n]\mathbf{y}^{\mathrm{H}}[n] + \mathbf{D}[m]\hat{\mathbf{H}}^{\mathrm{H}}[m])$$

$$= \zeta \boldsymbol{\theta}[n-1] + \mathbf{U}[n]\hat{\mathbf{H}}^{\mathrm{H}}[n-1] + \tilde{\mathbf{s}}[n]\tilde{\mathbf{e}}^{\mathrm{H}}[n]. \tag{7.107}$$

The resultant soft-decision-based MIMO-CTF modified RLS method is summarized in Algorithm 7.7.

Algorithm 7.7 MIMO-CTF a posteriori modified RLS tracking

$$\{\hat{\mathbf{s}}[n,k], \tilde{\mathbf{s}}[n,k]\} = \texttt{Detect}\{\mathbf{y}[n,k], \check{\mathbf{H}}[n,k]\} \tag{7.108a}$$

$$\boldsymbol{U}[n,k] = \mathrm{diag}(|\hat{\mathbf{s}}[n,k]|^2 - |\tilde{\mathbf{s}}[n,k]|^2) + \tilde{\mathbf{s}}[n,k]\hat{\mathbf{s}}[n,k]^{\mathrm{H}} \tag{7.108b}$$

$$\mathbf{d}[n,k] = \mathbf{y}[n,k] - \check{\mathbf{H}}[n,k]\tilde{\mathbf{s}}[n,k] \tag{7.108c}$$

$$\mathbf{\Phi}[n,k] = \zeta\mathbf{\Phi}[n,k] + \mathbf{U}[n,k] \tag{7.108d}$$

$$\boldsymbol{\theta}[n,k] = \zeta\boldsymbol{\theta}[n,k] + \mathbf{U}[n,k]\check{\mathbf{H}}[n,k] + \tilde{\mathbf{s}}[n,k]\mathbf{d}^{\mathrm{H}}[n,k] \tag{7.108e}$$

$$\hat{\mathbf{H}}[n,k] = (\mathbf{\Phi}^{-1}[n,k]\boldsymbol{\theta}[n,k])^{\mathrm{H}} \tag{7.108f}$$

7.8.1.5 MIMO-CTF Estimator Performance Analysis

The snapshots of the CTF estimation MSE exhibited by both hard- and soft-feedback-aided LMS and RLS MIMO-CTF tracking methods of Sections 7.8.1.1 and 7.8.1.2, respectively, as well as that of the modified RLS method of Section 7.8.1.4, are depicted in Figure 7.23. We considered the 4×4 MIMO-OFDM system characterized in Table 1.11. We assumed the transmission of a sequence of signal bursts comprising 24 OFDM symbols each. Furthermore, each signal burst was constituted by an 8 OFDM symbols pilot frame, followed by a 16 OFDM symbol data frame. Additionally, we assumed encountering an OFDM-symbol-normalized Doppler frequency of $f_D = 0.003$ and SNRs of 6.0 and 10.0 dB.

In Figure 7.23(a) we can see that at low SNRs, where the system suffers from frequent decision errors, the hard-feedback-aided LMS and RLS methods of Algorithms 7.5 and 7.6 exhibit a substantially worse performance than their soft-feedback-aided counterparts. On the other hand, Figure 7.23(b), which corresponds to the higher SNR value of 10 dB, where we have a relatively low probability of decision errors, demonstrates that the hard-feedback-aided RLS MIMO-CTF tracking method outperforms its soft-feedback-assisted counterpart. Nevertheless, the slightly lower performance of the soft-feedback-aided methods recorded at higher SNRs is a price worth paying for their significantly better robustness against error propagation at lower SNRs. Additionally, we can see in both Figures 7.23(a) and 7.23(b) that the modified RLS method of Algorithm 7.7 exhibits the best MSE performance among the soft-feedback-aided tracking methods considered.

Consequently, from the results of Figure 7.23 we may draw the conclusion that the soft-feedback-aided modified RLS MIMO-CTF tracking method of Algorithm 7.7 exhibits the best combination of attractive MSE performance and a high robustness against error propagation.

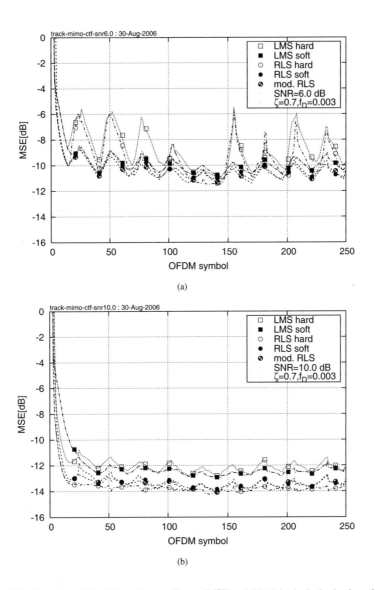

Figure 7.23: Snapshots of the **Mean Square Error (MSE)** exhibited by both the hard- and soft-feedback-aided recursive MIMO-CTF tracking methods of Sections 7.8.1.1, 7.8.1.2 and 7.8.1.4. We considered a 4×4 MIMO-OFDM system and a scenario of encountering an OFDM-symbol-normalized Doppler frequency of $f_D = 0.003$ as well as SNRs of (a) 6.0 dB and (b) 10.0 dB. The abscissa represents the index n of the received OFDM symbol.

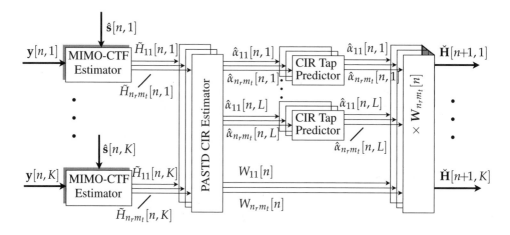

Figure 7.24: Detailed structure of the MIMO channel estimator corresponding to the DDCE module of Figure 7.1 in the context of the MIMO-OFDM system. The channel estimator comprises an array of PASTD modules, which performs recursive tracking of the MIMO-CIR. The resultant MIMO-CIR-related taps $\hat{\alpha}_{ij;l}[n]$ are filtered by an array of adaptive RLS prediction filters resulting in the a priori estimates of the MIMO-CIR-related taps $\check{\alpha}_{ij;l}[n+1]$. Finally, the a priori estimates of the subcarrier-related coefficients $\check{H}[n+1,k]$ are obtained by applying the array of transform matrices $W_{ij}[n]$ provided by the PASTD modules.

7.8.2   PASTD-Aided DDCE for MIMO-OFDM

As outlined in Section 7.8, we propose a MIMO channel estimation scheme which follows the DDCE philosophy of Figure 7.1. The detailed structure of our MIMO-DDCE channel estimator is illustrated in Figure 7.24. More specifically, our MIMO channel estimation scheme comprises an array of K per-subcarrier MIMO-CTF estimators, followed by an $(n_{\rm r} \times m_{\rm t})$-dimensional array of parametric CIR estimators and a corresponding array of $(n_{\rm r} \times m_{\rm t} \times L)$ CIR tap predictors, where L is the number of tracked CIR taps per link for the MIMO channel. The structure of both the parametric PASTD-aided MIMO-CIR tap estimators and that of the RLS MIMO-CIR tap predictors is to a large extent identical to those devised in Sections 7.5.2 and 7.6.4 in the context of our SISO channel estimation scheme advocated in Section 7.7. On the other hand, our MIMO-CTF estimators may employ one of the recursive MIMO-CTF tracking methods outlined in Sections 7.8.1.1, 7.8.1.2 or 7.8.1.4.

The resultant MIMO-DDCE scheme illustrated in Figure 7.24 and employing the modified RLS MIMO-CTF estimator of Algorithm 7.7, the PASTD-aided CIR estimator of Algorithm 7.2 as well as the RLS CIR tap predictor of Algorithm 7.3 is summarized in Algorithm 7.8.

7.8.2.1   PASTD-Aided MIMO-DDCE Performance Analysis

In this section we will characterize the achievable performance of the MIMO-DDCE scheme of Algorithm 7.8 in the context of the MIMO-OFDM system of Figure 1.17. More specifically, we consider a 2×2 MIMO-QPSK-OFDM system having $K = 128$ orthogonal QPSK-modulated subcarriers. The system parameters are outlined in Table 1.11. We employ an OFDM-frame-variant channel model having the time-variant seven-tap PDP characterized by the COST207 BU channel model of [347], as detailed in Section 1.7.2. Additionally, each individual propagation path undergoes fast Rayleigh fading at an OFDM-symbol-normalized Doppler frequency of $f_D = 0.001$ and $f_D = 0.005$. The resultant channel can be characterized as a multi-path Rayleigh-fading channel with slowly fluctuating PDP.

Firstly, Figure 7.25 characterizes the achievable MSE performance of the MIMO-DDCE method of Algorithm 7.8 for different values of the MIMO-CTF tracking scheme's forgetting factor ζ.

Algorithm 7.8 PASTD-aided MIMO-DDCE

MIMO-CTF tracking:

$$\text{for} \quad k = 1, \ldots, K \quad \text{do}$$

$$\boldsymbol{U}[n,k] = \text{diag}(|\hat{\mathbf{s}}[n,k]|^2 - |\tilde{\mathbf{s}}[n,k]|^2) + \tilde{\mathbf{s}}[n,k]\tilde{\mathbf{s}}[n,k]^{\text{H}} \tag{7.109a}$$

$$\mathbf{d}[n,k] = \mathbf{y}[n,k] - \tilde{\mathbf{H}}[n,k]\tilde{\mathbf{s}}[n,k] \tag{7.109b}$$

$$\boldsymbol{\Phi}[n,k] = \zeta\boldsymbol{\Phi}[n,k] + \mathbf{U}[n,k] \tag{7.109c}$$

$$\boldsymbol{\theta}[n,k] = \zeta\boldsymbol{\theta}[n,k] + \mathbf{U}[n,k]\check{\mathbf{H}}[n,k] + \tilde{\mathbf{s}}[n,k]\mathbf{d}^{\text{H}}[n,k] \tag{7.109d}$$

$$\tilde{\mathbf{H}}[n,k] = (\boldsymbol{\Phi}^{-1}[n,k]\boldsymbol{\theta}[n,k])^{\text{H}} \tag{7.109e}$$

$$\text{end for } k$$

CIR tracking:

$$\text{for} \quad i = 1, \ldots, n_{\text{r}} \quad \text{do}, \quad \text{for} \quad j = 1, \ldots, m_{\text{t}} \quad \text{do}$$

$$\boldsymbol{H}_1[n] = \hat{\boldsymbol{H}}_{ij}[n] \tag{7.109f}$$

$$\quad \text{for} \quad l = 1, 2, \ldots, L \quad \text{do}$$

$$\hat{\alpha}_{ij;l}[n] = \boldsymbol{w}_{ij;l}^{\text{H}}[n-1]\boldsymbol{H}_l[n] \tag{7.109g}$$

$$\lambda_{ij;l}[n] = \eta\lambda_{ij;l}[n-1] + |\hat{\alpha}_{ij;l}[n]|^2 \tag{7.109h}$$

$$\boldsymbol{e}_l[n] = \boldsymbol{H}_l[n] - \boldsymbol{w}_{ij;l}[n-1]\hat{\alpha}_{ij;l}[n] \tag{7.109i}$$

$$\boldsymbol{w}_{ij;l}[n] = \boldsymbol{w}_{ij;l}[n-1] + \boldsymbol{e}_l[n](\alpha_{ij;l}^*[n]/\lambda_{ij;l}[n]) \tag{7.109j}$$

$$\boldsymbol{H}_{l+1}[n] = \boldsymbol{H}_l[n] - \boldsymbol{w}_{ij;l}[n]\hat{\alpha}_{ij;l}[n] \tag{7.109k}$$

$$\quad \text{end for } l$$

$$\text{end for } i, \quad \text{end for } j$$

CIR prediction:

$$\text{for} \quad i = 1, \ldots, n_{\text{r}} \quad \text{do}, \quad \text{for} \quad j = 1, \ldots, m_{\text{t}} \quad \text{do}$$

$$\quad \text{for} \quad l = 1, 2, \ldots, L \quad \text{do}$$

$$e[n] = \hat{\alpha}_l[n] - \check{\alpha}_l[n] = \hat{\alpha}_l[n] - \boldsymbol{q}_l^{\text{H}}[n-1]\hat{\boldsymbol{\alpha}}_l[n-1] \tag{7.109l}$$

$$\boldsymbol{q}_l[n] = \boldsymbol{q}_l[n-1] + \boldsymbol{k}_l[n-1]e^*[n] \tag{7.109m}$$

$$\check{\alpha}_l[n+1] = \boldsymbol{q}_l^{\text{H}}[n]\hat{\boldsymbol{\alpha}}_l[n] \tag{7.109n}$$

$$\boldsymbol{g}[n] = \boldsymbol{P}_l[n-1]\hat{\boldsymbol{\alpha}}_l[n] \tag{7.109o}$$

$$\boldsymbol{k}_l[n] = \frac{\boldsymbol{g}[n]}{\beta + \hat{\boldsymbol{\alpha}}_l^{\text{H}}[n]\boldsymbol{g}[n]} \tag{7.109p}$$

$$\boldsymbol{P}_l[n] = \frac{1}{\beta}(\boldsymbol{I} - \boldsymbol{k}_l[n]\hat{\boldsymbol{\alpha}}_l^{\text{H}}[n])\boldsymbol{P}_l[n-1] \tag{7.109q}$$

$$\quad \text{end for } l$$

$$\text{end for } i, \quad \text{end for } j$$

CTF reconstruction:

$$\text{for} \quad i = 1, \ldots, n_{\text{r}} \quad \text{do}, \quad \text{for} \quad j = 1, \ldots, m_{\text{t}} \quad \text{do}$$

$$\check{\boldsymbol{H}}_{ij}[n+1] = \boldsymbol{W}_{ij}[n]\check{\boldsymbol{\alpha}}_{ij}[n+1] \tag{7.109r}$$

$$\text{end for } i, \quad \text{end for } j$$

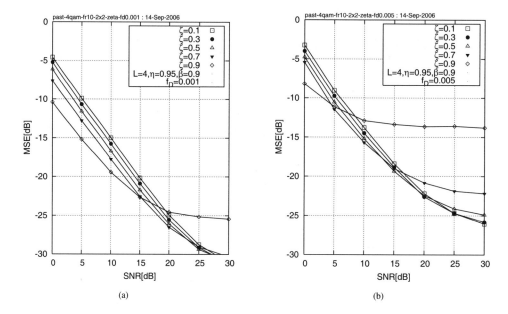

Figure 7.25: The **Mean Square Error (MSE)** exhibited by the 2×2 **SDM-4-QAM-OFDM** system employing the **SDM PASTD-aided DDCE** scheme of Algorithm 7.8. The PASTD-DDCE parameters are $\zeta = 0.1, 0.3, \dots, 0.9$ as well as $\eta = 0.95$, $\beta = 0.9$. We considered the scenarios of encountering Doppler frequencies of (a) $f_D = 0.001$ and (b) $f_D = 0.005$. The abscissa represents the average SNR recorded at the receive antenna elements.

Figures 7.25(a) and 7.25(b) correspond to encountering the Doppler frequencies of $f_D = 0.001$ and 0.005, respectively. As may be concluded from Figure 7.25, the optimum value of the forgetting factor ζ is largely dependent on the SNR as well as on the Doppler frequency encountered. Nevertheless, the compromise value of $\zeta = 0.7$ appears to constitute a relatively good choice in the practical range of SNR values and Doppler frequencies.

Secondly, Figure 7.26 characterizes the achievable MSE performance of the MIMO-DDCE method of Algorithm 7.8 for different values of the PASTD-aided CIR tracking scheme's forgetting factor η. Figures 7.26(a) and 7.26(b) correspond to encountering the Doppler frequencies of $f_D = 0.001$ and 0.005, respectively. Similar to the choice of the optimum MIMO-CTF tracking forgetting factor ζ, the optimum value of the PASTD-aided CIR tracking forgetting factor η is largely dependent on the SNR as well as on the Doppler frequency encountered and the compromise value of $\eta = 0.95$ appears to constitute a good choice across the practical range of SNR values and Doppler frequencies.

Furthermore, Figure 7.27 characterizes the achievable MSE performance of the MIMO-DDCE method of Algorithm 7.8 for different ranks L of the PASTD-aided CIR tracking-related estimated subspace, while assuming a constant value of the forgetting factors $\eta = 0.95$ and $\zeta = 0.7$. Figures 7.27(a) and 7.27(b) correspond to encountering the Doppler frequencies of $f_D = 0.001$ and 0.005, respectively. From Figure 7.27 we may conclude that a relatively high performance of the PASTD-aided CIR estimator may be achieved when assuming that the rank of the estimated CTF signal subspace is $L = 4$, regardless of the actual number of paths constituting the multi-path CIR encountered.

In order to characterize further the performance of the resultant MIMO channel estimation scheme, we will use the estimation efficiency criteria κ of Equation (7.85). In the case of a MIMO system, the

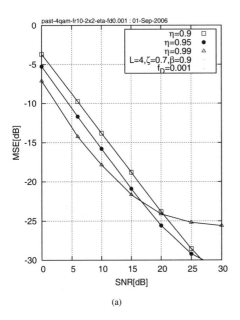
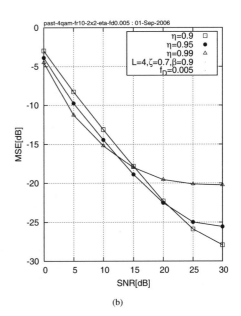

(a)   (b)

Figure 7.26: The **Mean Square Error (MSE)** exhibited by the 2×2 **SDM-4-QAM-OFDM** system employing the **SDM PASTD-aided DDCE** scheme of Algorithm 7.8. The PASTD-DDCE parameters are $\eta = 0.9, 0.95$ and 0.99 as well as $\zeta = 0.7$, $\beta = 0.9$. We considered the scenarios of encountering the Doppler frequencies of (a) $f_D = 0.001$ and (b) $f_D = 0.005$. The abscissa represents the average SNR recorded at the receive antenna elements.

channel estimation efficiency factor κ may be redefined as follows:

$$\kappa = \frac{1}{\sigma_e^2 \gamma} \frac{L m_t n_r}{K}, \tag{7.110}$$

where $L m_t n_r$ denotes the total number of the independent channel-related parameters estimated. The value of the channel estimation efficiency factor κ corresponding to the PAST-aided MIMO-DDCE scheme considered may be obtained empirically using the results depicted in Figure 7.27. Specifically we have $\kappa = -4\,\text{dB}$.

Finally, Figure 7.28 characterizes the achievable BER performance of the rate $\frac{1}{2}$ turbo-coded SDM-QPSK-OFDM system employing the MIMO-PASTD-DDCE method of Algorithm 7.8. The DDCE parameters are $\zeta = 0.7$, $L = 4$, $\eta = 0.95$ and $\beta = 0.9$. Furthermore, we assumed a pilot overhead of 10%. Figures 7.28(a) and 7.28(b) correspond to the 4×4 and 8×8 MIMO scenarios, respectively. We considered encountering the Doppler frequencies of $f_D = 0.001, 0.003$ and 0.005. Observe that the system proposed attains a virtually error-free performance of a rate $1/2$ turbo-coded 8×8 QPSK-OFDM system, exhibiting a total bit rate of 8 bps/Hz and having a pilot overhead of only 10%, at an SNR of 10 dB and normalized Doppler frequency of 0.003, which corresponds to the mobile terminal speed of roughly 65 km/h.[4]

Our future research is related to reducing the pilot overhead required, potentially leading to semi-blind channel estimation schemes.

---

[4]Additional system parameters are characterized in Table 1.11.

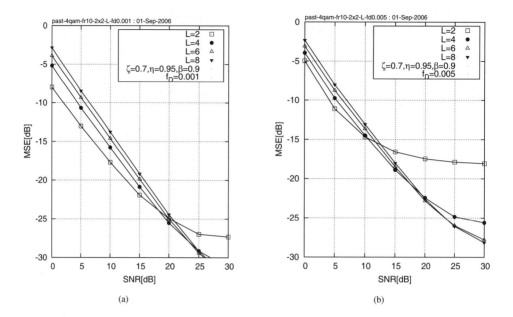

(a)   (b)

Figure 7.27: The **Mean Square Error (MSE)** exhibited by the 2×2 **SDM-4-QAM-OFDM** system employing the **SDM PASTD-aided DDCE** scheme of Algorithm 7.8 and tracking $L = 2, 4, 6$ and 8 CIR taps. The PASTD-DDCE parameters are $\zeta = 0.7, \eta = 0.95$ and $\beta = 0.9$. We considered the scenarios of encountering the Doppler frequencies of (a) $f_D = 0.001$ and (b) $f_D = 0.005$. The abscissa represents the average SNR recorded at the receive antenna elements.

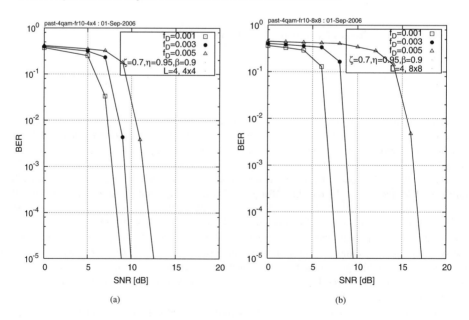

(a)   (b)

Figure 7.28: **BER** versus **SNR** performance exhibited by the rate $\frac{1}{2}$ turbo-coded (a) 4×4 and (b) 8×8 **SDM-QPSK-OFDM** system employing the **MIMO-PASTD-DDCE** method of Algorithm 7.8. The abscissa represents the average SNR recorded at the receive antenna elements.

7.9 Chapter Summary

The subject of this chapter was the design of efficient channel estimation schemes. We commenced our discourse by considering the family of conceptually simplest, low-complexity pilot symbol-assisted techniques. Then the decision-directed channel estimation philosophy was introduced, where the initial channel estimates are generated with the aid of pilots and a tentative data estimate is arrived at with the aid slicing, while exploiting the tentative channel estimate. Then, provided that all data decisions are error-free, we now have access to 100 % pilot information and hence a more accurate channel estimate may be generated. Since the channel fluctuations obey the Doppler frequency, the consecutive channel coefficients are correlated and can also be predicted into the future on the basis of the previous channel estimates. The chapter concluded by designing channel estimation schemes for MIMO-aided OFDM arrangements, which typically require the estimation of a high number of MIMO links.

Chapter 8

Iterative Joint Channel Estimation and MUD for SDMA-OFDM Systems

8.1 Introduction[1]

Multiple-Input, Multiple-Output (MIMO) Orthogonal Frequency-Division Multiplexing (OFDM) systems have recently attracted substantial research interest. On the one hand, the employment of multiple antennas offers an opportunity to exploit both transmitter and receiver diversity, thus significantly increasing the system's transmission integrity [107]. On the other hand, as a further benefit, OFDM exhibits robustness against both frequency-selective fading as well as the Inter-Symbol Interference (ISI) imposed by multi-path propagation. Specifically, intensive research efforts have been invested both in Bell Labs Layered Space–Time architecture (BLAST) [105, 108, 452] and in Space-Division Multiple Access (SDMA) based MIMO OFDM [3]. More specifically, in SDMA-OFDM systems the transmitted signals of L simultaneous uplink mobile users – each equipped with a single transmitter antenna – are received by the P different receiver antennas of the Base Station (BS). At the BS Multi-User Detection (MUD) [439] techniques are invoked for detecting the different users' transmitted signals with the aid of their unique user-specific spatial signature constituted by their Frequency-Domain CHannel Transfer Functions (FD-CHTFs) or, equivalently, Channel Impulse Responses (CIRs). Since the same time–frequency resource is shared by simultaneous users, a higher bandwidth efficiency can be achieved by SDMA systems in comparison with conventional multiplexing techniques, such as for example Time-Division Multiple Access (TDMA) or Frequency-Division Multiple Access (FDMA).

However, in these systems accurate channel estimation is required at the receiver to invoke both coherent demodulation and interference cancellation. Compared with Single-Input, Single-Output (SISO) systems, channel estimation in the MIMO scenario becomes more challenging, since a significantly increased number of independent transmitter–receiver channel links have to be estimated simultaneously for each subcarrier. Moreover, the interfering signals of the other transmitter antennas have to be suppressed. All these factors render channel estimation for MIMO-OFDM systems a new challenge.

---

[1]This chapter is partially based on ©IEEE Jiang, Akhtman & Hanzo 2007 [517].

MIMO-OFDM for LTE, Wi-Fi and WiMAX   Lajos Hanzo, Yosef Akhtman, Li Wang and Ming Jiang
© 2011 John Wiley & Sons, Ltd

In the literature, a number of blind channel estimation techniques have been proposed for MIMO-OFDM systems [83, 217, 228, 251, 518, 519]. However, most of these approaches suffer from either a slow convergence rate or a performance degradation due to the inherent limitations of blind search mechanisms. By contrast, the techniques benefiting from explicit training with the aid of known reference/pilot signals are typically capable of achieving a better performance at the cost of a reduced effective system throughput. For example, Li *et al.* [520] proposed an approach exploiting both transmitter diversity and the delay profile characteristics of typical mobile channels, which was further simplified and enhanced in [131, 521] and [134], respectively. Other schemes employed Minimum Mean Square Error (MMSE) [522], Constrained Least-Squares (CLS) [200], iterative Least-Squares (LS) [132, 173], Second-Order Statistics (SOS) based subspace [156] estimation algorithms as well as QR Decomposition combined with the M-algorithm (QRD-M) [172, 523] or techniques based on Time Of Arrival (TOA) [178] etc. Some researchers focused their attention on designing optimum training patterns or structures [131, 133, 524]. Furthermore, various joint approaches combining channel estimation with data symbol detection at the receiver were also proposed for Code-Division Multiple Access (CDMA) [519, 523], SISO OFDM [525] and MIMO OFDM [172, 526] systems. However, in the context of BLAST or SDMA types of multi-user MIMO-OFDM systems, all channel estimation techniques found in the literature were developed under the assumption that the number of users L is lower than [132, 156, 217, 228, 518, 522, 527] or equal to [133, 172, 178, 200, 251, 526, 528] the number of receiver antennas P. This assumption is critical for the following reasons. When $L > P$, which we refer to as an *overloaded* scenario, the $(P \times L)$-dimensional MIMO channel matrix representing the $(P \times L)$ number of channel links becomes singular, thus rendering the degree of freedom of the detector insufficient. This will catastrophically degrade the performance of numerous known detection approaches, such as for example the MMSE algorithm of [3, 439] and the QRD-M algorithm of [172]. Furthermore, the associated significant degradation of the MUD's performance in this overloaded scenario will inevitably result in severe error propagation in decision-directed types of channel estimators [3].

Against this background, in this chapter we propose a new Genetic Algorithm (GA) [328, 453–456] assisted iterative Joint Channel Estimation and Multi-User Detection (GA-JCEMUD) approach for multi-user MIMO SDMA-OFDM systems, which provides an effective solution to the multi-user MIMO channel estimation problem in the above-mentioned overloaded scenario. Our ambitious goal of supporting a high number of users is physically possible, because the proposed GA-based technique dispenses with any constraints concerning the rank of the channel matrix. In the literature, only a few channel estimation schemes have been proposed based on GAs. More specifically, Yen *et al.* [461] proposed a GA-aided multi-user CDMA single-antenna receiver, which jointly estimates the transmitted symbols and fading channel coefficients of all the users. A batch blind equalization scheme based on Maximum Likelihood (ML) concatenated channel and data estimation employing a Micro Genetic Algorithm (μGA) and the Viterbi Algorithm (VA) was proposed in [529]. In [530, 531], GA-based approaches were used to find optimum training sequences for channel estimation in OFDM systems. However, to our best knowledge, no techniques employing GAs for joint channel and data optimization can be found in the literature in the context of multi-user MIMO-OFDM. Furthermore, at the time of writing, the GAs invoked in the data detection literature [38, 459–461, 532, 533] can only provide a hard-decision output for the Forward Error Correction (FEC) or channel decoder, which inevitably limits the system's achievable performance. By contrast, our proposed GA is capable of providing 'soft' outputs and hence it becomes capable of achieving an improved performance with the aid of FEC decoders.

The structure of this chapter is as follows. Firstly an overview of the proposed scheme is provided in Section 8.2, followed by the introduction of the proposed iterative GA-JCEMUD in Section 8.3. More specifically, the methods used to generate initial FD-CHTF estimates and initial symbol estimates are discussed in Sections 8.3.1 and 8.3.2, respectively. The GA-aided joint optimization process is detailed in Section 8.3.3, commencing with a discussion of the GA individual's structure in Section 8.3.3.1, followed by a description of the initialization process in Section 8.3.3.2, the analysis of the genetically joint channel estimation and symbol detection in Section 8.3.3.3 and the derivation of a novel

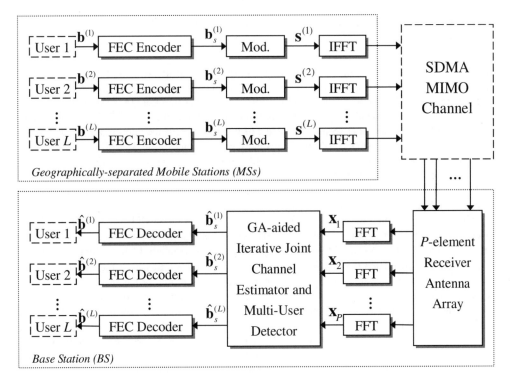

Figure 8.1: Schematic of the SDMA-OFDM uplink system employing the proposed GA-aided iterative joint channel estimator and MUD. ©IEEE Jiang, Akhtman & Hanzo 2007 [517]

soft-decoded GA in Section 8.3.3.4, respectively. Our simulation results are provided in Section 8.4, while Sections 8.5 and 8.6 conclude our findings.

8.2 System Overview

Figure 8.1 shows a schematic of the SDMA-OFDM uplink system using the proposed iterative GA-JCEMUD. As shown in the upper half of Figure 8.1, the information bit blocks $\mathbf{b}^{(l)}$ ($l = 1, \ldots, L$) of the L mobile users are first encoded by the L independent FEC encoders. The resultant coded bits $\mathbf{b}_s^{(l)}$ are then mapped to Quadrature Amplitude Modulation (QAM) or Phase-Shift Keying (PSK) symbols $\mathbf{s}^{(l)}$, which are modulated by the Inverse Fast Fourier Transform (IFFT) based OFDM modulators and transmitted over the SDMA MIMO channel described in Section 4.2.1.

At the BS illustrated in the lower half of Figure 8.1, the received signal constituted by the noise-contaminated superposition of all users' transmitted signals is OFDM demodulated at the P receiver antenna elements and forwarded to the iterative GA-JCEMUD for joint channel estimation and symbol detection, as will be detailed in Section 8.3. Then the detected soft bits $\hat{\mathbf{b}}_s^{(l)}$ are generated, which are forwarded to the L independent FEC decoders for channel decoding.

8.3 GA-Assisted Iterative Joint Channel Estimation and MUD

In this section, we will elaborate on the philosophy of the proposed iterative GA-JCEMUD, which is illustrated in Figure 8.2. We assume that each OFDM symbol consists of K subcarriers. All subcarriers of the first transmitted OFDM symbol of all the L users carry known pilot QAM symbols, which are spread by user-specific spreading codes before transmission.[2] Within the first OFDM symbol duration ($n = 0$), the BS pilot controller seen in the middle of Figure 8.2 feeds the pilots to the GA-JCEMUD printed in grey, which simultaneously processes the received signals $x_p[0, k]$ ($p = 1, \ldots, P$; $k = 1, \ldots, K$) at the P receiver antenna elements. In order to simplify the analysis, we focus our attention on the pth receiver. Based on the pilots and the corresponding received signals, the initial estimates of the FD-CHTFs $\tilde{H}_p^{(l)}[0, k]$ ($l = 1, \ldots, L$; $k = 1, \ldots, K$) can be generated, which will be subjected to time-domain filtering invoked at the pth receiver, as plotted at the top of Figure 8.2.

The time-domain filtering is invoked for each of the L users on an OFDM symbol basis. More specifically, for the lth user, the K initial FD-CHTF estimates $\tilde{H}_p^{(l)}[n, k]$ ($k = 1, \ldots, K$) associated with the current, i.e. the nth OFDM symbol, are processed by a K-length IFFT, resulting in the set of K uncorrelated CIR-related taps $\tilde{h}_p^{(l)}[n, k]$. Then, only the first K_0 CIR tap coefficients are retained with the rest set to zero. The value of K_0 depends on the delay profile of the channel, which is not known a priori at the receiver. However, in many application scenarios it is possible appropriately to overestimate K_0 on the basis of previous field experiments [228]. Generally speaking, the value of K_0 should be set to a sufficiently large number so that it exceeds the actual maximum delay spread of the channel. The reason for doing so is to ensure that the 'real' CIR taps are retained, while removing a significant part of the noise contaminating the high-delay, low-power CIR taps, and hence improving the initial CHTF estimates generated by the GA-JCEMUD. To elaborate a little further, on the one hand, the output of the GA-JCEMUD may be noisy, owing to low SNRs and/or a high number of users as well as receiver antennas (i.e. more channel links to estimate, rendering the estimation job more difficult). On the other hand, in practical applications the value of K is usually chosen to be significantly higher than the number of CIR taps corresponding to the channel's maximum delay spread. Furthermore, most of the channel's output power is contributed by the first several CIR taps. In other words, provided that K_0 is higher than the maximum channel delay spread, the last $(K - K_0)$ CIR taps from the set of the total K time-domain CIR taps generated by the IFFT from the CHTF estimates contain nothing but noise. Therefore, when the last $(K - K_0)$ CIR taps are set to zero, most of the noise contaminating the channel estimates can be removed without failing to capture the channel's 'real' CIR taps. Based on the above analysis, we point out that a good choice of K_0 is constituted by the number of samples in the OFDM cyclic prefix. Additionally, the 'significant-tap catching' approaches of [131, 520] can be employed for improving the channel estimator's performance by further removing the low-power taps – those that are likely to be constituted by noise, rather than by 'real' CIR taps – within the range of $[1, \ldots, K_0]$ according to a predefined amplitude threshold. Following the CIR tap filtering process discussed above, the retained CIR-related coefficients $\tilde{h}_p^{(l)}[n, k]$ ($k = 1, \ldots, K_0$) are converted to the improved a posteriori FD-CHTF estimates $\hat{H}_p^{(l)}[n, k]$ ($k = 1, \ldots, K$) by the Fast Fourier Transform (FFT). For more detailed discussions of these processes, refer to Chapters 15 and 16 of [3].

If in the $(n + 1)$th OFDM symbol duration a data symbol rather than a pilot symbol is transmitted, the pilot controller will carry out the following actions:

- enables a first-stage MUD for generating reference symbol estimates; and

- switches the operating mode of the GA-JCEMUD from *pilot-aided channel estimation* mode to *joint channel estimation and data detection* mode.

More specifically, the Optimized Hierarchy Reduced Search Algorithm (OHRSA) aided MUD of [534, 535] is employed as the first-stage MUD, as shown in Figure 8.2. It exploits the a posteriori

---

[2] More details of the process will be discussed in Section 8.3.1.

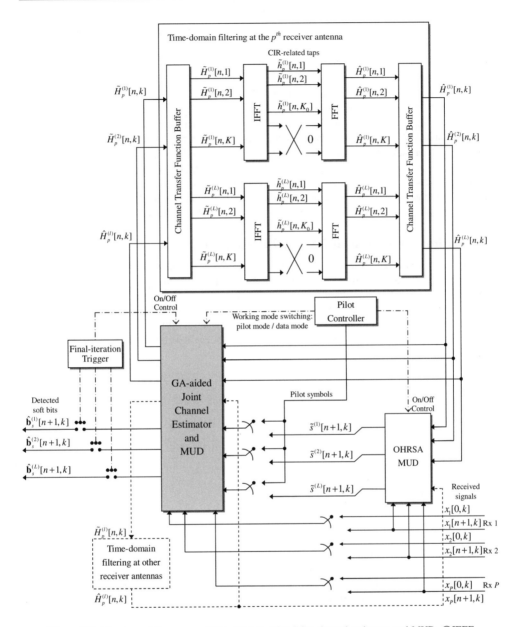

Figure 8.2: Structure of the proposed GA-aided iterative joint channel estimator and MUD. ©IEEE Jiang, Akhtman & Hanzo 2007 [517]

FD-CHTF estimates $\hat{H}_p^{(l)}[n,k]$ associated with the previous OFDM symbol to invoke subcarrier-by-subcarrier based detection, yielding an initial guess of the L users' transmitted symbols $s^{(l)}[n+1,k]$ ($l = 1,\ldots,L;\ k = 1,\ldots,K$). The resultant symbol estimates $\tilde{s}^{(l)}[n+1,k]$, the FD-CHTF estimates $\hat{H}_p^{(l)}[n,k]$ as well as the corresponding received signals $x_p[n+1,k]$ ($p = 1,\ldots,P;\ k = 1,\ldots,K$) are then forwarded to the GA-JCEMUD, where the FD-CHTFs and data symbols associated with the $(n+1)$th OFDM symbol are jointly optimized on a subcarrier-by-subcarrier basis.

The GA-optimized FD-CHTF estimates $\tilde{H}_p^{(l)}[n+1,k]$ are then forwarded to the time-domain filters for further enhancement, as mentioned earlier in this section. Now the cleansed a posteriori channel estimates $\hat{H}_p^{(l)}[n+1,k]$ are expected to be closer to their true values of $H_p^{(l)}[n+1,k]$ than the initially used estimates, i.e. $\hat{H}_p^{(l)}[n,k]$, which are associated with the previous OFDM symbol. Thus, based on the improved channel estimates, the OHRSA MUD is capable of providing a better initial guess of the transmitted symbols for the GA-JCEMUD. This decision-directed process can be invoked for a number of iterations to attain a further performance enhancement. After the final iteration, the final-iteration trigger portrayed on the left-hand side of Figure 8.2 terminates the GA-JCEMUD's operation and enables the output links, generating the L users' detected soft bits $\hat{b}_s^{(l)}[n+1,k]$ ($l=1,\ldots,L$; $k=1,\ldots,K$) corresponding to the $(n+1)$th OFDM symbol.

In the following sections, we will further detail the processes of obtaining initial FD-CHTF estimates with the aid of pilots, generating initial symbol estimates by the OHRSA MUD as well as jointly optimizing the FD-CHTFs and the data symbols using the GA, respectively.

8.3.1 Pilot-Aided Initial Channel Estimation

In order to obtain an initial estimate of the FD-CHTFs, each user's pilot OFDM symbol is multiplied by a user-specific spreading code before it is transmitted.[3] With the aid of the spread pilot OFDM symbols, an initial FD-CHTF estimate is attainable at the receivers, where the Multi-User Interference (MUI) is effectively reduced in proportion to the spreading factor.

More specifically, the orthogonal Walsh–Hadamard Transform (WHT) [3] codes of length L are chosen. The received symbol at the pth receiver antenna element associated with the kth subcarrier of the $n=0$th OFDM symbol duration can be formulated as

$$x_p[0,k] = \mathbf{c}\bar{\mathbf{H}}_p[0,k]\mathbf{s}[0,k] + n_p[0,k], \quad p=1,\ldots,P, \tag{8.1}$$

where the pilot signal vector $\mathbf{s}[0,k]$ and the diagonal FD-CHTF matrix $\bar{\mathbf{H}}_p[0,k]$ are given by

$$\mathbf{s}[0,k] = [s^{(1)}[0,k], s^{(2)}[0,k],\ldots,s^{(L)}[0,k]]^T, \tag{8.2}$$

$$\bar{\mathbf{H}}_p[0,k] = \mathrm{diag}[H_p^{(1)}[0,k], H_p^{(2)}[0,k],\ldots,H_p^{(L)}[0,k]], \tag{8.3}$$

respectively, and the user signature vector $\mathbf{c}$ is formulated as

$$\mathbf{c} = [\mathbf{c}^{(1)}, \mathbf{c}^{(2)},\ldots,\mathbf{c}^{(L)}], \tag{8.4}$$

where $\mathbf{c}^{(l)}$ ($l=1,\ldots,L$) represents the lth user's WHT code sequence, which is the lth row of the L-order recursive WHT matrix given by [3]

$$\mathbf{U}_{\mathrm{WHT}_L} = \frac{1}{\sqrt{2}}\begin{bmatrix} 1\cdot\mathbf{U}_{\mathrm{WHT}_{L/2}} & 1\cdot\mathbf{U}_{\mathrm{WHT}_{L/2}} \\ 1\cdot\mathbf{U}_{\mathrm{WHT}_{L/2}} & -1\cdot\mathbf{U}_{\mathrm{WHT}_{L/2}} \end{bmatrix}, \tag{8.5}$$

while the lowest-order WHT unitary matrix is defined as

$$\mathbf{U}_{\mathrm{WHT}_2} = \frac{1}{\sqrt{2}}\begin{bmatrix} 1 & 1 \\ 1 & -1 \end{bmatrix}. \tag{8.6}$$

Note that the pilot symbol vector $\mathbf{s}[0,k]$ of Equation (8.1) is known at the receivers. Furthermore, we can use the same unspread pilot QAM symbol for all users, i.e. $s^{(l)}[0,k] = s_0$ ($l=1,\ldots,L$). Hence, it directly follows from Equation (8.1) that

$$\check{x}_p[0,k] = \frac{s_0^*}{|s_0|^2}x_p[0,k]$$

$$= \mathbf{c}\mathbf{H}_p^T[0,k] + \frac{s_0^*}{|s_0|^2}n_p[0,k], \quad p=1,\ldots,P, \tag{8.7}$$

---

[3]Note that no spreading is applied to the data OFDM symbols.

where $(\cdot)^*$ denotes the complex conjugate and $\mathbf{H}_p[0, k]$ is the pth row of the FD-CHTF matrix $\mathbf{H}$, given by

$$\mathbf{H}_p[0, k] = [H_p^{(1)}[0, k], H_p^{(2)}[0, k], \ldots, H_p^{(L)}[0, k]]. \tag{8.8}$$

Thus, the initial FD-CHTF estimates can be obtained as follows:

$$\begin{aligned}
\tilde{\mathbf{H}}_p^T[0, k] &= \mathbf{c}^T \tilde{x}_p[0, k] \\
&= \mathbf{H}_p^T[0, k] + \underbrace{\frac{s_0^*}{|s_0|^2} n_p[0, k] \mathbf{c}^T}_{n_p'[0, k]}, \quad p = 1, \ldots, P.
\end{aligned} \tag{8.9}$$

After the time-domain CIR tap filtering operation shown in Figure 8.2, the refined channel estimates can then be used to assist the OHRSA MUD to detect the unknown transmitted symbols within the next OFDM symbol duration. Afterwards, the GA-JCEMUD will be set to joint channel estimation and data detection mode, providing the FD-CHTF estimates associated with the forthcoming OFDM symbols.

Depending on the specific performance–throughput design trade-off targeted, this process of generating initial channel estimates can be invoked at predefined time intervals. Here we denote the pilot overhead as ϵ, which is defined by the ratio of the number of pilot OFDM symbols to the total number of transmitted OFDM symbols. We will show in Section 8.4 that a good performance is achievable by the proposed scheme with a small value of ϵ.

8.3.2 Generating Initial Symbol Estimates

As mentioned earlier, for each subcarrier, an initial symbol estimate is first obtained with the aid of the first-stage OHRSA-assisted MUD [534, 535] shown in Figure 8.2, which exploits the a posteriori FD-CHTF estimates generated within the previous OFDM symbol duration. For the sake of notational convenience, in this section the index of $[n, k]$ is omitted. However, we note that the following analysis is conducted on a subcarrier basis.

As an extension of the Complex Sphere Decoder (CSD) method [448, 450], the OHRSA MUD is capable of achieving a near-optimum performance at a significantly reduced computational complexity. It is well known that the optimum ML MUD [3] employs an exhaustive search for finding the most likely transmitted signals. More explicitly, recall that the L-user symbol vector estimate $\hat{\mathbf{s}}_{\mathrm{ML}}$ can be obtained by minimizing the following metric:

$$\hat{\mathbf{s}}_{\mathrm{ML}} = \arg\left\{ \min_{\check{\mathbf{s}} \in \mathcal{M}^L} \|\mathbf{x} - \mathbf{H}\check{\mathbf{s}}\|^2 \right\}, \tag{8.10}$$

where $\check{\mathbf{s}}$ is an a priori candidate vector of the set $\mathcal{M}^L$, which consists of 2^{mL} trial vectors, where m denotes the number of Bits Per Symbol (BPS). More specifically, $\mathcal{M}^L$ is formulated as

$$\mathcal{M}^L = \{\check{\mathbf{s}} = [\check{s}^{(1)}, \check{s}^{(2)}, \ldots, \check{s}^{(L)}]^T \mid \check{s}^{(1)}, \check{s}^{(2)}, \ldots, \check{s}^{(L)} \in \mathcal{M}_c\}, \tag{8.11}$$

where $\mathcal{M}_c$ denotes the set containing the 2^m legitimate complex constellation points associated with the specific modulation scheme employed. Furthermore, it can be shown that Equation (8.10) is equivalent to [534, 535]

$$\hat{\mathbf{s}} = \arg\left\{ \min_{\check{\mathbf{s}} \in \mathcal{M}^L} \|\mathbf{V}(\check{\mathbf{s}} - \hat{\mathbf{s}}_{\mathrm{MMSE}})\|^2 \right\}, \tag{8.12}$$

where $\mathbf{V}$ is an upper-triangular matrix having positive real-valued elements on the main diagonal and satisfying

$$\mathbf{V}^H \mathbf{V} = \mathbf{H}^H \mathbf{H} + \sigma_n^2 \mathbf{I}, \tag{8.13}$$

while

$$\hat{\mathbf{s}}_{\mathrm{MMSE}} = (\mathbf{H}^H \mathbf{H} + \sigma_n^2 \mathbf{I})^{-1} \mathbf{H}^H \mathbf{x} \tag{8.14}$$

is the unconstrained MMSE-based estimate[4] of the transmitted signal vector s, with $\mathbf{I}$ and σ_n^2 the identity matrix and the AWGN noise variance, respectively. Since $\mathbf{V}$ is an upper-triangular matrix, a specific cost function can be derived:

$$
\begin{aligned}
\Phi(\check{\mathbf{s}}) &= \|\mathbf{V}(\check{\mathbf{s}} - \hat{\mathbf{s}}_{\mathrm{MMSE}})\|^2 \\
&= (\check{\mathbf{s}} - \hat{\mathbf{s}}_{\mathrm{MMSE}})^H \mathbf{V}^H \mathbf{V}(\check{\mathbf{s}} - \hat{\mathbf{s}}_{\mathrm{MMSE}}) \\
&= \sum_{i=1}^{L} \left| \sum_{j=i}^{L} v_{ij}(\check{s}^{(j)} - \hat{s}_{\mathrm{MMSE}}^{(j)}) \right|^2 \\
&= \sum_{i=1}^{L} \phi_i(\check{\mathbf{s}}_i),
\end{aligned}
\tag{8.15}
$$

where $\phi_i(\check{\mathbf{s}}_i)$ is a set of subcost functions. Note that the outputs of both $\Phi(\check{\mathbf{s}})$ and $\phi_i(\check{\mathbf{s}}_i)$ are real valued. Furthermore, we have

$$
\begin{aligned}
\phi_i(\check{\mathbf{s}}_i) &= \left| \sum_{j=i}^{L} v_{ij}(\check{s}^{(j)} - \hat{s}_{\mathrm{MMSE}}^{(j)}) \right|^2 \\
&= \left| v_{ii}(\check{s}^{(i)} - \hat{s}_{\mathrm{MMSE}}^{(i)}) + \sum_{j=i+1}^{L} v_{ij}(\check{s}^{(j)} - \hat{s}_{\mathrm{MMSE}}^{(j)}) \right|^2.
\end{aligned}
\tag{8.16}
$$

Based on Equation (8.16), $\Phi(\check{\mathbf{s}})$ can be redefined as the Cumulative Sub-Cost (CSC) function:

$$
\Phi_L(\check{\mathbf{s}}_L) = \phi_L(\check{\mathbf{s}}_L) = |v_{LL}(\check{s}^{(L)} - \hat{s}_{\mathrm{MMSE}}^{(L)})|^2, \quad i = L, \tag{8.17a}
$$

$$
\Phi_i(\check{\mathbf{s}}_i) = \Phi_{i+1}(\check{\mathbf{s}}_{i+1}) + \phi_i(\check{\mathbf{s}}_i), \quad i = L-1, \ldots, 1, \tag{8.17b}
$$

where $\check{\mathbf{s}}_i$ represents the subvectors of $\check{\mathbf{s}}$, formulated as

$$
\check{\mathbf{s}}_i = [\check{s}^{(i)}, \check{s}^{(i+1)}, \ldots, \check{s}^{(L)}], \quad i \in \{1, \ldots, L\}. \tag{8.18}
$$

In physically tangible terms, the Euclidean norm of Equation (8.17) can be interpreted as a weighted Euclidean distance between the candidate constellation point $\check{s}^{(l)}$ and the unconstrained MMSE estimate $\hat{s}_{\mathrm{MMSE}}^{(l)}$ of the transmitted signal component $s^{(l)}$. Explicitly, the CSC functions obey the property

$$
\Phi(\check{\mathbf{s}}) = \Phi_1(\check{\mathbf{s}}_1) > \Phi_2(\check{\mathbf{s}}_2) > \cdots > \Phi_L(\check{\mathbf{s}}_L) > 0 \tag{8.19}
$$

for all possible combinations of $\check{\mathbf{s}} \in \mathcal{M}^L$ and $\hat{\mathbf{s}}_{\mathrm{MMSE}} \in \mathbb{C}^{L \times 1}$, where the L-dimensional complex space $\mathbb{C}^{L \times 1}$ contains all possible unconstrained MMSE estimates $\hat{\mathbf{s}}_{\mathrm{MMSE}}$ of the transmitted signal vector s.

By exploiting the monotonously increasing nature of the non-binary, i.e. multi-bit, symbol-based CSC functions of Equation (8.19), a bit-based recursive search algorithm can be developed [534, 535], where the $(L_b = mL)$-dimensional bit vectors $\check{\mathbf{b}}$ constituting the L users' bits, rather than the symbol vectors $\check{\mathbf{s}}$, are used as the candidates for the CSC functions given by Equations (8.15) and (8.17). More specifically, two legitimate hypotheses of -1 and 1 are stipulated at each recursive step i of the search algorithm, concerning one of the bits of the bit-based trial vector $\check{\mathbf{b}}_i$. This allows us now to interpret physically the CSC functions as the Euclidian distance contribution of the specific $\check{\mathbf{b}}_i$, when considering a specific bit of a given symbol of a given user. The recursive search process commences with the evaluation of the CSC function of Equation (8.17a), followed by the calculation of the conditioned CSC function values of Equation (8.17b). Moreover, for each tentatively assumed value of $\check{\mathbf{b}}_i$ a successive recursive search step $(i-1)$ is invoked, which is conditioned on the hypotheses made in all preceding

---

[4]The unconstrained MMSE-based estimate denotes the resultant complex value calculated from Equation (8.14), as opposed to the constrained estimate, which is the hard-decoded version of the unconstrained estimate.

recursive steps $j = i, \ldots, L_b = mL$. Upon each arrival of the recursive process at the index $i = 1$, a complete bit-based candidate vector $\hat{\mathbf{b}}$ associated with a certain symbol vector $\check{\mathbf{s}}$ is hypothesized and the corresponding value of the cost function $\Phi(\hat{\mathbf{b}})$ formulated in Equation (8.15) is evaluated. Furthermore, with the aid of a carefully designed search strategy [534, 535], the OHRSA is capable of arriving at the optimum ML estimate at a significantly reduced complexity. For more details on the OHRSA MUD, the interested reader is referred to [534, 535].

8.3.3 GA-Aided Joint Optimization Providing Soft Outputs

With the aid of the initial FD-CHTF estimates of Section 8.3.1 and the initial symbol estimates of Section 8.3.2, the proposed GA-JCEMUD printed in the grey block of Figure 8.2 is employed to optimize jointly the estimates of the FD-CHTFs and multi-user data symbols.

8.3.3.1 Extended GA Individual Structure

In comparison with the pure GA-based MUDs [459, 460, 532], which optimize the multi-user data symbols only, the joint optimization work requires the FD-CHTFs to be simultaneously optimized along with the data symbols, as in [38, 461]. Furthermore, concerning the MIMO channel's structure, the GA *individuals'* representation of [38, 461] is extended to

$$
\begin{cases}
\check{\mathbf{s}}_{(y,x)}[n,k] = [\check{s}^{(1)}_{(y,x)}[n,k], \check{s}^{(2)}_{(y,x)}[n,k], \ldots, \check{s}^{(L)}_{(y,x)}[n,k]] \\[2mm]
\tilde{\mathbf{H}}_{(y,x)}[n,k] =
\begin{bmatrix}
\tilde{H}^{(1)}_{1,(y,x)}[n,k] & \tilde{H}^{(2)}_{1,(y,x)}[n,k] & \cdots & \tilde{H}^{(L)}_{1,(y,x)}[n,k] \\
\tilde{H}^{(1)}_{2,(y,x)}[n,k] & \tilde{H}^{(2)}_{2,(y,x)}[n,k] & \cdots & \tilde{H}^{(L)}_{2,(y,x)}[n,k] \\
\vdots & \vdots & \ddots & \vdots \\
\tilde{H}^{(1)}_{P,(y,x)}[n,k] & \tilde{H}^{(2)}_{P,(y,x)}[n,k] & \cdots & \tilde{H}^{(L)}_{P,(y,x)}[n,k]
\end{bmatrix}
\end{cases}
\tag{8.20}
$$

in the context of the kth subcarrier of the nth OFDM symbol, where the subscript (y, x) denotes the xth $(x = 1, \ldots, X)$ individual at the yth $(y = 1, \ldots, Y)$ *generation*. In compliance with classic GA terminology, any combination of a symbol vector $\check{\mathbf{s}}_{(y,x)}[n,k]$ and an FD-CHTF matrix $\tilde{\mathbf{H}}_{(y,x)}[n,k]$ represents a GA individual, where $\check{\mathbf{s}}_{(y,x)}[n,k]$ is referred to as the individual's *symbol chromosome* and $\tilde{\mathbf{H}}_{(y,x)}[n,k]$ as the associated *FD-CHTF chromosome*, respectively, while each element of a chromosome is termed a *gene*. Note that the symbol genes and the channel genes belong to different sets. More specifically, we have $\check{s}^{(l)}_{(y,x)}[n,k] \in \mathcal{M}_c$ and $\tilde{H}^{(l)}_{p,(y,x)}[n,k] \in \mathbb{C}$, where $\mathbb{C}$ denotes the set of all complex numbers. This particular association of a pair of hypothesized channel and data estimates is reminiscent of the so-called Per-Survivor Processing (PSP) based blind detection techniques of [536, 537]; however, it distinguished itself by invoking a genetically guided efficient search strategy.

8.3.3.2 Initialization

During the stage of initialization, the GA generates a *population* of X individuals represented by Equation (8.20), based on the initial FD-CHTF estimates of Section 8.3.1 and the initial symbol estimates of Section 8.3.2. More explicitly, at the kth subcarrier in the $(n+1)$th OFDM symbol duration, the genes of the $(y, x) = (1, 1)$st individual are generated as

$$
\begin{cases}
\check{s}^{(l)}_{(1,1)}[n+1,k] = \check{s}^{(l)}[n+1,k], \\
\tilde{H}^{(l)}_{p,(1,1)}[n+1,k] = \tilde{H}^{(l)}_{p}[n,k],
\end{cases}
\quad l = 1, \ldots, L; \; p = 1, \ldots, P,
\tag{8.21}
$$

where $\check{s}^{(l)}[n+1,k]$ represents the initial symbol estimate provided by the OHRSA MUD, while $\tilde{H}^{(l)}_{p}[n,k]$ denotes the initial FD-CHTF estimates associated with the previous, i.e. the nth, OFDM symbol. The other $(X - 1)$ individuals are then created by the GA's *mutation* operator:

$$\begin{cases} \tilde{s}^{(l)}_{(y,x)}[n+1,k] = MUTATION(\tilde{s}^{(l)}_{(1,1)}[n+1,k]), \\ \tilde{H}^{(l)}_{p,(y,x)}[n+1,k] = MUTATION(\tilde{H}^{(l)}_{p,(1,1)}[n+1,k]), \end{cases}$$
$$x = 2,\ldots,X; \; y = 1,\ldots,Y; \; l = 1,\ldots,L; \; p = 1,\ldots,P. \tag{8.22}$$

The details of the mutation process are discussed in Section 8.3.3.3.2. After the creation of the $y = 1$st generation, which consists of the population of the initially generated X individuals, the GA-based search process can be invoked for jointly optimizing the estimates of the multi-user symbols and FD-CHTFs.

8.3.3.3 Joint Genetic Optimization

The basic idea of the GA-based optimization is to find the optimum or a near-optimum solution according to a predefined *Objective Function (OF)*. In the context of the joint detection problem in SDMA-OFDM systems, the GA's OF can be based on the ML metric of Equation (8.10), formulated as

$$\Omega(\tilde{\mathbf{H}}[n,k], \tilde{\mathbf{s}}[n,k]) = \|\mathbf{x}[n,k] - \tilde{\mathbf{H}}[n,k] \cdot \tilde{\mathbf{s}}[n,k]\|^2, \tag{8.23}$$

where each combination of the trial data vector $\tilde{\mathbf{s}}[n,k]$ and trial FD-CHTF matrix $\tilde{\mathbf{H}}[n,k]$ constitutes a GA individual defined in Equation (8.20). The output of the OF is referred to as the *Objective Score (OS)*, and the individual having a lower OS is considered to have a higher *fitness* value. Explicitly, the GA's ultimate aim is to find the individual that has the highest fitness value. This is achieved with the aid of the genetic operators invoked during the evolution process, such as *cross-over* and *mutation* [38, 453], where specific genes of the different individuals are exchanged and mutated to produce the corresponding *offspring*. The number X of resultant offspring individuals then constitutes a new population, which forms the GA's next generation and is expected to have a statistically improved average fitness value in comparison with the parent population. Finally, the GA terminates when the generation index reaches a predefined value Y. Conventionally, the highest-fitness individual of the final population will be considered as the GA's final solution, which consists of the genetically improved FD-CHTF and data symbol estimates. For more details about the genetic optimization process, refer to Appendix A.1 and references [38, 459–461, 532].

Recall that the elements of the symbol chromosome $\tilde{\mathbf{s}}[n,k]$ of an individual belong to $\mathcal{M}_c$, i.e. to the legitimate constellation symbol set. Thus, the symbol mutation space is discrete and is limited by the 2^m constellation points in $\mathcal{M}_c$. By contrast, the mutation space of the FD-CHTFs is continuous and infinite, simply because the value of an FD-CHTF sample can be an arbitrary value on the complex plane $\mathbb{C}$. Therefore, different cross-over/mutation operators have to be employed for the symbol chromosome and channel chromosome, respectively.

8.3.3.3.1 Cross-over Operator

In our system, the uniform cross-over [38] is used to mutate the symbol chromosomes. This cross-over operator exchanges specific symbol genes of two parent individuals in order to generate the offspring. Theoretically, it may also be applied to the FD-CHTF mutation process, where the channel chromosomes of parent individuals are swapped. However, this 'exchange-only' mechanism neglects the continuous nature of the complex-valued FD-CHTFs, thus imposing a limitation during the cross-over process, when the genetic information is delivered from parents to offspring.

To improve the cross-over efficiency, the blend cross-over [538, 539] is invoked for combining the channel chromosomes from each of the two parent individuals into new channel chromosomes of the two offspring. More specifically, the offspring's channel chromosomes can be formulated as

$$\begin{cases} \tilde{H}^{(l)}_{p,(y+1,1)}[n,k] = \beta_1 \tilde{H}^{(l)}_{p,(y,1)}[n,k] + \beta_2 \tilde{H}^{(l)}_{p,(y,2)}[n,k], \\ \tilde{H}^{(l)}_{p,(y+1,2)}[n,k] = \beta_2 \tilde{H}^{(l)}_{p,(y,1)}[n,k] + \beta_1 \tilde{H}^{(l)}_{p,(y,2)}[n,k], \end{cases} \quad y = 1,\ldots,Y, \tag{8.24}$$

where the random weight factors $\beta_i \in [0, 1]$ $(i = 1, 2)$ satisfy

$$\beta_1 + \beta_2 = 1. \tag{8.25}$$

Similar to the uniform cross-over operator invoked for symbol chromosomes, in the blend cross-over operator a random binary cross-over mask is also created to identify the specific channel genes to be combined. Observing Equation (8.24), we can note that the genes of the new channel chromosomes are actually the linearly biased or weighted results averaged between the two parents in the context of the $(P \times L)$-dimensional complex space. When we have $\beta_1 = \beta_2 = 0.5$, the result becomes an unbiased average of the channel chromosomes of the two parents. Explicitly, the blend cross-over operator exploits the continuous and infinite nature of the FD-CHTFs and hence it is capable of amalgamating the parental genetic information in a more meritorious manner.

8.3.3.3.2 Mutation Operator

Similar to the cross-over operator, the mutation operator also requires the employment of different methods for the symbol and channel chromosomes, respectively. The mutation schemes based on discrete character sets, e.g. the classic Uniform Mutation (UM) [38], can be applied to the symbol chromosomes. On the other hand, the approach of [461], which is referred to here as Step Mutation (SM), can be employed to mutate the channel chromosomes. To elaborate a little further, the offspring's channel chromosomes can be generated by

$$\begin{cases} \Re(\tilde{H}^{(l)}_{p,(y+1,x)}[n, k]) = \Re(\tilde{H}^{(l)}_{p,(y,x)}[n, k]) + \theta_\Re \Delta^{(l)}_{p,(y,x)}[n, k], \\ \Im(\tilde{H}^{(l)}_{p,(y+1,x)}[n, k]) = \Im(\tilde{H}^{(l)}_{p,(y,x)}[n, k]) + \theta_\Im \Delta^{(l)}_{p,(y,x)}[n, k], \\ x = 1, \ldots, X; \ y = 1, \ldots, Y; \ l = 1, \ldots, L; \ p = 1, \ldots, P, \end{cases} \tag{8.26}$$

where $\Delta^{(l)}_{p,(y,x)}[n, k]$ is a random number within $(0, \lambda_{max}]$, while $\Re$ and $\Im$ respectively denote the real and imaginary components of the specific channel gene $\tilde{H}^{(l)}_{p,(y+1,x)}[n, k]$ to be mutated. The sign factors $\theta_\Re$ and $\theta_\Im$ of Equation (8.26) are uniformly and randomly generated, assuming values of $+1$ or -1.

We point out that the value of the maximum mutation step size λ_{max} is critical for the system's attainable performance, since it directly affects the convergence of the GA's optimization process. A low system performance will be expected when λ_{max} is insufficiently high, where the GA may get trapped in local rather than global minima. However, an excessively high value of λ_{max} will result in a slow convergence rate. Generally speaking, the value of λ_{max} should be adjusted according to the Doppler frequency encountered. More specifically, when we have a higher Doppler frequency, the consecutive channel fades in the time domain experienced by each of the subcarriers become faster, thus requiring a higher λ_{max} value to assist the GA in capturing the rapid changes of the channel fades. In low-Doppler scenarios, the situation is the inverse. Therefore, it is desirable that the value of λ_{max} is adjusted as a function of the Doppler frequency. However, in this case it is a challenging task to develop a closed-form function for quantifying the effects of the Doppler frequency, owing to the inherently nonlinear nature of the GA-based optimization process. Nonetheless, with the aid of computer simulations we can identify the appropriate values of λ_{max} for different Doppler frequencies, as will be discussed in Section 8.4.1.

8.3.3.3.3 Comments on the Joint Optimization Process

Note that since the FD-CHTFs and data symbols are jointly optimized within the same genetic process, the individuals having better FD-CHTF estimates will have a higher probability of producing better symbol estimates and vice versa. Thus, this joint optimization is a 'self-adaptive' process with its native intuition leading towards the optimum solution. Furthermore, compared with other techniques, e.g. that of [529], where the channel estimation and symbol detection are completed by the GA and the

VA separately, the proposed joint scheme is capable of reducing the associated complexity, since the channel estimation is simultaneously achieved with the aid of the same GA process, thus incurring no additional complexity.

Moreover, it is worth pointing out that at the time of writing the GA-aided detection schemes found in the literature [38,459–461,532] are only capable of providing single-individual hard-decoded symbol estimates, which inevitably limits the GA-aided system's attainable performance. In Section 8.3.3.4, we will introduce a method which enables the GA to provide soft outputs based on the entire population.

8.3.3.4 Generating the GA's Soft Outputs

In this section we derive an algorithm that enables the GA to output soft information. For the sake of simplicity, again, we omit the index $[n, k]$ in this section.

The soft-bit value or Log-Likelihood Ratio (LLR) associated with the (m_B)th bit position of the lth $(l = 1, \ldots, L)$ user's transmitted symbol $s^{(l)}$ can be formulated as [540]

$$\mathcal{L}_{l,m_B} = \ln \frac{P(b_{l,m_B} = 1|\mathbf{x}, \mathbf{H})}{P(b_{l,m_B} = 0|\mathbf{x}, \mathbf{H})}, \tag{8.27}$$

which is the natural logarithm of the quotient of probabilities that the bit considered has a value of $b_{l,m_B} = 1$ or $b_{l,m_B} = 0$. Note that the probability $P(b_{l,m_B} = b|\mathbf{x}, \mathbf{H})$ that the symbol transmitted by the lth user has the (m_B)th bit value of $b_{l,m_B} = b \in \{0, 1\}$ is given by the sum of all the probabilities of the symbol combinations which assume that $b_{l,m_B} = b$. Hence, Equation (8.27) can be equivalently rewritten as

$$\mathcal{L}_{l,m_B} = \ln \frac{\sum_{\check{s} \in \mathcal{M}_{l,m_B,1}^L} P(\check{s}|\mathbf{x}, \mathbf{H})}{\sum_{\check{s} \in \mathcal{M}_{l,m_B,0}^L} P(\check{s}|\mathbf{x}, \mathbf{H})}, \tag{8.28}$$

where $\mathcal{M}_{l,m_B,b}^L$ denotes the specific subset associated with the lth user, which is constituted by those specific trial vectors whose lth element's (m_B)th bit has a value of b, which is expressed as

$$\mathcal{M}_{l,m_B,b}^L = \{\check{s} = [\check{s}^{(1)}, \check{s}^{(2)}, \ldots, \check{s}^{(L)}]^T \mid \{\check{s}^{(1)}, \check{s}^{(2)}, \ldots, \check{s}^{(L)}\} \in \mathcal{M}_c\} \wedge \{b_{l,m_B} = b\}\}. \tag{8.29}$$

With the aid of Bayes' theorem [540], we have

$$P(\check{s}|\mathbf{x}, \mathbf{H}) = P(\mathbf{x}|\check{s}, \mathbf{H}) \frac{P(\check{s})}{P(\mathbf{x})}. \tag{8.30}$$

On substituting Equation (8.30) into Equation (8.28), we arrive at

$$\mathcal{L}_{l,m_B} = \ln \frac{\sum_{\check{s} \in \mathcal{M}_{l,m_B,1}^L} P(\mathbf{x}|\check{s}, \mathbf{H})}{\sum_{\check{s} \in \mathcal{M}_{l,m_B,0}^L} P(\mathbf{x}|\check{s}, \mathbf{H})}. \tag{8.31}$$

Note that here we have assumed that the different (2^m)-ary symbol combination vectors $\check{s}$ have the same probability, i.e. $P(\check{s})$, $\check{s} \in \mathcal{M}_c$, is a constant. On the other hand, recall that in SDMA-OFDM systems we have

$$\mathbf{x} = \mathbf{Hs} + \mathbf{n}. \tag{8.32}$$

It can be observed from Equation (8.32) that $\mathbf{x}$ is a random sample of the L-dimensional multivariate complex-Gaussian distribution, where the mean vector is $(\mathbf{Hs})$, while the $(P \times P)$-dimensional covariance matrix $\mathbf{R_n}$ is given by

$$\mathbf{R_n} = E\{\mathbf{nn}^H\} = \sigma_n^2 \mathbf{I}, \tag{8.33}$$

and the noise encountered at the P receiver antennas is assumed to be uncorrelated. Hence, the above-mentioned multi-variate complex-Gaussian distribution can be described by [541]

$$f(\mathbf{x}|\mathbf{s}, \mathbf{H}) = \frac{1}{\pi^P |\mathbf{R_n}|} \exp\{-(\mathbf{x} - \mathbf{Hs})^H \mathbf{R_n^{-1}} (\mathbf{x} - \mathbf{Hs})\}. \tag{8.34}$$

On substituting Equation (8.33) into Equation (8.34), we have

$$f(\mathbf{x}|\mathbf{s}, \mathbf{H}) = \frac{1}{\pi^P \sigma_n^2} \exp\left\{-\frac{1}{\sigma_n^2} \|\mathbf{x} - \mathbf{H}\mathbf{s}\|^2\right\}. \tag{8.35}$$

Note that $f(\mathbf{x}|\mathbf{s}, \mathbf{H}) = P(\mathbf{x}|\mathbf{s}, \mathbf{H})$ is the a priori probability that the vector $\mathbf{x}$ has been received under the condition that the vector $\mathbf{s}$ was transmitted over the MIMO channel characterized by the FD-CHTF matrix $\mathbf{H}$. Thus, Equation (8.31) can be further developed with the aid of Equation (8.35), yielding

$$\mathcal{L}_{l,m_B} = \ln \frac{\sum_{\breve{\mathbf{s}} \in \mathcal{M}_{l,m_B,1}^L} (1/\pi^P \sigma_n^2) \exp\{-(1/\sigma_n^2)\|\mathbf{x} - \mathbf{H}\breve{\mathbf{s}}\|^2\}}{\sum_{\breve{\mathbf{s}} \in \mathcal{M}_{l,m_B,0}^L} (1/\pi^P \sigma_n^2) \exp\{-(1/\sigma_n^2)\|\mathbf{x} - \mathbf{H}\breve{\mathbf{s}}\|^2\}}. \tag{8.36}$$

In order to avoid the exponential computation imposed by Equation (8.36), the maximum approximation [3] can be applied, yielding

$$\mathcal{L}_{l,m_B} \approx -\frac{1}{\sigma_n^2}[\|\mathbf{x} - \mathbf{H}\breve{\mathbf{s}}_{l,m_B,1}\|^2 - \|\mathbf{x} - \mathbf{H}\breve{\mathbf{s}}_{l,m_B,0}\|^2], \tag{8.37}$$

where

$$\breve{\mathbf{s}}_{l,m_B,b} = \arg\left\{\min_{\breve{\mathbf{s}} \in \mathcal{M}_{l,m_B,b}^L} [\|\mathbf{x} - \mathbf{H}\breve{\mathbf{s}}\|^2]\right\}, \quad b = 0, 1. \tag{8.38}$$

Furthermore, concerning the fact that the true FD-CHTF matrix $\mathbf{H}$ is unknown, and using Equation (8.23), Equation (8.37) can be represented as

$$\mathcal{L}_{l,m_B} \approx -\frac{1}{\sigma_n^2}[\overline{\Omega}_{l,m_B,1} - \overline{\Omega}_{l,m_B,0}], \tag{8.39}$$

where

$$\overline{\Omega}_{l,m_B,b} = \min[\Omega(\tilde{\mathbf{H}}, \breve{\mathbf{s}}_{l,m_B,b}), \omega], \quad b = 0, 1, \tag{8.40}$$

and $\omega = P \cdot L$ is a normalization factor. Equation (8.39) suggests that the LLRs can be obtained by evaluating the GA's OF. More explicitly, in order to calculate the LLR of the (m_B)th bit of the lth $(l = 1, \ldots, L)$ user at the specific subcarrier considered, the number X of individuals in the GA's final generation is divided into two groups, where the first (or second) group is constituted by those individuals that have a value of one (or zero) at the (m_B)th bit of the lth user's estimated transmitted symbol. The resultant lowest OS calculated in each of the two groups is then compared with ω, and the smaller of the two will be used in Equation (8.39) for calculating the corresponding LLR, which can therefore assist the channel decoder in improving the SDMA-OFDM system's performance.

It is worth pointing out that the proposed GA generating the above-mentioned population-based soft outputs only imposes a modest complexity increase in comparison with the conventional hard-decision individual-based GAs [38, 459–461, 532]. This is because the only additional operation required by the proposed scheme is to compare ω with the OSs which are already available, since the results of the OF evaluation carried out by the conventional GAs can be readily used.

8.4 Simulation Results

In this section, we will quantify the performance of the MIMO SDMA-OFDM system using the proposed GA-aided iterative joint channel estimation and MUD technique. Our attention was focused on overloaded scenarios which have not been investigated in the literature in this specific context at the time of writing. Nonetheless, we point out that the proposed GA-JCEMUD scheme performs equally well in conventional scenarios where the number of users L is less than or equal to the number of receiver antenna elements P.

More specifically, an overloaded scenario where $L = 4$ users were supported by $P = 2$ receiver antenna elements was considered. As an example, a simple two-path Rayleigh fading channel model

Table 8.1: Basic simulation parameters used in Section 8.4.

LDPC parameters	Modem		4-QAM
	Code rate		0.5
	Column weight		2.5
	Maximum iterations		10
	Block length of input bits		640 bits
GA-JCEMUD parameters	Symbol initialization		OHRSA [534, 535]
	Mating pool creation strategy		Pareto-Optimality [328]
	Selection method		Fitness-proportionate
	Cross-over scheme	FD-CHTF	Blend cross-over
		Symbol	Uniform cross-over
	Mutation scheme	FD-CHTF	Step mutation [38, 461]
		Symbol	Uniform mutation
	Mutation probability	FD-CHTF	0.20
		Symbol	0.15
	Maximum mutation step size λ_{max}		Varied (dependent on F_D)
	Elitism percentage		10%
	Incest prevention		Enabled
	Population size X		160
	Generations Y		5
	GA's output		Population-based
	Number of iterations		1 (unless specified)
	Pilot overhead ϵ		2.5% (unless specified)
Channel parameters	Paths		2
	Delay profile		$0.7740 \cdot z^0 + 0.6332 \cdot z^{-1}$
	K_0		8
	Subcarriers K		64
	Cyclic prefix		8
	F_D		0.003 (unless specified)

was employed, where the associated delay profile was $(0.7740 \cdot z^0 + 0.6332 \cdot z^{-1})$. The value of the parameter K_0 was set to $8 \gg 2$, which is a rather loose condition, potentially capable of tolerating an increase of the actual dispersion up to eight CIR taps. Each of the paths experienced independent Rayleigh fading having the same Doppler frequency of $F_D = f_d T_s$ normalized to the OFDM symbol rate, where f_d and T_s are the maximum Doppler frequency and the OFDM symbol duration including the cyclic prefix, respectively. The channel was assumed to be 'OFDM symbol invariant', implying that the CIR taps were assumed to be constant for the duration of one OFDM symbol, but they were faded at the beginning of each symbol. Furthermore, the fading envelope of the $(P \times L)$ number of user–receiver channel links were assumed to be uncorrelated. Each user's associated transmit power or signal variance was assumed to be unity.

Both scenarios with and without FEC coding were investigated. In the FEC-coded scenario, as an example, a half-rate binary Low-Density Parity Check (LDPC) [419, 428] code was employed. However, other FEC codes, e.g. Turbo Convolutional (TC) [314, 420, 421] codes, are also applicable to the proposed system. For the reader's convenience, the simulation parameters are summarized in Table 8.1. For more details on the GA's configuration, the interested reader is referred to Section 8.3.3.3 and Chapter 5.

8.4.1   Effects of the Maximum Mutation Step Size

As a preliminary investigation, in this section we attempt to identify the appropriate choices of the maximum mutation step size λ_{max} in conjunction with different Doppler frequencies. Firstly, we

LDPC-GA-JCEMUD-SDMA-OFDM, L4/P2, 4QAM, 2-path Rayleigh

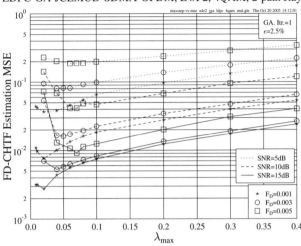

Figure 8.3: **FD-CHTF estimation MSE** versus λ_{max} performance of the **LDPC-coded iterative GA-JCEMUD-assisted SDMA-OFDM** system in the overloaded scenario, where $L = 4$ users were supported with the aid of $P = 2$ receiver antenna elements, while assuming different values of the OFDM-symbol-normalized Doppler frequency F_D. The basic simulation parameters are given in Table 8.1.

characterize the average FD-CHTF estimation Mean Square Error (MSE) performance of the GA-JCEMUD/SDMA-OFDM system using various values of λ_{max}, as shown in Figure 8.3, where the GA-JCEMUD invoked a single iteration and used a pilot overhead of $\epsilon = 2.5\%$. The average FD-CHTF estimation MSE is defined by

$$\overline{\text{MSE}} = \frac{1}{N_T} \sum_{n=1}^{N_T} \overline{\text{MSE}}[n], \tag{8.41}$$

where N_T is the total number of OFDM symbols transmitted, while $\overline{\text{MSE}}[n]$ is the average FD-CHTF estimation MSE associated with the nth OFDM symbol, given by

$$\overline{\text{MSE}}[n] = \frac{1}{PLK} \sum_{p=1}^{P} \sum_{l=1}^{L} \sum_{k=1}^{K} |\hat{H}_p^{(l)}[n,k] - H_p^{(l)}[n,k]|^2. \tag{8.42}$$

Here we point out that the FD-CHTF estimation MSE performance of the GA-JCEMUD is the same both with and without employing FEC coding. This is because the GA-aided joint optimization process has no direct interaction with the outer FEC code and thus it becomes independent of the codec. Explicitly, we can see in Figure 8.3 that the choice of λ_{max} has a substantial effect on the system's FD-CHTF estimation MSE performance, regardless of both F_D and the Signal-to-Noise Ratio (SNR).

Figure 8.4 shows the performance of the Bit Error Ratio (BER) versus λ_{max} of the LDPC-coded GA-JCEMUD-aided SDMA-OFDM system. It can be found that in scenarios associated with a higher SNR the effect of λ_{max} becomes more significant in terms of the achievable BER performance. The reason for this phenomenon is that at low SNRs the BER performance of the system is mainly dominated by the noise signal, and at high SNRs by the choice of λ_{max}, provided that the Doppler frequency is the same. Furthermore, a higher F_D value typically requires a higher λ_{max} to attain the best achievable BER performance, as we discussed in Section 8.3.3.3.2. According to Figure 8.4, the recommended values

LDPC-GA-JCEMUD-SDMA-OFDM, L4/P2, 4QAM, 2-path Rayleigh

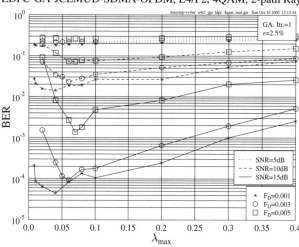

Figure 8.4: BER versus λ_{max} performance of the **LDPC-coded iterative GA-JCEMUD-assisted SDMA-OFDM** system in the overloaded scenario, where $L = 4$ users were supported with the aid of $P = 2$ receiver antenna elements, while assuming different values of the OFDM-symbol-normalized Doppler frequency F_D. The basic simulation parameters are given in Table 8.1. ©IEEE Jiang, Akhtman & Hanzo 2007 [517]

of λ_{max} in terms of achieving the best possible BER performance[5] are 0.4, 0.6 and 0.7 in conjunction with F_D values of 0.001, 0.003 and 0.005, respectively. These values of λ_{max} corresponding to the specific F_D values encountered were used to generate all the simulation results discussed in the rest of this chapter.

8.4.2 Effects of the Doppler Frequency

In Figure 8.5 we compare the BER versus SNR performance of both the uncoded and LDPC-coded GA-JCEMUD/SDMA-OFDM systems in conjunction with different values of F_D. The performances of the systems employing the linear MMSE MUD or the optimum ML MUD are also provided as references, both assuming perfect Channel State Information (CSI). A pilot overhead of $\epsilon = 2.5\%$ was assumed and the GA-JCEMUD used a single iteration. As shown in Figure 8.5, unsurprisingly, the performances of both the uncoded and coded GA-JCEMUD-aided systems degraded when F_D was increased, since a higher Doppler frequency implies that the channel fades more rapidly, which renders channel estimation more challenging. This is especially true for MIMO systems, even more so for overloaded MIMO systems, as discussed in Section 8.1. Nonetheless, with only a 2.5% pilot overhead, the proposed GA-JCEMUD/SDMA-OFDM system was capable of achieving a performance close to the perfect-CSI-aided optimum ML MUD at $F_D = 0.001$. By contrast, the system employing the MMSE MUD completely failed even with the aid of perfect CSI, owing to the insufficient degree of detection freedom experienced in overloaded scenarios. Furthermore, compared with the uncoded scenario, the performance degradation of the LDPC-coded GA-JCEMUD incurred by higher Doppler frequencies was much less dramatic, as observed in Figure 8.5.

---

[5]Note that the best choice of λ_{max} for achieving the best BER performance may be similar to, but may not necessarily be, the best choice for attaining the best FD-CHTF estimation MSE performance. This effect can be observed from Figures 8.3 and 8.4.

LDPC-GA-JCEMUD-SDMA-OFDM, L4/P2, 4QAM, 2-path Rayleigh

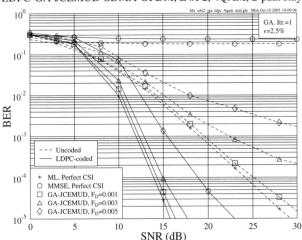

Figure 8.5: BER versus SNR performances of the **uncoded** and **LDPC-coded iterative GA-JCEMUD-assisted SDMA-OFDM** systems in the overloaded scenario, where $L = 4$ users were supported with the aid of $P = 2$ receiver antenna elements, while assuming different values of the OFDM-symbol-normalized Doppler frequency F_D. The basic simulation parameters are given in Table 8.1. ©IEEE Jiang, Akhtman & Hanzo 2007 [517]

8.4.3 Effects of the Number of GA-JCEMUD Iterations

Figure 8.6 shows the achievable performance of the proposed system invoking different numbers of GA-JCEMUD iterations. It was assumed that $\epsilon = 2.5\%$ and $F_D = 0.003$. As seen in Figure 8.6, a significant iteration gain was achieved when the GA-JCEMUD invoked additional iterations. Furthermore, the performance of the uncoded system consistently improved as the number of iterations was increased, while in the LDPC-coded system most of the gain was attained by the first GA-JCEMUD iteration. In Figure 8.7, the average FD-CHTF estimation MSE performance of the GA-JCEMUD/SDMA-OFDM system as well as the performance of the reference system employing $\epsilon = 100\%$ pilot overhead are compared. Observe from Figure 8.7 that, as expected, the FD-CHTF estimation MSE performance was improved when the number of GA-JCEMUD iterations was increased. Moreover, when the SNR exceeded about 13 dB, the GA-JCEMUD using $\epsilon = 2.5\%$ pilot overhead approached the best-case FD-CHTF estimation MSE performance associated with $\epsilon = 100\%$.

8.4.4 Effects of the Pilot Overhead

In Figure 8.8 the BER performance of the proposed system using different pilot overheads is investigated. In most cases, the GA-JCEMUD was capable of achieving a good performance using a pilot overhead as low as $\epsilon = 1.5\%$ to 2.5%. Furthermore, the increase of the pilot OFDM symbol overhead brings about more substantial benefits at the higher Doppler frequencies than at the lower ones, especially in the scenarios associated with higher SNRs, where an increasing fraction of the residual detection errors was inflicted by the inaccurate channel estimation.

8.4.5 Joint Optimization Versus Separate Optimization

In order to characterize further the advantages of the proposed GA-aided joint optimization scheme, in Figure 8.9 we compare the performances of the GA-JCEMUD and its counterpart, referred to as the

LDPC-GA-JCEMUD-SDMA-OFDM, L4/P2, 4QAM, 2-path Rayleigh

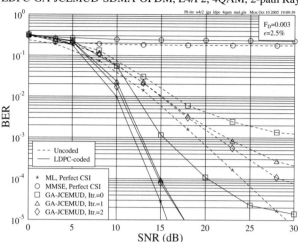

Figure 8.6: BER versus SNR performances of the **uncoded** and **LDPC-coded iterative GA-JCEMUD-assisted SDMA-OFDM** systems in the overloaded scenario, where $L = 4$ users were supported with the aid of $P = 2$ receiver antenna elements, while invoking different numbers of GA-JCEMUD iterations. The basic simulation parameters are given in Table 8.1. ©IEEE Jiang, Akhtman & Hanzo 2007 [517]

GA-JCEMUD-SDMA-OFDM, L4/P2, 4QAM, 2-path Rayleigh

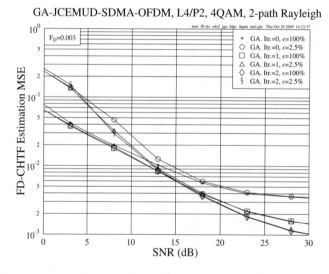

Figure 8.7: FD-CHTF estimation MSE versus SNR performance of the **iterative GA-JCEMUD-assisted SDMA-OFDM** system in the overloaded scenario, where $L = 4$ users were supported with the aid of $P = 2$ receiver antenna elements, while invoking different numbers of GA-JCEMUD iterations. The basic simulation parameters are given in Table 8.1. ©IEEE Jiang, Akhtman & Hanzo 2007 [517]

GA-based Channel-Estimator-assisted OHRSA MUD (GACE-OHRSA-MUD), where the OHRSA MUD is serially concatenated with the stand-alone GA-aided channel estimator. More specifically, in the GACE-OHRSA-MUD, channel estimation and symbol detection are separately accomplished

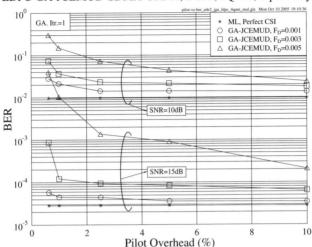

LDPC-GA-JCEMUD-SDMA-OFDM, L4/P2, 4QAM, 2-path Rayleigh

Figure 8.8: BER versus pilot overhead performance of the **LDPC-coded iterative GA-JCEMUD-assisted SDMA-OFDM** system in the overloaded scenario, where $L = 4$ users were supported with the aid of $P = 2$ receiver antenna elements, while assuming different values of the OFDM-symbol-normalized Doppler frequency F_D. The basic simulation parameters are given in Table 8.1. ©IEEE Jiang, Akhtman & Hanzo 2007 [517]

by the GA-aided channel estimator and the OHRSA MUD, respectively. In other words, the symbol estimates offered by the OHRSA MUD are fixed during the GA-aided optimization process of the FD-CHTF estimates. Explicitly, in this case the effect of error propagation due to inaccurate symbol and/or channel estimates will become more severe, resulting in a dramatic BER performance degradation in comparison with the proposed joint optimization scheme, as evidenced in Figure 8.9. Furthermore, the superiority of the GA-JCEMUD becomes even more conspicuous in high-Doppler scenarios.

8.4.6 Comparison of GA-JCEMUDs Having Soft and Hard Outputs

Figure 8.10 shows the performance comparison of the GA-JCEMUD providing either the conventional individual-based hard outputs [38, 459–461, 532] or the proposed population-based soft outputs. As expected, with the advent of FEC codes, the proposed soft GA is capable of significantly outperforming the conventional arrangement, especially when the channel fades more rapidly. This result implies that the proposed GA exhibited a higher robustness against fast fading channels than the conventional GAs [38, 459–461, 532].

8.4.7 MIMO Robustness

As a further investigation, a visual comparison of the true and estimated FD-CHTFs is portrayed in Figure 8.11. More specifically, the L users' FD-CHTFs associated with a specific receiver antenna element during a block of 40 consecutive OFDM symbols are plotted at an SNR value of 20 dB. Each dot of the curves plotted in Figure 8.11 represents a complex-valued FD-CHTF at a specific subcarrier. By observing the perfect channel-knowledge-based illustration at the top of Figure 8.11, we can see that the FD-CHTF at each subcarrier evolves over the duration of the 40 OFDM symbols, where the thickness of the ring-shaped formations indicates the amount of FD-CHTF change during this time interval. The full perimeter of the ring is constituted by the $K = 64$ spoke-like formations corresponding to the 64 OFDM subcarriers. Explicitly, the radii of the FD-CHTF rings associated with

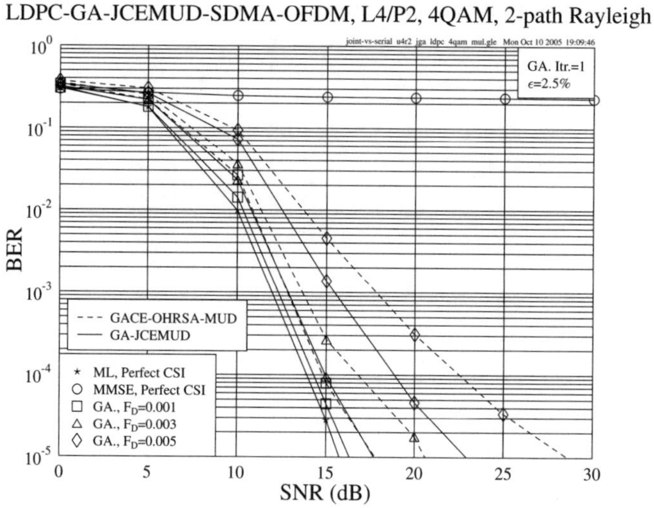

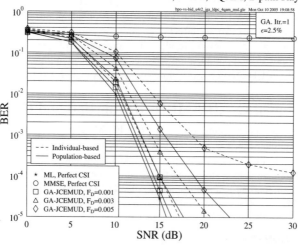

Figure 8.9: BER versus SNR performance of the **LDPC-coded SDMA-OFDM** system using either the **iterative GA-JCEMUD** or the **iterative GACE-OHRSA-MUD** in the overloaded scenario, where $L = 4$ users were supported with the aid of $P = 2$ receiver antenna elements, while assuming different values of the OFDM-symbol-normalized Doppler frequency F_D. The basic simulation parameters are given in Table 8.1. ©IEEE Jiang, Akhtman & Hanzo 2007 [517]

Figure 8.10: BER versus SNR performance of the **LDPC-coded SDMA-OFDM** system using either the conventional **individual-based** or the proposed **population-based** GA-JCEMUD in the overloaded scenario, where $L = 4$ users were supported with the aid of $P = 2$ receiver antenna elements, while assuming different values of the OFDM-symbol-normalized Doppler frequency F_D. The basic simulation parameters are given in Table 8.1. ©IEEE Jiang, Akhtman & Hanzo 2007 [517]

the four user–receiver channel links are significantly different. This is because each individual link is subjected to independent fading, and although the Doppler frequencies encountered at the four links

True Channel Transfer Functions, 2-path Rayleigh

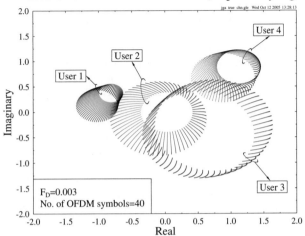

Estimated Channel Transfer Functions, 2-path Rayleigh

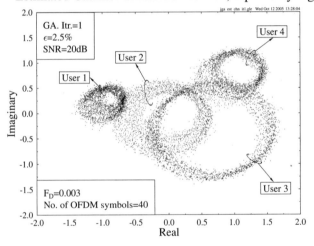

Figure 8.11: Channel estimation performance of the **iterative GA-JCEMUD-assisted SDMA-OFDM** system in the overloaded scenario, where $L = 4$ users were supported with the aid of $P = 2$ receiver antenna elements. The estimated FD-CHTFs $\hat{H}_1^{(l)}[n, k]$ $(l = 1, \ldots, L; \ k = 1, \ldots, K)$ associated with $n = 1, \ldots, 40$ consecutive OFDM symbols at the $p = 1$st receiver antenna are plotted at an SNR value of 20 dB, and compared with the true FD-CHTFs. The basic simulation parameters are given in Table 8.1. ©IEEE Jiang, Akhtman & Hanzo 2007 [517]

are identical, their short-term envelope fluctuation observed over the 40 OFDM symbol durations is different. However, by comparing the subfigures at the top and bottom of Figure 8.11, we can see that the FD-CHTF estimates closely match their true values, resulting in a similar FD-CHTF contour for each of the four channel links. This implies that the proposed GA-JCEMUD is capable of *simultaneously* capturing the fading envelope changes of *each* individual user–receiver link, regardless of its instant variety of fading. Since an equally good performance was attained over all the user–receiver links, this demonstrates the global robustness of the proposed approach in MIMO scenarios.

8.5   Conclusions

From our discussions in the previous sections, we conclude that the proposed GA-aided iterative joint channel estimation and MUD scheme generating soft outputs constitutes an effective solution to the channel estimation problem in multi-user MIMO SDMA-OFDM systems. Furthermore, the GA-JCEMUD is capable of exhibiting a robust performance in overloaded scenarios, where the number of users is higher than the number of receiver antenna elements, either with or without FEC coding. This attractive property enables the SDMA-OFDM system potentially to support an increased number of users.

Note that in this chapter the GA-JCEMUD was used to estimate the CHTFs of a two-tap channel model, while the value of the parameter K_0 was set to 8, which is four times higher than the actual number of CIR taps. However, we point out that as long as the condition of $K_0 > 2$ is satisfied, the performance of the proposed scheme can be improved upon reducing the value of K_0, since more noise will be removed during the CIR tap filtering process. Furthermore, this implies that, given a fixed value of K_0 which has to be higher than the maximum channel delay spread encountered, the more taps the channel has, the better the performance of the GA-JCEMUD, as a benefit of having an increased ratio of the 'real' number of CIR taps in comparison with K_0.

8.6   Chapter Summary

In MIMO-OFDM systems accurate channel estimation is required at the receiver in order to carry out coherent demodulation and interference cancellation. However, channel estimation is more challenging in the MIMO scenario than in a SISO scenario, owing to an increased number of independent transmitter–receiver channel links as well as the interference imposed by multiple transmitter antennas. Our discussions commenced in Section 8.1 with an extensive review of the family of channel estimation techniques found in the MIMO-OFDM literature, where a number of channel estimation approaches including blind [83, 217, 228, 251, 518, 519], pilot-aided [131, 132, 134, 200, 520, 521] and joint [172, 526] estimation schemes have been proposed. However, in the context of BLAST or SDMA types of multi-user MIMO-OFDM systems, none of the channel estimation techniques found in the literature [133, 156, 172, 178, 200, 217, 228, 251, 518, 522, 526, 527] allows the number of users to exceed that of the receiver antennas. We refer to this as an overloaded scenario, where we have an insufficient detection degree of freedom, rendering the channel estimation an even more challenging task.

In an effort to overcome the channel estimation problem in overloaded MIMO-OFDM systems, in this chapter a GA-assisted iterative Joint Channel Estimation and Multi-User Detection (GA-JCEMUD) approach was proposed. We commenced with an overview of the proposed scheme in Section 8.2, where we outlined our system schematic in Figure 8.1, followed by a detailed discussion in Section 8.3. The philosophy of the proposed iterative GA-JCEMUD was portrayed in Figure 8.2, accompanied by our corresponding elaborations. In Section 8.3.1, we generated initial estimates of the FD-CHTFs with the aid of pilot OFDM symbols spread by WHT-based spreading codes. On the other hand, the Optimized Hierarchy Reduced Search Algorithm (OHRSA) MUD [534, 535] was employed as a first-stage detector in order to generate initial symbol estimates, as discussed in Section 8.3.2.

With the aid of the initial FD-CHTF and symbol estimates, the proposed GA-JCEMUD printed in the grey block of Figure 8.2 can be employed for jointly optimizing the estimates of the FD-CHTFs and multi-user data symbols. This process was detailed in Section 8.3.3. More specifically, concerning the MIMO channel's structure, in Section 8.3.3.1 we introduced the extended structure of the GA individuals. Following a description of the initialization process in Section 8.3.3.2, the GA-based joint optimization was presented in Section 8.3.3.3, which consists of a discussion of the cross-over operators in Section 8.3.3.3.1, the mutation operators in Section 8.3.3.3.2 and further discussions in Section 8.3.3.3.3. Furthermore, in order to overcome the limitations imposed by the conventional GA-aided MUDs, which can only provide single-individual hard-decoded symbol estimates, a new algorithm was derived in Section 8.3.3.4 to enable the GA to provide soft outputs based on the entire population. Since the only additional operation required by the proposed scheme is to compare the normalization factor $\omega = P \cdot L$ with the OSs, which have already been calculated during the OF evaluation process carried out by the conventional GAs, the proposed GA generating the above-mentioned population-based soft outputs imposes only a modest complexity increase in comparison with its conventional hard-decision individual-based counterpart.

The numerical results of the proposed uncoded and LDPC-coded iterative GA-JCEMUD-aided SDMA-OFDM system were provided in Section 8.4. As a preliminary investigation, the effect of the maximum mutation step size λ_{max} used in the GA-JCEMUD was first identified in Section 8.4.1, where it was found that the choice of λ_{max} has a significant impact on the system's MSE as well as BER performance, and these effects became more substantial when the SNR was increased. Furthermore, to attain the best achievable BER performance, typically a higher λ_{max} is required in the scenario associated with a higher OFDM-symbol-normalized Doppler frequency F_D. This is because, when we have a higher Doppler frequency, the consecutive time-domain channel fades experienced by each of the subcarriers fluctuate more rapidly, thus requiring a higher λ_{max} value to assist the GA in capturing the rapid changes of the channel fades. In addition, the values of λ_{max} required to attain the best BER performance were identified for different values of F_D with the aid of Figure 8.4.

In Section 8.4.2, our research was dedicated to probing the effect of Doppler frequency. As expected, the performances of both the uncoded and coded GA-JCEMUD-aided systems degraded when F_D was increased. However, when using a low pilot overhead of $\epsilon = 2.5\%$, the proposed GA-JCEMUD/SDMA-OFDM system was capable of achieving a performance close to the perfect-CSI-aided optimum ML MUD at $F_D = 0.001$, as evidenced in Figure 8.5, while the system employing the MMSE MUD completely failed even with the aid of perfect CSI. In Section 8.4.3, we showed that the proposed system's performance can be improved by increasing the number of GA-JCEMUD iterations, both with and without channel coding. At an SNR value of about 13 dB, the GA-JCEMUD using $\epsilon = 2.5\%$ pilot overhead approached the best-case FD-CHTF estimation MSE performance associated with $\epsilon = 100\%$. When examining the effect of the pilot overhead ϵ, we observed that the GA-JCEMUD was capable of achieving a good performance using as low a pilot overhead as $\epsilon = 1.5\%$ to 2.5% in most of the scenarios considered, as shown in Figure 8.8 of Section 8.4.4.

In order to characterize further the advantages of the proposed joint optimization mechanism, in Section 8.4.5 we compared the performances of the GA-JCEMUD and its counterpart, namely the GACE-OHRSA-MUD, which serially concatenates the OHRSA MUD with the stand-alone GA-aided channel estimator. It was shown that the former outperformed the latter, especially in high-Doppler scenarios. This demonstrated the superiority of the joint optimization mechanism over the conventionally combined detection architecture. Moreover, Section 8.4.6 exhibits the further benefits of the proposed GA-JCEMUD scheme owing to its ability to provide soft outputs. With the advent of FEC codes, the proposed population-based soft-decoded GA of Section 8.3.3.4 was capable of significantly outperforming the conventional individual-based hard-decoded GAs [38, 459–461, 532], especially when the channel fades rapidly. This result implies that the proposed GA exhibited a higher robustness against fast fading channels than the conventional GAs. Finally, in Section 8.4.7 the GA-JCEMUD's robustness recorded in MIMO scenarios was verified. As shown in Figure 8.11, the proposed iterative GA-JCEMUD was capable of *simultaneously* capturing the fading envelope changes of *each* individual user–receiver link, regardless of its instant variety of fading, and thus achieving an equally good performance over all the user–receiver links. This result potently demonstrates the robustness of the proposed approach.

Part II

Coherent versus Non-coherent and Cooperative OFDM Systems

Part 3

Cooperative OFDM System

List of Symbols in Part II

General notation

- The superscript $*$ is used to indicate complex conjugation. Therefore, a^* represents the complex conjugate of the variable a.

- The superscript T is used to indicate matrix transpose operation. Therefore, $\mathbf{A}^T$ represents the transpose of the matrix $\mathbf{A}$.

- The superscript H is used to indicate complex conjugate transpose operation. Therefore, $\mathbf{A}^H$ represents the complex conjugate transpose of the matrix $\mathbf{A}$.

- The notation b_i represents the ith entry of the column vector $\mathbf{b}$.

- The notation $\overrightarrow{b_i}$ represents the ith entry of the row vector $\overrightarrow{\mathbf{b}}$.

- The notation $a_{i,j}$ represents the entry located in the ith row and jth column of the matrix $\mathbf{A}$.

- The notation $*$ denotes the convolutional process. Therefore, $a * b$ represents the convolution between variables a and b.

- The notation $\hat{x}$ represents the estimate of x.

- The notation $\breve{x}$ represents the trial candidate of x.

- The notation $X(f)$ is the Fourier transform of $x(t)$.

Special symbols

- **Matrices and vectors**:

$\mathbf{H}$	The Channel State Information (CSI) matrix
$\mathbf{W}$	The AWGN (CSI) matrix
$\mathbf{S}_n$	The nth transmitted space–time signal matrix
$\mathbf{I}_n$	The $(n \times n)$-element identity matrix
$\mathbf{s}$	The signal column vector transmitted from a mobile station
$\mathbf{y}$	The received signal column vector at the base station
$\mathbf{w}$	The AWGN column vector at the base station
$\hat{\mathbf{c}}$	The search centre of hypersphere search space employed by the Sphere Detection (SD)

- **Variables**:

U	The number of users within a system
M	The number of transmit antenna elements employed at a terminal
N	The number of receive antenna elements employed at a terminal
M_c	The modulation constellation size
$\mathcal{M}_c$	The set of modulation constellation points

C	The search radius of the hypersphere, in which the SD carries out the search for the solution
$\mathcal{C}_i$	The decoupled search centre for $\check{s}_i$
$\mathcal{D}_i$	The accumulated Partial Euclidean Distance (PED) between $\check{\mathbf{s}}_i = [\check{s}_i \check{s}_{i+1} \dots \check{s}_U]$ and the centre $\hat{\mathbf{c}}_i = [\hat{c}_i \hat{c}_{i+1} \dots \hat{c}_U]$ of the hyper-sphere search space
K	The number of candidates having the lowest accumulated PEDs retained at each tree search level by the K-best SD
T_b	The length of the correlated-envelope block-fading intervals
η	The bandwidth efficiency in bps/Hz
$\mathcal{L}$	Transmitted MIMO symbol candidate list generated by the List Sphere Decoder (LSD)
$\mathcal{N}_{cand}$	The size of the candidate list $\mathcal{L}$ generated by the LSD
L_f	The transmission frame length
M_r	The number of available cooperating MSs within a cell
f_d	The normalized Doppler frequency

- **Mathematical operations**:

$\sum$	Sum operation
$\otimes$	Kronecker product
$log[\cdot]$	Logarithm operation
$max(\cdot)$	The maximum value of a matrix/vector
$min(\cdot)$	The minimum value of a matrix/vector
$vec(\cdot)$	Vertical stacking of the columns of a matrix
$row(\cdot)$	Vertical stacking of the rows of a matrix
$tr(\cdot)$	Trace operation of a matrix
$QR(\cdot)$	The QR decomposition of a square matrix
$det(\cdot)$	The determinant operation
$p(\cdot)$	The probability density function
$\|\cdot\|^2$	The second-order norm
$\Re\{\cdot\}$	Real part of a complex value
$\Im\{\cdot\}$	Imaginary part of a complex value
$E\{\cdot\}$	Expectation of a random variable
$\mathbf{A}^H$	Matrix/vector Hermitian adjoint
$\mathbf{A}^{-1}$	Matrix inverse
$\mathbf{A}^T$	Matrix/vector transpose
$\mathbf{A}^*$	Matrix/vector/variable complex conjugate

- **Symbols**:

$\mathbb{R}^n$	n-dimensional real-valued Euclidean space
$\mathbb{C}^n$	n-dimensional complex-valued Euclidean space
Π	Interleaver
Π^{-1}	Deinterleaver
E_{total}	The total power of a sphere packing symbol

Chapter 9

Reduced-Complexity Sphere Detection for Uncoded SDMA-OFDM Systems

9.1 Introduction

9.1.1 System Model

Figure 9.1 shows an SDMA/OFDM Uplink (UL) transmission scenario, where each of the U users is equipped with a single transmit antenna, while the BS has N receive antenna elements. Based on our rudimentary discourse on MIMO-OFDM in Section 1.1.1, for each subcarrier the link between each pair of transmit and receiver antennas may be characterized with the aid of a unique user-specific FDCTF denoted as h_{nu} in Figure 9.1. The subscripts of h, i.e. u and n, represent the user and receive antenna element index at the BS, respectively. For example, the FD-CTF or the spatial signature of the uth user can be expressed as a column vector:

$$\mathbf{h}_u = [h_{1u}, h_{2u}, \ldots, h_{Nu}]^T, \tag{9.1}$$

with $u \in 1, \ldots, U$. If the transmitted signal of the uth user is denoted by s_u and the received signal plus the Additive White Gaussian Noise (AWGN) at the nth receive antenna element is represented by y_n and w_n, respectively, the entire SDMA/OFDM system can be described on a per-subcarrier basis by a matrix equation written as

$$\mathbf{y} = \mathbf{Hs} + \mathbf{w}, \tag{9.2}$$

where the received signal's column vector is $\mathbf{y} \in \mathbb{C}^{N \times 1}$, the transmitted signal's column vector is $\mathbf{s} \in \mathbb{C}^{U \times 1}$ and the noise's column vector is $\mathbf{w} \in \mathbb{C}^{N \times 1}$, which are given by the following equations, respectively:

$$\mathbf{y} = [y_1, y_2, \ldots, y_N]^T, \tag{9.3}$$

$$\mathbf{s} = [s_1, s_2, \ldots, s_U]^T, \tag{9.4}$$

$$\mathbf{w} = [w_1, w_2, \ldots, w_N]^T. \tag{9.5}$$

The FD-CTF matrix $\mathbf{H} \in \mathbb{C}^{N \times U}$ is constituted by the number U of user-specific CTF vectors defined by Equation (9.1), with $\mathbf{h}_u \in \mathbb{C}^{N \times 1}$, where $u = 1, 2, \ldots, U$. Explicitly, the FD-CTF matrix $\mathbf{H}$ can be

MIMO-OFDM for LTE, Wi-Fi and WiMAX   Lajos Hanzo, Yosef Akhtman, Li Wang and Ming Jiang

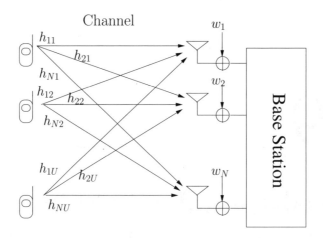

Figure 9.1: Schematic of an SDMA uplink MIMO channel scenario.

expressed as

$$\mathbf{H} = [\mathbf{h}_1 \ \mathbf{h}_2 \ \dots \ \mathbf{h}_U], \tag{9.6}$$

where each column represents a user's unique spatial signature. Here, we assume that the FD-CTF H_{nu} between user $u \in 1, 2, \dots, U$ and receive antenna element $n \in 1, 2, \dots, N$ are independent, stationary, complex-valued Gaussian-distributed processes with a zero mean and a unit variance [542]. Furthermore, both the transmitted signal of each of the U users and the AWGN encountered at each of the N antenna elements exhibit a zero mean and a variance of $2\sigma_s^2$ and $2\sigma_w^2$, respectively.

9.1.2　Maximum Likelihood Detection

The Maximum Likelihood (ML) detector jointly detects the U different users' complex symbols that are most likely to have been transmitted. The stylized schematic of the ML detector is shown in Figure 9.2, where M_c is the constellation size of a specific modulation scheme. Observe that the received signal's column vector $\mathbf{y}$ of Equation (9.2) possesses a U-dimensional multi-variate complex-Gaussian distribution, with a vector of mean values of $\mathbf{Hs}$ and a covariance matrix given by $\mathbf{R_w} \in \mathbb{C}^{N \times N}$. The latter is given by

$$\mathbf{R_w} = E\{\mathbf{w}\mathbf{w}^H\} \tag{9.7}$$

$$= 2\sigma_w^2 \mathbf{I}, \tag{9.8}$$

under the assumption that the noise contributions added at each receive antenna element are uncorrelated. Consequently, the a priori probability function of the received signal vector $\mathbf{y}$ is equivalent to the complex Gaussian distribution function, which can be written as [542]

$$P(\mathbf{y}|\mathbf{s}, \mathbf{H}) = f(\mathbf{y}|\mathbf{s}, \mathbf{H}) = \frac{1}{\pi^N |\mathbf{R_w}|} \exp[-(\mathbf{y} - \mathbf{Hs})^H \mathbf{R_w}^{-1} (\mathbf{y} - \mathbf{Hs})] \tag{9.9}$$

$$= \frac{1}{(2\sigma_w^2 \pi)^N} \exp\left(-\frac{1}{2\sigma_w^2} \|\mathbf{y} - \mathbf{Hs}\|^2\right). \tag{9.10}$$

On the other hand, the basic idea behind the ML detector is to maximize the a posteriori probability $P(\check{\mathbf{s}}|\mathbf{y}, \mathbf{H})$, where the candidate vector $\check{\mathbf{s}} \in \mathbb{C}^{U \times 1}$ is an element of the set M_c^U of trial vectors which was transmitted over the channel characterized by the channel matrix $\mathbf{H} \in \mathbb{C}^{N \times U}$, and under the condition that the received signal vector is $\mathbf{y}$. Importantly, the relationship between

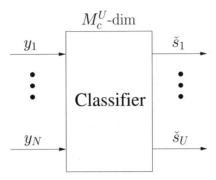

Figure 9.2: Representation of the optimum ML detector.

the a posteriori probability and the a priori probability can be formulated with the aid of Bayes' theorem [542] as follows:

$$P(\check{\mathbf{s}}|\mathbf{y}, \mathbf{H}) = P(\mathbf{y}|\check{\mathbf{s}}, \mathbf{H})\frac{P(\check{\mathbf{s}})}{P(\mathbf{y})}, \tag{9.11}$$

where $P(\check{\mathbf{s}}) = 1/M_c^U$ is a constant, since it is assumed that all symbol vector probabilities are identical. Furthermore, since all probabilities have to sum to unity, we have

$$\sum_{\check{\mathbf{s}} \in M_c^U} P(\check{\mathbf{s}}|\mathbf{y}, \mathbf{H}) = 1. \tag{9.12}$$

Additionally, the total probability $P(\mathbf{y})$ can be expressed by

$$P(\mathbf{y}) = \sum_{\hat{\mathbf{s}} \in M_c^U} P(\mathbf{y}|\check{\mathbf{s}}, \mathbf{H})P(\check{\mathbf{s}}), \tag{9.13}$$

which is also a constant. Consequently, we have

$$\frac{P(\check{\mathbf{s}})}{P(\mathbf{y})} = const, \tag{9.14}$$

which leads to the conclusion that, for the ML detector, the problem of finding the optimum solution $\hat{\mathbf{s}}_{ML}$ which maximizes the a posteriori probability of $P(\check{\mathbf{s}}|\mathbf{y}, \mathbf{H})$ is equivalent to maximizing the a priori probability of $P(\mathbf{y}|\check{\mathbf{s}}, \mathbf{H})$. Hence, according to Equation (9.10), the problem is also equivalent to minimizing the Euclidean distance metric $\|\mathbf{y} - \mathbf{H}\check{\mathbf{s}}\|^2$, i.e.

$$\hat{\mathbf{s}}_{ML} = \arg \min_{\check{\mathbf{s}} \in M_c^U} \|\mathbf{y} - \mathbf{H}\check{\mathbf{s}}\|^2. \tag{9.15}$$

The ML detector is capable of achieving the optimum BER performance by jointly detecting all the U different users' symbols at the cost of a potentially excessive computational complexity, which depends on the size of the modulation constellation and/or the number of users supported by the system, since the ML detector evaluates the Euclidean distance metric of Equation (9.15) for all the possible transmitted symbol vectors. For example, if an SDMA-OFDM system employs 16-QAM and supports $U = 8$ users, a full-search of 2^{32} possibilities will be encountered in order to find the optimum solution, imposing an excessive computational complexity.

9.1.3   Chapter Contributions and Outline

The motivation of finding a low-complexity solution while achieving a near-ML performance has driven researchers to develop new algorithms. Recently, inspired by the Sphere Detection (SD) algorithm originally introduced by Porst and Finke [543] to calculate efficiently a vector of short length in a lattice, Viterbo and Boutros have applied the original SD algorithm in communication systems [544] in order to approach the ML performance at a complexity which is polynomially, rather than exponentially, dependent on the number of unknowns, which opened up a whole new research area. Different types of SDs and complexity reduction schemes have been proposed, for example, in [545–548] for the depth-first SD. By contrast, the schemes proposed in [549–552] were designed for the breadth-first SD. As a benefit of the superior performance of the SD algorithm, it will serve as a key mechanism to reduce the complexity of diverse MIMO-OFDM scenarios throughout this treatise. Hence, for the sake of further developing the SD algorithm and applying it to various problems, a comprehensive understanding of the SD's operating principle is a vital prerequisite. Thus, the main objective of this chapter is to review the fundamentals of both the depth-first and the breadth-first tree search SDs and to carry out in-depth comparative studies in terms of their corresponding complexity reduction schemes as well as their achievable performance. More specifically, the main contributions of this chapter are as follows:

- *Compare and analyse the most influential complexity reduction schemes proposed in the literature for the conventional depth-first SD, the breadth-first SD as well as for the recently proposed OHRSA detector, which may be regarded as an advanced extension of the depth-first SD.*

- *Extend the performance versus complexity studies of the above-mentioned SD algorithms to challenging rank-deficient MIMO scenarios.*

The outline of this chapter is as follows. In Section 9.2 the SD fundamentals are reviewed, followed by a discourse on Generalized Sphere Detection (GSD), which are capable of operating in rank-deficient MIMO systems. The most influential complexity reduction schemes proposed for the depth-first and breadth-first SDs are discussed in Sections 9.3.1 and 9.3.2, respectively. Then, Section 9.3.3 introduces the recently proposed OHRSA detector and analyses both its hierarchical search structure and its optimization strategies in comparison with the complexity-reduction schemes of its conventional SD counterparts. The achievable BER performance versus complexity imposed by the above-mentioned SDs is characterized in Section 9.4 for both full-rank and rank-deficient MIMO systems. Finally, our concluding remarks are provided in Section 9.5.

9.2   Principle of SD

9.2.1   Transformation of the ML Metric

As discussed in Section 9.1.2, the ML solution for an SDMA system of Equation 9.2 can be written as

$$\hat{\mathbf{s}}_{ML} = \arg\min_{\check{\mathbf{s}} \in \mathcal{M}_c^U} \| \mathbf{y} - \mathbf{H}\check{\mathbf{s}} \|^2, \tag{9.16}$$

where $\mathcal{M}_c$ is the set of M_c legitimate symbol points in the modulation constellation and U is the number of users supported by the system. Thus, a potentially excessive-complexity search is likely to be encountered, depending on the value of M_c and/or U, which prevents the application of the full-search-based ML detectors in most practical high-throughput scenarios. Fortunately, Equation (9.16) can be extended as [553]:

$$\hat{\mathbf{s}}_{ML} = \arg\min_{\check{\mathbf{s}} \in \mathcal{M}_c^U} \| \mathbf{y} - \mathbf{H}\check{\mathbf{s}} \|^2 \tag{9.17}$$

$$= \arg\min_{\check{\mathbf{s}} \in \mathcal{M}_c^U} \left\{ (\check{\mathbf{s}} - \hat{\mathbf{c}})^H \mathbf{H}^H \mathbf{H} (\check{\mathbf{s}} - \hat{\mathbf{c}}) + \underbrace{\mathbf{y}^H (\mathbf{I} - \mathbf{H}(\mathbf{H}^H \mathbf{H})^{-1} \mathbf{H}^H) \mathbf{y}}_{\varphi} \right\}, \tag{9.18}$$

where

$$\hat{\mathbf{c}} = (\mathbf{H}^H \mathbf{H})^{-1} \mathbf{H}^H \mathbf{y} \qquad (9.19)$$

which is the unconstrained ML estimate of $\mathbf{s}$ or the LS solution of Equation (9.2). Importantly, the value of φ in Equation (9.18) is independent of the argument $\check{\mathbf{s}}$, when minimizing the Objective Function (OF) of Equation (9.16). Hence, the trial candidate $\check{\mathbf{s}}$ minimizing $\|\mathbf{y} - \mathbf{H}\check{\mathbf{s}}\|^2$ also minimizes $(\check{\mathbf{s}} - \hat{\mathbf{c}})^H \mathbf{H}^H \mathbf{H}(\check{\mathbf{s}} - \hat{\mathbf{c}})$, Thus, we have:

$$\hat{\mathbf{s}}_{ML} = \arg\min_{\check{\mathbf{s}} \in \mathcal{M}_c^U} (\check{\mathbf{s}} - \hat{\mathbf{c}})^H \mathbf{H}^H \mathbf{H}(\check{\mathbf{s}} - \hat{\mathbf{c}}). \qquad (9.20)$$

In fact, the well-known SD algorithm was derived from the mathematical problem of finding the shortest vector in a lattice, which was originally described in [554] and refined in [555]. Even when exploiting the above-mentioned simplifications, finding the ML solution $\hat{\mathbf{s}}$ still has to be carried out on an exhaustive search basis for the entire M_c^U number of legitimate transmitted signal vector combinations.

Therefore, in the following sections, two different types of SD algorithms will be introduced and compared, which are capable of significantly reducing the associated search complexity, namely the original SD algorithm of [544] which is also referred to as a *Depth-First SD* and the *K-Best SD* of [550], which can be regarded as a *Breadth-First SD*.

9.2.2 Depth-First Tree Search [544]

For the depth-first SD scheme, a search radius C is set in order to limit the search range. Specifically, we limit the search according to

$$(\check{\mathbf{s}} - \hat{\mathbf{c}})^H \mathbf{H}^H \mathbf{H}(\check{\mathbf{s}} - \hat{\mathbf{c}}) \le C, \qquad (9.21)$$

where C is the Initial Search Radius (ISR), which has to be sufficiently high in order to contain the ML solution of Equation (9.16). Let

$$\mathbf{G} = \mathbf{H}^H \mathbf{H}, \qquad (9.22)$$

which is a $(U \times U)$ Grammian matrix [544]. Thus, we can obtain the $(U \times U)$ upper-triangular matrix $\mathbf{U}$ which satisfies $\mathbf{U}^H \mathbf{U} = \mathbf{H}^H \mathbf{H}$ with the aid of, for example, the ubiquitous Cholesky factorization [544]. Thus, the entries of the upper-triangular matrix $\mathbf{U}$ are denoted by $u_{i,j}$, satisfying $u_{i,j} = 0$ if $i > j$ for $i, j = 1, 2, \ldots, U$. Furthermore, the entries on the diagonal of $\mathbf{U}$ are denoted by $u_{i,i}$, which are assumed to be of positive real value without loss of generality [553]. Consequently, bearing in mind that the matrix $\mathbf{U}$ is upper triangular, we can rewrite Equation (9.21) as

$$(\check{\mathbf{s}} - \hat{\mathbf{c}})^H \mathbf{H}^H \mathbf{H}(\check{\mathbf{s}} - \hat{\mathbf{c}}) = (\check{\mathbf{s}} - \hat{\mathbf{c}})^H \mathbf{U}^H \mathbf{U}(\check{\mathbf{s}} - \hat{\mathbf{c}}) \qquad (9.23)$$

$$= \sum_{i=1}^{U} \left| \sum_{j=i}^{U} u_{i,j}(\check{s}_j - \hat{c}_j) \right|^2 \qquad (9.24)$$

$$= \sum_{i=1}^{U} \left| u_{i,i}(\check{s}_i - \hat{c}_i) + \sum_{j=i+1}^{U} u_{i,j}(\check{s}_j - \hat{c}_j) \right|^2 \le C. \qquad (9.25)$$

Hence, we can recursively calculate the bound for each $\check{s}_i$ value with the aid of Equation (9.25), if we start from $i = U$. Specifically, in light of Equation (9.25), we can enumerate legitimate values for $\check{s}_U$ based on the following derived criterion as

$$|\check{s}_U - \hat{c}_U| \le \frac{\sqrt{C}}{u_{U,U}}. \qquad (9.26)$$

Then, as indicated by Equation (9.26), after choosing a legitimate symbol value for $\check{s}_U$ around $\hat{c}_U$ within a radius of $\sqrt{C}/u_{U,U}$, we can continue to choose a trial legitimate value for $\check{s}_{U-1}$ satisfying the criterion derived from Equation (9.25), which can be expressed as

$$|u_{U-1,U-1}(\check{s}_{U-1} - \hat{c}_{U-1}) + u_{U-1,U}(\check{s}_U - \hat{c}_U)|^2 + |u_{U,U}(\check{s}_U - \hat{c}_U)|^2 \leq C, \tag{9.27}$$

or equivalently

$$\left| \check{s}_{U-1} - \left(\hat{c}_{U-1} - \frac{u_{U-1,U}}{u_{U-1,U-1}} \xi_U \right) \right| \leq \frac{\sqrt{C - |u_{U,U}\xi_U|^2}}{u_{U-1,U-1}}, \tag{9.28}$$

where

$$\xi_i \triangleq \check{s}_i - \hat{c}_i. \tag{9.29}$$

Now a trial value can be chosen for s_{U-1} around $\hat{c}_{U-1} - (u_{U-1,U}/u_{U-1,U-1})\xi_U$ within a radius of $\sqrt{C - |u_{U,U}\xi_U|^2}/u_{U-1,U-1}$ in light of Equation (9.28). The recursive process continues by choosing a trial candidate for s_{U-2} based on its corresponding criterion. Following the rationale of Equation (9.25), the decoupled search space for the ith component $\check{s}_i$ can be evaluated by

$$|\check{s}_i - \mathcal{C}_i| \leq \frac{\sqrt{C - \mathcal{D}_{i+1}}}{u_{i,i}}, \tag{9.30}$$

where

$$\mathcal{C}_i \triangleq \left(\hat{c}_i - \sum_{j=i+1}^{U} \frac{u_{i,j}}{u_{i,i}} \xi_j \right) \tag{9.31}$$

and

$$\mathcal{D}_i \triangleq \left| \sum_{l=i}^{U} \sum_{j=l}^{U} u_{l,j} \xi_j \right|^2 \tag{9.32}$$

are defined as the decoupled search centre for $\check{s}_i$ and the accumulated Partial Euclidean Distance (PED) between $\check{\mathbf{s}}_i = [\check{s}_i \; \check{s}_{i+1} \; \ldots \; \check{s}_U]$ and the centre $\hat{\mathbf{c}}_i = [\hat{c}_i \; \hat{c}_{i+1} \; \ldots \; \hat{c}_U]$ of the hypersphere, respectively. Thus, this recursive process can be continued, until i reaches 1. Then the search radius C is updated by calculating the Euclidean distance between the newly obtained signal point $\check{\mathbf{s}}$ and the centre $\hat{\mathbf{c}}$ of the hypersphere, namely the unconstrained ML solution. Equivalently, we have

$$C = \mathcal{D}_1. \tag{9.33}$$

Following this a new search is carried out within a smaller compound confined by the newly obtained search radius. The search then proceeds in the same way until no more legitimate signal points can be found in the increasingly reduced search space. Consequently, the last found legitimate signal point $\check{\mathbf{s}}$ is regarded as the ML solution.

To elaborate a little further, the search radius $\sqrt{C - \mathcal{D}_{i+1}}/u_{i,i}$ for $\check{s}_i$ in Equation (9.30) provides information on how large the remaining search space is that has to be scoured for identifying s_i. Moreover, in light of Equations (9.31) and (9.32), the relationship between the decouple search centre for $\check{s}_i$ and its corresponding accumulated PED for $\check{\mathbf{s}}_i$ can be expressed as

$$\mathcal{D}_i = \mathcal{D}_{i+1} + u_{i,i}|\check{s}_i - \mathcal{C}_i|^2, \tag{9.34}$$

which indicates that, given a specific $\mathcal{D}_{i+1}$, the value of $\mathcal{D}_i$ only depends on the tentative choice for the current s_i value.

Intuitively, an astutely selected ISR C can substantially speed up the search process, since the employment of a small radius excludes a high proportion of the low-probability lattice points at the very beginning. However, the radius must not be set too small either, since that would jeopardise finding the ML solution of Equation (9.16). Hence, the appropriate choice of the ISR is a key factor in determining both the performance and the complexity imposed by the SD discussed in this chapter. In practice, the ISR has to be set according to the noise level by obeying [553]

$$C^2 = 2\sigma_w^2 JN - \mathbf{y}^T(\mathbf{I} - \mathbf{H}(\mathbf{H}^T\mathbf{H})^{-1}\mathbf{H}^T)\mathbf{y} \tag{9.35}$$

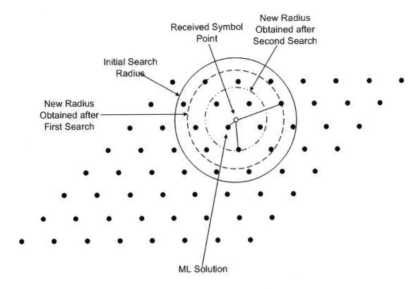

Figure 9.3: Geometric representation of the SD algorithm.

to ensure that the probability of detection failure becomes negligible [544], where N is the number of receive antenna elements, while $J \geq 1$ is a parameter appropriately selected to ensure that the detector will indeed capture the true transmitted signal vector $\mathbf{s}$.

The SD algorithm can be interpreted as a geometric problem, which is shown in Figure 9.3, where the depth-first SD is applied to a one-dimensional case, i.e. to a single-user system, for the sake of convenience. In the example shown in Figure 9.3, the employment of 64-QAM was assumed. At the receiver, the shape of the constellation is assumed to be distorted to a diamond shape instead of the original square shape, owing to the routinely encountered multi-path channel-induced phase rotation and magnitude attenuation. Instead of carrying out a full search over the entire 64-point constellation, as the ML detection would in order to find the statistically optimum solution, the SD initializes the search radius depending on the estimated SNR, which confines the search area to the outermost circle centred at the reconstructed received symbol point $\mathbf{y}_{reconstr} = \mathbf{H}\hat{\mathbf{c}}$, where $\hat{\mathbf{c}}$ is the unconstrained ML solution. As seen from Figure 9.3, the search area is significantly reduced in comparison with the ML detector. It is indeed intuitive that only the trial lattice points in the immediate neighbourhood of the received point are worth examining. Inside the search area confined by the radius, all the symbols are deemed to be tentative candidates for the transmitted symbol. Now the core operation of the SD algorithm is activated. Specifically, a new radius is calculated by measuring the distance between the candidate and the reconstructed received symbol point $\mathbf{y}_{reconstr}$, which should be no higher than the original radius. Then another arbitrary symbol point is chosen from the newly obtained search area as the trial transmitted point. Again, the search radius is updated with the value of the distance between the newly obtained trial point and the reconstructed received symbol point $\mathbf{y}_{reconstr}$. These operations continue, until the detector finds the specific legitimate constellation point which is nearest to $\mathbf{y}_{reconstr}$. At the end of the search, we assume that the last trial point that was found is the ML solution. In the example shown in Figure 9.3, the detector reaches the optimum ML solution after two radius updates. Hence, only three trial points are examined in terms of their Euclidean distance with respect to the reconstructed received symbol point $\mathbf{y}_{reconstr}$. Therefore, the potentially full search carried out by the ML detector is avoided by the SD.

A better way of illustrating the depth-first SD algorithm's philosophy, when it is applied to multi-dimensional scenarios, i.e. to multi-user systems, is constituted by the search tree example provided for

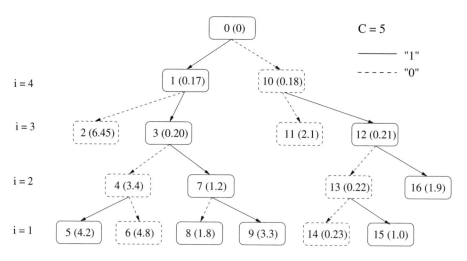

Figure 9.4: Illustration of the depth-first SD algorithm with the aid of the classic tree searching. The figure in brackets indicates the PED of a specific node for the trial point in the modulated constellation, while the number outside represents the order in which the points are visited.

the scenario of the (4×4) BPSK-modulated SDMA-OFDM system characterized in Figure 9.4. Before we elaborate further on the original depth-first SD with the aid of the search tree of Figure 9.4, it is important to note that the SD detector earmarks a legitimate symbol point as the tentative decision for $\check{s}_i$ only if the resultant $\mathcal{D}_i$ of Equation (9.34) is no higher than the search radius C, implying that the earmarked symbol point for $\check{s}_i$ is located inside the circle of Equation (9.30) centred at $\mathcal{C}_i$. Otherwise, this point is not earmarked. As shown in Figure 9.4, the depth-first SD commences its search procedure using an ISR of $C = 5$ from the top level ($i = 4$). For each tree node, the number within brackets denotes the corresponding accumulated PED of that node, while the number outside the brackets indicates the order in which the node is visited. The dashed line represents a binary zero, whereas the solid line denotes a binary one. As we can see in Figure 9.4, the search is carried out from left to right, but in both downward and upward directions along the tree. Specifically, there are two scenarios that may be encountered during the tree search portrayed in Figure 9.4. Firstly, the search may reach a leaf node at the bottom, i.e. the lowest level corresponding to s_1 in Figure 9.4. The other possible scenario is that the detector cannot find any point inside the circle of Equation (9.30) for the ith element s_i, or, equivalently, the accumulated PEDs of all the candidates for s_i are higher than the current search radius C. In the first case, once the search reaches a leaf node, e.g. at its fifth step the detector reaches a tree leaf having a Euclidean distance of 4.2 as shown in Figure 9.4, which is smaller than the current search radius of $C = 5$, then the detector starts the search process again with the reduced radius $C = 4.2$. In the second case, the detector must have made at least one erroneous tentative point selection for the previous $(U - i)$ lattice coordinates. In this scenario, the detector goes back to the $(i + 1)$th search tree level and selects another tentative point for s_{i+1} within the circle formulated by Equation (9.30), and proceeds downwards along the tree again to try and find a legitimate decision for s_i. If all the available tentative points for s_{i+1} fail to lead to a legitimate decision, the search backtracks to s_{i+2} with the same objective, and so on. For example, at the ninth step seen in Figure 9.4, the detector is unable to find a legitimate point within the new smaller hypersphere having the radius of 1.8, which was obtained at the previous step, hence the search backtracks to level $i = 4$, since no more available candidates can be found within the corresponding search area for s_2, and s_3. In the end, after visiting a total of 15 tree nodes and leaves in Figure 9.4, the SD chooses the tree leaf having a minimum Euclidean distance of 0.23 and backtracks to the level $i = 4$ to yield the final ML solution $\hat{s}_{ML}$.

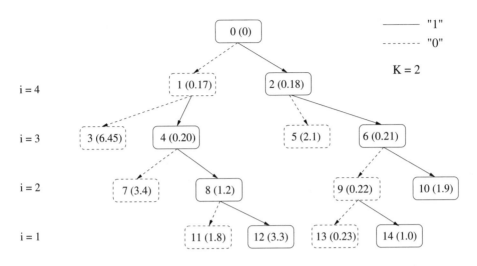

Figure 9.5: Illustration of breadth-first SD algorithm by the corresponding tree searching.

9.2.3 Breadth-First Tree Search [550]

Based on our discussions on the depth-first SD algorithm in Section 9.2.2, we can observe that the tree search is carried out in a depth-first manner, with the goal of reaching a leaf node to ensure that the newly calculated Euclidean distance allows us rapidly to shrink the search hypersphere. However, as we will see in Section 9.2.5, the computational complexity of the depth-first SD depends very much on the ISR C of Equation (9.35), and the appropriate choice of C constitutes a design challenge. Therefore, another tree search scheme was proposed to circumvent this problem based on the idea of searching the tree in a *breadth-first* manner by limiting the number of tree nodes to be expanded to K, where K denotes the maximum number of nodes having the K lowest accumulated PEDs at every level of the tree. Hence, the computational complexity of the tree search is reduced, while circumventing the problem of finding an appropriate choice of the ISR. More importantly, an SNR-independent computational complexity is expected and the search is guaranteed to be carried out in the downward direction along the tree.

The search tree of the K-best SD algorithm using $K = 2$ is shown in Figure 9.5, which was applied to the same example of Figure 9.4, where the depth-first SD algorithm was employed. Since we use $K = 2$, following the evaluation of the PEDs of all nodes at a certain level, only the two nodes having the lowest PEDs are expanded or pursued further at each level. Consequently, the detector successfully finds the ML solution to a high probability, which has a Euclidean distance of $\mathcal{D}_1 = 0.23$ in Figure 9.5 with respect to the centre $\hat{c}$ of the search hypersphere. Comparing the two expanded search trees portrayed in Figure 9.4 and Figure 9.5, we can see that a higher complexity reduction was achieved by the K-best SD detector. However, we cannot simply conclude that the K-best SD is always better than its depth-first counterpart, since upon reducing the ISR of the latter, a higher complexity reduction may be expected to be attained. On the other hand, if K is set to an excessively low value, such as $K = 1$ for example, the K-best SD becomes unable to find the true ML solution owing to the fact that the detector discontinues the search along the true ML branch as early as the fourth level in Figure 9.5 by choosing to expand and pursue a node having a PED of 0.17. Therefore, the K-best SD does not necessarily arrive at the ML solution, while the depth-first SD does. More discussions on the comparison of these two SDs in terms of achievable performance and imposed complexity will be carried out based on the simulation results in Sections 9.2.5 and 9.4.

9.2.4   Generalized Sphere Detection (GSD) for Rank-Deficient Systems

Our discussions in the previous sections implied the assumption that the number of users U, or the number of the transmit antennas M, is no more than that of the receive antennas N, i.e. we have $U, M \leq N$. However, this is not always the case in practice, e.g. when SD is implemented in a typical downlink of an SDM-OFDM system, where the number of antenna elements employed by the BS exceeds that used at the MS. In this scenario the channel matrix $\mathbf{H}$ of Equation (9.6) becomes non-invertible and hence the system is referred to as rank deficient, where the SDs discussed in Section 9.2 fail to work. Recall that the SD applied in a MIMO system, where the number of transmit antennas M is no higher than the number of receive antennas N, i.e. $M \leq N$, the QR decomposition or the Cholesky factorization has to be invoked for decomposing the Grammian matrix $\mathbf{G} = \mathbf{H}^T \mathbf{H}$ in order to obtain the upper-triangular matrix $\mathbf{U}$ having a rank of M, which is identical to the length of the transmitted MIMO symbol vector $\mathbf{s}$. However, for rank-deficient systems the rank of the matrix $\mathbf{H}$ is lower than the number of transmitted symbols to be estimated, which in turn results in zero elements along the diagonal of the upper-triangular matrix $\mathbf{U}$. Recall that in the decoupled search space of Equation (9.30) for the ith component $\check{s}_i$ in SD, written here as

$$|\check{s}_i - \mathcal{C}_i| \leq \frac{\sqrt{C - \mathcal{D}_{i+1}}}{u_{i,i}}, \tag{9.36}$$

all the diagonal elements $u_{i,i}$ have to be non-zero integers. Similarly, the Cholesky decomposition will also fail since the matrix $\mathbf{G} = \mathbf{H}^T \mathbf{H}$ is no longer positive definite. Hence two different techniques of circumventing this problem will be briefly introduced in Sections 9.2.4.1 and 9.2.4.2.

9.2.4.1   GSD [556]

After examining the resultant upper-triangular matrix $\mathbf{U}$ evaluated by the QR decomposition in the context of a rank-deficient system where $M > N$, it may be readily shown that the diagonal elements in the first N rows of the $(M \times M)$ matrix $\mathbf{U}$ are non-zero, while the diagonal elements in the remaining $(M - N)$ rows are zero. Hence, if $\mathbf{U}$ is partitioned so that the first N rows and the remaining $(M - N)$ rows are separated, we can use the resultant $(N \times M)$ matrix that has non-zero diagonal elements for the SD of the first N transmitted symbols based on one of the $M_c^{(M-N)}$ possible combinations of the remaining $(M - N)$ symbols. Essentially, this GSD algorithm [556] can be considered as the combination of the SD for the first N transmitted symbols in $\mathbf{s}$ and the full ML detection of the remaining $(M - N)$ symbols, which is a conceptually straightforward method that eliminates the problem of having zero diagonal elements in the upper-triangular matrix $\mathbf{U}$ of a rank-deficient MIMO system. The SD scheme invoked in the GSD can be any of the SDs discussed in Sections 9.2.2 and 9.2.3.

Owing to the fact that only N symbols are detected by using low-complexity SD, while all possible $M_c^{(M-N)}$ combinations of the remaining $(M - N)$ symbols have to be tested by the ML detector, the complexity of this GSD scheme is expected to be high, especially when the number of the transmit antennas is significantly higher than that of the receive antennas, i.e. when $M \gg N$. More quantitatively, the resultant complexity is an exponential function of $(M - N)$ [557], potentially preventing its practical application. Thus, our forthcoming discussions will be focused on the design of more efficient SDs applicable to rank-deficient systems.

9.2.4.2   GSD Using a Modified Grammian Matrix [557]

In Section 9.2.4.1, a particular partitioning of the matrix $\mathbf{U}$ is conducted in order to circumvent the problem of having zero diagonal elements. In this section, a different GSD scheme will be discussed, which carries out the Cholesky factorization of a modified Grammian matrix $\tilde{\mathbf{G}}$ in order to obtain an upper-triangular matrix $\mathbf{U}$ having non-zero diagonal elements. The basic idea behind the GSD algorithm of [557] is that, under the assumption of using a constant modulus modulation scheme, such as BPSK and QPSK, which implies that every element in the signal vector $\check{s}$ has a constant modulus, the product $\alpha \check{s}_i^* \check{s}_i$ becomes a constant value of α under the assumption of unitary transmit power. Consequently, we have an equivalent ML solution for the corresponding SD formulated as [557]

$$\hat{s}_{ML} = \arg\min_{\check{s} \in \mathcal{M}_c^M} \{\|\mathbf{y} - \mathbf{H}\check{s}\|_2^2 + \alpha \check{s}^H \check{s}\} \tag{9.37}$$

$$= \arg\min_{\check{s} \in \mathcal{M}_c^M} \{(\check{s} - \hat{c})^H (\mathbf{H}^H \mathbf{H} + \alpha\mathbf{I})(\check{s} - \hat{c})$$

$$+ \underbrace{\mathbf{y}^H (\mathbf{I} - \mathbf{H}(\mathbf{H}^H \mathbf{H} + \alpha^2 \mathbf{I})^{-1} \mathbf{H}^H)\mathbf{y}}_{\varphi}\}, \tag{9.38}$$

where

$$\hat{c} = \underbrace{(\mathbf{H}^H \mathbf{H} + \alpha\mathbf{I})^{-1}}_{\triangleq \tilde{\mathbf{G}}} \mathbf{H}^H \mathbf{y}, \tag{9.39}$$

and $\mathbf{I}$ represents the identity matrix. Normally, α is set to be the noise variance $2\sigma_w^2$, i.e. we arrive at

$$\hat{c} = (\mathbf{H}^H \mathbf{H} + 2\sigma_w^2 \mathbf{I})^{-1} \mathbf{H}^H \mathbf{y}, \tag{9.40}$$

which is the MMSE solution of Equation (9.2).

Since the last term denoted by the φ portion of Equation (9.38) is independent of the value of $\check{s}$, Equation (9.38) can be simplified as

$$\hat{s}_{ML} = \arg\min_{\check{s} \in \mathcal{M}_c^M} \{(\check{s} - \hat{c})^H \tilde{\mathbf{G}}(\check{s} - \hat{c})\}. \tag{9.41}$$

Furthermore, the modified Grammian matrix, $\tilde{\mathbf{G}}$, is always Hermitian and positive definite in contrast to the original Grammian matrix $\mathbf{G} = \mathbf{H}^H \mathbf{H}$. Hence, the modified Grammian matrix $\tilde{\mathbf{G}}$ can be Cholesky factorized in order to attain an upper-triangular matrix $\mathbf{U}$ having non-zero diagonal elements, regardless of the rank of the matrix $\mathbf{H}$, i.e. we have $\tilde{\mathbf{G}} = \mathbf{U}^H \mathbf{U}$. Consequently, the metric of the GSD can be expressed based on Equation (9.41) as

$$\hat{s}_{ML} = \arg\min_{\check{s} \in \mathcal{M}_c^M} \{(\check{s} - \hat{c})^H \mathbf{U}^H \mathbf{U}(\check{s} - \hat{c})\}, \tag{9.42}$$

which is in a form identical to that of Equation (9.20) for the full-rank scenario. Finally, owing to the fact that all diagonal elements in $\mathbf{U}$ are now non-zero, the standard SD tree search algorithm of Sections 9.2.2 and 9.2.3 can be applied to Equation (9.42).

9.2.5 Simulation Results

In this section, the achievable performance versus the complexity imposed by the SD is discussed and analysed in comparison with conventional ML detection based on our simulation results. The system parameters used in all of our simulations throughout the chapter are shown in Table 9.1. Note that the power delay profile of the three-path frequency-selective channel is given by $P(\tau) = \sum_{k=0}^{2} P(\tau_k)\delta(t - k\tau)$, where τ is the delay spread and we have $P(\tau_k) = [0.5\ 0.3\ 0.2]$ for $k = 0, 1, 2$. It is assumed that each user has a single transmit antenna and perfect FD-CHTF estimation is available in all the simulations. The ISR of the depth-first SD was adjusted according to the SNR level [553]. Specifically, we used the setting of $C = 2\sigma_w^2 JN$, where the parameter J was chosen to satisfy $J \geq 1$, while N is the number of receive antennas.

- **BER Performance and Computational Complexity Versus SNR**
  Both the BER performance achieved and the computational complexity imposed by the ML as well as by the aforementioned two types of SD algorithms are shown in Figure 9.6 for the fully

Table 9.1: Summary of system parameters.

System parameters	Choice
System	SDMA-OFDM
Uplink/downlink	Uplink
Number of subcarriers	128
CIR model	Three-path frequency-selective channel
CIR tap fading	OFDM symbol invariant
Channel estimation	Ideal
Transmit antennas per user	1
Initial squared search radius	SNR based

loaded (4×4) antenna SDMA-OFDM scenario, where 16-QAM transmissions were employed. The BER curves of both the depth-first SD and the K-best SD ($K = 16$) virtually coincide with that of the ML detector. The y-axis on the right quantifies the algorithm's complexity expressed in terms of the number of real-valued additions and multiplications versus E_b/N_0, as shown by the dashed line. As seen from Figure 9.6, both SD algorithms are capable of approaching the ML performance at a significantly lower complexity compared with the ML detector. More importantly, upon comparing the depth-first SD and the K-best SD detectors, we found that the former, which carries out the tree search in a depth-first manner, exhibits a complexity dependent on E_b/N_0. Specifically, the higher the received signal power, the lower the computational complexity. Since the complexity of the depth-first SD is variable, it is less suitable for real-time implementation [550]. This phenomenon can be explained as follows. When the signal $\mathbf{y}$ is received at a higher SNR, the ML solution is typically closer to the search centre $\hat{\mathbf{c}}$ of the hypersphere search space, which is obtained either by the LS algorithm of Equation (9.19) or by the MMSE algorithm of Equation (9.40). Hence the ISR can be set to a smaller value, in order to avoid a time-consuming search within a large hypersphere. Therefore, in our simulations, the ISR C was set according to the noise level, as mentioned previously. On the other hand, the K-best SD detector exhibits a constant computational complexity, since its complexity depends only on the maximum number of nodes K to be considered for each search tree level, on the modulation scheme used and on the number of transmit antennas employed. Hence, when all these parameters are fixed, the complexity of the K-best SD remains constant. It is observed from Figure 9.6 that the complexity imposed by the K-best SD is significantly lower than that imposed by its depth-first counterpart, when the SNR is low, while the former becomes slightly higher than the latter when the SNR encountered is high.

- **Complexity Versus the Number of Transmit Antennas or Users**
  Figure 9.7 portrays the complexity of both the ML and the SDs versus the number of users U in the scenario of a fully loaded 4-QAM SDMA-OFDM system. Observe in Figure 9.7 that the ML detector's complexity increases exponentially with U, which is independent of the value of the SNR, since the ML detector jointly detects the number U of users, imposing a potentially excessive computational complexity of M_c^U Euclidean distance metric evaluations between all possible tentative transmitted signal vectors $\check{\mathbf{s}}$ and the received signal vector $\mathbf{y}$. As shown in Figure 9.7, a significant complexity gain is achieved by both types of SDs over the ML detector, which further escalates as the number of transmit antennas increases. Again, the complexity of the depth-first SD is dependent on the SNR, while the K-best SD exhibits an SNR-independent complexity, as observed in Figure 9.7. According to [558], the order of SD complexity in the context of an m-dimensional lattice is at most $O(m^{4.5})$ at low SNRs, and $O(m^3)$ at high SNRs. Again, we can observe from Figure 9.7 that the K-best SD ($K = 16$) exhibits a significantly

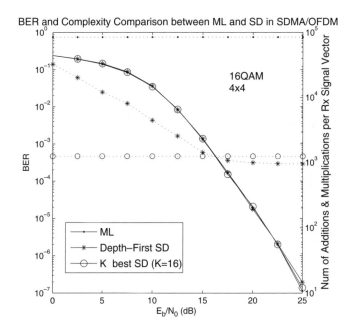

Figure 9.6: Comparison of the ML and SD algorithm. The y-axis on the left quantifies the BER performance of the ML and SD algorithms using solid lines, while the right y-axis quantifies the complexity versus E_b/N_0, which is plotted using dashed lines. All system parameters were summarized in Table 9.1.

lower complexity than its depth-first counterpart at an SNR of 4 dB, while it exhibits a complexity slightly higher when the SNR is 20 dB.

- **Effects of K on the BER Performance and the Complexity of K-best SD**
  Figure 9.8 reveals the effects of the parameter K on both the achievable BER performance and the computational complexity of the K-best SD detector. Observe in Figure 9.8 that K has to be set to at least 16 for the SD to approach the ML detector's performance. However, setting K to be lower than that would reduce the computational complexity imposed, as the dashed line representing the complexity versus E_b/N_0 trend indicates, which is achieved at the cost of a degradation in BER performance. The same conclusion can be drawn from Figure 9.8 as deduced from Figure 9.6 earlier, i.e. the complexity of the K-best SD algorithm is independent of the received signal power. Thus, for a given scenario, the trade-off between the achievable BER performance and the computational complexity imposed is effectively controlled by the choice of K.

- **Effects of the ISR on the Complexity of Depth-First SD**
  From our previous results shown in Figure 9.6 we infer that the complexity of the SD may vary as the received signal's SNR changes. Essentially, the complexity of the SD is dependent on the specific choice of the ISR C that confines the search area, which in turn determines the efficiency of the search. Figure 9.9 offers an insight into the dependence of the SD's complexity on the ISR C. The associated complexity increases significantly as C is increased. Therefore, a judicious choice of the ISR plays a vital role in determining both the performance and the complexity of the SD scheme. If it is set too small, the resultant initial search space may not contain the ML solution. On the other hand, it should not be set too high, otherwise a near-exhaustive search may be encountered.

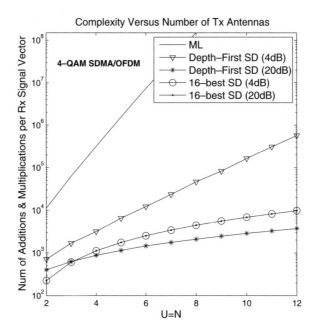

Figure 9.7: Complexity versus the number of transmit antennas. All system parameters were summarized in Table 9.1.

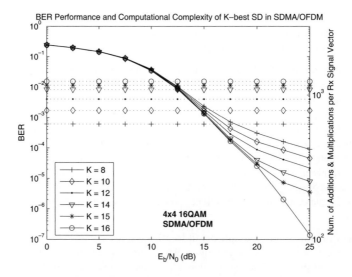

Figure 9.8: Effects of K on the BER performance and complexity of the K-best SD. The y-axis on the right represents the scale for the dashed lines, indicating the complexity versus E_b/N_0 trends, while the y-axis on the left indicates the solid lines showing the BER performance versus E_b/N_0. All system parameters were summarized in Table 9.1.

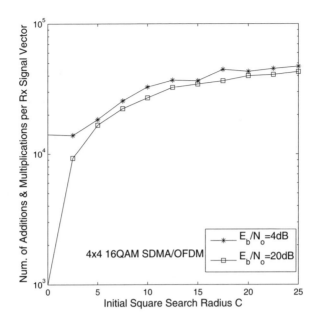

Figure 9.9: The complexity of depth-first SD versus ISR. All system parameters were summarized in Table 9.1.

9.3   Complexity-Reduction Schemes for SD

9.3.1   Complexity-Reduction Schemes for Depth-First SD

9.3.1.1   ISR Selection Optimization [548]

From our previous discussions based on the simulation results of Figure 9.9, we know that the choice of the ISR is crucial as regards the performance of the depth-first SD. Hence, the key to further reducing the associated complexity is to optimize the ISR selection. All our simulations characterized so far have employed an experimentally motivated ISR scheme, where the ISR C is defined as $C = 2\sigma_w^2 JN$, where the parameter J is chosen to satisfy $J \geq 1$, while N represents the number of receive antennas. However, this ISR scheme is suboptimal, since it is unable to guarantee that there is always at least one legitimate signal point within the initial hypersphere, potentially leading to a decoding failure. The failure may require a second tentative decoding using a larger ISR and hence wastes valuable computational resources. Two other ISR selection schemes are investigated in this section, namely the MMSE-based ISR selection and a hybrid scheme, which is constituted by a contribution from the previous two schemes:

- **MMSE-Based ISR Selection Scheme**
  The idea behind this ISR selection scheme is appealingly simple. In order to guarantee successful decoding, the ISR is set to the Euclidean distance between the received signal point $\mathbf{y}$ and the MMSE-solution-based reconstructed received signal $\mathbf{y}_{mmse}$, which can be expressed as [548]

$$\hat{\mathbf{y}}_{mmse} = \mathbf{H}\hat{\mathbf{s}}_{mmse}, \qquad (9.43)$$

  where $\hat{\mathbf{s}}_{mmse}$ is the hard-decision-based MMSE solution, which can be written as

$$\hat{\mathbf{s}}_{mmse} = (\mathbf{H}^H\mathbf{H} + 2\sigma_w^2\mathbf{I})^{-1}\mathbf{H}^H\mathbf{y}. \qquad (9.44)$$

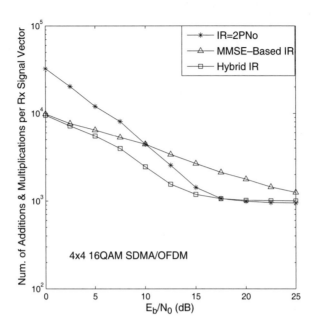

Figure 9.10: Comparison of different ISR selection schemes for depth-first SD. All system parameters were summarized in Table 9.1.

As expected, the ISR C can be formulated as

$$C = \|\mathbf{y} - \mathbf{y}_{mmse}\|^2. \tag{9.45}$$

- **Hybrid ISR Selection Scheme**
  The hybrid ISR selection scheme obtains its ISR based on a combination of the above-mentioned experimentally adjusted solution and the MMSE-based solution. Specifically, we assume that C_1 and C_2 are the ISRs calculated by the aforementioned two ISR schemes, respectively. In order to achieve a reduced complexity, the hybrid ISR scheme opts for the smaller of the two, i.e. for

$$C = \min(C_1, C_2). \tag{9.46}$$

Our comparison of the three previously discussed ISR schemes is provided in Figure 9.10, which suggests that the hybrid ISR scheme achieves the lowest complexity over the entire SNR range of interest. However, it suffers from the same problem of potential decoding failure as the pure experimentally adjusted ISR scheme. On the other hand, the MMSE-based ISR is the most reliable one in terms of guaranteeing successful sphere decoding [548]. In terms of complexity, the MMSE-based scheme outperforms the experimentally motivated arrangement at low SNRs while imposing a higher complexity at high SNRs.

9.3.1.2   Optimal Detection Ordering [559]

In the context of an SDMA system supporting U transmitted data streams, the original SD algorithm of [560] commences the detection of symbols from the Uth signal component to the first one, without considering any specifically beneficial detection order. However, if we expand the ML error formula of

Equation (9.25), we get

$$Err_{ml} = \sum_{i=1}^{U} \left| u_{ii}(\check{s}_i - \hat{c}_i) + \sum_{j=i+1}^{U} u_{ij}(\check{s}_j - \hat{c}_j) \right|^2 \tag{9.47}$$

$$= |u_{U,U}\xi_U|^2 + |u_{U-1,U-1}\xi_{U-1} + u_{U-1,U}\xi_U|^2 \tag{9.48}$$

$$+ |u_{U-2,U-2}\xi_{U-2} + u_{U-2,U-1}\xi_{U-1} + u_{U-2,U}\xi_U|^2 + \dots, \tag{9.49}$$

where $\xi_i = \check{s}_i - \hat{c}_i$. Then we can observe in Equation (9.49) that the transmitted symbol $\check{s}_U$ appears U times in the above summation, $\check{s}_{U-1}$ appears $(U-1)$ times, ..., and $\check{s}_1$ appears only once. Based on this observation, we infer that the correct detection probability of the first detected symbol $\check{s}_U$ has an impact on all of the following $(U-1)$ detection steps, while the weight of $\check{s}_{U-1}$ is somewhat lower, since it has an impact only on the next $(U-2)$ steps etc. In other words, the highest-quality signal in terms of SNR should be detected first. This philosophy is the essence of the detection ordering technique, which is a key advance applied for example in the context of the V-BLAST system [561].

Under the assumption that each transmitted stream has an identical transmit power and that each signal experiences the same amount of noise after passing through the channel, the received signal x_m of the mth transmitted signal component can be written as

$$y_u = \sum_{n=1}^{N} h_{nu} \cdot s_u, \tag{9.50}$$

where the noise term is omitted here for convenience and h_{nu} represents the FD-CHTF between the uth user and the nth receive antenna, while $\mathbf{h}_u$ is the uth column of the FD-CHTF matrix $\mathbf{H}$. Hence, we can see that the SNR of the uth signal component is proportional to the norm of its corresponding column $\mathbf{h}_u$ in the FD-CHTF matrix $\mathbf{H}$. Bearing in mind the above-mentioned rationale of detection ordering, the norm of the column vector $\mathbf{h}_u$ $(u = 1, 2, \dots, U)$ is ordered as

$$\|\mathbf{h}_1\| \le \|\mathbf{h}_2\| \le \dots \le \|\mathbf{h}_U\|. \tag{9.51}$$

Consequently, when the SD is applied to this reordered FD-CHTF matrix $\mathbf{H}$, the detection of $\check{s}$ proceeds in a descending order of the channel SNR, which may be estimated with the aid of frequency-domain pilots. After finding the ML solution, the resultant vector $\hat{s}$ of modulated symbols has to be reordered again, according to the symbol positions of the original sequence. The complexity reduction facilitated by the most beneficial detection ordering scheme is revealed in Figure 9.11.

9.3.1.3 Search Algorithm Optimization

9.3.1.3.1 Sorted Sphere Detection (SSD)

Although the depth-first SD scheme [544] of Section 9.2.2 is capable of approaching the ML performance at a significantly reduced complexity, it does not operate efficiently at every search step. In fact, the search commences from the surface of the sphere towards the centre. The search carried out in this order does not take into account the definition of the ML solution, which is defined by that specific valid lattice point which is closest to the centre $\hat{c}$ of the search sphere [545]. Therefore, the SD follows a zigzag-shaped search trajectory from the surface of the sphere towards its centre $\hat{c}$ as the search for the ML solution proceeds, which is not as efficient as it could be.

Thus, modifications can be introduced in the search order of the SD algorithm in order to reduce its complexity further. Bearing in mind the aforementioned definition of the ML solution, the modified SD should commence its search near the centre of the sphere. Consequently, a reduced-complexity SD was proposed in [545], where the elements in the candidate set $\mathcal{B}_i$ for the ith signal component $\check{s}_i$ are first sorted in ascending order according to the metric

$$|\check{s}_i - \mathcal{C}_i|, \tag{9.52}$$

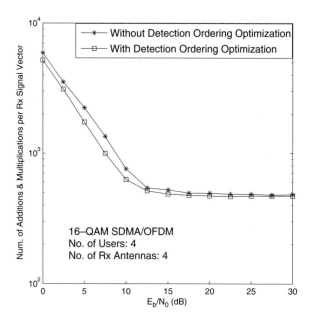

Figure 9.11: The computational complexity benefit of detection order optimization for depth-first SD. All system parameters were summarized in Table 9.1.

in which C_i given by Equation (9.31) represents the decoupled centre of the search area of Equation (9.30) for $\check{s}_i$. Essentially, upon sorting the legitimate candidates for $\check{s}_i$ according to their distance from the decoupled search centre C_i of Equation (9.31), the modified algorithm commences its search from the most promising lattice point. Thus, the SD complexity is expected to be significantly reduced by the rapid reduced search radius. We refer to this modified SD scheme as the Sorted Sphere Detection (SSD) algorithm, which is expected to exhibit a reduced complexity.

9.3.1.3.2 SSD Using Updated Bounds

Another SD method operating on the basis of SSD was proposed in [545], which is capable of achieving an even lower complexity. Specifically, when a new candidate lattice point is found within the search hypersphere, in addition to updating the search radius, the following three modifications are introduced:

1. The decouple search areas of Equation (9.30) recorded for all candidate basis sets $\mathcal{B}_i$ ($i = 1, 2, \ldots, U$) are also updated immediately with the aid of the most recently obtained lattice point.

2. The next round of the search is carried out commencing from $\check{s}_1$, instead of $\check{s}_U$.

3. The new search for $\check{s}_i$ is carried out without going back to start from the first component in the newly obtained smaller candidate set $\mathcal{B}_i$.

Note that the immediate update of the decoupled search area of Equation (9.30) for each tree search level actually eliminates some of the search candidates at the rightmost end of the sorted set $\mathcal{B}_i$, but with its leftmost end unchanged [545]. This facilitates the above-mentioned third action, which in turn allows the SD to avoid searching candidates already identified during the last round of the search. Finally, we refer to this modified SD scheme as the Updated-Bound-aided SSD (SSD-UB).

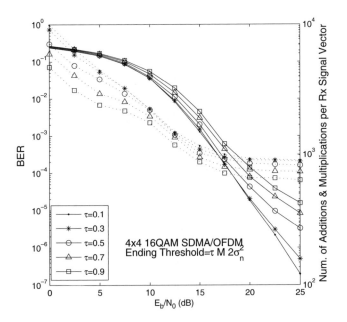

Figure 9.12: BER performance and computational complexity of termination-threshold-aided SSD. The y-axis on the left quantifies the BER performance of the ML and SD algorithms using solid lines, while the right y-axis quantifies the complexity versus E_b/N_0, which is plotted using dashed lines. All system parameters were summarized in Table 9.1.

9.3.1.3.3 SSD Using Termination Threshold

A more intuitive approach that retains most of the benefits of the SSD reduces the complexity further by introducing a search-termination threshold t [559] informing the SD to curtail the search when the ML error term of Equation (9.20) becomes lower than t, where the ML error refers to the newly obtained squared search radius of Equation (9.33). This procedure aims to avoid testing all possible tentative ML solution points one by one, which is time consuming. Recall that the SSD reorders the components in the ith basis set $\mathcal{B}_i$, which contains all the tentative points within the search hypersphere for the ith signal component $\breve{s}_i$, in ascending order according to the metric given by Equation (9.52). Therefore, the point considered first in the set $\mathcal{B}_i$ is the most promising one. Thus, with the aid of the termination threshold t, the search procedure may be curtailed, provided that the newly obtained lattice point is sufficiently close to the received signal. Hence, the appropriate choice of the termination threshold is the key point for ensuring the efficiency of this reduced-complexity SD. Specifically, if the termination threshold t is set too small, it does not have any effect, since it is unlikely that the ML error would be smaller than t. On the other hand, if t is too large, the search for the ML solution may be curtailed when it tests a non-ML point, whose distance from the received symbol point is less than t. In this scenario, the complexity imposed can be further reduced at the cost of a degradation in performance. A judicious choice of the termination threshold t is given by [559]

$$t = \tau \cdot U \cdot 2\sigma_w^2, \tag{9.53}$$

where U is the number of users, σ_w^2 is the noise level and τ is a parameter typically set to 0.1, 0.3, etc. Consequently, the termination threshold should be set proportional to the number of transmit antennas as well as to the noise power. In this treatise, we refer to this reduced-complexity SD as the Termination-Threshold-aided SSD (SSD-TT). As shown in Figure 9.12, there is a trade-off between the achievable

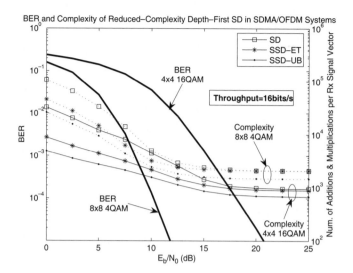

Figure 9.13: BER performance and computational complexity of reduced-complexity depth-first SDs. The y-axis on the left quantifies the BER performance of the ML and SD algorithms using solid lines, while the right y-axis quantifies the complexity versus E_b/N_0, which is plotted using dashed lines. All system parameters were summarized in Table 9.1.

performance and the complexity imposed by the SD, which is controlled by the appropriate choice of the termination threshold.

Let us now compare the search algorithm optimization schemes discussed in this section in Figure 9.13. Our comparisons are carried out in the scenarios of both (8×8)-element 4-QAM and (4×4)-element 16-QAM SDMA-OFDM systems, which have an identical throughput of 16 bits per symbol. In both cases, the updated-bound-assisted SD achieves a significantly lower computational complexity than the termination-threshold-assisted arrangement, rendering it a more effective complexity-reduction scheme. The termination-threshold-assisted scheme is capable of attaining an evident complexity reduction when the SNR is relatively low, while imposing only a slightly lower complexity than the original SD of [560] when the SNR is in excess of 17.5 dB. On the other hand, when comparing two different SDMA-OFDM systems, we found that the (8×8)-antenna 4-QAM system substantially outperforms the (4×4)-element 16-QAM system in terms of achievable BER, as a benefit of its higher diversity gain and its lower-density modulation constellation, while imposing an acceptable computational complexity. More specifically, for a given target BER of 10^{-5}, we have an SNR gain of about 9 dB if the (8×8)-antenna 4-QAM scheme is employed, rather than the (4×4)-element 16-QAM arrangement. This is achieved at the cost of less than three times increased computational complexity, as quantified in terms of the number of real-valued additions and multiplications per received signal vector, when the updated-bound-assisted scheme is employed.

In addition to their reduced complexity, the search algorithm optimization schemes discussed in this section have a further benefit of rendering the complexity of the SD less sensitive to the specific choice of the ISR, which can be observed from Figure 9.14.

9.3.2   Complexity-Reduction Schemes for K-Best SD

9.3.2.1   Optimal Detection Ordering

Having discussed various complexity-reduction schemes designed for the depth-first SD, let us now consider a range of complexity-reduction schemes applicable to the K-best SD. The detection ordering

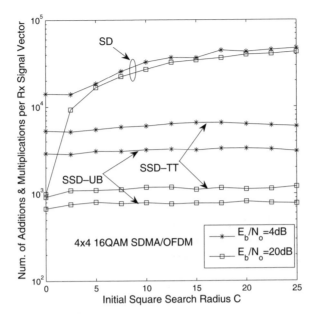

Figure 9.14: Complexity versus the square ISR of reduced-complexity depth-first SDs. All system parameters were summarized in Table 9.1.

optimization scheme introduced in Section 9.3.1.2, which is capable of effectively reducing the complexity of the depth-first SD, was found suitable also for the K-best SD, which achieved a similar performance to that shown in Figure 9.11. For a rudimentary introduction to this scheme, refer to Section 9.3.1.2.

9.3.2.2 Search-Radius-Aided K-Best SD

Based on the portrayal of the K-best SD in Sections 9.2.3 and 9.2.5, it becomes explicit that its computational complexity is controlled by the parameter K, for a certain modulation scheme and a certain number of transmit antennas or users. This is in contrast to its depth-first counterpart, which achieves a low complexity, despite approaching the ML performance with the aid of rapid shrinking of the original search radius. Intuitively, if we can introduce a search radius for employment in the K-best SD, its complexity can be further reduced by discarding the unlikely ML candidate nodes which are located outside the sphere confined by the search radius, thus reducing the number of tentative nodes at each level. Consequently, since the partial Euclidean distances evaluated for some of the nodes exceed the radius, there may be less than K nodes that have to be considered for each level, resulting in an additional complexity reduction. In contrast to the gradually reduced radius of the depth-first SD algorithm, the radius used for the K-best SD remains unchanged during the entire search process, since it carries out the tree search in the downwards direction only and the search is ceased whenever it reaches tree leaf level, i.e. the lowest level of the tree exemplified in Figure 9.5. Hence, exactly the same search radius selection problem is encountered by the K-best SD, as faced by the depth-first SD. In order to avoid having no lattice points inside the sphere, which in turn results in a repeated search using an increased radius, the radius selection schemes used for the K-best SD should guarantee that at least one lattice point is located in the search sphere. In this report, two radius selection schemes for K-best SD will be examined, namely the *LS-Criterion-Based* and the *MMSE-Criterion-Based* radius calculation schemes; the latter has already been discussed in the context of depth-first SD in Section 9.3.1.1.

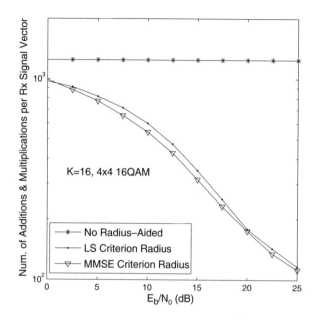

Figure 9.15: Complexity versus SNR of the radius-based K-best SD.

In Figure 9.15 we characterize these two radius-based K-best SDs and the original K-best SD of Section 9.2.3 [550], where we find that a significantly lower complexity can be achieved by both of the radius-based K-best SDs, compared with the original K-best SD of [550]. Hence, the radius-based K-best SD no longer exhibits an SNR-independent complexity as characterized in Figure 9.15, because a higher complexity reduction can be attained when the SNR increases. On the other hand, the complexity of the MMSE-criterion-based radius scheme of Section 9.3.1.1 is evidently lower than that of the LS-criterion-based radius scheme, owing to the fact that the former scheme is expected to operate using a smaller search radius which is capable of reducing the number of nodes at each level that would be expanded.

9.3.2.3 Complexity-Reduction Parameter δ for Low SNRs

Although the complexity of K-best SD can be significantly reduced by introducing a search radius, it still exhibits a relatively high complexity when the SNR is low, as we can observe from Figure 9.15. Intuitively, when the noise level is high, i.e. at low SNRs, investing excessive detection efforts in terms of a large search space becomes futile. This will become more explicit by considering the ML detector which has a high computational complexity and yet hardly achieves any performance gain in comparison with the MMSE detector, for example, when the SNR is low. In order to mitigate the problem, we introduce a complexity-reduction parameter δ, which allows us to reduce the complexity of the K-best SD when the SNR is low. A similar parameter γ was employed in the OHRSA detector of [562] in order to control its complexity, which will be discussed in Section 9.3.3.

The parameter δ is used as follows. When the SNR corresponding to the currently detected ith signal component is lower than δ, i.e. we have $\|\mathbf{h}_i\|^2/\sigma_w^2 < \delta$, only the tentative constellation point yielding the smallest value of $|s_i - \mathcal{C}_i|$ is considered, rather than testing all the original K candidates. Moreover, owing to the employment of the detection ordering optimization scheme of Section 9.3.1.2, the SNRs associated with the signal components about to be detected, i.e. $\check{s}_{i-1}, \check{s}_{i-2}, \ldots, \check{s}_1$, will also be lower than δ. Thus, only a single tentative point will be enumerated, which in fact represents the final decision for the corresponding signal components.

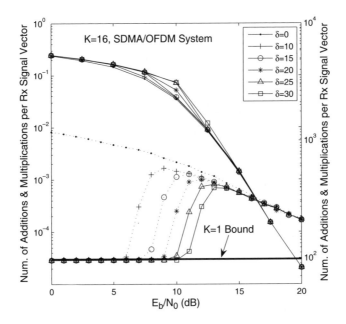

Figure 9.16: Effect of Parameter δ on the BER and complexity of the K-best SD. The solid lines scaled on the y-axis on the left represent the BER performance, while the dashed lines scaled on the y-axis on the right show the corresponding complexity.

Consequently, the complexity associated with a low SNR is significantly reduced at the cost of a modest BER performance degradation, provided that we choose the value of δ appropriately, as observed in Figure 9.16. Specifically, if δ is chosen to be 10 for the K-best SD ($K = 16$) in the scenario of a (4×4)-antenna 16-QAM SDMA-OFDM system, the corresponding BER curve is almost the same as the ML curve, but the corresponding complexity curve indicates a significantly lower complexity which coincides with the $K = 1$ complexity bound for the SNR range spanning from 0 to 6 dB. In other words, with the aid of δ, the original computationally demanding low-SNR range imposes a computational complexity associated with the case of $K = 1$.

9.3.3   OHRSA – An Advanced Extension of SD

9.3.3.1   Hierarchical Search Structure

Recently, another advanced tree search detection method, referred to as the OHRSA, was proposed in [563] as an extension of the conventional depth-first SD, which is capable of further reducing the detection complexity. Since its pre-processing stage actually employs exactly the same strategy as the GSD discussed in Section 9.2.4.2, the OHRSA may also be readily applied to rank-deficient MIMO systems, where the number of transmit antennas or users exceeds that of the receive antennas. Hence, the emphasis of this section will be on the search philosophy.

In order to enable the OHRSA detector to handle rank-deficient scenarios, the Grammian matrix $\tilde{\mathbf{G}}$ of Equation (9.39) is employed, which can be Cholesky factorized to obtain the upper-triangular matrix $\mathbf{U}$. Thus, the OF of the OHRSA may be formulated in a similar manner to the conventional SDs by rewriting Equation (9.25) based on the fact that the diagonal elements u_{ii} are positive real values [563]:

$$J(\check{s}) = \sum_{i=1}^{U} \left| u_{i,i}(\check{s}_i - \hat{c}_i) + \sum_{j=i+1}^{M} u_{i,j}(\check{s}_j - \hat{c}_j) \right|^2 \tag{9.54}$$

$$= \sum_{i=1}^{U} \left| \sum_{j=i}^{U} u_{i,j}(\check{s}_j - \hat{c}_j) \right|^2 \tag{9.55}$$

$$= \sum_{i=1}^{U} \phi_i(\check{s}_i), \tag{9.56}$$

where the Sub-Cost Function (SCF) $\phi_i(\check{s}_i)$ can be written as

$$\phi_i(\check{s}_i) = \left| \sum_{j=i}^{U} u_{i,j}(\check{s}_j - \hat{c}_j) \right|^2 \tag{9.57}$$

$$= \left| u_{i,i}(\check{s}_i - \hat{c}_i) + \sum_{j=i+1}^{U} u_{i,j}(\check{s}_j - \hat{c}_j) \right|^2 \tag{9.58}$$

$$= \left| u_{i,i}(\check{s}_i - \hat{c}_i) + a_i \right|^2, \tag{9.59}$$

where

$$a_i \triangleq \sum_{j=i+1}^{U} u_{i,j}(\check{s}_j - \hat{x}_j), \tag{9.60}$$

which is a complex-valued scalar independent of the ith element $\check{s}_i$ of the a priori candidate signal vector $\check{s}$.

Furthermore, according to [563], the so-called Cumulative Sub-Cost Function (CSCF) $J_i(\check{s}_i)$ is defined recursively as

$$J_U(\check{s}_U) = \phi_U(\check{s}_U) = \left| u_{U,U}(\check{s}_U - \hat{c}_U) \right|^2 \tag{9.61}$$

$$J_i(\check{s}_i) = \sum_{i}^{U} \phi_i(\check{s}_i) \tag{9.62}$$

$$= \sum_{j=i+1}^{U} \phi_j(\check{s}_j) + \phi_i(\check{s}_i) \tag{9.63}$$

$$= J_{i+1}(\check{s}_{i+1}) + \phi_i(\check{s}_i), \quad i = 1, \ldots, U-1, \tag{9.64}$$

where $\check{s}_i$ is defined as the candidate subvector given by $\check{s}_i = [\check{s}_i, \ldots, \check{s}_U]$. According to Equation (9.64), a recursive search can be carried out starting from the calculation of $J_M(\check{s}_M)$. At the ith recursive step, a tentative candidate $\check{s}_i$ is selected from the set of M_c possible hypotheses for the transmitted signal s_i associated with the ith user. Then, based on the value of the tentative candidate $\check{s}_i$, $J_i(\check{s}_i)$ is evaluated, which depends only on the tentative values of $\check{s}_j$, where $j = i, i+1, \ldots, U$. The recursive calculation of the SCF $J_i(\check{s})$ proceeds until i reaches 1. The resultant OF of $J(\check{s})$ is equal to the value of the CSC function $J_1(\check{s}_i)$, i.e.

$$J(\check{s}) = J_1(\check{s}_1), \tag{9.65}$$

which can be derived from Equation (9.56) and Equation (9.64). Hence, a recursive search process may be formulated on the basis of Equation (9.64) for testing all legitimate tentative signal vectors $\check{s}$ and then the value of its corresponding OF $J(\check{s})$ is stored. Then i is reset to U, and according to Equation (9.64) a new recursive process is commenced from the calculation of $J_U(\check{s}_U)$. Finally, after an exhaustive computation of all the M_c^U number of values for $J(\check{s})$ corresponding to all possible hypothesized signal vectors š, the ML solution is guaranteed to be found as the one associated with the lowest value of $J(š)$. The recursive hierarchical search formulated in Equation (9.64) is in fact also carried out in the conventional depth-first SD algorithms of Section 9.2.2, but with a significantly small search space (i.e. within the search hypersphere) given by the search radius, which is updated once a hypothesized signal vector š is obtained. Essentially, the recursive hierarchical search discussed so far in this section is the same as the full search technique employed in conventional ML detectors, which exhibits a potentially excessive complexity if a high-throughput modulation scheme is employed or a high number of users are supported by the system. Instead of introducing a search radius to confine the search area of the SD, the OHRSA invokes several optimization rules on the basis of exploiting the properties of the CSCF $J_i(š_i)$ of Equation (9.64). Note that the SCF ϕ_i given by Equation (9.57) is always positive, therefore the value of the CSCF $J_i(š_i)$ monotonically increases as the hierarchical search continues. Specifically, we have

$$J(š) = J_1(š_1) > J_2(š_2) > \cdots > J_M(š_M) > 0. \tag{9.66}$$

The hierarchical search structure combined with the property given by Equation (9.66) allow the search process to achieve a significant complexity reduction, which will be considered in the next section in comparison with the complexity-reduction techniques discussed for the depth-first SD in Section 9.3.1, since the OHRSA detector also falls into the category of depth-first SDs.

9.3.3.2 Optimization Strategies for the OHRSA Versus Complexity-Reduction Techniques for the Depth-First SD

In Section 9.3.3.1 we argued that the conventional depth-first SD of Section 9.2.2 and the OHRSA algorithms share the same recursive hierarchical search structure. Given the aim of decreasing the number of OF evaluations required for finding the ML solution, the optimization strategy of OHRSA will be contrasted to the complexity-reduction techniques of SD in our following discourse.

9.3.3.2.1 Best-First Detection Strategy

This strategy is identical to the detection ordering optimization technique discussed in Section 9.3.1.2 for the depth-first SD. Briefly, the best-first detection strategy entails detecting the received signals in a descending order according to their received signal quality expressed in terms of the SNR encountered, which is proportional to the norm of its corresponding column vector in the channel transfer function matrix $\mathbf{H}$ of Equation (9.6). The corresponding mathematical proof was provided in Section 9.3.1.2 and will not be restated here.

9.3.3.2.2 Sorting Criterion

Recall that in the SSD technique of Section 9.3.1.3, the elements in the resultant tentative candidate set $\mathcal{B}_i$ delimited by the decoupled search area of Equation (9.30) for the specific signal component $š_i$ are sorted in an ascending order according to their distance from the decoupled search centre $\mathcal{C}_i$ of Equation (9.31). The rationale of this was based on the idea that the ML solution is likely to be located near the centre of the decoupled search area. Thus the SD becomes capable of promptly finding the ML solution, avoiding a 'zigzagging' search from the surface of the sphere to the ML solution, which is closest to the centre $\hat{\mathbf{c}}$ of the hypersphere.

    The rationale of the SSD of Section 9.3.1.3.1 can be transplanted into the OHRSA, despite the fact that their mathematical sorting criteria are quite different from each other. To expound a little further, for the OHRSA, there is no need for the concept of a search radius and corresponding search sphere, which is the basic difference between the OHRSA and the conventional SD of Section 9.2.2. However, bearing in mind the definition of the ML solution and the specific property of the OHRSA formulated in Equation (9.66), another scheme capable of achieving the same objective of avoiding futile search steps may be devised. Specifically, in the context of the OHRSA, the ML solution $\mathbf{s}_{ML}$, can be interpreted as the tentative signal vector $\check{s}$ whose corresponding OF $J(\check{s})$ is the smallest one. On the other hand, the CSCF of Equation (9.64) is increased cumulatively as the recursive search proceeds from $\check{s}_U$ to $\check{s}_1$ and hence we arrive at the final value of the OF $J(\check{s})$ formulated in Equation (9.66) which is repeated here for convenience:

$$J(\check{s}) = J_1(\check{s}_1) > J_2(\check{s}_2) > \cdots > J_U(\check{s}_U) > 0. \tag{9.67}$$

Let us now rewrite Equation (9.64) as follows:

$$J_i(\check{s}_i) = J_{i+1}(\check{s}_{i+1}) + \phi_i(\check{s}_i), \quad i = 1, \ldots, U - 1. \tag{9.68}$$

Based on the above two equations, it is intuitive that in order to arrive at the lowest possible OF value $J(\check{s})$ after a single cycle of the recursive search loop is completed, the increment $\phi_i(\check{s}_i)$ seen in Equation (9.68) should be as small as possible at each recursive step. If we denote the set of M_c tentative candidate values of the transmitted signal component s_i at each recursive step $i = U, \ldots, 1$ as $\{\tilde{s}_m\}_{m=1,\ldots,M_c} \in \mathcal{M}_c$, the set of potential candidates $\{\tilde{s}_m\}_{m=1,\ldots,M_c}$ should be tested in an ascending order according to their corresponding values of $\phi_i(\check{s}_i) = \phi_i(\tilde{s}_m, \check{s}_{i+1})$, as formulated in Equation (9.57). As a consequence, we have

$$\phi_i(\tilde{s}_1, \check{s}_{i+1}) < \cdots < \phi_i(\tilde{s}_m, \check{s}_{i+1}) < \cdots < \phi_i(\tilde{s}_U, \check{s}_{i+1}), \tag{9.69}$$

where according to Equation (9.59) we have

$$\phi_i(\tilde{s}_m, \check{s}_{i+1}) = |u_{i,i}(\tilde{s}_m - \check{c}_i) + a_i|^2. \tag{9.70}$$

Therefore, with the aid of this sorting criterion, the more likely candidates for ML solution are tested earlier.

9.3.3.2.3 Local Termination Threshold

In contrast to the sorting technique employed in the conventional SD algorithms of Sections 9.3.1 and 9.3.2, the computational complexity of the OHRSA can only be further reduced if it is combined with other surrogate techniques, since no radius reduction is used to confine the search area. As an example, a local Termination Threshold (TT) can be introduced to control the operation of the OHRSA, e.g. to curtail operation based on the OF value computed at the current level search. Recall that the *global* TT technique of Section 9.3.1.3 instructs the SD to curtail its search and output the most recently found signal vector $\check{s}$ as the ML solution, when the Euclidean distance between the newly obtained signal vector and the search centre $\hat{c}$ is equal to or smaller than the preset termination threshold. The TT technique used in the OHRSA is a *local* one, which is invoked to curtail the current recursive search loop instead of discontinuing the search altogether. Therefore, the local TT employed in the OHRSA is reminiscent of the search bound formulated in Equation (9.30) for the depth-first SD algorithm, which confines the decoupled search area for a specific signal component $\check{s}_i$.

The local TT of the OHRSA may be formulated as

$$J_{\min} = \min\{J_{\min}, J(\check{s})\}, \tag{9.71}$$

which is updated every time a new OF value $J(\check{s})$ is obtained and hence the recursive search reaches the decision for deciding upon signal component $\check{s}_1$. Therefore, with the aid of the sorting criterion of Equation (9.69), the search loop is discontinued at the ith recursive search step aimed at deciding upon the signal component $\check{s}_i$, whenever the search satisfies $J_i(\tilde{s}_m, \check{s}_{i+1}) > J_{\min}$. And the search steps back to the $(i + 1)$th detection step, where another tentative candidate $\tilde{s}_m$ is chosen for $\check{s}_{i+1}$. By contrast, if the most recently obtained $J_{i+1}(\tilde{s}_m, \check{s}_{i+2}) < J_{\min}$, then the algorithm returns to the ith detection step. In the worst-case scenario, when the detection loop returns to $i = M$ and all the potential candidates for $\check{s}_M$ have been tested but the algorithm still fails to find a new search path to reach $J_1(\tilde{s}_m, \check{s}_2)$, the detector outputs the currently available tentative signal vector $\check{s}$, whose corresponding OF $J(\check{s})$ has the minimum value, as the ML solution.

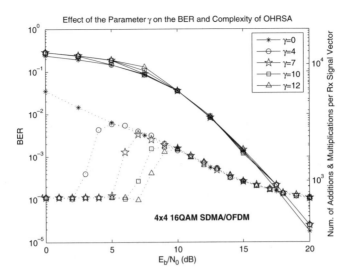

Figure 9.17: BER and complexity of the OHRSA detector. The real lines together with the left y-axis show the BER trends versus the SNR, while the dashed lines with the aid of the right y-axis exhibit the complexity trend versus the SNR.

9.3.3.2.4 Performance Evaluation

In Figure 9.17 both the BER performance achieved and the complexity imposed by the OHRSA detector are portrayed in conjunction with different complexity-reduction parameter values γ. As argued in Section 9.3.2.3, the appropriate SNR-dependent choice of the complexity-reduction parameter allows us to avoid the computationally demanding and yet inefficient detection of the specific signal components which have their signal energy well below the noise floor [562]. Following on from our previous discussion of the parameter δ employed by the K-best SD in Section 9.3.2.3, recall that δ has a similar role to that of the parameter γ in the context of the OHRSA detector. Suffice it to state here that the introduction of the parameter γ reduces the complexity of the OHRSA at low SNRs as we can see from the results of Figure 9.17, which is achieved at the cost of a slight BER performance degradation. By comparing Figures 9.17 and 9.16 we found that the BER performance degradation suffered by the OHRSA detector occurs in an SNR range which is different from that of the K-best SD detector of Section 9.3.2.3. More specifically, the BER performance degradation of the OHRSA detector takes place in the SNR range associated with the highest complexity reduction, i.e. in the low-SNR range. By contrast, the performance degradation of the K-best SD becomes most pronounced in the moderate SNR range.

9.4 Comparison of the Depth-First, K-Best and OHRSA Detectors

9.4.1 Full-Rank Systems

In this section, we compare the depth-first and K-best SDs of Sections 9.2.2 and 9.2.3 and the OHRSA detector of Section 9.3.3, which can be regarded as an advanced extension of the depth-first SD in the specified scenario of full-rank systems. Figures 9.18(a) and 9.18(b) show both the BER performance and the computational complexity of these three detectors in the scenarios of (4×4) 16-QAM and (8×8) 4-QAM SDMA-OFDM systems, respectively. Both systems had an effective throughput of $4 \cdot 4 =$

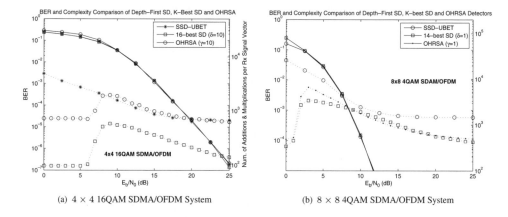

(a) 4×4 16QAM SDMA/OFDM System   (b) 8×8 4QAM SDMA/OFDM System

Figure 9.18: BER and complexity comparison of depth-first SD, K-best SD and OHRSA detectors. The solid lines scaled on the left y-axis show the BER trends versus the SNR, while the dashed lines scaled on the right y-axis exhibit the complexity trends versus the SNR. All the remaining system parameters were summarized in Table 9.1.

16 and $8 \cdot 2 = 16$ bits per symbol. By choosing an appropriate K and δ value for the K-best SD of Section 9.2.3, it was ensured in Figure 9.18 that it was capable of maintaining a near-ML BER performance while exhibiting the lowest complexity of the three in both scenarios. When comparing our identical throughput of 16 bits per symbol, the system which employs an antenna arrangement of (4×4) elements and the 16-QAM scheme has a significantly worse BER performance at a commensurately reduced complexity.

9.4.2 Rank-Deficient Systems

In this section, we compare the three types of SDs in terms of their BER performance and computational complexity in the context of rank-deficient 4-QAM SDMA/OFDM systems in conjunction with different antenna arrangements. In Figure 9.19 the BER curves associated with the depth-first SD, the K-best SD and the OHRSA detectors are portrayed, demonstrating that all of them achieve a near-ML performance in the different rank-deficient scenarios considered. However, unlike the other two detectors, the K-best SD does not guarantee an ML performance without an appropriate choice of K. More specifically, setting $K = 32$, which ensures that the K-best SD does exhibit an ML BER performance in an (8×5)-element system, does not necessarily guarantee an ML performance if the rank-deficient system becomes more asymmetrical in terms of having an excessive number of transmitters. For example, for an antenna arrangement of (8×4) elements, we can see this phenomenon in Figure 9.19. In other words, more computational efforts are required to approach the ML performance as the difference between the number of transmit and receive antennas increases. This will become more explicit by considering Figure 9.20.

To expound a little further, Figure 9.20 compares the complexity of these three detectors in both (8×4)-element and (8×7)-element 4-QAM systems. We observe that all of these detectors exhibit a significantly lower complexity in the context of the latter system than in the former one, since, in the latter, the number of receive antennas increases to approach that of the transmit antennas, making the system less rank deficient. In these 4-QAM scenarios, we found that the OHRSA detector has the lowest computational complexity, while the depth-first SD and the K-best SD typically exhibit a similar complexity, although their specific relationship depends on the SNR encountered. An interesting observation from Figure 9.20 is that, instead of decreasing, the complexity of the OHRSA detector increases as the SNR increases in the high-SNR region, i.e. in the SNR range spanning from 12 to

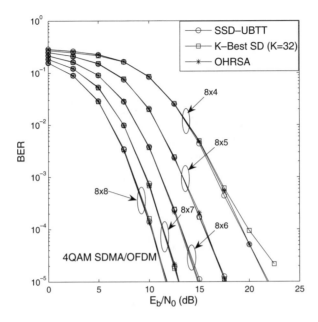

Figure 9.19: BER performance comparison of the depth-first SD, the K-best SD and the OHRSA detectors in rank-deficient systems. All the remaining system parameters were summarized in Table 9.1.

25 dB. The reason behind this phenomenon can be explained as follows. In the heavily loaded system, the interference between the different antenna elements becomes much more significant, while in the high-SNR region, the noise variance becomes low and hence a well-shaped decision lattice is created, which suggests that the OHRSA requires a sufficiently high complexity budget in order to approach the ML solution. Furthermore, owing to the specific search strategy of the OHRSA detector of Section 9.3.3, an erroneous decision is more likely to be made at the higher level of the search tree. Therefore, instead of decreasing, the complexity of the OHRSA detector increases as the SNR increases in the high-SNR region.

9.5 Chapter Conclusions

In this chapter, one of the most promising low-complexity near-ML detectors, namely the SD, has been investigated. Specifically, the derivation of the SD's objective function from the conventional ML metric was performed in Section 9.2.1, followed by a discourse on the SD's tree search process in Sections 9.2.2 and 9.2.3. In particular, depending on whether the tree search was carried out in both the downward and upward directions of Figure 9.4 or solely in the downward direction of Figure 9.5, SDs were classified into two categories, namely the families of depth-first and breadth-first SDs. The search space of the former, which is a hypersphere, initially confined by the ISR C of Figure 9.3, rapidly shrinks upon regularly updating the search radius, as soon as the depth-first tree search reaches a leaf node. In contrast to the former, the breadth-first SD or the so-called K-best SD confined the search space by introducing a parameter K which indicates the number of best candidates retained for each search tree level, rather than employing a search radius C. Hence, it was found in Figures 9.6 and 9.9 of Section 9.2.5 that the complexity imposed by the depth-first SD may vary depending on the received SNR and on the choice of the ISR C, whereas the K-best SD may exhibit a constant complexity, regardless of the received SNR. As for the achievable performance, both types of SDs are capable of attaining the exact ML performance, provided that the ISR C of Figure 9.3 derived for the depth-first SD

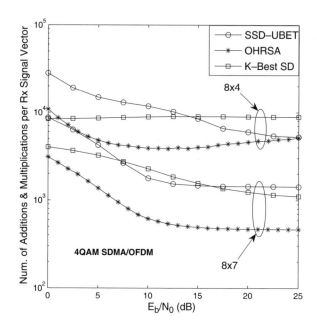

Figure 9.20: The complexity comparison of the depth-first SD, K-best SD and the OHRSA detectors in rank-deficient systems. The complexity curve corresponding to the K-best SD in the scenario of (8×4)-antenna arrangement is obtained by setting $K = 32$, whereas under the antenna arrangement of (8×7) elements, we set $K = 18$, since K is expected to have a larger value as the rank-deficient system becomes more asymmetrical in terms of having an excessive number of transmitters, in order to maintain a near-ML BER performance. All the remaining system parameters were summarized in Table 9.1.

or the parameter K for the breadth-first SD is chosen to be sufficiently high. Additionally, owing to the SNR-independent computational complexity, the K-best SD is more suitable for real-time applications and it may be readily implemented in a pipelined fashion.

In the scenario of rank-deficient MIMO systems, where the number of transmit antennas M is higher than that of receive antennas N, the Grammian matrix $\mathbf{G}$ of Equation (9.22) has $(M - N)$ diagonal elements of zero. Hence, Cholesky factorization of $\mathbf{G}$ cannot be directly applied, so the conventional SD has to be modified in order to apply it in rank-deficient situations, which results in the so-called GSD of Section 9.2.4. Two SD methods have been introduced in Section 9.2.4 to handle the challenging rank-deficient scenarios. Essentially, the first scheme of Section 9.2.4.1 may be regarded as a combination of the standard SD for the first N transmitted symbols and ML detection for the remaining $(M - N)$ symbols, since only the diagonal elements of the last $(M - N)$ rows of the Grammian matrix $\mathbf{G}$ are zero. The other technique of Section 9.2.4.2, dealing with the problem of having $(M - N)$ zero diagonal elements in the matrix $\mathbf{G}$, is to generate a modified Grammian matrix $\tilde{\mathbf{G}}$ which becomes $\mathbf{G} + \alpha \mathbf{I}$, where a judicious choice of the parameter α is required in order to achieve a sufficiently low computational complexity. As detailed in Section 9.2.4.2, the parameter α is chosen to be the noise variance $2\sigma_w^2$. Thus, the GSD commences its search for the ML solution within a search space centred around the MMSE solution, rather than the LS solution of the conventional SD.

The OHRSA detector, which was developed as an extension of the GSD detector of Section 9.2.4, was introduced in Section 9.3.3. It was studied in comparison with the most influential complexity reduction schemes invoked for the SDs, which were detailed in Sections 9.3.1 and 9.3.2. More specifically, the OHRSA invokes exactly the same pre-processing operations as the GSD, which were shown to be capable of dealing with rank-deficient scenarios in Section 9.2.4.2, where the number of transmit antennas is higher than that of receive antennas. Furthermore, a comprehensive discussion on the search techniques used by the OHRSA algorithm was provided in Section 9.3.3 in comparison with the classic SDs. Essentially, both the OHRSA and the SD rely on a hierarchical search structure, and they both rely on identical ML metric equations. On the other hand, although the search strategy of the OHRSA is quite different from that of the conventional SD, their basic philosophy may be deemed as being reminiscent of each other.

Simulation results were provided in Section 9.4 to investigate the achievable performance versus the complexity imposed by the OHRSA detector in comparison with those of several reduced-complexity SDs. It was shown in Figure 9.19 that all these low-complexity near-ML detectors are capable of approaching the ML performance. As for the complexity, the OHRSA detector does not always exhibit a lower complexity than its classic counterparts. For example, it was observed in Figure 9.20 that in a rank-deficient system using 4-QAM the OHRSA detector may indeed impose a significantly lower complexity compared with the conventional SDs. However, it was demonstrated in Figure 9.18 that when 16-QAM or even higher throughput modulation schemes are employed, or when the number of transmit antennas is not higher than that of e receive antennas, the complexity of the OHRSA detector may in fact becomes higher than that of its conventional SD counterparts. On the other hand, recall from Figure 9.18 that the K-best SD, which is assisted by the complexity-reduction techniques of Section 9.3.2, exhibits a modest complexity in comparison with the depth-first SD.

Chapter **10**

Reduced-Complexity Iterative Sphere Detection for Channel-Coded SDMA-OFDM Systems

10.1 Introduction[1]

The radio spectrum is a scarce resource. Therefore, one of the most important objectives in the design of future communications systems is the efficient exploitation of the available spectrum, in order to accommodate the ever-increasing traffic demands. Any effort to achieve bandwidth-efficient transmissions over hostile wireless channels typically requires advanced channel coding. Powerful turbo codes were introduced by Berrou in [565, 566] in the context of iteratively decoding two parallel concatenated convolutional codes. His work was later extended to serially concatenated codes [567] and then found its way gradually into iterative detector designs, such as for example iterative multi-user detectors [568]. Despite their modest complexity, iterative detection and decoding mechanisms are capable of approaching the capacity limits for transmission over wireless MIMO channels.

10.1.1 Iterative Detection and Decoding Fundamentals

10.1.1.1 System Model

Before introducing the channel coding blocks in our MIMO system model, let us briefly review the mathematical model of an SDMA system supporting U users and having N receive antennas at the BS, which is formulated as

$$ \mathbf{y} = \mathbf{Hs} + \mathbf{n}, \tag{10.1} $$

where $\mathbf{y}$, $\mathbf{H}$, $\mathbf{s}$ and $\mathbf{n}$ are the $(N \times 1)$-element received signal column vector, the $(N \times U)$-element FD-CTF matrix, the $(U \times 1)$-element transmitted signal column vector, and the $(N \times 1)$-element AWGN column vector, respectively. Each element s_m of the transmitted signal vector $\mathbf{s}$ can be further represented as $s_u = map\langle x^u \rangle$, $u = 1, 2, \ldots, U$, where $x^{\langle u \rangle}$ is a $(\log_2 M_c \times 1)$ block of raw

---

[1]This chapter is partially based on ©IEEE Wang & Hanzo 2007 [564].

MIMO-OFDM for LTE, Wi-Fi and WiMAX   Lajos Hanzo, Yosef Akhtman, Li Wang and Ming Jiang
© 2011 John Wiley & Sons, Ltd

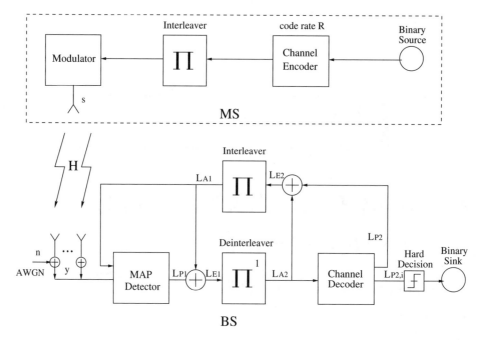

Figure 10.1: Schematic diagram of iterative detection and decoding. ©IEEE Wang & Hanzo 2007 [564]

bits. In other words, each element of the transmitted signal vector **s**, i.e. a constellation symbol, contains $\log_2 M_c$ information bits.

When no channel encoder is employed at the transmitter, the estimates of the transmitted signal **s** can be obtained by the low-complexity near-ML detectors of Chapter 9. Note that all the low-complexity near-ML SDs we encountered in Chapter 9 are HIHO (Hard-In, Hard-Out) detectors.

Owing to the employment of channel coding, the SNR required to achieve a desirable BER may be further reduced. Hence in Figure 10.1 a MIMO system employing a channel encoder and an iterative receiver is portrayed. The interleaver and deinterleaver pair seen at the receiver side of Figure 10.1 divides the receiver into two parts, namely the inner MAP detector and the outer decoder. Note that in Figure 10.1, the subscript '1' denotes variables associated with the inner detector, while the subscript '2' represents variables associated with the outer channel decoder. It was detailed throughout [314] and [569] that the iterative exchange of extrinsic information between these serially concatenated receiver blocks results in substantial performance improvements. In this treatise we assume familiarity with the classic turbo detection principles [314]. Naturally, the inner MIMO detector has to be capable of processing the soft-bit information provided by the soft-output channel decoder. On the other hand, the outer channel decoder also has to be capable of processing the soft reliability information provided by the soft-output inner MIMO detector. The resultant soft-bit information is iteratively exchanged between the inner MIMO detector and the outer channel decoder.

10.1.1.2 MAP Bit Detection

In contrast to the conventional HIHO detector, which outputs hard-symbol decisions, and hence results in hard-bit decisions also at the output of the demodulator, the inner MIMO detector of Figure 10.1 has to be capable of providing soft-bit reliability information for further processing by the outer channel decoder. The advantage of providing soft-bit information is that the channel decoder benefits from exploiting the reliability information provided by the detector and returns to the detector its improved-confidence soft information in the interest of iteratively improving the resultant A Posteriori Probability (APP). Hence, the probability of bit errors is minimized. This SISO scheme may be referred to as an MAP detector. Conventionally, the APP is quantified in terms of the Log-Likelihood Ratio (LLR) as [314]

$$L_D(x_k|\mathbf{y}) = \ln \frac{P[x_k = +1|\mathbf{y}]}{P[x_k = -1|\mathbf{y}]}, \tag{10.2}$$

where $\mathbf{y}$ is the received symbol vector and x_k, $k = 0, 1, \ldots, U \cdot \log_2 M_c - 1$, is the kth element of the corresponding transmitted bit vector $\mathbf{x}$. Since the bits in the vector $\mathbf{x}$ have been channel encoded and scrambled by the interleaver, we may assume that they are statistically independent of each other. With the aid of Bayes' theorem, the LLRs of Equation (10.2) can be rewritten as [553, 570]

$$L_D(x_k|\mathbf{y}) = \ln \frac{p(\mathbf{y}|x_k = +1)P[x_k = +1]/p(\mathbf{y})}{p(\mathbf{y}|x_k = -1)P[x_k = -1]/p(\mathbf{y})} \tag{10.3}$$

$$= \ln \frac{P[x_k = +1]}{P[x_k = -1]} + \ln \frac{p(\mathbf{y}|x_k = +1)}{p(\mathbf{y}|x_k = -1)} \tag{10.4}$$

$$= L_A(x_k) + \underbrace{\ln \frac{\sum_{\mathbf{x} \in \mathbb{X}_{k,+1}} p(\mathbf{y}|\mathbf{x}) \cdot \exp \sum_{j \in \mathbb{J}_{k,x}} L_A(x_j)}{\sum_{\mathbf{x} \in \mathbb{X}_{k,-1}} p(\mathbf{y}|\mathbf{x}) \cdot \exp \sum_{j \in \mathbb{J}_{k,x}} L_A(x_j)}}_{L_E(x_k|\mathbf{y})}, \tag{10.5}$$

where $\mathbb{X}_{k,+1}$ represents the set of $M_c^U/2$ legitimate transmitted bit vectors $\mathbf{x}$ associated with $x_k = +1$, and, similarly, $\mathbb{X}_{k,-1}$ is defined as the set corresponding to $x_k = -1$. Specifically, we have

$$\mathbb{X}_{k,+1} = \{\mathbf{x}|x_k = +1\}, \quad \mathbb{X}_{k,-1} = \{\mathbf{x}|x_k = -1\}. \tag{10.6}$$

Note here that the value of $x_k = -1$ represents a logical value of 0, while $x_k = 1$ represents a logical value of 1. Furthermore $\mathbb{J}_{k,\mathbf{x}}$ is the set of indices j, which is defined as

$$\mathbb{J}_{k,\mathbf{x}} = \{j|j = 0, 1, \ldots, U \cdot \log_2 M_c - 1, j \neq k\}. \tag{10.7}$$

The a priori LLR value L_A defined for the jth bit is given by [314]

$$L_A(x_j) = \ln \frac{P[x_j = +1]}{P[x_j = -1]}. \tag{10.8}$$

According to [553], following a number of manipulations, the a posteriori LLR value can be expressed with the aid of the a priori LLRs as

$$L_D(x_k|\mathbf{y}) = L_A(x_k) + \underbrace{\ln \frac{\sum_{\mathbf{x} \in \mathbb{X}_{k,+1}} p(\mathbf{y}|\mathbf{x}) \cdot \exp(\frac{1}{2}\mathbf{x}_{[k]}^T \cdot \mathbf{L}_{A,[k]})}{\sum_{\mathbf{x} \in \mathbb{X}_{k,-1}} p(\mathbf{y}|\mathbf{x}) \cdot \exp(\frac{1}{2}\mathbf{x}_{[k]}^T \cdot \mathbf{L}_{A,[k]})}}_{L_E(x_k|\mathbf{y})}, \tag{10.9}$$

where the subscript $[k]$ denotes the exclusion of the kth element of a vector. Hence, $\mathbf{x}_{[k]}$ represents a specific subvector of the bit vector $\mathbf{x}$ obtained by omitting the kth component and retaining the rest of them. Similarly, $\mathbf{L}_{A,[k]}$ represents the specific subvector of the a priori LLR vector $\mathbf{L}_A$ obtained by excluding the kth element, where $\mathbf{L}_A$ is the vector containing the a priori LLR value of all the bits in $\mathbf{x}$.

Observe from Equation (10.9) that the a posteriori LLR is equal to the sum of the a priori LLR and the so-called *extrinsic* LLR, which is the second component in the equation. Note that, although the above derivation of the soft reliability information is valid for the bit vector $\mathbf{x}_1$ which is associated with the inner MIMO detector, the subscript '1' is omitted, since Equation (10.9) also holds for the bit vector $\mathbf{x}_2$ associated with the outer channel code. Assuming that an AWGN channel is encountered, the conditional probability of receiving the MIMO output signal $\mathbf{y}$, provided that $\mathbf{x}$ was transmitted, namely $p(\mathbf{y}|\mathbf{x})$, can be computed as

$$p(\mathbf{y}|\mathbf{s} = map(\mathbf{x})) = \frac{\exp[-(1/2\sigma_w^2) \cdot \|\mathbf{y} - \mathbf{Hs}\|^2]}{(2\pi\sigma_w^2)^N}, \tag{10.10}$$

where the denominator is a constant when the noise variance $2\sigma_w^2$ is constant, hence it can be omitted in the calculation of the LLR values. In order to reduce the computational complexity imposed, the *Jacobian logarithm* [314] may be employed to approximate the *extrinsic* LLRs as follows:

$$jac\ln(a_1, a_2) = \ln(e^{a_1} + e^{a_2}), \tag{10.11}$$

$$= \max(a_1, a_2) + \ln(1 + e^{-|a_1 - a_2|}), \tag{10.12}$$

where the second term may be omitted in order to approximate further the original log value, since $\ln(1 + e^{-|a_1 - a_2|})$ can be regarded as a refinement of the coarse approximation provided by the maximum. Consequently, when using the above-mentioned Jacobian approximation, the extrinsic LLR, i.e. the second term of Equation (10.9), can be rewritten as

$$L_e(x_k|\mathbf{y}) = \frac{1}{2} \max_{\mathbf{x} \in \mathbb{X}_{k,+1}} \left\{ -\frac{1}{\sigma_w^2} \|\mathbf{y} - \mathbf{Hs}\|^2 + \mathbf{x}_{[k]}^T \cdot \mathbf{L}_{A,[k]} \right\}$$

$$- \frac{1}{2} \max_{\mathbf{x} \in \mathbb{X}_{k,-1}} \left\{ -\frac{1}{\sigma_w^2} \|\mathbf{y} - \mathbf{Hs}\|^2 + \mathbf{x}_{[k]}^T \cdot \mathbf{L}_{A,[k]} \right\}, \tag{10.13}$$

which represents the information exchanged between the inner MIMO detector and the outer channel decoder, as seen in Figure 10.1.

10.1.2   Chapter Contributions and Outline

Even with the aid of the Jacobian approximation of Equation (10.12), the calculation of the *extrinsic* LLR value using Equation (10.13) may still impose an excessive computational complexity, depending on the number of users U and on the constellation size M_c of the modulation scheme employed, since a brute-force full search has to be carried out by the MAP detector in order to find the joint maximum of the two terms of Equation (10.13). From our discourse on the SD scheme provided in Section 9.2 as well as in light of the corresponding complexity reduction techniques of Section 9.3, we may argue that the HIHO SD constitutes a computationally efficient solution to the ML detection problem in uncoded MIMO systems. For the sake of approaching the channel capacity at a low complexity, the SISO SD algorithm was contrived by Hochwald and ten Brink in [553], where a list of the best hypothesized transmitted MIMO symbol candidates was generated, which was representative of the entire lattice in computing the soft-bit information, resulting in the concept of the List Sphere Decoder (LSD) of Section 10.2.1. However, in order to achieve a good performance, when the LSD is employed in an iterative detection-aided channel-coded system, the list size has to remain sufficiently large, resulting in a potentially excessive complexity. Hence, to reduce further the complexity imposed by the LSD of Section 10.2.1, we propose various solutions to the problem of how to maintain a near-MAP performance with the aid of a small candidate list size. More specifically, the novel contributions of this chapter are as follows:

- Our discovery is that in contrast to the conventional SD, it is plausible to set the search centre of the SD to a point which is typically closer to the real ML solution than the conventional LS or MMSE solution. Commencing the search from a more accurate search centre may be considered as a process of search-complexity reduction.

- A generic centre-shifting SD scheme is proposed for channel-coded iterative receivers based on the above-mentioned perception, which substantially reduces the detection complexity by decomposing it into two stages, namely the iterative search-centre-update phase and the reduced-complexity search around it. Three search-centre-update algorithms are devised in order to shift the search centre iteratively to a point closer to the true ML point with the aid of the soft-bit information delivered by the outer channel decoder.

- We propose a novel complexity-reduction scheme, referred to as the A priori LLR Threshold (ALT) technique, for the LSD, which is also based on exploitation of the soft-bit information, i.e. the a priori LLRs provided by the outer channel decoder in the context of iterative detection-aided channel-coded systems.

- We significantly improve the performance of the conventional two-stage SD-aided turbo receiver by intrinsically amalgamating our proposed centre-shifting-assisted SD with the decoder of a Unity-Rate Code (URC) having an Infinite Impulse Response (IIR), both of which are embedded in a channel-coded SDMA-OFDM transceiver, thus creating a powerful three-stage serially concatenated scheme. Moreover, to achieve a near-capacity performance, Irregular Convolutional Codes (IrCCs) are used as the outer code for the proposed iterative centre-shifting SD-aided three-stage system.

- The convergence characteristics of the proposed schemes are visualized and analysed with the aid of EXIT charts. Furthermore, performance versus complexity comparisons are carried out among the above-mentioned novel schemes.

The remainder of this chapter is organized as follows. The fundamentals of the conventional LSD are briefly reviewed in Section 10.2.1, followed by a discussion on the centre-shifting theory in the context of the SD in Section 10.2.2, which partitions the SD into two parts, i.e. the search-centre-update phase and the search around it. Then, three search-centre-update algorithms are contrived in Section 10.2.3 in order to update the search centre iteratively to a point which is expected to be increasingly closer to the true ML MIMO symbol point. This search-centre update is achieved by exploiting the soft-bit information delivered to the outer channel decoder in the iterative receiver. The ALT-based SD scheme is devised in Section 10.3 in the interest of achieving a complexity reduction, which also relies on exploitation of the soft-bit information gleaned from the outer channel decoder, but in a different manner in comparison with the centre-shifting SD scheme of Section 10.2. In Section 10.4 we demonstrate that the iterative decoding convergence of the conventional two-stage system may be improved by constructing a three-stage system with the aid of the URC encoder/decoder pair of Figure 10.33. Furthermore, IrCCs are employed as the outer code for the proposed iterative centre-shifting SD-aided three-stage system to achieve a near-capacity performance. Finally, we summarize the findings of this chapter and provide our concluding remarks in Section 10.5.

10.2 Channel-Coded Iterative Centre-Shifting SD

10.2.1 Generation of the Candidate List

10.2.1.1 List Generation and Extrinsic LLR Calculation

The inner MIMO detector in Figure 10.1 was chosen to be one of the SDs detailed in Chapter 9, in order to approach the MAP performance while avoiding a potentially excessive computational complexity, which is likely to be encountered by the employment of the conventional MAP detector. However, when calculating the soft information generated by the HIHO SD of Section 9.2, finding the ML solution of $\hat{s}_{ML} = \arg\min_{\hat{s} \in M_c^U} \|y - Hs\|^2$ does not necessarily solve the problem of maximizing the two terms in Equation (10.13), because here the search for $s_{ML} = \arg\min_{\hat{s} \in M_c^U} \|y - Hs\|^2$ in each term is carried out in the bit domain having $x_k = 1$ or $x_k = -1$, rather than in the original MIMO symbol domain in the scenario of HIHO SD. Therefore, conventional SDs cannot be directly employed in the iterative detection scheme shown in Figure 10.1, because the ML solution s_{ML} provides us with a single hard-decision-based MIMO symbol value, rather than the required bit-based soft information. Fortunately, based on the idea that although the MIMO bit vector for maximizing the two terms in Equation (10.13) is not necessarily the ML MIMO symbol solution s_{ML}, the bit vector is typically located near the ML MIMO symbol solution $\mathbf{s}_{ML}$. Hence, finding the MIMO bit vector which maximizes the two terms of Equation (10.13) does not require a full search of the entire lattice. Similar to the conventional SD, the search can be carried out in a significantly smaller hypersphere containing the ML solution $\mathbf{s}_{ML}$, but instead of simply finding the ML solution, the SD has to output a list $\mathcal{L}$ which contains the ML solution as well as its neighbours, which might constitute the MIMO bit vector maximizing the two terms of Equation (10.13) with a high probability. Finally, by doing the subtraction between the two obtained values of the OFs corresponding to the two terms of Equation (10.13), we can get the *extrinsic* LLR required.

Based on the above discussions, simple modifications of the conventional depth-first SD of Section 9.2.2 may be carried out by appropriately modifying: (1) the search radius update strategy; and (2) the output stack for storing the aforementioned list $\mathcal{L}$. As for the search radius, it has to be constant all the time during the search regardless of whether a new signal point was found. However, this does not mean that there is no need to calculate the Euclidean distance between the newly obtained signal point and the received signal point, because their distance is used as the metric controlling the update of the output stack. Again, the output stack was introduced for storing the aforementioned list $\mathcal{L}$. Let us assume that the size of $\mathcal{L}$ is preset to be $\mathcal{N}_{cand}$. When a new signal point is found inside the sphere, two possible actions may be taken: (1) the newly obtained signal point is added directly to the output stack $\mathcal{L}$, provided that it is not full; (2) if the stack is already full, the new signal point is compared with the element having the largest distance from the received signal point, and replaces it if the new signal point has a smaller distance. Consequently, the resultant list $\mathcal{L}$ contains the ML solution as well as $(\mathcal{N}_{cand} - 1)$ candidates which are close to the former. According to [571], during generation of the candidate list $\mathcal{L}$, the search radius can only be reduced to the value of the maximum distance metric found in the list $\mathcal{L}$, if the output stack is full. Based on this intuition, if there are more signal vectors having $x_k = 1$, the resultant soft reliability information indicates with a high probability that the kth bit is a logical one. On the other hand, if there are more signal vectors having $x_k = -1$, a reasonable decision can be made implying that the kth bit is a logical zero. Hence, we can finally rewrite Equation (10.13) for the LSD as

$$
\begin{aligned}
L_e(x_k|\mathbf{y}) \approx &\frac{1}{2} \max_{\mathbf{x} \in \mathcal{L} \cap \mathbb{X}_{k,+1}} \left\{ -\frac{1}{\sigma_w^2} \|\mathbf{y} - \mathbf{Hs}\|^2 + \mathbf{x}_{[k]}^T \cdot \mathbf{L}_{A,[k]} \right\} \\
&- \frac{1}{2} \max_{\mathbf{x} \in \mathcal{L} \cap \mathbb{X}_{k,-1}} \left\{ -\frac{1}{\sigma_w^2} \|\mathbf{y} - \mathbf{Hs}\|^2 + \mathbf{x}_{[k]}^T \cdot \mathbf{L}_{A,[k]} \right\}.
\end{aligned}
\tag{10.14}
$$

The above approximation becomes an equality when the output stack $\mathcal{L}$ contains the entire lattice, i.e. $\mathcal{N}_{cand} = M_c^U$. However, as mentioned earlier, the maximizer of both terms of Equation (10.14) is located near the ML solution, hence the size of the list $\mathcal{L}$ required to achieve a desired performance is typically far smaller than M_c^U.

As for the application of the K-best SD of Section 9.2.3 in our channel-coded system, the list generation is more straightforward than for its depth-first counterpart discussed previously in this section. Specifically, instead of generating a single signal vector after the breadth-first tree search, which is expected to be the near-ML solution, the K-best SD retains $\mathcal{N}_{cand}$ best tree leaf candidates having the lowest accumulated Euclidean distances from the received signal point $\mathbf{y}$. Eventually, after backtracking from these tree leaves, $\mathcal{N}_{cand}$ signal vectors can be generated, constituting the list $\mathcal{L}$.

10.2.1.2 Computational Complexity of LSDs

Let us now quantify the computational complexity of both the soft-output LSD and the exact MAP detectors in terms of the number of OF evaluations, which corresponds to the two terms in Equation (10.14). As mentioned previously, the approximation in Equation (10.14) becomes an equality when $\mathcal{L}$ represents the entire search space, constituted by $\mathcal{N}_{cand} = M_c^U = 2^{U \cdot BPS}$ OF evaluations, where U represents the number of users in SDMA systems, M_c is the size of the specific constellation and BPS is the number of bits per symbol. Therefore, the complexity of the exact MAP detector can be calculated as the total number of OF evaluations given by

$$\mathcal{C}_{MAP} = U \cdot BPS \cdot 2^{(U \cdot BPS)}. \tag{10.15}$$

Clearly, the complexity grows exponentially with the product of the number of users U and the number of bits per symbol BPS. Let us consider an eight-user 4-QAM SDMA system as an example. It corresponds to a complexity of $\mathcal{C}_{MAP} = 1\,048\,576$ OF evaluations, which is excessive. If a 16-QAM scheme is employed, the complexity is increased to 1.3744×10^{11} OF evaluations, which is implementationally infeasible.

As for the computational complexity imposed by the LSD of Section 10.2.1.1, it may be significantly reduced by generating a list of candidates having a length of $\mathcal{N}_{cand}$, where $2^{U \cdot BPS} \geq \mathcal{N}_{cand} \geq 1$, since the corresponding complexity can be expressed as

$$\mathcal{C}_{MAP} = U \cdot BPS \cdot \mathcal{N}_{cand}. \tag{10.16}$$

Consequently, the complexity has become linearly proportional to the length of the list $\mathcal{L}$. In the following sections, we can observe that the value of $\mathcal{N}_{cand}$ can be set to a small fraction of $2^{U \cdot BPS}$, especially when a high-throughput modulation scheme, e.g. 64-QAM, is employed and/or a high number of users are supported by the system.

10.2.1.3 Simulation Results and 2D EXIT-Chart Analysis

Our forthcoming EXIT-chart analysis and Monte Carlo simulations, if not stated otherwise, will be carried out in the scenario of (8×4)-element rank-deficient 4-QAM SDMA-OFDM systems, under the simplifying assumptions that perfect channel estimation is available at the BS and that the channel is time invariant. Note that the power delay profile of the three-path frequency-selective channel considered is given by $P(\tau) = \sum_{k=0}^{2} P(\tau_k)\delta(t - k\tau)$, where τ is the delay spread and $P(\tau_k) = [0.5\ 0.3\ 0.2]$ for $k = 0, 1, 2$. We employ a constraint-length $K_c = 3$, half-rate Recursive Systematic Convolutional (RSC) code RSC(2,1,3) having the octally represented generator polynomials of (6/13). The length of the interleaver between the channel encoder and the modulator/mapper is $10\,240$ bits. It is reasonable to set the length of the list to be the same as the parameter K of the K-best SD, which represents the maximum number of candidates to be retained at each search tree level. Our system parameters are summarized in Table 10.1.

Figure 10.2 depicts the EXIT functions of both the K-best LSD and the outer convolutional decoder. Observe in Figure 10.2 that the EXIT curve corresponding to the SD, which we refer to as the inner decoder, intersects that of the outer decoder before reaching the convergence point of $[I_A(MUD) = 1, I_E(MUD) = 1]$. Therefore, regardless of the number of iterations invoked and the length of the interleaver, residual errors may persist at this specific SNR $= 8$ dB. More importantly, as seen in Figure 10.2, the shape of the EXIT curve of the inner decoder depends significantly on the size of the list $\mathcal{N}_{cand}$ employed, which is equal to K in all forthcoming simulations. Specifically, having a longer list leads to a steeper and hence more beneficial slope of the EXIT curve. In other words, the EXIT curves of the inner decoder and the outer decoder will intersect at a higher $[I_A, I_E]$ value, when the list is extended. The phenomenon where the inner decoder's EXIT curve may even decay as the a priori information fed back by the outer decoder increases can be explained by the fact that the inner and outer decoders exchange flawed information owing to a shortage of candidate solutions, more particularly the absence of the ML solution in the candidate list, which is not long enough. Consequently, the maximum achievable iteration gain may be significantly reduced when employing a very small list, although, as expected, the overall computational complexity imposed by the soft-bit-information calculation is substantially reduced. Furthermore, we can infer from Figure 10.2 that the BER performances corresponding to different list sizes do not dramatically differ from each other at low SNRs, when the open tunnel between the EXIT curves of the inner and outer decoders closes at low $[I_A, I_E]$ values. This is because all inner EXIT curves corresponding to different list sizes have similar

Table 10.1: Summary of system parameters for the K-best SD-aided coded SDMA-OFDM system.

System parameters	Choice
System	SDMA-OFDM
Number of subcarriers	128
Uplink/downlink	Uplink
Modulation	4-QAM
Number of users/transmit antennas	8
Number of receive antennas	4
Transmit antennas per user	1
Block length	10 240 bits
CIR model	$P(\tau_k) = [0.5\ 0.3\ 0.2]$, for $k = 0,\ 1,\ 2$
CIR tap fading	OFDM symbol invariant
Channel estimation	Ideal
Detector/MAP	K-best LSD
List length $\mathcal{N}_{cand}$	$= K$
Channel encoder	RSC(2,1,3)
	Generator polynomials (6/13)
	Code termination (Off)
No. of iterations (Variable)	Iterations terminate as soon as
	the resultant trajectory line
	reaches the convergence point

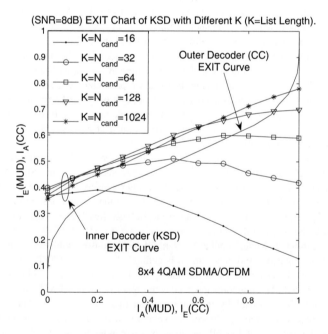

Figure 10.2: The 2D EXIT chart of the K-best SD using different list lengths in the scenario of an (8×4)-antenna 4-QAM SDMA-OFDM system at SNR $= 8$ dB. All other system parameters are listed in Table 10.1.

$[I_A, I_E]$ starting points for a given SNR. On the other hand, a higher iteration gain can be achieved by a longer list at high SNRs. These inferences can be verified by the BER results depicted in Figure 10.3.

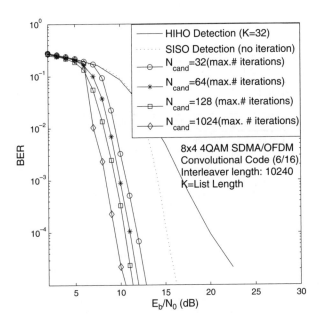

Figure 10.3: The achievable BER performance of the conventional K-best SD-aided iterative detection in the scenario of an (8×4)-antenna 4-QAM SDMA-OFDM system with different K. In all cases, the maximum iteration gain has been achieved.

Figure 10.3 compares the achievable BER performances of the K-best LSD-aided iterative detector having different list sizes in the scenario of the (8×4) rank-deficient SDMA-OFDM system. It can be seen that, compared with the uncoded system, a significant performance gain is achieved by employing the channel encoder/decoder. Moreover, the attainable performance can be further improved by invoking the iterative detection scheme of Figure 10.1 which exchanges soft information between the inner decoder, i.e. the soft-output K-best SD and the convolutional decoder. The difference between the attainable iteration gains exhibited by the inner decoder using different list sizes remains insignificant until the SNR increases to about $5\,dB$, which is also the convergence threshold of the inner decoder having the list length of $K = 128$. The convergence threshold associated with the list length of $K = 32$, on the other hand, is about $7\,dB$. In other words, useful iteration gain can only be observed for relatively high SNRs, provided that a sufficiently high list length is employed. Hence, the BER performance suffers from having an insufficiently long list size. On the other hand, the computational complexity imposed and the memory required by the LSD may be substantially reduced with the aid of iterative detection, as quantified in Table 10.2.

More explicitly, Table 10.2 shows the trade-off between the SNR required and the computational complexity imposed by the K-best LSD/MAP detector at the target BER of 10^{-5}. Note that we quantify the computational complexity of the list generation in the K-best LSD in terms of the total number of PED evaluations according to Equation (9.32) in Section 9.2.2, whereas we calculate the complexity of the soft-information generation at the output of the K-best LSD/MAP detector using Equation (10.16) in terms of the total number of OF evaluations corresponding to the two terms of Equation (10.14).

Therefore, we can observe from both Figure 10.3 and Table 10.2 that in order to achieve a near-MAP BER performance, we have to set both K and the list size $\mathcal{N}_{cand}$ to at least 1024. In other words, for a given target BER of 10^{-5}, to achieve a performance gain of $3\,dB$ over the system where both K and $\mathcal{N}_{cand}$ are set to 32 in the scenario of an (8×4)-element overloaded 4-QAM SDMA-OFDM system, substantial computational and memory investments have to be made, which require

Table 10.2: Simulation results of the conventional K-best LSD-aided iterative detection in the scenario of an (8×4)-element 4-QAM rank-deficient SDMA-OFDM system as depicted in Figure 10.1. Note that the computational complexity of the list generation by the LSD is calculated in terms of the total number of PED evaluations, while that of the soft-information generation by the LSD/MAP detector is computed using Equation (10.16) in terms of the total number of OF evaluations corresponding to the two terms in Equation (10.14).

	(8×4) 4-QAM SDMA-OFDM rank-deficient system				
BER	List $(= K)$	Memory	SNR(dB)	SD compl.	MAP compl.
10^{-5}	32	256	14	724	1 024 (2 iter.)
	64	512	13.2	1 364	2 048 (2 iter.)
	128	1024	11.2	2 388	4 096 (2 iter.)
	1024	8196	10.5	13 652	32 768 (2 iter.)

nearly 19 times more PED evaluations per channel use for the candidate list generation, 32 times more OF evaluations per channel use for the LLR calculation and 32 times more memory requirements per channel use. Although the computational complexity imposed is only a small fraction of that required by the EXACT MAP detector (which requires, for example, more than 10^6 OF evaluations for the LLR calculation in this particular scenario), it is still substantially higher than desirable, especially in heavily rank-deficient systems.

10.2.2   Centre-Shifting Theory for SDs

Recall from Section 9.2 that the philosophy of various types of SD is that of finding the ML solution which minimizes the ML error term of Equation (9.16), which is then transformed into the problem of finding the specific MIMO symbol which minimizes the first term of Equation (9.18) or the first term of Equation (9.38). More explicitly, according to Equation (9.20) we have

$$\hat{\mathbf{s}}_{ML} = \arg\min_{\check{\mathbf{s}} \in M_c^U} (\check{\mathbf{s}} - \hat{\mathbf{c}})^H \mathbf{H}^H \mathbf{H} (\check{\mathbf{s}} - \hat{\mathbf{c}}), \qquad (10.17)$$

where $\hat{\mathbf{c}} = (\mathbf{H}^H \mathbf{H})^{-1} \mathbf{H}^H \mathbf{y}$ is the unconstrained ML estimate of $\mathbf{s}$, i.e. the LS solution. In addition, according to Equation (9.41) we get

$$\hat{\mathbf{s}}_{ML} = \arg\min_{\check{\mathbf{s}} \in M_c^U} (\check{\mathbf{s}} - \hat{\mathbf{c}})^H (\mathbf{H}^H \mathbf{H} + 2\sigma_w^2 \mathbf{I}) (\check{\mathbf{s}} - \hat{\mathbf{c}}), \qquad (10.18)$$

where $2\sigma_w^2$ represents the noise variance and hence $\hat{\mathbf{c}} = (\mathbf{H}^H \mathbf{H} + 2\sigma_w^2 \mathbf{I}) \mathbf{H}^H \mathbf{y}$ corresponds to the MMSE solution.

Therefore, when using SD, the ML solution can be found by creating a reduced-size search hypersphere centred around the LS solution or the MMSE solution and then reducing the search radius when possible. During our investigations of SD, we suggested the plausible idea of setting the search centre to a MIMO signal constellation point, which is typically closer to the real ML solution than the conventional LS or MMSE solution. To some extent, extending the search from a more accurate search centre can be considered as a process of search-complexity reduction. In fact, the computational complexity reduction achieved by the MMSE-based centre over the LS-aided one was quantified in Figure 9.15 in Section 9.3.2. Hence, it is plausible that the closer the search centre to the real ML solution, the lower the computational complexity, as has been verified by all of our simulations in the context of the SD-aided uncoded SDMA-OFDM systems considered.

Consequently, the SD can be split into two independent functional blocks, namely the centre calculation or centre update block and the SD's hypersphere search block, as shown in Figure 10.4.

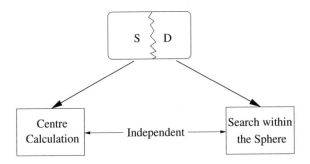

Figure 10.4: Independent SD blocks: the search-centre calculation and the search.

Table 10.3: Performance versus complexity characterization of self-iterative K-best SD in full-rank 16-QAM systems.

(4×4) 16-QAM SDMA-OFDM full-rank system				
SNR (dB)	K	No. of iterations	BER	Complexity
20	16	None	1.481×10^{-5}	2048
	8	1	0.0188	2048
	4	2	0.0069	8192
	2	4	0.0063	16 384
	1	16	0.0001	1230

Table 10.4: Performance versus complexity characterization of iterative K-best SD in rank-deficient systems.

(8×4) 4-QAM SDMA-OFDM rank-deficient system				
SNR (dB)	K	No. of iterations	BER	Complexity
16	8	1 (upper)	0.150 78	2048
		4 (upper)	0.018 8	2048
		4 (upper), 3 (both)	0.006 9	8192
	8	6 (upper)	0.006 3	16 384

Hence, the search can be carried out independently of the search-centre calculation. Thus, the search centre can be obtained by arbitrary detection schemes, not just by the conventional LS or MMSE detection scheme. This observation turns the SD into a high-flexibility detector, which can be readily combined with other well-established linear or nonlinear detectors. As a result, the total computational complexity imposed by the SD consists of that of the detector which provides the search centre for consecutive search operation of Figure 10.4. In other words, the affordable computational complexity can be flexibly split between the centre calculation phase and the search phase of Figure 10.4. The simple schematic of Figure 10.4 is further detailed in Figure 10.5, where the triangularization of the channel matrix $\mathbf{H}$ and the PED calculation previously detailed in Section 9.2 are portrayed more explicitly. It is also plausible that an improved performance versus complexity trade-off emerges as the search-centre calculation is regularly updated, before further triangularization and PED calculation are carried out as seen in Figure 10.5.

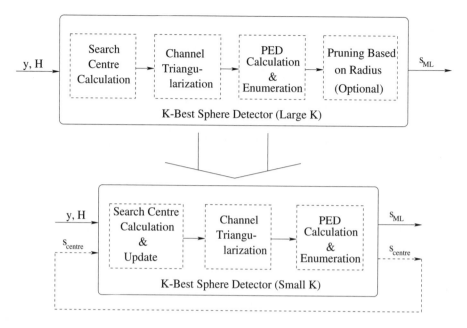

Figure 10.5: The structure of the iterative K-best SD using the centre-shifting scheme. The search centre of the SD can be set to a more accurate centre than the original LS or MMSE solution to reduce the computational complexity required by each iteration, while maintaining the performance. A benefit of this is that the memory requirements can be largely reduced, although the overall computational complexity remains almost the same as proved by our simulations.

10.2.3 Centre-Shifting K-Best SD-Aided Iterative Receiver Architectures

The novel idea of centre shifting, which was proposed in the context of an uncoded system in Section 10.2.2, has the benefit of less memory requirements imposed by the K-best SD, since K can be set to a small value. However, the overall computational complexity reduction may still remain modest if the iterative scheme shown in Figure 10.5 is employed, since a fraction of the original computational complexity imposed by the search process is in fact transferred to the centre calculation phase. Hence, the overall computational complexity may remain similar to that of the non-iterative SD.

In order to provide further insights, the performance versus complexity characterization of the K-best SD is provided in Tables 10.3 and 10.4 for both a full-rank and a rank-deficient system, respectively.

On the other hand, the centre-shifting scheme applied for the K-best SD is expected to become significantly more powerful, if it is employed in the scenario of the iterative detection-aided channel-coded system of Figure 10.6, since the process of obtaining a more accurate search centre is further aided by the channel decoder, which substantially contributes towards the total error-correction capability of the iterative receiver. Beneficially, no additional computational complexity is imposed by calculating the search centre based on the output of the channel decoder. Note that although the SD process is repeated according to the number of iterations, the overall computational complexity imposed by the iterative receiver may be substantially reduced while maintaining a high BER performance, since K and N_{cand} can be set to substantially lower values when combined with the centre-shifting scheme than that required without it.

In our forthcoming discourse on the centre-shifting K-best SD-aided iterative receiver, first of all we propose three different receiver architectures employing different centre-calculation schemes. Then we will opt for using the best of the three centre-calculation schemes in a Unity-Rate Code (URC)

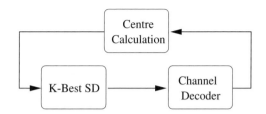

Figure 10.6: Centre-shifting-aided K-best SD in coded system.

assisted three-stage iterative receiver in Section 10.4. More explicitly, the schematic of Figure 10.1 is extended in Figure 10.33 of Section 10.4 with a URC decoder. Accordingly, the receiver incorporates the URC's decoder, as seen in Figure 10.33. *During our EXIT-chart-assisted receiver design, our aim is to construct a low-complexity near-MAP detector which is capable of supporting high-throughput modulation schemes operating in heavily rank-deficient systems.*

10.2.3.1 Direct Hard-Decision Centre-Update-Based Two-Stage Iterative Architecture

10.2.3.1.1 Receiver Architecture and EXIT-Chart-Aided Analysis

Our first proposed centre-calculation scheme is the Direct Hard-Decision Centre-Shifting (DHDC) scheme portrayed in Figure 10.7, which calculates the search centre for the forthcoming detection iteration by imposing hard decisions on the interleaved a posteriori LLRs at the output of the channel decoder. Then it remodulates the resultant bit streams of all the SDMA users, in order to generate the mapped symbol matrix, where each column corresponds to the most recently obtained search centre.

The main purpose of invoking the centre-shifting scheme for the K-best SD in the context of the iterative detection-aided system of Figure 10.1 is to increase the maximum attainable iterative gain while maintaining an affordable complexity. The list size $\mathcal{N}_{cand}$ is equal to the number of tentative MIMO symbol candidates retained at each tree search level, which is set to the lowest possible value in order to reduce the computational complexity imposed. Naturally, additional computational efforts are imposed by the SD based on the updated search centre since after a certain number of iterations the candidate list used for the LLR calculation is regenerated. But again, as a whole, the total memory requirements of the K-best SD and the overall computational complexity imposed by the list generation and the LLR calculation of the K-best SD is expected to be substantially reduced with the aid of the iterative information exchange between the centre-shifting scheme and the channel decoder. In order to investigate the benefits of invoking the centre-shifting scheme, EXIT charts are used to analyse the modified SD block, which has two inputs and one output, as shown in Figure 10.8. The centre-calculation phase portrayed in Figure 10.4 is transplanted into the SD block of Figure 10.8. The two inputs seen in Figure 10.8 are the a priori LLRs and the interleaved a posteriori LLRs provided by the channel decoder, whereas the output is the resultant *extrinsic* LLR. As a consequence, we have to employ the 3D EXIT chart first and then project it to two dimensions, in order to obtain the 2D EXIT chart of the iterative receiver, as will be detailed in the context of Figure 10.9.

More explicitly, Figure 10.9 depicts the 3D EXIT chart of the DHDC-aided K-best SD iterative receiver, where $K = 256$ is used in our (4×4)-element 4-QAM SDMA-OFDM system. Since the total number of MIMO symbols is $4^4 = 256$, the SD is actually the exact MAP detector, which computes the LLRs by conducting the totally $M \cdot BPS \cdot \mathcal{N}_{cand} = 4 \cdot 2 \cdot 256 = 2048$ OF evaluations, which correspond to the evaluations of the two terms in Equation (10.14). We evaluate the *extrinsic* Mutual Information (MI), I_E, at the output of the SD, which is quantified on the vertical axis of Figure 10.9(a), after providing the SD with the two inputs required, which correspond to the a priori LLRs and the a posteriori LLRs gleaned from the channel decoder, respectively. The MI associated with the two inputs, namely I_A and I_D, is quantified on the two abscissas, namely on the x-axis and

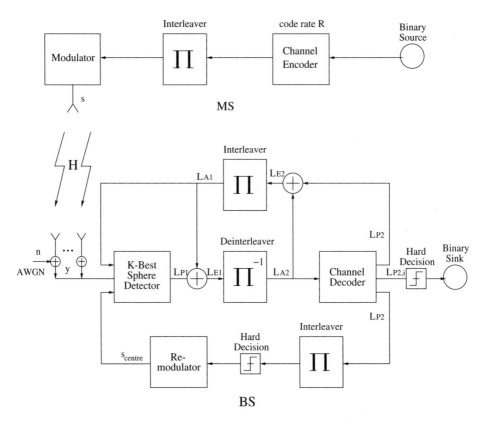

Figure 10.7: DHDC-assisted K-best SD-aided iterative detection scheme.

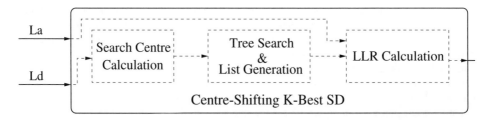

Figure 10.8: Structure of the centre-shifting K-best SD: a two-input, one-output block.

y-axis, respectively. The two parallel EXIT planes of the inner decoder, i.e. the K-best SD, recorded for different SNRs in Figure 10.9(a), indicate that the *extrinsic* output I_E is actually independent of the input I_D, implying that the DHDC scheme is unable to glean any benefits for the exact MAP detector. This is not unexpected, since the inner EXIT curve seen in Figure 10.9(b), which was obtained by projecting on the 3D EXIT chart of Figure 10.9(a) to the 2D EXIT chart on the plane given by the two axes, which quantify the *extrinsic* MI and the a priori MI at the output and input of the MUD, i.e. the SD, respectively, does not encounter the problem of going down as the input MI I_A increases as shown in Figure 10.2 of Section 10.2.1.3. Thus, it depends only on the SNR of the received signal and the a priori LLRs. Therefore, we can infer that when the list size is sufficiently high, the EXIT curve

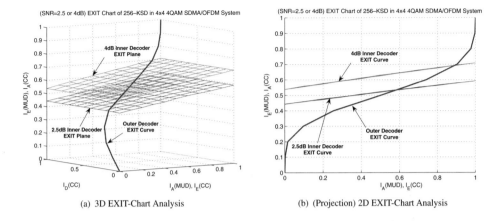

(a) 3D EXIT-Chart Analysis   (b) (Projection) 2D EXIT-Chart Analysis

Figure 10.9: EXIT-chart analysis of the DHDC-based K-best SD-aided iterative receiver. Since K is set to 256 in a (4×4) 4-QAM system, the K-best SD is in fact the exact ML detector. All other system parameters are listed in Table 10.1.

of the inner decoder is proportional to the input a priori LLRs, whereas the DHDC scheme provides hardly any performance improvement, since it fails to provide an iterative gain.

However, if we reduce the list size $\mathcal{N}_{cand}$ of the SD to a relatively small value, the output *extrinsic* LLRs are no longer independent of the input I_D LLRs, based on which the iterative detection-aided DHDC scheme of Figure 10.7 updates the search centre of the K-best SD as seen in Figure 10.10, where the inner and outer EXIT surfaces are depicted for $K = \mathcal{N}_{cand} = 32$ and $K = \mathcal{N}_{cand} = 128$, respectively. More specifically, we can observe from Figure 10.10 that for a given input a priori MI I_A, the output *extrinsic* MI I_E is proportional to the input a posteriori MI I_D. On the other hand, the relationship between the output I_E and the input I_A represented by the corresponding LLR in Figure 10.8 is quite different when the other input I_D is fixed. More explicitly, when I_D is not high enough, because the search centre is insufficiently accurate, the output I_E seen in Figure 10.10 may not be proportional to the input I_A. However, when the I_D is sufficient high, since the SD carries out detection using an accurate centre, which is close to the ML solution, the output I_E is expected to increase proportionally as the input I_A approaches unity, even for a small value of $\mathcal{N}_{cand}$. In fact, the observations based on Figure 10.10 coincide with the simulation results shown in Figure 10.2, where the EXIT curve of the K-best SD starts to decrease, despite having an increasing value of I_A when K and $\mathcal{N}_{cand}$ are insufficiently high. In this scenario, having small K and $\mathcal{N}_{cand}$ values may yield a candidate list which may not contain the ML solution with a high probability, which in turn leads to flawed information being exchanged between the inner decoder and the outer decoder during iterative detection. Consequently, instead of increasing the iterative gain, using more iterations results in a reduced output I_E value for the K-best SD. However, as a benefit of the centre-shifting scheme, we can improve the resultant candidate list without increasing K and $\mathcal{N}_{cand}$, by simply updating the search centre to a more accurate one. Hence, the quality of the output soft-bit information, i.e. L_E, is improved without increasing the list size $\mathcal{N}_{cand}$ or K. Finally, based on the above discussions, we summarize the aforementioned relationships in Figure 10.11.

Note that the two inputs of the K-best SD shown in Figure 10.8, namely the a priori LLRs and the a posteriori LLRs, are not perfectly independent. By contrast, the 3D EXIT chart of Figure 10.10 was obtained by providing the K-best SD with two perfectly independent Gaussian-distributed LLRs. Hence it is anticipated that the actual decoding trajectory will deviate from the EXIT-chart-based performance predictions. As a result, the iterative detector may not be able to achieve an infinitesimally low BER at

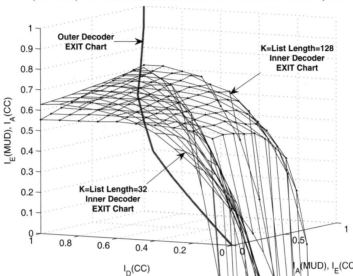

Figure 10.10: The 3D EXIT-chart analysis of the DHDC-aided K-best SD using $K = 32$ or 128 in the (4×4)-element 4-QAM SDMA-OFDM system operating at $SNR = 4\,dB$. All other system parameters are listed in Table 10.1.

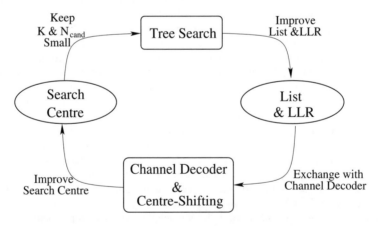

Figure 10.11: Relationship between the search centre and the resultant *extrinsic* LLRs when the centre-shifting scheme is invoked and K and $\mathcal{N}_{cand}$ are fixed to a relatively small value.

the same channel SNR where the EXIT-chart analysis succeeded in creating a narrow, but marginally open tunnel.

10.2.3.1.2 Simulation Results

Let us now characterize the achievable performance of the DHDC-aided K-best SD iterative receiver in the scenario of an (8×4)-element rank-deficient 4-QAM SDMA-OFDM system. Since the values of K and $\mathcal{N}_{cand}$ were set relatively low, we know from the EXIT chart of Figure 10.2 and the BER curve

Table 10.5: Summary of system parameters for the K-best SD-aided coded SDMA-OFDM System.

System parameters	Choice
System	SDMA-OFDM
Number of subcarriers	128
Uplink/downlink	Uplink
Modulation	4-QAM
Number of users/transmit antennas	8
Number of receive antennas	4
Transmit antennas per user	1
Block length	10 240
CIR model	Three-path frequency-selective channel
CIR tap fading	OFDM symbol invariant
Channel estimation	Ideal
Detector/MAP	Centre-shifting-aided K-best LSD
List length $\mathcal{N}_{cand}$	$= K = 128$
Channel encoder	RSC(2,1,3)
	Generator polynomials (6/13)
	Code termination (Off)
Iteration mode	Once no more iterative gain can
	be achieved by the conventional iterative
	receiver, the centre-shifting function is switched on

of Figure 10.3 that the conventional K-best SD iterative receiver dispensing with centre shifting suffers from a performance degradation compared with the more complex system using $K = \mathcal{N}_{cand} = 1024$. Therefore, it is beneficial to switch off the DHDC scheme of Section 10.2.3.1 during the first few iterations. However, when the maximum attainable iterative gain is achieved with the DHDC scheme being switched off, the DHDC scheme is activated again in order to update the search centre of the K-best SD. This centre-update action may be expected to create a wider EXIT tunnel between the EXIT curves of the inner and outer decoder, potentially facilitating an easier passage of the decoding trajectory through the tunnel. Our system parameters are summarized in Table 10.5.

Figure 10.12 reveals the BER performance improvement brought about by the DHDC-aided K-best SD iterative receiver over that of the conventional SD iterative receiver using no centre shifting in the scenario of an (8×4)-element rank-deficient SDMA-OFDM system. The BER curves of the iterative receiver correspond to a variable number of iterations, which were enabled to iterate until perceivable iterative gains were achieved. Specifically, within the SNR range of 6–13 dB, where useful performance improvements can be observed, a maximum performance gain of 2 dB can be achieved by the DHDC scheme over the system using no centre shifting, if we fix the values of K and $\mathcal{N}_{cand}$ to 64. By contrast, a slightly lower performance gain of about 1.5 dB can be attained if K and $\mathcal{N}_{cand}$ are set to 32, and hence the complexity is reduced by about a factor of four. It is worth emphasizing that the DHDC-aided system associated with $K = \mathcal{N}_{cand} = 64$ is capable of achieving a near-MAP performance, which can only be attained by setting K and $\mathcal{N}_{cand}$ to at least 1024 for the system using non-centre shifting in the heavily rank-deficient scenario considered. Hence, the memory required by the K-best SD was significantly reduced.

More importantly, the associated computational complexity is also expected to be substantially reduced if we consider $SNR = 8$ dB in Figure 10.13, where the corresponding EXIT chart is portrayed. Figure 10.13(a) depicts the 3D EXIT chart of the DHDC-aided K-best SD iterative receiver in the scenario of an (8×4)-element rank-deficient system. Since the function of the DHDC centre-shifting scheme is only switched on when the maximum iterative gain of the scheme using no centre shifting is achieved, i.e. when the resultant trajectory reaches the crossing point of the EXIT curves of the inner

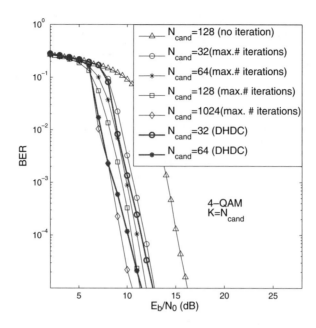

Figure 10.12: BER performance improvement brought about by the DHDC scheme in the context of an (8×4)-element rank-deficient SDMA-OFDM system: significant BER performance improvement can be achieved by the employment of the DHDC scheme. All other system parameters are listed in Table 10.1.

and the outer decoder, the staircase-shaped decoding trajectory follows exactly the same path as with the DHDC scheme disabled, until it reaches the intersection. Then, with the aid of the increasingly accurate search centre provided by the DHDC centre-shifting scheme, the decoding trajectory continues to evolve through the tunnel of Figure 10.13(a) between the 3D EXIT surface of the inner decoder and the EXIT curve of the outer decoder, both of which are obtained by considering the a posteriori LLR values. The resultant additional iterative gain brought about by the DHDC scheme may be more explicitly observed if we refer to the projection of the 3D EXIT chart depicted in Figure 10.13(b). As we can see, the maximum MI measured at the output of the channel decoder of the iterative receiver using no centre shifting is about $I_E = 0.85$ after four iterations exchanging *extrinsic* information between the inner and outer decoder. By contrast, the maximum achievable MI approaches about $I_E = 0.95$ with the aid of the DHDC scheme when activating three additional iterations, thus resulting in a further reduced BER.

10.2.3.2 Two-Stage Iterative Architecture Using a Direct Soft-Decision Centre Update

In Section 10.2.3.1 we updated the search centre of the SD by making hard decisions at the output of the channel decoder when generating the a posteriori LLRs. Given this simple centre-update strategy, the centre-shifting scheme was capable of achieving evident performance gains as we can see in Figure 10.12 of Section 10.2.3.1.2. However, the attainable performance improvements are expected to be increased by exploiting the slightly more sophisticated centre-calculation technique of theDirect Soft-Decision Centre-Shifting (DSDC) scheme, to be introduced in our forthcoming discourse. These further improvements are expected, because the action of subjecting the LLRs to hard decisions discards the useful soft information contained in the LLRs, which indicates how reliable our estimate of the most recently obtained centre is. Consequently, in the DSDC scheme we calculate the soft LLRs of the symbols based on the interleaved soft-bit information. Then the SD carries out the detection again with this newly obtained search centre during the next iteration.

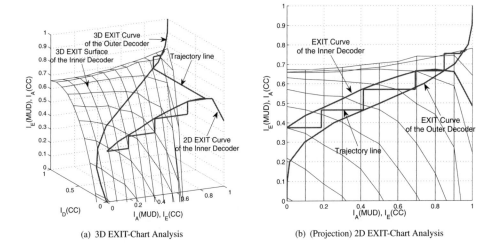

(a) 3D EXIT-Chart Analysis

(b) (Projection) 2D EXIT-Chart Analysis

Figure 10.13: EXIT-chart analysis of DHDC-shifting K-best SD-aided iterative receiver in the scenario of an (8×4) rank-deficient 4-QAM SDMA-OFDM system (SNR $= 8$ dB, $K = \mathcal{N}_{cand} = 128$). All other system parameters are listed in Table 10.1.

10.2.3.2.1 Soft-Symbol Calculation

Recall that a posteriori soft-bit information delivered from the channel decoder to the SD is defined to be the logarithm of the bit-probability ratios of its two legitimate values [314], namely $+1$ and -1. Then, given the received signal vector $\mathbf{y}$ as formulated in Equation (10.2), we can rewrite Equation (10.2) in a more convenient form as follows:

$$L(x_k|\mathbf{y}) = \ln \frac{P[x_k = +1|\mathbf{y}]}{P[x_k = -1|\mathbf{y}]}. \tag{10.19}$$

Therefore, bearing in mind that we have $P[x_k = +1|\mathbf{y}] = 1 - P[x_k = -1|\mathbf{y}]$, and taking the exponent of both sides in Equation (10.19), it is possible to derive the probability that $x_k = +1$ or $x_k = -1$ was transmitted in terms of their LLRs as follows:

$$e^{L(x_k|\mathbf{y})} = \frac{P[x_k = +1|\mathbf{y}]}{1 - P[x_k = +1|\mathbf{y}]}. \tag{10.20}$$

From Equation (10.20) we arrive at

$$P[x_k = +1|\mathbf{y}] = \frac{e^{L(x_k|\mathbf{y})}}{1 + e^{L(x_k|\mathbf{y})}}$$
$$= \frac{1}{1 + e^{-L(x_k|\mathbf{y})}}. \tag{10.21}$$

Similarly, we have

$$P[x_k = -1|\mathbf{y}] = \frac{1}{1 + e^{+L(x_k|\mathbf{y})}}. \tag{10.22}$$

In the following, we consider 4-QAM as an example to discuss briefly the soft-symbol calculation process with the aid of the LLR-to-probability conversion formula of Equation (10.21) and Equation (10.22). The symbol alphabet of the 4-QAM scheme is given in Table 10.6, which indicates that a 4-QAM symbol consists of 2 bits, the first of which determines the imaginary part of the symbol while the second controls the real part. Specifically, given the probabilities of two successive bits which

Table 10.6: The 4-QAM symbol alphabets over the complex numbers (i denotes $\sqrt{-1}$).

4-QAM symbol alphabets over the complex numbers			
j	1	2	3	4
$x_{j,1}\, x_{j,2}$	00	01	10	11
s_j	$(+1+i)/\sqrt{2}$	$(-1+i)/\sqrt{2}$	$(+1-i)/\sqrt{2}$	$(-1-i)/\sqrt{2}$

constitute a 4-QAM symbol, from their two legitimate values of $+1$ and -1, we can calculate the jth user's soft-symbol, s_j, as follows:

$$
\begin{aligned}
s_j &= [\Re(s_j);\ \Im(s_j)] \\
&= [P[x_{j,2} = -1|\mathbf{y}] \cdot (+1) + P[x_{j,2} = +1|\mathbf{y}] \cdot (-1); \\
&\quad P[x_{j,1} = -1|\mathbf{y}] \cdot (+1) + P[x_{j,1} = +1|\mathbf{y}] \cdot (-1)]/\sqrt{2},
\end{aligned}
\tag{10.23}
$$

where we assumed that the 2 bits are independent of each other, which is not entirely true owing to their correlation imposed by the Gray mapping to the 4-QAM symbols. The probabilities $P[x_{j,k} = \pm 1|\mathbf{y}]$ can be calculated from Equation (10.21) and Equation (10.22) based on the a posteriori LLR values received from the outer channel decoder.

10.2.3.2.2 Receiver Architecture and EXIT-Chart-Aided Analysis

Based on the idea of retaining the soft-bit information contained in the a posteriori LLRs, we propose the iterative DSDC-aided K-best SD receiver portrayed in Figure 10.14, where the soft-decision block substitutes the hard-decision and remodulation functionality of the DHDC-aided iterative receiver shown in Figure 10.7. The scheme of Figure 10.14 provides a soft search centre for the K-best SD and, based on the soft centres, the SD is expected to generate a better candidate list for the following LLR calculation, which is then delivered to the outer channel decoder. Although the soft-centre calculation imposes a slightly higher computational complexity than its hard-decision-based counterpart, the iterative DSDC-aided K-best SD receiver is capable of attaining a higher performance gain over the conventional iterative receiver, as observed throughout our forthcoming EXIT-chart analysis and BER results.

Figure 10.15 compares the 3D EXIT charts of the DHDC-aided and DSDC-aided iterative receivers, at $\mathrm{SNR} = 8\,\mathrm{dB}$, in the scenario of an (8×4)-element rank-deficient SDMA-OFDM system. We can observe in Figure 10.15(a) that the DSDC scheme's EXIT surface is distinctly higher than that of the DHDC-aided SD at the same values of K or N_{cand} ($K = N_{cand} = 64$ in this case). Thus, for a given number of iterations, a higher iterative gain is expected. On the other hand, since the tunnel between the EXIT surface of the SD and the EXIT curve of the outer convolutional channel decoder opens at a lower $[I_A, I_E]$ point, when the DSDC scheme is invoked instead of the DHDC arrangement, the centre-shifting scheme may provide performance benefits at lower SNRs. Specifically, when the SNR is too low, the EXIT surface of the SD may still be beneath the EXIT curve of the outer decoder, even though the centre-shifting scheme was switched on after the maximum attainable iterative gain has been achieved by the iterative receiver dispensing with centre shifting. Therefore, the higher the corresponding EXIT surface, the better the achievable performance of the centre-shifting scheme. We can observe in Figure 10.15(b) that the EXIT surface of the DSDC-aided receiver is still slightly higher for $K = 64$ than that of the DHDC-aided one using $K = 128$ and hence potentially doubling the associated complexity. In other words, the computational complexity imposed by the SD can be substantially reduced with the aid of the DSDC scheme, without sacrificing the attainable iterative gain.

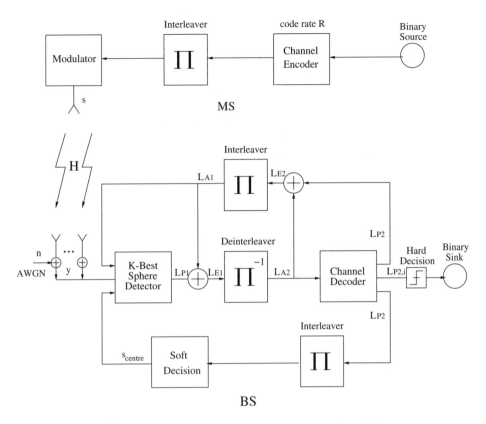

Figure 10.14: DSDC-shifting K-best SD-aided iterative detection scheme.

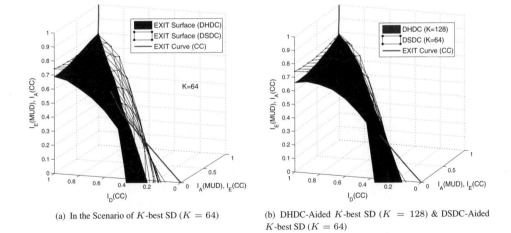

(a) In the Scenario of K-best SD ($K = 64$)

(b) DHDC-Aided K-best SD ($K = 128$) & DSDC-Aided K-best SD ($K = 64$)

Figure 10.15: EXIT-chart comparison of the DHDC and DSDC centre-shifting schemes in the scenario of a 4-QAM (8×4)-element SDMA-OFDM system at $\mathrm{SNR} = 8\,\mathrm{dB}$. All other system parameters are listed in Table 10.1.

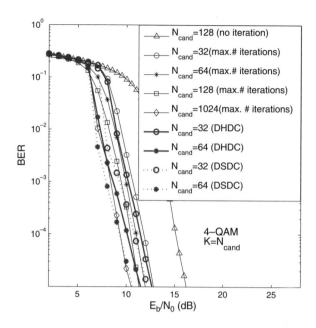

Figure 10.16: BER performance improvements provided by the DSDC scheme in the context of an (8×4)-element rank-deficient SDMA-OFDM system. Compared with the DHDC-aided receiver, the DSDC-aided K-best SD iterative receiver is capable of achieving a better BER performance at a slightly higher computational complexity imposed by the search-centre calculation process. All other system parameters are listed in Table 10.1.

10.2.3.2.3 Simulation Results

Figure 10.16 depicts the BER curves of the DSDC-aided K-best SD iterative receiver in comparison with those of the conventional iterative receiver dispensing with centre shifting and the DHDC-aided iterative receiver in the scenario of an (8×4)-element rank-deficient 4-QAM SDMA-OFDM system. The system parameters used and the iteration mode control remain the same as listed in Table 10.5, except that the centre-shifting scheme is changed to the DSDC. As seen in Figure 10.16, a better BER performance can be achieved in both scenarios where $K = N_{cand} = 32$ and $K = N_{cand} = 64$ were employed by the DSDC-aided iterative receiver than that by the DHDC-aided one. These observations confirm our discussions based on the EXIT-chart analysis of Figure 10.15. Remarkably, by having a list length of $K = N_{cand} = 64$, the DSDC-aided receiver outperforms the conventional iterative receiver using no centre shifting having a high complexity associated with a list size of $K = N_{cand} = 1024$. This remarkable performance improvement is achieved while simultaneously approaching the performance of the exact MAP detector, which may be implementationally infeasible, especially in such a heavily rank-deficient system. The additional iterative gain attained by the DSDC scheme can be observed from the EXIT charts plotted in Figure 10.17. More specifically, since the value of K and N_{cand} are as low as 64, no additional iterative gains can be achieved by the system using the DHDC scheme beyond a few iterations. By contrast, as soon as the DSDC scheme is activated, a substantially higher iteration gain is attained.

10.2.3.3 Two-Stage Iterative Architecture Using an Iterative SIC-MMSE-Aided Centre Update

As evidenced by our simulation results shown in Figure 10.16, upon exploiting the soft-bit information contained in the a posteriori LLRs gleaned from the channel decoder, the DSDC centre-shifting scheme brings about a higher performance gain than its hard-decision-based counterpart, i.e. the DHDC scheme.

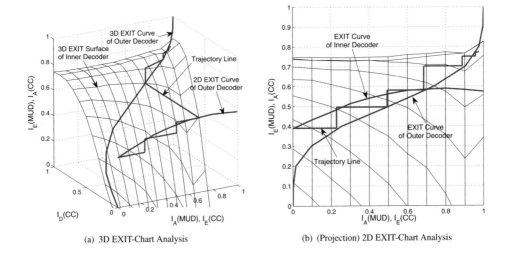

(a) 3D EXIT-Chart Analysis   (b) (Projection) 2D EXIT-Chart Analysis

Figure 10.17: EXIT-chart analysis of the DSDC-based K-best SD-aided iterative receiver in the scenario of an (8×4)-element rank-deficient 4-QAM SDMA-OFDM system (SNR $= 8\,$dB, $K = \mathcal{N}_{cand} = 64$). All other system parameters are listed in Table 10.1.

In order to exploit the soft information further so that the K-best SD iterative receiver benefits more substantially from the centre-shifting scheme at the cost of a slightly higher computational complexity, we take advantage of the iterative Soft-Interference-Cancellation-aided MMSE (SIC-MMSE) [572,573] algorithm in order to generate the search centre for the SD.

10.2.3.3.1 SIC-Aided MMSE Algorithm [572, 573]

As discussed in Section 10.2.3.2, given the a posteriori LLRs, we can calculate the corresponding soft symbol for a specific modulation scheme, i.e. 4-QAM, using Equation (10.21), Equation (10.22) and Equation (10.23). Similarly, given the a priori LLRs, we can also define the jth user's soft symbol, more precisely the mean of the jth user's symbol, as [572]

$$\bar{s}_j = E[s_j] = \sum_q s_j^{(q)} \cdot P[s_j = s_j^{(q)}], \tag{10.24}$$

where q is the number of points in the modulation constellation, e.g. $q = 4$ for 4-QAM or QPSK, while $s_j^{(q)}$ represents the qth legitimate value of the symbol s_j. Consequently, for 4-QAM, we arrive at

$$\begin{aligned}
\bar{s}_j &= (\Re(s_j); \Im(s_j)) \\
&= [P[x_{j,2} = -1] \cdot (+1) + P[x_{j,2} = +1] \cdot (-1); \\
&\quad P[x_{j,1} = -1] \cdot (+1) + P[x_{j,1} = +1] \cdot (-1)]/\sqrt{2}, \tag{10.25}
\end{aligned}$$

where $P[x_k = \pm 1]$ can be computed according to [314]:

$$\begin{aligned}
P[x_k = +1] &= \frac{e^{L(x_k)}}{1 + e^{L(x_k)}} \\
&= \frac{1}{1 + e^{-L(x_k)}} \tag{10.26}
\end{aligned}$$

and

$$P[x_k = -1] = \frac{1}{1 + e^{+L(x_k)}}, \tag{10.27}$$

respectively. On the other hand, we define the covariance of the jth user's symbol as [572, 574]

$$v_j = Cov[s_j, s_j]$$
$$= E[s_j s_j^*] - E[\bar{s}_j]E[\bar{s}_j^*] \tag{10.28}$$
$$= 1 - |\bar{s}_j|^2, \tag{10.29}$$

for constant modulus modulation schemes, such as BPSK, QPSK and 4-QAM.

The estimated symbol of the jth user generated by the MMSE algorithm can be expressed with the aid of the SIC principle as [572, 573]

$$\hat{s}_j = \bar{s}_j + v_j \mathbf{w}_j^H (\mathbf{y} - \mathbf{H}\bar{\mathbf{s}}), \tag{10.30}$$

where the jth column of the MMSE weight matrix $\mathbf{W}_{MMSE}$ can be expressed as [572, 573]

$$\mathbf{w}_{j,MMSE} = (\mathbf{H}\mathbf{V}\mathbf{H}^H + 2\sigma_w^2 \mathbf{I}_P)^{-1} \mathbf{h}_j, \tag{10.31}$$

where $\mathbf{I}_P$ represents the $(P \times P)$-element identity matrix and $\mathbf{V} = \text{diag}[v_1, v_2, \ldots, v_J]$.

As we may notice for the first iteration, the a priori LLRs gleaned from the outer decoder are not available, i.e. we have $L_A(CC) = 0$, which in turn leads to $\bar{s}_j = 0$ and $v_j = 1, j = 1, 2, \ldots, J$. In the sequel, the resultant search centre computed in Equation (10.30) is actually the conventional MMSE solution, where Equation (10.30) converges to the non-SIC-aided MMSE algorithm expressed as

$$\check{s}_j = \mathbf{w}_j^H \mathbf{y}, \tag{10.32}$$

where

$$\mathbf{w}_{j,MMSE} = (\mathbf{H}\mathbf{H}^H + 2\sigma_n^2 \mathbf{I}_P)^{-1} \mathbf{h}_j. \tag{10.33}$$

However, the SIC-MMSE starts to take effect from the second iteration onward, which is expected to provide a more accurate search centre for the SD than both the previously investigated DHDC and DSDC schemes, since in addition to retaining the soft-bit information during the soft-symbol generation, it carries out the soft interference cancellation at each iteration.

10.2.3.3.2 Receiver Architecture and EXIT-Chart Analysis

In this section we investigate the SIC-MMSE-aided iterative centre-shifting K-best SD receiver depicted in Figure 10.18, where the a posteriori LLR feedback-based DHDC and DSDC schemes are replaced by the SIC-MMSE-aided search-centre calculation, which is carried out on the a priori LLRs gleaned from the channel decoder. Therefore, as portrayed in Figure 10.19, we may consider the modified SD as a single input component fed with the a priori LLRs and producing a single output, namely the *extrinsic* LLRs, which contains both the centre-calculation part and the original SD part. Then the EXIT chart used to analyse the system becomes two dimensional.

The benefits of the SIC-MMSE-aided centre-shifting scheme become clearer if we refer to the EXIT charts obtained in the (8×4)-element rank-deficient scenario of the 4-QAM SDMA-OFDM system as seen in Figure 10.20. Recall from Figure 10.2 that the inner decoder's EXIT curve decayed upon increasing the a priori information, owing to the flawed information exchange between the inner and outer decoders, which was caused by the employment of an insufficiently large candidate list size N_{cand} and by the number of candidates K retained at each search level. The comparisons in Figure 10.20(a) indicate that this problem was effectively solved by the application of the SIC-MMSE-aided centre-shifting scheme. More explicitly, when using the SIC-MMSE scheme, the inner decoder's EXIT curve no longer decays when the a priori MI increases, even when using a limited list size of $K = N_{cand} = 16$. On the other hand, as shown in Figure 10.20(b), when K and N_{cand} are increased to 128 and 1024 for the K-best SD using no centre shifting, both of the resultant inner decoder's EXIT curves increase. As expected, the EXIT curve corresponding to $K = N_{cand} = 1024$ reaches a higher end point than that

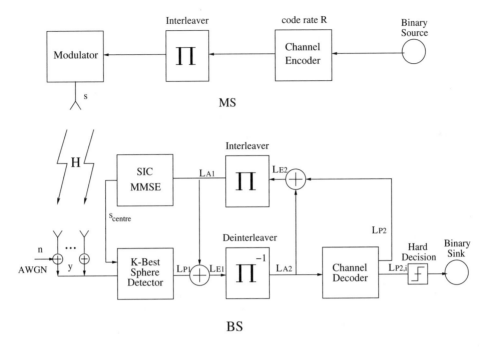

Figure 10.18: SIC-MMSE-aided centre-shifting K-best SD scheme.

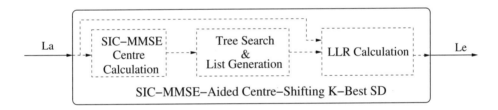

Figure 10.19: Structure of the SIC-MMSE-aided centre-shifting K-best SD: a SISO block.

associated with $K = N_{cand} = 128$. However, as a benefit of the SIC-MMSE centre-shifting scheme, the EXIT curve of the inner decoder may arrive at an even higher end point, despite using smaller K and N_{cand} values than that of the SD dispensing with centre shifting and high values of K and N_{cand}, such as 1024. Hence, we can infer from the above observations that the SIC-MMSE-aided receiver is capable of achieving a near-MAP BER performance conjunction with small values of K and N_{cand}.

10.2.3.3.3 Simulation Results

Both subfigures of Figure 10.21 show a significant performance gain, which was facilitated by the SIC-MMSE-aided centre-shifting K-best SD iterative receiver. Specifically, as seen in Figure 10.20(a), the SIC-MMSE-aided centre-shifting K-best SD is capable of approaching almost the same iterative gain by setting $K = 16$ as iterative SD using no centre shifting does in conjunction with $K = 1024$, at a BER of 10^{-5}. Hence, both the associated memory requirements and the computational complexity imposed are substantially reduced. Explicitly, for a fixed value of K, such as for example $K = 32$ and for the same target BER of 10^{-5}, we can observe that the iterative gain over the non-iterative receiver

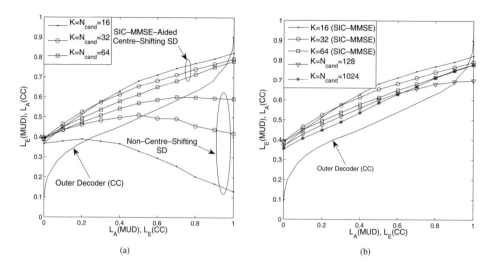

Figure 10.20: EXIT-chart comparison of SIC-MMSE-aided centre-shifting and the non-centre-shifting K-best SD iterative receiver in the scenario of the (8×4)-element rank-deficient 4-QAM SDMA-OFDM system at $SNR = 8\,dB$. All other system parameters are listed in Table 10.1.

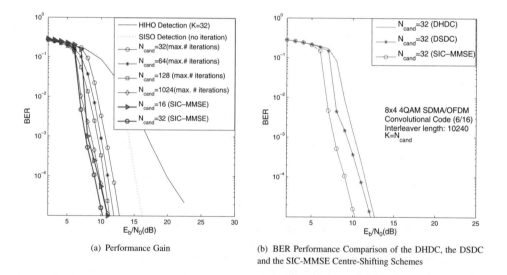

(a) Performance Gain

(b) BER Performance Comparison of the DHDC, the DSDC and the SIC-MMSE Centre-Shifting Schemes

Figure 10.21: BER performance of the SIC-MMSE-aided centre-shifting K-best SD iterative receiver. All other system parameters are listed in Table 10.1.

was doubled by the SIC-MMSE-aided centre-shifting scheme, when compared with that achieved by the iterative SD using no centre shifting, corresponding to about 6 dB.

On the other hand, in Figure 10.21(b) we compare the achievable performance of the three proposed centre-shifting schemes, namely the DHDC, the DSDC and the SIC-MMSE, in the context of K-best SD in the scenario of an (8×4)-element rank-deficient SDMA-OFDM system. We classify the centre-shifting schemes into two categories, namely the hard-decision-based methods, such as the DHDC scheme, and the soft-decision-based techniques, which include the two other centre-shifting schemes,

Table 10.7: Performance comparison of the conventional non-centre-shifting K-best SD and the SIC-MMSE-aided centre-shifting K-best SD iterative detection in the scenario of an (8×4) rank-deficient SDMA-OFDM system. Note that the computational complexity of the SD, i.e. the list generation by the SD, is calculated in terms of the total number of PED evaluations, while that of the soft-information generation by the SD/MAP detector is computed on the basis of Equation (10.16) in terms of the total number of OF evaluations corresponding to the two terms in Equation (10.14).

Performance gain and computational complexity reduction achieved by the SIC-MMSE scheme in an (8×4)-element 4-QAM SDMA-OFDM rank-deficient system							
BER	Centre shifting	$N_{cand}(K)$	Iterations	SNR	Memory	SD compl.	MAP compl.
10^{-5}	None	1024	3	10.5	8196	13652	49152
		128	3	11.2	1024	2388	6144
		64	2	12	512	1364	2048
		32	2	12.8	256	724	1024
		16	2	15	128	404	512
	SIC-MMSE	64	3	10.2	512	4092	3072
		32	3	10.2	256	2172	1536
		16	3	11	128	1212	768

exploiting the soft-bit information that arrives at the SD from the outer channel decoder. As argued earlier, the better the soft information exploited by the centre-shifting scheme, the higher the achievable performance improvement or the higher the attainable complexity reduction facilitated by the SISO SD-aided iterative receiver. As seen in Figure 10.21(b), performance gains of about 2.5 and 2 dB are attained by the SIC-MMSE centre-shifting scheme over the DHDC and the DSDC schemes at the cost of a slightly higher computational complexity, respectively, at the target BER of 10^{-5}. Therefore, the SIC-MMSE-aided centre-shifting scheme significantly outperforms the other two by invoking the idea of SIC.

We quantify the achievable performance gain and the complexity reduction facilitated by the SIC-MMSE-aided centre-shifting scheme in Table 10.7, in comparison with the conventional non-centre-shifting SD-aided iterative receiver. Again, we view the SD module as being constituted by two serially concatenated parts, namely the SD and the MAP decoder, which are responsible for carrying out the list generation and the soft-bit-information calculation, respectively. Table 10.7 quantifies the computational complexity imposed by the SD section in terms of the total number of PED evaluations, and that associated with the MAP part in terms of the total number of OF evaluations corresponding to the two terms in Equation (10.14). Thus, as explicitly indicated in Table 10.7, in order to achieve a near-MAP performance, i.e. to achieve a BER of 10^{-5} at an SNR below 11 dB in the context of an (8×4)-element rank-deficient SDMA-OFDM system, we have to use at least $K = N_{cand} = 1024$ for the non-centre-shifting SD-aided iterative receiver. However, using the SIC-MMSE-aided centre-shifting scheme, we can achieve the same goal by setting $K = N_{cand} = 16$ while imposing a factor of 11 lower computational complexity than that associated with the list-generation part and imposing a factor of 64 lower computational efforts by the soft-bit-information calculation of the SD receiver using no centre shifting. A further additional performance gain of 0.8 dB can be obtained by setting $K = 32$, at the cost of a modestly increased computational complexity. Furthermore, our extensive simulation results indicate that in a heavily rank-deficient system, setting K to a value higher than 32 can hardly improve the achievable performance gain further, if our target BER is below 10^{-2}, since a near-MAP performance has already been achieved.

10.3   A Priori LLR-Threshold-Assisted Low-Complexity SD

It transpires from Section 9.2.3 that having an insufficiently large candidate list, $N_{cand} = K$, does not guarantee the K-best SD of Section 9.2.3 and that its candidate list includes the ML point, while its depth-first counterpart of Section 9.2.2 does. When the value of K was kept low to maintain a low computational complexity, this resulted in a considerable performance degradation in Figure 9.8. In order to circumvent this problem, in this section another novel complexity-reduction scheme, referred to as the A priori LLR Threshold (ALT) technique, is designed for the K-best SD. Similar to the centre-shifting scheme, its philosophy is also based on exploitation of the a priori LLRs provided by the outer channel decoder, albeit in a rather different way. First of all, in Section 10.3.1, the operating principle of this novel complexity-reduction technique is highlighted. The analysis of this technique in terms of its achievable performance and the computational complexity imposed is carried out with the aid of our simulation results in Section 10.3.2.

10.3.1   Principle of the ALT-Aided Detector

Firstly, let us review the definition of the a priori LLRs, which is the logarithm of the ratio of the bit probabilities associated with $+1$ and -1 [314], which can be expressed as follows:

$$L_A(x_j) = \ln \frac{P[x_j = +1]}{P[x_j = -1]}. \tag{10.34}$$

Therefore, the sign of the resultant LLRs indicates whether the current bit is more likely to be $+1$ or -1, whereas the magnitude reflects how reliable the decision concerning the current bit is. For example, given a large positive a priori LLR delivered by the outer channel decoder of Figure 10.1, it is implied that the corresponding transmitted bit is likely to have been $+1$. In light of this, the search tree of the depth-first SD of Section 9.2.2 may be significantly simplified by invoking an ALT. To be specific, first we consider BPSK modulation as an example. If the a priori LLR of the mth user's BPSK symbol is sufficiently high (higher than the ALT), there is no need to carry out the detection for that particular user during the SD process. In other words, at the $(m + 1)$th search tree level, all the resultant tree nodes are expanded by a single branch instead of retaining both legitimate detection options. Therefore, for the depth-first SD of Section 9.2.2, the computational complexity is expected to be significantly reduced as we can observe from the search tree portrayed in Figure 10.22 in the scenario of a BPSK SDMA system where four users are supported. The application of the ALT scheme in the context of the example shown in Figure 9.4 of Section 9.2.2 generates a more simple tree structure, which imposes a reduced detection complexity. Since ALT $= 7$, which is lower than the absolute values of both the $(m = 4)$th user's and $(m = 2)$th user's LLRs arriving from the outer decoder of Figure 10.1 after a certain number of iterations, the SD will discard all the branches corresponding to $s_4 = 0$ at the $(m = 4)$th level and $s_2 = 1$ at the $(m = 2)$th level. Consequently, the final ML solution is attained after visiting only nine tree nodes and leaves in Figure 10.22. Hence, as long as the ALT is not too low, the computational complexity imposed can be substantially reduced by invoking the ALT scheme without any BER performance degradation, which becomes explicit by comparing the search trees as shown in Figures 9.4 and 10.22.

The depth-first SD of Section 9.2.2 was briefly revisited in the previous section when invoking the ALT technique, which is also applicable in the context of the breadth-first-style K-best SD. More explicitly, the main benefit of employing the ALT scheme, for the breadth-first SD, such as the K-best SD of Section 9.2.3, is not the achievable complexity reduction, but rather the potential performance improvement attained, since although there are still only K candidates expected to be retained at the mth search tree level of Figure 10.23, the affordable search complexity is assigned to the candidates having a specific bit value at the mth position, which is determined by the specific LLR-based decision. In other words, the LLR-based search tree pruning portrayed in Figure 10.23(b) decreases the probability

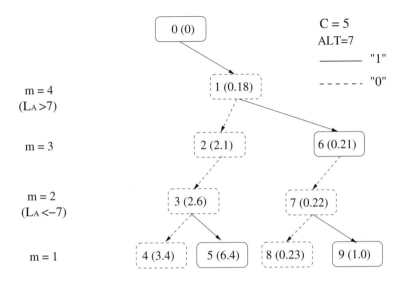

Figure 10.22: Illustration of the depth-first SD algorithm with the aid of the ALT scheme where ALT = 7. The number in brackets indicates the PED of a specific node for the trial point in the modulated constellation, while the number outside represents the order in which the points are visited.

of discarding a potentially correct path at an early search stage, especially when the value of K is set relatively small. When applying the ALT-assisted K-best SD to a four-user BPSK-modulated SDMA system, the resultant search tree is as portrayed in Figure 10.23. As can be seen in Figure 10.23(a), when dispensing with the ALT scheme, after 15 PED evaluations the SD opts for the specific tree leaf having a Euclidean distance of 0.56. Then the search portrayed in Figure 10.23(a) backtracks to the $(m = 4)$th level, yielding the hypothesized ML solution. However, as the ALT scheme is invoked, a better pruning search tree is obtained for the K-best SD, which is shown in Figure 10.23(b), if we assume that the absolute values of the a priori LLRs of both the $(m = 4)$th and the $(m = 2)$th users exceed the preset ALT value. Only nine PED evaluations have been carried out for this particular example, indicating a considerable reduction of the computational complexity imposed. However, even more importantly, the SD successfully identifies the true ML solution, which is different from the one generated by the SD characterized in Figure 10.23(a), where the latter dispenses with the ALT technique. This is achieved by backtracking during the search from a different tree leaf having a smaller Euclidean distance of 0.39 to the $(m = 4)$th level. Hence the incorrect search branch corresponding to $s_4 = 0$ is truncated as early as at the $(m = 4)$th level, thus reducing the computational complexity imposed while simultaneously avoiding the situation of discarding a potential path leading to the true ML solution, which may be the case for the non-ALT-assisted K-best SD at the $(m = 2)$th level owing to the fact that the true ML path may have a temporarily larger PED.

10.3.2 Features of the ALT-Assisted K-Best SD Receiver

10.3.2.1 BER Performance Gain

In this section, we concentrate our investigations on the novel ALT scheme in the context of the K-best SD, which is based on our simulation results. Figure 10.24 depicts the BER performance of the ALT-assisted K-best SD in the scenario of the (8×4)-element rank-deficient 4-QAM SDMA-OFDM system in comparison with the system dispensing with the ALT technique. Given a target BER of 10^{-5} and a fixed list length $\mathcal{N}_{cand} = K = 16$, a performance gain of 2.5 dB can be achieved by setting the ALT to 7, whereas a performance gain of about 1 dB can be obtained for the system using $K = 128$.

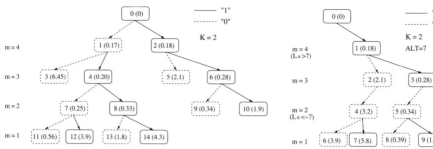

(a) The Search Tree of the Non-ALT-Assisted K-best SD in the Scenario of a 4-User BPSK SDMA System, where $K = 2$

(b) The Search Tree of the ALT-Assisted K-best SD in the Scenario of a 4-User BPSK SDMA System, where $K = 2$ and the ALT $= 7$

Figure 10.23: Illustration of the K-best SD algorithm with the aid of ALT scheme where ALT $= 7$. The number in brackets indicates the PED of a specific node for the trial point in the modulated constellation, while the number outside represents the order in which the points are visited.

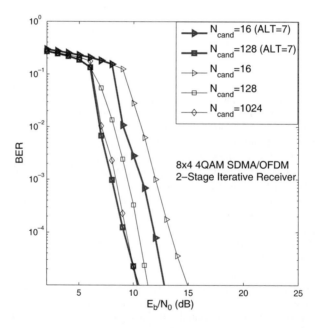

Figure 10.24: BER performance of the two-stage LLR-threshold-aided K-best SD iterative receiver in an (8×4)-element 4-QAM SDMA-OFDM system.

Actually, the ALT-aided receiver associated with $K = 128$ has already attained the MAP performance, which required at least $K = 1024$ for the equivalent system dispensing with the ALT technique. The ratio of these system complexities is as high as eight.

10.3.2.2 Computational Complexity

Under the assumption that the conventional K-best SD iterative receiver dispensing with the ALT technique generates the candidate list only once at the first iteration, which is stored in the memory for the *extrinsic* LLR calculation of the forthcoming iterations, the performance gains attained by both

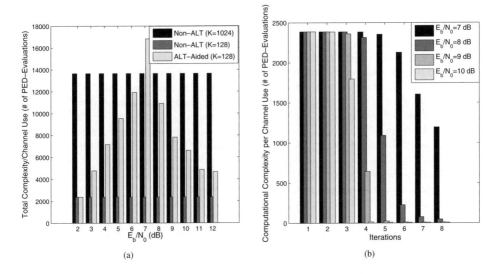

Figure 10.25: Histogram of the computational complexity of candidate list generation imposed by the ALT-aided K-best SD in the (8×4)-element rank-deficient 4-QAM SDMA-OFDM system: (a) the overall computational complexity per channel use for different E_b/N_0 values (ALT = 7); (b) the computational complexity per channel use of each iteration ($K = 128$, ALT = 7). Note that the maximum number of iterations for all different E_b/N_0 values was fixed to 8 and the iterative detection was terminated as soon as no more iteration gain can be achieved, i.e. the resultant EXIT trajectory line reached the convergence point of $[I_A, I_E] = [1, 1]$.

the centre-shifting scheme introduced in Section 10.2.3 and the ALT scheme are achieved at the cost of an acceptable computational complexity investment, since the candidate list has to be regenerated at each iteration. However, the memory requirement imposed is expected to be reduced, since there is no need to store the resultant candidate list. The complexity imposed by invoking the ALT scheme can be viewed in Figure 10.25(a), where the overall computational complexity quantified in terms of the number of PED evaluations per channel use imposed by the system operating both with and without the aid of the ALT scheme (ALT = 7) is plotted for the (8×4)-element rank-deficient 4-QAM SDMA-OFDM system. Specifically, since SD has to be carried out only once per channel use, regardless of how many iterations have been carried out, the receiver not benefitting from the ALT technique exhibits the same computational complexity of 2388 PED evaluations per channel use for $K = 128$, regardless of the channel SNR. By contrast, the number of PED evaluations required by the ALT-assisted receiver differs for different SNRs. To be specific, observe in Figure 10.25(a) that the complexity increases steadily as the SNR increases from 2 to 7 dB, peaking at about 17 000 PED evaluations per channel use. Beyond 7 dB, the complexity decays steadily as the SNR increases further, but levels out around 5000 PED evaluations at an SNR of 12 dB. Upon inspecting both Figure 10.24 and Figure 10.25(a), we can see that no performance gain is achieved by the ALT-aided receiver when the SNR is lower than 6 dB, despite the additional computational efforts of regenerating the candidate list at each iteration. This is not unexpected, since it is unlikely that the a priori LLRs gleaned by the outer channel decoder become higher than the threshold of the ALT scheme, because the intersection of the inner and outer EXIT curve occurs at a low I_A value in Figure 10.38(a). Hence, it is unwise to activate the ALT scheme when the SNR is low, since it may impose an increased complexity without any performance improvements.

On the other hand, as seen in Figure 10.24, with the advent of the ALT scheme the K-best SD becomes capable of achieving a near-MAP performance by setting $K = N_{cand} = 128$ instead of 1024. Recall our arguments on the complexity of the *extrinsic* LLR calculation for the LSD outlined in Section 10.2.1.2 that the corresponding complexity is linearly proportional to N_{cand}, as explicitly expressed in Equation (10.16). From this perspective, given a fixed target BER performance, the computational complexity imposed by the *extrinsic* LLR calculation of the K-best SD can be considerably reduced by employing the ALT scheme. Furthermore, with reference to Figure 10.25(a), the candidate list generation complexity of the ALT-aided receiver is well below that of its 'non-ALT-aided' counterpart for the SNR range spanning from 2 to 12 dB, except for SNRs in the immediate vicinity of 7 dB, if our aim is to achieve the near-MAP BER performance quantified in Figure 10.24, which can be attained by having $K = N_{cand} = 1024$ for the system operating without the ALT technique or by setting $K = N_{cand} = 128$ in the presence of the ALT scheme. More specifically, the number of PED evaluations per channel use carried out by the non-ALT-aided system using $N_{cand} = 1024$ remains as high as 13 652, regardless of the SNR and the number of iterations. On the other hand, in the presence of the ALT scheme, the candidate list has to be regenerated at each iteration; nonetheless, the total complexity imposed is substantially reduced, except for SNRs in the immediate vicinity of 7 dB. There are two reasons for this: (1) when the SNR is low, the number of iterations providing a useful gain is low, because there is no open tunnel between the EXIT curves of the inner and the outer decoder, unless the SNR is sufficiently high; (2) by contrast, when the SNR is high, the resultant staircase-shaped decoding trajectory can readily pass through the widely open EXIT tunnel and reach the point of perfect convergence at $[I_A, I_E] = [1, 1]$ after a low number of iterations. Furthermore, when the SNR is high, the number of PED evaluations carried out at each iteration is expected to decrease, as the iterations continues, as observed in Figure 10.25(b). This is due to the fact that the a priori LLRs fed back from the outer decoder of Figure 10.1 to the SD are likely to become higher than the LLR threshold after the first few iterations, and this allows the ALT-assisted SD directly to truncate the low-probability branches, leading to a reduced constellation size, which in turn results in a reduced complexity. More specifically, the complexity histogram of Figure 10.25(b) indicates that the higher the SNR, the more sharply the complexity drops as the iterations continue. Actually, when the SNR is relatively high, the complexity imposed becomes more modest after a few iterations, since the majority of the a priori LLRs fed back from the outer decoder to the SD of Figure 10.1 becomes higher than the LLR threshold. From a different perspective, this observation also explains the reason why we experience a complexity peak at the moderate SNR of 7 dB, where the ALT-related complexity does not decrease sufficiently substantially as the iterations continue and hence a high number of iterations are required to attain the maximum achievable iteration gain, since only a rather narrow EXIT tunnel was created between the EXIT curves of the inner and the outer decoder.

10.3.2.3 Choice of LLR Threshold

In the previous ALT-related simulations of Sections 10.3.2.1 and 10.3.2.2, we maintained an LLR threshold of ALT = 7, which ensured that the proposed ALT scheme performed well. However, it is intuitive that the LLR threshold cannot be set arbitrarily, since it plays a vital role in determining the system's performance. To be more specific, if the threshold is set too high, the ALT scheme can hardly affect the system's operation, since the a priori LLRs provided by the outer decoder are unlikely to be higher than the threshold, even after several iterations. By contrast, if the threshold is set to an excessively low value, although the computational complexity can be substantially reduced, naturally, a BER performance degradation is imposed. The above conjectures are verified by our simulation results shown in Figure 10.26, where the bars in the histogram of Figure 10.26(a) represent the computational complexity imposed, which was quantified in terms of the number of the PED evaluations per channel use. The LLR thresholds employed by the ALT-assisted K-best SD iterative receiver were set to values of ALT = 4, 7 and 10. Observe in Figure 10.26(a) that the lower the LLR threshold, the higher the complexity reduction attained. The corresponding BER curves plotted in Figure 10.26(b), however, demonstrate that when the threshold is set to an excessively low value, this may be expected to impose a performance degradation, as the SNR increases. This is not unexpected because, when the SNR becomes high, the a priori LLRs fed back by the outer decoder to the SD become predominantly higher than the

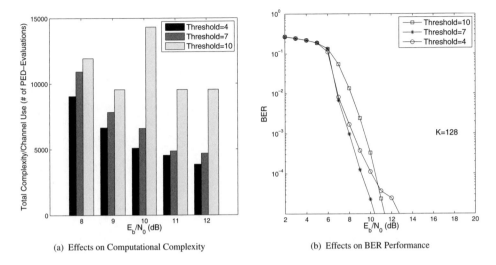

(a) Effects on Computational Complexity   (b) Effects on BER Performance

Figure 10.26: Effects of the LLR threshold on both the BER performance and the computational complexity of the K-best SD iterative receiver in an (8×4)-element 4-QAM SDMA-OFDM system. Note that the maximum number of iterations for all different E_b/N_0 values was 8 and the iterative detection was terminated as soon as no more iteration gain was achieved, i.e. the resultant decoding trajectory either reached the convergence point, or became trapped.

LLR threshold set at the very beginning of the iterative detection process. This may trigger an aggressive search tree truncation, which in turn results in discarding the true ML branch. In other words, in this scenario the truncation introduced by the ALT technique was activated too early, before the receiver attained a sufficiently high iterative gain. For example, given a target BER of 10^{-5}, a performance gain of about 1.5 dB was observed in Figure 10.26(b) over that of the receiver operating without the ALT technique, with the aid of a threshold of ALT = 7, while a performance degradation of about 1.5 dB was imposed by setting the threshold to ALT = 4. Note in Figure 10.26(b) that the BER curve corresponding to the threshold of ALT = 10 actually coincides with that of the 'non-ALT-assisted' system, as shown in Figure 10.24, implying that the ALT scheme does not have any beneficial effect with the aid of such a high threshold value. In conclusion, the threshold has to be carefully adjusted to achieve the target performance as a function of the SNR encountered.

10.3.2.4 Non-Gaussian-Distributed LLRs Caused by the ALT Scheme

Although the proposed ALT scheme is capable of providing useful performance improvements, a vital problem, which limits its capacity, can be observed from Figure 10.27, where the EXIT charts of the ALT-assisted receiver are plotted for four and six iterations. The decoding trajectories seen in both Figure 10.27(a) and Figure 10.27(b), which indicate the practically achievable mutual information improvements at the outputs of the inner and the outer decoders during the iterative process as a benefit of exploiting the a priori information available, do not match the corresponding theoretical EXIT curves very well, leading to an achievable maximum iteration gain which is significantly lower than that implied by the theoretical EXIT curves. In comparison with the EXIT chart depicted in Figure 10.27(b), where we have $K = N_{cand} = 128$, this EXIT-chart mismatch becomes even worse when K and N_{cand} are as low as 16, as shown in Figure 10.27(a). More specifically, even though a widely open EXIT tunnel was created between the EXIT curves of the inner and the outer decoders in Figure 10.27(a) with the aid of the ALT scheme, the decoding trajectory fails to reach the point of perfect convergence at $(1, 1)$ since it becomes trapped at the point $(0.96, 0.96)$, regardless of the number of iterations. This results in

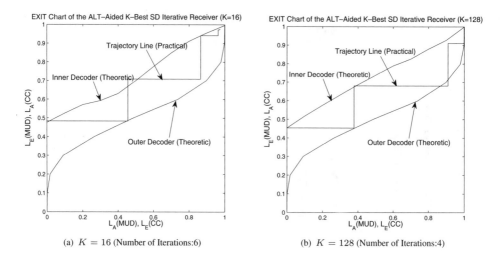

(a) $K = 16$ (Number of Iterations:6)   (b) $K = 128$ (Number of Iterations:4)

Figure 10.27: EXIT chart of the ALT-aided K-best SD iterative receiver in the scenario of an (8×4)-element 4-QAM SDMA-OFDM system at SNR $= 10$ dB.

a significantly worse performance in comparison with the situation when we use $K = N_{cand} = 128$, as observed in Figure 10.24. Actually, the ALT-aided receiver remains unable to achieve a near-error-free performance at SNR $= 10$ dB even for $K = 128$, regardless of the number of iterations, and despite having an open tunnel between the EXIT curves of the inner and outer decoder.

Recall that the EXIT-chart analysis of an iterative receiver is sufficiently accurate only on condition, when the a priori LLRs at the input and the a posteriori LLRs at the output of a constituent module of the iterative scheme exhibit a Gaussian distribution. That is the reason why the length of the interleaver between the inner and outer decoders has to be sufficiently high, in order to maintain an approximate Gaussian distribution. Otherwise, a mismatch may occur between the predicted and the practically achievable gains, yielding a smaller iterative gain and difficulties in system performance prediction. However, we found from our previous discussions and simulation results presented in Figure 10.27 for LSDs that, apart from the interleaver length, the maximum iteration gain is also substantially affected by the value of K and N_{cand}, as evidenced by the EXIT chart of Figure 10.2 presented in Section 10.2.1.3. Again, a non-Gaussian distribution exhibited by the resultant LLRs at the input and output of the SD is the cause of this phenomenon, as indicated by the simulation-based histograms of both the a priori LLRs and of the *extrinsic* LLRs of the SD module after each iteration in Figure 10.28(a) and Figure 10.28(b). When K and N_{cand} are sufficiently high, such as 1024, as shown in Figure 10.28(a), an approximate Gaussian distribution is recorded for the LLRs upon increasing the number of iterations while also exhibiting an increasingly higher variance. However, when K and N_{cand} are set to an excessively low value, such as 32 as shown in Figure 10.28(b), after two iterations the majority of the resultant LLRs at both the input and the output of the SD have values which are close to the LLR truncation value of 32 used in our case, leading to a distinctively non-Gaussian distribution. Hence, we cannot expect the EXIT-chart analysis, which is based on the premise of experiencing a Gaussian LLR distribution, to produce an accurate performance prediction.

In order to investigate further the reason behind the EXIT-chart mismatch seen in Figure 10.27, we compare the histograms of both the a priori and the *extrinsic* LLRs of the SD with and without the ALT scheme in Figure 10.29(a) and Figure 10.29(b). Consequently, we found that the application of the ALT scheme actually degrades the accuracy of the approximate Gaussian distribution exhibited by the LLRs at an earlier stage of the iterations, resulting in a more severe EXIT-chart mismatch problem. To be specific, in the absence of the ALT scheme, we observe in Figure 10.29(a) that at the fifth iteration the

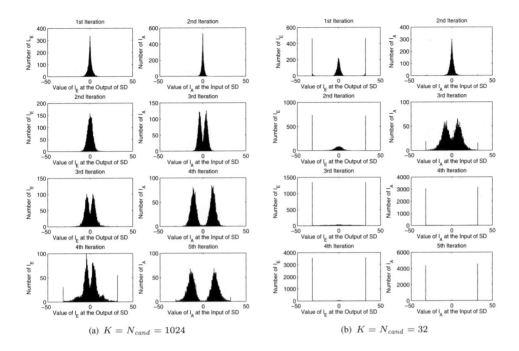

Figure 10.28: Histograms of the LLRs at both the input and the output of the K-best SD during the iterative process in the scenario of an (8×4)-element 4-QAM SDMA-OFDM system for $K = N_{cand} = 1024$ and for $K = N_{cand} = 32$ at SNR $= 10$ dB.

Gaussian-like distribution is eliminated. However, in the presence of the ALT scheme, a non-Gaussian distribution appears even earlier after the fourth iteration, as seen in Figure 10.29(b). Hence, although the theoretical EXIT curve of the ALT-aided receiver obtained under the assumption of having a near-Gaussian-distributed LLR all the time, as previously shown in Figure 10.27, can indeed reach the $(1, 1)$ point, the problem of EXIT-chart mismatch imposed by the non-Gaussian distribution of the LLRs at both the input and output of the SD is aggravated by the application of the ALT scheme. This leads to a more limited iterative gain than that expected from the theoretical EXIT curves of Figure 10.27.

10.3.3 ALT-Assisted Centre-Shifting Hybrid SD

10.3.3.1 Comparison of the Centre-Shifting and the ALT Schemes

Both the proposed centre-shifting and ALT schemes require the repeated generation of candidate lists throughout the iterative detection process. In this section, we first compare the ALT scheme with the SIC-MMSE-aided centre-shifting scheme, which was formed to be the most efficient of all the three centre-shifting schemes proposed in Section 10.2.3.

From the BER curves depicted in Figures 10.21 and 10.24, we observe that in order to achieve the near-MAP performance exhibited by the SIC-MMSE-aided centre-shifting K-best SD iterative receiver using $K = 32$, we have to set $K = 128$ for the ALT-assisted receiver. In other words, given a target BER, the SIC-MMSE-aided centre-shifting scheme imposes a significantly lower complexity than the ALT scheme, as quantified in Figure 10.30(a), where their corresponding computational complexity is characterized versus the SNR quantified in terms of the total number of PED evaluations per channel use. Specifically, although a fairly sharp drop can be seen in the complexity imposed by the ALT-assisted receiver as the SNR increased from a moderate level to a relatively high value, the ALT-assisted

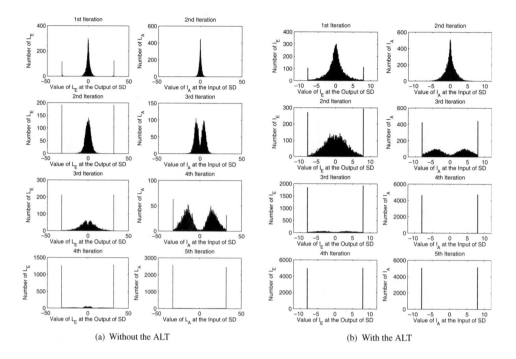

(a) Without the ALT   (b) With the ALT

Figure 10.29: Histogram of the LLRs at both the input and the output of the K-best SD during the iterative process in the scenario of an (8×4)-element 4-QAM SDMA-OFDM system for $K = N_{cand} = 128$, $\mathrm{SNR} = 10\,\mathrm{dB}$ and with as well as without the ALT technique.

receiver still requires a considerably higher computational effort to match the BER performance of its centre-shifting-aided counterpart. On the other hand, the above-mentioned sharp drop in the complexity imposed by the ALT-aided K-best SD when the SNR is increased relatively high is caused by the fact that the complexity imposed per iteration decreases as the iterative detection proceeds, as can be observed in Figure 10.30(b), where we have a relatively high of $\mathrm{SNR} = 8\,\mathrm{dB}$.

10.3.3.2 ALT-Assisted Centre-Shifting Hybrid SD

Since the computational complexity imposed by the ALT scheme per iteration is expected to decrease as the iterations proceed as observed in Figure 10.30(b), in this section we propose a hybrid SD-aided iterative receiver, which combines the benefits of the ALT scheme and the SIC-MMSE centre-shifting scheme, in attempting to reduce the associated complexity further. In comparison with the centre-shifting SD receiver dispensing with the ALT technique, a small performance degradation is imposed if the ALT scheme is employed, as seen in Figure 10.31. This is not unexpected, since a non-Gaussian distribution was exhibited by the soft-bit information, i.e. by the LLRs, which are exchanged between the inner and outer decoders, leading to a limited iterative gain. On the other hand, the computational complexity imposed by the candidate list generation phase of the SD is significantly reduced, as seen in Figure 10.32. More specifically, Figure 10.32(a) depicts the overall computational complexity of the hybrid receiver per channel use for different SNRs in contrast to that of the pure SIC-MMSE-assisted centre-shifting SD receiver. By contrast, Figure 10.32(b) shows the computational complexity per channel use at each iteration, i.e. as a function of the iteration index. Both of the two receivers we compared here have to carry out the candidate list regeneration at each iteration. As shown in Figure 10.32(b), when invoking the ALT scheme, the hybrid system exhibits a gradually reduced

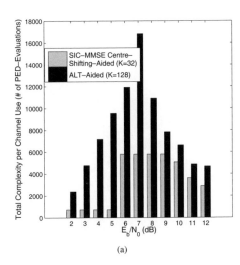
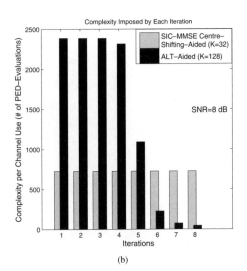

(a)   (b)

Figure 10.30: The candidate-list-generation-related computational complexity comparison of the SIC-MMSE centre-shifting-aided and the ALT-aided K-best SD iterative receiver in the (8×4)-element SDMA-OFDM system: (a) the overall computational complexity per channel use for different E_b/N_0 values; (b) computational complexity per channel use of each iteration. Note that the maximum number of iterations for all different E_b/N_0 values is 8 and the iterative detection will be terminated as soon as no more iteration gain can be achieved, i.e. the resultant trajectory line reaches the convergence point.

complexity as the iterations proceed, while the pure centre-shifting-aided receiver imposes a constant complexity at each iteration. Hence, the resultant overall computational complexity of the hybrid receiver is significantly reduced. To be specific, for the candidate list generation phase, only about half the computational efforts are required by the hybrid receiver at high SNRs, as seen in Figure 10.32(a).

10.4 URC-Aided Three-Stage Iterative Receiver Employing SD

Recently, a URC three-stage serially concatenated system was proposed [575] in the context of SISO MMSE turbo equalization. A rate 1 encoder and its corresponding decoder are amalgamated with the transmitter and the receiver, respectively. Therefore, the *extrinsic* LLRs are exchanged between three blocks, i.e. the MMSE equalizer, the URC decoder and the convolutional decoder at the receiver, resulting in a significant performance gain which was explicitly indicated by the resultant EXIT charts shown in [575, 576]. In this section, we transplant the URC-aided three-stage concept into our SD-aided MIMO system. The investigation of the resultant system's performance has been carried out using both EXIT-chart analysis and Monte Carlo simulations. Finally, the performance of the centre-shifting scheme in this scenario will also be studied.

10.4.1 URC-Aided Three-Stage Iterative Receiver

Figure 10.33 depicts the system model of the SD-aided three-stage serially concatenated transceiver in the context of an Uplink (UL) SDMA-OFDM system, where each user has a single transmit antenna, as we have always assumed so far. At the transmitter, a block of L information bits u_1 is first encoded by the convolutional channel encoder I in order to generate the coded bits c_1, which are interleaved by the interleaver Π_1 of Figure 10.33. Then the resultant permuted bits u_2 are successively fed through

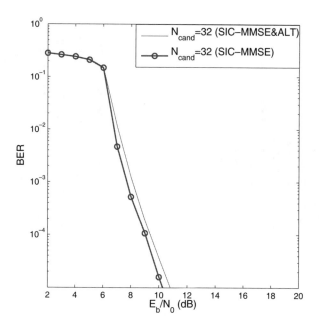

Figure 10.31: BER performance of the two-stage K-best SD iterative receiver using the combined SIC-MMSE centre-shifting and the ALT schemes in an (8×4)-element 4-QAM SDMA-OFDM system. All other system parameters are listed in Table 10.1.

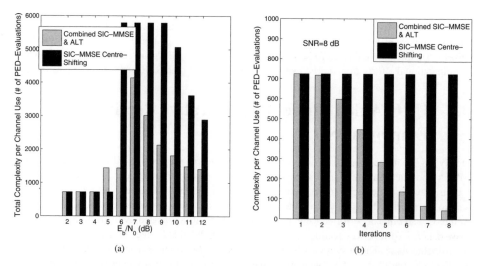

Figure 10.32: The candidate-list-generation-related computational complexity comparison of the hybrid K-best SD iterative receiver which combines the SIC-MMSE centre-shifting scheme and the ALT technique: (a) the overall computational complexity per channel use for different E_b/N_0 values; (b) computational complexity per channel use for each iteration. Note that the maximum number of iterations for all different E_b/N_0 values is 8 and the iterative detection will be terminated as soon as no more iteration gain can be achieved, i.e. the resultant trajectory line reaches the convergence point.

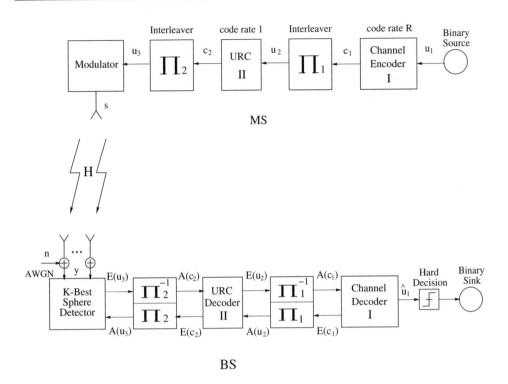

Figure 10.33: URC-aided three-stage iterative detection scheme.

the URC encoder II and the interleaver Π_2, yielding the interleaved double-encoded bits u_3, which are delivered to the bit-to-symbol modulator/mapper. Note that the labels u and c represent the uncoded and coded bits, respectively, corresponding to a specific module as indicated by the subscript. For example, u_2 and c_2 denote the uncoded bits and the coded bits at the input and the output of the URC encoder II of Figure 10.33, respectively. At the receiver of Figure 10.33, which comprises three modules, namely the SD, the URC decoder II and the convolutional channel decoder II, the extrinsic information is exchanged among the blocks in a number of consecutive iterations. Specifically, as shown in Figure 10.33, $A(\cdot)$ represents the a priori information expressed in terms of the LLRs, while $E(\cdot)$ denotes the corresponding *extrinsic* information. Hence, the URC decoder generates two *extrinsic* outputs by processing two a priori inputs delivered from both the SD and the convolutional decoder II. After completing the last iteration, the estimates $\hat{u}_1$ of the original transmitted information bit u_1 are produced by the convolutional channel decoder I.

We denote the MI between the a priori value $A(s)$ and the symbol s as $I_{A(s)}$, and the MI between the *extrinsic* value $E(s)$ and the symbol s by $I_{E(s)}$. Hence, the MI of the two outputs of the URC decoder, namely $I_{E(u_2)}$ and $I_{E(c_2)}$, are functions of the two a priori MI inputs, namely $I_{A(u_2)}$ and $I_{A(c_2)}$. Explicitly, we have [575]

$$I_{E(u_2)} = T_{u_2}(I_{A(u_2)}, I_{A(c_2)}), \tag{10.35}$$

$$I_{E(c_2)} = T_{c_2}(I_{A(u_2)}, I_{A(c_2)}). \tag{10.36}$$

Therefore, two 3D EXIT charts corresponding to the above two equations are needed in order to describe fully the EXIT characteristics of the URC decoder. In contrast to the double-input, double-output URC module of Figure 10.33, both the SD and the convolutional decoder can be viewed as SISO modules, for a given received signal vector. Thus, a single 2D EXIT chart is sufficient for characterizing each of

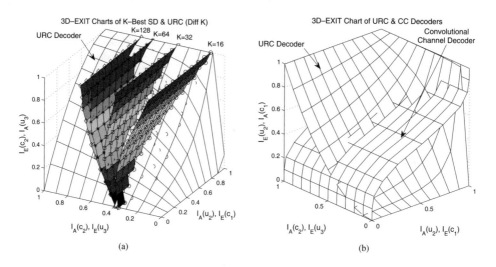

Figure 10.34: The (SNR $= 8$ dB) 3D EXIT charts of the K-best SD-aided three-stage iterative receiver in the scenario of the (8×4) rank-deficient 4-QAM SDMA-OFDM system.

them. Similarly, we have the corresponding EXIT functions expressed as [575]

$$I_{E(u_3)} = T_{u_3}(I_{A(u_3)}, E_b/N_0) \tag{10.37}$$

for the SD and

$$I_{E(c_1)} = T_{c_1}(I_{A(c_1)}) \tag{10.38}$$

for the convolutional channel decoder. We note that since the MI $I_{E(u_3)}$ of the SD's output is independent of $I_{A(u_2)}$, the traditional EXIT curve of the SD portrayed in the 2D space of Figure 10.2 can be extended to the 3D space by sliding the EXIT curve along the $I_{A(u_2)}$-axis. That is to say, the EXIT characteristics of the SD can be portrayed as an EXIT surface in one of the two 3D EXIT charts of the URC decoder, namely Figure 10.34(a). Similarly, the EXIT surface of the outer convolutional decoder can be generated as depicted in Figure 10.34(b) together with the other 3D EXIT chart of the URC decoder, since $I_{E(c1)}$ of Equation (10.38) is independent of $I_{A(c2)}$. Consequently, in total two 3D EXIT charts are required for plotting all the EXIT functions. To be specific, Figure 10.34(a) for Equation (10.36) and Equation (10.37), and Figure 10.34(b) for Equation (10.35) and Equation (10.38).

The intersection of the surfaces of the SD and the URC decoder characterizes the best possible achievable performance for different fixed values of $I_{A(u_2)}$ as the iterations between the SD and the URC decoder are carried out, during which the soft-bit information is exchanged. More importantly, according to Equation (10.35), for each point $(I_{A(u_2)}, I_{A(c_2)}, I_{E(c_2)})$ of the intersection line as seen in Figure 10.34(a), there is a specific point $(I_{A(u_2)}, I_{A(c_2)}, I_{E(u_2)})$ determined by the two a priori inputs of the URC decoder on the surface of the URC decoder in Figure 10.34(b). Hence, there must be a line (not plotted) on the surface of the URC decoder in Figure 10.34(b) corresponding to the intersection line in Figure 10.34(a). In order to simplify the complicated 3D EXIT-chart representation, we view the SD and the URC decoder as a joint module with single input $I_{E(u_2)}$ and single output $I_{A(u_2)}$. As a result, a classical 2D EXIT chart can be plotted, which can also be obtained by projecting the aforementioned line on the surface of the URC decoder in Figure 10.34(b) on the $I_{E(u_2)}-I_{A(u_2)}$ plane, as seen in Figure 10.35(a).

Figure 10.35(a) shows the 2D EXIT chart of decoder I and the combined module of decoder II and the SD, in comparison with that of the conventional two-stage iterative receiver. As observed in Figure 10.2, due to the insufficient length of the candidate list, the maximum achievable iterative gain

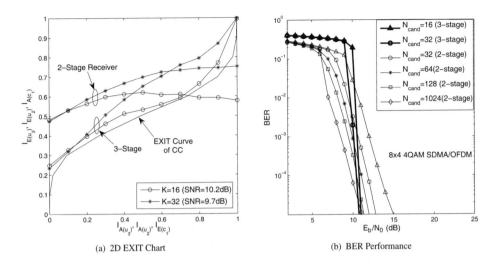

(a) 2D EXIT Chart

(b) BER Performance

Figure 10.35: EXIT analysis and BER performance of the three-stage iterative receiver using the K-best SD.

becomes rather limited, since the EXIT curve of the SD intersects that of the channel decoder at an earlier stage, if we have K or N_{cand} values of 16 or 32. In other words, when the resultant decoding trajectory gets trapped at the intersection point of the EXIT chart, where the decoding convergence point, after a certain number of iterations, typically a residual error floor, persists. However, with the aid of the URC decoder II seen in Figure 10.33, the point of the EXIT curve intersection of the joint decoder II and SD module and that of decoder I moves close to the $(1, 1)$ point, resulting in a near-error-free performance, as long as there is an open tunnel between the two EXIT curves. More specifically, as observed in Figure 10.35(a), for the SD-aided iterative receiver using $K = 16$, an open tunnel is created between the EXIT curve of the joint SD and decoder II module and that of decoder I at an SNR of 10.2 dB. Thus, the corresponding BER curve plotted in Figure 10.35(b) confirmed the predictions of the EXIT-chart analysis seen in Figure 10.35(a), indicating that the BER decreases sharply once the SNR is in excess of about 10.2 dB. Similarly, when we have $K = 32$, a lower convergence threshold of 9.7 dB is associated with an even earlier decrease of the BER curve, as shown in Figure 10.35(b). Consequently, given a target BER of 10^{-5}, a performance gain of nearly 4 dB and 2 dB can be attained over the conventional two-stage iterative receiver when employing the URC decoder II in conjunction with $K = 16$ and $K = 32$, respectively. However, as a price, the BER of the three-stage scheme is expected to be higher than that of the two-stage receiver at low SNRs. The reason behind this phenomenon becomes clearer if we refer to the EXIT-chart comparison of the two-stage and three-stage iterative receivers characterized in Figure 10.35(a), where we observe that the EXIT curve of the inner decoder of the conventional two-stage receiver has a significantly higher starting point than that of its three-stage counterpart, resulting in a lower convergence threshold, which in turn leads to a potential higher iterative gain at relatively low SNRs. Although the employment of the URC encoder/decoder pair at the transmitter/receiver is capable of moving the EXIT curve intercept point closer to $(1, 1)$, an open tunnel can only be formed if the value of K or N_{cand} as well as that of the SNR is sufficiently high. This explains why the BER curve of the SD using $(K = 32)$ drops sharply at a lower SNR than that of the SD employing $(K = 16)$, as seen in Figure 10.35(b).

The reason why a URC will make the slope of the EXIT-chart curve steeper, hence resulting in a lower error floor and a higher BER waterfall threshold, can be interpreted as follows. Since the URC has an Infinite Impulse Response (IIR) due to its recursive coding structure, the corresponding EXIT-chart curve is capable of reaching the highest point of perfect convergence to an infinitesimally low

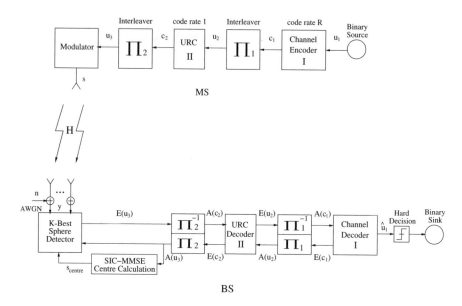

Figure 10.36: SIC-MMSE-aided centre-shifting K-best SD-aided three-stage iterative detection scheme.

BER $(1, 1)$, provided that the interleaver length is sufficiently large [577]. On the other hand, since the URC decoder employs the MAP decoding scheme, the extrinsic probability computed at the output of the URC decoder contains the same amount of information as the sequence at the input of the URC decoder. In other words, the area under the inner EXIT curve remains unchanged regardless of the employment of the URC [578, 579]. Hence, a higher end point of the EXIT curve leads to a lower starting point, implying a steeper slope of the EXIT curve.

10.4.2 Performance of the Three-Stage Receiver Employing the Centre-Shifting SD

The decay observed in Figure 10.2 for the combined SD and URC decoder II module's EXIT curve observed when K and N_{cand} are set to an insufficiently high value is caused by the corresponding EXIT surface of the SD as plotted in Figure 10.34(a). Our previous investigations for the centre-shifting scheme indicated that the SIC-MMSE-aided scheme is capable of ensuring that the EXIT curve of the inner decoder, namely that of the SD, monotonically increases upon increasing I_A, as seen in Figure 10.20(a). Hence, we apply the SIC-MMSE centre-shifting scheme in the context of the URC-aided three-stage iterative receiver, in order to improve the shape of the EXIT curve seen in Figure 10.35(a), which may result in a relatively high convergence threshold. The three-stage SIC-MMSE-aided centre-shifting K-best SD-assisted iterative receiver is portrayed in Figure 10.36, where the SIC-MMSE-aided centre calculation is applied. Thus, redetection using an updated search centre has to be carried out during each iteration that invokes the SD.

Figure 10.37 shows the resultant 3D EXIT chart of the three-stage scheme, where we observe that the EXIT surfaces do not suffer from as severe a bending as those of the non-centre-shifting-aided receiver characterized in Figure 10.34(a), even when K or N_{cand} is relatively small. The resultant EXIT curve of the combined SD and URC decoder II module is plotted in Figure 10.38(a), for $K = N_{cand} = 16$, which does not touch the EXIT curve of decoder I. To be specific, the original convergence threshold of the three-stage receiver using no centre shifting is about 10.2 dB, since an open tunnel is just formed

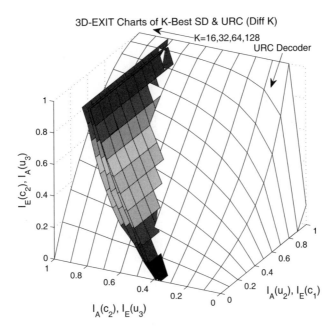

Figure 10.37: The 3D EXIT chart of the three-stage SIC-MMSE-aided centre-shifting K-best SD-assisted iterative receiver.

for SNRs in excess of this level. For SNRs below this level, the EXIT curve of the combined module would fall below that of decoder I, leading to a consistently closed EXIT tunnel, as exemplified by the situation characterized by $SNR = 9.6$ dB, as also portrayed in Figure 10.38(a). However, owing to the employment of the SIC-MMSE-aided centre-shifting scheme, a wide open tunnel has been created between the EXIT curves. The convergence threshold of the SIC-MMSE centre-shifting-aided three-stage scheme was reduced to about 9.6 dB and the resultant BER curve is plotted in Figure 10.38(b) in comparison with that of the three-stage receiver dispensing with centre shifting. Indeed, with the aid of the centre-shifting scheme, the BER curve starts to drop more sharply at a slightly lower SNR, which is similar to the convergence threshold observed in Figure 10.38(a), yielding a performance gain of 0.5 dB for the target BER of 10^{-5}. It is not unexpected that the attainable performance improvement is insignificant, since the SIC-MMSE centre-shifting scheme fails to increase the relatively low starting point of the EXIT curve, which is brought about by the employment of the URC decoder II.

10.4.3 Irregular Convolutional Codes for Three-Stage Iterative Receivers

The so-called Irregular Convolutional Codes (IrCCs) [580, 581], proposed by Tüchler and Hagenauer, encode appropriately chosen 'fractions' of the input stream using punctured constituent convolutional codes having different code rates. The appropriate 'fractions' are specifically designed with the aid of EXIT charts to improve the convergence behaviour of iteratively decoded systems. Thus, with the aid of IrCCs, we are able to solve the mismatch between the EXIT curve of the inner decoder in the three-stage receiver and the EXIT curve of the RSC(2,1,3) code marked by crosses in Figure 10.39. Our goal is to achieve an improved convergence behaviour for the three-stage concatenated system by minimizing the area between the EXIT curve of the amalgamated two-compound inner code and that of the outer code. The resultant EXIT curve of the optimized IrCC having a code rate of 0.5 is represented by the dotted line in Fig. 10.39. Hence, a narrow but still open EXIT-chart tunnel is created, which implies having

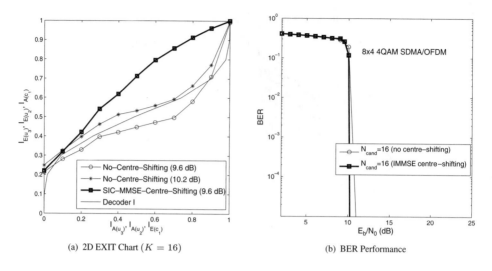

(a) 2D EXIT Chart ($K = 16$)   (b) BER Performance

Figure 10.38: EXIT analysis and BER performance of SIC-MMSE-aided centre-shifting K-best SD three-stage iterative receiver.

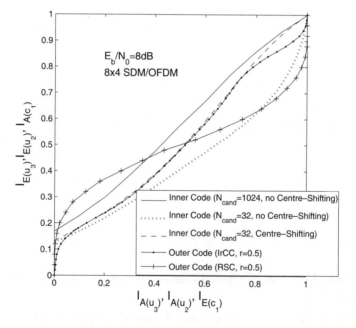

Figure 10.39: EXIT charts of the URC-aided three-stage receiver in the scenario of an (8×4)-element SDMA-OFDM system at $E_b/N_0 = 8$ dB.

a near-capacity performance attained at the cost of a potentially high number of decoding iterations, although the 'per-iteration' complexity may be low.

Monte Carlo simulations were performed to characterize the decoding convergence prediction of the IrCC design in the high-throughput overloaded (8×4) SDMA-OFDM system. As our benchmark system, the half-rate RSC(2,1,3) code's EXIT curve marked by crosses in Figure 10.39 is employed

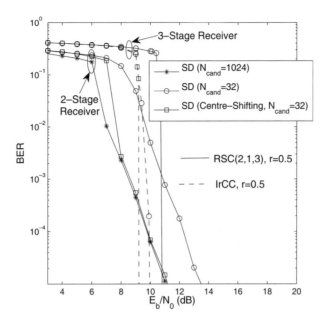

Figure 10.40: BER performance of the three-stage iterative receiver using the K-best SD in the scenario of an (8×4)-element SDMA-OFDM system.

as the outer code of the traditional two-stage receiver. As our proposed scheme, the half-rate IrCC corresponding to the EXIT curve represented by the dotted line in Figure 10.39 is used as the outer code in the URC-assisted three-stage receiver. Figure 10.40 compares the BER performance of both systems, where we can see that at relatively high SNRs, both three-stage concatenated receivers – namely the one using the SD employing the classic RSC code as well as the one employing the optimized IrCC code – are capable of outperforming the traditional two-stage receiver equipped with the SD. Specifically, given the target BER of 10^{-5}, a performance gain of 2.5 dB can be attained by the three-stage receiver over its two-stage counterpart, when both of them employ the SD ($N_{cand} = 32$) and the regular RSC. Remarkably, when amalgamated with the URC encoder/decoder, the three-stage receiver using the SD and $N_{cand} = 32$ becomes capable of outperforming the two-stage receiver using the high-complexity near-MAP SD having $N_{cand} = 1024$, provided that the SNR is in excess of about 11 dB. Furthermore, an additional performance gain of 1 dB can be attained by employing the optimized IrCC in comparison with the classic RSC-aided three-stage system. Moreover, in order to enhance the achievable performance further, when the SIC-MMSE-aided iterative centre-shifting SD is invoked, another approximately 1 dB additional performance gain is attained. Consequently, as observed in Figure 10.40, given the target BER of 10^{-5}, overall performance gains of 4.5 dB and 2 dB are attained by our proposed system in comparison with its SD ($N_{cand} = 32$) aided and SD ($N_{cand} = 1024$) assisted two-stage counterparts, respectively.

In line with the EXIT-chart-based predictions of Figure 10.39, a sharp BER improvement is achieved by the three-stage receiver, as seen in Figure 10.40, since the EXIT curve of the inner code will rise above that of the outer code for SNRs in excess of a certain level, resulting in a consistently open EXIT tunnel leading to the point of convergence at $(1, 1)$, which is exemplified in Figure 10.41 by the curve recorded at SNR $= 9.5$ dB when using the half-rate IrCC as the outer code. Also shown in Figure 10.41 is the staircase-shaped decoding trajectory evolving through the open tunnel to the point of convergence at $(1, 1)$, as recorded during our Monte Carlo simulations. The activation order of the three SISO modules used is [3 2 1 2 1 2], where the integers represent the index (I) of the three SISO

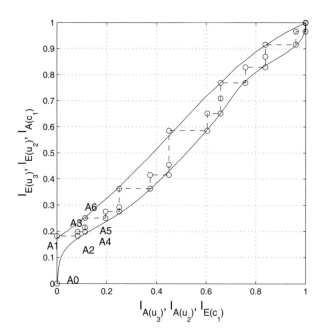

Figure 10.41: EXIT charts and recorded decoding trajectory for the three-stage receiver using IrCC at $E_b/N_0 = 9.5\,\text{dB}$ in the scenario of an (8×4)-element SDMA-OFDM system.

modules. Specifically, $I = 3$ denotes the SD, $I = 2$ represents the URC decoder II and $I = 1$ denotes the channel decoder I of Figure 10.36. Hence, the vertical coordinates of the points A_1, A_3 and A_5 in Figure 10.41 quantify the $I_{E(u_2)}$ value measured at the output of the URC decoder II corresponding to its three successive activations during the first iteration, respectively, while the segments between A_1 and A_2 as well as between A_3 and A_4 represent two successive activations of the channel decoder I during the first iteration, respectively. The segment between A_5 and A_6 in Figure 10.41 denotes the beginning of a new iteration associated with similar decoding activations.

Figure 10.42 depicts the computational complexity – which is quantified in terms of the number of PED evaluations corresponding to the term ϕ of Equation (9.25) – imposed by the SD versus E_b/N_0 plot for the above-mentioned receivers. Note that the computational complexity imposed by the K-best SD dispensing with the centre-shifting scheme remains constant for both two-stage and three-stage receivers regardless of the SNR and the number of iterations, under the assumption that the buffer size is sufficiently large to store the resultant candidate list $\mathcal{L}$, which is generated by the SD just once during the first iteration between the SD and the channel decoder. On the other hand, every time the search centre $\mathbf{x}_c$ in the transmit domain is updated, the SD is required to regenerate the candidate list. However, as observed in Figure 10.42, the candidate list size N_{cand} can be substantially reduced with the aid of the centre-shifting scheme, hence the resultant overall complexity imposed by the SD becomes significantly lower than that of the receiver using no centre shifting. Explicitly, the candidate list generation complexity of the SIC-MMSE centre-shifting-aided two-stage receiver is well below that of the receiver using no centre shifting right across the SNR range spanning from 2 to 12 dB. This statement is valid if our aim is to achieve the near-MAP BER performance quantified in Figure 10.40, which can be attained by having $K = N_{cand} = 1024$ for the system operating without the centre-shifting scheme or by setting $K = N_{cand} = 32$ in the presence of the centre-shifting scheme. Actually, the number of PED evaluations carried out per channel use by the system dispensing with the centre-shifting scheme remains as high as 13 652, regardless of the SNR and the

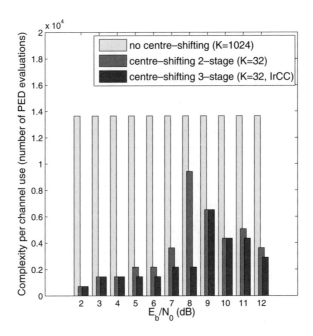

Figure 10.42: Complexity reduction achieved by the three-stage iterative receiver using the K-best SD in the scenario of an (8×4)-element SDMA-OFDM system.

number of iterations. On the other hand, in the presence of the centre-shifting scheme, the candidate list has to be regenerated at each iteration; nonetheless, the total complexity imposed is substantially reduced. We can also observe from Figure 10.42 that the centre-shifting K-best SD employed by the URC-aided three-stage system imposes a computational complexity which is even below that of its centre-shifting-aided two-stage counterpart, while achieving a performance gain of 2 dB for the target BER of 10^{-5}, as seen in Figure 10.40. Hence, the significant complexity reduction facilitated by the proposed SD scheme in the context of the three-stage receiver outweighs the relatively small additional complexity cost imposed by the URC, which only employs a two-state trellis, leading to an overall reduced complexity. Furthermore, in addition to the complexity reduction achieved by the proposed scheme, another benefit is the attainable memory reduction, since there is no need to store the resultant candidate list for the forthcoming iterations. As a result, the memory size required can be substantially reduced by having a significantly reduced value of K.

10.5 Chapter Conclusions

In this chapter, our main objective was to reduce the complexity encountered by the conventional LSD in the channel-coded iterative receiver and to contrive a near-capacity design for the SD-aided MIMO system. To be specific, although the conventional LSD was capable of achieving a significant complexity reduction in comparison with the exact MAP detector, it may still impose a potentially excessive complexity, since the LSD has to generate soft information for every transmitted bit, which requires the observation of a high number of hypotheses about the transmitted MIMO symbol, thus generating a large candidate list to represent the entire lattice. This complexity problem may be aggravated by supporting an increased number of users and/or using a high-order modulation scheme, especially in high-dimensional rank-deficient MIMO systems. Therefore, in order to maintain a near-MAP performance, while relying on a small set of symbol hypotheses, we proposed two complexity-

Table 10.8: Summary of the SD-aided receiver investigations of Chapter 10. Note that the computational complexity of the SD, i.e. the list generation by the SD, is calculated in terms of the total number of PED evaluations, while that of the soft-information generation by the SD/MAP detector is computed on the basis of Equation (10.16) in terms of the total number of OF evaluations corresponding to the two terms in Equation (10.14).

Performance and computational complexity of various SD-aided receivers in an (8×4)-element 4-QAM SDMA-OFDM rank-deficient system (target BER: 10^{-5})							
SD-aided receiver		$N_{cand}(K)$	Iterations	SNR	Memory	SD compl.	MAP compl.
Two-stage receiver (RSC)	Conventional SD (no centre shifting, no ALT)	1024	3	10.5	8196	13652	49152
		128	3	11.2	1024	2388	6144
		64	2	12	512	1364	2048
		32	2	12.8	256	724	1024
		16	2	15	128	404	512
	SIC-MMSE-aided centre-shifting SD (no ALT)	64	3	10.2	512	4092	3072
		32	3	10.2	256	2172	1536
		16	3	11	128	1212	768
	ALT-aided SD (no centre shifting)	128	4	10.2	1024	5070	8192
		16	6	12.8	128	N/A	1536
	SIC-MMSE-aided centre-shifting SD using ALT	32	4	10.8	256	1490	2048
Three-stage receiver (IrCC)	SIC-MMSE aided centre-shifting SD IrCC, $r = 0.5$	32	9	9.2	256	6520	4608

reduction techniques, namely the iterative centre-shifting-based SD scheme of Section 10.2 and the ALT-assisted SD scheme of Section 10.3, both of which rely on exploitation of the soft-bit information delivered by the outer channel decoder in the iterative receiver.

More specifically, in Section 10.2.3 three different algorithms were proposed for the iterative centre-shifting SD, namely the DHDC, the DSDC and the SIC-MMSE- assisted search-centre calculation schemes. It was shown in Figure 10.21(b) that the SIC-MMSE-aided scheme outperforms the other two. This is not unexpected, since although the SIC-MMSE-aided SD scheme imposes a slightly higher complexity in order to attain a more accurate search centre, a significant complexity reduction may be achieved, which is associated with the list generation and soft-bit-information calculation carried out by the SD. Our proposition in Section 10.2.2 was that the search centre may be generated by a more sophisticated detector than the LS or MMSE detector of conventional SDs. This generic proposition turned the SD into a high-flexibility detector, which may be beneficially combined with highly sophisticated or low-complexity linear or nonlinear detectors. In other words, the total affordable computational complexity may be flexibly split between the SD's search-centre calculation phase and the search phase.

Based on the exploitation of the soft-bit information, namely the a priori LLRs gleaned from the outer channel decoder, in Section 10.3 we proposed another reduced-complexity technique termed the ALT-aided SD scheme. Given the definition of the a priori LLRs, the sign of the LLR indicates whether the current bit is more likely to be $+1$ or -1, whereas the magnitude reflects how reliable the decision concerning the current bit is. Hence, the basic idea behind the ALT-aided SD scheme of Section 10.3 is as follows: when the absolute value of the a priori LLR of a specific bit is larger than the preset ALT threshold, we assume that we have reliable knowledge of this bit being 0 or 1. As a result, the tree search of the SD may be significantly simplified, since the number of tentative candidates for the corresponding tree search level may be reduced. As evidenced by Figure 10.26(b), the threshold has to be carefully adjusted to achieve the target performance as a function of the SNR encountered. As seen in Figure 10.24, the proposed ALT scheme is capable of providing useful performance improvements, although these are slightly less significant than those achieved by the SIC-MMSE-based centre-shifting-assisted SD scheme in Figure 10.21(a). This is because the non-Gaussian distribution of the LLRs recorded at the output of the ALT-aided SD in Figure 10.28(b) during the iterative detection process limits the efficiency of the iterative detection process and imposes difficulties on the EXIT-chart-assisted performance prediction.

Finally, motivated by the URC-aided three-stage SISO turbo equalizer of [575], in Section 10.4 we significantly improved the performance of the conventional two-stage SD-aided turbo receiver of Figure 10.1. We achieved this improvement by intrinsically amalgamating the SD with the decoder of a URC having an IIR, both of which were embedded in a channel-coded SDMA-OFDM transceiver, thereby creating the powerful three-stage serially concatenated scheme of Figure 10.36. To achieve a near-capacity performance, observed in Figure 10.36, IrCCs were used as the outer code for the proposed iterative centre-shifting SD-aided three-stage system. Consequently, we demonstrated in Figure 10.40 that at the target BER of 10^{-5}, an E_b/N_0 performance gain as high as 4.5 dB was attained by the system of Figure 10.36 using the MMSE-based centre-shifting SD relying on a low-complexity candidate list size of $N_{cand} = 32$, in comparison with its two-stage counterpart of Figure 10.18 in the challenging scenario of an (8×4)-element rank-deficient 4-QAM SDMA-OFDM uplink system.

In Table 10.8 we quantitatively summarize and compare the different SD-aided receivers' performance as well as complexity in the context of an (8×4)-element 4-QAM SDMA-OFDM rank-deficient system. As observed in Table 10.8, the combination of the SIC-MMSE-aided centre-shifting and the ALT schemes is capable of achieving a near-MAP performance at the lowest complexity for the conventional two-stage receiver. A further performance gain of 1.6 dB may be attained by our near-capacity design using the URC-aided three-stage receiver.

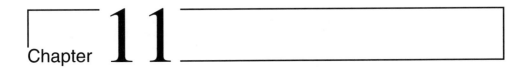

Chapter 11

Sphere-Packing Modulated STBC-OFDM and its Sphere Detection

11.1 Introduction

11.1.1 System Model

In previous chapters, the SDMA-OFDM system supporting U single-antenna-assisted MSs was considered, where the multi-user data streams sharing the same time/frequency channel can be distinguished at the BS with the aid of their unique user-specific spatial signature constituted by their CIRs, resulting in a potentially significant increase in spectral efficiency. The separability of the individual MIMO links relies on the presence of rich multi-path propagation, which requires a sufficiently high antenna spacing, in order to ensure that the individual channels between pairs of the transmit and receive antennas exhibit an independent Rayleigh distribution and the absence of a strong Line-Of-Sight (LOS) path. In light of this interpretation, SDMA transmissions can be viewed as MIMO schemes maximizing the system's overall throughput.

On the other hand, instead of maximizing the system's overall throughput, another powerful family of MIMO schemes was designed for achieving spatial diversity, which is usually quantified in terms of the number of decorrelated spatial branches available at the transmitter or receiver. This number is referred to as the *diversity order*. An effective and practical way of achieving MIMO-aided diversity is to employ space–time coding (STC), which is a specific coding technique designed for MIMO systems equipped with multiple spatially separated transmit antennas. When the signals of a specific user are launched from different transmit antennas during different time slots, the independently fading channels are less likely to encounter deep fades simultaneously. Hence, the MIMO system is capable of exploiting the independently fading paths of multi-path propagation environments, provided that the transmit antennas are sufficiently far from each other so that they experience independent fading. In practice the separation of the antennas has to be of the order of at least a few wavelengths, otherwise the maximum achievable diversity gain may erode. At a carrier frequency of 2 GHz the wavelength is 15 cm and hence the required antenna separation would be in excess of, say, half a metre – a distance which is affordable at the BS, but not at the MS. Fortunately, single-antenna-aided MSs can share their antennas with the aid of cooperative communication principles, as we will detail later in the book.

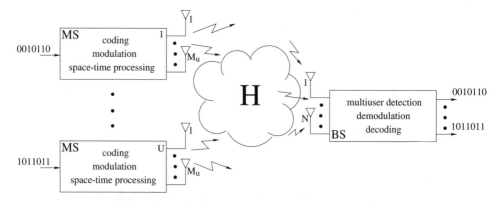

Figure 11.1: Schematic of a generalized SDMA-OFDM wireless uplink transmission system.

As argued above, the family of MIMO transmission schemes may be subdivided into two main categories: those designed for achieving spatial multiplexing and spatial diversity, which aim to maximize the data rate and to minimize the transmission error rate, respectively. It is worth noting, however, that there have been proposals in the literature which are capable of striking a more flexible compromise between the achievable rate and reliability [582]. Therefore, in this chapter we consider a UL Multi-User MIMO (MU-MIMO) system, where the SDMA technique used as a spatial multiple-access scheme significantly increases the system's overall throughput without requiring extra spectrum. Meanwhile, in order to attain the spatial diversity gains, multiple antennas are assumed to be employed by each user. Thus, the multiple-transmit-antenna-assisted STC scheme may be employed by each uplink user to provide diversity gains in addition to the spatial multiplexing gain, which renders the system more robust to hostile wireless fading channels.

More specifically, the generalized SDMA-OFDM UL transmission scenario is as depicted in Figure 11.1, where each of the U synchronous co-channel users/MSs is equipped with M_u transmit antennas and employs a specific STC scheme, while the BS has N receive antenna elements. Hence, under the assumption that each MS is equipped with the same number of transmit antenna elements, the total number of transmit antennas is $M = M_1 + M_2 + \cdots + M_U = U \cdot M_u$. The space–time modulator of the uth MS maps a *sequence* of bits $b(k)_{k=0}^{B-1}$ to a *sequence* of $M_u \times 1$ symbol vectors $s(t)_{t=0}^{J-1}$, which are transmitted with the aid of its own M_u transmit antenna elements during J successive symbol periods of duration T. Generally, each space–time signal of the uth MS can be expressed as a $(M_u \times J)$-element matrix given by

$$\mathbf{S} = [\mathbf{s}^{(u)}(T), \mathbf{s}^{(u)}(2T), \dots, \mathbf{s}^{(u)}(jT)] \tag{11.1}$$

$$= \begin{bmatrix} s_1^{(u)}(T) & s_1^{(u)}(2T) & \cdots & s_1^{(u)}(JT) \\ s_2^{(u)}(T) & s_2^{(u)}(2T) & \cdots & s_2^{(u)}(JT) \\ \vdots & \vdots & \ddots & \vdots \\ s_{M_u}^{(u)}(T) & s_{M_u}^{(u)}(2T) & \cdots & s_{M_u}^{(u)}(JT), \end{bmatrix}_{M_u \times J}, \tag{11.2}$$

where $s_m^{(u)}(jT)$ denotes the symbol transmitted by the transmit antenna element m $(m = 1, 2, \dots, M_u)$ of user u $(u = 1, 2, \dots, U)$ during the jth symbol period $(j = 1, 2, \dots, J)$. Hence, the resultant average rate, or throughput, of a specific space–time code is B/J bps/Hz. Note that for a pure spatial multiplexing system, we have $J = 1$, since there is no spatial or temporal correlation introduced among the transmitted signals.

The multiple transmit antennas of each MS are assumed to be positioned as far apart as possible to ensure that signals launched from the different transmit antennas experience independent or spatially uncorrelated fading channels.[1] Hence, at an arbitrary time, the link between each pair of transmit and receiver antennas may be characterized with the aid of a unique transmit-antenna-specific FD-CTF, which is denoted as $h_{nm}^{(u)}$. The superscript and subscript on h represent the user and the antenna element index, respectively. For example, the FD-CTF or the spatial signature of transmit antenna element m of the uth user can be expressed as a column vector:

$$\mathbf{h}_m^{(u)} = [h_{1m}^{(u)}, h_{2m}^{(u)}, \ldots, h_{Nm}^{(u)}]^T, \tag{11.3}$$

with $m \in 1, \ldots, M_u$ and $u \in 1, \ldots, U$. If the mth transmit antenna element's signal is denoted by $s_m^{(u)}$, and the received signal plus the additive white Gaussian noise (AWGN) at receive antenna element n is represented by y_n and w_n, respectively, the entire SDMA system can be described by a matrix equation written as

$$\mathbf{y} = \mathbf{H}\mathbf{s} + \mathbf{w}, \tag{11.4}$$

where the received signal vector is $\mathbf{y} \in \mathbb{C}^{N \times 1}$, the transmitted signal vector is $\mathbf{s} \in \mathbb{C}^{M \times 1}$, and the noise vector is $\mathbf{w} \in \mathbb{C}^{N \times 1}$, which are given by the following equations, respectively:

$$\mathbf{y} = [y_1, y_2, \ldots, y_N]^T, \tag{11.5}$$

$$\mathbf{s} = [s_1^{(1)}, s_2^{(1)}, \ldots, s_{M_1}^{(1)}, \ldots, s_1^{(U)}, s_2^{(U)}, \ldots, s_{M_U}^{(U)}]^T, \tag{11.6}$$

$$= [s_1, s_2, \ldots, s_M]^T, \tag{11.7}$$

$$\mathbf{w} = [w_1, w_2, \ldots, w_N]^T. \tag{11.8}$$

The system's overall FD-CTF matrix $\mathbf{H} \in \mathbb{C}^{N \times M}$ consists of M FD-CTF column vectors $\mathbf{h}_m^{(u)} \in \mathbb{C}^{N \times 1}$, which correspond to the spatial signature of the mth transmit antenna element of the uth user, as defined by Equation (11.3). Hence, the FD-CTF matrix $\mathbf{H}$ can be expressed as

$$\mathbf{H} = \underbrace{[\mathbf{h}_1^{(1)}, \mathbf{h}_2^{(1)}, \ldots, \mathbf{h}_{M_1}^{(1)}, \ldots, \mathbf{h}_1^{(U)}, \mathbf{h}_2^{(U)}, \ldots, \mathbf{h}_{M_U}^{(U)}]}_{M \ columns}, \tag{11.9}$$

where each column represents a certain transmit antenna's unique spatial signature of a specific user. Here, we assume that the FD-CTF $h_{nm}^{(u)}$ between the uth user's transmit antenna element $m \in 1, 2, \ldots, M_u$ and receive antenna element $n \in 1, 2, \ldots, N$ are independent, stationary, complex-valued Gaussian-distributed processes with a zero mean and a unit variance [542]. Furthermore, both the mth transmit antenna element's signal $s_m^{(u)}$ of user u and the AWGN noise, w_n, at the nth antenna element exhibit a zero mean and a variance of σ_s^2 and $2\sigma_w^2$, respectively. Note that the elements of the matrix $\mathbf{H}$ represent the FD-CTF, since our SDMA systems are considered to be combined with OFDM systems [542] as briefly discussed in Section 1.1.1.

Finally, although the above-mentioned system model describes a generalized MU-MIMO transmission scheme, it is also applicable to a Single-User MIMO (SU-MIMO) scenario, when setting $U = 1$.

11.1.2 Chapter Contributions and Outline

The concept of combining orthogonal transmit diversity designs with the principle of Sphere Packing (SP) [583] was introduced by Su *et al.* in [584] to maximize the achievable coding advantage, demonstrating that the proposed SP-aided STBC (STBC-SP) scheme was capable of outperforming the conventional orthogonal designed STBC schemes of [585–587] in the SU-MIMO DL scenario. Against this background, our main contribution in this chapter is the challenging design of the K-best

---

[1]In the case of compact handheld communicators, the affordable separation may not be sufficiently high to ensure independent fading in the uplink, since the wavelength at a carrier frequency of 2 GHz is 15 cm, hence perhaps a separation of half the wavelength is affordable. A laptop backplane is more suitable for providing a higher antenna separation, but even in this case the associated correlation of the signals would erode the achievable transmit diversity gain. More independent fading may be ensured by forming a virtual MIMO from the single antennas of cooperating users.

SD for SP-modulated systems, which extends the employment of STBC-SP schemes to MU-MIMO scenarios, while approaching the MAP performance at a moderate complexity. More specifically, the novel contributions of this chapter are listed as follows:

- We improve the STBC performance by jointly designing the space–time signals of the two time slots of an SDMA UL scheme using SP modulation, while existing orthogonal designs make no attempt to do so owing to its potentially complex detection.

- We solve this potential complexity problem by further developing the K-best SD for detection of SP modulation, because SP offers a substantial SNR reduction at the cost of increased complexity, which is reduced by the new SD.

The remainder of this chapter is organized as follows. The fundamentals of orthogonal STBC schemes are briefly reviewed in Section 11.2.1, followed by a discourse on the orthogonal design of STBC schemes using SP modulation in Section 11.2.2. Then, in Section 11.3 our SD design contrived for the STBC-SP-assisted MU-MIMO system is detailed. More specifically, based on the bit-by-bit MAP detection scheme designed for the STBC-SP-aided MU-MIMO system derived in Section 11.3.1, a multi-layer tree search referred to as the user-based tree search is proposed in Section 11.3.2 in order to render the conventional SD applicable to the above-mentioned SP-modulated scenario. Finally, we provide our concluding remarks in Section 11.4.

11.2 Orthogonal Transmit Diversity Design with SP Modulation

11.2.1 STBCs

11.2.1.1 STBC Encoding

STBCs describe the relationship between the original symbol stream stored in the column vector $\mathbf{x}$ and the redundant signal replicas artificially constructed at the transmitter for transmission from the different antennas during different time slots or symbol periods. Generally, an STBC can be described by an $(M_u \times J)$-dimensional transmission matrix as defined earlier in Equation (11.2) of Section 11.1.1.

A simple but elegant Orthogonal STBC (OSTBC) scheme employing two transmit antennas was discovered by Alamouti [585] and was later generalized by Tarokh *et al.* in [586] to an arbitrary number of antennas. This remarkable scheme enables the receiver to perform ML detection based on low-complexity linear processing, yet achieving the maximum attainable transmit diversity by imposing a low extra encoding complexity at the transmitter. The corresponding block diagram of Alamouti's STBC-aided transmitter employing a constellation size of M_c symbols is shown in Figure 11.2, where $2 \log_2 M_c$ bits of the information source are fed into the constellation mapper in order to generate the modulated symbols x_1 and x_2. Instead of using spatial multiplexing to double the throughput in comparison with its single-antenna-based counterpart, the two-transmit-antenna-aided MS launches the signals x_1 and x_2 as well as their conjugates simultaneously from the two antennas during two successive symbol periods or time slots. To be more specific, during the first symbol period, x_1 and x_2 are transmitted from antennas Tx1 and Tx2, respectively. Then, in the forthcoming symbol period, $-x_2^*$ is assigned to antenna Tx1 and x_1^* is assigned to antenna Tx2, so that correlation is introduced in both time and spatial domains. Hence, according to the generalized STBC transmission matrix defined in Equation (11.2), during the two consecutive symbol periods of jT and $(j+1)T$ the uth MS associated with the transmitted codeword of Alamouti's scheme – also known as the $\mathbf{G}_2$ STBC scheme – can be

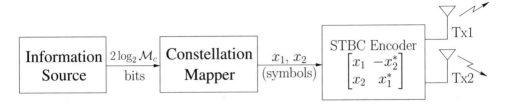

Figure 11.2: Transmit diversity using Alamouti's G2 STBC.

represented with the aid of the following matrix:

$$
\begin{aligned}
\mathbf{G}_2 &= \begin{bmatrix} s_1^{(u)}(jT) & s_1^{(u)}((j+1)T) \\ s_2^{(u)}(jT) & s_2^{(u)}((j+1)T) \end{bmatrix} \\
&= \begin{bmatrix} x_1^{(u)} & -x_2^{(u)*} \\ x_2^{(u)} & x_1^{(u)*} \end{bmatrix},
\end{aligned} \tag{11.10}
$$

where the column index of the matrix denotes the symbol period index, while the row index represents the transmit antenna index.

11.2.1.2 Equivalent STBC Channel Matrix

A key assumption when employing the above $\mathbf{G}_2$ scheme is that the channel magnitude and phase are quasi-static, as defined for $h_{nm}^{(u)}(t)$ in Section 11.1.1, implying that the FD-CTF observed for the path between the uth MS's mth transmit antenna and the nth receive antenna of the BS at time instant t remains constant during two consecutive symbol periods. Explicitly, in the context of the $\mathbf{G}_2$ scheme we arrive at

$$
h_{nm}^{(u)}(jT) = h_{nm}^{(u)}[(j+1)T], \tag{11.11}
$$

where $m = 1, 2, n = 1, 2, \ldots, N$ and $u = 1, 2, \ldots, U$ denote the indices of the transmit antennas of a specific MS, of the receive antenna and of the MS, respectively. Furthermore, it is also assumed throughout this chapter that the FD-CTF $h_{nm}^{(u)}(t)$ is perfectly known at the receiver. Therefore the noise-contaminated signals of the uth MS received by the nth antenna at the BS during the jth symbol period can be expressed as

$$
y_n^{(u)}(jT) = \sum_{m=1}^{M_u} h_{nm}^{(u)}(jT)s_m^{(u)}(jT) + w_n(jT), \tag{11.12}
$$

where M_u is the number of transmit antennas employed by the uth MS. Furthermore, with the aid of Equation (11.10) and Equation (11.11), we can expand Equation (11.12) to obtain an expression for the signal of the uth MS received by the nth antenna in two consecutive symbol periods, respectively, as follows:

$$
y_n^{(u)}(jT) = h_{n1}^{(u)}(jT)x_1^{(u)} + h_{n2}^{(u)}(jT)x_2^{(u)} + w_n(jT), \tag{11.13}
$$

$$
y_n^{(u)}((j+1)T) = -h_{n1}^{(u)}(jT)x_2^{(u)*} + h_{n2}^{(u)}(jT)x_1^{(u)*} + w_n((j+1)T). \tag{11.14}
$$

For notational simplicity, the time index can be omitted. Consequently, we have

$$
y_{1,n}^{(u)} = h_{n1}^{(u)}x_1^{(u)} + h_{n2}^{(u)}x_2^{(u)} + w_{1,n}, \tag{11.15}
$$

$$
y_{2,n}^{(u)} = -h_{n1}^{(u)}x_2^{(u)*} + h_{n2}^{(u)}x_1^{(u)*} + w_{2,n}. \tag{11.16}
$$

According to [314, 588], Equation (11.15) and Equation (11.16) together can be rewritten in a more compact matrix form as

$$\tilde{\mathbf{y}}_n^{(u)} = \tilde{\mathbf{H}}_n^{(u)} \cdot \mathbf{x}^{(u)} + \tilde{\mathbf{w}}_n, \tag{11.17}$$

where

$$\mathbf{x}^{(u)} = [x_1^{(u)} \quad x_2^{(u)}]^T \tag{11.18}$$

represents the transmitted symbols of the uth MS during two consecutive symbol periods, and $\tilde{\mathbf{H}}_n^{(u)}$ is defined as the *equivalent STBC channel matrix* between the M_u transmit antennas of the uth MS and the nth receive antenna at the BS, which can be expressed for the $\mathbf{G}_2$ scheme as [314, 588]

$$\tilde{\mathbf{H}}_n^{(u)} = \begin{bmatrix} h_{n1}^{(u)} & h_{n2}^{(u)} \\ h_{n2}^{(u)*} & -h_{n1}^{(u)*} \end{bmatrix}. \tag{11.19}$$

Moreover, $\tilde{\mathbf{y}}$ of Equation (11.17) is defined as the *equivalent received signal vector*, which is given by

$$\tilde{\mathbf{y}}_n^{(u)} = [y_{1,n}^{(u)} \quad y_{2,n}^{(u)*}]^T, \tag{11.20}$$

where the first element $y_{1,n}$ corresponds to the signal received by the nth antenna during the first symbol period and the second element $y_{2,n}^*$ is the conjugate of the signal received at the same antenna during the second symbol period, while $\tilde{\mathbf{w}}$ of Equation (11.17) is referred to as the *equivalent noise vector*, which is written as

$$\tilde{\mathbf{w}}_n = [w_{1,n} \quad w_{2,n}^*]^T, \tag{11.21}$$

where, again, $w_{1,n}$ denotes the AWGN imposed on the nth receive antenna during the first symbol period and $w_{2,n}^*$ is the conjugate of the AWGN inflicted during the second symbol period. The AWGN encountered during each symbol period has a zero mean and a variance of $2\sigma_w^2$.

11.2.1.3 STBC Diversity Combining and Maximum Likelihood Detection

Without loss of generality, let us now consider a $\mathbf{G}_2$-assisted SU-MIMO system supporting only the uth MS with the aid of a single receive antenna at the BS, whose equivalent system model may be expressed as

$$\tilde{\mathbf{y}} = \tilde{\mathbf{H}} \cdot \mathbf{x} + \tilde{\mathbf{w}} \tag{11.22}$$

$$= \begin{bmatrix} h_{11}^{(u)} & h_{12}^{(u)} \\ h_{12}^{(u)*} & -h_{11}^{(u)*} \end{bmatrix} \begin{bmatrix} x_1^{(u)} \\ x_2^{(u)} \end{bmatrix} + \begin{bmatrix} w_{1,1} \\ w_{2,1}^* \end{bmatrix}. \tag{11.23}$$

For notational simplicity, both the receive antenna index and the user index are omitted, resulting in

$$\tilde{\mathbf{y}} = \begin{bmatrix} h_1 & h_2 \\ h_2^* & -h_1^* \end{bmatrix} \begin{bmatrix} x_1 \\ x_2 \end{bmatrix} + \begin{bmatrix} w_1 \\ w_2^* \end{bmatrix}. \tag{11.24}$$

In light of the orthogonality of $\tilde{\mathbf{H}}_n^{(u)}$ of Equation (11.19), we multiply both sides of Equation (11.24) by the conjugate transpose of $\tilde{\mathbf{H}}$, yielding

$$\breve{\mathbf{y}} = \tilde{\mathbf{H}}^H \cdot \tilde{\mathbf{y}} \tag{11.25}$$

$$= \begin{bmatrix} |h_1|^2 + |h_2|^2 & 0 \\ 0 & |h_1|^2 + |h_2|^2 \end{bmatrix} \begin{bmatrix} x_1 \\ x_2 \end{bmatrix} + \breve{\mathbf{w}} \tag{11.26}$$

$$= \begin{bmatrix} (|h_1|^2 + |h_2|^2)x_1 + h_1^* w_1 + h_2 w_2^* \\ (|h_1|^2 + |h_2|^2)x_2 + h_2^* w_1 - h_1 w_2^* \end{bmatrix}, \tag{11.27}$$

where $\breve{\mathbf{w}} = \tilde{\mathbf{H}}^* \cdot \tilde{\mathbf{w}}$ has a zero mean and a covariance of $(|h_1|^2 + |h_2|^2) \cdot \mathbf{I}_2$, while the elements of $\breve{\mathbf{w}}$ are i.i.d. [588]. The process of obtaining the estimates of the transmitted symbol vector as outlined in

Equation (11.25) is also referred to as *STBC-aided diversity combining*. Then the estimated vector $\check{\mathbf{y}}$ is forwarded to the ML detector, which uses the detection rule outlined in [588]:

$$\hat{\mathbf{x}} = \arg\min_{\check{\mathbf{x}} \in \mathcal{M}_c^M} \|\check{\mathbf{y}} - (|h_1|^2 + |h_2|^2) \cdot \check{\mathbf{x}})\|^2, \tag{11.28}$$

where $\mathcal{M}_c$ is the constellation set of the modulation scheme. Therefore, $\mathcal{M}_c^M$ denotes the M-dimensional legitimate constellation set.

According to Equation (11.27), the x_1 and x_2 transmitted from the two transmit antennas during the two successive symbol periods do not interfere with each other's estimates, which correspond to the first and second elements of $\check{\mathbf{y}}$. As a result, the observation enables us to 'decompose' the ML detection rule into two independent low-complexity detection operations for x_1 and x_2, as follows [585]:

$$\hat{x}_1 = \arg\min_{\check{x}_1 \in \mathcal{M}_c}(|h_1|^2 + |h_2|^2 - 1)|\check{x}_1|^2 + d^2(\check{y}_1, \check{x}_1), \tag{11.29}$$

$$\hat{x}_2 = \arg\min_{\check{x}_2 \in \mathcal{M}_c}(|h_1|^2 + |h_2|^2 - 1)|\check{x}_2|^2 + d^2(\check{y}_2, \check{x}_2), \tag{11.30}$$

where $d^2(\mathbf{x}, \mathbf{y})$ is the *squared Euclidean distance* between the signal vector $\mathbf{x}$ and $\mathbf{y}$. Typically, for constant modulus modulation schemes, such as BPSK or QPSK, the detection criteria described by Equation (11.29) and Equation (11.30) can be further simplified as [585]

$$\hat{x}_1 = \arg\min_{\check{x}_1 \in \mathcal{M}_c} d^2(\check{y}_1, \check{x}_1), \tag{11.31}$$

$$\hat{x}_2 = \arg\min_{\check{x}_2 \in \mathcal{M}_c} d^2(\check{y}_2, \check{x}_2). \tag{11.32}$$

Consequently, the original ML detector's search space is substantially reduced from $\mathcal{M}_c^M$ to $(M \cdot \mathcal{M}_c)$, resulting in a reduced detection complexity, while maintaining the ML performance as well as the maximum achievable transmit diversity gain.

For multiple-receive-antenna-aided scenarios, i.e. for $N \geq 2$, the same STBC decoding process, namely that of Equation (11.25), can be invoked for each receive antenna and then the outputs of the antennas are combined, before passing them through the ML detector. Assuming that there are N receive antennas, the STBC decoding process is represented as

$$\check{\mathbf{y}} = \tilde{\mathbf{H}}^H \cdot \tilde{\mathbf{y}} \tag{11.33}$$

$$= (|h_{11}|^2 + |h_{12}|^2 + \cdots + |h_{N1}|^2 + |h_{N2}|^2) \cdot \mathbf{I}_2 \cdot \begin{bmatrix} x_1 \\ x_2 \end{bmatrix} + \check{\mathbf{w}} \tag{11.34}$$

$$= \begin{bmatrix} \sum_{n=1}^N (|h_{n1}|^2 + |h_{n2}|^2)x_1 + \check{w}_1 \\ \sum_{n=1}^N (|h_{n1}|^2 + |h_{n2}|^2)x_2 + \check{w}_2 \end{bmatrix}, \tag{11.35}$$

where

$$\tilde{\mathbf{H}} = \begin{bmatrix} \tilde{\mathbf{H}}_1^{(1)} \\ \tilde{\mathbf{H}}_2^{(1)} \\ \vdots \\ \tilde{\mathbf{H}}_N^{(1)} \end{bmatrix} = \begin{bmatrix} h_{11} & h_{12} \\ h_{12}^* & -h_{11}^* \\ \vdots & \\ h_{N1} & h_{N2} \\ h_{N2}^* & -h_{N1}^* \end{bmatrix} \tag{11.36}$$

and $\check{w}_1$ and $\check{w}_2$ are the noise terms corresponding to the first and second components of $\check{\mathbf{w}} = \tilde{\mathbf{H}}^H \cdot \tilde{\mathbf{w}}$. Then the estimated vector $\check{\mathbf{y}}$ is fed into the ML detector, which invokes the detection rules described by Equation (11.31) and Equation (11.32) in order to recover the transmitted symbol, if a constant modulus constellation is employed. Otherwise, we have the generalized detection criterion for our $\mathbf{G}_2$-aided system having N receive antennas, which are given by

$$\hat{x}_1 = \arg\min_{\breve{x}_1 \in \mathcal{M}_c} \left(\sum_{n=1}^{N} (|h_{n1}|^2 + |h_{n2}|^2) - N \right) |\breve{x}_1|^2 + d^2(\breve{y}_1, \breve{x}_1), \tag{11.37}$$

$$\hat{x}_2 = \arg\min_{\breve{x}_2 \in \mathcal{M}_c} \left(\sum_{n=1}^{N} |(h_{n1}|^2 + |h_{n2}|^2) - N \right) |\breve{x}_2|^2 + d^2(\breve{y}_2, \breve{x}_2). \tag{11.38}$$

Hence, according to Equation (11.35), when employing N receive antennas, a total transmit and receive diversity associated with the diversity order of $M \cdot N = 2N$ can be achieved by the $\mathbf{G}_2$-aided system without having a transmission rate less than unity in comparison with the single-antenna-aided multi-user system, since the coding rate of the $\mathbf{G}_2$ STBC is unity. Additionally, as observed from Equation (11.35) for a single-user system, after the STBC decoding operation of Equation (11.33), there is no Multi-Stream Interference (MSI), as a benefit of the orthogonality of the equivalent channel matrix $\tilde{\mathbf{H}}_n^{(u)}$ of Equation (11.19).

11.2.1.4   Other STBCs and Orthogonal Designs

Again, orthogonal STBC designs have recently attracted considerable interests in multiple-antenna-aided wireless systems, as motivated by the STBC scheme proposed by Alamouti in [585] for a two-transmit-antenna scenario, which was further generalized for an arbitrary number of transmit antennas by Tarokh *et al.* in [586]. In [586], Tarokh *et al.* also showed that the maximum achievable rate of OSTBC schemes designed for complex-valued constellations cannot exceed one, i.e. we have $R \leq 1$. Later it was shown in [589] by Liang and Xia that this rate is in fact always smaller than unity, i.e. we have $R < 1$, when the number of transmit antennas exceeds two. Recently, Su and Xia in [587] proved that the rate cannot exceed $3/4$ for more than two transmit antennas.

According to [587], the process of square-shaped orthogonal encoder-matrix design can be carried out by commencing from $\mathbf{G}_1(x_1) = x_1\mathbf{I}_1$, and then recursively invoking the OSTBC construction equation as follows:

$$\mathbf{G}_{2^k}(x_1, \ldots, x_{k+1}) = \begin{bmatrix} \mathbf{G}_{2^{k-1}}(x_1, \ldots, x_k) & -x_{k+1}^*\mathbf{I}_{2^{k-1}} \\ x_{k+1}\mathbf{I}_{2^{k-1}} & \mathbf{G}_{2^{k-1}}^H(x_1, \ldots, x_k) \end{bmatrix} \tag{11.39}$$

for $k = 1, 2, 3, \ldots$, where $(\cdot)^*$ represents the conjugate of a complex symbol, $(\cdot)^H$ denotes the Hermitian of a complex matrix, and $\mathbf{I}_n$ is a $(n \times n)$-element identity matrix. Again, the rows and columns respectively represent the spatial and temporal dimensions. Therefore, $\mathbf{G}_{2^k}$ is an orthogonal design of $(2^k \times 2^k)$ elements, which determines how to transmit $(k + 1)$ complex modulated symbols, i.e. $x_1, x_2, \ldots, x_{k+1}$, from 2^k transmit antennas during 2^k consecutive symbol periods. Hence, the resultant symbol rate of $\mathbf{G}_{2^k}$ is equal to $(k + 1)/2^k$. Considering the OSTBC design contrived for four transmit antennas for example, where we have $k = 2$, we arrive at

$$\mathbf{G}_4(x_1, x_2, x_3, x_4) = \begin{bmatrix} x_1 & -x_2^* & -x_3^* & 0 \\ x_2 & x_1^* & 0 & -x_3^* \\ x_3 & 0 & x_1^* & x_2^* \\ 0 & x_3 & -x_2 & x_1 \end{bmatrix}. \tag{11.40}$$

11.2.2   Orthogonal Design of STBC Using SP Modulation

11.2.2.1   Joint Orthogonal Space–Time Signal Design for Two Antennas Using SP

Conventionally, the orthogonal design of STBCs [585–587] discussed in Section 11.2.1 is based on conventional PSK/QAM modulated symbols. In other words, the inputs $(x_1, x_2, \ldots, x_{k+1})$ of the STBC encoder are chosen independently from the constellation corresponding to a specific modulation scheme, then mapped to 2^k transmit antennas using for example Equation (11.39), which are then transmitted during 2^k consecutive symbol periods. Therefore, no effort was made by Alamouti's scheme to design jointly the input symbols $(x_1, x_2, \ldots, x_{k+1})$. However, it was shown by Su *et al.* in [584] that combining the orthogonal design with SP [583] is capable of attaining extra coding gains by maximizing the diversity product[2] of the STBC signals in the presence of temporally correlated fading channels. The diversity product expression for the square-shaped OSTBC matrix $\mathbf{G}_{2^k}$ expressed in the context of time-correlated fading channels is given by [584]

$$\zeta_{\mathbf{G}_{2^k}} = \frac{1}{2\sqrt{k+1}} \min_{(x_1,\ldots,x_{k+1}) \neq (\tilde{x}_1,\ldots,\tilde{x}_{k+1})} \left(\sum_{i=1}^{k+1} |x_i - \tilde{x}_i|^2 \right)^{1/2}, \tag{11.41}$$

where x_i and $\tilde{x}_i$ are the elements of two distinct space–time signalling matrices C and $\tilde{C}$, respectively. From Equation (11.41) we can observe that the diversity product is actually determined by the Minimum Euclidean Distance (MED) among all the possible ST signal vectors $(x_1, x_2, \ldots, x_{k+1})$. Thus, the idea of combining the individual antenna signals into a joint ST design using SP is both straightforward and desirable, since the SP modulated symbols have the best known MED in the $2(k+1)$-dimensional real-valued Euclidean space $\mathbb{R}^{2(k+1)}$ [583]. Hence the system becomes capable of maximizing the achievable diversity product of STBC codes, which in turn minimizes the transmission error probability.

Without loss of generality, we consider the $\mathbf{G}_2$ scheme again as an simple example, where the corresponding space–time signalling matrix of Equation (11.10) is rewritten here for convenience:

$$\mathbf{G}_2(x_1, x_2) = \begin{bmatrix} x_1 & -x_2^* \\ x_2 & x_1^* \end{bmatrix}, \tag{11.42}$$

where the elements of the input vector (x_1, x_2) of the STBC encoder are chosen independently from PSK/QAM modulation constellations conventionally, e.g. BSPK or 4-QAM, as shown in Figure 11.3. Let us now define the lattice D_4 as an SP having the best MED from all other $(L-1)$ legitimate phasor points in four-dimensional real-valued Euclidean space $\mathbb{R}^4$ [583], which may also be defined as a lattice that consists of all legitimate SP constellation points having integer coordinates $[a_1, a_2, a_3, a_4]$. These coordinates uniquely and unambiguously describe the legitimate combinations of the two time slots' modulated symbols in the $\mathbf{G}_2$ scheme, while obeying the SP constraint of

$$a_1 + a_2 + a_3 + a_4 = p, \tag{11.43}$$

where p is an even integer [590]. Furthermore, each two-dimensional complex-valued input vector of the $\mathbf{G}_2$ scheme, i.e. (x_1, x_2), can be represented in the following way:

$$(x_1, x_2) = (\Re\{x_1\} + j\Im\{x_1\}, \Re\{x_2\} + j\Im\{x_2\}), \tag{11.44}$$

where $\Re\{\cdot\}$ and $\Im\{\cdot\}$ denote the real and imaginary components of a complex number. In other words, any two-dimensional complex-valued vector, i.e. (x_1, x_2), in the two-dimensional complex-valued space $\mathbb{C}^2$, can be represented by four real numbers, which as a whole correspond to the coordinates of a four-dimensional real-valued phasor in the $\mathbb{R}^4$ space represented in the following way:

$$(x_1, x_2) \Longleftrightarrow (\Re\{x_1\}, \Im\{x_1\}, \Re\{x_2\}, \Im\{x_2\}). \tag{11.45}$$

Hence, the joint design of (x_1, x_2) in the two-dimensional complex-valued space $\mathbb{C}^2$ is readily transformed into the four-dimensional real-valued Euclidean space $\mathbb{R}^4$. Explicitly, the procedure of joint signal design over two individual transmit antennas is portrayed in Figure 11.4, from which we can see that with the aid of the above-mentioned SP scheme, the joint ST signal design of the individual transmit antennas can be achieved by maximizing the coding advantage of $\mathbf{G}_2$ by maximizing the

---

[2]Defined as the estimated SNR gain over an uncoded system having the same diversity order as the coded system [586].

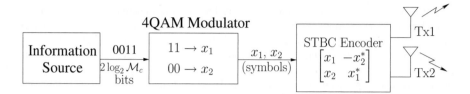

Figure 11.3: Alamouti's G2 STBC scheme using 4-QAM modulation.

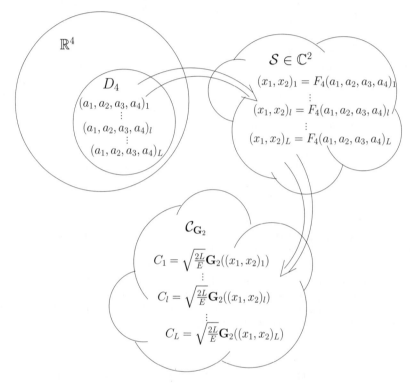

Figure 11.4: Procedure of joint ST signal design using SP for $\mathbf{G}_2$ STBC scheme.

Euclidean distance of the four-tuples [590]

$$(x_{l,1}, x_{l,2}) = F_4(a_{l,1}, a_{l,2}, a_{l,3}, a_{l,4})$$
$$= (a_{l,1} + ja_{l,2}, a_{l,3} + ja_{l,4}). \tag{11.46}$$

On choosing L legitimate constellation points from the lattice D_4 to construct a set denoted by $\mathcal{A} = \{\boldsymbol{a}_l = [a_{l,1}, a_{l,2}, a_{l,3}, a_{l,4}]^T \in \mathbb{R}^4 : 0 \le l \le L - 1\}$, the L resultant energy-normalized codewords given by

$$\mathbf{C}_l = \sqrt{\frac{2L}{E_{total}}} \mathbf{G}_2(F_4(\boldsymbol{a}_l)), \quad l = 0, 1, \ldots, L - 1, \tag{11.47}$$

where $E_{total} \triangleq \sum_{l=1}^{L}(|a_{l,1}|^2 + |a_{l,2}|^2 + |a_{l,3}|^2 + |a_{l,4}|^2)$, constitute the ST signal space $\mathcal{C}_{\mathbf{G}_2}$, whose diversity product $\zeta_{\mathbf{G}_2}$ is determined by the MED of the set $\mathcal{A}$ formulated in Equation (11.41).

Consequently, we arrive at the SP-aided $\mathbf{G}_2$ scheme portrayed in Figure 11.5, where an SP symbol represented by the two-dimensional complex-valued phasor points of $(x_{l,1}, x_{l,2})$ is transmitted over two

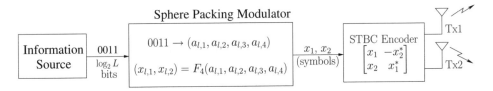

Figure 11.5: Alamouti's G2 STBC scheme using SP modulation.

symbol periods, resulting in a coding rate of $R = 0.5$ in comparison with the conventionally modulated system, where the coding rate of $\mathbf{G}_2$ is unity. For example, if we assume that an $L = 16$-point SP (16-SP) scheme is employed, the effective throughput per channel use can be calculated as

$$T_{eff} = (\log_2 L) \cdot R = 2 \text{ bps}, \tag{11.48}$$

which is equal to that of the 4-QAM modulated system depicted in Figure 11.3.

11.2.2.2   SP Constellation Construction [583, 591]

According to Equation (11.47) describing the $\mathbf{G}_2$ codeword construction based on the SP scheme, a power normalization factor of $\sqrt{2L/E_{total}}$ is used. Thus, it is desirable to choose a specific subset of L SP constellation points from the entire set of legitimate SP constellation points hosted by D_4 based on the criterion of maintaining the minimum total energy E_{total}, while having a certain MED among the selected SP symbols. However, a potentially excessive computer search has to be carried out to attain the best subset of the L SP symbols out of all possible choices, when searching for the SP symbols having the best MED, hence maximizing the coding advantage of the resultant STBC scheme, if there are more than L SP symbols satisfying the above-mentioned minimum total energy condition.

Furthermore, the lattice D_4 can be divided into layers or shells which classify all the legitimate constellation points into a layer according to their Euclidean distances from the origin, i.e. their norms or energy. As an example, the first 10 SP layers in the D_4 SP constellation are provided in Table 11.1, where we view the four-integer-element phasor of a specific layer as the basis vector of an SP constellation and any choice of signs and any ordering of the coordinates are legitimate [583]. To be more explicit, all legitimate permutations and signs for the corresponding constellation points listed in Table 11.1 have to be applied in order to generate the full list of SP constellation points for a specific layer.

11.2.3   System Model for STBC-SP-Aided MU-MIMO Systems

Let us now construct the generalized system equations for an STBC-aided UL MU-MIMO scenario, where the SDMA-OFDM system supports a total of U UL users and employs N receive antennas at the BS. For the sake of simplicity, the $\mathbf{G}_2$ scheme using two transmit antennas is employed by each user. The overall equivalent MU-MIMO system equation can be derived in complete analogy to the case of STBC-assisted SU-MIMO systems as discussed in Section 11.2.1.2 with the aid of the so-called equivalent channel matrix. After straightforward manipulations, under the assumption that the CIR taps between each of the two transmit antennas of a specific user and the nth receive antenna at the BS

Table 11.1: The first 10 layers of the lattice D_4 © [591].

Layer	Constellation points				Norm	Number of combinations
0	0	0	0	0	0	1
1	±1	±1	0	0	2	24
2	±2	0	0	0	4	8
	±1	±1	±1	±1	4	16
3	±2	±1	±1	0	6	96
4	±2	±2	0	0	8	24
5	±2	±2	±1	±1	10	96
	±3	±1	0	0	10	48
6	±3	±1	±1	±1	12	64
	±2	±2	±2	0	12	32
7	±3	±2	±1	0	14	192
8	±2	±2	±2	±2	16	16
	±4	0	0	0	16	8
9	±4	±1	±1	0	18	96
	±3	±2	±2	±1	18	192
	±3	±3	0	0	18	24
10	±4	±2	0	0	20	48
	±3	±3	±1	±1	20	96

remain constant during two consecutive symbol periods, we have

$$
\tilde{\mathbf{y}} = \begin{bmatrix} \tilde{\mathbf{y}}_1 \\ \tilde{\mathbf{y}}_2 \\ \vdots \\ \tilde{\mathbf{y}}_N \end{bmatrix}_{2N \times 1} = \begin{bmatrix} \sum_{u=1}^{U} \tilde{\mathbf{H}}_1^{(u)} \cdot \mathbf{x}^{(u)} \\ \sum_{u=1}^{U} \tilde{\mathbf{H}}_2^{(u)} \cdot \mathbf{x}^{(u)} \\ \vdots \\ \sum_{u=1}^{U} \tilde{\mathbf{H}}_N^{(u)} \cdot \mathbf{x}^{(u)} \end{bmatrix}_{2N \times 1} + \begin{bmatrix} \tilde{\mathbf{w}}_1 \\ \tilde{\mathbf{w}}_2 \\ \vdots \\ \tilde{\mathbf{w}}_N \end{bmatrix}_{2N \times 1} \tag{11.49}
$$

$$
= \tilde{\mathbf{H}} \cdot \mathbf{x} + \tilde{\mathbf{w}}, \tag{11.50}
$$

where the overall equivalent channel matrix $\tilde{\mathbf{H}}$ can be expressed as

$$
\tilde{\mathbf{H}} = \begin{bmatrix} \tilde{\mathbf{H}}_1^{(1)} & \tilde{\mathbf{H}}_1^{(2)} & \cdots & \tilde{\mathbf{H}}_1^{(U)} \\ \tilde{\mathbf{H}}_2^{(1)} & \tilde{\mathbf{H}}_2^{(2)} & \cdots & \tilde{\mathbf{H}}_2^{(U)} \\ \vdots & \vdots & \ddots & \vdots \\ \tilde{\mathbf{H}}_N^{(1)} & \tilde{\mathbf{H}}_N^{(2)} & \cdots & \tilde{\mathbf{H}}_N^{(U)} \end{bmatrix}, \tag{11.51}
$$

with each submatrix $\tilde{\mathbf{H}}_n^{(u)}$ being defined as Equation (11.19). Additionally, the transmitted symbol vector $\mathbf{x}$ of the entire MU-MIMO system is a column vector created by concatenating each user's transmitted symbol vector $\mathbf{x}^{(u)}$ which is given by

$$
\mathbf{x}^{(u)} = F_4(\mathbf{a}^{(u)}) = [x_1^{(u)} \ x_2^{(u)}]^T. \tag{11.52}
$$

Thus the transmitted symbol vector $\mathbf{x}$ may be expressed as

$$
\mathbf{x} = [F_4((\mathbf{a}^{(1)})^T) \quad F_4((\mathbf{a}^{(2)})^T) \quad \cdots \quad F_4((\mathbf{a}^{(U)})^T)]^T \tag{11.53}
$$

$$
= [F_4((\mathbf{a}^{(1)})^T \quad (\mathbf{a}^{(2)})^T \quad \cdots \quad (\mathbf{a}^{(U)})^T)]^T \tag{11.54}
$$

$$
= [(\mathbf{x}^{(1)})^T \quad (\mathbf{x}^{(2)})^T \quad \cdots \quad (\mathbf{x}^{(U)})^T]^T. \tag{11.55}
$$

Thus, by defining $\mathbf{a} = [(\boldsymbol{a}^{(1)})^T \; (\boldsymbol{a}^{(2)})^T \; \ldots \; (\boldsymbol{a}^{(u)})^T]^T$, we have

$$\mathbf{x} = F_4(\mathbf{a}). \tag{11.56}$$

Moreover, as observed in Equation (11.49), the *equivalent received noise-contaminated signal vector* $\tilde{\mathbf{y}}$ and the *equivalent noise vector* $\tilde{\mathbf{w}}$ are formed by concatenating the N two-elements subvectors $\tilde{\mathbf{y}}_n$ and $\tilde{\mathbf{w}}_n$, respectively, which can be written as

$$\tilde{\mathbf{y}}_n = [y_{1,n} \; y_{2,n}^*]^T, \tag{11.57}$$

where the first element $y_{1,n}$ corresponds to the signal received by the nth antenna during the first symbol period and the second element $y_{2,n}^*$ is the conjugate of the signal received at the same antenna during the second symbol period, while

$$\tilde{\mathbf{w}}_n = [w_{1,n} \; w_{2,n}^*]^T, \tag{11.58}$$

where again $w_{1,n}$ denotes the AWGN imposed on the nth receive antenna during the first symbol period and $w_{2,n}^*$ is the conjugate of the AWGN inflicted during the second symbol period. The AWGN encountered during each symbol period has a zero mean and a variance of $2\sigma_w^2$.

11.3 Sphere Detection Design for SP Modulation

According to our discussions in Section 11.2.1.3, an OSTBC scheme eliminates the MSI among the MIMO elements of a specific user, owing to the orthogonality of the equivalent channel matrix $\tilde{\mathbf{H}}_n^{(u)}$ formulated in Equation (11.19). Therefore, the receiver is capable of performing ML detection based on low-complexity linear processing in order to achieve full transmit diversity by imposing a negligible extra encoding complexity at the MS in the STBC-SP-assisted SU-MIMO UL scenario considered in Section 11.2.1.3. However, in the context of an MU-MIMO system, the resultant overall equivalent channel matrix $\tilde{\mathbf{H}}$ of Equation (11.51) is no longer orthogonal, since we have

$$
\tilde{\mathbf{H}}^H \tilde{\mathbf{H}} =
\begin{bmatrix}
(\tilde{\mathbf{H}}_1^{(1)})^* & (\tilde{\mathbf{H}}_2^{(1)})^* & \cdots & (\tilde{\mathbf{H}}_N^{(1)})^* \\
(\tilde{\mathbf{H}}_1^{(2)})^* & (\tilde{\mathbf{H}}_2^{(2)})^* & \cdots & (\tilde{\mathbf{H}}_N^{(2)})^* \\
\vdots & \vdots & \ddots & \vdots \\
(\tilde{\mathbf{H}}_1^{(U)})^* & (\tilde{\mathbf{H}}_2^{(U)})^* & \cdots & (\tilde{\mathbf{H}}_N^{(U)})^*
\end{bmatrix}
\begin{bmatrix}
\tilde{\mathbf{H}}_1^{(1)} & \tilde{\mathbf{H}}_1^{(2)} & \cdots & \tilde{\mathbf{H}}_1^{(U)} \\
\tilde{\mathbf{H}}_2^{(1)} & \tilde{\mathbf{H}}_2^{(2)} & \cdots & \tilde{\mathbf{H}}_2^{(U)} \\
\vdots & \vdots & \ddots & \vdots \\
\tilde{\mathbf{H}}_N^{(1)} & \tilde{\mathbf{H}}_N^{(2)} & \cdots & \tilde{\mathbf{H}}_N^{(U)}
\end{bmatrix}
$$

$$
=
\begin{bmatrix}
\sum_{n=1}^{N}(|h_{n1}^{(1)}|^2 + |h_{n2}^{(1)}|^2)\mathbf{I}_2 & \mathbf{MUI} & \cdots & \mathbf{MUI} \\
\mathbf{MUI} & \sum_{n=1}^{N}(|h_{n1}^{(2)}|^2 + |h_{n2}^{(2)}|^2)\mathbf{I}_2 & \cdots & \mathbf{MUI} \\
\vdots & \vdots & \ddots & \vdots \\
\mathbf{MUI} & \mathbf{MUI} & \cdots & \sum_{n=1}^{N}(|h_{n1}^{(U)}|^2 + |h_{n2}^{(U)}|^2)\mathbf{I}_2
\end{bmatrix},
\tag{11.59}
$$

where $\mathbf{I}_2$ denotes a (2×2)-element identity matrix and the term $\mathbf{MUI}$ refers to the (2×2)-element MUltiple-access Interference (MUI) submatrix, which contains non-zero elements imposed by the co-channel users. Hence, although as a benefit of having a diagonal submatrix $\sum_{n=1}^{N}(|h_{n1}^{(u)}|^2 + |h_{n2}^{(u)}|^2)\mathbf{I}_2$ in Equation (11.59) there is no MSI between the two transmit antennas of a specific MS, a significant performance loss will be caused by the MUI in the context of a multi-user system in comparison with that of the single-user scenario considered in Section 11.2.1.3, provided that we still simply apply the detection criterion formulated in Equation (11.37) and Equation (11.38) of

Section 11.2.1.3, i.e.

$$\hat{x}_1 = \underset{\check{x}_1 \in \mathcal{M}_c}{\arg\min} \left(\sum_{n=1}^{N} (|h_{n1}|^2 + |h_{n2}|^2) - N \right) |\check{x}_1|^2 + d^2(\check{y}_1, \check{x}_1), \tag{11.60}$$

$$\hat{x}_2 = \underset{\check{x}_2 \in \mathcal{M}_c}{\arg\min} \left(\sum_{n=1}^{N} |(h_{n1}|^2 + |h_{n2}|^2) - N \right) |\check{x}_2|^2 + d^2(\check{y}_2, \check{x}_2), \tag{11.61}$$

in order separately to carry out signal detection for each user without considering the effects of MUI produced by the co-channel users. To mitigate the effects of MUI imposed in the multi-user scenario considered, a successive interference cancellation (SIC) scheme was proposed in [588], which significantly improves the achievable BER performance of the STBC-aided multi-user system. On the other hand, the powerful near-ML SD techniques of Chapter 9 designed for classic modulation schemes are also readily applicable to STBC-aided multi-user systems, at a potentially reduced complexity. Hence, in order to avoid using the traditional brute-force ML detector, we intend to develop further the K-best SD of Section 9.2.3 to be used at the BS in the STBC-SP-assisted SDMA-OFDM UL scenario, in order to achieve a near-MAP performance at a moderate complexity.

11.3.1 Bit-Based MAP Detection for SP-Modulated MU-MIMO Systems

According to Equation (11.50) and Equation (11.56), the conditional PDF $p(\tilde{\mathbf{y}}|\mathbf{a})$ for MU-MIMO systems using $N_D = 4$-dimensional real-valued SP modulation is given by

$$p(\tilde{\mathbf{y}}|\mathbf{a}) = \frac{1}{(2\pi\sigma_w^2)^{N_D/2}} e^{-(1/2\sigma_w^2)\|\tilde{\mathbf{y}} - \tilde{\mathbf{H}} \cdot F_4(\mathbf{a})\|^2}. \tag{11.62}$$

Then, using Bayes' theorem, and exploiting the independence of the bits in the vector $\mathbf{b} = [b_1, b_2, \ldots b_{B \cdot U}]$ carried by the received symbol vector $\tilde{\mathbf{y}}$, we can factorize the joint bit probabilities into their products [553], hence the LLR of bit k for $k = 1, \ldots, B \cdot U$ can be written as

$$L_D(\mathbf{b}_k|\tilde{\mathbf{y}}) = L_A(\mathbf{b}_k) + \ln \underbrace{\frac{\sum_{\mathbf{a} \in \mathcal{A}_{k=1}^U} p(\tilde{\mathbf{y}}|\mathbf{a}) \cdot e^{\sum_{j=1, j \neq k}^{B \cdot U} \mathbf{b}_j L_A(\mathbf{b}_j)}}{\sum_{\mathbf{a} \in \mathcal{A}_{k=0}^U} p(\tilde{\mathbf{y}}|\mathbf{a}) \cdot e^{\sum_{j=1, j \neq k}^{B \cdot U} \mathbf{b}_j L_A(\mathbf{b}_j)}}}_{L_E(\mathbf{b}_k|\tilde{\mathbf{y}})}, \tag{11.63}$$

where $\mathcal{A}_{k=1}^U$ and $\mathcal{A}_{k=0}^U$ are subsets of the multi-user SP symbol constellation $\mathcal{A}^U$ where we have $\mathcal{A}_{k=1}^U \triangleq \{\mathbf{a} \in \mathcal{A}^U : b_k = 1\}$, and in a similar fashion, $\mathcal{A}_{k=0}^U \triangleq \{\mathbf{a} \in \mathcal{A}^U : b_k = 0\}$. Using Equation (11.62), we arrive at

$$L_D(\mathbf{b}_k|\tilde{\mathbf{y}}) = L_A(\mathbf{b}_k) + \ln \underbrace{\frac{\sum_{\mathbf{a} \in \mathcal{A}_{k=1}^U} e^{[-(1/2\sigma_w^2)\|\tilde{\mathbf{y}} - \tilde{\mathbf{H}} \cdot F_4(\mathbf{a})\|^2 + \sum_{j=1, j \neq k}^{B \cdot U} \mathbf{b}_j L_A(\mathbf{b}_j)]}}{\sum_{\mathbf{a} \in \mathcal{A}_{k=0}^U} e^{[-(1/2\sigma_w^2)\|\tilde{\mathbf{y}} - \tilde{\mathbf{H}} \cdot F_4(\mathbf{a})\|^2 + \sum_{j=1, j \neq k}^{B \cdot U} \mathbf{b}_j L_A(\mathbf{b}_j)]}}}_{L_E(\mathbf{b}_k|\tilde{\mathbf{y}})}. \tag{11.64}$$

11.3.2 SD Design for SP Modulation

11.3.2.1 Transformation of the ML Metric

Although the basic idea behind the ML detector is to maximize the a posteriori probability of the received signal vector $\tilde{\mathbf{y}}$, this problem can be readily transformed into an issue of maximizing the a priori probability of Equation (11.62) with the aid of Bayes' theorem [542]. Consequently, maximizing the a priori probability of Equation (11.62) is equivalent to minimizing the Euclidean distance $\|\tilde{\mathbf{y}} - \tilde{\mathbf{H}}F_4(\mathbf{a})\|^2$. Therefore, the ML solution can be written as

$$\hat{\mathbf{a}}_{ML} = \arg\min_{\breve{\mathbf{a}} \in \mathcal{A}^U} \|\tilde{\mathbf{y}} - \tilde{\mathbf{H}} \cdot F_4(\breve{\mathbf{a}})\|_2^2, \tag{11.65}$$

where $F_4(\cdot)$ is defined in Equation (11.56) in the context of our multi-user system. Observe from Equation (11.65) that a potentially excessive complexity search may be encountered, depending on the size of the set $\mathcal{A}^U$, which prevents the application of the full-search-based ML detectors in high-throughput scenarios. By comparing the unconstrained LS solution of $\hat{\mathbf{a}}_{ls} = F_4^{-1}(\hat{\mathbf{x}}_{ls}) = F_4^{-1}((\tilde{\mathbf{H}}^H\tilde{\mathbf{H}})\tilde{\mathbf{H}}^H\tilde{\mathbf{y}})$ with all legitimate constrained/sliced solutions, namely $\breve{\mathbf{a}} \in \mathcal{A}^U$, the ML solution of Equation (11.65) can be transformed into

$$\hat{\mathbf{a}}_{ML} = \arg\min_{\breve{\mathbf{a}} \in \mathcal{A}^U} F_4(\breve{\mathbf{a}} - \hat{\mathbf{a}}_{ls})^H (\tilde{\mathbf{H}}^H\tilde{\mathbf{H}})F_4(\breve{\mathbf{a}} - \hat{\mathbf{a}}_{ls}). \tag{11.66}$$

11.3.2.2 Channel Matrix Triangularization

Let us now generate the $(2U \times 2U)$-dimensional upper-triangular matrix $\mathbf{U}$, which satisfies $\mathbf{U}^H\mathbf{U} = \tilde{\mathbf{H}}^H\tilde{\mathbf{H}}$ with the aid of Cholesky factorization [544]. Then, upon defining a matrix $\mathbf{Q}$ with elements $q_{i,i} \triangleq u_{i,i}^2$ and $q_{i,j} \triangleq u_{i,j}/u_{i,i}$, we can rewrite Equation (11.66) as

$$\hat{\mathbf{a}}_{ML} = \arg\min_{\breve{\mathbf{a}} \in \mathcal{A}^U} F_4(\breve{\mathbf{a}} - \hat{\mathbf{a}}_{ls})^H \mathbf{U}^H \mathbf{U} F_4(\breve{\mathbf{a}} - \hat{\mathbf{a}}_{ls})$$

$$= \arg\min_{\breve{\mathbf{a}} \in \mathcal{A}^U} \left\{ \sum_{u=1}^{U} q_{2u-1,2u-1} \left[\mathbf{e}_1^{(u)} + \sum_{v=u+1}^{U} q_{2u-1,2v-1}\mathbf{e}_1^{(v)} + \sum_{v=u}^{U} q_{2u-1,2v}\mathbf{e}_2^{(v)} \right]^2 \right.$$

$$\left. + \sum_{u=1}^{U} q_{2u,2u} \left[\mathbf{e}_2^{(u)} + \sum_{v=u+1}^{U} q_{2u,2v-1}\mathbf{e}_1^{(v)} + \sum_{v=u+1}^{U} q_{2u,2v}\mathbf{e}_2^{(v)} \right]^2 \right\}, \tag{11.67}$$

where $\mathbf{e}^{(u)}$ is the uth two-element subvector of the multi-user vector $\mathbf{e} = [(\mathbf{e}^{(1)})^T \ldots (\mathbf{e}^{(u)})^T \ldots (\mathbf{e}^{(U)})^T]^T$, corresponding to the uth user, and is given by

$$\mathbf{e}^{(u)} = \breve{\mathbf{x}}^{(u)} - \hat{\mathbf{x}}_{ls}^{(u)}, \tag{11.68}$$

where $\breve{\mathbf{x}}^{(u)} = [\breve{x}_1^{(u)}, \breve{x}_2^{(u)}]^T = F_4(\mathbf{a}^{(u)})$, $\mathbf{a}^{(u)} \in \mathcal{A}$ and $\hat{\mathbf{x}}_{ls}^{(u)} = [\hat{x}_{ls\,1}^{(u)}, \hat{x}_{ls\,2}^{(u)}]^T = F_4(\hat{\mathbf{a}}_{ls}^{(u)})$. Hence, the sum in $\{\cdot\}$ of Equation (11.67) is the *user-based* accumulated PED between the tentative symbol vector $\breve{\mathbf{x}} = [(\breve{\mathbf{x}}^{(1)})^T, (\breve{\mathbf{x}}^{(2)})^T, \ldots, (\breve{\mathbf{x}}^{(U)})^T]^T$ and the search centre $\hat{\mathbf{x}}_{ls} = [(\hat{\mathbf{x}}_{ls}^{(1)})^T, (\hat{\mathbf{x}}_{ls}^{(2)})^T, \ldots, (\hat{\mathbf{x}}_{ls}^{(U)})^T]^T$.

11.3.2.3 User-Based Tree Search

Let us now recall the tree search carried out by the K-best SD of Section 9.2.3 for conventional modulation schemes, such as BPSK, where each tree level represents an independent data stream corresponding to a certain transmit antenna of a specific user. Each tree node corresponds to a legitimate BPSK symbol[3] in the constellation of domain $\mathbb{C}^1$. Consequently, in the absence of joint ST signal design for the $M_u = 2$ transmit antennas, the BPSK constellations of the two adjacent tree levels corresponding to a specific user are independent and identical. Figure 11.6 shows the search tree of the K-best SD when it is applied to a four-transmit-antenna scenario, where K is set to two, which means that a maximum number of two candidates are retained at each level. Consequently, the search is carried out in the downward direction only along the search tree. The number in brackets indicates the corresponding PED of the current tree node accumulated from the top level of the tree down to the current tree node, while the number outside the brackets represents the order in which the tree nodes are visited. At the (m = 4) level, both the candidates are retained, which results in four candidates at

---

[3]In this treatise, we consider complex-valued BPSK symbols having zero imaginary parts.

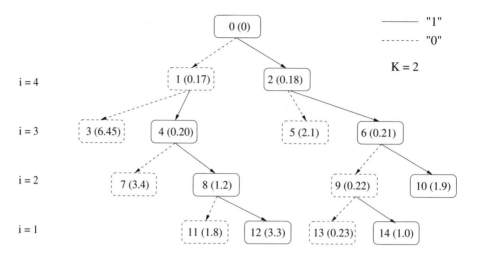

Figure 11.6: The search tree of K-best SD in the scenario of a four-transmit-antenna BPSK SDMA system, where $K = 2$.

the (m = 3) level. Then we choose two candidates having the lowest two accumulated PEDs from the search centre $\hat{\mathbf{x}}_c$, i.e. 0.2 and 0.21, out of four, which again generate four candidates at the (m = 2) level. The search goes on in a similar way, until it reaches the tree leaf point having the lowest Euclidean distance of 0.23 from $\hat{\mathbf{x}}_c$. Then the estimated signal vector can be obtained by doing the backtracking, which is assumed to be the ML solution. However, the K-best SD does not necessarily find the ML solution, unless the value of K is large enough. An extreme example is when $K = 1$, when the resultant K-best SD degenerates into the linear LS detector.

On the other hand, when the joint ST signals are transmitted from the $M_u = 2$ transmit antennas of the uth user, they are combined into a joint ST design with the aid of the SP scheme as discussed in Section 11.2.2. The corresponding SP-symbol-based search tree structure is depicted in Figure 11.7, where the search tree of the modified K-best SD is exemplified in the context of a UL SDMA system supporting $U = 2$ $\mathbf{G}_2$-SP-aided users, where $K = 2$ and a four-point-SP constellation is employed. Explicitly, the two adjacent tree levels corresponding to the SP symbols of the jointly designed STBC-SP data streams of a specific user should be considered together in the tree search process, resulting in multi-dimensional/multi-layer tree nodes in the $\mathbb{C}^2$ SP symbol domain, which we refer to as a *user-wise* tree search. The resultant 2D complex-valued tree node consists of two complex-valued BPSK symbols, which are the constituent components of a transformed SP symbol $F_4(\mathbf{a})$. On the other hand, due to the joint consideration of the two adjacent BPSK tree levels, the number of effective search tree levels is reduced by a factor of two, while each symbol becomes quaternary, instead of being binary.

As observed in Figure 11.7, since a 4-SP scheme is employed and the number of candidate tree nodes retained at each tree level is $K = 2$, each of the two selected tree nodes having the smallest two accumulated PED values at the previous search tree level of the survivor path has to be expanded into four child nodes at the current level. Consequently, in analogy to the conventional K-best algorithm [550], both the calculation of the user-based accumulated PEDs and the tree pruning process continue in the downward direction of Figure 11.7 all the way along the tree, until they reach the tree leaf level, producing a candidate list of $\mathcal{L}_{cand} \in \mathcal{A}^U$. This list contains $N_{cand} = K$ SP symbol candidate solutions, which are then used for the *extrinsic* bit LLR calculation using Equation (11.64). Having a reduced candidate list size assists us in achieving a substantial complexity reduction. Explicitly, after

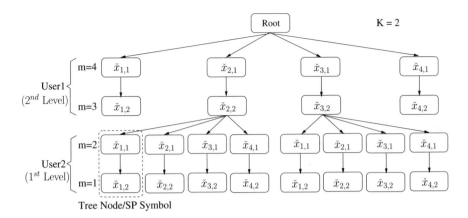

Figure 11.7: The search tree of K-best SD in the scenario of an STBC-SP-aided uplink SDMA system: the number of users is $U = 2$, the number of transmit antennas per user is $M_u = 2$ and the number of candidates retained at each search tree level is $K = 2$.

the max-log approximation, we arrive at

$$
\begin{aligned}
L_E(\mathbf{b}_k|\tilde{\mathbf{y}}) = {} & \max_{\mathbf{a} \in \mathcal{L}_{cand} \cap \mathcal{A}_{k=1}^U} \left[-\frac{1}{2\sigma_w^2} \|\tilde{\mathbf{y}} - \tilde{\mathbf{H}} \cdot F_4(\mathbf{a})\|^2 + \sum_{j=1,j\neq k}^{B\cdot U} \mathbf{b}_j L_A(\mathbf{b}_j) \right] \\
& - \max_{\mathbf{a} \in \mathcal{L}_{cand} \cap \mathcal{A}_{k=0}^U} \left[-\frac{1}{2\sigma_w^2} \|\tilde{\mathbf{y}} - \tilde{\mathbf{H}} \cdot F_4(\mathbf{a})\|^2 + \sum_{j=1,j\neq k}^{B\cdot U} \mathbf{b}_j L_A(\mathbf{b}_j) \right].
\end{aligned}
\tag{11.69}
$$

Finally, the K-best SD algorithm designed for the $N_D = 4$-dimensional SP modulation scheme is summarized as follows:

The pre-processing phase:

1. Obtain the upper-triangular matrix $\mathbf{U}$ via Cholesky factorization of the Grammian matrix $\mathbf{G} = \tilde{\mathbf{H}}^H\tilde{\mathbf{H}}$, i.e. $\mathbf{U} = Chol(\mathbf{G})$.

2. Calculate the search centre $\hat{\mathbf{x}}_{ls}$ by

$$
\hat{\mathbf{x}}_{ls} = \mathbf{G}^{-1}\tilde{\mathbf{H}}^H\mathbf{y}.
\tag{11.70}
$$

The tree search phase:
The first stage:

1. $m = M$, $d_M = \hat{x}_{lsM}$, where M is the total number of transmit antennas supported by the system.

2. Calculate the corresponding PED for each SP symbol $(\check{x}_{l,1}, \check{x}_{l,2})$, $l = 1, 2, \ldots, L$, in the constellation of $\mathbb{C}^2$ domain as follows:

$$
e_M = \hat{x}_{lsM} - \check{x}_{l,1},
\tag{11.71}
$$

$$
d_{M-1} = \hat{x}_{lsM-1} + \frac{u_{M-1,M}}{u_{M-1,M-1}} e_M,
\tag{11.72}
$$

$$
PED = u_{M-1,M-1}^2(d_{M-1} - \check{x}_{l,2}).
\tag{11.73}
$$

3. Choose K SP symbols $(\check{x}_{k,1}, \check{x}_{k,2})$, $k = 1, 2, \ldots, K$, that have the K smallest PEDs.

4. For each chosen SP symbol, compute

$$e_{M-1} = \hat{x}_{lsM-1} - \check{x}_{k,2},\tag{11.74}$$

$$d_{M-2} = \hat{x}_{lsM-1} + \frac{u_{M-2,M-1}}{u_{M-2,M-2}}e_{M-1} + \frac{u_{M-1,M}}{u_{M-1,M-1}}e_M.\tag{11.75}$$

The mth stage:

1. $m = m - 2$.

2. For each survived search tree path from the previous tree level, calculate the corresponding PED for each SP symbol $(\check{x}_{l,1}, \check{x}_{l,2})$ in the constellation of the $\mathbb{C}^2$ domain:

$$e_m = \hat{x}_{ls\ m} - \check{x}_{l,1},\tag{11.76}$$

$$d_{m-1} = \hat{x}_{ls\ m-1} + \sum_{j=m}^{M} \frac{u_{m-1,j}}{u_{m-1,m-1}}e_j,\tag{11.77}$$

$$PED = u_{m-1,m-1}^2(d_{m-1} - \check{x}_{l,2}).\tag{11.78}$$

3. Choose K SP symbols $(\check{x}_{k,1}, \check{x}_{k,2})$, $k = 1, 2, \ldots, K$, that have the K smallest PEDs.

4. For each chosen SP symbol, compute

$$e_{m-1} = \hat{x}_{ls\ m-1} - \check{x}_{k,2},\tag{11.79}$$

$$d_{m-2} = \hat{x}_{ls\ m-1} + \frac{u_{m-2,m-1}}{u_{m-2,m-2}}e_{m-1} + \frac{u_{m-1,m}}{u_{m-1,m-1}}e_m.\tag{11.80}$$

5. If $m - 1 = 1$, obtain the solution by backtracking from the tree leaf having the largest accumulated PED to the tree root. Otherwise, go to step 1 of the mth stage.

11.3.3 Simulation Results and Discussion

The schematic of the system is depicted in Figure 11.8, where the transmitted source bits of the uth user are channel encoded and then interleaved by a random bit interleaver. The B interleaved bits $\mathbf{b}^{(u)} = b_{0,\ldots,B-1}^{(u)} \in \{0, 1\}$ are mapped to an SP-modulated symbol $\mathbf{a}^{(u)} \in \mathcal{A}$ by the SP modulator/mapper of Figure 11.8, where $B = \log_2 L$. The $\mathbf{G}_2$ encoder then maps the SP-modulated symbol $\mathbf{a}^{(u)}$ to an ST signal $\mathbf{C}^{(u)} = \sqrt{(2L/E_{total})}\mathbf{G}_2(F_4(\mathbf{a}^{(u)})) \in \mathcal{C}_{\mathbf{G}_2}$ by invoking Equation (11.42) and Equation (11.46). Finally, the ST signal $\mathbf{C}^{(u)}$ is transmitted from the two transmit antennas of the uth MS during two consecutive time slots.

In Figure 11.8 the interleaver and deinterleaver pair seen at the BS divide the receiver into two parts, namely the SD (*inner decoder*) and the channel decoder (*outer decoder*). Note that in Figure 11.8, L_A, L_E and L_D denote the a priori, the *extrinsic* and the a posteriori LLRs, while the subscripts '1' and '2' represent the bit LLRs associated with the inner decoder and the outer decoder, respectively. It was detailed throughout [314] and [569] that the iterative exchange of extrinsic information between these serially concatenated receiver blocks results in substantial performance improvements. In this treatise we assume familiarity with the classic turbo detection principles [314, 569].

To investigate the performance of the STBC-SP-assisted multi-user SDMA-OFDM UL system, we compare the SP-modulated system with its conventionally modulated counterpart in the two scenarios using the system parameters summarized in Table 11.2. Figures 11.9(a) and 11.9(b) depict, respectively, the corresponding EXIT charts [592] used as a convenient visualization technique for analysing the convergence characteristics of turbo receivers. This technique computes the MI of the output *extrinsic* and input a priori components, which are denoted by I_E and I_A respectively, corresponding to the associated bits for each of the iterative SISO blocks of Figure 11.8, i.e. to the SD and the RSC(2,1,3) channel decoder. As observed in Figure 11.9, the maximum achievable iterative gains of traditional

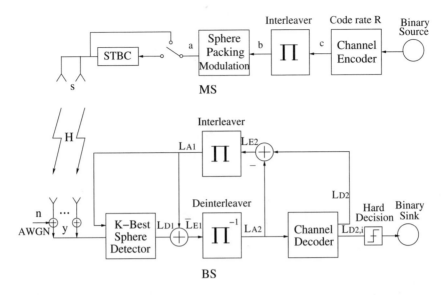

Figure 11.8: Schematic of the uplink SP-modulated multi-user MIMO system using K-best SD.

Table 11.2: Summary of system parameters.

	Scenario I	Scenario II
Modulation	4-QAM/16-SP	16-QAM/256-SP
Users supported	4	2
Doppler frequency		0.1
System		SDMA-OFDM
Subcarriers		1024
STBC		$\mathbf{G}_2$
Rx at BS		4
CIR model		$P(\tau_k) = [0.5\ 0.3\ 0.2]$
Detector/MAP		K-best LSD
List length		$\mathcal{N}_{cand} = K$
Channel code	Half-rate RSC(2,1,3) (5/7)	
BW efficiency		4 bps/Hz

QAM-modulated systems employing the conventional K-best SD using $N_{cand} = K = 128$ are rather limited in comparison with our SP-aided K-best SD specifically designed for SP signals having the same list size of $N_{cand} = 128$. This is because the EXIT curve of the SD using the conventional 4- and 16-QAM-based system has a relatively low I_E value at $I_A = 1$, in contrast to the corresponding EXIT curve of its SP-modulated counterpart. Nonetheless, we also observe from Figure 11.9 that the SD's EXIT curve in the QAM-modulated system emerges from a higher starting point at $I_A = 0$ than that of its SP-modulated counterpart. This leads to a potentially lower BER at relatively low SNRs, where I_A is also low, although the exact detection-convergence behaviour is determined by the SD's complexity as well as by the SNR. Observe in Figure 11.9 that in principle the employment of SP modulation is capable of eliminating the EXIT curve intercept point at a lower SNR, thus leading to an infinitesimally low BER. However, an open EXIT tunnel can only be formed if the value of $K = N_{cand}$ as well as that of the SNR are sufficiently high.

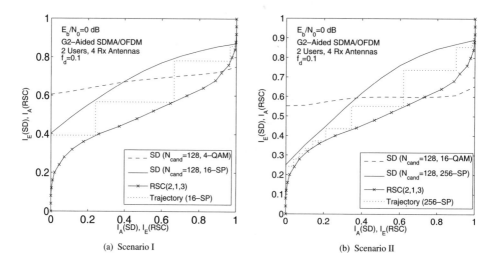

(a) Scenario I

(b) Scenario II

Figure 11.9: EXIT charts of STBC-SP-aided iterative receiver of Figure 11.8 employing the modified K-best SD and the parameters of Table 11.2. The overall system throughput is 8 bits per symbol.

Monte Carlo simulations were performed for characterizing the above-mentioned decoding convergence prediction in both Scenario I and Scenario II of Table 11.2. Figures 11.10(a) and 11.10(b) suggest that the SP-modulated system exhibits a relatively higher BER at low SNRs in both scenarios, which matches the predictions of the EXIT charts seen in Figure 11.9. On the other hand, as a benefit of employing the SP modulation, performance gains of 1.5 and 3.5 dB can be achieved by 16-SP- and 256-SP- modulated systems in Scenario I and Scenario II of Table 11.2, respectively, in comparison with their identical-throughput QAM-based counterparts, given a target BER of 10^{-4} and $N_{cand} = 128$. Furthermore, as observed from Figures 11.10(a) and 11.10(b), an attractive compromise can be achieved between the achievable performance and the complexity imposed by adjusting the list size N_{cand} employed by the K-best SD.

11.4 Chapter Conclusions

In comparison with the SDMA-OFDM system models employed in previous chapters, where only a single transmit antenna is employed by each MS, in this chapter a looser constraint is assumed for the sake of allowing the employment of multiple antennas at each MS, in order to enhance the system's robustness to the hostile wireless fading channel with the aid of transmit diversity gains. Therefore, the simple but elegant OSTBC scheme, which was initially devised by Alamouti [585] for two-transmit-antenna-aided transmission, may be employed in the MU-MIMO scenario. As discussed in Section 11.2.1, the OSTBC scheme is capable of enabling the receiver to perform ML detection based on low-complexity linear processing, yet achieving the maximum attainable transmit diversity by imposing a low encoding complexity at the transmitter. Furthermore, in comparison with the conventional orthogonal design of STBCs based on PSK/QAM modulated symbols, in Section 11.2.2 we proposed an orthogonal transmit diversity design using SP modulation, which is capable of attaining extra coding gains by maximizing the diversity product of the STBC signals in the presence of temporally correlated fading channels.

On the other hand, although the resultant STBC-SP scheme has recently been investigated by researchers in the context of SU-MIMO systems, existing designs make no attempt to employ it in MU-MIMO systems owing to its relatively complex detection scheme. More specifically, despite having no

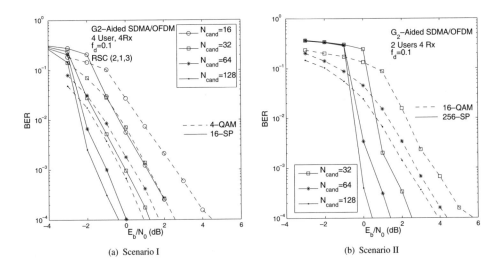

Figure 11.10: BER performance of the system of Figure 11.8 in Scenario I and Scenario II of Table 11.2. The overall system throughput is 8 bits per symbol.

MSI between the transmit antennas of a specific user, a significant performance loss will be inflicted by the MUI imposed by the co-channel users if we insist on using low-complexity linear detection schemes, as in the SU-MIMO scenario. Although SIC-based nonlinear detection schemes [593] are capable of enhancing the achievable performance, these improvements erode if the number of users increases, especially when the system becomes rank deficient, potentially resulting in an inadequate performance. Based on this background, we proposed the so-called multi-layer tree search mechanism in order to render the powerful low-complexity near-ML SD scheme applicable to the STBC-SP-assisted MU-MIMO system. Consequently, with the aid of our K-best SD, a significant performance gain can be attained by the SP-modulated system over its conventionally modulated identical-throughput counterpart in MU-MIMO scenarios. For example, a performance gain of 3.5 dB was achieved in Figure 11.10 over a 16-QAM benchmark by the 256-SP scheme in the scenario of a four-receive-antenna SDMA UL system supporting $U = 2$ $\mathbf{G}_2$-assisted users, given a target BER of 10^{-4}.

Chapter 12

Multiple-Symbol Differential Sphere Detection for Differentially Modulated Cooperative OFDM Systems

12.1 Introduction[1]

Multiple-antenna-aided transmit diversity arrangements [594] constitute powerful techniques of mitigating the deleterious effects of fading, hence improving the end-to-end system performance, which is usually achieved by multiple co-located antenna elements at the transmitter and/or receiver, as discussed in Chapter 11. However, in cellular communication systems, it is often impractical for the mobile to employ several antennas for the sake of achieving a diversity gain owing to its limited size. Furthermore, owing to the limited separation of the antenna elements, they rarely experience independent fading, which limits the achievable diversity gain and may be further compromised by the detrimental effects of the shadow fading, imposing further signal correlation among the antennas in their vicinity. Fortunately, as depicted in Figure 12.1, in multi-user wireless systems mobiles may cooperatively share their antennas in order to achieve uplink transmit diversity by forming a Virtual Antenna Array (VAA) in a distributed fashion. Thus, so-called cooperative diversity relying on the cooperation among multiple terminals may be achieved [595, 596].

On the other hand, in order to carry out classic coherent detection, channel estimation is required at the receiver, which relies on using training pilot signals or tones and exploits the fact that in general the consecutive CIR taps are correlated in both the time and frequency domains of the OFDM subchannels. However, channel estimation for an M-transmitter, N-receiver MIMO system requires the estimation of $(M \times N)$ CIRs, which imposes both an excessive complexity and a high pilot overhead, especially in mobile environments associated with relatively rapidly fluctuating channel conditions. Therefore, in such situations, differential encoded transmissions combined with non-coherent detection and hence requiring no CSI at the receiver become an attractive design alternative, leading to differential-modulation-assisted cooperative communications [596]. Three different channel models corresponding

---

[1]This chapter is partially based on ⓒIEEE Wang & Hanzo 2007 [39].

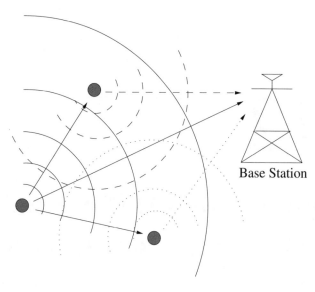

Figure 12.1: Cooperative diversity exploiting cooperation among multiple terminals.

Table 12.1: Channel models considered: sampling frequency $f_s = 10\,\text{MHz}$ and the unit of the power profile is decibels.

Channel models	
Typical urban	Parameter
No. of taps	$N_{taps} = 6$
Power profile	$\boldsymbol{\sigma} = [-7.219\,04 - 4.219\,04 - 6.219\,04 - 10.219 - 12.219 - 14.219]$
Delay profile	$\boldsymbol{\tau} = [0\ 2\ 6\ 16\ 24\ 50]$
Rural area	Parameter
No. of taps	$N_{taps} = 4$
Power profile	$\boldsymbol{\sigma} = [-2.407\,88 - 4.407\,88 - 12.4079 - 22.4079]$
Delay profile	$\boldsymbol{\tau} = [0\ 2\ 4\ 6]$
Hilly terrain	Parameter
No. of taps	$N_{taps} = 6$
Power profile	$\boldsymbol{\sigma} = [-4.053\,25 - 6.053\,25 - 8.053\,25 - 11.0533 - 10.0533 - 16.0533]$
Delay profile	$\boldsymbol{\tau} = [0\ 2\ 4\ 6\ 150\ 172]$

to three distinct communication environments will be considered in this chapter, namely the Typical Urban (TH), the Rural Area (RA) and the Hilly Terrain (HT) scenarios summarized in Table 12.1.

12.1.1 Differential Phase-Shift Keying and Detection

12.1.1.1 Conventional Differential Signalling and Detection

In this section, we briefly review the conventional differential encoding and detection process. Let $\mathcal{M}_c$ denote an M_c-ary PSK constellation which is defined as the set $\{2\pi m/M_c; m = 0, 1, \ldots, M_c - 1\}$, where $v[n] \in \mathcal{M}_c$ represent the data to be transmitted over a slow-fading frequency-flat channel. The differential signalling process commences by transmitting a single reference symbol $s[0]$, which is

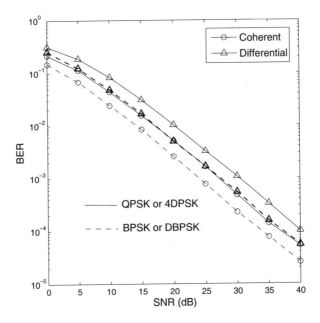

Figure 12.2: BER performance comparison between conventional coherent and differential detection in an SISO system.

normally set to unity, followed by a differential encoding process, which can be expressed as

$$s[n] = s[n-1]v[n], \qquad (12.1)$$

where $s[n-1]$ and $s[n]$ represent the symbols transmitted during the $(n-1)$th and nth time slots, respectively.

By representing the signals arriving at the receiver corresponding to the $(n-1)$th and nth transmitted symbols as

$$y[n-1] = h[n-1]s[n-1] + w[n-1], \qquad (12.2)$$

$$y[n] = [n]s[n] + w[n], \qquad (12.3)$$

respectively, and assuming a slow-fading channel, i.e. $h[n-1] = h[n]$, we arrive at

$$y[n] = h[n-1]s[n-1]v[n] + w[n] \qquad (12.4)$$

$$= y[n-1]v[n] + \underbrace{w[n] - w[n-1]v[n]}_{w'[n]}, \qquad (12.5)$$

where $w[n-1]$ and $w[n]$ denote the AWGN with a variance of $2\sigma_w^2$ added at the receiver during the two consecutive time slots. Consequently, the differentially encoded data $v[n]$ can be recovered in the same manner as in the conventional coherent detection scheme in a single-input, single-output context by using $y[n-1]$ as the reference signal of the differential detector. This is achieved without any CSI at the expense of a 3 dB performance loss in comparison with its coherent counterpart caused by the doubled noise $w'[n]$ at the decision device, which has a variance of $4\sigma_w^2$. This can be verified by the BER curves of the single-antenna-aided OFDM system characterized in Figure 12.2 for two different throughputs, i.e. for 1 bits per symbol and for 2 bits per symbol, respectively. The other system parameters are summarized in Table 12.2.

Table 12.2: Summary of system parameters for differential-modulation-aided OFDM system.

System parameters	Choice
System	OFDM
Subcarrier BW	$\Delta f = 10\,\text{kHz}$
Number of subcarriers	$D = 1024$
Modulation	DPSK in time domain
Normalized Doppler freq.	$f_d = 0.001$
Channel model	Typical urban, refer to Table 12.1

12.1.1.2 Effects of Time-Selective Channels on Differential Detection

Apart from the 3 dB performance loss suffered by Conventional Differential Detection (CDD) in slow-fading scenarios as discussed in Section 12.1.1, an error floor may be encountered by the CDD in fast-fading channels, if DPSK modulation is carried out in the time direction, i.e. for the same subcarrier of consecutive OFDM symbols, since the fading channel is deemed to be more correlated between the same subcarrier of consecutive OFDM symbols than between adjacent subcarriers of a given OFDM symbol. In other words, the assumption that $h[n-1] = h[n]$ no longer holds, leading to unrecoverable phase information between consecutive transmitted DPSK symbols even in the absence of noise. Furthermore, all the channel models considered in Table 12.1 exhibit temporally Rayleigh-distributed fading for each of the D subcarriers employed by the OFDM system with the autocorrelation function expressed as

$$\varphi_{hh}^{t}[\kappa] \triangleq \mathcal{E}\{h[n+\kappa]h^{*}[n]\} \tag{12.6}$$

$$= J_0(2\pi f_d \kappa), \tag{12.7}$$

where $J_0(\cdot)$ denotes the zeroth-order Bessel function of the first kind and f_d is the normalized Doppler frequency.

Figure 12.3(a) depicts the magnitude of the temporal correlation function for various normalized Doppler frequencies f_d, while Figure 12.3(b) plots the corresponding BER curves of the DQPSK-modulated CDD-aided OFDM system with the system parameter summarized in Table 12.2. Given a Doppler frequency of $f_d = 0.001$, the BER curves decrease continuously as the SNR increases. However, the BER curve tends to create an error floor when f_d becomes high, which is caused by the relative mobility between the transmitter and the receiver. For example, with a relatively high Doppler frequency of $f_d = 0.03$, the magnitude of the temporal correlation function of the typical urban channel model of Table 12.1 decreases rapidly as κ increases. Therefore, the CDD, which is capable of achieving a desirable performance in slow-fading channels, suffers from a considerable performance loss when the transmit terminal is moving at a high speed relative to the receiver.

12.1.1.3 Effects of Frequency-Selective Channels on Differential Detection

Our discussions in Section 12.1.1.2 were focused on the CDD employing differentially encoded modulation along the Time Domain (TD) – which is referred to here as T-DQPSK modulation – for each of the D subcarriers of an OFDM system. In general, the time- and frequency-domain differential encoding have their own merits. Specifically, the T-DQPSK-modulated OFDM system is advantageous for employment in continuous transmissions, because the effective throughput remains high, since the overhead constituted by the reference symbol $s[0]$ tends to zero in conjunction with a relatively large transmission block/frame size, namely with an increasing transmission frame duration. However, T-DQPSK-aided OFDM is less suitable for burst transmission, when the consecutive OFDM symbols may experience fairly uncorrelated fading. Hence, employment of frequency-domain differentially encoded modulation – which is referred to here as F-DPSK – is preferable for the above-mentioned scenario. Before investigating the impact of the channel's frequency selectivity for the channel models

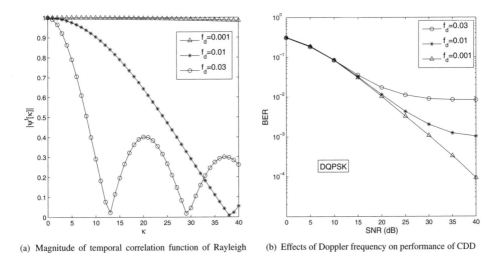

(a) Magnitude of temporal correlation function of Rayleigh fading channels

(b) Effects of Doppler frequency on performance of CDD

Figure 12.3: Impact of mobility on the performance of CDD.

summarized in Table 12.1 on performance of the CDD, we review the frequency-domain (FD) autocorrelation function of OFDM having D active subcarriers and a subcarrier frequency spacing of Δf, which can be expressed as

$$\varphi_{hh}^{f}[\mu] \triangleq \mathcal{E}\{h[k+\mu]h^{*}[k]\}, \tag{12.8}$$

$$= \sum_{l-1}^{N_{taps}} \sigma_{l}^{2} e^{-j2\pi\mu\Delta f\tau_{l}}, \tag{12.9}$$

where N_{taps}, σ_{l} and τ_{l} represent the number of paths, the elements of the power profile $\boldsymbol{\sigma}$ and the delay profile $\boldsymbol{\tau}$ of the channel models given by Table 12.1, respectively.

Accordingly, Figure 12.4(a) depicts the magnitude of the FD autocorrelation function for the three different channel models of Table 12.1, namely the TU, RA and HT channel models, assuming that we have $D = 1024$ and $\Delta f = 10\,\text{kHz}$. Note that the OFDM symbol duration is

$$T_{f} = DT_{s} + T_{g}, \tag{12.10}$$

where $T_{s} = 1/(\Delta fD)$ is the OFDM symbol duration and T_{g} denotes the guard interval. We observe that the magnitude of the spectral correlation of the RA channel model decreases slowly as μ increases, since the maximum path delay τ_{max} is as small as $6T_{s}$. Thus, a moderately frequency-selective channel is expected, resulting in a gracefully decreasing BER curve, as observed in Figure 12.4(b), where the BER curves corresponding to the TU and HT channel models were also plotted. The latter two BER curves exhibit an error floor as the SNR increases, especially the one corresponding to the HT scenario. This is not unexpected, since we observe a sharp decay in $|\varphi_{hh}^{f}[\mu]|$ during the interval ($\mu = 0, 1, \ldots, 4$) and a 'strong non-concave' behaviour for $|\varphi_{hh}^{f}[\mu]|$, as seen in Figure 12.4(a). This is caused by the large maximum path delay of $\tau_{max} = 172T_{s}$.

12.1.2 Chapter Contributions and Outline

As observed in Sections 12.1.1.2 and 12.1.1.3, significant channel-induced performance degradations suffered by the CDD-aided direct-transmission-based OFDM system simply imply that the cooperative

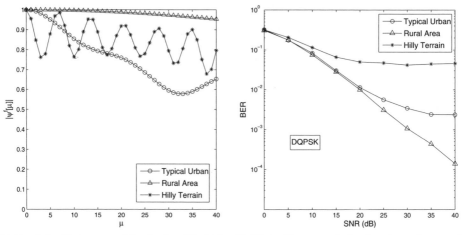

(a) Magnitude of frequency correlation function of Rayleigh fading channels

(b) Effects of frequency-selective channel on performance of CDD

Figure 12.4: Impact of frequency-selective channels on the performance of CDD.

diversity gains achieved by the cooperative system may erode as the relative mobile velocities of the cooperating users with respect to both each other and the BS increase. The detrimental effects of highly time-selective channels imposed on the T-DQPSK-modulated scenario were characterized in Figure 12.3(b), while those of heavily frequency-selective channels on the F-DPSK-modulated system were quantified in Figure 12.4(b). In order to eliminate this performance erosion and still achieve full cooperative diversity in conjunction with differential detection in wideband OFDM-based cooperative systems, in Section 12.2 we will invoke the Multiple-Symbol Differential Sphere Detection (MSDSD) technique, which was proposed by Lutz *et al.* in [597] in order to cope with fast-fading channels in SISO narrowband scenarios. We will demonstrate in Section 12.3 that, although a simple MSDSD scheme may be implemented at the relay, more powerful detection schemes are required at the BS of both the DAF- and DDF-aided cooperative systems in order to achieve a desirable end-to-end performance. Hence, the novel contributions of this chapter are as follows:

- A generalized equivalent multiple-symbol-based system model is constructed for the differentially encoded cooperative system using either the Differential Amplify-and-Forward (DAF) or Differential Decode-and-Forward (DDF) scheme.

- With the aid of the multi-layer search tree mechanism proposed for the SD in Chapter 11 in the context of the SP-modulated MIMO system, the MSDSD is specifically designed for both the DAF- and DDF-aided cooperative systems based on the above-mentioned generalized equivalent multiple-symbol system model. Our design objective is to retain the maximum achievable diversity gains at high mobile velocities, e.g. when T-DQPSK is employed, while imposing a low complexity.

The remainder of this chapter is organized as follows. The principle of the single-path MSDSD, which was proposed for employment in SISO systems, is reviewed in Section 12.2, where we will demonstrate that the MSDSD is capable of significantly mitigating the channel-induced error floor for both T-DQPSK- and F-DPSK-modulated OFDM systems, provided that the second-order statistics of the fading and noise are known at the receiver. Given the duality of the time and frequency dimensions, we will only consider the T-DQPSK-modulated system in Section 12.3, where we focus our attention on the multi-path MSDSD design, which is detailed for both the DAF- and DDF-aided cooperative cellular UL. The construction of the generalized equivalent multiple-symbol cooperative system model

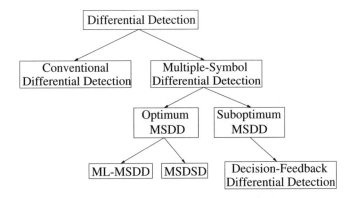

Figure 12.5: Differential detection classification.

is detailed in Section 12.3.3.1. Finally, we provide our concluding remarks in Section 12.4 based on the simulation results of Section 12.3.4.

12.2 The Principle of Single-Path MSDSD [597]

Differential detection schemes may be broadly divided into two categories, namely CDD and Multiple-Symbol Differential Detection (MSDD), as seen in Figure 12.5. Since a data symbol is mapped to the phase difference between the successive transmitted PSK symbols, CDD estimates the data symbol by directly calculating the phase difference of the two successive received symbols. In contrast to CDD having an observation window size of $N_{wind} = 2$, the MSDD collects $N_{wind} > 2$ consecutively received symbols for joint detection of the $(N_{wind} - 1)$ data symbols. This family may be further divided into two subgroups, namely the optimum maximum-likelihood (ML)-MSDD and suboptimum MSDD schemes, as seen in Figure 12.5. The ML-MSDD is the optimum scheme in terms of performance, but it exhibits a potentially excessive computational complexity in conjunction with a large observation window size N_{wind}. One of the suboptimum approaches that may be employed to achieve a low-complexity near-ML-MSDD is the linear-prediction-based Decision-Feedback Differential Detection (DFDD). Recently, the SD algorithm [556] was also used to resolve the complexity problem imposed by the ML-MSDD without sacrificing the achievable performance [597–600], leading to the so-called MSDSD arrangement, which will be introduced in the forthcoming sections.

12.2.1 ML Metric for MSDD

The basic idea behind ML-MSDD is the exploitation of the correlation between the phase distortions experienced by the consecutive N_{wind} transmitted DPSK symbols [601]. In other words, the receiver makes a decision about a block of $(N_{wind} - 1)$ consecutive symbols based on N_{wind} received symbols, enabling the detector to exploit the statistics of the fading channels. Ideally, the error floor encountered when performing CDD as observed in Figure 12.3 and Figure 12.4 can be essentially eliminated, provided that the value of N_{wind} is sufficiently high.

More explicitly, the MSDD at the receiver jointly processes the ith received symbol vector consisting of N_{wind} consecutively received symbols

$$\mathbf{y}[i_{N_{wind}}] \triangleq [y[(N_{wind} - 1)i - (N_{wind} - 1)], \ldots, y[(N_{wind} - 1)i]]^T, \tag{12.11}$$

where $i_{N_{wind}}$ is the symbol vector index, in order to generate the ML estimate vector $\hat{\mathbf{s}}[i_{N_{wind}}]$ of the corresponding N_{wind} transmitted symbols

$$\mathbf{s}[i_{N_{wind}}] \triangleq [s[(N_{wind}-1)i-(N_{wind}-1)], \ldots, s[(N_{wind}-1)i]]^T. \tag{12.12}$$

Then, when using differential decoding by carrying out the inverse of the differential encoding process of Equation (12.1), the estimated vector $\hat{\mathbf{v}}[i_{N_{wind}}]$ of the corresponding $(N_{wind}-1)$ differentially encoded data symbols

$$\mathbf{v}[i_{N_{wind}}] \triangleq [v[(N_{wind}-1)i-(N_{wind}-2)], \ldots, v[(N_{wind}-1)i]]^T \tag{12.13}$$

can be attained. Note that, due to differential encoding, consecutive blocks $\mathbf{y}[i_{N_{wind}}]$ overlap by one scalar received symbol [602]. For the sake of representational simplicity, we omit the symbol block index $i_{N_{wind}}$ without any loss of generality.

Under the assumption that the fading is a complex-valued zero-mean Gaussian process with a variance of σ_h^2 and that the channel noise has a variance of $2\sigma_w^2$, the PDF of the received symbol vector $\mathbf{y} = [y_0, y_1, \ldots, y_{N_{wind}-1}]^T$ conditioned on the transmitted symbol vector $\mathbf{s} = [s_0, s_1, \ldots, s_{N_{wind}-1}]^T$ spanning N_{wind} symbol periods is expressed as [597]

$$p(\mathbf{y}|\mathbf{s}) = \frac{\exp(-Tr\{\mathbf{y}^H\Psi^{-1}\mathbf{y}\})}{(\pi^{N_{wind}} det\,\Psi)}, \tag{12.14}$$

where

$$\Psi = \mathcal{E}\{\mathbf{y}\mathbf{y}^H|\mathbf{s}\} \tag{12.15}$$

denotes the conditional autocorrelation matrix of the Rayleigh fading channel. Then, the ML solution which maximizes the probability of Equation (12.14) can be obtained by exhaustively searching the entire transmitted symbol vector space. Thus, the ML metric of the MSDD can be expressed as [602]

$$\hat{\mathbf{s}}_{ML} = \underset{\check{\mathbf{s}} \in \mathcal{M}_c^{N_{wind}}}{\arg\max}\, p(\mathbf{y}|\check{\mathbf{s}}) \tag{12.16}$$

$$= \underset{\check{\mathbf{s}} \in \mathcal{M}_c^{N_{wind}}}{\arg\min}\, Tr\{\mathbf{y}^H\Psi^{-1}\mathbf{y}\}. \tag{12.17}$$

12.2.2   Metric Transformation

In order to elaborate further on the ML metric of Equation (12.16), we extended the conditional autocorrelation matrix Ψ as [597]:

$$\Psi = \mathcal{E}\{\mathbf{y}\mathbf{y}^H|\mathbf{s}\}, \tag{12.18}$$

$$= \mathrm{diag}(\mathbf{s})\mathcal{E}\{\mathbf{h}\mathbf{h}^H\}\mathrm{diag}(\mathbf{s}^H) + \mathcal{E}\{\mathbf{n}\mathbf{n}^H\}, \tag{12.19}$$

$$= \mathrm{diag}(\mathbf{s})(\mathcal{E}\{\mathbf{h}\mathbf{h}^H\} + 2\sigma_w^2\mathbf{I}_{N_{wind}})\mathrm{diag}(\mathbf{s}^H), \tag{12.20}$$

$$= \mathrm{diag}(\mathbf{s})\,\mathbf{C}\,\mathrm{diag}(\mathbf{s}^H), \tag{12.21}$$

where $\mathcal{E}\{\mathbf{h}\mathbf{h}^H\}$ is the channel's covariance matrix in either the time or the frequency domain, which is determined by the specific domain of the differential encoding. More explicitly, the elements of the covariance matrix $\mathcal{E}\{\mathbf{h}\mathbf{h}^H\}$ can be computed by Equation (12.7), when differential encoding at the transmitter is carried out along the TD. Otherwise, we employ Equation (12.9) to obtain the covariance matrix, when differential encoding is employed in the FD. Furthermore, in the context of Equation (12.21) we define

$$\mathbf{C} \triangleq (\mathcal{E}\{\mathbf{h}\mathbf{h}^H\} + 2\sigma_w^2\mathbf{I}_{N_{wind}}) \tag{12.22}$$

in order to simplify Equation (12.20).

Since we have $\mathrm{diag}(\mathbf{s})^{-1} = \mathrm{diag}(\mathbf{s})^H = \mathrm{diag}(\mathbf{s}^*)$, the ML decision rule of Equation (12.17) can be reformulated as

$$\hat{\mathbf{s}}_{ML} = \underset{\mathbf{s} \in \mathcal{M}_c^{N_{wind}}}{\arg\min} \{\mathbf{y}^H \mathbf{\Psi}^{-1} \mathbf{y}\} \tag{12.23}$$

$$= \underset{\mathbf{s} \in \mathcal{M}_c^{N_{wind}}}{\arg\min} \{\mathbf{y}^H \mathrm{diag}(\mathbf{s}) \, \mathbf{C}^{-1} \, \mathrm{diag}(\mathbf{s})^H \mathbf{y}\} \tag{12.24}$$

$$= \underset{\mathbf{s} \in \mathcal{M}_c^{N_{wind}}}{\arg\min} \{\mathbf{s} \, \mathrm{diag}(\mathbf{y})^H \mathbf{C}^{-1} \mathrm{diag}(\mathbf{y})\mathbf{s}^*\} \tag{12.25}$$

$$= \underset{\mathbf{s} \in \mathcal{M}_c^{N_{wind}}}{\arg\min} \{\mathbf{s} \, \mathrm{diag}(\mathbf{y})^H \mathbf{F}^H \mathbf{F} \, \mathrm{diag}(\mathbf{y})\mathbf{s}^*\}, \tag{12.26}$$

where $\mathbf{F}$ is an upper-triangular matrix obtained using the Cholesky factorization of the inverse matrix $\mathbf{C}^{-1}$, i.e. we have

$$\mathbf{C}^{-1} = \mathbf{F}^H \mathbf{F}. \tag{12.27}$$

Then, by further defining an upper-triangular matrix as

$$\mathbf{U} \triangleq (\mathbf{F} \, \mathrm{diag}(\mathbf{y}))^*, \tag{12.28}$$

we finally arrive at [597]

$$\hat{\mathbf{s}}_{ML} = \underset{\mathbf{s} \in \mathcal{M}_c^{N_{wind}}}{\arg\min} \{\|\mathbf{U}\mathbf{s}\|^2\}, \tag{12.29}$$

which completes the process of transforming the ML-MSDD metric of Equation (12.17) to a *shortest-vector* problem [597].

12.2.3   Complexity Reduction Using SD

While the performance of the MSDD improves steadily as N_{wind} is increased, the drawback is its potentially excessive computational complexity, which increases exponentially with N_{wind}. On the other hand, SD algorithms [556,560,603] are well known for their efficiency when solving the so-called shortest-vector problem in the context of multi-user, multi-stream detection in MIMO systems. Thus, due to the upper-triangular structure of the $\mathbf{U}$ matrix, the traditional SD algorithm can be employed to solve the shortest-vector problem as indicated by Equation (12.29). Consequently, the ML solution of the ML-MSDD metric of Equation (12.17) can be obtained on a component-by-component basis at a significantly lower complexity. Note that all the SD algorithms discussed in Chapter 9 can be employed to solve the shortest-vector problem of Equation (12.29).

12.2.4   Simulation Results

Monte Carlo simulations are provided in this section in order to characterize the achievable performance and the complexity imposed by the MSDSD for both TD and FD differentially encoded OFDM systems. The simulation parameters are summarized in Table 12.3.

12.2.4.1   Time-Differential-Encoded OFDM System

Let us now consider the application of the MSDSD in the TD differentially encoded OFDM system for three different normalized Doppler frequencies in the presence of the typical urban channel given by Table 12.1. The T-DQPSK modulation scheme is employed at the transmitter, while the MSDSD employing three different observation window sizes N_{wind} is used at the receiver, namely $N_{wind} = 2, 6, 9$. Note that, as mentioned in Section 12.1.1, when we have $N_{wind} = 2$ the MSDSD actually degenerates to the CDD. Additionally, since T-DQPSK is employed, a relatively short transmission frame length of 101 OFDM symbols is utilized in order to reduce the detection delay imposed by

Table 12.3: Simulation parameters for time-differential-modulation-aided OFDM system.

System parameters	Choice
System	OFDM
Subcarrier BW	$\Delta f = 10\,\text{kHz}$
Number of subcarriers	$D = 1024$
Modulation	T-DQPSK/F-DQPSK
Frame length	101 OFDM symbols
Normalized Doppler freq.	$f_d = 0.001, 0.01, 0.03$
Channel model	Typical urban if not specified

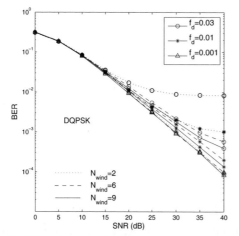

(a) BER performance of a T-DQPSK-modulated OFDM system using MSDSD in Rayleigh fading channels having different normalized Doppler frequencies

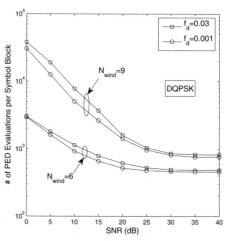

(b) Complexity imposed by the MSDSD versus the SNR

Figure 12.6: The application of MSDSD in the time-differential-modulated OFDM system.

the MSDSD. Figure 12.6(a) depicts the BER performance of the MSDSD for normalized Doppler frequencies $f_d = 0.03, 0.01, 0.001$, where we observe that for the slow-fading channel associated with $f_d = 0.001$, there is no need to employ an observation window size of more than $N_{wind} = 2$, since the CDD does not suffer from an error floor. In other words, the MSDSD is unable to improve the CDD's performance further by increasing N_{wind}. However, when the channel becomes more uncorrelated, i.e. when we have $f_d = 0.03$ or 0.01, the BER curve is shifted downwards by employing an N_{wind} value larger than 2, approaching that observed for $f_d = 0.001$, at the expense of imposing a higher computational complexity. The complexity imposed by the MSDSD versus the SNR is plotted in Figure 12.6(b), where the complexity curves corresponding to $N_{wind} = 9$ are evidently above those corresponding to $N_{wind} = 6$. Moreover, the complexity imposed by the MSDSD decreases steadily as the SNR increases and finally levels out in the high-SNR range. This is not unexpected, since under the assumption of having a reduced noise contamination, it is more likely that the ML solution point $\hat{s}_{ML}$ is located near the search centre (the origin in this case) of the SD used for finding the MSDD solution. As a result, the SD's search process may converge much more rapidly, imposing a reduced complexity. Again, for more details about the characteristics of SDs, refer to Chapter 9. Furthermore, we can also observe from Figure 12.6(b) that the Doppler frequency has a crucial effect not only on the performance achieved by the MSDSD, but also on its complexity.

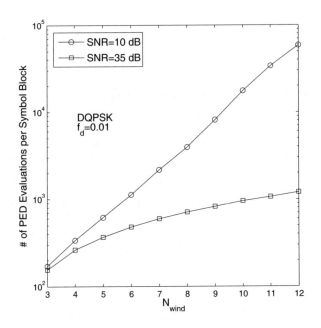

Figure 12.7: Complexity imposed by the MSDSD versus the observation window size N_{wind}.

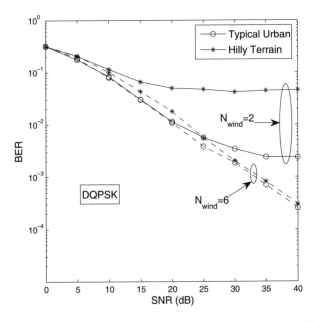

Figure 12.8: BER performance of F-DQPSK-modulated OFDM system using the MSDSD for the different channel models of Table 12.1.

Given a Doppler frequency of $f_d = 0.01$, let us now investigate the complexity of the MSDSD from a different angle by plotting the complexity versus N_{wind} in Figure 12.7, where complexity curves are drawn for two different SNRs. Although both of the curves exhibit an increase upon increasing the value of N_{wind}, the one corresponding to the relatively low SNR of 10 dB rises more sharply than the other one recorded for an SNR of 35 dB.

12.2.4.2   Frequency-Differential-Encoded OFDM System

As discussed in Section 12.1.1.3 for the scenario of burst transmissions or detection-delay-sensitive communications, F-DPSK is preferable to its TD counterpart. However, the channels experienced by the OFDM modem may exhibit a moderate time but a significant frequency selectivity, as exemplified by the TU and HT channel models given in Table 12.1. Therefore, the BER curves corresponding to the TU and HT channel models exhibit an error floor when using the CDD associated with $N_{wind} = 2$, as observed in Figure 12.8, due to the channel's frequency selectivity. Other simulation parameters are summarized in Table 12.3. Similar to the results obtained in the T-DPSK scenario, the error floor can be eliminated with the aid of the MSDSD, where the observation window size was $N_{wind} = 6$. Remarkably, a significant performance improvement is achieved by the MSDSD for the severely frequency-selective HT environment as seen in Figure 12.8. The BER curve associated with the CDD levels out as soon as the SNR increases beyond 20 dB, while the MSDSD using $N_{wind} = 6$ completely removes the error floor, resulting in a steadily decreasing BER curve as the SNR increases.

12.3   Multi-path MSDSD Design for Cooperative Communication

12.3.1   System Model

After the brief review on the principle of the MSDSD designed for single-path channels in Section 12.2, we continue by specifically designing an MSDSD scheme for the cooperative system discussed in Section 12.1. As depicted in Figure 12.9, we consider a U-user cooperation-aided system, where signal transmission involves two transmission phases, namely the broadcast phase and the relay phase, which are also referred to as phase I and II. A user who directly sends his/her own information to the destination is regarded as a *source* node, while the other users who assist in forwarding the information received from the source node are considered as *relay* nodes. In both phases, any of the well-established multiple-access schemes can be employed by the users to guarantee an orthogonal transmission among them, such as Time-Division Multiple Access (TDMA), Frequency-Division Multiple Access (FDMA) or Code-Division Multiple Access (CDMA). In this discussion, TDMA is considered for the sake of simplicity. Furthermore, due to the symmetry of channel allocation among users, as indicated in Figure 12.9, we focus our attention on the information transmission of source terminal T_S seen in Figure 12.10, which potentially employs $(U - 1)$ relay terminals $T_{R_1}, T_{R_2}, \ldots, T_{R_{U-1}}$ in order to achieve cooperative diversity by forming a VAA. Without loss of generality, we simply assume the employment of a single antenna for each of the collaborating MSs and that of N receive antennas for the BS. Additionally, a unitary total power P shared by the collaborating MSs for transmitting a symbol is assumed.

Owing to the potential transmission inefficiency and implementational difficulty imposed by the channel estimation in cooperation-aided systems, differential encoding and detection without acquisition of the CSI is preferable to the employment of substantially more complex coherent transmission techniques, as we discussed in Section 12.1. Hence, we assume that in phase I, the source broadcasts its differentially encoded signals, while the destination as well as the relay terminals are also capable of receiving the signal transmitted by the source. In the forthcoming phase II, we consider two possible cooperation protocols which can be employed by the relay nodes: the relay node may either directly forward the received signal to the destination after signal amplification

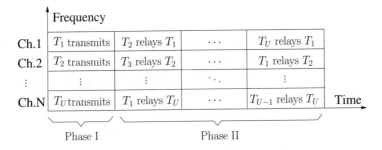

Figure 12.9: Repetition-based channel allocation scheme.

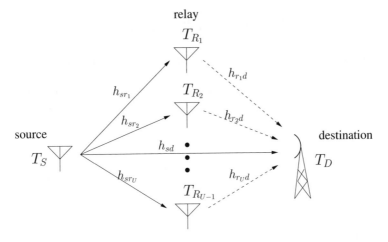

Figure 12.10: Cooperative communication schematic of multiple-relay nodes. ©IEEE Wang & Hanzo 2007 [39]

(the Amplify-and-Forward (AF) scheme) or differentially decode and re-encode the received signal before its retransmission (the Decode-and-Forward (DF) scheme).

Recall from Section 12.1.1.1 that the information is conveyed in the difference of the phases of two consecutive PSK symbols for differentially encoded transmission. In the context of the user cooperation-aided system of Figure 12.10, the source terminal T_S broadcasts the lth differentially encoded frame $\mathbf{s}^l$ during phase I, which consists of L_f DMPSK symbols $s[n]$ $(n = 1, 2, \ldots, L_f)$ given by Equation (12.1). According to Equation (12.1), the differential encoding process of the source node may be expressed as

$$s_s[n] = s_s[n-1]v_s[n], \tag{12.30}$$

where $v_{sd}[n] \in \mathcal{M}_c = \{e^{j2\pi m/M_c}; m = 0, 1, \ldots, M_c - 1\}$ is the information symbol obtained after bit-to-symbol mapping, and $s_{sd}[n] \in \mathcal{M}_c = \{e^{j2\pi m/M_c}; m = 0, 1, \ldots, M_c - 1\}$ represents the differentially encoded symbols during the nth time slot. We assume a total power of unity, i.e. $P = 1$, for transmitting a DMPSK symbol of the source over the entire user cooperation period and introduce the broadcast transmit power ratio η which is equal to the transmit power P_s employed by the source. Hence, during the forthcoming phase II, the total power consumed by all the $(U - 1)$ relay nodes used for transmitting the signal received from the source is $\sum_{u=1}^{U-1} P_{r_u} = 1 - \eta$, where P_{r_u} is the power consumed by the relay terminal T_{R_u} for conveying the signal of the source node. To mitigate the impairments imposed by the time-selective channels on the T-DPSK-modulated transmission, frame-based rather than symbol-based user cooperation is carried out, which is achieved at the expense of both a higher detection delay and increased memory requirements.

Furthermore, according to the cooperative strategy of Figure 12.9, where each of the $(U - 1)$ spatially dispersed relay nodes helps forward the signal from the source node to the destination node in $(U - 1)$ successive time slots, we construct a *single-symbol system model* for the source node's nth transmit symbol in the context of the TDMA-based user-cooperation-aided system of Figure 12.10 as

$$\mathbf{Y}_n = \mathbf{P}\mathbf{S}_n\mathbf{H}_n + \mathbf{W}_n, \tag{12.31}$$

where the diagonal matrix $\mathbf{P}$ is introduced to describe the transmit power allocation among the collaborating MSs and is defined as

$$\mathbf{P} \triangleq \text{diag}([\sqrt{P_s} \ \sqrt{P_{r_1}} \ \dots \ \sqrt{P_{r_{U-1}}}]). \tag{12.32}$$

Additionally, in Equation (12.31) $\mathbf{S}_n$ and $\mathbf{Y}_n$ represent the *transmitted user-cooperation-based signal matrix* and the received signal matrix at the destination, respectively, during both phase I and phase II. Additionally, $\mathbf{H}_n$ and $\mathbf{W}_n$ denote the channel matrix and the AWGN matrix, respectively. Upon further elaborating of Equation (12.31), we get

$$\begin{bmatrix} y_{sd_1}[n] & \cdots & y_{sd_N}[n] \\ y_{r_1d_1}[n+1\cdot L_f] & \cdots & y_{r_1d_N}[n+1\cdot L_f] \\ \vdots & \cdots & \vdots \\ y_{r_{U-1}d_1}[n+(U-1)L_f] & \cdots & y_{r_{U-1}d_N}[n+(U-1)L_f] \end{bmatrix}_{U \times N}$$

$$= \mathbf{P}\begin{bmatrix} s_s[n] & 0 & \cdots & 0 \\ 0 & s_{r_1}[n+1\cdot L_f] & \cdots & 0 \\ \vdots & \vdots & \ddots & \vdots \\ 0 & 0 & \cdots & s_{r_{U-1}}[n+(U-1)L_f] \end{bmatrix}_{U \times U}$$

$$\times \begin{bmatrix} h_{sd_1}[n] & \cdots & h_{sd_N}[n] \\ h_{r_1d_1}[n+1\cdot L_f] & \cdots & h_{r_1d_N}[n+1\cdot L_f] \\ \vdots & \cdots & \vdots \\ h_{r_{U-1}d_1}[n+(U-1)L_f] & \cdots & h_{r_{U-1}d_N}[n+(U-1)L_f] \end{bmatrix}_{U \times N}$$

$$+ \begin{bmatrix} w_{sd_1}[n] & \cdots & w_{sd_N}[n] \\ w_{r_1d_1}[n+1\cdot L_f] & \cdots & w_{r_1d_N}[n+1\cdot L_f] \\ \vdots & \cdots & \vdots \\ w_{r_{U-1}d_1}[n+(U-1)L_f] & \cdots & w_{r_{U-1}d_N}[n+(U-1)L_f] \end{bmatrix}_{U \times N}, \tag{12.33}$$

where the rows and columns of the transmitted user-cooperation-based signal matrix $\mathbf{S}_n$ denote the spatial and temporal dimensions, respectively. Moreover, since the source and multiple relay terminals are assumed to be far apart, the elements of the channel matrix $\mathbf{H}_n$, corresponding to the CIRs between the source and the destination nodes as well as those between the relay node and the destination node, are mutually uncorrelated, but each of them may be correlated along the TD according to the time-selective characteristics of the channel. Additionally, the elements of the AWGN matrix are modelled as independent complex-valued Gaussian random variables with zero mean and a variance of $N_0 = 2\sigma_w^2$.

More specifically, since we have the transmitted symbol $s_s[n] \in \mathcal{M}_c = \{e^{j2\pi m/M_c}; m_s = 0, 1, \dots, M_c - 1\}$ at the source node, the $(U \times U)$-element transmitted signal matrix $\mathbf{P}\mathbf{S}_n$ in the general system model of Equation (12.33) can be reformatted for the DAF-aided cooperative system as

$$\mathbf{PS}_n = \begin{bmatrix} \sqrt{P_s} \cdot e^{j2\pi m_s/M_c} & 0 & \cdots & 0 \\ 0 & f_{AM_{r_1}} y_{sr_1}[n] & \cdots & 0 \\ \vdots & \vdots & \ddots & \vdots \\ 0 & 0 & \cdots & f_{AM_{r_{U-1}}} y_{sr_{U-1}}[n] \end{bmatrix}, \tag{12.34}$$

where $f_{AM_{r_u}}$ is the signal gain employed by the uth relay node to make sure that the average transmitted power of the uth relay is P_{r_u} and

$$y_{sr_u}[n]$$
$$= \sqrt{P_s} \cdot s_s[n] h_{sr_u}[n] + w_{sr_u}[n] \tag{12.35}$$
$$= \sqrt{P_s} \cdot e^{j2\pi m_s/M_c} h_{sr_u}[n] + w_{sr_u}[n] \quad (m_s = 0, 1, \ldots, M_c - 1) \tag{12.36}$$

represents the signal received at the uth relay node during the broadcast phase I.

As for the DDF-aided user cooperation system, where the relay node differentially detects and re-encodes the signal received from the source node before forwarding it to the destination, the $(U \times U)$-element transmitted signal matrix $\mathbf{PS}_n$ in the general system model of Equation (12.33) can be rewritten as follows under the assumption that the output of the differentially detected relay is error-free:

$$\mathbf{PS}_n = \begin{bmatrix} \sqrt{P_s} \cdot e^{j2\pi m_s/M_c} & 0 & \cdots & 0 \\ 0 & \sqrt{P_{r_1}} \cdot e^{j2\pi m_s/M_c} & \cdots & 0 \\ \vdots & \vdots & \ddots & \vdots \\ 0 & 0 & \cdots & \sqrt{P_{r_{U-1}}} \cdot e^{j2\pi m_s/M_c} \end{bmatrix}. \tag{12.37}$$

12.3.2 Differentially Encoded Cooperative Communication Using CDD

In this section, for the sake of simplicity, we consider two differential-modulation-based two-user cooperative schemes, namely, the DAF and DDF. Both these schemes are amenable to the CDD in fading channels after a linear signal combination process, which will be discussed in our forthcoming discourse.

12.3.2.1 Signal Combining at the Destination for DAF Relaying

For the DAF scheme, the $(U - 1)$ relay nodes of Figure 12.10 amplify the signal received from the source node and forward it to the destination node in a preset order over $(U - 1)$ successive time slots during phase II. In order to ensure that the average transmit power of the uth relay node remains P_{r_u}, the corresponding amplification factor $f_{AM_{r_u}}$ in Equation (12.34) employed by the uth relay node can be specified as [604]

$$f_{AM_{r_u}} = \sqrt{\frac{P_{r_u}}{P_s \sigma_{sr_u}^2 + N_0}}, \tag{12.38}$$

where $\sigma_{sr_u}^2$ is the variance of the channel's envelope spanning between the source and the uth relay node, which can be obtained by long-term averaging of the received signals. Therefore, the signal received at the destination from the uth relay node $y_{r_u d}[n + uL_f]$ in Equation (12.33) can be represented as follows [604]:

$$y_{r_u d}[n + uL_f] = f_{AM_{r_u}} y_{sr_u}[n] h_{r_u d}[n + uL_f] + w_{r_u d}[n + uL_f], \tag{12.39}$$

where $y_{sr_u}[n]$ is the signal received from the source node at the uth relay node during the broadcast phase I, which was given by Equation (12.35).

The destination BS linearly combines the signal at each of the N receive antennas received from the source through the direct link during the broadcast, namely phase I and those at each receive antenna received from all the relay nodes during phase II, followed by the CDD process operating without acquiring any CSI. Based on the multi-channel differential detection principle of [467], we combine the multi-path signal of the U-user cooperation system of Figure 12.10 prior to the CDD process as

$$y = \sum_{i=1}^{N} \left[a_0 (y_{sd_i}[n-1])^* y_{sd_i}[n] + \sum_{u=1}^{U-1} a_u (y_{r_u d_i}[n+uL_f-1])^* y_{r_u d_i}[n+uL_f] \right], \quad (12.40)$$

where L_f is the length of the frame, while the coefficients a_0 and a_u $(u = 1, 2, \ldots)$ are respectively given by

$$a_0 = \frac{1}{N_0}, \quad (12.41)$$

$$a_u = \frac{P_s \sigma_{sr_u}^2 + N_0}{N_0 (P_s \sigma_{sr_u}^2 + P_{r_u} \sigma_{r_u d}^2 + N_0)}, \quad (12.42)$$

where $\sigma_{sr_u}^2$ and $\sigma_{r_u d}^2$ are the variances of the link between the source and relay nodes as well as of the link between the relay node and the BS, respectively. By assuming that the CIRs h_{sr_u} as well as $h_{r_u d}$ are almost constant over two successive symbol periods, the destination node carries out the CDD based on the combined signal y of Equation (12.40) as

$$e^{j2\pi \tilde{m}/M_c} = \underset{\tilde{m}=0,1,\ldots,M_c-1}{\arg\max} \Re\{e^{-j2\pi \tilde{m}/M_c} y\}, \quad (12.43)$$

where $\Re\{\cdot\}$ denotes the real component of a complex number.

12.3.2.2 Signal Combining at Destination for DDF Relaying

For the DDF-aided U-user cooperation system of Figure 12.10, each relay node differentially decodes and re-encodes the signal received from the source node, before forwarding it to the BS. Similarly, based on the multi-channel differential detection techniques of [467, 605], the combined signal prior to differential detection by the DDF scheme can be expressed in exactly the same form as that of Equation (12.40) for the DAF scheme, which is repeated here for convenience:

$$y = \sum_{i=1}^{N} \left[a_0 (y_{sd_i}[n-1])^* y_{sd_i}[n] + \sum_{u=1}^{U-1} a_u (y_{r_u d_i}[n+uL_f-1])^* y_{r_u d_i}[n+uL_f] \right], \quad (12.44)$$

noting that different diversity combining weights of a_0 and a_u $(u = 1, 2, \ldots, U-1)$ are used. Note also that the choice of diversity combining weights may affect the achievable system performance. For example, when the normalized total power of $P = 1$ used for transmitting a symbol during the entire user-cooperation-aided process is equally divided among the source and relay nodes, i.e. when we have $P_s = P_{r_u} = 1/U$ $(u = 1, 2, \ldots, U-1)$, the SNR of the combiner output is maximized by opting for [605]

$$a_0 = a_u = \frac{1}{N_0}, \quad (12.45)$$

provided that the corresponding channel variances are identical. Although the choice of the diversity combining weights is not optimal in general, it is optimal for the case when the SNR of the source–destination link and those of the multiple relay–destination links are the same. Again, by assuming that the CIRs taps h_{sr_u} as well as $h_{r_u d}$ are constant during two successive symbol periods, the CDD process of Equation (12.43) can be carried out by the destination after combining the multi-path signals.

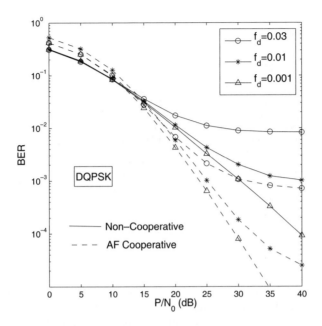

Figure 12.11: BER performance of the DAF-aided DQPSK-modulated two-user cooperative OFDM system in Rayleigh fading channels at different normalized Doppler frequencies. The system parameters were summarized in Table 12.4.

12.3.2.3 Simulation Results

Figure 12.11 depicts the BER performance versus P/N_0 for both the single-user non-cooperative system and the two-user DAF-aided cooperative system, using the simulation parameters summarized in Table 12.4. Note that we consider a scenario where the total power P used for transmitting a differentially encoded symbol during an entire user cooperation process is equally shared between the source and relay nodes, and the SNRs at the receiver of the relay and destination nodes are identical. Additionally, in order to carry out a fair comparison between the non-cooperative and cooperative systems, we assume that the power consumed by the single-user non-cooperative system when transmitting a single T-DQPSK symbol is also equal to $P = 1$, which is identical to that consumed by its user-cooperation-aided counterpart. As observed from Figure 12.11, in the presence of the slowly fading TU channel of Table 12.1 associated with $f_d = 0.001$, the DDF-aided two-user cooperative system is capable of achieving the maximum attainable spatial diversity order of two, resulting in a significant performance gain of 10 dB, given a target BER of 10^{-4}. This high gain is not unexpected, since it is unlikely that both the direct and relay links suffer from a deep fade. However, since the T-DQPSK modulation scheme is employed, the performance achieved by the CDD at the destination node degrades significantly as the normalized Doppler frequency f_d becomes higher. This is due to, for example, the relative mobility of the source and relay nodes with respect to the BS. For the sake of simplicity, here we assume the same normalized Doppler frequency exhibited by all the three links of the two-user cooperative system, namely the source–relay, relay–destination and source–destination links. As shown in Figure 12.11, an error floor is formed by the BER curves corresponding to the more time-selective scenarios associated with an increased normalized Doppler frequency f_d ranging from 0.001 to 0.03, which is an undesirable situation encountered also by the classic single-user non-cooperative benchmark system. However, the lowest achievable end-to-end BER of 10^{-3} exhibited by the CDD operating with the aid of the DAF-aided cooperation scheme is still lower than the BER of 10^{-2} achieved by the non-cooperative system under the assumption of $f_d = 0.03$.

Table 12.4: Summary of system parameters for a T-DQPSK-modulated two-user cooperative OFDM system.

System parameters	Choice
System	Two-user cooperative OFDM
Number of relay nodes	1
Subcarrier BW	$\Delta f = 10\,\text{kHz}$
Number of subcarriers	$D = 1024$
Modulation	T-DQPSK
Frame length L_f	101
CRC	CCITT-6
Normalized Doppler freq.	$f_d = 0.03, 0.01, 0.001$
Channel model	Typical urban, refer to Table 12.1
Channel variances	$\sigma_{sd}^2 = \sigma_{sr}^2 = \sigma_{rd}^2 = 1$
Power allocation	$P_s = P_{r_1} = 0.5P = 0.5$
SNR at relay and destination	$P_s/N_0 = P_{r_1}/N_0$

In comparison with the DAF-aided cooperative system, where the relay node directly forwards the amplified signal to the destination, the differential decoding and re-encoding of the DDF-aided system are carried out by the relay node before forwarding, as discussed in Section 12.3.2.2. The simulation parameters are summarized in Table 12.4, where we can see that a Cyclic Redundancy Check (CRC) code is employed by the relay node in order to determine whether the current decoded signal is correct or not and only the error-free decoded signal is forwarded to the destination. Otherwise, the relay remains silent during phase II. Figure 12.12 plots the BER curves of the DDF-aided two-user cooperative system using the CDD at both the relay and destination nodes in contrast to those of its non-cooperative counterpart. Again, the DDF-aided cooperative scheme is capable of achieving the maximum attainable diversity order of two, leading to a significant performance gain for transmission over a slow-fading channel associated with $f_d = 0.001$. Furthermore, observe by comparing Figure 12.12 that a similarly negative impact is imposed on the end-to-end BER performance by the relative mobility of the source, relay and destination nodes for the DDF scheme as that imposed for the DAF scheme. Moreover, also note in Figure 12.12 that although the DDF-aided cooperative system outperforms its non-cooperative counterpart at the three different values of the normalized Doppler frequency considered, the achievable performance gain becomes more negligible as f_d increases. Specifically, only a slightly lower error floor is exhibited in Figure 12.12 by the DDF-aided system associated with $f_d = 0.03$ than that presented by the classic single-user non-cooperative arrangement. In addition, as observed from both Figure 12.11 and Figure 12.12, both the DAF- and DDF-aided cooperative systems exhibit a worse BER performance than the classic non-cooperative one in the relatively low-SNR range spanning from 0 to 15 dB, which can also be observed for the co-located multiple-transmit-antenna-assisted system. This trend is not unexpected, since the effective SNR experienced at the receiver is halved for the two-transmit-antenna-aided system, and the benefit of diversity is overwhelmed by the deleterious effects of the noise when the SNR is low.

Let us now investigate the benefit of the CRC-based error-detection capability of the relay node on the end-to-end BER performance of a DDF-aided two-user cooperative system in Figure 12.13, where the BER curves corresponding to different CRC codes are plotted in contrast to those of the so-called fixed-relay-based cooperative system as well as to that of the single-user non-cooperative one. Note that, as summarized in Table 12.4, the frame length L_f employed is 101 DQPSK symbols, whereas CCITT-6 was used by the relay node similarly to the previously simulated DDF-aided cooperative system of Figure 12.12, which exhibits a desirable error-detection capability for this relatively short frame length, since a full diversity order of two can be achieved. To improve the achievable transmission efficiency, a CRC code using as few parity bits as possible is preferable, such as CCITT-4. However,

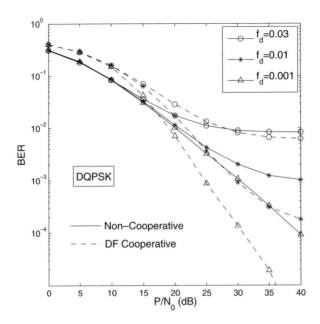

Figure 12.12: BER performance of the DDF-aided DQPSK-modulated cooperative system in Rayleigh fading channels at different normalized Doppler frequencies. The system parameters were summarized in Table 12.4.

as observed in Figure 12.13, the achievable BER performance of the DDF-aided cooperative system gradually degrades as the SNR increases, leading to an approximately 4 dB performance gain reduction at a target BER of 10^{-5} in comparison with the system employing the CCITT-6. Another extreme example worth considering is a fixed-relay-based cooperative system, where the relay forwards the re-encoded differential signal to the destination without checking whether the differentially decoded bits are correct or not. Hence, the achievable transmission efficiency is improved by sacrificing the maximum achievable diversity gain. Specifically, without the aid of the CRC, no spatial diversity gain can be achieved, although an additional transmit antenna provided by the relay node further assists the source by forwarding the signal to the BS. The reason for this trend is that without CRC checking the original diversity gain is eroded by the flawed information delivered by the relay node, which is further combined with the signal received via the direct link at the destination. Hence, a flexible compromise between maintaining a high transmission efficiency and the maximum achievable diversity gain can be struck by employing an appropriate CRC code.

In comparison with the classic co-located multiple-transmit-antenna-assisted system, the performance of the user-cooperation-aided system is affected both by the channel quality of the source–destination and relay–destination links and by that of the source–relay link. This statement is true for both the DAF- and DDF-aided cooperative systems as evidenced by our forthcoming discussions. Figure 12.14 compares the BER performance achieved by the two-user cooperative system employing either the DAF or the DDF scheme in two different scenarios, namely for a noisy source–relay link, as assumed in the scenarios characterized in Figures 12.12 and 12.13, and for a perfect noise-free source–relay link. In other words, the relay is assumed to have perfect knowledge of the source node's transmitted signal in the latter scenario, which can also be regarded as the conventional co-located multiple-transmit-antenna-aided system, if the DDF scheme is employed. Additionally, recall from Figures 12.11 and 12.12 that the maximum diversity order of two can indeed be achieved by the T-DQPSK-modulated two-user cooperative system using the CDD when a quasi-static scenario of a

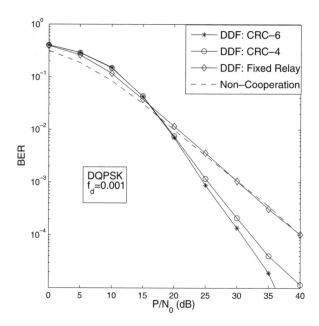

Figure 12.13: Benefits of the CRC-based error-detection capability at the relay node on the end-to-end BER performance of a DDF-aided DQPSK-modulated cooperative system. The system parameters were summarized in Table 12.4.

normalized Doppler frequency $f_d = 0.001$ is assumed. Although the maximum achievable diversity gain cannot be increased by having a perfect source–relay link, observe in Figure 12.14 that the system's BER performance was indeed improved. To be more specific, a performance gain as high as 5 dB was attained in Figure 12.14 for the system using the DDF scheme by having a perfect source–relay link, whereas only a negligible performance gain was attained in Figure 12.14 by its DAF-aided counterpart. Furthermore, by comparing the performance achieved by the DAF and DDF schemes in Figure 12.14, we observe that the latter is slightly outperformed by the former if the transmissions between the source and relay nodes are carried out over a noisy link having an SNR at the relay node which is equal to that measured at the destination node. However, it is expected that the latter will outperform the former as a benefit of having a better-quality source–relay link, as indicated by the extreme example of having a noise-free source–relay link, which was characterized in Figure 12.14. Therefore, when the source–relay link is of poor quality, it is preferable to employ the DAF scheme, which outperforms the DDF scheme despite its lower complexity, since there is no need to carry out any differential decoding and re-encoding.

12.3.3   Multi-path MSDSD Design for Cooperative Communication

In order to mitigate the potential negative impact induced by strongly time-selective or frequency-selective channels on the conventional T-DQPSK or F-DQPSK scenarios of Section 12.1.1, the single-path MSDSD introduced in Section 12.2 constitutes an attractive scheme for employment by the relay nodes, when differential decoding is carried out at relay nodes using the DF protocol. Figure 12.15 characterizes the achievable performance improvements of the DDF-aided two-user cooperative system attained by the single-path MSDSD scheme at the relay node in time-selective Rayleigh fading channels at different normalized Doppler frequencies. When employing the MSDSD scheme using $N_{wind} = 6$ at the relay node, observe in Figure 12.15 that the error floors encountered in time-selective channels corresponding to $f_d = 0.01$ and $f_d = 0.03$ are significantly mitigated, resulting in a substantial

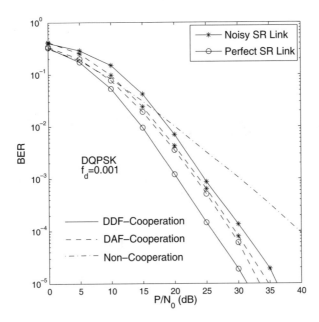

Figure 12.14: Impact of the source–relay link's quality on the end-to-end BER performance of a T-DQPSK-modulated two-user cooperative system. The system parameters were summarized in Table 12.4.

performance gain. For example, given a target BER of 10^{-4}, a performance gain in excess of 5 dB can be achieved for $f_d = 0.01$ as seen in Figure 12.15. However, since the end-to-end performance of the user cooperative system of Figure 12.10 is determined by the robustness of the differential detection schemes employed at both the relay and destination nodes, the single-path MSDSD-aided relay terminals alone are unable to guarantee a desirable end-to-end performance. Hence, although a significant performance gain can be attained by improving the detection capability at the relay node, there is still a substantial performance gap between the BER curve obtained at $f_d = 0.01$ or $f_d = 0.03$ and that corresponding to $f_d = 0.001$. The maximum diversity order of two is not achieved at $f_d = 0.03$ or $f_d = 0.01$, as indicated by the slope of the BER curve seen in Figure 12.15. Hence, for further improving the performance of the DDF-aided cooperative system or that of the DAF-aided one, a powerful differential detector has to be applied at the destination node, which is robust to the impairments imposed by time-selective channels. Unfortunately, the single-path MSDSD scheme cannot be directly employed by the destination node in order jointly to decode differentially the multi-path signals received from the source and relay nodes. Thus, a potential channel-induced performance degradation may still occur when carrying out conventional differential detection of signals received over the multi-path channel, which is discussed in Section 12.3.2. In the forthcoming sections, based on the principle of the single-path MSDSD, we will propose an MSDSD scheme specifically designed for user-cooperation-aided communication systems, which is capable of jointly detecting differentially the multi-path signals delivered by the source and relay nodes.

12.3.3.1 Derivation of the Metric for Optimum Detection

12.3.3.1.1 Equivalent System Model for the DDF-Aided Cooperative Systems

Following on from the principle of the single-path MSDSD discussed in Section 12.2, the receiver operating without knowledge of the CSI at the destination node collects N_{wind} consecutive

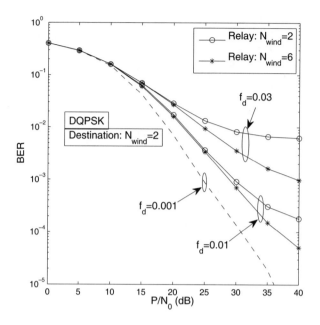

Figure 12.15: BER performance of DDF-aided DQPSK-modulated cooperative system using MSDSD-aided relays in Rayleigh fading channels.

user-cooperation-based space–time symbols $\mathbf{S}_n$ $(n = 0, 1, \ldots, N_{wind} - 1)$. These samples are then used jointly to detect a block of $(N_{wind} - 1)$ consecutive symbols $v_s[n]$ $(n = 0, 1, \ldots, N_{wind} - 2)$, which were differentially encoded by the source during phase I by exploiting the correlation between the phase distortions experienced by the adjacent samples $\mathbf{S}_n$ $(n = 0, 1, \ldots, N_{wind} - 1)$. The nth user-cooperation-based space–time symbol $\mathbf{S}_n$ was defined specifically for the DDF-aided cooperative system in Equation (12.37), which is rewritten here as

$$
\mathbf{S}_n = \begin{bmatrix} e^{j2\pi m_s/M_c} & 0 & \cdots & 0 \\ 0 & e^{j2\pi m_s/M_c} & \cdots & 0 \\ \vdots & \vdots & \ddots & \vdots \\ 0 & 0 & \cdots & e^{j2\pi m_s/M_c} \end{bmatrix},
\tag{12.46}
$$

where we have $m_s = 0, 1, \ldots, M_c - 1$. Since the total power used for transmitting a single symbol $\mathbf{S}_n$ during the entire user-cooperation process is normalized, we have

$$
P_s + \sum_{u=1}^{U-1} P_{r_u} = 1,
\tag{12.47}
$$

where U is the number of users in the user-cooperation-aided system of Figure 12.10. Moreover, with the aid of Equations (12.33) and (12.37), we can rewrite the generalized single-symbol-based cooperative system model of Equation (12.31) for the DDF-aided cooperative transmission, resulting in the *equivalent single-symbol-based system model* as follows:

$$
\mathbf{Y}_n = \mathbf{P}\mathbf{S}_n\mathbf{H}_n + \mathbf{W}_n
\tag{12.48}
$$

$$
= \mathbf{S}_n\mathbf{P}\mathbf{H}_n + \mathbf{W}_n
\tag{12.49}
$$

$$
= \tilde{\mathbf{S}}_n\tilde{\mathbf{H}}_n + \tilde{\mathbf{W}}_n,
\tag{12.50}
$$

where the equivalent user-cooperation transmitted signal's unitary matrix $\tilde{\mathbf{S}}_n$ is represented by

$$\tilde{\mathbf{S}}_n = \mathbf{S}_n = \begin{bmatrix} e^{j2\pi m_s/M_c} & 0 & \cdots & 0 \\ 0 & e^{j2\pi m_s/M_c} & \cdots & 0 \\ \vdots & \vdots & \ddots & \vdots \\ 0 & 0 & \cdots & e^{j2\pi m_s/M_c} \end{bmatrix}, \quad m_s = 0, 1, \ldots, M_c - 1, \quad (12.51)$$

and the equivalent channel matrix $\tilde{\mathbf{H}}_n$ can be expressed as

$$\tilde{\mathbf{H}}_n = \mathbf{P}\mathbf{H}_n \tag{12.52}$$

$$= \begin{bmatrix} \sqrt{P_s} \cdot h_{sd_1}[n] & \cdots & \sqrt{P_s} \cdot h_{sd_N}[n] \\ \sqrt{P_{r_1}} \cdot h_{r_1 d_1}[n+1\cdot L_f] & \cdots & \sqrt{P_{r_1}} \cdot h_{r_1 d_N}[n+1\cdot L_f] \\ \vdots & \cdots & \vdots \\ \sqrt{P_{r_{U-1}d}} \cdot h_{r_{U-1}d_1}[n+(U-1)L_f] & \cdots & \sqrt{P_{r_{U-1}d}} \cdot h_{r_{U-1}d_N}[n+(U-1)L_f] \end{bmatrix}. \tag{12.53}$$

In addition, according to Equation (12.33) the received signal matrix $\mathbf{Y}_n$ and the equivalent noise matrix $\tilde{\mathbf{W}}_n$ may be written as

$$\mathbf{Y}_n = \begin{bmatrix} y_{sd_1}[n] & \cdots & y_{sd_N}[n] \\ y_{r_1 d_1}[n+1\cdot L_f] & \cdots & y_{r_1 d_N}[n+1\cdot L_f] \\ \vdots & \cdots & \vdots \\ y_{r_{U-1}d_1}[n+(U-1)L_f] & \cdots & y_{r_{U-1}d_N}[n+(U-1)L_f] \end{bmatrix} \tag{12.54}$$

and

$$\tilde{\mathbf{W}}_n = \mathbf{W}_n = \begin{bmatrix} w_{sd_1}[n] & \cdots & w_{sd_N}[n] \\ w_{r_1 d_1}[n+1\cdot L_f] & \cdots & w_{r_1 d_N}[n+1\cdot L_f] \\ \vdots & \cdots & \vdots \\ w_{r_{U-1}d_1}[n+(U-1)L_f] & \cdots & w_{r_{U-1}d_N}[n+(U-1)L_f] \end{bmatrix}, \tag{12.55}$$

respectively.

12.3.3.1.2 Equivalent System Model for the DAF-Aided Cooperative System

Similarly, with the aid of Equations (12.33), (12.34) as well as (12.35) and following a number of straightforward manipulations left out here for compactness, we arrive at the *equivalent single-symbol system model* for the DAF-aided cooperation system based on the generalized single-symbol cooperative system model of Equation (12.31) as follows:

$$\mathbf{Y}_n = \tilde{\mathbf{S}}_n \tilde{\mathbf{H}}_n + \tilde{\mathbf{W}}_n, \tag{12.56}$$

where the received signal matrix $\mathbf{Y}_n$ at the BS is expressed identically to that of the DDF-aided system as

$$\mathbf{Y}_n = \begin{bmatrix} y_{sd_1}[n] & \cdots & y_{sd_N}[n] \\ y_{r_1 d_1}[n+1\cdot L_f] & \cdots & y_{r_1 d_N}[n+1\cdot L_f] \\ \vdots & \cdots & \vdots \\ y_{r_{U-1}d_1}[n+(U-1)L_f] & \cdots & y_{r_{U-1}d_N}[n+(U-1)L_f] \end{bmatrix}, \tag{12.57}$$

and the equivalent user-cooperation transmitted signal matrix $\tilde{\mathbf{S}}_n$ can be written as

$$
\tilde{\mathbf{S}}_n = \begin{bmatrix}
e^{j2\pi m_s/M_c} & 0 & \cdots & 0 \\
0 & e^{j2\pi m_s/M_c} & \cdots & 0 \\
\vdots & \vdots & \ddots & \vdots \\
0 & 0 & \cdots & e^{j2\pi m_s/M_c}
\end{bmatrix}, \quad m_s = 0, 1, \ldots, M_c - 1, \quad (12.58)
$$

which is identical to the transmitted signal matrix given in Equation (12.51) for the DDF-aided system. However, the resultant equivalent channel matrix $\tilde{\mathbf{H}}_n$ of the DAF-aided system is different from that obtained for its DDF-aided counterpart of Equation (12.52), which is expressed as

$$
\tilde{\mathbf{H}}_n = [\tilde{\mathbf{h}}_1 \ \tilde{\mathbf{h}}_2 \ldots \tilde{\mathbf{h}}_N], \quad (12.59)
$$

where the ith column vector $\tilde{\mathbf{h}}_i$ may be written as

$$
\mathbf{h}_i = \begin{bmatrix}
\sqrt{P_s} \cdot h_{sd_i}[n] \\
\sqrt{\frac{P_{r_1}}{\sigma_{sr_1}^2 + (N_0/P_s)}} h_{sr_1}[n] h_{r_1 d_i}[n + 1 \cdot L_f] \\
\vdots \\
\sqrt{\frac{P_{r_{U-1}}}{\sigma_{sr_{U-1}}^2 + (N_0/P_s)}} h_{sr_{U-1}}[n] h_{r_{U-1} d_i}[n + (U-1) \cdot L_f]
\end{bmatrix}. \quad (12.60)
$$

Moreover, the resultant equivalent noise term $\tilde{\mathbf{W}}_n$ can be represented as

$$
\tilde{\mathbf{W}}_n = [\tilde{\mathbf{w}}_1 \ \tilde{\mathbf{w}}_2 \ldots \tilde{\mathbf{w}}_N], \quad (12.61)
$$

where the ith column vector $\tilde{\mathbf{w}}_i$ may be expressed as

$$
\tilde{\mathbf{w}}_i = \begin{bmatrix}
w_{sd}[n] \\
\sqrt{\frac{P_{r_1}}{P_s \sigma_{sr_1}^2 + N_0}} w_{sr_1}[n] h_{r_1 d_i}[n + 1 \cdot L_f] + w_{r_1 d_i}[n + 1 \cdot L_f] \\
\vdots \\
\sqrt{\frac{P_{r_{U-1}}}{P_s \sigma_{sr_{U-1}}^2 + N_0}} w_{sr_{U-1}}[n] h_{r_{U-1} d_i}[n + (U-1) \cdot L_f] + w_{r_{U-1} d_i}[n + (U-1) \cdot L_f]
\end{bmatrix}.
$$
$$(12.62)$$

12.3.3.1.3   Optimum Detection Metric

Then, based on Equation (12.50) and Equation (12.56), we can construct the general input–output relation of the channel for multiple differential symbol transmissions for both DAF- and DDF-aided user-cooperative systems, where we have the *equivalent multiple-symbol-based system model* as

$$
\underline{\mathbf{Y}} = \underline{\tilde{\mathbf{S}}_d} \underline{\tilde{\mathbf{H}}} + \underline{\tilde{\mathbf{W}}}. \quad (12.63)
$$

Note that if $\mathbf{A}$ represents a matrix, then $\underline{\mathbf{A}}$ is a block matrix, $\mathbf{A}_d$ denotes a diagonal matrix, and $\underline{\mathbf{A}}_d$ represents a block diagonal matrix. The block matrix $\underline{\mathbf{Y}}$ hosting the received signal, which contains signals received during N_{wind} successive user-cooperation-based symbol durations corresponding to N_{wind} consecutively transmitted differential symbols $s_s[n]$ ($n = 0, 1, \ldots, N_{wind} - 1$) of the source node, is defined as

$$
\underline{\mathbf{Y}} = [\mathbf{Y}_n^T \ \mathbf{Y}_{n+1}^T \cdots \mathbf{Y}_{n+N_{wind}-1}^T]^T, \quad (12.64)
$$

and the block matrix $\underline{\tilde{\mathbf{H}}}$ representing the channel as well as the block matrix $\underline{\tilde{\mathbf{W}}}$ of the AWGN are defined likewise by vertically concatenating N_{wind} matrices $\tilde{\mathbf{H}}_n$ ($n = 0, 1, \ldots, N_{wind} - 1$) and $\tilde{\mathbf{W}}_n$

$(n = 0, 1, \ldots, N_{wind} - 1)$, respectively. Therefore, we can represent $\underline{\tilde{\mathbf{H}}}$ as

$$\underline{\tilde{\mathbf{H}}} = [\tilde{\mathbf{H}}_n^T \ \tilde{\mathbf{H}}_{n+1}^T \ldots \tilde{\mathbf{H}}_{n+N_{wind}-1}^T]^T, \tag{12.65}$$

and express $\underline{\tilde{\mathbf{W}}}$ as

$$\underline{\tilde{\mathbf{W}}} = [\tilde{\mathbf{W}}_n^T \ \tilde{\mathbf{W}}_{n+1}^T \ldots \tilde{\mathbf{W}}_{n+N_{wind}-1}^T]^T. \tag{12.66}$$

Furthermore, the diagonal block matrix of the transmitted signal is constructed as

$$\underline{\tilde{\mathbf{S}}}_d = \mathrm{diag}(\tilde{\mathbf{S}}_n, \ \tilde{\mathbf{S}}_{n+1}, \ \ldots, \ \tilde{\mathbf{S}}_{n+N_{wind}-1}) \tag{12.67}$$

$$= \begin{bmatrix} \tilde{\mathbf{S}}_n & 0 & \cdots & 0 \\ 0 & \tilde{\mathbf{S}}_{n+1} & \cdots & 0 \\ \vdots & \vdots & \ddots & \vdots \\ 0 & 0 & \cdots & \tilde{\mathbf{S}}_{n+N_{wind}-1} \end{bmatrix}, \tag{12.68}$$

where $\tilde{\mathbf{S}}_n$ $(n = 0, 1, \ldots, N_{wind} - 1)$ was given by Equation (12.51) or Equation (12.58).

Note that all the elements in $\tilde{\mathbf{H}}_n$ and $\tilde{\mathbf{W}}_n$ of (12.52) and (12.55) possess a standard Gaussian distribution for the DDF-aided cooperative system, whereas most terms in $\tilde{\mathbf{H}}_n$ and $\tilde{\mathbf{W}}_n$ of (12.59) and (12.61) do not for its DAF-aided counterpart. However, our informal simulation-based investigations suggest that the resultant noise processes are near-Gaussian distributed in the DAF-aided scenario. As a result, the PDF of the corresponding received signal in (12.63) is also near-Gaussian, especially for low SNRs, as seen in Figure 12.16. Hence, under the simplifying assumption that the *equivalent* fading and noise are zero-mean complex-Gaussian processes in the DAF-aided cooperative system, the PDF of the non-coherent receiver's output $\underline{\mathbf{Y}}$ at the BS for both the DAF- and DDF-aided cooperative systems can be obtained based on its counterpart of Equation (12.14) derived for the single-transmit-antenna scenario in Section 12.2 as

$$Pr(\underline{\mathbf{Y}}|\underline{\tilde{\mathbf{S}}}_d) = \frac{\exp(-Tr\{\underline{\mathbf{Y}}^H \underline{\mathbf{\Psi}}^{-1} \underline{\mathbf{Y}}\})}{(\pi^{UN_{wind}} \det(\underline{\mathbf{\Psi}}))^N}, \tag{12.69}$$

where the conditional autocorrelation matrix is given by

$$\underline{\mathbf{\Psi}} = \mathcal{E}\{\underline{\mathbf{Y}}\underline{\mathbf{Y}}^H | \underline{\tilde{\mathbf{S}}}_d\}, \tag{12.70}$$

$$= \underline{\tilde{\mathbf{S}}}_d \mathcal{E}\{\underline{\tilde{\mathbf{H}}}\,\underline{\tilde{\mathbf{H}}}^H\} \underline{\tilde{\mathbf{S}}}_d^H + \mathcal{E}\{\underline{\tilde{\mathbf{W}}}\,\underline{\tilde{\mathbf{W}}}^H\}. \tag{12.71}$$

Specifically, for the DDF-aided cooperative system having an equivalent channel matrix $\tilde{\mathbf{H}}_n$ given by Equation (12.52) and a noise matrix given by Equation (12.55), the channel's autocorrelation matrix $\mathcal{E}\{\underline{\tilde{\mathbf{H}}}\,\underline{\tilde{\mathbf{H}}}^H\}$ formulated in Equation (12.71) can be further extended as

$$\mathcal{E}\{\underline{\tilde{\mathbf{H}}}\,\underline{\tilde{\mathbf{H}}}^H\} = \mathcal{E}\left\{ \begin{bmatrix} \tilde{\mathbf{H}}_n \\ \vdots \\ \tilde{\mathbf{H}}_{n+N_{wind}-1} \end{bmatrix} [\tilde{\mathbf{H}}_n^* \ \cdots \ \tilde{\mathbf{H}}_{n+N_{wind}-1}^*] \right\} \tag{12.72}$$

$$= N \begin{bmatrix} \Gamma_{DF}(0) & \Gamma_{DF}(1) & \cdots & \Gamma_{DF}(N_{wind} - 1) \\ \Gamma_{DF}(-1) & \Gamma_{DF}(0) & \cdots & \Gamma_{DF}(N_{wind} - 2) \\ \vdots & \vdots & \ddots & \vdots \\ \Gamma_{DF}(1 - N_{wind}) & \Gamma_{DF}(2 - N_{wind}) & \cdots & \Gamma_{DF}(0) \end{bmatrix}, \tag{12.73}$$

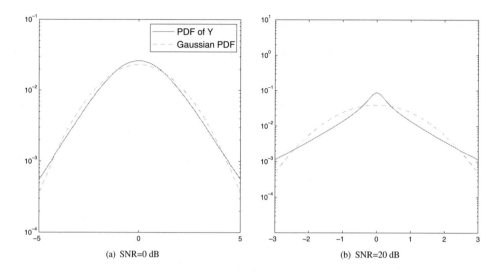

(a) SNR=0 dB   (b) SNR=20 dB

Figure 12.16: PDF of the received signal $\underline{\mathbf{Y}}$ of Equation (12.64) in the DAF-aided cooperative system. ©IEEE Wang & Hanzo 2007 [39]

by defining

$$\Gamma_{DF}(\kappa) \triangleq \begin{bmatrix} \varphi_{sd}^t[\kappa] & 0 & \cdots & 0 \\ 0 & \varphi_{r_1 d}^t[\kappa] & \cdots & 0 \\ \vdots & \vdots & \ddots & \vdots \\ 0 & 0 & \cdots & \varphi_{r_{U-1} d}^t[\kappa] \end{bmatrix} \mathbf{P}^2 \tag{12.74}$$

$$= \begin{bmatrix} P_s \varphi_{sd}^t[\kappa] & 0 & \cdots & 0 \\ 0 & P_{r_1} \varphi_{r_1 d}^t[\kappa] & \cdots & 0 \\ \vdots & \vdots & \ddots & \vdots \\ 0 & 0 & \cdots & P_{r_{U-1}} \varphi_{r_{U-1} d}^t[\kappa] \end{bmatrix}, \tag{12.75}$$

where $\mathbf{P}$ is the transmit power allocation matrix given by Equation (12.32), while $\varphi_{sd}^t[\kappa]$ and $\varphi_{r_u d}^t[\kappa]$ respectively represent the channel's autocorrelation function for the direct link and relay–destination link between the uth relay node and the destination BS. Under the assumption of Rayleigh fading channels, the channel's autocorrelation function can be expressed as

$$\varphi^t[\kappa] \triangleq \mathcal{E}\{h[n + \kappa]h^*[n]\} \tag{12.76}$$
$$= J_0(2\pi f_d \kappa), \tag{12.77}$$

with $J_0(\cdot)$ denoting the zeroth-order Bessel function of the first kind and as usual f_d representing the normalized Doppler frequency. Furthermore, under the assumption of an identical noise variance observed at each terminal, $\mathcal{E}\{\tilde{\underline{\mathbf{W}}}\tilde{\underline{\mathbf{W}}}^H\}$ of the DDF-aided system can be expressed with the aid of the equivalent noise matrix given by Equation (12.55) as

$$\mathcal{E}\{\tilde{\underline{\mathbf{W}}}\tilde{\underline{\mathbf{W}}}^H\} = N_0 N \mathbf{I}_{UN_{wind}}, \tag{12.78}$$

where N and N_0 respectively denote the number of receive antennas employed at the BS and the Gaussian noise variance, while $I_{UN_{wind}}$ is a $(UN_{wind} \times UN_{wind})$-element identity matrix.

On the other hand, when considering the DAF-aided user-cooperative system having an equivalent channel matrix $\tilde{\mathbf{H}}_n$ given by Equation (12.59) and a noise matrix given by Equation (12.61), the channel's autocorrelation matrix $\mathcal{E}\{\tilde{\mathbf{H}}\tilde{\mathbf{H}}^H\}$ can be expressed as

$$
\mathcal{E}\{\tilde{\mathbf{H}}\tilde{\mathbf{H}}^H\} = \mathcal{E}\left\{ \begin{bmatrix} \tilde{\mathbf{H}}_n \\ \vdots \\ \tilde{\mathbf{H}}_{n+N_{wind}-1} \end{bmatrix} \begin{bmatrix} \tilde{\mathbf{H}}_n^* & \cdots & \tilde{\mathbf{H}}_{n+N_{wind}-1}^* \end{bmatrix} \right\} \tag{12.79}
$$

$$
= N \begin{bmatrix} \Gamma_{AF}(0) & \Gamma_{AF}(1) & \cdots & \Gamma_{AF}(N_{wind}-1) \\ \Gamma_{AF}(-1) & \Gamma_{AF}(0) & \cdots & \Gamma_{AF}(N_{wind}-2) \\ \vdots & \vdots & \ddots & \vdots \\ \Gamma_{AF}(1-N_{wind}) & \Gamma_{AF}(2-N_{wind}) & \cdots & \Gamma_{AF}(0) \end{bmatrix}, \tag{12.80}
$$

where

$$
\Gamma_{AF}(\kappa) \triangleq \begin{bmatrix} \varphi_{sd}^t[\kappa] & 0 & \cdots & 0 \\ 0 & \varphi_{sr_1}^t[\kappa]\varphi_{r_1d}^t[\kappa] & \cdots & 0 \\ \vdots & \vdots & \ddots & \vdots \\ 0 & 0 & \cdots & \varphi_{sr_{U-1}}^t[\kappa]\varphi_{r_{U-1}d}^t[\kappa] \end{bmatrix} \mathbf{P}^2 \mathbf{F}_{AM}^2 \tag{12.81}
$$

$$
= \begin{bmatrix} P_s\varphi_{sd}^t[\kappa] & 0 & \cdots & 0 \\ 0 & \dfrac{P_{r_1}\varphi_{sr_1}^t[\kappa]\varphi_{r_1d}^t[\kappa]}{\sigma_{sr_1}^2+(N_0/P_s)} & \cdots & 0 \\ \vdots & \vdots & \ddots & \vdots \\ 0 & 0 & \cdots & \dfrac{P_{r_{U-1}}\varphi_{sr_{U-1}}^t[\kappa]\varphi_{r_{U-1}d}^t[\kappa]}{\sigma_{sr_{U-1}}^2+(N_0/P_s)} \end{bmatrix} \tag{12.82}
$$

with the diagonal matrix $\mathbf{F}_{AM}$ is defined as

$$
\mathbf{F}_{AM} = \begin{bmatrix} 1 & 0 & \cdots & 0 \\ 0 & f_{AM_{r_1}} & \cdots & 0 \\ \vdots & \vdots & \ddots & \vdots \\ 0 & 0 & \cdots & f_{AM_{r_{U-1}}} \end{bmatrix}, \tag{12.83}
$$

which contains all the signal gain factors $f_{AM_{r_u}}$ ($u = 1, 2, \ldots, N_{wind} - 1$) of Equation (12.38) employed by the $(U - 1)$ relay nodes, respectively, in the U-user-cooperation-aided communication system of Figure 12.10. Moreover, with the aid of the equivalent noise matrix given by Equation (12.61) for the DAF-aided system, we can express $\mathcal{E}\{\tilde{\mathbf{W}}\tilde{\mathbf{W}}^H\}$ as

$$
\mathcal{E}\{\tilde{\mathbf{W}}\tilde{\mathbf{W}}^H\} = N\mathbf{I}_{N_{wind}} \otimes \begin{bmatrix} N_0 & 0 & \cdots & 0 \\ 0 & \left(\dfrac{P_{r_1}\sigma_{r_1d}^2}{P_s\sigma_{sr_1}^2 + N_0}+1\right)N_0 & \cdots & 0 \\ \vdots & \vdots & \ddots & \vdots \\ 0 & 0 & \cdots & \left(\dfrac{P_{r_{U-1}}\sigma_{r_{U-1}d}^2}{P_s\sigma_{sr_{U-1}}^2 + N_0}+1\right)N_0 \end{bmatrix}, \tag{12.84}
$$

where N represents the number of receive antennas employed at the BS, while $\mathbf{I}_{N_{wind}}$ denotes an $(N_{wind} \times N_{wind})$-element identity matrix. Note that $\otimes$ denotes the Kronecker product. Hence, the noise autocorrelation matrices $\mathcal{E}\{\tilde{\mathbf{W}}\tilde{\mathbf{W}}^H\}$, which were given by Equations (12.78) and (12.84) for the DDF-

and DAF-aided systems, respectively, are diagonal due to the temporally and spatially uncorrelated nature of the AWGN.

Although the basic idea behind the ML detector is that of maximizing the a posteriori probability of the received signal block matrix $\mathbf{Y}$, this problem can be readily shown to be equivalent to maximizing the a priori probability of Equation (12.69) with the aid of Bayes' theorem [542]. Thus, based on the ML detection rule, an exhaustive search has to be carried out over the entire transmitted signal vector space in order to find the specific solution which maximizes the a priori probability of Equation (12.69). Thus, the ML metric of the multi-path MSDD can be expressed as

$$\hat{\mathbf{S}}_{ML} = \underset{\tilde{\mathbf{S}}_d \to \tilde{\mathbf{s}} \in \mathcal{M}_c^{N_{wind}}}{\arg\max} \quad Pr(\mathbf{Y}|\tilde{\mathbf{S}}_d) \tag{12.85}$$

$$= \underset{\tilde{\mathbf{S}}_d \to \tilde{\mathbf{s}} \in \mathcal{M}_c^{N_{wind}}}{\arg\min} \quad Tr\{\mathbf{Y}^H \mathbf{\Psi}^{-1} \mathbf{Y}\}, \tag{12.86}$$

where s is a column vector hosting all the diagonal elements of the diagonal matrix $\tilde{\mathbf{S}}_d$. Note that although s has UN_{wind} elements, each of which is chosen from an identical constellation set of $\mathcal{M}_c$, we have $\mathbf{s} \in \mathcal{M}_c^{N_{wind}}$ instead of $\mathbf{s} \in \mathcal{M}_c^{UN_{wind}}$, since all the U diagonal elements of our derived equivalent U-user-cooperation transmitted signal $\tilde{\mathbf{S}}_n$ of Equation (12.51) or (12.58) have the same symbol value as that of the nth signal transmitted from the source in the broadcast phase I. More specifically, $\tilde{\mathbf{s}}$ may be expressed as

$$\tilde{\mathbf{s}} = [\underbrace{\tilde{s}_1 \tilde{s}_2 \dots \tilde{s}_U}_{\tilde{s}_1} \dots \underbrace{\tilde{s}_{(n-1)U+1} \dots \tilde{s}_{nU}}_{\tilde{s}_n} \dots \underbrace{\tilde{s}_{N_{wind}U+1} \dots \tilde{s}_{N_{wind}U}}_{\tilde{s}_{N_{wind}}}]^T, \tag{12.87}$$

where the subvector $\tilde{s}_n$ is a column vector containing all the diagonal elements of the matrix $\tilde{\mathbf{S}}_n$.

12.3.3.2 Transformation of the ML Metric

Again, in a user-cooperation-aided system, the noise contributions imposed at the relay and destination nodes are both temporally and spatially uncorrelated, thus we have diagonal noise autocorrelation matrices for both the DDF-aided and DAF-aided systems, as observed in Equations (12.78) and (12.84), respectively. Additionally, the equivalent transmitted signal matrix $\tilde{\mathbf{S}}_d$ of the user-cooperation-aided system as constructed in either Equation (12.51) or Equation (12.58) for the above-mentioned two systems is a unitary matrix, hence we have

$$\tilde{\mathbf{S}}_d^{-1} = \tilde{\mathbf{S}}_d^{H}. \tag{12.88}$$

Then, we can further extend Equation (12.71) as

$$\mathbf{\Psi} = \tilde{\mathbf{S}}_d \mathcal{E}\{\tilde{\mathbf{H}}\tilde{\mathbf{H}}^H\}\tilde{\mathbf{S}}_d^H + \mathcal{E}\{\tilde{\mathbf{W}}\tilde{\mathbf{W}}^H\} \tag{12.89}$$

$$= \tilde{\mathbf{S}}_d(\mathcal{E}\{\tilde{\mathbf{H}}\tilde{\mathbf{H}}^H\} + \mathcal{E}\{\tilde{\mathbf{W}}\tilde{\mathbf{W}}^H\})\tilde{\mathbf{S}}_d^H \tag{12.90}$$

$$= \tilde{\mathbf{S}}_d \mathbf{C} \tilde{\mathbf{S}}_d^H, \tag{12.91}$$

where we have

$$\mathbf{C} \triangleq \mathcal{E}\{\tilde{\mathbf{H}}\tilde{\mathbf{H}}^H\} + \mathcal{E}\{\tilde{\mathbf{W}}\tilde{\mathbf{W}}^H\}, \tag{12.92}$$

which is defined as the ($UN_{wind} \times UN_{wind}$)-element *channel-noise autocorrelation* matrix. Now, the ML metric of Equation (12.86) generated for the multi-path MSDD can be reformulated by substituting Equation (12.91) characterizing Ψ into Equation (12.86) as

$$\hat{\mathbf{S}}_{ML} = \underset{\tilde{\mathbf{S}}_d \to \tilde{\mathbf{s}} \in \mathcal{M}_c^{N_{wind}}}{\arg\min} \quad Tr\{\mathbf{Y}^H \mathbf{\Psi}^{-1} \mathbf{Y}\} \tag{12.93}$$

$$= \underset{\tilde{\mathbf{S}}_d \to \tilde{\mathbf{s}} \in \mathcal{M}_c^{N_{wind}}}{\arg\min} \quad Tr\{\mathbf{Y}^H (\tilde{\mathbf{S}}_d \mathbf{C} \tilde{\mathbf{S}}_d^H)^{-1} \mathbf{Y}\}. \tag{12.94}$$

Furthermore, since the $\tilde{\underline{\mathbf{S}}}_d$ is unitary, we get

$$\hat{\underline{\mathbf{S}}}_{ML} = \underset{\tilde{\underline{\mathbf{S}}}_d \to \tilde{\mathbf{s}} \in \mathcal{M}_c^{N_{wind}}}{\arg\min} \quad Tr\{\underline{\mathbf{Y}}^H \tilde{\underline{\mathbf{S}}}_d \mathbf{C}^{-1} \tilde{\underline{\mathbf{S}}}_d^{\,H} \underline{\mathbf{Y}}\}. \tag{12.95}$$

Now we define two matrix transformation operators, namely $\mathcal{F}_y(\cdot)$ and $\mathcal{F}_s(\cdot)$, for the received signal matrix $\underline{\mathbf{Y}}$ of Equation (12.54) or (12.57) and the transmitted signal matrix $\tilde{\underline{\mathbf{S}}}_d$ of Equation (12.51) or (12.58), respectively, in the scenario of a differentially modulated U-user cooperative system employing N receive antennas at the BS and jointly detecting differentially N_{wind} received symbols. Specifically, the operator $\mathcal{F}_y(\cdot)$ is defined as follows:

$$\mathcal{F}_y(\underline{\mathbf{Y}}) \triangleq \begin{bmatrix} \vec{\mathbf{y}}_1 & \mathbf{0} & \cdots & \mathbf{0} \\ \mathbf{0} & \vec{\mathbf{y}}_2 & \cdots & \mathbf{0} \\ \vdots & \vdots & \ddots & \vdots \\ \mathbf{0} & \mathbf{0} & \cdots & \vec{\mathbf{y}}_{UN_{wind}} \end{bmatrix}, \tag{12.96}$$

where $\vec{\mathbf{y}}_i$ is the ith row of the matrix $\underline{\mathbf{Y}}$ and the resultant matrix is a $(UN_{wind} \times UNN_{wind})$-element matrix. On the other hand, the operator $\mathcal{F}_s(\cdot)$, which is applied to the diagonal transmitted signal matrix $\underline{\mathbf{S}}_d$, is defined as

$$\mathcal{F}_s(\tilde{\underline{\mathbf{S}}}_d) \triangleq \begin{bmatrix} \tilde{s}_1 \mathbf{I}_N \\ \tilde{s}_2 \mathbf{I}_N \\ \vdots \\ \tilde{s}_{UN_{wind}} \mathbf{I}_N \end{bmatrix}, \tag{12.97}$$

where $\tilde{s}_i$ is the ith element of the column vector $\tilde{\mathbf{s}}$ of Equation (12.87) hosting all the UN_{wind} diagonal elements of the diagonal matrix $\tilde{\underline{\mathbf{S}}}_d$. Thus, the resultant matrix is of $(UNN_{wind} \times N)$ dimension.

Consequently, we exploit the transformation operators $\mathcal{F}_y(\cdot)$ defined in Equation (12.96) and $\mathcal{F}_s(\cdot)$ defined in Equation (12.97), which allow us further to reformulate the ML solution expression of Equation (12.95) as

$$\hat{\underline{\mathbf{S}}}_{ML} = \underset{\tilde{\underline{\mathbf{S}}}_d \to \tilde{\mathbf{s}} \in \mathcal{M}_c^{N_{wind}}}{\arg\min} \quad Tr\{\underline{\mathbf{Y}}^H \tilde{\underline{\mathbf{S}}}_d \mathbf{C}^{-1} \tilde{\underline{\mathbf{S}}}_d^{\,H} \underline{\mathbf{Y}}\} \tag{12.98}$$

$$= \underset{\mathbf{S}_{\mathcal{F}} \to \tilde{\mathbf{s}} \in \mathcal{M}_c^{N_{wind}}}{\arg\min} \quad Tr\{\mathbf{S}_{\mathcal{F}}^T \mathbf{Y}_{\mathcal{F}}^H \mathbf{C}^{-1} \mathbf{Y}_{\mathcal{F}} \mathbf{S}_{\mathcal{F}}^*\}, \tag{12.99}$$

where we have

$$\underline{\mathbf{Y}}_{\mathcal{F}} = \mathcal{F}_y(\underline{\mathbf{Y}}) \tag{12.100}$$

and

$$\underline{\mathbf{S}}_{\mathcal{F}} = \mathcal{F}_s(\tilde{\underline{\mathbf{S}}}_d) = \begin{bmatrix} \tilde{s}_1 \mathbf{I}_N \\ \tilde{s}_2 \mathbf{I}_N \\ \vdots \\ \tilde{s}_{UN_{wind}} \mathbf{I}_N \end{bmatrix} = \begin{bmatrix} \mathbf{S}_{\mathcal{F}_1} \\ \mathbf{S}_{\mathcal{F}_2} \\ \vdots \\ \mathbf{S}_{\mathcal{F}_{N_{wind}}} \end{bmatrix}, \tag{12.101}$$

where the $(UN \times N)$-dimensional matrix $\underline{\mathbf{S}}_{\mathcal{F}_i}$ represents the ith submatrix of the block matrix $\underline{\mathbf{S}}_{\mathcal{F}}$, which may be expressed as

$$\underline{\mathbf{S}}_{\mathcal{F}_i} = \begin{bmatrix} \tilde{s}_{U(i-1)+1} \mathbf{I}_N \\ \tilde{s}_{U(i-1)+2} \mathbf{I}_N \\ \vdots \\ \tilde{s}_{Ui} \mathbf{I}_N \end{bmatrix}_{UN \times N}, \tag{12.102}$$

where all the non-zero elements have an identical symbol value, which corresponds to the ith symbol transmitted from the source during the broadcast phase I.

12.3.3.3 Channel-Noise Autocorrelation Matrix Triangularization

Let us now generate the $(UN_{wind} \times UN_{wind})$-element upper-triangular matrix $\mathbf{F}$, which satisfies $\mathbf{F}^H \mathbf{F} = \mathbf{C}^{-1}$ with the aid of Cholesky factorization. Then we arrive at

$$\hat{\underline{\mathbf{S}}}_{ML} = \arg\min_{\underline{\mathbf{S}}_{\mathcal{F}} \to \tilde{s} \in \mathcal{M}_c^{N_{wind}}} Tr\{\underline{\mathbf{S}}_{\mathcal{F}}^T \mathbf{Y}_{\mathcal{F}}^H \mathbf{F}^H \mathbf{F} \underline{\mathbf{Y}}_{\mathcal{F}} \underline{\mathbf{S}}_{\mathcal{F}}^*\}. \tag{12.103}$$

Then, by further defining a $(UN_{wind} \times UNN_{wind})$-element matrix $\mathbf{U}$ as

$$\mathbf{U} \triangleq (\mathbf{F}\underline{\mathbf{Y}}_{\mathcal{F}})^* \tag{12.104}$$

$$= \begin{bmatrix} \mathbf{U}_{1,1} & \mathbf{U}_{1,2} & \cdots & \mathbf{U}_{1,N_{wind}} \\ \mathbf{0} & \mathbf{U}_{2,2} & \cdots & \mathbf{U}_{2,N_{wind}} \\ \vdots & \vdots & \ddots & \vdots \\ \mathbf{0} & \mathbf{0} & \cdots & \mathbf{U}_{N_{wind},N_{wind}} \end{bmatrix}, \tag{12.105}$$

where we have

$$\mathbf{U}_{i,j} \triangleq \begin{bmatrix} u_{U(i-1)+1, UN(j-1)+1} & u_{U(i-1)+1, UN(j-1)+2} & \cdots & u_{U(i-1)+1, UNj} \\ u_{U(i-1)+2, UN(j-1)+1} & u_{U(i-1)+2, UN(j-1)+2} & \cdots & u_{U(i-1)+2, UNj} \\ \vdots & \vdots & \ddots & \vdots \\ u_{Ui, UN(j-1)+1} & u_{Ui, UN(j-1)+2} & \cdots & u_{Ui, UNj} \end{bmatrix}_{U \times UN}, \tag{12.106}$$

we finally arrive at

$$\hat{\underline{\mathbf{S}}}_{ML} = \arg\min_{\underline{\mathbf{S}}_{\mathcal{F}} \to \tilde{s} \in \mathcal{M}_c^{N_{wind}}} \|\mathbf{U}\underline{\mathbf{S}}_{\mathcal{F}}\|^2, \tag{12.107}$$

which completes the process of transforming the multi-path ML-MSDD metric of Equation (12.86) to a *shortest-vector* problem.

12.3.3.4 Multi-dimensional Tree-Search-Aided MSDSD Algorithm

Although the problem of finding an optimum solution for the ML-MSDD has been transformed into the so-called *shortest-vector* problem of Equation (12.107), the multi-path ML-MSDD designed for user-cooperation-aided systems may impose a potentially excessive computational complexity when aiming at finding the solution which minimizes Equation (12.107), especially when a high-order differential modulation scheme and/or a high observation window size N_{wind} are employed. Fortunately, in light of the SD algorithms discussed in Chapter 9, the computational complexity imposed may be significantly reduced by carrying out a tree search within a reduced-size hypersphere confined by either the search radius C for the depth-first SD or the maximum number of candidates K retained at each search tree level for the breadth-first SD. In our following discourse, we consider the depth-first SD algorithm as an example and demonstrate how to reduce the complexity imposed by the ML-MSSD.

In order to search for the ML solution of Equation (12.107) in a confined hypersphere, an initial search radius C is introduced. Thus, we obtain the metric relevant for the multi-path MSDSD scheme as

$$\hat{\underline{\mathbf{S}}}_{ML} = \arg\min_{\underline{\mathbf{S}}_{\mathcal{F}} \to \tilde{s} \in \mathcal{M}_c^{N_{wind}}} \|\mathbf{U}\tilde{\mathbf{s}}\|^2 \leq C^2 \tag{12.108}$$

$$= \arg\min_{\underline{\mathbf{S}}_{\mathcal{F}} \to \tilde{s} \in \mathcal{M}_c^{N_{wind}}} \left\| \begin{bmatrix} \mathbf{U}_{1,1} & \mathbf{U}_{1,2} & \cdots & \mathbf{U}_{1,N_{wind}} \\ \mathbf{0} & \mathbf{U}_{2,2} & \cdots & \mathbf{U}_{2,N_{wind}} \\ \vdots & \vdots & \ddots & \vdots \\ \mathbf{0} & \mathbf{0} & \cdots & \mathbf{U}_{N_{wind},N_{wind}} \end{bmatrix} \begin{bmatrix} \underline{\mathbf{S}}_{\mathcal{F}_1} \\ \underline{\mathbf{S}}_{\mathcal{F}_2} \\ \vdots \\ \underline{\mathbf{S}}_{\mathcal{F}_{N_{wind}}} \end{bmatrix} \right\|^2 \leq C^2 \tag{12.109}$$

$$= \arg\min_{\underline{\mathbf{S}}_{\mathcal{F}} \to \tilde{s} \in \mathcal{M}_c^{N_{wind}}} \left\| \sum_{n=1}^{N_{wind}} \left(\sum_{m=n}^{N_{wind}} \mathbf{U}_{n,m} \underline{\mathbf{S}}_{\mathcal{F}_m} \right) \right\|^2 \leq C^2. \tag{12.110}$$

Since the tree search is carried out commencing from $n = N_{wind}$ to $n = 1$, the accumulated PED between the candidate $\underline{\mathbf{S}}_{\mathcal{F}}$ and the origin can be expressed as

$$\mathcal{D}_n = \underbrace{\left\| \mathbf{U}_{n,n}\underline{\mathbf{S}}_{\mathcal{F}_n} + \sum_{m=n+1}^{N_{wind}} \mathbf{U}_{n,m}\underline{\mathbf{S}}_{\mathcal{F}_m} \right\|^2}_{\delta_n} + \underbrace{\left\| \sum_{l=n+1}^{N_{wind}} \left(\sum_{m=l}^{N_{wind}} \mathbf{U}_{l,m}\underline{\mathbf{S}}_{\mathcal{F}_m} \right) \right\|^2}_{\mathcal{D}_{n+1}} \leq C^2. \quad (12.111)$$

Furthermore, due to the employment of a differential modulation scheme, the information is encoded as the phase difference between the consecutively transmitted symbols. Hence, in light of the multi-layer tree search proposed for the SD in Section 11.3.2.3, the MSDSD scheme can start the search from $n = (N_{wind} - 1)$ by choosing a trial submatrix for $\underline{\mathbf{S}}_{\mathcal{F}_{N_{wind}-1}}$ satisfying

$$\mathcal{D}_{N_{wind}-1} \leq C^2 \quad (12.112)$$

from the legitimate candidate pool, after simply assuming that the N_{wind}th symbol transmitted by the source is $s_s = 1$. That is, according to Equation (12.102) we have

$$\underline{\mathbf{S}}_{\mathcal{F}_{N_{wind}}} = \underbrace{[\mathbf{I}_N \ \mathbf{I}_N \ldots \mathbf{I}_N]^T}_{U \text{ identity submatrices}}. \quad (12.113)$$

Given the trial submatrix $\underline{\mathbf{S}}_{\mathcal{F}_{N_{wind}-1}}$ satisfying Equation (12.112), the search continues and a candidate matrix is selected for $\underline{\mathbf{S}}_{\mathcal{F}_{N_{wind}-2}}$ based on the criterion that the value of the resultant PED computed using Equation (12.111) does not exceed the squared radius, i.e.

$$\mathcal{D}_{N_{wind}-2} \leq C^2. \quad (12.114)$$

This recursive process will continue until n reaches 1, i.e. when we choose a trial value for $\tilde{s}_1$ within the computed range. Then the search radius C is updated by calculating the Euclidean distance between the newly obtained signal point $\underline{\mathbf{S}}_{\mathcal{F}}$ and the origin and a new search is carried out within a reduced compound confined by the newly obtained search radius. The search then proceeds in the same way, until no more legitimate signal points can be found in the increasingly reduced search area. Consequently, the last legitimate signal point $\underline{\mathbf{S}}_{\mathcal{F}}$ found this way is regarded as the ML solution of Equation (12.107). Therefore, in comparison with the multi-path ML-MSDD algorithm of Equation (12.107), the MSDSD algorithm may achieve a significant computational complexity reduction, as does its single-path counterpart, as observed in Section 12.2. For more details on the principle of SD algorithms refer to Chapter 9 and on the idea of multi-layer tree search to Chapter 11.

12.3.4 Simulation Results

12.3.4.1 Performance of the MSDSD-Aided DAF-User-Cooperation System

As discussed in Section 12.3.2.3, the relative mobility among users imposes a performance degradation on the user-cooperation-aided system. Thus, the multi-path MSDSD scheme proposed in Section 12.3.3, which relies on the exploitation of the correlation between the phase distortions experienced by the N_{wind} consecutive transmitted DPSK symbols, is employed in order to mitigate the channel-induced error floor encountered by the CDD characterized in Figure 12.17. The system parameters used in our simulations are summarized in Table 12.5.

Figure 12.17 depicts the BER performance improvement achieved by the MSDSD employed at the destination node for the DAF-aided two-user cooperative system in the presence of three different normalized Doppler frequencies, namely $f_d = 0.03$, 0.01 and 0.001. With the aid of the MSDSD employing $N_{wind} = 6$ at the destination node, both the error floors experienced in Rayleigh channels having normalized frequencies of $f_d = 0.03$ and 0.01 are significantly mitigated. Specifically, the

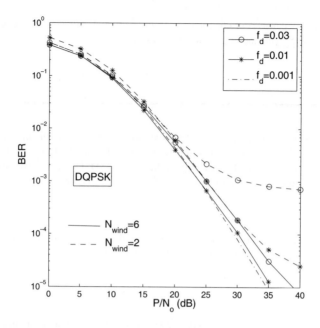

Figure 12.17: BER performance improvement achieved by the MSDSD employing $N_{wind} = 6$ for the DAF-aided T-DQPSK-modulated cooperative system in time-selective Rayleigh fading channels. All other system parameters are summarized in Table 12.5.

Table 12.5: Summary of system parameters used for the T-DQPSK-modulated two-user cooperative OFDM system.

System parameters	Choice
System	Two-user cooperative OFDM
Number of relay nodes	1
Subcarrier BW	$\Delta f = 10\,\text{kHz}$
Number of subcarriers	$D = 1024$
Modulation	T-DQPSK
Frame length L_f	101
CRC	CCITT-6
Normalized	If it is not specified,
Doppler frequency	$f_{d,sd} = f_{d,sr} = f_{d,rd} = f_d$
Channel model	Typical urban, refer to Table 12.1
Channel variances	$\sigma_{sd}^2 = \sigma_{sr}^2 = \sigma_{rd}^2 = 1$
Power allocation	$P_s = P_{r_1} = 0.5P = 0.5$
SNR at relay and destination	$P_s/N_0 = P_{r_1}/N_0$

BER curve corresponding to the normalized Doppler frequency $f_d = 0.01$ almost coincides with that associated with $f_d = 0.001$, indicating a performance gain of about 10 dB over the system dispensing with the MSDSD. Remarkably, in the scenario of a fast-fading channel having $f_d = 0.03$, the BER curve obtained when the CDD is employed at the destination node levels out just below 10^{-3}, as the SNR increases. By contrast, with the aid of the MSDSD the resultant BER curve decreases steadily,

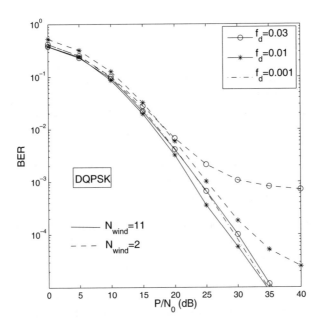

Figure 12.18: BER performance improvement achieved by the MSDSD scheme employing $N_{wind} = 11$ for the DAF-aided T-DQPSK-modulated cooperative system in time-selective Rayleigh fading channels. All other system parameters are summarized in Table 12.5.

suffering a modest performance loss of only about 4 dB at the target BER of 10^{-5} in comparison with the curve associated with $f_d = 0.001$. Hence, the more time selective the channel, the more significant the performance improvement achieved by the proposed MSDSD scheme.

For further reducing the detrimental impact induced by the time-selective channel on the DAF-aided user-cooperative system, an observation window size of $N_{cand} = 11$ is employed by the MSDSD arrangement at the destination node at the expense of a higher detection complexity. As seen in Figure 12.18, the MSDSD using $N_{wind} = 11$ is capable of eliminating the error floor encountered by the system employing the CDD, even when the channel is severely time selective, i.e. for $f_d = 0.03$. In other words, the BER curve corresponding to the MSDSD-aided system in Figure 12.18 and obtained for $f_d = 0.03$ coincides with that of its CDD-aided counterpart recorded for $f_d = 0.001$. Furthermore, the MSDSD-aided system with $N_{wind} = 11$ in a fast-fading channel associated with $f_d = 0.01$ is able to outperform the system employing $N_{wind} = 2$, even if the latter is operating in a slow-fading channel having $f_d = 0.001$. Therefore, even in the presence of a severely time-selective channel, the DAF-aided user-cooperative system employing the MSDSD is capable of achieving an attractive performance by jointly detecting differentially a sufficiently high number of consecutively received user-cooperation-based joint symbols $\mathbf{S}_n$ $(n = 0, 1, \ldots, N_{wind} - 1)$ of Equation (12.58) by exploiting knowledge of the equivalent channel autocorrelation matrix $\mathcal{E}\{\tilde{\mathbf{H}}\tilde{\mathbf{H}}^H\}$ of Equation (12.79), which characterizes the CIR statistics of both the direct and relay links.

All the previously described simulations were carried out under the assumption that an identical normalized Doppler frequency is exhibited by each link of the user-cooperation system, i.e. that we have $f_{d,sd} = f_{d,sr} = f_{d,rd} = f_d$. However, a more realistic scenario is the one where the relative speeds of all the cooperative users as well as of the destination terminal are different from each other, leading to a different Doppler frequency for each link. Thus, in order to investigate the impact of different relative speeds among all the nodes on the attainable end-to-end performance of the DAF-aided system, Monte Carlo simulations were carried out for the three different scenarios summarized in Table 12.6. In all the

Table 12.6: Normalized Doppler frequency of three different scenarios.

	$f_{d,sd}$	$f_{d,sr}$	$f_{d,rd}$
Scenario I (S moves, R&D relatively immobile)	0.03	0.03	0.001
Scenario II (R moves, S&D relatively immobile)	0.001	0.03	0.03
Scenario III (D moves, S&R relatively immobile)	0.03	0.001	0.03

three situations, only one of the three nodes in the two-user cooperation-aided system is supposed to move relative to the other two nodes at a speed resulting in a normalized Doppler frequency of 0.03, while the latter two remain stationary relative to each other, yielding a normalized Doppler frequency of 0.001.

In Figure 12.19 the BER curves corresponding to the three different scenarios of Table 12.6 are bounded by the two dashed–dotted BER curves having no legends, which were obtained by assuming an identical normalized Doppler frequency of $f_d = 0.03$ and $f_d = 0.001$ for each link in the user-cooperation-aided system, respectively. This is not unexpected, since the two above-mentioned BER bounds correspond to the least and most desirable time-selective channel conditions considered in this chapter, respectively. The channel quality of the direct link characterized in terms of its grade of time selectivity predetermines the achievable performance of the DAF-aided user-cooperation-assisted system employing the MSDSD. Hence, it is observed in Figure 12.19 that the system is capable of attaining a better BER performance in Scenario II ($f_{d,sd} = 0.001$) than in the other two scenarios ($f_{d,sd} = 0.03$). However, as seen in Figure 12.19, due to the high speed of the relay node observed in Scenario II relative to the source and destination nodes, the MSDSD employing $N_{wind} = 6$ remains unable to eliminate completely the impairments induced by the time-selective channel, unless a higher N_{wind} value is employed. Therefore, a modest performance degradation occurs in comparison with the $f_d = 0.001$ scenario. On the other hand, the MSDSD-aided system exhibits a similar performance in Scenario I and Scenario III, since the source–relay and relay–destination links are symmetric and thus they are exchangeable in the context of the DAF scheme, as observed in Equation (12.81).

12.3.4.2 Performance of the MSDSD-Aided DDF User-Cooperation System

Despite the fact that the performance degradation experienced by the conventional DDF-aided user-cooperation system employing the CDD in severely time-selective channels can be mitigated by utilizing the single-path MSDSD at the relay node, a significant performance loss remains unavoidable due to the absence of a detection technique at the destination node, which is robust to the time-selective channel, as previously seen in Figure 12.15. Fortunately, the multi-path-based MSDSD designed for the user-cooperation-aided system devised in Section 12.3.3 can be employed at the destination node in order to mitigate further the channel-induced performance degradation of the DDF-aided system.

Figure 12.20 demonstrates a significant performance improvement attained by the multi-path-based MSDSD design employing $N_{wind} = 6$ at the destination node of the DDF-aided two-user cooperative system over its counterpart dispensing with MSDSD at the destination at both $f_d = 0.03$ and $f_d = 0.001$ for each link, respectively. The more severely time selective the channel, the higher the end-to-end performance gain that can be achieved by the MSDSD-assisted DDF-aided system. Specifically, for a given target BER of 10^{-3}, a performance gain as high as 9 dB is achieved at $f_d = 0.03$, whereas only negligible performance improvement is attained at $f_d = 0.01$. On the other hand, by comparing the simulation results of Figure 12.17 and Figure 12.20, we observe that the performance gains achieved by the MSDSD employed at the destination node of the DDF-aided system is significantly lower than those recorded for its DAF-aided counterpart. Even though $N_{wind} = 11$ is employed, there is still a conspicuous gap between the BER curves corresponding to high values of f_d and the one obtained at $f_d = 0.001$ in the context of the DDF-aided system, as shown in Figure 12.21. This trend is not

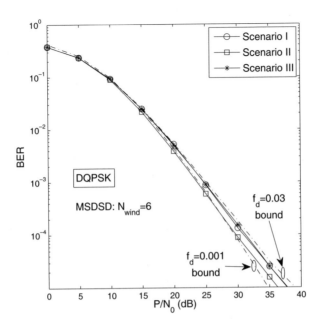

Figure 12.19: The impact of the relative mobility among the source, relay and destination nodes on the BER performance of the DAF-aided T-DQPSK-modulated cooperative system employing MSDSD at the destination node in Rayleigh fading channels. All other system parameters are summarized in Tables 12.5 and 12.6.

unexpected owing to the fact that the design of the multi-path MSDSD used in the DDF-aided user-cooperation-assisted system is carried out under the assumption of an idealized perfect reception-and-forward process at the relay node, while actually the relay will keep silent when it fails to detect the received signal correctly, as detected by the CRC check. In other words, the MSDSD employed at the destination simply assumes that the relay node has knowledge of the signal transmitted by the source node as implied by the system model of Equation (12.37) describing the DDF-aided system, operating without realizing that sometimes only noise is presented to the receive antenna during the relay phase II.

In comparison with its DAF-aided counterpart, the end-to-end performance of the DDF-aided system is jointly determined by the robustness of the differential detection technique to time-selective channels at the destination node, as well as by that at the relay node. Previously, we employed the same observation window size N_{wind} for the MSDSDs used at both the relay and destination nodes. However, in reality there exist situations where the affordable overall system complexity is limited and hence a low value of N_{wind} has to be used at both the relay and destination nodes. Thus, it is beneficial to characterize the importance of the detection technique employed at the relay and destination nodes to determine the system's required complexity. Figure 12.22 plots the BER curve of the DDF-aided two-user cooperative system for $N_{wind} = 6$ at the relay node and for $N_{wind} = 2$ at the destination node versus that generated by reversing the N_{wind} allocation, i.e. by having $N_{wind} = 2$ and $N_{wind} = 6$ at the relay and destination nodes, respectively. Observe in Figure 12.22 that the system having a more robust differential detector at the relay node slightly outperforms the other in the high-SNR range at both $f_d = 0.03$ and $f_d = 0.01$. This is because a less robust detection scheme employed at the relay node may erode the benefits of relaying in the DDF-aided user-cooperation-assisted system. Naturally, this degrades the achievable performance of the MSDSD at the destination, which carries out the detection based on the assumption of a reliable relayed signal. Hence, in the context of the DDF-aided

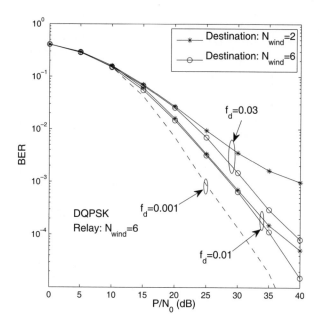

Figure 12.20: BER performance improvement achieved by the multi-path-based MSDSD scheme employing $N_{wind} = 6$ at the destination node of the DDF-aided T-DQPSK-modulated cooperative system in Rayleigh fading channels. All other system parameters are summarized in Table 12.5.

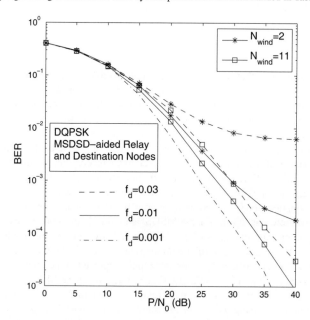

Figure 12.21: BER performance improvement achieved by the multi-path MSDSD employing $N_{wind} = 11$ at the destination node of the DDF-aided T-DQPSK-modulated cooperative system in Rayleigh fading channels. All other system parameters are summarized in Table 12.5.

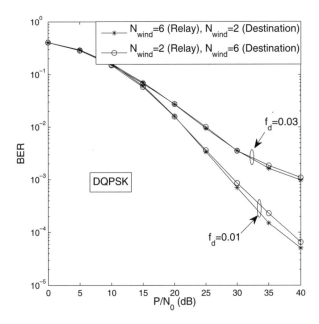

Figure 12.22: BER performance of the DDF-aided T-DQPSK-modulated cooperative system employing MSDSD in conjunction with different detection-complexity allocations in Rayleigh fading channels. All other system parameters are summarized in Table 12.5.

user-cooperation-assisted system employing the MSDSD, a higher complexity should be invested at the relay node in the interest of achieving an enhanced end-to-end performance.

Let us now investigate the effect of the relative mobility of the source, relay and destination nodes on the achievable BER performance of the DDF-aided two-user cooperative system by considering the BER curves corresponding to the three scenarios of Table 12.6, in Figure 12.23. Based on our previous discussions, we understand that the performances of the detection schemes employed at both the relay and destination nodes are equally important factors in determining the achievable end-to-end system performance, which are mainly affected by the Doppler frequency characteristics of both the source–relay link and the source–destination link in the DDF-aided user-cooperation-assisted system. In Scenario I of Table 12.6 the system exhibits the worst BER performance, which is roughly the same as the $f_d = 0.03$ performance bound, since the benefits brought about by a high-quality, near-stationary relay–destination link may be eroded by a low-quality, high-Doppler source–relay link dominating the achievable performance of the MSDSD scheme at the relay node, which in turn substantially degrades the achievable end-to-end system performance. In Scenario II of Table 12.6, we assumed that the source and destination nodes experience a low Doppler frequency in the direct link ($f_{d,sd} = 0.001$), which is one of the two above-mentioned dominant links in the DDF-aided system. Thus, for a given target BER of 10^{-4}, the system achieves a performance gain as high as 5 dB in Scenario II over that attained in the benchmark scenario having an identical Doppler frequency of $f_d = 0.03$ for each link, as observed in Figure 12.23. Moreover, the achievable performance gain can be almost doubled if the system is operating in Scenario III, where in turn the other important link, namely the source–relay link, becomes a slow-fading channel associated with $f_d = 0.001$. Remarkably, the performance achieved in Scenario III is comparable with that attained by the same system in the benchmark scenario, where we have $f_d = 0.001$ for each of the three links. More specifically, the system operating in Scenario III only suffers a performance loss of about 1 dB at a target BER of 10^{-4} in comparison with that associated with the slow-fading benchmark scenario.

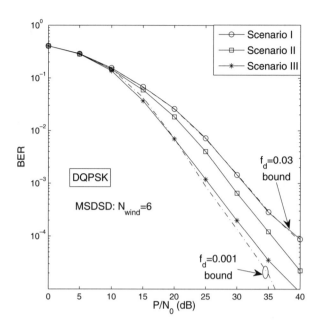

Figure 12.23: The impact of the relative mobility among the source, relay and destination nodes on the BER performance of the DDF-aided T-DQPSK-modulated cooperative system employing MSDSD at both the relay and destination nodes in Rayleigh fading channels. All other system parameters are summarized in Tables 12.5 and 12.6.

12.4   Chapter Conclusions

Cooperative diversity, emerging as an attractive diversity-aided technique to circumvent the cost and size constraints of implementing multiple antennas on a pocket-sized mobile device with the aid of antenna sharing among multiple cooperating single-antenna-aided users, is capable of effectively combating the effects of channel fading and hence improving the attainable performance of the network. However, the user-cooperation mechanism may result in a complex system when using coherent detection, where not only the BS but also the cooperating MSs would require channel estimation. Channel estimation would impose both an excessive complexity and a high pilot overhead. This situation may be further aggravated in mobile environments associated with relatively rapidly fluctuating channel conditions. Therefore, the consideration of cooperative system design without assuming knowledge of the CSI at transceivers becomes more realistic, which inspires the employment of differentially encoded modulation at the transmitter and that of non-coherent detection dispensing with both the pilots and channel estimation at the receiver. However, as discussed in Section 12.1.1, the performance of the low-complexity CDD-aided direct-transmission-based OFDM system may substantially degrade in highly time-selective or frequency-selective channels, depending on the domain in which the differential encoding is carried out. Fortunately, as argued in Section 12.2, the single-path MSDSD, which has been contrived to mitigate the channel-induced error floor encountered by differentially encoded single-input, single-output transmission, jointly detects differentially multiple consecutively received signals by exploiting the correlation among their phase distortions. Hence, inspired by the proposal of the single-path MSDSD, our main objective in this chapter is specifically to design a multi-path MSDSD which is applicable to the differentially encoded cooperative systems in order to make the overall system robust to the effects of the hostile wireless channel. To this end, in Section 12.3.3.1 we constructed a generalized equivalent multiple-symbol system model for the cooperative system employing either the DAF or DDF scheme, which facilitated the process of

Table 12.7: Performance summary of the MSDSD investigated in Chapter 12. The system parameters were given by Table 12.5. Note that 'N/C' means the target BER is not achievable, regardless of the SNR, while 'N/A' means the data are not available.

			BER			
			P/N_0 (dB)		Gain (dB)	
	f_d	N_{wind}	10^{-3}	10^{-4}	10^{-3}	10^{-4}
	Performance of the single-relay-aided cooperative system					
Non-cooperative system	$f_{d,sd} = f_{d,sr} = f_{d,rd} = 0.001$	2	30	40	—	—
		6	30	40	0.0	0.0
	$f_{d,sd} = f_{d,sr} = f_{d,rd} = 0.01$	2	40	N/C	—	—
		6	32	N/A	8	N/A
	$f_{d,sd} = f_{d,sr} = f_{d,rd} = 0.03$	2	N/C	N/C	—	—
		6	35	N/A	∞	N/A
DAF cooperative system	$f_{d,sd} = f_{d,sr} = f_{d,rd} = 0.001$	2	23.5	29	—	—
		6	23.5	29	0.0	0.0
	$f_{d,sd} = f_{d,sr} = f_{d,rd} = 0.01$	2	25	33	—	—
		6	23.5	30	1.5	3
	$f_{d,sd} = f_{d,sr} = f_{d,rd} = 0.03$	2	32.5	N/C	—	—
		6	25	32	7.5	∞
	$f_{d,sd} = f_{d,sr} = 0.03, f_{d,rd} = 0.001$	6	24	31	—	—
	$f_{d,sr} = f_{d,rd} = 0.03, f_{d,sd} = 0.001$	6	23	30	1	1
	$f_{d,sd} = f_{d,rd} = 0.03, f_{d,sr} = 0.001$	6	24	31	0.0	0.0
DDF cooperative system	$f_{d,sd} = f_{d,sr} = f_{d,rd} = 0.001$	R: 2, D: 2	24.5	31	—	—
		R: 6, D: 2	24.5	31	0.0	0.0
		R: 2, D: 6	24.5	31	0.0	0.0
		R: 6, D: 6	24.5	31	0.0	0.0
	$f_{d,sd} = f_{d,sr} = f_{d,rd} = 0.01$	R: 2, D: 2	30	58	—	—
		R: 6, D: 2	29	37	1	21
		R: 2, D: 6	30	38	0	20
		R: 6, D: 6	29	35.5	1	22.5
	$f_{d,sd} = f_{d,sr} = f_{d,rd} = 0.03$	R: 2, D: 2	N/C	N/C	—	—
		R: 6, D: 2	40	N/C	∞	0
		R: 2, D: 6	41	N/C	∞	0
		R: 6, D: 6	31.3	41	∞	∞
	$f_{d,sd} = f_{d,sr} = 0.03, f_{d,rd} = 0.001$	R, D: 6	31	40	—	—
	$f_{d,sr} = f_{d,rd} = 0.03, f_{d,sd} = 0.001$	R, D: 6	29	36	2	4
	$f_{d,sd} = f_{d,rd} = 0.03, f_{d,sr} = 0.001$	R, D: 6	25.5	32	5.5	8

transforming the optimum detection metric to a shortest-vector problem, as detailed in Section 12.3.3.2. Then, it was shown in Section 12.3.3.4 that the resultant shortest-vector problem may be efficiently solved by a multi-layer tree search scheme, which is similar to that proposed in Section 11.3.2.3. This procedure relies on the channel-noise autocorrelation matrix triangularization procedure of Section 12.3.3.3.

Our Monte Carlo simulation results provided in Section 12.3.4.1 demonstrated that the resultant multi-path MSDSD employed at the BS is capable of completely eliminating the performance loss encountered by the DAF-aided cooperative system, provided that a sufficiently high value of N_{wind} is used. For example, observe in Figure 12.18 that, given a target BER of 10^{-3}, a performance gain of about 10 dB can be attained by the proposed MSDSD employing $N_{wind} = 11$ for a DQPSK-modulated two-user cooperative system in a relatively fast-fading channel associated with a normalized Doppler frequency of 0.03. In contrast to the DAF-aided cooperative system, it was shown in Figure 12.21 of

Section 12.3.4.2 that, although a significant performance improvement can also be achieved by the multi-path MSDSD at the BS in highly time-selective channels for the DDF-aided system, the channel-induced performance loss was not completely eliminated, even when $N_{wind} = 11$ was employed. This was because the MSDSD employed at the BS simply assumed a guaranteed perfect decoding at the relay, operating without taking into account that sometimes only noise is presented to the receive antenna during the relay's phase II, i.e. when the relay keeps silent owing to the failure of recovering the source's signal. Furthermore, our investigation of the proposed MSDSD in the practical Rayleigh fading scenario, where a different Doppler frequency is assumed for each link, demonstrated that the channel quality of the direct source–destination link characterized in terms of its grade of time selectivity predetermines the achievable performance of the DAF-aided cooperative system. By contrast, the source–relay and relay–destination links are symmetric and thus they may be interchanged without affecting the end-to-end performance. By contrast, observe in Figure 12.23 that the achievable performance of the DDF-aided system employing the MSDSD is dominated by the source–relay link. This is not unexpected, since a high-quality, near-stationary source–relay link enhances the performance of the MSDSD at the BS, making its assumption of a perfect decoding at the relay more realistic. Finally, based on the simulation results obtained in this chapter, we quantitatively summarize the performance gains achieved by the MSDSD for the direct-transmission-based non-cooperative system as well as for both the DAF- and DDF-aided cooperative systems in Table 12.7.

Chapter **13**

Resource Allocation for the Differentially Modulated Cooperation-Aided Cellular Uplink in Fast Rayleigh Fading Channels

13.1 Introduction[1]

13.1.1 Chapter Contributions and Outline

It was observed in Chapter 12 that the differentially modulated user-cooperative uplink systems employing either the DAF scheme of Section 12.3.2.1 or the DDF scheme of Section 12.3.2.2 were capable of achieving cooperative diversity gain while circumventing the cost and size constraints of implementing multiple antennas in a pocket device. Additionally, by avoiding the challenging task of estimating all the $(N_t \times N_r)$ CIRs of multi-antenna-aided systems, the differentially encoded cooperative system may exhibit a better performance than its coherently detected, but non-cooperative, counterpart, since the CIRs cannot be perfectly estimated by the terminals. The CIR estimation becomes even more challenging when the MS travels at a relatively high speed, resulting in a rapidly fading environment. On the other hand, although it was shown in Chapter 12 that a full spatial diversity can usually be achieved by the differentially modulated user-cooperative uplink system, the achievable end-to-end BER performance may significantly depend on the specific choice of the cooperative protocol employed and/or on the quality of the relay channel. Therefore, in the scenario of differentially modulated cooperative uplink systems, where multiple cooperating MSs are roaming in the area between a specific MS and the BS seen in Figure 13.1, an appropriate Cooperative-Protocol Selection (CPS) as well as a matching Cooperating-User Selection (CUS) becomes necessary in order to maintain a desirable end-to-end performance. Motivated by the above-mentioned observations, the novel contributions of this chapter are as follows:

---

[1]This chapter is partially based on ©IEEE Wang & Hanzo 2007 [606].

MIMO-OFDM for LTE, Wi-Fi and WiMAX   Lajos Hanzo, Yosef Akhtman, Li Wang and Ming Jiang
© 2011 John Wiley & Sons, Ltd

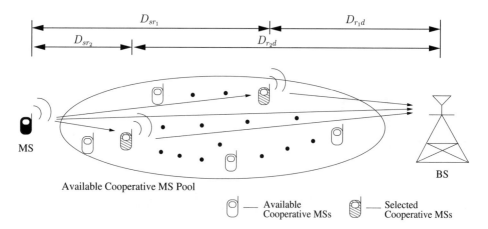

Figure 13.1: Cooperation-aided uplink systems using relay selection. ©IEEE Wang & Hanzo 2007 [606]

- The achievable end-to-end performance is theoretically analysed for both the DAF- and DDF-aided cooperative systems.

- Based on the above-mentioned analytical results, both CUS schemes and Adaptive Power Control (APC) schemes are proposed for the above two types of cooperative system in the interest of achieving the best possible performance.

- Intensive comparative studies of the most appropriate resource allocation in the context of both DAF- and DDF-assisted cooperative systems are carried out.

- In order to make the most of the complementarity of the DAF- and DDF-aided cooperative systems, a more flexible resource-optimized adaptive hybrid cooperation-aided system is proposed, yielding a further improved performance.

The remainder of this chapter is organized as follows. In Section 13.2 we first theoretically analyse the achievable end-to-end performance of both the DAF- and DDF-assisted cooperative systems. Then, based on the BER performance analysis of Section 13.2, in Sections 13.3.1 and 13.3.2 we will propose appropriate CUS schemes for both the above-mentioned two types of cooperative systems, along with an optimized power control arrangement. Additionally, in order to improve further the achievable end-to-end performance of the cooperation-aided UL of Figure 13.1 and to create a flexible cooperative mechanism, in Section 13.4 we will also investigate the CPS of the UL in conjunction with the CUS as well as the power control, leading to a resource-optimized adaptive cooperation-aided system. Finally, our concluding remarks will be provided in Section 13.5.

13.1.2   System Model

To be consistent with the system model employed in Chapter 12, the U-user TDMA UL is considered for the sake of simplicity. Again, due to the symmetry of channel allocation among users, as indicated in Figure 12.9, we focus our attention on the information transmission of a specific source MS seen in Figure 13.1, which potentially employs M_r out of the $\mathcal{P}_{cand} = (U-1)$ available relay stations in order to achieve cooperation-aided diversity by forming a VAA. Without loss of generality, we simply assume the employment of a single antenna for each terminal. For simple analytical tractability, we assume that the sum of the distances D_{sr_u} between the source MS and the uth RS, and that between the uth RS and the destination BS, which is represented by D_{r_ud}, is equal to the distance D_{sd} between the source MS and the BS. Equivalently, as indicated by Figure 13.1, we have

$$D_{sr_u} + D_{r_ud} = D_{sd}, \quad u = 1, 2, \ldots, U - 1. \tag{13.1}$$

Furthermore, by considering a path-loss exponent of v [607], the average power $\sigma^2_{i,j}$ at the output of the channel can be computed according to the internode distance $D_{i,j}$ as follows:

$$\sigma^2_{i,j} = C \cdot D^{-v}_{i,j}, \quad i,j \in \{s, r_u, d\}, \tag{13.2}$$

where C is a constant which can be normalized to unity without loss of generality and the subscripts s, r_u and d represent the source, the uth relay and the destination, respectively. Thus, Equation (13.2) can be expressed as

$$\sigma^2_{i,j} = D^{-v}_{i,j}, \quad i,j \in \{s, r_u, d\}. \tag{13.3}$$

Additionally, under the assumption of having a total transmit power of P and assuming that M_r cooperating MSs are activated out of a total of $\mathcal{P}_{cand}$, we can express the associated power constraint as

$$P = P_s + \sum_{m=1}^{M_r} P_{r_m}, \tag{13.4}$$

where P_s and P_{r_m} $(m = 1, 2, \ldots, M_r)$ are the transmit power employed by the source MS and the mth RS, respectively.

13.2 Performance Analysis of the Cooperation-Aided UL

In this section, we commence analysing the error probability performance of both the DAF-aided and DDF-aided systems, where the MSDSD devised in Chapter 12 is employed in order to combat the effects of fast fadings caused by the relative mobility of the MSs and BS in the cell. Recall from Chapter 12 that the Doppler-frequency-induced error floor encountered by the CDD (or equivalently by the MSDSD using $N_{wind} = 2$) is expected to be significantly eliminated by jointly detecting $N_{wind} > 2$ consecutive received symbols with the aid of the MSDSD, provided that N_{wind} is sufficiently high. Therefore, under the assumption that the associated performance degradation can be mitigated by the MSDSD in both the DAF-aided and DDF-aided cooperative systems, it is reasonable to expect that the BER performance exhibited by the cooperation-assisted system employing the MSDSD in a relatively rapidly fading environment can be closely approximated by that achieved by the CDD in slow-fading channels. Hence, in the ensuing two sections our performance analysis is carried out without considering the detrimental effects imposed by the mobility of the MSs, since these effects are expected to be mitigated by employment of the MSDSD of Section 12.3. Consequently, our task may be interpreted as the performance analysis of a CDD-assisted differentially modulated cooperative system operating in slow-fading channels.

13.2.1 Theoretical Analysis of Differential Amplify-and-Forward Systems

13.2.1.1 Performance Analysis

First of all, without loss of accuracy, we drop the time index n and rewrite the signal of Equation (12.35) received at the mth cooperating MS and that of Equation (12.39) from the mth RS at the BS as follows:

$$y_{sr_m} = \sqrt{P_s} s_s h_{sr_m} + w_{sr_m}, \tag{13.5}$$

$$y_{r_m d} = f_{AM_{r_m}} y_{sr_m} h_{r_m d} + w_{r_m d}, \tag{13.6}$$

where the amplification factor $f_{AM_{r_m}}$ employed by the mth relay node can be specified as [604]

$$f_{AM_{r_m}} = \sqrt{\frac{P_{r_m}}{P_s \sigma^2_{sr_m} + N_0}}, \tag{13.7}$$

with N_0 being the variance of the AWGN imposed at all cooperating MSs as well as at the BS. Then, we can further reformat Equation (13.6) with the aid of Equation (13.5) in order to express the signal received at the destination BS from the RS as

$$y_{r_m d} = f_{AM_{r_m}} h_{r_m d}(\sqrt{P_s} h_{sr_m} s_s + w_{sr_m}) + w_{r_m d} \tag{13.8}$$

$$= f_{AM_{r_m}} \sqrt{P_s} h_{r_m d} h_{sr_m} s_s + f_{AM_{r_m}} h_{r_m d} w_{sr_m} + w_{r_m d}. \tag{13.9}$$

Hence, we can calculate the received SNR per symbol at the BS for both the direct and the relaying links, respectively, as

$$\gamma_{sd}^s = \frac{P_s |h_{sd}|^2}{N_0}, \tag{13.10}$$

$$\gamma_{r_m d}^s = \frac{P_s P_{r_m} |h_{sr_m}|^2 |h_{r_m d}|^2}{N_0 (P_s \sigma_{sr_m}^2 + P_{r_m} |h_{r_m d}|^2 + N_0)}. \tag{13.11}$$

Furthermore, MRC is assumed to be employed at the BS prior to the CDD scheme for the system using the DAF arrangement characterized in Equation (12.40) of Section 12.3.2.1, which is rewritten here for convenience:

$$y = a_0 (y_{sd}[n-1])^* y_{sd}[n] + \sum_{m=1}^{M_r} a_m (y_{r_m d}[n + mL_f - 1])^* y_{r_m d}[n + mL_f], \tag{13.12}$$

where L_f is the length of the transmission packet, while the coefficients a_0 and a_m $(m = 1, 2, \ldots, M_r)$ are given by

$$a_0 = \frac{1}{N_0}, \tag{13.13}$$

$$a_m = \frac{P_s \sigma_{sr_m}^2 + N_0}{N_0 (P_s \sigma_{sr_m}^2 + P_{r_m} |h_{r_m d}|^2 + N_0)}. \tag{13.14}$$

According to the basic property of the MRC scheme, the SNR at the MRC's output can be expressed as

$$\gamma^s = \gamma_{sd}^s + \sum_{m=1}^{M_r} \gamma_{r_m d}^s. \tag{13.15}$$

Equivalently, we can express the SNR per bit at the output of the MRC as

$$\gamma^b = \frac{\gamma_{sd}^s}{\log_2 M_c} + \sum_{m=1}^{M_r} \frac{\gamma_{r_m d}^s}{\log_2 M_c}$$

$$= \gamma_{sd}^b + \sum_{m=1}^{M_r} \gamma_{r_m d}^b, \tag{13.16}$$

where M_c is the constellation size of a specific modulation scheme.

On the other hand, the end-to-end BER expression conditioned on the SNR per bit at the combiner's output, namely γ^b of Equation (13.16), for the DAF-aided system activating M_r RSs for a specific source MS can be expressed as [608]

$$P_{BER|\gamma^b}^{DAF}(a, b, M_r) = \frac{1}{2^{2(M_r+1)} \pi} \int_{-\pi}^{\pi} f(a, b, M_r + 1, \theta) e^{-\alpha(\theta)\gamma^b} d\theta, \tag{13.17}$$

where [608]

$$f(a, b, L, \theta) = \frac{b^2}{2\alpha(\theta)} \sum_{l=1}^{L} \binom{2L-1}{L-l} [(\beta^{-l+1} - \beta^{l+1}) \cos((l-1)(\phi + \pi/2))$$

$$- (\beta^{-l+2} - \beta^l) \cos(l(\phi + \pi/2))], \tag{13.18}$$

$$\alpha(\theta) = \frac{b^2(1 + 2\beta \sin\theta + \beta^2)}{2} \tag{13.19}$$

and

$$\beta = a/b. \tag{13.20}$$

In Equation (13.17) the parameters a and b are the modulation-dependent factors defined in [467]. Specifically, $a = 10^{-3}$ and $b = \sqrt{2}$ for DBPSK modulation, while $a = \sqrt{2 - \sqrt{2}}$ and $b = \sqrt{2 + \sqrt{2}}$ for DQPSK modulation using Gray coding. Additionally, the parameter β, which is defined as Equation (13.20), can be calculated according to the specific modulation scheme employed [467]. Moreover, the parameter L of Equation (13.18) denotes the number of diversity paths. For example, when M_r cooperating MSs are activated, we have $L = M_r + 1$, assuming that the BS combines the signals received from all the M_r RSs as well as that from the direct link.

On the other hand, since a non-dispersive Rayleigh fading channel is considered here, the PDF of the channel's fading amplitude r can be expressed as [607]

$$p_r(r) = \begin{cases} \frac{2r}{\Omega} e^{-r^2/\Omega}, & 0 \le r \le \infty \\ 0, & r < 0, \end{cases} \tag{13.21}$$

where $\Omega = \overline{r^2}$ represents the mean square value of the fading amplitude. Hence, the PDF of the instantaneous received SNR per bit at the output of the Rayleigh fading channel is given by the so-called Γ distribution [607]

$$p_{\gamma^b}(\gamma) = \begin{cases} \frac{1}{\overline{\gamma^b}} e^{-\gamma/\overline{\gamma^b}}, & \gamma \ge 0 \\ 0, & \gamma < 0 \end{cases} \tag{13.22}$$

where $\overline{\gamma^b}$ denotes the average received SNR per bit, which can be expressed as

$$\overline{\gamma^b} = \frac{P_{t,bit} \cdot \Omega}{N_0} \tag{13.23}$$

$$= \frac{P_{t,symbol} \cdot \Omega}{N_0 \cdot \log_2 \mathcal{M}_c}, \tag{13.24}$$

with $P_{t,bit}$ and $P_{t,symbol}$ representing the transmit power per bit and per symbol, respectively.

Now, the unconditional end-to-end BER of the DAF-aided cooperative system can be calculated by averaging the conditional BER expression of Equation (13.17) over the entire range of received SNR per bit values by weighting it according to its probability of occurrence represented with the aid of its PDF in Equation (13.22) as follows [608, 609]:

$$P_{BER}^{DAF}(a, b, M_r) = \int_{-\infty}^{+\infty} P_{BER|\gamma^b}^{DAF} \cdot p_{\gamma^b}(\gamma) \, d\gamma \tag{13.25}$$

$$= \frac{1}{2^{2(M_r+1)}\pi} \int_{-\pi}^{\pi} f(a, b, M_r + 1, \theta) \int_{-\infty}^{+\infty} e^{-\alpha(\theta)\gamma} p_{\gamma^b}(\gamma) \, d\gamma \, d\theta \tag{13.26}$$

$$= \frac{1}{2^{2(M_r+1)}\pi} \int_{-\pi}^{\pi} f(a, b, M_r + 1, \theta) \mathcal{M}_{\gamma^b}(\theta) \, d\theta, \tag{13.27}$$

where the joint Moment Generating Function (MGF) [609] of the received SNR per bit γ^b given by Equation (13.16) is defined as

$$\mathcal{M}_{\gamma^b}(\theta) = \int_{-\infty}^{+\infty} e^{-\alpha(\theta)\gamma} p_{\gamma^b}(\gamma)\, d\gamma \tag{13.28}$$

$$= \underbrace{\int_{-\infty}^{+\infty} \cdots \int_{-\infty}^{+\infty}}_{(M_r+1)\text{-fold}} e^{-\alpha(\theta)(\gamma_{sd} + \sum_{m=1}^{M_r} \gamma_{r_m d})} p_{\gamma_{sd}^b}(\gamma_{sd})$$

$$\times \prod_{m=1}^{M_r} p_{\gamma_{r_m d}^b}(\gamma_{r_m d})\, d\gamma_{sd}\, d\gamma_{r_1 d} \cdots d\gamma_{r_{M_r} d} \tag{13.29}$$

$$= \mathcal{M}_{\gamma_{sd}^b}(\theta) \prod_{m=1}^{M_r} \mathcal{M}_{\gamma_{r_m d}^b}(\theta), \tag{13.30}$$

with $\mathcal{M}_{\gamma_{sd}^b}(\theta)$ and $\mathcal{M}_{\gamma_{r_m d}^b}(\theta)$ representing the MGF of the received SNR per bit γ_{sd}^b of the direct link and that of the received SNR per bit $\gamma_{r_m d}^b$ of the mth relay link. Specifically, with the aid of Equation (13.22) we have [604, 609]

$$\mathcal{M}_{\gamma_{sd}^b}(\theta) = \frac{1}{1 + k_{sd}(\theta)}, \tag{13.31}$$

$$\mathcal{M}_{\gamma_{r_m d}^b}(\theta) = \frac{1}{1 + k_{sr_m}(\theta)} \left(1 + \frac{k_{sr_m}(\theta)}{1 + k_{sr_m}(\theta)} \frac{P_s \sigma_{sr_m}^2 + N_0}{P_{r_m}} \frac{1}{\sigma_{r_m d}^2} Z_{r_m}(\theta) \right), \tag{13.32}$$

where

$$k_{sd}(\theta) \triangleq \frac{\alpha(\theta) P_s \sigma_{sd}^2}{N_0}, \tag{13.33}$$

$$k_{sr_m}(\theta) \triangleq \frac{\alpha(\theta) P_s \sigma_{sr_m}^2}{N_0} \tag{13.34}$$

and

$$Z_{r_m}(\theta) \triangleq \int_0^\infty \frac{e^{-(u/\sigma_{r_m d}^2)}}{u + R_{r_m}(\theta)}\, du, \tag{13.35}$$

with

$$R_{r_m}(\theta) \triangleq \frac{P_s \sigma_{sr_m}^2 + N_0}{P_{r_m}[1 + k_{sr_m}(\theta)]}. \tag{13.36}$$

According to Equations (3.352.2) and (8.212.1) of [610], Equation (13.35) can be further extended as

$$Z_{r_m}(\theta) = -e^{R_{r_m}(\theta)/\sigma_{r_m d}^2} \left(\zeta + \ln \frac{R_{r_m}(\theta)}{\sigma_{r_m d}^2} + \int_0^{R_{r_m}(\theta)/\sigma_{r_m d}^2} \frac{e^{-t} - 1}{t}\, dt \right), \tag{13.37}$$

where $\zeta \triangleq 0.577\,215\,664\,90\ldots$ denotes the Euler constant. In order to circumvent the integration, Equation (13.37) can be expressed with aid of the Taylor series as

$$Z_{r_m}(\theta) = -e^{R_{r_m}(\theta)/\sigma_{r_m d}^2} \left(\zeta + \ln \frac{R_{r_m}(\theta)}{\sigma_{r_m d}^2} + \sum_{n=1}^\infty \frac{(-R_{r_m}(\theta)/\sigma_{r_m d}^2)^n}{n \cdot n!} \right) \tag{13.38}$$

$$\approx -e^{R_{r_m}(\theta)/\sigma_{r_m d}^2} \left(\zeta + \ln \frac{R_{r_m}(\theta)}{\sigma_{r_m d}^2} + \sum_{n=1}^{N_n} \frac{(-R_{r_m}(\theta)/\sigma_{r_m d}^2)^n}{n \cdot n!} \right), \tag{13.39}$$

where the parameter N_n is introduced to control the accuracy of Equation (13.39). Since the Taylor series in Equation (13.38) converges quickly, the integration in Equation (13.37) can be approximated by the sum of the first N_n elements in Equation 13.39. Consequently, the average BER of the DAF-aided cooperative system where the desired source MS relies on M_r cooperating MSs activated in order to form a VAA can be expressed as

$$P_{BER}^{DAF}(a, b, M_r) = \frac{1}{2^{2(M_r+1)}\pi} \int_{-\pi}^{\pi} \frac{f(a, b, M_r + 1, \theta)}{1 + k_{sd}(\theta)} \prod_{m=1}^{M_r} \frac{1}{1 + k_{sr_m}(\theta)}$$
$$\times \left(1 + \frac{k_{sr_m}(\theta)Z_{r_m}(\theta)}{1 + k_{sr_m}(\theta)} \frac{P_s\sigma_{sr_m}^2 + N_0}{P_{r_m}\sigma_{r_m d}^2}\right) d\theta. \tag{13.40}$$

Using the same technique as in [604], the BER expression of Equation (13.40) can be upper-bounded by bounding $Z_{r_m}(\theta)$ of Equation (13.35), to simplify the exact BER expression of Equation (13.40). Specifically, $R_{r_m}(\theta)$ of Equation (13.36) reaches its minimum value when $\alpha(\theta)$ of Equation (13.19) is maximized at $\theta = \pi/2$, which in turn maximizes $Z_{r_m}(\theta)$ of Equation (13.35). Thus, the error probability of Equation (13.40) may be upper-bounded as

$$P_{BER}^{DAF}(a, b, M_r) \lessapprox \frac{1}{2^{2(M_r+1)}\pi} \int_{-\pi}^{\pi} \frac{f(a, b, M_r + 1, \theta)}{1 + k_{sd}(\theta)} \prod_{m=1}^{M_r} \frac{1}{1 + k_{sr_m}(\theta)}$$
$$\times \left(1 + \frac{k_{sr_m}(\theta)Z_{r_m,max}}{1 + k_{sr_m}(\theta)} \frac{P_s\sigma_{sr_m}^2 + N_0}{P_{r_m}\sigma_{r_m d}^2}\right) d\theta, \tag{13.41}$$

where

$$Z_{r_m,max} \triangleq -e^{R_{r_m,min}/\sigma_{r_m d}^2}\left(\zeta + \ln \frac{R_{r_m,min}}{\sigma_{r_m d}^2} + \sum_{n=1}^{N_n} \frac{(-R_{r_m,min}/\sigma_{r_m d}^2)^n}{n \cdot n!}\right), \tag{13.42}$$

in which

$$R_{r_m,min} \triangleq \frac{P_s\sigma_{sr_m}^2 + N_0}{P_{r_m}[1 + P_s\sigma_{sr_m}^2 b^2(1 + \beta)^2/2N_0]}. \tag{13.43}$$

Similarly, the average BER of Equation (13.40) can be lower-bounded by minimizing $Z_{r_m}(\theta)$ of Equation (13.35) at $\theta = -\pi/2$. Specifically, from Equation (13.40) we arrive at the error probability expression of

$$P_{BER}^{DAF}(a, b, M_r) \gtrapprox \frac{1}{2^{2(M_r+1)}\pi} \int_{-\pi}^{\pi} \frac{f(a, b, M_r + 1, \theta)}{1 + k_{sd}(\theta)} \prod_{m=1}^{M_r} \frac{1}{1 + k_{sr_m}(\theta)}$$
$$\times \left(1 + \frac{k_{sr_m}(\theta)Z_{r_m,min}}{1 + k_{sr_m}(\theta)} \frac{P_s\sigma_{sr_m}^2 + N_0}{P_{r_m}\sigma_{r_m d}^2}\right) d\theta, \tag{13.44}$$

where

$$Z_{r_m,min} \triangleq -e^{R_{r_m,max}/\sigma_{r_m d}^2}\left(\zeta + \ln \frac{R_{r_m,max}}{\sigma_{r_m d}^2} + \sum_{n=1}^{N_n} \frac{(-R_{r_m,max}/\sigma_{r_m d}^2)^n}{n \cdot n!}\right), \tag{13.45}$$

in which

$$R_{r_m,max} \triangleq \frac{P_s\sigma_{sr_m}^2 + N_0}{P_{r_m}[1 + P_s\sigma_{sr_m}^2 b^2(1 - \beta)^2/2N_0]}. \tag{13.46}$$

For further simplifying the BER expressions of Equations (13.41) and (13.44), we can neglect all the additive terms of '1' in the denominators of both of the above-mentioned BER expressions by considering the relatively high-SNR region. Consequently, after some further manipulations, the

Table 13.1: Summary of system parameters.

System parameters	Choice
System	User-cooperative cellular uplink
Cooperative protocol	DAF
Number of relay nodes	M_r
Number of subcarriers	$D = 1024$
Modulation	DPSK
Packet length	$L_f = 128$
Normalized Doppler frequency	$f_d = 0.008$
Path-loss exponent	Typical urban area, $v = 3$ [607]
Channel model	Typical urban, refer to Table 12.1
Relay location	$D_{sr_m} = D_{sd}/2, m = 1, 2, \ldots, M_r$
Power control	$P_s = P_{r_m} = P/(M_r + 1), m = 1, 2, \ldots, M_r$
Noise variance at MS and BS	N_0

approximated high-SNR BER upper bound and its lower-bound counterpart respectively can be expressed as follows:

$$P_{BER,high-snr}^{DAF}(a, b, M_r) \lessapprox \frac{F(a, b, M_r + 1)N_0^{M_r+1}}{P_s\sigma_{sd}^2} \prod_{m=1}^{M_r} \frac{P_{r_m}\sigma_{r_m,d}^2 + P_s\sigma_{sr_m}^2 Z_{r_m,max}}{P_s P_{r_m}\sigma_{sr_m}^2\sigma_{r_m d}^2} \quad (13.47)$$

$$P_{BER,high-snr}^{DAF}(a, b, M_r) \gtrapprox \frac{F(a, b, M_r + 1)N_0^{M_r+1}}{P_s\sigma_{sd}^2} \prod_{m=1}^{M_r} \frac{P_{r_m}\sigma_{r_m,d}^2 + P_s\sigma_{sr_m}^2 Z_{r_m,min}}{P_s P_{r_m}\sigma_{sr_m}^2\sigma_{r_m d}^2}, \quad (13.48)$$

where

$$F(a, b, L) = \frac{1}{2^{2L}\pi} \int_{-\pi}^{\pi} \frac{f(a, b, L, \theta)}{\alpha^L(\theta)} d\theta. \quad (13.49)$$

Then $R_{r_m,min}$ of Equation (13.43) and $R_{r_m,max}$ of Equation (13.46) can be approximated as

$$R_{r_m,min} \approx \frac{2N_0}{b^2(1 + \beta)^2 P_{r_m}}, \quad (13.50)$$

$$R_{r_m,max} \approx \frac{2N_0}{b^2(1 - \beta)^2 P_{r_m}}, \quad (13.51)$$

respectively. Importantly, both the BER upper and lower bounds of Equations (13.47) and (13.48) imply that a DAF-aided cooperative system having M_r selected cooperating users is capable of achieving a diversity order of $L = (M_r + 1)$, as indicated by the exponent L of the noise variance N_0.

13.2.1.2   Simulation Results and Discussion

Let us now consider a DAF-aided cooperative cellular uplink system using M_r relaying MSs in an urban area having a path-loss exponent of $v = 3$. Without loss of generality, all the activated relaying MSs are assumed to be located about half-way between the source MS and the BS, while the total power used for transmitting a single modulated symbol is equally shared among the source MS and the M_r RSs. To be more specific, we have $D_{sr_m} = D_{sd}/2, P_s = P_{r_m} = P/(M_r + 1), m = 1, 2, \ldots, M_r$. Moreover, the normalized Doppler frequency is set to $f_d = 0.01$ under the assumption that multiple MSs are randomly moving around in the same cell. The system parameters considered in this section are summarized in Table 13.1.

The theoretical BER curves of Equation (13.40) versus the SNR received for slow-fading channels are depicted in Figure 13.2 in comparison with the results obtained by our Monte Carlo simulations, where the MSDSD of Section 12.3 using $N_{wind} = 8$ is employed at the BS to eliminate the performance

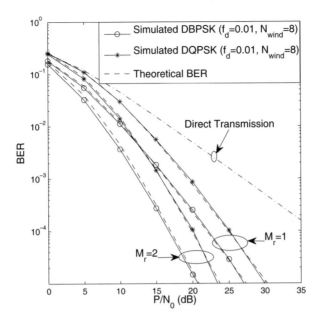

Figure 13.2: BER performance versus SNR for DAF-aided cooperative cellular systems, where there are M_r activated cooperating MSs, each having fixed transmit power and location. The MSDSD using $N_{wind} = 8$ is employed at the BS. All other system parameters are summarized in Table 13.1.

loss imposed by the relative mobility of the cooperating MSs, which is again modelled by a normalized Doppler frequency of $f_d = 0.01$. As suggested previously in Section 13.2.1.1, the Taylor series in Equation (13.38) converges rapidly and hence we employ $N_n = 5$ in Equation (13.39) to reduce the computational complexity, while maintaining the required accuracy. Observe in Figure 13.2 that all theoretical BER curves, corresponding to different numbers of activated cooperating MSs and to DBPSK and DQPSK modulation schemes, match well with the BER curves obtained by our Monte Carlo simulations. Therefore, with the aid of the MSDSD of Section 12.3 employed at the BS, a full diversity order of $L = (M_r + 1)$ can be achieved by the DAF-aided cooperative system in rapidly fading channels, where the achievable BER performance can be accurately predicted using Equation (13.40).

Additionally, the BER upper and lower bounds of Equations (13.41) and (13.44) derived for both DBPSK- and DQPSK-modulated DAF-aided cooperative systems are plotted in Figures 13.3(a) and 13.3(b), respectively, versus the theoretical BER curve of Equation (13.40). Both the lower and upper bounds are tight in comparison with the exact BER curve of Equation (13.40) when the DBPSK modulation scheme is used, as observed in Figure 13.3(a). On the other hand, a relatively loose upper bound is obtained by invoking Equation (13.41) for the DQPSK-modulated system, while the lower bound associated with Equation (13.44) still remains very tight. Therefore, it is sufficiently accurate to approximate the BER performance of the DAF-aided cooperative system using the lower bound of Equation (13.44).

Furthermore, in order to simplify the lower-bound expression of Equation (13.41), the integration term of Equation (13.37) is omitted completely, assuming that we have $N_n = 0$ in Equation (13.45). The corresponding BER curves are depicted in Figure 13.4 versus those obtained when $N_n = 5$. It can be seen that the lower bound obtained after discarding the integration term in Equation (13.37) still remains accurate and tight in the relatively high-SNR region. More specifically, the resultant BER lower bound remains tight over a wide span of SNRs and only becomes inaccurate when the SNR of P/N_0 dips below 5 dB and 10 dB for the DBPSK- and DQPSK-modulated cooperative systems, respectively.

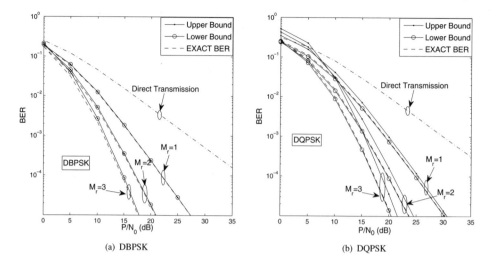

Figure 13.3: BER lower and upper bounds versus SNR for DAF-aided cooperative cellular systems where there are M_r activated cooperating MSs, each having fixed transmit power and location. All other system parameters are summarized in Table 13.1.

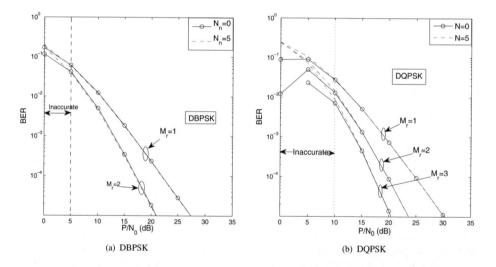

Figure 13.4: Impact of N_n of Equation (13.45) on the BER lower bounds versus SNR for DAF-aided cooperative cellular systems, where there are M_r activated cooperating MSs, each having fixed transmit power and location. All other system parameters are summarized in Table 13.1.

When the SNR is sufficiently high and hence employment of the high-SNR-based lower bound of Equation (13.48) can be justified, its validity is verified by the BER curves of Figures 13.5(a) and 13.5(b) for the DBPSK- and DQPSK-modulated systems, respectively. Specifically, the simplified high-SNR-based BER lower bound of Equation (13.48) having $N_n = 0$ in Equation (13.45) is capable of accurately predicting the BER performance achieved by the DAF-aided cooperative cellular uplink, provided that the transmitted SNR expressed in terms of P/N_0 is in excess of 15 dB for both the DBPSK and DQPSK modulation schemes considered.

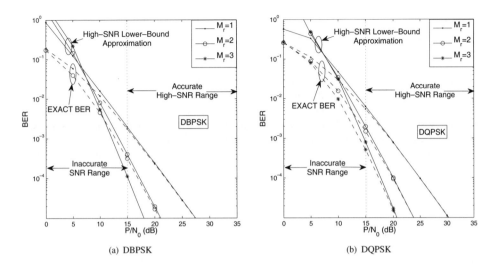

Figure 13.5: High-SNR-based BER lower bounds versus SNR for DAF-aided cooperative cellular systems, where there are M_r activated cooperating MSs, each having transmit power and location. All other system parameters are summarized in Table 13.1.

13.2.2 Theoretical Analysis of DDF Systems

13.2.2.1 Performance Analysis

In the following discourse, the analytical BER performance expressions will be derived for a DDF-aided cooperative cellular system in order to facilitate our resource allocation to be outlined in Section 13.3.2. In contrast to its DAF-aided counterpart of Section 13.2.1, the M_r cooperating MSs selected will make sure that the information contained in the frame or packet received from the source MS can be correctly recovered by differentially decoding the received signal with the aid of CRC checking, prior to forwarding it to the BS. In other words, some of the M_r cooperating MSs selected may not participate during the relaying phase, to avoid potential error propagation due to the imperfect signal recovery. By simply assuming that the packet length is sufficiently high with respect to the channel's coherent time, the worst-case Packet Loss Ratio (PLR) at the mth cooperating MS can be expressed as

$$P_{PLR_m, upper} = 1 - (1 - P_{SER_m})^{L_f},\qquad(13.52)$$

for a given packet length L_f, where P_{SER_m} represents the symbol error rate at the mth cooperating MS, which can be calculated as [611]

$$P_{SER_m} = \frac{M_c - 1}{M_c} + \frac{|\rho_m|\tan(\pi/M_c)}{\xi(\rho_m)}\left[\frac{1}{\pi}\arctan\left(\frac{\xi(\rho_m)}{|\rho_m|}\right) - 1\right],\qquad(13.53)$$

where ρ_m and the function $\xi(x)$, respectively, can be written as follows:

$$\rho_m = \frac{P_s\sigma_{sr_m}^2/N_0}{1 + (P_s\sigma_{sr_m}^2/N_0)},\qquad(13.54)$$

$$\xi(x) = \sqrt{1 - |x|^2 + \tan^2(\pi/M_c)}.\qquad(13.55)$$

Then, based on the $P_{PLR_m, upper}$ expression of Equation (13.52), the average end-to-end BER upper bound of a DDF-aided cooperative system can be obtained. Explicitly, in the context of a system where only $M_r = 1$ cooperating user is selected to participate in relaying the signal from the source MS to the

BS, the average end-to-end BER upper bound $P_{BER,upper}^{DDF}$ is obtained by the summation of the average BERs of two scenarios as

$$P_{BER,upper}^{DDF} = (1 - P_{PLR_1,upper})P_{BER}^{\Phi_1} + P_{PLR_1,upper}P_{BER}^{\Phi_2}, \tag{13.56}$$

where Φ_1 is defined as the first scenario when the cooperating MS perfectly recovers the information received from the source MS and thus transmits the differentially remodulated signal to the BS. By contrast, Φ_2 is defined as the second scenario, when the cooperating MS fails to decode correctly the signal received from the source MS and hence remains silent during the relaying phase. Therefore, the scenarios Φ_1 and Φ_2 can be simply represented as follows, depending on whether the transmit power P_{r1} of the cooperating MS is zero or not during the relaying phase. Thus, we can represent Φ_1 and Φ_2 as

$$\Phi_1 \triangleq \{P_{r_1} \neq 0\}, \tag{13.57}$$

$$\Phi_2 \triangleq \{P_{r_1} = 0\}, \tag{13.58}$$

respectively. Recall our BER analysis carried out for the DAF-aided system in Section 13.2.1.1, where the end-to-end BER expression of a cooperative system conditioned on the received SNR per bit γ^b can be written as

$$P_{BER|\gamma^b}(a, b, L) = \frac{1}{2^{2L}\pi} \int_{-\pi}^{\pi} f(a, b, L, \theta)e^{-\alpha(\theta)\gamma^b} d\theta, \tag{13.59}$$

where $f(a, b, L, \theta)$ given by Equation (13.18) is a function of the number of multi-path components L and is independent of the received SNR per bit γ^b. The parameters a and b are modulation dependent, as defined in [467]. Consequently, the unconditional end-to-end BER, $P_{BER}^{\Phi_i}$, corresponding to the scenario Φ_i can be expressed as

$$P_{BER}^{\Phi_i} = \int_{-\infty}^{\infty} P_{BER|\gamma_{\Phi_i}^b} \cdot p_{\gamma_{\Phi_i}^b}(\gamma) \, d\gamma, \tag{13.60}$$

where $p_{\gamma_{\Phi_i}^b}(\gamma)$ represents the PDF of the received SNR per bit after diversity combining at the BS in the scenario Φ_i of Equations (13.57) and (13.58).

On the other hand, since the MRC scheme is employed at the BS to combine the signals potentially forwarded by multiple cooperating MSs and the signal transmitted from the source MS as characterized by Equation (12.44) using the combining weights of Equation (12.45), the received SNR per bit after MRC combining is simply the sum of that of each combined path, which is expressed as

$$\gamma_{\Phi_1}^b = \gamma_{sd}^b + \gamma_{r_1 d}^b, \tag{13.61}$$

$$\gamma_{\Phi_2}^b = \gamma_{sd}^b. \tag{13.62}$$

Therefore, the unconditional BER of the scenario Φ_1 can be computed as

$$P_{BER}^{\Phi_1} = \frac{1}{2^{2L}\pi} \int_{-\pi}^{\pi} f(a, b, L = 2, \theta) \int_{-\infty}^{\infty} e^{-\alpha(\theta)\gamma_{\Phi_1}^b} p_{\gamma_{\Phi_1}^b}(\gamma) \, d\gamma \, d\theta \tag{13.63}$$

$$= \frac{1}{2^{2L}\pi} \int_{-\pi}^{\pi} f(a, b, L = 2, \theta)\mathcal{M}_{\gamma_{\Phi_1}^b}(\theta) \, d\theta, \tag{13.64}$$

where the joint MGF of the received SNR per bit recorded at the BS for the scenario Φ_1 is expressed as

$$\mathcal{M}_{\gamma_{\Phi_1}^b}(\theta) = \int_{-\infty}^{\infty} e^{-\alpha(\theta)\gamma_{\Phi_1}^b} p_{\gamma_{\Phi_1}^b}(\gamma) \, d\gamma \tag{13.65}$$

$$= \int_{-\infty}^{\infty} \int_{-\infty}^{\infty} e^{-\alpha(\theta)(\gamma_{sd}^b + \gamma_{r_1 d}^b)} p_{\gamma_{sd}^b}(\gamma_{sd}) p_{\gamma_{r_1 d}^b}(\gamma_{r_1 d}) \, d\gamma_{sd} \, d\gamma_{r_1 d} \tag{13.66}$$

$$= \mathcal{M}_{\gamma_{sd}^b}(\theta)\mathcal{M}_{\gamma_{r_1 d}^b}(\theta) \tag{13.67}$$

$$= \frac{N_0^2}{(N_0 + \alpha(\theta)P_s\sigma_{sd}^2)(N_0 + \alpha(\theta)P_{r_1}\sigma_{r_1 d}^2)}, \tag{13.68}$$

with $p_{\gamma_{sd}^b}(\gamma_{sd})$ and $p_{\gamma_{r_1 d}^b}(\gamma_{r_1 d})$, respectively, denoting the PDF of the received SNR per bit for the direct link and for the RD relay link. Both of these expressions were given by Equation (13.22). In parallel, the unconditional BER corresponding to the scenario Φ_2 can be obtained as

$$P_{BER}^{\Phi_2} = \frac{1}{2^{2L}\pi} \int_{-\pi}^{\pi} f(a, b, L = 1, \theta) \int_{-\infty}^{\infty} e^{-\alpha(\theta)\gamma_{\Phi_2}^b} p_{\gamma_{\Phi_2}^b}(\gamma)\, d\gamma\, d\theta \tag{13.69}$$

$$= \frac{1}{2^{2L}\pi} \int_{-\pi}^{\pi} f(a, b, L = 1, \theta) \mathcal{M}_{\gamma_{\Phi_2}^b}(\theta)\, d\theta, \tag{13.70}$$

where the MGF of the received SNR per bit recorded at the BS for the scenario Φ_2 is written as

$$\mathcal{M}_{\gamma_{\Phi_2}^b}(\theta) = \int_{-\infty}^{\infty} e^{-\alpha(\theta)\gamma_{\Phi_2}^b} p_{\gamma_{\Phi_2}^b}(\gamma)\, d\gamma \tag{13.71}$$

$$= \int_{-\infty}^{\infty} e^{-\alpha(\theta)\gamma_{sd}^b} p_{\gamma_{sd}^b}(\gamma_{sd})\, d\gamma_{sd} \tag{13.72}$$

$$= \frac{N_0}{N_0 + \alpha(\theta)P_s\sigma_{sd}^2}. \tag{13.73}$$

Similarly, the BER upper bound can also be attained for cooperative systems relying on $M_r > 1$ cooperating users. For example, when $M_r = 2$, the average end-to-end BER upper bound $P_{BER,upper}^{DDF}$ becomes the sum of the average BERs of four scenarios expressed as

$$P_{BER,upper}^{DDF} = (1 - P_{PLR_1,upper})(1 - P_{PLR_2,upper})P_{BER}^{\Phi_1} + P_{PLR_1,upper}(1 - P_{PLR_2,upper})P_{BER}^{\Phi_2}$$
$$+ (1 - P_{PLR_1,upper})P_{PLR_2,upper}P_{BER}^{\Phi_3} + P_{PLR_1,upper}P_{PLR_2,upper}P_{BER}^{\Phi_4}, \tag{13.74}$$

where the four scenarios are defined as follows:

$$\Phi_1 = \{P_{r_1} \neq 0, P_{r_2} \neq 0\}, \tag{13.75}$$
$$\Phi_2 = \{P_{r_1} = 0, P_{r_2} \neq 0\}, \tag{13.76}$$
$$\Phi_3 = \{P_{r_1} \neq 0, P_{r_2} = 0\}, \tag{13.77}$$
$$\Phi_4 = \{P_{r_1} = 0, P_{r_2} = 0\}. \tag{13.78}$$

13.2.2.2 Simulation Results and Discussion

Under the assumption of a relatively rapidly Rayleigh fading channel associated with a normalized Doppler frequency of $f_d = 0.008$ and a packet length of $L_f = 16$ DQPSK-modulated symbols, the BER curves corresponding to DDF-aided cooperative systems with $M_r = 1$ and $M_r = 2$ cooperating MSs are plotted in comparison with the worst-case theoretical BERs of Equations (13.56) and (13.74) in Figure 13.6(a). Since the worst-case BER expression derived in Section 13.2.2.1 for the DDF-aided system does not take into account the negative impact of the time-selective channel, the resultant asymptotic line may not be capable of accurately approximating the true achievable BER performance of a DDF-aided system employing the CDD in the context of a rapidly fading environment. However, with the aid of the MSDSD of Section 12.3 using $N_{wind} > 2$, the performance loss induced by the relative mobility of the cooperating terminals and the BS can be significantly eliminated. Thus, as revealed by Figure 13.6(a), the worst-case BER bound closely captures the dependency of the system's BER on the P/N_0 ratio. On the other hand, the BER curves of DDF-aided cooperative systems employing the MSDSD using different packet lengths L_f are plotted together with the corresponding worst-case theoretical BER bound in Figure 13.6(b). Likewise, the theoretical BER bound based on Equation (13.56) closely captures the dependency of the MSDSD-aided system's BER on the packet length L_f employed in the scenario of a rapidly fading channel associated with a normalized Doppler frequency of $f_d = 0.008$.

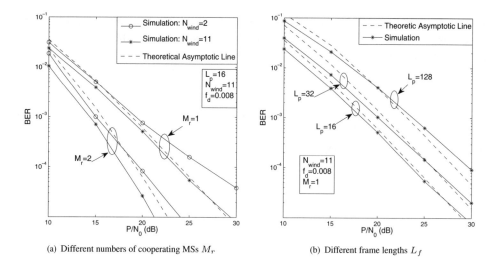

(a) Different numbers of cooperating MSs M_r

(b) Different frame lengths L_f

Figure 13.6: BER performance versus SNR for DDF-aided cooperative cellular systems, where there are M_r activated cooperating MSs, each having fixed transmit power and location. The MSDSD using $N_{wind} = 11$ is employed at the BS. All other system parameters are summarized in Table 13.2.

Table 13.2: Summary of system parameters.

System parameters	Choice
System	User-cooperative cellular uplink
Cooperative protocol	DDF
Number of relay nodes	M_r
Number of subcarriers	$D = 1024$
Modulation	DQPSK
Packet length	L_f
CRC	CCITT-4
Normalized Doppler freq.	$f_d = 0.008$
Path-loss exponent	Typical urban area, $v = 3$ [607]
Channel model	Typical urban, refer to Table 12.1
Relay location	$D_{sr_m} = D_{sd}/2, m = 1, 2, \ldots, M_r$
Power control	$P_s = P_{r_m} = P/(M_r + 1), m = 1, 2, \ldots, M_r$
Noise variance at MS and BS	N_0

13.3   CUS for the Uplink

User-cooperation-aided cellular systems are capable of achieving substantial diversity gains by forming VAAs constituted by the concerted action of distributed mobile users, while eliminating the space and cost limitations of the shirt-pocket-sized mobile phones. Hence, the cost of implementing user cooperation in cellular systems is significantly reduced, since there is no need specifically to set up additional RSs. On the other hand, it is challenging to realize user cooperation in a typical coherently detected cellular system, since $(N_t \times N_r)$ CIRs have to be estimated. For eliminating the implementationally complex channel estimation, in particular at the RSs, it is desirable to employ differentially detected modulation schemes in conjunction with the MSDSD scheme of Section 12.3. Furthermore, even if the Doppler-frequency-induced degradations are eliminated by employing the MSDSD, another major problem is how to choose the required number of cooperating users from the pool of P_{cand} available candidates, which may significantly affect the end-to-end performance of the cooperative system. These effects have been observed in our previous simulation results shown in Figure 12.14 in Section 12.3.2.3, where we indicated that the quality of the source–relay link quantified in terms of the SNR, which is dominated by the specific location of the cooperating users, plays a vital role in determining the achievable end-to-end performance of a cooperative system. Moreover, the employment of Adaptive Power Control (APC) among the cooperating users is also important in order to maximize the achievable transmission efficiency. Hence, we will commence our discourse on the above-mentioned two schemes, namely the CUS and the APC schemes, in the context of the cooperative uplink, which will be based on the end-to-end performance analysis carried out in Section 13.2. More specifically, we will propose a CUS scheme combined with APC for the DAF-aided cooperative system employing the MSDSD of Section 12.3 and its DDF-aided counterpart in Sections 13.3.1 and 13.3.2, respectively.

13.3.1 CUS Scheme for DAF Systems with APC

13.3.1.1 APC for DAF-Aided Systems [609]

As discussed in Section 13.1.2, for the sake of simplicity and analytical tractability, we assume that the source MS is sufficiently far away from the BS and the available cooperating MSs can be considered to be moving along the direct Line-Of-Sight (LOS) path between them, as specified by Equation (13.1) of Section 13.1.2. Explicitly, Equation (13.1) can be rewritten by normalizing D_{sd} to 1, as follows:

$$D_{sr_u} + D_{r_u d} = D_{sd} = 1, \quad u = 1, 2, \ldots, P_{cand}, \tag{13.79}$$

where P_{cand} is the RS pool size. This simplified model is readily generalized to a more realistic geography by taking into account the angle between the direct link and the relaying links. Furthermore, given a path-loss exponent of v, the average power $\sigma_{i,j}^2$ of the channel fading coefficient can be computed according to Equation (13.3), which is repeated here for convenience:

$$\sigma_{i,j}^2 = D_{i,j}^{-v}, \quad i, j \in \{s, r_u, d\}. \tag{13.80}$$

Then, by defining

$$d_m \triangleq \frac{D_{sr_u}}{D_{sd}}, \tag{13.81}$$

we can represent $\sigma_{sr_u}^2$ and $\sigma_{r_u d}^2$ respectively as

$$\sigma_{sr_u}^2 = \sigma_{sd}^2 \cdot d_m^v = d_m^v, \tag{13.82}$$

$$\sigma_{r_u d}^2 = \sigma_{sd}^2 \cdot (1 - d_m)^v = (1 - d_m)^v. \tag{13.83}$$

It was found in Section 13.2.1.2 that the simpler high-SNR-based BER lower-bound expression of Equation (13.48) associated with $N_n = 0$ in Equation (13.45) is tight over a wide range of SNRs of interest, e.g. for SNRs in excess of 15 dB for both the uncoded DBPSK- and DQPSK-modulated DAF-aided cooperative systems, as observed in Figure 13.5. Therefore, a power control scheme taking into account the location of the selected cooperating mobile users can be formulated, in order to minimize the BER of Equation (13.48) under the total transmit power constraint of Equation (13.4), i.e. when we have $P = P_s + \sum_{m=1}^{M_r} P_{r_m}$.[2] Thus, we arrive at

$$
\begin{aligned}
&[\hat{P}_s, \{\hat{P}_{r_m}\}_{m=1}^{M_r} \mid \{d_m\}_{m=1}^{M_r}] \\[2mm]
&= \operatorname*{arg\,min}_{\check{P}_s, \{\check{P}_{r_m}\}_{m=1}^{M_r}} \left\{ \frac{F(a, b, M_r + 1) N_0^{M_r+1}}{\check{P}_s \sigma_{sd}^2} \prod_{m=1}^{M_r} \frac{\check{P}_{r_m} \sigma_{r_m,d}^2 + \check{P}_s \sigma_{sr_m}^2 Z_{r_m,min}}{\check{P}_s \check{P}_{r_m} \sigma_{sr_m}^2 \sigma_{r_m d}^2} \right\}
\end{aligned} \tag{13.84}
$$

$$
= \operatorname*{arg\,min}_{\check{P}_s, \{\check{P}_{r_m}\}_{m=1}^{M_r}} \left\{ \frac{F(a, b, M_r + 1) N_0^{M_r+1}}{\check{P}_s \sigma_{sd}^2} \prod_{m=1}^{M_r} \frac{\check{P}_{r_m} \sigma_{sd}^2 (1 - d_m)^v + \check{P}_s \sigma_{sd}^2 d_m^v Z_{r_m,min}}{\check{P}_s \check{P}_{r_m} \sigma_{sd}^4 d_m^v (1 - d_m)^v} \right\} \tag{13.85}
$$

$$
= \operatorname*{arg\,min}_{\check{P}_s, \{\check{P}_{r_m}\}_{m=1}^{M_r}} \left\{ \frac{1}{\check{P}_s^{M_r+1}} \prod_{m=1}^{M_r} \frac{\check{P}_{r_m} (1 - d_m)^v + \check{P}_s d_m^v \tilde{Z}_{r_m,min}}{\check{P}_{r_m}} \right\}, \tag{13.86}
$$

which is subjected to the power constraint of $P = P_s + \sum_{m=1}^{M_r} P_{r_m}$ and $P_{r_m} > 0$ $(m = 1, 2, \ldots, M_r)$. The variable $\tilde{Z}_{r_m,min}$ in Equation (13.86) is defined as

$$
\tilde{Z}_{r_m,min} \triangleq -e^{\tilde{R}_{c_m}} (\zeta + \ln \tilde{R}_{c_m}), \tag{13.87}
$$

where we have

$$
\tilde{R}_{c_m} \triangleq \frac{R_{r_m,max}}{(1 - d_m)^v} \tag{13.88}
$$

$$
= \frac{2N_0}{(1 - d_m)^v b^2 (1 - \beta)^2 P_s c_m}. \tag{13.89}
$$

In order to find the solution of the minimization problem formulated in Equation (13.86) with the aid of the Lagrangian method, we first define the function $f(P_s, c_m)$ by taking the logarithm of the right-hand side of Equation (13.86) as

$$
f(P_s, c_m) \triangleq \ln \left(\frac{1}{P_s^{M_r+1}} \prod_{m=1}^{M_r} \frac{c_m (1 - d_m)^v + d_m^v \tilde{Z}_{r_m,min}}{c_m} \right) \tag{13.90}
$$

$$
= -(M_r + 1) \ln P_s - \sum_{m=1}^{M_r} \ln c_m + \sum_{m=1}^{M_r} \ln(c_m (1 - d_m)^v - d_m^v \tilde{Z}_{r_m,min}), \tag{13.91}
$$

where

$$
c_m \triangleq \frac{P_{r_m}}{P_s}. \tag{13.92}
$$

Furthermore, we define the function $g(P_s, c_m)$ based on the transmit power constraint of Equation (13.4) as follows:

$$
g(P_s, c_m) \triangleq \mathbf{c}^T \mathbf{1} - \frac{P}{P_s}, \tag{13.93}
$$

where

$$
\mathbf{c} \triangleq [1, c_1, \ldots, c_{M_r}]^T, \tag{13.94}
$$

---

[2]In this context we note that here we effectively assume that perfect power control is used when a specific mobile is transmitting its own data as well when it is acting as an RS. Naturally, the associated transmit power may be rather different in these two modes.

and **1** represents an $(M_r \times 1)$-element column vector containing all ones. Then, the Lagrangian function Λ can be defined as

$$\Lambda(P_s, c_m, \lambda) \triangleq f(P_s, c_m) + \lambda g(P_s, c_m) \tag{13.95}$$

$$= -(M_r + 1)\ln P_s - \sum_{m=1}^{M_r} \ln c_m + \sum_{m=1}^{M_r} \ln(c_m(1 - d_m)^v - d_m^v \tilde{Z}_{r_m, min})$$

$$+ \lambda\left(\mathbf{c}^T \mathbf{1} - \frac{P}{P_s}\right), \tag{13.96}$$

where λ is the Lagrangian multiplier. Hence, the first-order conditions for the optimum solution can be found by setting the partial derivatives of Equation (13.96) with respect to both P_s and c_m to zero:

$$\frac{\partial \Lambda(P_s, c_m, \lambda)}{\partial P_s} = -\frac{M_r + 1}{P_s} + \lambda \frac{P}{P_s^2} + \sum_{m=1}^{M_r} \frac{d_m^v \frac{e^{\tilde{R}_{cm}}}{P_s}[\tilde{R}_{cm}(\zeta + \ln \tilde{R}_{cm}) + 1]}{c_m(1 - d_m)^v - d_m^v e^{\tilde{R}_{cm}}(\zeta + \ln \tilde{R}_{cm})} = 0, \tag{13.97}$$

$$\frac{\partial \Lambda(P_s, c_m, \lambda)}{\partial c_m} = \lambda - \frac{1}{c_m} + \frac{(1 - d_m)^v + d_m^v\left[\frac{\tilde{R}_{cm} e^{\tilde{R}_{cm}}}{c_m}(\zeta + \ln \tilde{R}_{cm}) + \frac{e^{\tilde{R}_{cm}}}{c_m}\right]}{c_m(1 - d_m)^v - d_m^v e^{\tilde{R}_{cm}}(\zeta + \ln \tilde{R}_{cm})} = 0, \tag{13.98}$$

$$\frac{\partial \Lambda(P_s, c_m, \lambda)}{\partial \lambda} = \mathbf{c}^T \mathbf{1} - \frac{P}{P_s} = 0. \tag{13.99}$$

Consequently, by combining Equations (13.97) and (13.98), after a few further manipulations we obtain

$$\frac{(M_r + 1)P_s}{P} - \frac{1}{c_m} + \frac{(1 - d_m)^v + d_m^v\left[\frac{2N_0}{b^2(1-\beta^2)(1-d_m)^v P_s c_m^2} e^{\tilde{R}_{cm}(\zeta + \ln \tilde{R}_{cm}) + \frac{e^{\tilde{R}_{cm}}}{c_m}}\right]}{c_m(1 - d_m)^v - d_m^v e^{\tilde{R}_{cm}}(\zeta + \ln \tilde{R}_{cm})}$$

$$-\frac{1}{P}\sum_{m=1}^{M_r} \frac{\frac{2N_0 d_m^v e^{\tilde{R}_{cm}}}{b^2(1-\beta)^2(1-d_m)^v}\left(\zeta + \ln \tilde{R}_{cm} + \frac{1}{\tilde{R}_{cm}}\right)}{c_m[c_m(1 - d_m)^v - d_m^v e^{\tilde{R}_{cm}}(\zeta + \ln \tilde{R}_{cm})]} = 0. \tag{13.100}$$

Therefore, the optimum power control can be obtained by finding the specific values of c_m ($m = 1, 2, \ldots, M_r$) that satisfy both Equation (13.99) and (13.100), which involves an $L = (M_r + 1)$-dimensional search as specified in the summation of Equation (13.100) containing the power control of each of the M_r cooperating users. Hence, a potentially excessive computational complexity may be imposed by the search for the optimum power control solution. To reduce the search space significantly, the summation in the last term of Equation (13.100) may be removed, leading to

$$\frac{(M_r + 1)P_s}{P} - \frac{1}{c_m} + \frac{(1 - d_m)^v + d_m^v\left[\frac{2N_0}{b^2(1-\beta^2)(1-d_m)^v P_s c_m^2} e^{\tilde{R}_{cm}(\zeta + \ln \tilde{R}_{cm}) + \frac{e^{\tilde{R}_{cm}}}{c_m}}\right]}{c_m(1 - d_m)^v - d_m^v e^{\tilde{R}_{cm}}(\zeta + \ln \tilde{R}_{cm})}$$

$$-\frac{1}{P}\frac{\frac{2N_0 d_m^v e^{\tilde{R}_{cm}}}{b^2(1-\beta)^2(1-d_m)^v}\left(\zeta + \ln \tilde{R}_{cm} + \frac{1}{\tilde{R}_{cm}}\right)}{c_m[c_m(1 - d_m)^v - d_m^v e^{\tilde{R}_{cm}}(\zeta + \ln \tilde{R}_{cm})]} = 0, \tag{13.101}$$

so that the resultant Equation (13.101) depends only on the specific c_m value of interest. In other words, the original $(M_r + 1)$-dimensional search is reduced to a single-dimensional search, resulting in a substantially reduced power control complexity, while the resultant power control is close to that corresponding to Equation (13.100).

13.3.1.2 CUS Scheme for DAF-Aided Systems

Since the quality of the relay-related channels, namely the source–relay and the relay–destination links, dominates the achievable end-to-end performance of a DAF-aided cooperative system, the appropriate choice of cooperating users from the candidate pool of MSs roaming between the source MS and the BS as depicted in Figure 13.1 appears to be important in the scenario of cellular systems. In parallel with the APC scheme designed for the DAF-aided cooperative system discussed in Section 13.3.1.1, the CUS scheme is devised based on the minimization problem of Equation (13.84), which can be further simplified as

$$[\{\hat{d}_m\}_{m=1}^{M_r} \mid P_s, \{P_{r_m}\}_{m=1}^{M_r}] = \underset{\{\check{d}_m\}_{m=1}^{M_r}}{\arg\min} \left\{ \prod_{m=1}^{M_r} \frac{P_{r_m}\sigma_{sd}^2(1-\check{d}_m)^v + P_s\sigma_{sd}^2\check{d}_m^v Z_{r_m,min}}{\sigma_{sd}^4\check{d}_m^v(1-\check{d}_m)^v} \right\}$$

(13.102)

$$= \underset{\{\check{d}_m\}_{m=1}^{M_r}}{\arg\min} \left\{ \prod_{m=1}^{M_r} \frac{P_{r_m}(1-\check{d}_m)^v + P_s\check{d}_m^v \tilde{Z}_{r_m,min}}{\check{d}_m^v(1-\check{d}_m)^v} \right\},$$

(13.103)

which is subjected to the physical constraint of having a normalized relay location of $0 < d_m < 1$ $(m = 1, 2, \ldots, M_r)$ measured from the source.

Although Equation (13.103) can be directly solved numerically, it is difficult to get physically tangible insights from a numerical solution. To simplify further the minimization problem of Equation (13.103), we define the function $f(d_m)$ by taking the logarithm of the right-hand side of Equation (13.103), leading to

$$f(d_m) \triangleq \ln\left(\prod_{m=1}^{M_r} \frac{P_{r_m}(1-d_m)^v + P_s d_m^v \tilde{Z}_{r_m,min}}{d_m^v(1-d_m)^v} \right)$$

(13.104)

$$= -v \sum_{m=1}^{M_r} \ln(d_m(1-d_m)) + \sum_{m=1}^{M_r} \ln(P_{r_m}(1-d_m)^v + P_s d_m^v \tilde{Z}_{r_m,min}).$$

(13.105)

Then, by differentiating Equation (13.105) with respect to the normalized relay locations d_m $(m = 1, 2, \ldots, M_r)$ and equating the results to zero, we get

$$\frac{\partial f_{d_m}}{\partial d_m} = \frac{v(2d_m - 1)}{d_m(1 - d_m)}$$

$$+ \frac{-P_{r_m}v(1-d_m)^{v-1} + P_s v d_m^{v-1}\tilde{Z}_{r_m,min} + P_s d_m^v \frac{v(e^{\check{R}_{cm}} - \tilde{Z}_{r_m,min}\check{R}_{cm})}{1-d_m}}{P_{r_m}(1-d_m)^v + P_s d_m^v \tilde{Z}_{r_m,min}} = 0.$$

(13.106)

Hence, the optimum normalized relay distance of d_m for a specific power control can be obtained by finding the specific d_m values which satisfy Equation (13.106). Consequently, the original M_r-dimensional search of Equation (13.103) is broken down into M_r single-dimensional search processes.

Although the optimized location of the cooperating users can be calculated for a given power control, the resultant location may not be the global optimum in terms of the best achievable BER performance. In other words, to attain the globally optimum location and then activate the available cooperating candidates that happen to be closest to the optimum location, an iterative power versus RS location optimization process has to be performed. To be more specific, the resultant global optimization steps are as follows:

Step 1: Initialize the starting point $(\{c_m\}_{m=1}^{M_r}, \{d_m\}_{m=1}^{M_r})$ for the search in the $2M_r$-dimensional space, hosting the M_r powers and RS locations.

Step 2: Calculate the locally optimum location $\{d_{m,local}\}_{m=1}^{M_r}$ of the cooperating users for the current power control, $\{c_m\}_{m=1}^{M_r}$.

Step 3: If we have $\{d_{m,local}\}_{m=1}^{M_r} \neq \{d_m\}_{m=1}^{M_r}$, then let $\{d_m\}_{m=1}^{M_r} = \{d_{m,local}\}_{m=1}^{M_r}$. Otherwise, stop the search, since the globally optimum solution has been found: $\{d_{m,global}\}_{m=1}^{M_r} = \{d_{m,local}\}_{m=1}^{M_r}$ and $\{c_{m,global}\}_{m=1}^{M_r} = \{c_m\}_{m=1}^{M_r}$.

Step 4: Calculate the locally optimum power control $\{c_{m,local}\}_{m=1}^{M_r}$ of the cooperating RSs for the current location, $\{d_m\}_{m=1}^{M_r}$.

Step 5: If we have $\{c_{m,local}\}_{m=1}^{M_r} \neq \{c_m\}_{m=1}^{M_r}$, then let $\{c_m\}_{m=1}^{M_r} = \{c_{m,local}\}_{m=1}^{M_r}$ and continue to **Step 1**. Otherwise, stop the search, since the globally optimum solution has been found: $\{d_{m,global}\}_{m=1}^{M_r} = \{d_{m,local}\}_{m=1}^{M_r}$ and $\{c_{m,global}\}_{m=1}^{M_r} = \{c_m\}_{m=1}^{M_r}$.

Furthermore, it is worth emphasizing that the above optimization process requires an 'offline' operation. Hence, its complexity does not contribute to the complexity of the real-time CUS scheme. As mentioned previously in this section, since it is likely that no available cooperating MS candidate is situated in the exact optimum location found by the offline optimization, the proposed CUS scheme simply chooses the available MS that roams closest to the optimum location and then adaptively adjusts the power control. The rationale of the CUS scheme is based on the observation that the achievable BER is proportional to the distance between the cooperating MS and the optimum location, as will be seen in Section 13.3.1.3.

13.3.1.3 Simulation Results and Discussion

Both the APC and CUS schemes designed for the DAF-aided cooperative system, which were devised in Sections 13.3.1.1 and 13.3.1.2, respectively, are based on the high-SNR-related BER lower bound of Equation (13.48), which was shown to be a tight bound for a wide range of SNRs in Figure 13.5. In order to characterize further the proposed APC and CUS schemes and to gain insights into the impact of power control as well as that of the cooperating user's location on the end-to-end BER performance of the DAF-aided uplink supporting different number of cooperating users, the BER lower bounds are plotted versus P_s/P and d_m in Figures 13.7(a) and 13.7(b), respectively, in comparison with the exact BER of Equation (13.40) and with its upper bound of Equation (13.47). DQPSK modulation is assumed to be used here. Furthermore, in order to cope with the effects of the rapidly fluctuating fading channel, the MSDSD scheme of Section 12.3 is employed at the BS. For the sake of simplicity, we assume that an equal power is allocated to all activated cooperating MSs, which are also assumed to be located at the same distance from the source MS. All the other system parameters are summarized in Table 13.3. Observe from both Figures 13.7(a) and 13.7(b) that at a moderate SNR of 15 dB the lower bounds remain tight across the entire horizontal axes, i.e. regardless of the specific values of P_s/P and d_m. By contrast, the upper bound of Equation (13.47) fails to predict accurately the associated BER trends, especially when the number of activated cooperating MSs, M_r, is high. Therefore, despite using the much simpler optimization metrics of Equations (13.86) and (13.103), which are based on the high-SNR-related BER lower bound of Equation (13.48), the APC and CUS schemes of Sections 13.3.1.1 and 13.3.1.2 are expected to remain accurate for quite a wide range of SNRs.

Furthermore, both the power control strategy and the specific location of the cooperating MSs play a vital role in determining the achievable BER performance of the DAF-aided cooperative system. Specifically, as shown in Figure 13.7(a), under the assumption that all the activated cooperating users are located about half-way between the source MS and the BS, i.e. for $d_m = 0.5$ ($m = 1, 2, \ldots, M_r$), and for an equal power allocation among the cooperating users, i.e. for $P_{r_m} = (P - P_s)/M_r$ $m = 1, 2, \ldots, M_r$), the minimum of the BER curve is shifted to the left when an increased number of cooperating MSs participate in signal relaying. This indicates that the transmit power employed by the source MS should be decreased in order to attain the best achievable end-to-end BER performance. On the other hand, under the assumption of an equal power allocation among the source MS and all the cooperating MSs, i.e. where we have $P_s = P_{r_m} = P/M_r$ ($m = 1, 2, \ldots, M_r$), we observe from Figure 13.7(b) that the shape of the BER curves indicates a stronger sensitivity of the system's performance to the location of the cooperating users. This trend becomes even more dominant as the number of cooperating MSs, M_r, increases. However, in contrast to the phenomenon observed in Figure 13.7(a), the position of the BER minimum remains nearly unchanged, as observed in Figure 13.7(b), indicating that the optimum location of the cooperating users remains unaffected for this specific system arrangement, regardless of M_r.

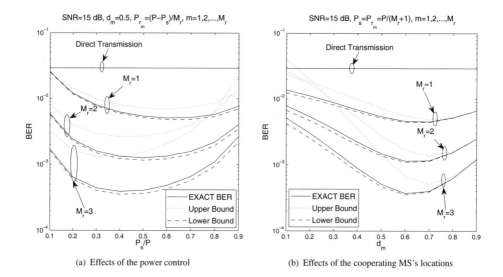

(a) Effects of the power control   (b) Effects of the cooperating MS's locations

Figure 13.7: Effects of the power control and of the cooperating MS's location on the BER performance of DQPSK-modulated DAF-aided cooperative cellular systems having M_r activated cooperating MSs. All other system parameters are summarized in Table 13.3.

Table 13.3: Summary of system parameters.

System parameters	Choice
System	User-cooperative cellular uplink
Cooperative protocol	DAF
Number of relay nodes	M_r
Number of subcarriers	$D = 1024$
Modulation	DQPSK
Detection	MSDSD ($N_{wind} = 11$)
Packet length	$L_f = 128$
Normalized Doppler freq.	$f_d = 0.008$
Path-loss exponent	Typical urban area, $v = 3$ [607]
Channel model	Typical urban, refer to Table 12.1
Noise variance at MS and BS	N_0

Importantly, the horizontal coordinate of the BER minimum represents the optimum MS location for the equal power allocation arrangement employed. Therefore, the achievable BER seen in Figure 13.7(b) is proportional to the distance between the RS and the optimum location, which provides the rationale for our distance-based CUS scheme.

In order to examine the tightness of the high-SNR-based BER lower bound of Equation (13.48) for the DAF-aided cooperative system at different transmit SNRs of P/N_0, the BER lower bounds corresponding to three distinct values of P/N_0 versus different power controls and relay locations are depicted in Figures 13.8(a) and 13.8(b), respectively. Let us assume that $M_r = 2$ cooperating MSs are activated. With an SNR as high as 20 dB, the lower bound is tight, as seen in both Figures 13.8(a) and 13.8(b). As the SNR decreases, the lower bound becomes increasingly loose, but remains capable of accurately predicting the BER trends and the best achievable performance in the vicinity of a moderate SNR level of 15 dB. However, when the SNR falls to as low a value as 10 dB, the lower bound remains

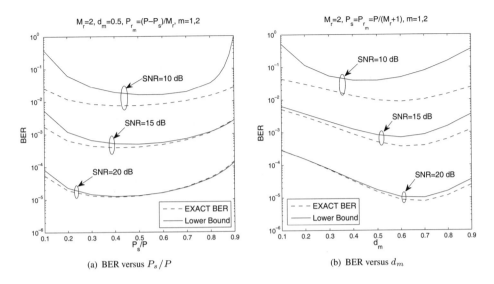

(a) BER versus P_s/P (b) BER versus d_m

Figure 13.8: Effects of the SNR on the tightness of the high-SNR-based BER lower bound for the DQPSK-modulated DAF-aided cooperative cellular uplink having two activated cooperating MSs. All other system parameters are summarized in Table 13.3.

no longer tight to approximate the exact BER, thus the APC and CUS schemes devised under the assumption of a high SNR may not hold the promise of an accurate solution. Nevertheless, since the low SNR range corresponding to high BER levels, such as for example 10^{-2}, is not within our range of interest, the proposed APC and CUS schemes of Sections 13.3.1.1 and 13.3.1.2 are expected to work appropriately for a wide range of SNRs.

Let us now continue by investigating the performance improvements achieved by the optimization of the power control and the cooperating user's location. In Figure 13.9(a) the BER performance of the DAF-aided cooperative system employing the APC scheme of Section 13.3.1.1 is depicted versus the cooperating user's location, d_m, in comparison with that of the system dispensing with the APC scheme. Again, we simply assume that multiple activated cooperating users are located at the same distance from the source user. Observe in Figure 13.9 that significant performance improvements can be achieved by the APC scheme when the cooperating user is situated closer to the BS than to the source MS. Hence the attainable BER is expected to be improved as the cooperating user moves increasingly closer to the BS. For example, the single cooperating user ($M_r = 1$) DAF-aided cooperative system using the APC scheme is capable of attaining its lowest possible BER at SNR $= 15$ dB, when we have $d_1 = D_{sr_1}/D_{sd} = 0.8$. Therefore, the performance improvement achieved by the APC scheme largely depends on the specific location of the cooperating users. Furthermore, the performance gains attained by the APC scheme for a specific arrangement of d_m is also dependent on the number of activated cooperating MSs, M_r. More specifically, when we have $M_r = 3$, a substantially larger gap is created between the BER curve of the system dispensing with the APC scheme and that of its APC-aided counterpart than that observed for $M_r = 1$, as seen in Figure 13.9(a).

At the same time, the BER performance of the DAF-aided system using relay location optimization is plotted in Figure 13.9(b) in comparison with that of the cooperative system, where the multiple activated cooperating users roam midway between the source MS and the BS. Similarly, a potentially substantial performance gain can be achieved by optimizing the location of the cooperating users, although, naturally, this gain depends on the specific power control regime employed as well as on the number of activated cooperating users. To be specific, observe in Figure 13.9(b) that it is desirable to assign the majority of the total transmit power to the source MS in favour of maximizing the achievable

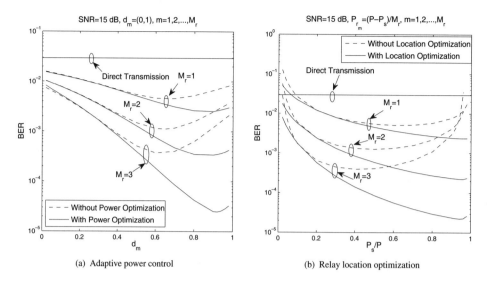

(a) Adaptive power control   (b) Relay location optimization

Figure 13.9: Power and relay location optimization for DQPSK-modulated DAF-aided cooperative cellular systems having M_r activated cooperating MSs. All other system parameters are summarized in Table 13.3. ©IEEE Wang & Hanzo 2007 [606]

performance gain by location optimization. Moreover, the more the cooperating users are activated, the higher the performance enhancement attained. Importantly, in the presence of a deficient power control regime, e.g. when less than 10% of the overall transmit power is assigned to the source MS, the DAF-aided system may suffer from a severe performance loss, regardless of the location of the cooperating users. This scenario results in an even worse performance than that of the non-cooperative system. Therefore, by observing Figures 13.9(a) and 13.9(b) we infer that for the DAF-aided cooperative uplink, it is beneficial to assign the majority of the total transmit power to the source MS and choose the specific cooperating users roaming in the vicinity of the BS in order to enhance the achievable end-to-end BER performance.

The above observations concerning the cooperative resource allocation of the DAF-aided system can also be inferred by depicting the three-dimensional BER surface versus both the power control and the cooperating MS's location in Figure 13.10(a) for a single-RS-aided cooperative system ($M_r = 1$). Indeed, the optimum solution is located in the area where both P_s/P and d_1 have high values. In order to reach the optimum operating point, the iterative optimization process discussed in Section 13.3.1.2 has to be invoked. The resultant optimization trajectory is depicted in Figure 13.10(b) together with the individual power-optimization- and location-optimization-based curves. The intersection point of the latter two lines represents the globally optimum joint power–location solution. As seen in Figure 13.10(b), by commencing the search from the centre of the two-dimensional power–location plane, the optimization process converges after four iterations between the power and location optimization phases, as the corresponding trajectory converges on the above-mentioned point of intersection.

Let us now consider a DAF-aided DQPSK-modulated cooperative cellular system employing both the CUS and APC schemes of Sections 13.3.1.1 and 13.3.1.2, where $M_r = 3$ cooperating MSs are activated in order to amplify and forward the signal received from the source MS to the BS, which are selected from $\mathcal{P}_{cand} = 9$ candidates roaming between the latter two. Without loss of generality, we simply assume that the locations of all the cooperating candidates are independent and uniformly distributed along the direct LOS link connecting the source MS and the BS, which are expected to change from time to time. Figure 13.11 depicts the performance of the DAF-aided cooperative

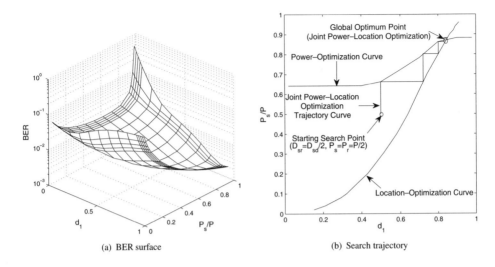

(a) BER surface   (b) Search trajectory

Figure 13.10: Optimum cooperative resource allocation for DQPSK-modulated DAF-aided cooperative cellular systems having a single activated cooperating MS at SNR = 15 dB. All other system parameters are summarized in Table 13.3.

system employing the CUS and APC schemes of Sections 13.3.1.1 and 13.3.1.2 in comparison with both that exhibited by its counterpart dispensing with the above-mentioned techniques and that of the direct-transmission-based system operating without user cooperation in Rayleigh fading channels associated with different normalized Doppler frequencies. Figure 13.11 demonstrates that the DAF-aided cooperative system is capable of achieving a significantly better performance than the non-cooperative system. Observe in Figure 13.11 that a further significant performance gain of 10 dB can be attained by invoking the CUS and APC schemes for a cooperative system employing the CDD of Section 12.1.1 ($N_{wind} = 2$), at a BER target of 10^{-5} and a normalized Doppler frequency of $f_d = 0.008$. Furthermore, employment of the CUS combined with the APC makes the cooperative cellular system more robust to the deleterious effects of time-selective channels. Indeed, observe in Figure 13.11 that an error floor is induced by a normalized Doppler frequency of $f_d = 0.03$ at a BER of 10^{-3} for the cooperative system dispensing with the CUS and APC arrangements, while the BER curve corresponding to the system carrying out cooperative resource allocation only starts to level out at a BER of 10^{-5}. For the sake of further eliminating the BER degradation caused by severely time-selective channels, the MSDSD employing $N_{wind} > 2$ can be utilized at the BS. As observed in Figure 13.11, for a target BER level of 10^{-5}, a P/N_0 degradation of about 7 dB was induced by increasing f_d from 0.008 to 0.03 for the CDD-aided system, while it was reduced to 1 dB by activating the MSDSD scheme of Section 12.3 using $N_{wind} = 11$.

Let us now consider the BER performance of DAF-aided cooperative systems dispensing with at least one of the two above-mentioned schemes, which is plotted in Figure 13.12(a). To be more specific, given a target BER of 10^{-5}, performance gains of 6 and 2.5 dB can be achieved respectively by employment of the CUS and APC over the benchmark system, where three cooperating users are randomly selected from the available nine RS candidates and the total transmit power is equally divided between the source and the relaying MSs. Hence, the distance-based CUS scheme of Section 13.3.1.2 performs well as a benefit of activating the RS candidates closest to the predetermined optimum locations, even in conjunction with a relatively small cooperating RS candidate pool, where it is more likely that none of the available RS candidates is situated in the optimum locations. In order further to enhance the achievable end-to-end performance, the APC is carried out based on the cooperating users' location as activated by the CUS and results in a performance gain as high as about 9.5 dB over

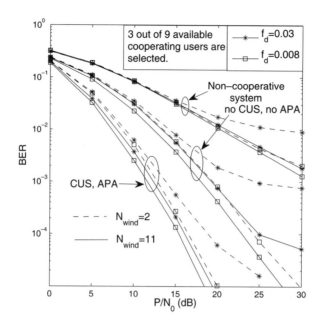

Figure 13.11: Performance improvements achieved by the CUS and APC schemes for a DAF-aided DQPSK-modulated user-cooperative cellular system employing the MSDSD of Section 12.3, where three out of nine cooperating user candidates are activated. All other system parameters are summarized in Table 13.3.

the benchmark system, as demonstrated in Figure 13.12(a). Moreover, besides providing performance gains, the CUS and APC schemes are also capable of achieving a significant complexity reduction in the context of the MSDSD employed by the BS, as seen in Figure 13.12(b), where the complexity imposed by the MSDSD using $N_{wind} = 11$ expressed in terms of the number of the PED evaluations versus P/N_0 is portrayed correspondingly to the four BER curves of Figure 13.12(a). Although the complexity imposed by the MSDSD in all of the four scenarios considered decreases steadily, the transmit SNR increases and then levels out at a certain SNR value around 20 dB. Observe in Figure 13.12(b) that a reduced complexity is imposed when either the CUS or the APC scheme is employed. Remarkably, the complexity imposed by the MSDSD at the BS can be reduced by a factor of about 10 for a wide range of transmit SNRs, when the CUS and APC are amalgamated. By carefully comparing the simulation results of Figures 13.12(a) and 13.12(b), it may be readily observed that the transmit SNR level, which guarantees the BER of 10^{-5}, is roughly the SNR level at which the complexity imposed by the MSDSD starts to level out. Therefore, it is inferred from the above observations that an appropriate cooperative resource allocation expressed in terms of the transmit power control and the appropriate cooperating user selection may significantly enhance the achievable end-to-end BER performance of the DAF-aided cooperative cellular uplink, while substantially reducing the computing power required by the MSDSD at the BS.

In a typical cellular system, the number of users roaming in a cell may also be referred to as the size of the cooperating user candidate pool denoted by $\mathcal{P}_{cand}$ in the scenario of the user-cooperative uplink. In order to investigate its impacts on the end-to-end BER performance of the DAF-aided cooperative system employing the CUS and APC schemes, the BER curves corresponding to different values of $\mathcal{P}_{cand}$ are plotted versus the transmit SNR, P/N_0, against that of the idealized scenario used as a benchmark, where the activated RSs are situated exactly in the optimum locations and have the optimum power control. Again, we assume that $M_r = 3$ RSs are activated, which are selected from the $\mathcal{P}_{cand}$ MSs roaming in the same cell. Interestingly, despite having a fixed number of activated

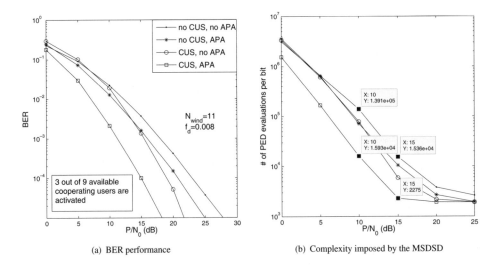

(a) BER performance   (b) Complexity imposed by the MSDSD

Figure 13.12: BER performance and the MSDSD complexity reductions achieved by the CUS and APC schemes for DAF-aided DQPSK-modulated user-cooperative cellular uplink, where three out of nine cooperating RS candidates are activated. All other system parameters are summarized in Table 13.3.

cooperating MSs, the end-to-end BER performance of the DAF-aided system steadily improves and approaches that of the idealized benchmark system upon increasing the value of $\mathcal{P}_{cand}$, as observed in Figure 13.13(a). On the other hand, it can be seen in Figure 13.13(b) that the higher the number of cooperating candidates, the lower the computational complexity imposed by the MSDSD at the BS. Specifically, by increasing the size of the candidate pool from $\mathcal{P}_{cand} = 3$ to 9, a performance gain of about 7 dB can be attained, while simultaneously achieving a detection complexity reduction factor of 6.5 at the target BER of 10^{-5}. In comparison with the idealized scenario, where an infinite number of cooperating candidates are assumed to be independently and uniformly distributed between the source MS and the BS, the DAF-aided cooperative system using both the CUS and APC schemes only suffers a negligible performance loss when $\mathcal{P}_{cand} = 9$ cooperating candidates. Therefore, the benefits brought about by the employment of the CUS and APC schemes may be deemed substantial in a typical cellular uplink.

13.3.2 CUS Scheme for DDF Systems with APC

In contrast to the process of obtaining the optimum power and location allocation arrangements discussed in Section 13.3.1 for DAF-aided cooperative systems, the first-order conditions obtained by differentiating the BER bound of a DDF-aided cooperative system formulated in Equations (13.56) and (13.74) for the $M_r = 1$ and $M_r = 2$ scenarios have complicated forms which are impervious to analytical solution. However, their numerical solution is feasible, instead of resorting to Monte Carlo simulations. Explicitly, by taking $M_r = 1$ as an example, the optimum power control can be obtained for a given RS location arrangement by minimizing the worst-case BER of Equation (13.56), yielding

$$[\hat{P}_s, \{\hat{P}_{r_m}\}_{m=1}^{M_r} \mid \{d_m\}_{m=1}^{M_r}]$$
$$= \underset{\check{P}_s, \{\check{P}_{r_m}\}_{m=1}^{M_r}}{\arg\min} \ \{(1 - P_{PLR_1,upper})P_{BER}^{\Phi_1} + P_{PLR_1,upper}P_{BER}^{\Phi_2}\}, \qquad (13.107)$$

where $P_{PLR_1,upper}$ is the worst-case packet loss ratio at the cooperating MS, which is given by Equation (13.52), while $P_{BER}^{\Phi_1}$ and $P_{BER}^{\Phi_2}$ are given by Equations (13.64) and (13.70), respectively,

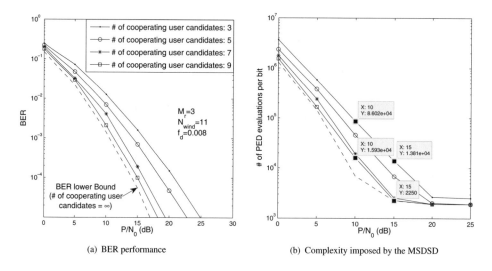

(a) BER performance (b) Complexity imposed by the MSDSD

Figure 13.13: The effects of the size of the cooperating RS pool on the DAF-aided DQPSK-modulated user-cooperative cellular uplink employing the CUS and APC schemes, where $M_r = 3$ cooperating users are activated. All other system parameters are summarized in Table 13.3.

corresponding to the average BER measured at the BS both with and without the signal forwarded by the RS. In parallel, the optimum location allocation can be obtained for a specific power control arrangement as

$$[\{\hat{d}_m\}_{m=1}^{M_r} \mid P_s, \{P_{r_m}\}_{m=1}^{M_r}]$$
$$= \arg\min_{\{\hat{d}_m\}_{m=1}^{M_r}} \left\{ (1 - P_{PLR_1,upper}) P_{BER}^{\Phi_1} + P_{PLR_1,upper} P_{BER}^{\Phi_2} \right\}. \tag{13.108}$$

Then, to attain the globally optimum location and activate the available cooperating candidates that happen to be closest to the optimum location, an iterative power versus RS location optimization process identical to that discussed in Section 13.3.1.2 in the context of an AF scheme has to be performed. Again, the rationale of the proposed CUS scheme for the DDF-aided system is based on the observation that the achievable BER is proportional to the distance between the cooperating MS and the optimum location, as will be demonstrated in Section 13.3.2.1.

13.3.2.1 Simulation Results and Discussion

The beneficial effects of cooperative resource allocation, in terms of the transmit power and the cooperating user's location on the achievable BER performance of the DDF-aided cooperative system, are investigated in Figure 13.14. Under the assumption that the channel fluctuates extremely slowly, e.g. for $f_d = 0.0001$, the worst-case BER performances corresponding to Equation (13.56) for $M_r = 1$ and to Equation (13.74) for $M_r = 2$, for the DQPSK-modulated DDF-aided cooperative systems employing either equal power allocation or the optimized power control, are plotted versus the different cooperating users' locations in Figure 13.14(a). The information bit stream is CCITT-4 coded by the source MS in order to carry out the CRC checking at the cooperating MS with the aid of a 32-bit CRC sequence. Hence, to maintain a relatively high effective throughput, two different transmission packet lengths are used, namely $L_f = 128$ and $L_f = 64$ DQPSK symbols. All other system parameters are summarized in Table 13.3. Observe in Figure 13.14(a) that the end-to-end BER performance can be substantially enhanced by employing the optimized power control, if the cooperating MS is not roaming in the neighbourhood of the source MS. Similar to the observation obtained for its DAF-aided counterparts

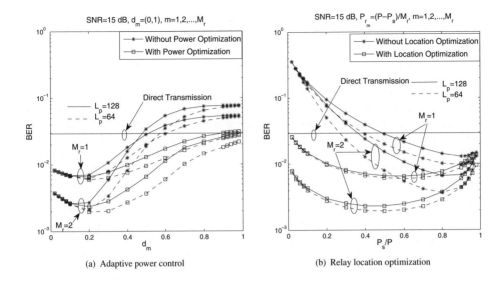

Figure 13.14: Power and relay location optimization for the DQPSK-modulated DDF-aided cooperative cellular systems having M_r activated cooperating MSs. All other system parameters are summarized in Table 13.4. ©IEEE Wang & Hanzo 2007 [606]

Table 13.4: Summary of system parameters.

System parameters	Choice
System	User-cooperative cellular uplink
Cooperative protocol	DDF
Number of relay nodes	M_r
Number of subcarriers	$D = 1024$
Modulation	DQPSK
CRC code	CCITT-4
Detection	MSDSD ($N_{wind} = 11$)
Packet length	L_f
Normalized Doppler freq.	f_d
Path-loss exponent	Typical urban area, $v = 3$ [607]
Channel model	Typical urban, refer to Table 12.1
Noise variance at MS and BS	N_0

characterized in Figure 13.9(a) of Section 13.3.1.3, the higher the number of active cooperating MSs, M_r, the more significant the performance gain attained by optimizing the power control for the DDF-aided system. However, due to the difference between the relaying mechanisms employed by the two above-mentioned cooperative systems, it is interesting to observe that the trends seen in Figure 13.14(a) are quite different from those emerging from Figure 13.9(a). Specifically, recall from the results depicted in Figure 13.9(a) that it is desirable to choose multiple cooperating users closer to the BS than to the source MS in a DAF-aided cooperative system, especially when employing the optimized power control for sharing the power among the cooperating users. By contrast, Figure 13.14(a) demonstrates that the cooperating MSs roaming in the vicinity of the source MS are preferred for a DDF-aided system in the interest of maintaining a better BER performance. Furthermore, the performance gap between the DAF-aided systems employing both the equal and optimized power allocations becomes wider as the cooperating MS moves closer to the optimum location corresponding to the horizontal coordinate of the lowest-BER point in Figure 13.9(a). By contrast, only a negligible performance improvement can be achieved by optimizing the power control, if the cooperating MS is close to the optimum location corresponding also to the horizontal coordinate of the lowest-BER point in Figure 13.14(a). In other words, the DDF-aided system suffers a relatively modest performance loss by employing the simple equal power allocation, if the multiple cooperating MSs are closer to their desired locations. Additionally, recall from Figure 13.9(a) recorded for the DAF-aided system that the worst-case BER performance was encountered by having no cooperating user closer to the optimum locations, regardless of whether the optimum power control is used or not, but the performance of this RS-aided DAF system was still slightly better than that of the conventional direct transmission system. By contrast, the DDF-aided system employing equal power allocation may unfortunately be outperformed by the direct-transmission-based non-cooperative system, if the cooperating MSs are located nearer to the BS than to the source MS. Finally, in contrast to the DAF-assisted system, the performance achieved by the DDF-aided system is dependent on the specific packet length, L_f, due to the potential relaying deactivation controlled by the CRC check carried out at the cooperating MS. To be specific, the shorter the packet length L_f, the lower the resultant BER.

In parallel, the BER performance of the above-mentioned DDF-aided systems is depicted versus P_s/P in Figure 13.14(b). Here, the transmit power of $(P - P_s)$ is assumed to be equally shared across multiple cooperating users. Again, similar to the results recorded for the DAF-aided system in Figure 13.9(b), a significant performance gain can be attained by locating the cooperating MS at the optimum position rather than in the middle of the source MS and BS path. This performance gain is expected to become even higher as the number of actively cooperating MSs, M_r, increases, as seen in Figure 13.14(b). By contrast, for optimum cooperating user location, instead of allocating the majority of the total transmit power to the source MS – as was suggested by Figure 13.9(b) for the DAF-aided system in the interest of achieving an improved BER performance – the results of Figure 13.14(b) suggest that only about half of the total power has to be assigned to the source MS, if the DDF scheme is used. Furthermore, the mild sensitivity of the BER performance observed in Figure 13.14(b) for the DDF-aided system benefiting from the optimum cooperating user location as far as the power control is concerned coincides with the trends seen in Figure 13.14(a), i.e. a desirable BER performance can still be achieved without optimizing the power control, provided that all the cooperating MSs roam in the vicinity of their optimum locations. Interestingly, in contrast to the conclusions inferred from Figure 13.14(a) for the DAF-aided system, the originally significant performance differences caused by the different packet lengths of $L_f = 128$ and $L_f = 64$ can be substantially reduced for the DDF-aided system, provided that the cooperating user is situated at the optimum location. Finally, as observed in Figure 13.14(b), when no active RS can be found in the vicinity of the optimum cooperating user locations, the DDF-aided system might be outperformed by its more simple direct transmission counterpart in the presence of deficient power control imposed by high power control errors.

Observe for the $M_r = 1$ scenario by merging Figures 13.14(a) and 13.14(b) that the globally optimum cooperative resource allocation characterized in terms of the transmit power control and RS selection regime can be visualized as the horizontal coordinates of the lowest point of the resultant 3D BER surface portrayed in Figure 13.15(a), where the 3D BER surface corresponding to different L_f values is plotted versus P_s/P and $d_1 = D_{sr_1}/D_{sd}$ for the DDF-aided cooperative system. The smaller the packet length L_f, the lower the BER. This is because the likelihood that the activated cooperating MS improves the signal relaying is inversely proportional to the packet length L_f. However, observe in Figure 13.15(a) that the gap between the different BER curves of 3D surface becomes relatively small in the vicinity of the globally optimum BER point, as predicted by Figures 13.14(a) and 13.14(b). On the other hand, similar to the results of Figure 13.10(b) recorded for the DAF-aided cooperative system, we plot the power-optimized curve versus d_1, while drawing the location-optimized curve versus P_s/P for the DDF-aided system associated with $M_r = 1$ in Figure 13.15(b), where the intersection of the two curves is the globally optimum solution corresponding to the projection of the lowest BER point onto the horizontal plane in Figure 13.15(a). The globally optimum solution can be found by the joint

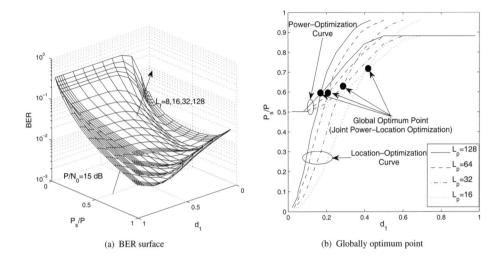

(a) BER surface   (b) Globally optimum point

Figure 13.15: Optimum cooperative resource allocation for the DQPSK-modulated DDF-aided cooperative cellular systems having a single activated cooperating MS at $SNR = 15$ dB. All other system parameters are summarized in Table 13.4.

power–location iterative optimization process discussed in Section 13.3.1.2. Furthermore, the globally optimum resource allocation, denoted by the black dot in Figure 13.15(b), changes as the packet length L_f varies. To be more specific, by increasing the packet length L_f, the optimum cooperating user location moves increasingly closer to the source MS, while the percentage of the total transmit power assigned to the source MS gradually decreases. This is not unexpected, since the probability of perfectly recovering all the symbols of the source MS by the cooperating MS is reduced on employing a higher packet length L_f, which has to be increased by choosing a cooperating MS closer to the source MS in the interest of increasing the received SNR at the cooperating MS.

Let us now continue by examining the BER performance improvement achieved by optimizing the resource allocation for the DDF-aided cooperative system in Figure 13.16, where the four subfigures depict the BER performance of the systems both with and without optimized cooperative resource allocation in terms of the transmit power and relay locations, while varying the packet length L_f. As seen in Figure 13.16, significant performance gains can be attained by using an optimum power control among the M_r cooperating users and the source user, as well as by assuming that all the M_r actively cooperating users are situated in their optimum locations, especially when we have a relatively large packet length L_f. Although a better PLR performance is attained when using short packets, the achievable performance gain is reduced, as indicated by the increasingly narrower gap between the BER curves obtained with and without the optimized resource allocation. Consider the $M_r = 2$ scenario as an example, where the originally achievable performance gain of 5 dB recorded for $L_f = 128$ is reduced to about 0.5 dB for $L_f = 16$ at a BER of 10^{-5}. In fact, this phenomenon coincides with the observation inferred from our previous simulation results, such as for example the 3D BER surface shown in Figure 13.15(a), which can be explained by the fact that the BER and PLR performance loss induced by a high packet length L_f may be significantly reduced by optimizing the cooperative resource allocation. Again for the scenario of $M_r = 2$, a performance loss of 5 dB is endured when employing $L_f = 64$ instead of $L_f = 16$ in the absence of resource allocation optimization, whereas the performance loss is reduced to 1.5 dB when the cooperative resource allocation is optimized. Furthermore, we also found that, interestingly, the asymptotic theoretical curves based on the worst-case BER expressions of Equation (13.56) and Equation (13.74) for $M_r = 1$ and $M_r = 2$, respectively, become tighter for the DDF-aided system using optimized resource allocation.

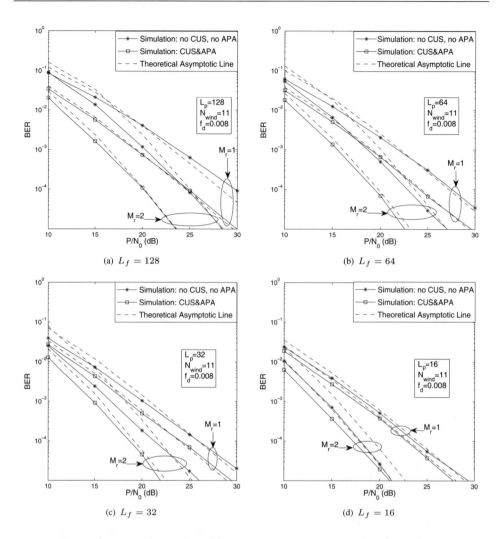

Figure 13.16: Performance improvement achieved by optimizing the cooperative resources for the DQPSK-modulated DDF-aided cooperative cellular systems employing the MSDSD in a relatively fast Rayleigh fading channel, where the M_r activated cooperating users are assumed to be situated at their optimum location. All other system parameters are summarized in Table 13.4.

Figure 13.17 separately investigates the impact of the CUS and that of the APC on the end-to-end BER performance of a DDF-aided cooperative system employing the MSDSD in a relatively rapidly Rayleigh fading channel associated with $f_d = 0.008$, where $N_{wind} = 8$ is employed to combat the performance degradation induced by the time-selective fading channel. Similar to the results of Figure 13.12(a) recorded for the DAF-aided system, a more significant performance improvement can be attained by invoking CUS than APC. However, in contrast to the DAF-aided system, the joint employment of the CUS and APC schemes for the DDF-aided system only leads to a negligible additional performance gain over the scenario where only the CUS is carried out. This is not unexpected, if we recall the observations inferred from Figure 13.14(a), i.e. the additional performance improvement achieved by optimizing the power control gradually erodes as the activated cooperating MS approaches the optimum location. Furthermore, unlike the CUS scheme, which simply selects the cooperating

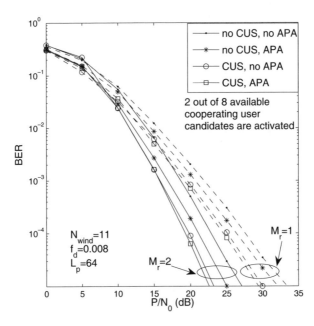

Figure 13.17: Performance improvements achieved by the CUS and APC schemes for the DDF-aided DQPSK-modulated user-cooperative cellular system employing the MSDSD in a relatively fast Rayleigh fading channel, where two out of eight cooperating users are activated. All other system parameters are summarized in Table 13.4.

MS that is closest to the optimum location calculated in an offline manner, the APC scheme, which conducts a real-time search for the optimum power control based on the actual location of the activated cooperating MS, may impose an excessive complexity. Hence, for reducing the complexity, the DDF-aided cooperative system may simply employ equal power allocation, while still being capable of achieving a desirable performance with the aid of the CUS scheme.

13.4 Joint CPS and CUS for the Differential Cooperative Cellular UL Using APC

From our discussions on the performance of the DAF- and DDF-aided cooperative cellular uplink in Sections 13.3.1.3 and 13.3.2.1, respectively, we may conclude that the above-mentioned two scenarios exhibit numerous distinct characteristics due to the employment of different relaying mechanisms. Therefore, the comparison of these two cooperative schemes will be further detailed in Section 13.4.1. Based on the initial comparison of the DAF and DDF schemes, a novel hybrid CPS scheme will be proposed in Section 13.4.2. In conjunction with the CUS and APC arrangements, we will then create a more flexible cooperative system, where the multiple cooperating MSs roaming in different areas might employ different relaying mechanisms to assist in forwarding the source MS's message to the BS to achieve the best possible BER performance. This system may be viewed as a sophisticated hybrid of a BS-aided ad hoc network or – alternatively – as an ad hoc network-assisted cellular network.

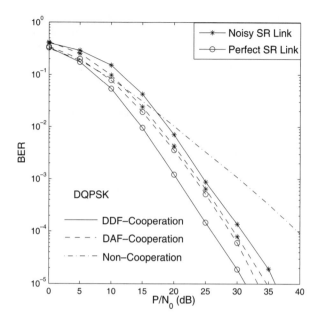

Figure 13.18: Impact of the source–relay link's quality on the end-to-end BER performance of a DQPSK-modulated cooperative system employing $M_r = 1$ cooperating RS roaming about half-way between the source MS and the BS. The CDD is employed by both the RS and the BS in a Rayleigh fading channel having a Doppler frequency of $f_d = 0.001$.

13.4.1   Comparison Between the DAF- and DDF-Aided Cooperative Cellular UL

13.4.1.1   Sensitivity to the Source–Relay Link Quality

The fundamental difference between the DAF and DDF schemes is whether decoding and re-encoding operations are required at the RS or not. Thus, generally speaking, the overall complexity imposed by the DDF-aided cooperative system is expected to be higher than that of its DAF-aided counterpart. However, as a benefit of preventing error propagation by the RS, the DDF-aided system is expected to outperform the DAF-aided one, provided that a sufficiently high source–relay link quality guarantees a near-error-free transmission between the source MS and the RS, as previously indicated by Figure 12.14 of Section 12.3.2.3. For convenience, we repeat these results here in Figure 13.18, where we observe that the sensitivity of the DDF-aided system to the source–relay link quality is significantly higher than that of the DAF-aided system. This is because the CRC employed may suggest to the RS to refrain from participating in forwarding the signal to the BS with a high probability, when the source–relay link is of low quality, which in turn leads to a rapid performance degradation. In practice, a high performance can be achieved for the DDF-aided system by activating the cooperating MSs roaming in the vicinity of the source MS and/or by invoking channel encoding.

13.4.1.2   Effect of the Packet Length

In contrast to the DAF-aided system, where the achievable performance is independent of the packet length L_f employed in the absence of the channel encoding, the DDF-aided system's performance is sensitive to the packet length L_f, as was previously demonstrated for example by Figure 13.16 of Section 13.3.2.1. This trend is not unexpected, since in the absence of the channel coding the PLR increases proportional to the value of L_f. This in turn may precipitate errors in the context of a DDF-

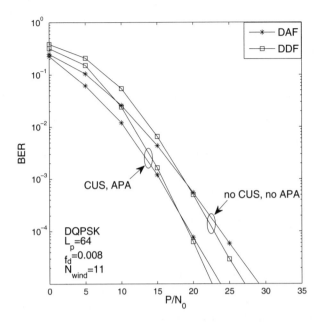

Figure 13.19: Performance comparison between the DAF- and DDF-aided DQPSK-modulated user-cooperative cellular systems employing the MSDSD, where two out of eight cooperating user candidates are activated. All other system parameters are summarized in Table 13.3.

aided system. However, this performance degradation can be substantially reduced by invoking the CUS of Section 13.3.2, as evidenced by Figure 13.14.

13.4.1.3 Cooperative Resource Allocation

As demonstrated by the simulation results of Sections 13.3.1.3 and 13.3.2.1, significant performance gains can be attained for both the DAF- and DDF-aided cellular uplink by optimizing the associated cooperative resource allocation with the aid of the CUS and APC schemes of Section 13.3. More explicitly, the BER performance of both the above-mentioned systems operating with and without the CUS and APC schemes is contrasted in Figure 13.19, where it is assumed that the $M_r = 2$ out of the $\mathcal{P}_{cand} = 8$ available cooperating MS candidates are activated and the MSDSD of Section 12.3 using $N_{wind} = 11$ is employed in order to eliminate the detrimental effects of the fading having a Doppler frequency of $f_d = 0.008$. Moreover, the variance of the noise added at each terminal of the cooperative system is assumed to be identical, namely N_0. Indeed, as seen in Figure 13.19, the performance of both the DAF and DDF systems is significantly enhanced by the employment of the CUS and APC schemes. We also note that the DAF-assisted system exhibits a better performance than the DDF-aided one, when the SNR of P/N_0 is relatively low, while the former is expected to be outperformed by the latter, as the SNR of P/N_0 is in excess of 20 dB. Again, this trend is not unexpected, since the sensitivity of the BER performance to the source–relay link's quality leads to a more rapid BER decrease upon increasing the SNR of P/N_0.

On the other hand, we also observed in Table 13.5 that, due to the distinct relaying mechanisms which lead to different levels of sensitivity to the quality of the source–relay link, the desirable cooperative resource allocation arrangement for the DAF-aided system may be quite different from that of its DDF-aided counterpart. As indicated by the RS's location arrangement of $[d_1, d_2, \ldots, d_{M_r}]$ seen in Table 13.5, the cooperating MSs roaming in the area near the BS are expected to be activated for the DAF-aided cooperative uplink, while those roaming in the neighbourhood of the source MS should be

Table 13.5: Cooperative resource allocation for DAF- and DDF-aided uplinks.

		DAF-aided uplink		DDF-aided uplink ($L_f = 64$)	
M_r	P/N_0 (dB)	$[P_s, P_{r_1}, \ldots, P_{r_{M_r}}]$	$[d_1, d_2 \ldots, d_{M_r}]$	$[P_s, P_{r_1}, \ldots, P_{r_{M_r}}]$	$[d_1, d_2, \ldots, d_{M_r}]$
1	10	$[0.882, 0.118]$	$[0.811]$	$[0.582, 0.418]$	$[0.192]$
	20	$[0.882, 0.118]$	$[0.871]$	$[0.622, 0.378]$	$[0.231]$
	30	$[0.882, 0.118]$	$[0.891]$	$[0.622, 0.378]$	$[0.231]$
2	10	$[0.76, 0.2, 0.04]$	$[0.74, 0.88]$	$[0.602, 0.202, 0.196]$	$[0.26, 0.26]$
	20	$[0.76, 0.2, 0.04]$	$[0.82, 0.91]$	$[0.602, 0.202, 0.196]$	$[0.31, 0.31]$
	30	$[0.78, 0.2, 0.02]$	$[0.85, 0.94]$	$[0.602, 0.202, 0.196]$	$[0.31, 0.31]$
3	10	$[0.88, 0.04, 0.04, 0.04]$	$[0.89, 0.89, 0.89]$	$[0.502, 0.102, 0.202, 0.194]$	$[0.31, 0.21, 0.26]$
	20	$[0.88, 0.04, 0.04, 0.04]$	$[0.92, 0.92, 0.92]$	$[0.502, 0.102, 0.202, 0.194]$	$[0.36, 0.26, 0.26]$
	30	$[0.88, 0.04, 0.04, 0.04]$	$[0.93, 0.93, 0.93]$	$[0.702, 0.102, 0.102, 0.094]$	$[0.41, 0.41, 0.41]$

selected for its DDF-aided counterpart in the interest of achieving the best possible BER performance. It is also indicated in Table 13.5 that the increase of the SNR, P/N_0, or the number of activated cooperating MSs, M_r, will move the desirable RS's location slightly further away from the source MS towards the BS for both the DAF- and DDF-aided scenarios. As for the optimized power control, the majority of the total transmit power P, i.e. about 88%, should be allocated to the source MS for the DAF-aided system, as revealed by the optimized power control arrangement of $[P_s, P_{r_1}, \ldots, P_{r_{M_r}}]$ seen in Table 13.5. By contrast, only about 60% of the power should be assigned to the source MS for the DDF-aided system. It is noteworthy that the optimized transmit power assigned to the M_r RSs as well as their optimum locations are not expected to be identical in both the DAF- and DDF-aided scenarios, as revealed in Table 13.5.

Furthermore, by comparing Figure 13.12(a) of Section 13.3.1.3 and Figure 13.17 of Section 13.3.2.1, we observe that a significant performance degradation may occur if the DAF-aided system dispenses with either the CUS or the APC scheme. By contrast, only a negligible performance loss is imposed when the DDF-aided system dispenses with the APC scheme rather than with the CUS scheme. Additionally, the CUS scheme of Section 13.3.2 is carried out by selecting the cooperating MSs roaming in the area closest to the optimum locations which may be determined offline, i.e. before initiating a voice call or data session. By contrast, the APC scheme of Section 13.3.2 may impose a relatively high real-time complexity, when calculating the optimum power control arrangement based on the current location of the activated RS. Hence, to minimize the complexity imposed by the cooperative resource allocation process, the DDF-aided system employing the CUS scheme may dispense with APC, simply opting for the equal power allocation arrangement at the expense of a moderate performance loss. In contrast to the DDF scheme, the DAF-aided system has to tolerate a high BER performance degradation if it dispenses with the APC scheme. It is also noteworthy that in contrast to the DAF-aided cooperative system, the DDF-assisted scheme employing neither the CUS nor the APC may be outperformed by the classic non-cooperative system, as observed in Figure 13.14, which is a consequence of its sensitivity to the quality of the source–relay link.

13.4.2　Joint CPS and CUS Scheme for the Cellular UL Using APC

Each cooperative cellular uplink considered so far in the book employed either DAF or DDF principles. As argued in the context of Figure 13.20, they both have their desirable RS area, when the CUS is employed. Generally speaking, the neighbourhood of the BS and that of the source MS are the specific areas where the RS should be activated for the DAF- and DDF-aided scenarios, respectively, again as discussed in Sections 13.3.1 and 13.3.2. Thus, often no available cooperating MS is roaming in the desirable RS location area, and hence a performance loss may be imposed by selecting a cooperating MS roaming far away from the optimum RS location. Furthermore, although the DDF-aided system exhibits a better performance than its DAF-aided counterpart in the presence of a high source–relay link quality, the former may be outperformed by the latter, as the quality of the source–relay link degrades despite imposing a higher overall system complexity. On the other hand, from our comparison of the DAF- and DDF-aided cooperative systems in Section 13.4.1, we realized that the two above-mentioned relaying mechanisms have complementary characteristics, reflected, for example, by their distinct optimum cooperative resource allocations. In light of the complementarity of the two relaying schemes, a more flexible cooperative scenario can be created, where either the DAF or DDF scheme is activated in the interest of enhancing the achievable performance of the cooperative system, while maintaining a moderate complexity. In contrast to the conventional cooperative system employing a single cooperative mechanism, the cooperating MSs roaming in different areas between the source MS and the BS may be activated and the relay schemes employed by each activated RS may be adaptively selected, to achieve the best possible performance.

For the sake of simplicity, let us now consider the hybrid cooperative cellular uplink employing the joint CPS and CUS scheme, as portrayed in Figure 13.20, where $M_r = 2$ cooperating MSs roaming in the preferred DDF- and the DAF-RS-area are activated, in order to forward the source MS's information to the BS. The particular cooperative protocol employed by the activated RSs is determined according to the specific area in which they happen to be situated. In order to make the most of the complementarity of the DAF and DDF schemes, it may be assumed that one of the cooperating MSs is activated in the preferred area of the DAF-RS, while the other is from the 'DDF-area', although, naturally, there may be more than one cooperating MS roaming within a specific desirable area. Finally, under the assumption that the first selected cooperating MS is roaming in the 'DDF-area', while the second one is roaming in the 'DAF-area', the MRC scheme employed by the BS, which combines the signals received from the source MS and the cooperating MSs, can be expressed as

$$y = a_0(y_{sd}[n-1])^* y_{sd}[n] + \sum_{m=1}^{2} a_m (y_{r_m d}[n + mL_f - 1])^* y_{r_m d}[n + mL_f], \qquad (13.109)$$

where L_f is the length of the transmission packet, while the coefficients a_0 and a_m $(m = 1, 2)$ are given by

$$a_0 = a_1 = \frac{1}{N_0} \qquad (13.110)$$

and

$$a_2 = \frac{P_s \sigma_{sr_2}^2 + N_0}{N_0 (P_s \sigma_{sr_2}^2 + P_{r_2} \sigma_{r_2 d}^2 + N_0)}. \qquad (13.111)$$

In order to determine the optimum RS areas for the hybrid cooperative system employing $M_r = 2$ cooperating users, the worst-case BER expression will first be derived in a similar manner to that derived for the DDF-aided system of Section 13.2.1 in our following discourse.

First of all, let us define the scenario Φ_1 as the situation when the cooperating MS employing the DDF scheme perfectly recovers the information from the source MS and then transmits the differentially remodulated signal to the BS, which is formulated as

$$\Phi_1 \triangleq \{P_{r_1} \neq 0\}. \qquad (13.112)$$

By contrast, the scenario Φ_2 is defined as the situation, when the cooperating MS using the DDF scheme fails to correctly decode the signal received from the source MS and keeps silent during the relay phase, which can be formulated as

$$\Phi_2 \triangleq \{P_{r_1} = 0\}. \qquad (13.113)$$

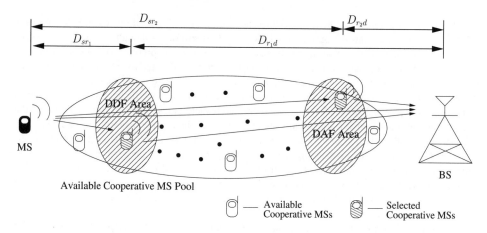

Figure 13.20: Cooperation-aided uplink systems using the joint CPS and CUS scheme.

Then, based on the differentially encoded conditional BER of Equation (13.59) invoked in Section 13.2.2, the unconditional BER observed at the BS is the scenario of Φ_1 can be expressed as

$$P_{BER}^{\Phi_1} = \frac{1}{2^{2L}\pi} \int_{-\pi}^{\pi} f(a, b, L = 3, \theta) \int_{-\infty}^{\infty} e^{-\alpha(\theta)\gamma_{\Phi_1}^b} p_{\gamma_{\Phi_1}^b}(\gamma)\, d\gamma\, d\theta \tag{13.114}$$

$$= \frac{1}{2^{2L}\pi} \int_{-\pi}^{\pi} f(a, b, L = 3, \theta) \mathcal{M}_{\gamma_{\Phi_1}^b}(\theta)\, d\theta, \tag{13.115}$$

where $\gamma_{\Phi_1}^b$ denotes the received SNR per bit after MRC combining, which can be written as

$$\gamma_{\Phi_1}^b = \gamma_{sd}^b + \gamma_{r_1 d}^b + \gamma_{r_2 d}^d. \tag{13.116}$$

Then, the joint MGF, $\mathcal{M}_{\gamma_{\Phi_1}^b}(\theta)$, of the received SNR per bit experienced at the BS in the scenario Φ_1 is expressed as

$$\mathcal{M}_{\gamma_{\Phi_1}^b}(\theta) = \int_{-\infty}^{\infty} e^{-\alpha(\theta)\gamma_{\Phi_1}^b} p_{\gamma_{\Phi_1}^b}(\gamma)\, d\gamma \tag{13.117}$$

$$= \int_{-\infty}^{\infty} \int_{-\infty}^{\infty} \int_{-\infty}^{\infty} e^{-\alpha(\theta)(\gamma_{sd}^b + \gamma_{r_1 d}^b + \gamma_{r_2 d}^b)}$$

$$\times p_{\gamma_{sd}^b}(\gamma_{sd}) p_{\gamma_{r_1 d}^b}(\gamma_{r_1 d}) p_{\gamma_{r_2 d}^b}(\gamma_{r_2 d})\, d\gamma_{sd}\, d\gamma_{r_1 d}\, d\gamma_{r_2 d} \tag{13.118}$$

$$= \mathcal{M}_{\gamma_{sd}^b}(\theta) \mathcal{M}_{\gamma_{r_1 d}^b}(\theta) \mathcal{M}_{\gamma_{r_2 d}^b}(\theta), \tag{13.119}$$

where

$$\mathcal{M}_{\gamma_{sd}^b}(\theta) = \frac{N_0}{N_0 + \alpha(\theta)P_s \sigma_{sd}^2}, \tag{13.120}$$

$$\mathcal{M}_{\gamma_{r_1 d}^b}(\theta) = \frac{N_0}{N_0 + \alpha(\theta)P_{r_1} \sigma_{r_1 d}^2}, \tag{13.121}$$

$$\mathcal{M}_{\gamma_{r_2 d}^b}(\theta) = \frac{1}{1 + k_{sr_2}(\theta)}\left(1 + \frac{k_{sr_2}(\theta)}{1 + k_{sr_2}(\theta)} \frac{P_s \sigma_{sr_2}^2 + N_0}{P_{r_2}} \frac{1}{\sigma_{r_2 d}^2} Z_{r_2}(\theta)\right), \tag{13.122}$$

and $k_{r_2 d}(\theta)$ and $Z_{r_2}(\theta)$ are given by Equations (13.34) and (13.35), respectively.

Table 13.6: Cooperative resource allocation for the hybrid cooperative uplink.

M_r	P/N_0 (dB)	$[P_s, P_{r_1}, \ldots, P_{r_{M_r}}]$	$[d_1, d_2, \ldots, d_{M_r}]$
2	10	$[0.702, 0.202, 0.096]$	$[0.26, 0.86]$
	20	$[0.702, 0.202, 0.096]$	$[0.31, 0.86]$
	30	$[0.702, 0.202, 0.096]$	$[0.31, 0.91]$

In parallel, the unconditional BER corresponding to the scenario Φ_2 can be formulated as

$$P_{BER}^{\Phi_2} = \frac{1}{2^{2L}\pi} \int_{-\pi}^{\pi} f(a, b, L = 2, \theta) \int_{-\infty}^{\infty} e^{-\alpha(\theta)\gamma_{\Phi_2}^b} p_{\gamma_{\Phi_2}^b}(\gamma)\, d\gamma\, d\theta \tag{13.123}$$

$$= \frac{1}{2^{2L}\pi} \int_{-\pi}^{\pi} f(a, b, L = 2, \theta) \mathcal{M}_{\gamma_{\Phi_2}^b}(\theta)\, d\theta, \tag{13.124}$$

where $\gamma_{\Phi_2}^b$ denotes the received SNR per bit after MRC combining, which can be expressed as

$$\gamma_{\Phi_2}^b = \gamma_{sd}^b + \gamma_{r_2 d}^d, \tag{13.125}$$

and hence the MGF of the received SNR per bit recorded at the BS for the scenario Φ_2 is written as

$$\mathcal{M}_{\gamma_{\Phi_2}^b}(\theta) = \int_{-\infty}^{\infty} e^{-\alpha(\theta)\gamma_{\Phi_2}^b} p_{\gamma_{\Phi_2}^b}(\gamma)\, d\gamma \tag{13.126}$$

$$= \int_{-\infty}^{\infty} \int_{-\infty}^{\infty} e^{-\alpha(\theta)(\gamma_{sd}^b + \gamma_{r_2 d}^b)} p_{\gamma_{sd}^b}(\gamma_{sd}) p_{\gamma_{r_2 d}^b}(\gamma_{r_2 d})\, d\gamma_{sd}\, d\gamma_{r_2 d}, \tag{13.127}$$

$$= \mathcal{M}_{\gamma_{sd}^b}(\theta)\mathcal{M}_{\gamma_{r_2 d}^b}(\theta), \tag{13.128}$$

where $\mathcal{M}_{\gamma_{sd}^b}(\theta)$ and $\mathcal{M}_{\gamma_{r_2 d}^b}(\theta)$ are given by Equations (13.120) and (13.122), respectively.

Finally, based on the worst-case packet loss ratio of $P_{PLR_1, upper}$ given by Equation (13.52), the average end-to-end BER upper bound, $P_{BER, upper}^{CPS}$, is obtained by the summation of the average BERs of two scenarios as

$$P_{BER, upper}^{CPS} = (1 - P_{PLR_1, upper})P_{BER}^{\Phi_1} + P_{PLR_1, upper}P_{BER}^{\Phi_2}. \tag{13.129}$$

Hence, when using the minimum BER criterion, the desirable RS area can be located by finding the globally optimum RS locations using the iterative power versus RS location optimization process of Sections 13.3.2 or 13.3.2. Considering the $M_r = 2$ scenario as an example, the globally optimum power and distance allocation arrangements are summarized in Table 13.6 under the assumption that the first cooperating MS is activated in the DDF mode. As expected, the figures shown in Table 13.6 reveal that the 'DDF-area' and the 'DAF-area' are still located in the vicinity of the source MS and the BS, respectively. Additionally, the majority of the total transmit power, i.e. about 70%, should be allocated to the source MS, while two-thirds of the remaining power should be assigned to the cooperating MS roaming in the 'DDF-area'.

The BER performance of the hybrid cooperative cellular uplink, where $M_r = 2$ out of $\mathcal{P}_{cand} = 8$ cooperating MSs are activated, is portrayed in comparison with that of its DAF- and DDF-aided counterparts in Figure 13.21. Remarkably, as demonstrated by Figure 13.21, the hybrid cooperative system outperforms both the DAF- and DDF-aided systems, regardless of whether the joint CPS–CUS–APC scheme is employed. These conclusions remain valid across a wide SNR range of our interest, although the performance advantage of the hybrid scheme over the latter two systems decreases in the context of the joint CPS–CUS–APC scheme. Furthermore, as the SNR increases, the DDF-aided system is expected to become superior to the other two systems, since the DDF-aided system

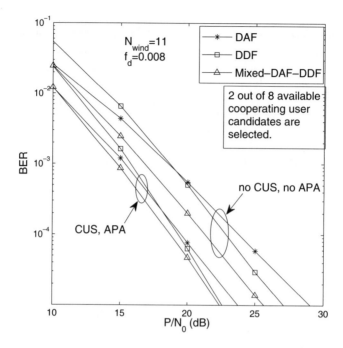

Figure 13.21: Performance improvement by the joint CPS and CUS for the DQPSK-modulated user-cooperative cellular uplink employing the MSDSD, where two out of eight cooperating user candidates are activated. All other system parameters are summarized in Table 13.3.

performs best when error-free transmissions can be assumed between the source MS and the RS. By contrast, if the SNR is low, the DAF-aided system performs best among the three. In addition to the performance advantage of the joint CPS–CUS–APC hybrid cooperative system, the overall system complexity becomes moderate in comparison with that of the DDF-aided system, since only half of the activated MSs have to decode and re-encode the received signal prior to forwarding it. Therefore, the proposed hybrid cooperative system employing the joint CPS–CUS–APC scheme is capable of achieving an attractive performance, despite maintaining a moderate overall system complexity.

13.5 Chapter Conclusions

In this chapter, CUS schemes and APC schemes designed for both the DAF- and DDF-aided cooperative systems were investigated based on our theoretical performance analysis. Significant performance gains can be achieved with the aid of the optimized resource allocation arrangements for both the DAF- and DDF-aided systems. Owing to the different levels of sensitivity to the quality of the source–relay link, the optimum resource allocation arrangements corresponding to the two above-mentioned systems were shown to be quite different. Specifically, it is desirable that the activated cooperating MSs are roaming in the vicinity of the source MS for the DDF-aided system, while the cooperating MSs roaming in the neighbourhood of the BS are preferred for the DAF-aided counterpart. In comparison with the former system, a larger portion of the total transmit power should be allocated to the source MS in the context of a DAF-aided system. Apart from achieving an enhanced BER performance, the complexity imposed by the MSDSD of Chapter 12 may also be significantly reduced by employing the CUS and APC schemes, even in the context of rapidly fading channels. Based on the simulation results throughout this chapter, the natures of the DAF- and DDF-aided systems are summarized and compared in Table 13.7.

Table 13.7: Comparison between the DAF- and DDF-aided cooperative cellular uplinks.

	DAF-aided uplink	DDF-aided uplink	References
Overall performance	Better when SR link quality is poor	Better when SR link quality is good	Figure 13.19
Overall complexity	Relatively low, no decoding at RSs	Relatively high, decoding and re-encoding at RSs	
Performance's sensitivity to source–relay link quality	Relatively moderate	Strong	Figures 12.19 12.23, 13.19
Performance's sensitivity to packet length L_f	Insensitive	Strong without CUS, minor with CUS	Figures 13.14, 13.16
Desirable RS locations	Near the BS	Near the source MS	Table 13.5
Desirable transmit power for the source MS	About 88% of the total power	About 60% of the total power	Table 13.5
Worst-case performance (Inappropriate resource allocation)	Slightly better than the non-cooperative system	Significantly worse than the non-cooperative system	Figures 13.9, 13.14
Importance of CUS and APC	Equally important	CUS is significantly more important	Figures 13.12(a), 13.17

Table 13.8: Summary of the resource-optimized cooperative systems investigated in Chapter 12.

		Performance gains achieved by various two-relay-aided differential cooperative systems with and without cooperative resource optimization			
Target BER	System type	Power control $[P_s, P_{r_1}, P_{r_2}]$	Relay selection $[d_{r_1}, d_{r_2}]$	SNR (dB)	Gain (dB)
10^{-3}	Direct transmission	N/A	N/A	27.3	—
	DAF-aided	[0.33, 0.33, 0.33]	[0.5,0.5]	18.8	8.5
	Cooperative System	[0.76, 0.2, 0.04]	[0.81, 0.9]	15.4	11.9
	DDF-aided	[0.33, 0.33, 0.33]	[0.5,0.5]	18.9	8.4
	Cooperative system	[0.602, 0.202, 0.196]	[0.29, 0.29]	15.8	11.5
	Hybrid DAF/DDF	[0.33, 0.33, 0.33]	[0.5,0.5]	16.9	10.4
	Cooperative system	[0.702, 0.202, 0.096]	[0.28, 0.86]	14.9	12.4
10^{-5}	Direct transmission	N/A	N/A	50	—
	DAF-aided	[0.33, 0.33, 0.33]	[0.5,0.5]	29	21
	Cooperative system	[0.76, 0.2, 0.04]	[0.82, 0.91]	23.7	26.3
	DDF-aided	[0.33, 0.33, 0.33]	[0.5,0.5]	27	23
	Cooperative system	[0.602, 0.202, 0.196]	[0.31, 0.31]	22.5	27.5
	Hybrid DAF/DDF	[0.33, 0.33, 0.33]	[0.5,0.5]	25.7	24.3
	Cooperative system	[0.702, 0.202, 0.096]	[0.31, 0.86]	22.3	27.7

Furthermore, in order to make the most of the complementarity of the two above-mentioned cooperative systems, a more flexible resource-optimized adaptive hybrid cooperation-aided system was proposed in this chapter, where the cooperative protocol employed by a specific cooperating MS may also be adaptively selected in the interest of achieving the best possible BER performance.

Finally, we quantitatively summarize and compare the performance gains achieved by the DAF-aided, the DDF-aided as well as the hybrid cooperative systems over the direct-transmission-based system in Table 13.8, based on the simulation results obtained throughout the chapter. Observe in Table 13.8 that, given a target BER of 10^{-3}, the DAF-aided cooperative system is capable of achieving a slightly higher performance gain than that attained by its DDF-aided counterpart, regardless of the employment of the optimized resource allocation. However, given a target BER of 10^{-5}, the latter becomes capable of achieving performance gains of 2 and 1.2 dB over the former for the non-optimized and optimized resource allocation arrangements, respectively, as seen in Table 13.8. Furthermore, among the three types of cooperative systems investigated in this chapter, the adaptive hybrid DAF/DDF cooperative system performs the best for a wide range of SNRs. Remarkably, as observed in Table 13.8, the hybrid cooperative system is capable of achieving performance gains over its direct-transmission-based counterpart, which are as high as 12.4 and 27.7 dB for the BER targets of 10^{-3} and 10^{-5}, respectively, when the optimized resource allocation is employed.

Chapter 14

The Near-Capacity Differentially Modulated Cooperative Cellular Uplink

14.1 Introduction

In point-to-point communication systems using a single antenna or co-located multiple antennas, it is feasible to achieve a high spectral efficiency by approaching Shannon's capacity limit with the aid of channel coding, as argued in Chapter 10. However, in contrast to the well-understood limitations of point-to-point single-user transmissions, researchers are only beginning to understand the fundamental performance limits of wireless multi-user networks, such as, for example, the cooperative cellular uplink considered in Chapters 12 and 13. To be more specific, in the scenarios of the uncoded DAF as well as DDF cooperative cellular uplinks, the best achievable BER performance can be approached by optimizing both the power control and the cooperating user selection, as discussed in Chapter 13. Naturally, the resultant cooperative system's performance is expected to be better than that of non-cooperative transmission. The attainable transmit diversity gains as well as path-loss reduction achieved by the cooperative relay-aided system were considered in Chapter 13, which translate into substantially enhanced robustness against fading for a given transmit power, or into a significantly reduced transmit power requirement for the same BER performance. However, the transmit diversity gains or cooperative diversity gains promised by the cooperative system considered are actually achieved at the cost of suffering a significant multiplexing loss compared with direct transmissions, which is imposed by the half-duplex communications of practical transceivers. More explicitly, realistic transceivers cannot transmit and receive simultaneously, because at a typical transmit power of, say, 0 dB and receiver sensitivity of -100 dBm the transmit power leakage imposed by the slightest power amplifier nonlinearity would leak into the receiver's Automatic Gain Control (AGC) circuit and would saturate it. Hence, the saturated AGC would become desensitized against low-power received signals. Furthermore, the cooperative diversity gains achieved by the relay-aided system over its direct-transmission-based counterpart may become modest in practical channel-coded scenarios, where the interleaving and channel coding gains dominate. Therefore, when a cooperative wireless communication system is designed to approach the maximum achievable spectral efficiency by taking the cooperation-induced multiplexing loss into account, it is not obvious whether or not the repetition-based relay-aided system becomes superior to its direct-transmission-based counterpart, especially when advanced channel coding techniques are employed. In other words, in the interest of achieving a high spectral efficiency, we have to answer the grave fundamental question: is it worth introducing cooperative mechanisms into the development of wireless networks?

14.1.1   System Architecture and Channel Model

14.1.1.1   System Model

Since the realistic condition of having an imperfect source–relay communication link is taken into account, the predominant DF as well as AF protocols employed may suffer from potential error propagation and noise enhancement, respectively, as observed in Chapters 12 and 13 where no channel coding was used. Fortunately, due to the advances of channel coding, well-designed channel-coded DF schemes are capable of guaranteeing near-error-free SR transmissions without noise enhancement, which in turn typically results in a superior performance in contrast to their AF-aided counterparts. In this context, only the differentially encoded and non-coherently detected DF-aided cooperative system dispensing with channel estimation will be investigated in the context of channel coding in this chapter. Naturally, in this scenario a more advanced channel-coded RS is required. Examples of channel-coded cooperative system designs may be found in [612, 613], although these contributions used coherent detection. To be consistent with the system model employed in Chapters 12 and 13, the differentially modulated TDMA cellular uplink is considered without any loss of generality, where no ICR estimation is required. For the sake of simplicity, we consider a single-relay-assisted scenario, where only one cooperating MS is activated in order to decode and re-encode the signal received from the source MS prior to forwarding the signal to the BS. Again, we simply assume the employment of a single antenna for each terminal, owing to the cost and size constraints of portable transceivers. Although we revealed in Chapter 13 that an optimized transmit power control and RS selection scheme may result in an enhanced end-to-end BER performance for the uncoded DDF-aided system, we simply assume here that the total transmit power is equally divided between the source MS and the single cooperating RS, which is assumed to be located half-way between the source MS and the BS, as depicted in Figure 14.1. This is because the emphasis in this chapter is on investigating the achievable network capacity of a general repetition-coded cooperative scenario and on techniques of approaching it. More specifically, for the sake of analytical tractability, we simply assume that the sum of the normalized distances D_{sr} between the source MS and the RS, and that between the RS and the destination BS, which is represented by D_{rd}, is equal to the normalized distance D_{sd} between the source MS and the BS. Naturally, the normalized SD distance is equal to unity. As a result, observe in Figure 14.1 that we have

$$D_{sr} = D_{rd} = \frac{1}{2}D_{sd} = 1. \tag{14.1}$$

Furthermore, as seen in Equation (13.3), the normalized average power $\sigma_{i,j}^2$ at the output of the channel is inversely proportional to the internode distance $D_{i,j}$, which is rewritten as follows:

$$\sigma_{i,j}^2 = D_{i,j}^{-v}, \quad i,j \in \{s,r,d\}, \tag{14.2}$$

where v denotes the path-loss exponent [607] and the subscripts s, r and d represent the source, relay and destination, respectively. Additionally, under the assumption of having a total transmit power of P and an equal power allocation among the source and cooperating MSs, we may express the associated power constraint as

$$P_s = P_r = \frac{1}{2}P, \tag{14.3}$$

where P_s and P_r are the transmit power of the source and cooperating MSs, respectively.

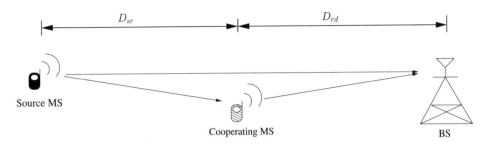

Figure 14.1: Single-relay-aided cooperative cellular uplink.

14.1.1.2 Channel Model

Throughout this chapter we assume that the complex-valued baseband signals undergo Rayleigh fading, which is modelled by multiplying the transmitted signal by a complex-valued Gaussian random variable. In order to provide a good approximation for TDMA-based cooperative systems and to facilitate the study of the non-coherent detection-based channel capacity, we consider a block-fading Rayleigh channel, where the fading coefficients are assumed to change in an i.i.d. manner from block to block. This assumption allows us to focus our attention on a single fading block [614, 615]. On the other hand, instead of employing the *standard block-fading channel*, where the fading coefficient remains constant over the duration of several consecutive symbol periods, we consider here a *time-selective block-fading model* [616], which includes the former as a special case. In the time-selective block-fading channel model considered, the channel's envelope exhibits correlation within a transmission block according to the Doppler frequency induced by the relative movement of the transceivers.

Consider a single-antenna-assisted point-to-point transmission scheme communicating over a block-fading channel which exhibits a correlated envelope for the duration of T_b consecutive symbols. Then, the received signal may be formulated as

$$\mathbf{y} = \mathbf{S}_d \mathbf{h} + \mathbf{w}, \tag{14.4}$$

where

$$\mathbf{y} = [y_1, y_2, \ldots, y_{T_b}]^T, \tag{14.5}$$

$$\mathbf{h} = [h_1, h_2, \ldots, h_{T_b}]^T \tag{14.6}$$

and

$$\mathbf{w} = [w_1, w_2, \ldots, w_{T_b}]^T \tag{14.7}$$

representing the received signal column vector, the fading coefficient column vector obeying a complex-valued Gaussian distribution $\mathcal{CN}(0, \sigma_h^2)$ and the Gaussian noise column vector having a distribution of $\mathcal{CN}(0, 2\sigma_w^2)$, respectively. The diagonal elements of the matrix $\mathbf{S}_d$ in Equation (14.4) may be expressed as

$$\mathbf{S}_d = \text{diag}\{\mathbf{s}\} = \begin{bmatrix} s_1 & 0 & \cdots & 0 \\ 0 & s_2 & \cdots & 0 \\ \vdots & \vdots & \ddots & \vdots \\ 0 & 0 & \cdots & s_{T_b} \end{bmatrix}, \tag{14.8}$$

where

$$\mathbf{s} = [s_1, s_2, \ldots, s_{T_b}]^T, \tag{14.9}$$

which represents the T_b consecutively transmitted signals within a fading block. Furthermore, in the cooperative communication scenario of Figure 14.1, the normalized channel fading variance σ_h^2 of each link was formulated in Equation (14.2) by taking the path loss into account. Given the assumption of

Rayleigh fading, **h** is a zero-mean complex-valued Gaussian vector with a $(T_b \times T_b)$-element covariance matrix Σ_h, which may be written as

$$\Sigma_h = \mathcal{E}\{\mathbf{h}\mathbf{h}^H\}$$

$$= \sigma_h^2 \cdot \begin{bmatrix} \varphi^t[0] & \varphi^t[1] & \cdots & \varphi^t[T_b - 1] \\ \varphi^t[-1] & \varphi^t[0] & \cdots & \varphi^t[T_b - 2] \\ \vdots & \vdots & \ddots & \vdots \\ \varphi^t[1 - T_b] & \varphi^t[2 - T_b] & \cdots & \varphi^t[0] \end{bmatrix}, \quad (14.10)$$

where $\varphi^t[\kappa]$ represents the channel's autocorrelation function, which can be expressed as

$$\varphi^t[\kappa] \triangleq \mathcal{E}\{h[n + \kappa]h^*[n]\} \quad (14.11)$$

$$= J_0(2\pi f_d \kappa), \quad (14.12)$$

with $J_0(\cdot)$ denoting the zeroth-order Bessel function of the first kind and, as usual, f_d represents the normalized Doppler frequency.

14.1.2   Chapter Contributions and Outline

Against the aforementioned background, the main objectives of this chapter are to investigate the necessity of introducing the cooperative mechanisms of Figure 14.1 into wireless networks, such as cellular voice and data networks. This design dilemma may be approached both from a pure capacity perspective and from the practical perspective of approaching the Discrete-input, Continuous-output Memoryless Channel (DCMC) capacity of the cooperative network. More specifically, the novel contributions of this chapter are as follows:

- From a pure capacity perspective, we answer the grave fundamental dilemma of whether it is worth incorporating cooperative mechanisms into wireless networks.

- A novel Irregular Distributed Hybrid Concatenated Differential (Ir-DHCD) coding scheme is proposed for the DDF cooperative system, in order to maximize the system's spectral efficiency.

- Based on our low-complexity near-capacity design criterion, we propose a practical framework of designing an Ir-DHCD-assisted cooperative system which is capable of performing close to the network's corresponding non-coherent DCMC capacity.

- In order to reduce further the complexity imposed, while approaching the cooperative network's DCMC capacity, the so-called adaptive-window-duration-based SISO iterative MSDSD scheme is proposed.

The remainder of this chapter is organized as follows. The fundamental performance limits of the non-coherent detection-aided direct-transmission-based system will be first studied in Section 14.2, followed by a review of the MAP-based SISO MSDSD in Section 14.3, which is capable of achieving a near-capacity performance at a low complexity. In order to answer the previously mentioned question related to the ultimate spectral efficiency of the repetition-based cooperative relay-aided system, the fundamental performance limits of the DDF-aided cooperative system will be investigated in Section 14.4.1 in comparison with those of its direct-transmission-based counterpart. Then, based on the novel Ir-DHCD coding scheme of Section 14.4.2 contrived for the DDF-aided cooperative system, a practical framework designed for approaching the DCMC capacity of the cooperative network will be proposed in Section 14.4.3, which is – naturally – different from that of point-to-point links. Hence, Section 14.4.4 will demonstrate that the cooperative scheme designed is capable of performing close to the corresponding network's DCMC capacity. Finally, our concluding remarks will be provided in Section 14.5.

14.2 Channel Capacity of Non-coherent Detectors

Since one of our goals in this chapter is to compare the maximum achievable spectral efficiency of the DDF-aided cooperative system and that of its differentially modulated direct-transmission-based counterpart as discussed in Section 14.1, the corresponding fundamental performance limits have to be investigated in the first place. Hence, in this section we first focus our attention on the non-coherent DCMC capacity of the classic single-antenna-assisted point-to-point communication scenario, based on which the non-coherent DCMC network capacity of the DDF-aided cooperative system will be studied in Section 14.4.1.

Recall the conditional PDF of the received signal vector from Equation (12.14), which was used for the derivation of the ML metric of the multiple-symbol differential detection (MSDD) scheme discussed in Section 12.2.1. The PDF of the received signal vector $\mathbf{y}$ in Equation (14.5) was conditioned on the transmitted signal vector $\mathbf{s}$ of Equation (14.9), which may be readily expressed as

$$p(\mathbf{y}|\mathbf{s}) = \frac{\exp(-\mathbf{y}^H \Psi^{-1} \mathbf{y})}{\det(\pi\Psi)}, \tag{14.13}$$

where

$$\Psi = \mathcal{E}\{\mathbf{y}\mathbf{y}^H|\mathbf{s}\} \tag{14.14}$$

$$= \mathbf{S}_d \Sigma_h \mathbf{S}_d^H + 2\sigma_w^2 \mathbf{I}_{T_b} \tag{14.15}$$

with $\mathbf{I}_{T_b}$ denoting the $(T_b \times T_b)$-element identity matrix.

Since differentially encoded modulation schemes, such as DQPSK, are assumed to be employed at MSs, and each element s_i of the transmitted signal vector $\mathbf{s}$ is chosen independently from a finite constellation set $\mathcal{M}_c$ with equal probabilities, the non-coherent DCMC capacity can be expressed as a function of the SNR as follows:

$$C(\text{SNR}) = H(\mathbf{y}) - H(\mathbf{y}|\mathbf{s}), \tag{14.16}$$

where H represents the differential entropy [617] of a random variable $\mathbf{x}$ defined as $H(\mathbf{x}) = -\int p(\mathbf{x}) \log_2 p(\mathbf{x}) \, d\mathbf{x}$, with $p(\cdot)$ denoting the corresponding PDF. According to [617], the differential entropy $H(\mathbf{y}|\mathbf{s})$ may be readily calculated as follows:

$$H(\mathbf{y}|\mathbf{s}) = -\int p(\mathbf{y}|\mathbf{s}) \ln p(\mathbf{y}|\mathbf{s}) \, d\mathbf{y} \tag{14.17}$$

$$= -\int p(\mathbf{y}|\mathbf{s})[-\mathbf{y}^H \Psi^{-1} \mathbf{y} - \ln \det(\pi\Psi)] \, d\mathbf{y} \tag{14.18}$$

$$= \mathcal{E}\left\{\sum_{i,j} \mathbf{y}_i^* (\Psi^{-1})_{i,j} \mathbf{y}_j\right\} + \ln \det(\pi\Psi) \tag{14.19}$$

$$= \mathcal{E}\left\{\sum_{i,j} \mathbf{y}_i^* \mathbf{y}_j (\Psi^{-1})_{i,j}\right\} + \ln \det(\pi\Psi) \tag{14.20}$$

$$= \sum_{i,j} \mathcal{E}\{\mathbf{y}_j \mathbf{y}_i^*\} (\Psi^{-1})_{i,j} + \ln \det(\pi\Psi) \tag{14.21}$$

$$= \sum_j \sum_i \Psi_{j,i} (\Psi^{-1})_{i,j} + \ln \det(\pi\Psi) \tag{14.22}$$

$$= \sum_j (\Psi\Psi^{-1})_{jj} + \ln \det(\pi\Psi) \tag{14.23}$$

$$= T_b + \ln \det(\pi\Psi) \tag{14.24}$$

$$= \ln \det(\pi e\Psi) \quad nats \tag{14.25}$$

$$= \log \det(\pi e\Psi) \quad bits. \tag{14.26}$$

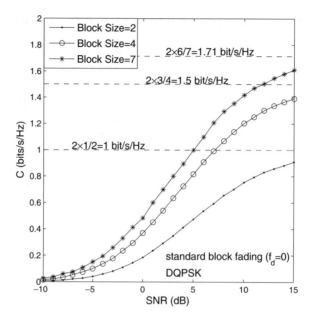

Figure 14.2: Non-coherent DCMC capacity of the SISO standard block-fading channel ($f_d = 0$) for different block sizes of $T_b = 2, 4$ and 7.

On the other hand, the entropy $H(\mathbf{y})$ of the continuous-valued faded and noise-contaminated received signal vector $\mathbf{y}$ cannot be evaluated in a closed form. When the fading block size T_b over which the fading envelope is assumed to be correlated is limited, a practical approach to the numerical evaluation of $H(\mathbf{y})$ is to carry out Monte Carlo integration as follows [618]:

$$H(\mathbf{y}) = - \int p(\mathbf{y}) \log p(\mathbf{y}) \, d\mathbf{y} \tag{14.27}$$

$$= -\mathcal{E}\left\{\log_2\left(\frac{1}{M_c^{T_b} \det(\pi \Psi)} \sum_{\check{s} \in \chi} \exp(-\mathbf{y}^H \Psi^{-1} \mathbf{y})\right)\right\}, \tag{14.28}$$

where χ is the set of all $M_c^{T_b}$ hypothetically transmitted symbol vectors š. The expectation value in Equation (14.28) is taken with respect to both different CIR realizations and to the noise.

The non-coherent DCMC capacity of the standard block-fading channel computed using the DQPSK modulation scheme is plotted in Figure 14.2 for various fading block sizes of $T_b = 2, 4$ and 7. As observed in Figure 14.2, all the three DCMC capacity curves associated with different values of T_b rise as the SNR increases and are expected to saturate when the SNR reaches a certain value, although this is not explicitly shown in the figure. Moreover, a larger fading block size T_b results in a higher DCMC capacity over a wide range of SNRs, while the capacity difference between the scenarios of $T_b = 2, 4$ and 7 becomes increasingly wider as the SNR increases. Nonetheless, the capacity difference above the saturation SNR remains constant. In other words, although an identical differential modulation scheme (DQPK) is employed, the maximum achievable spectral efficiency associated with a sufficiently high SNR is dependent on the fading block size T_b. This is not unexpected, since the differential signalling process commences by transmitting a reference symbol for each fading block, as argued earlier in Section 12.1.1.1, which does not contain any information. This reference symbol constitutes unexploited transmission overhead, i.e. redundancy, which thus imposes a diminishing capacity erosion as T_b is increased. Thus, given a sufficiently high SNR, the maximum achievable

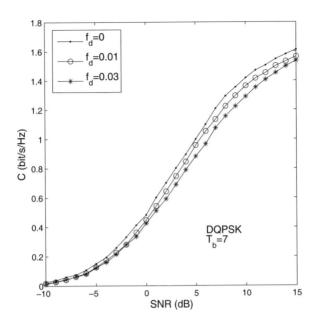

Figure 14.3: Non-coherent DCMC capacity of the SISO time-selective block-fading channel for various normalized Doppler frequencies ($f_d = 0, 0.01, 0.03$) and for $T_b = 7$. The fading envelope was correlated over each fading block, but changed in a random independent manner between fading blocks.

bandwidth efficiency η_{max}, which can be calculated as

$$\eta_{max} = \log_2 M_c \times (T_b - 1)/T_b \quad \text{bps/Hz}, \tag{14.29}$$

approaches that of the coherent-detection-aided transmission scheme, as the fading block size T_b increases towards infinity.

On the other hand, according to [616], the predictability of the channel is characterized by the rank Q of the channel's covariance matrix Σ_h formulated in Equation (14.10). For example, the block-fading channel, where the fading envelope remains constant over the entire fading block, is associated with the most predictable fading envelope when the channel's covariance matrix has a rank of $Q = 1$. By contrast, the fading process has a finite differential entropy and becomes less predictable when $Q = T_b$. Figure 14.3 compares the non-coherent DCMC capacity of the time-selective block-fading channel computed from Equation 14.16 for the DQPSK modulation scheme and for various normalized Doppler frequencies characterizing the correlation of the fading envelope exhibited over each fading block period. When we have an increased channel unpredictability due to increased Doppler frequency, a capacity loss is observed in Figure 14.3. In addition, it is also shown in [616] that for a fixed value of Q, the non-coherent capacity approaches the coherent capacity as T_b increases towards infinity. Hence, the observation of Figures 14.2 and 14.3 suggests that the non-coherent DCMC capacity of a time-selective block-fading channel is dependent on both the fading block size T_b and the fading correlation over blocks characterized by the corresponding covariance matrix Σ_h.

14.3   SISO MSDSD

For the sake of creating a near-capacity system design, the hard-input, hard-output multiple-symbol differential sphere detector (MSDSD) of Chapter 12 is invoked here. The MSDSD is capable of approaching the optimum ML performance in a channel-uncoded scenario at a significantly lower

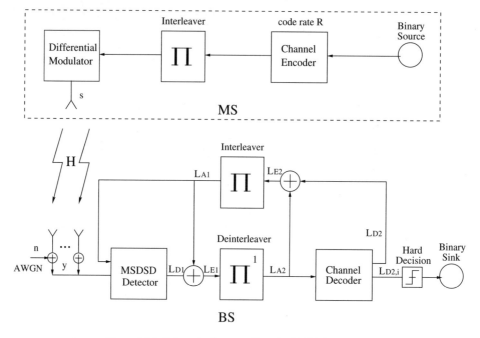

Figure 14.4: Schematic diagram of iterative MSDSD detection.

complexity than the brute-force full-search-based maximum likelihood MSDD (ML-MSDD). The MSDSD will be employed in the context of the well-known bit-interleaved coded modulation scheme using iterative detection (BICM-ID) [314], as portrayed in Figure 14.4. Hence, the MSDSD of Chapter 12 has to be modified in order to be able to process as well as to generate soft-bit information at its input and output, respectively, enabling the exchange of soft information between the outer channel decoder and itself.

14.3.1   Soft-Input Processing [598]

Recall from Section 12.2 that the principle of the ML-MSDD or the MSDSD is to maximize the a posteriori probability $Pr(\mathbf{s}|\mathbf{y})$, which can be expressed as

$$\hat{\mathbf{s}}_{ML} = \max_{\check{\mathbf{s}}\in\chi} Pr(\check{\mathbf{s}}|\mathbf{y}) \tag{14.30}$$

$$= \max_{\check{\mathbf{s}}\in\chi} \frac{p(\mathbf{y}|\check{\mathbf{s}})\,Pr(\check{\mathbf{s}})}{p(\mathbf{y})} \tag{14.31}$$

using Bayes' theorem [467, 542]. Owing to the equiprobable nature of the transmitted symbol vectors **s** and the independence of $p(\mathbf{y})$ from **s**, Equation (14.31) may be further simplified, yielding the ML metric as

$$\hat{\mathbf{s}}_{ML} = \max_{\check{\mathbf{s}}\in\chi} p(\mathbf{y}|\check{\mathbf{s}}). \tag{14.32}$$

When coupled with the outer channel decoder of Figure 14.4 in order to construct the iterative detection-aided receiver, the inner detector is provided with soft-bit information, i.e. with the LLRs, output by the outer decoder, which is regarded as a priori knowledge of the transmitted symbol vector **s**. For convenience, the above-mentioned a priori LLR $L_{A1}(x_k)$ of the kth bit of the bit vector **x**, which was defined in Equation (10.8) of Section 10.1.1.2, is rewritten here as

$$L_{A1}(x_k) = \ln \frac{Pr(x_k = +1)}{Pr(x_k = -1)}. \tag{14.33}$$

Thus, the transmitted symbol vectors $\mathbf{s} \in \chi$ can no longer be considered as equiprobable by the inner MSDD detector of Figure 14.4. Consequently, bearing in mind the aim of maximizing the a posteriori probability $Pr(\mathbf{s}|\mathbf{y})$ of the transmitted symbol vector $\mathbf{s}$, we refer to the proposed detector as the MAP-MSDD scheme, which can be implemented by incorporating the a priori soft information delivered by the channel decoder of Figure 14.4, based on the MAP metric, expressed as follows:

$$\hat{\mathbf{s}}_{MAP} = \max_{\check{\mathbf{s}} \in \chi} Pr(\check{\mathbf{s}}|\mathbf{y}) \tag{14.34}$$

$$= \max_{\check{\mathbf{s}} \in \chi} p(\mathbf{y}|\check{\mathbf{s}}) Pr(\check{\mathbf{s}}). \tag{14.35}$$

The a priori probability $Pr(\mathbf{s})$ of the transmitted symbol vector may be readily computed from the a priori LLRs of Equation (14.33) by taking into account the binary-to-M_c-ary bit-to-symbol mapping scheme characterized as $\mathbf{s} = \text{map}(\mathbf{x})$, under the assumption that, due to interleaving, the coded symbols may indeed be considered to be independent.

Furthermore, the MAP metric of Equation (14.35) may be reformulated by taking the logarithm of the right-hand side as

$$\hat{\mathbf{s}}_{MAP} = \max_{\check{\mathbf{s}} \in \chi} \{\ln(p(\mathbf{y}|\check{\mathbf{s}}) Pr(\check{\mathbf{s}}))\} \tag{14.36}$$

$$= \max_{\check{\mathbf{s}} \in \chi} \{\ln(p(\mathbf{y}|\check{\mathbf{s}})) + \ln(Pr(\check{\mathbf{s}}))\}. \tag{14.37}$$

Then, based on the conditional PDF $p(\mathbf{y}|\mathbf{s})$ of Equation (14.13), the transformation of the MAP metric of Equation (14.35) to the so-called *shortest-vector* problem can be completed by further reformatting Equation (14.37) as

$$\hat{\mathbf{s}}_{MAP} = \max_{\check{\mathbf{s}} \in \chi} \{-\mathbf{y}^H \Psi^{-1} \mathbf{y} + \ln(Pr(\check{\mathbf{s}}))\} \tag{14.38}$$

$$= \min_{\check{\mathbf{s}} \in \chi} \{\|\mathbf{U}\check{\mathbf{s}}\|^2 - \ln(Pr(\check{\mathbf{s}}))\}, \tag{14.39}$$

where $\mathbf{U}$ is an upper-triangular matrix defined in Equation (12.28) of Section 12.2.2. Additionally, according to the principles of differentially encoded modulation discussed in Section 12.1.1.1, a phase shift common to all components of the transmitted symbol vector $\mathbf{s}$ does not alter the MAP metric of Equation (14.39), which in turn yields the same data symbol vector $\mathbf{v}$. This is not unexpected, since the nth data symbol v_n of the symbol vector $\mathbf{v}$ to be transmitted is differentially encoded as the phase difference between the two consecutive transmitted symbols of s_{n-1} and s_n. Hence, for the sake of convenience, we may assume that the phase of the last element of the transmitted symbol vector $\mathbf{s}$ is fixed and set to zero. In the sequel, the other elements of the vector $\mathbf{s}$ may be obtained cumulatively as

$$s_n = \begin{cases} \prod_{m=n}^{N_{wind}-1} v_m^*, & 1 \leq n \leq N_{wind} - 1 \\ 1, & n = N_{wind}, \end{cases} \tag{14.40}$$

where N_{wind} is the observation window size employed by the MSDSD. Owing to the unique relations among the data bit vector $\mathbf{x}$, the data symbol vector $\mathbf{v}$ and the differentially encoded signalling symbol vector $\mathbf{s}$, they are treated interchangeably in our forthcoming discourse. Moreover, owing to the independence of the $(N_{wind} - 1)$ symbols v_n ($n = 1, 2, \ldots, N_{wind} - 1$) from each other, we have

$$\ln[Pr(\mathbf{s})] = \ln[Pr(\mathbf{x})] = \ln[Pr(\mathbf{v})] = \sum_{n=1}^{N_{wind}-1} \ln[Pr(v_n)]. \tag{14.41}$$

Then, by exploiting the upper-triangular structure of $\mathbf{U}$, Equation (14.39) can be rewritten as

$$\hat{\mathbf{v}}_{MAP} = \min_{\check{\mathbf{v}} \to \check{\mathbf{s}} \in \chi} \left\{ \| \mathbf{U}\check{\mathbf{s}} \|^2 - \ln(Pr(\check{\mathbf{v}})) \right\} \tag{14.42}$$

$$= \min_{\check{\mathbf{v}} \to \check{\mathbf{s}} \in \chi} \left\{ \sum_{n=1}^{N_{wind}} \left(\left| \sum_{m=n}^{N_{wind}} u_{n,m}\check{s}_m \right|^2 \right) - \sum_{n=1}^{N_{wind}-1} \ln(Pr(\check{v}_n)) \right\} \tag{14.43}$$

$$= \min_{\check{\mathbf{v}} \to \check{\mathbf{s}} \in \chi} \left\{ \sum_{n=1}^{N_{wind}-1} \left(\left| \sum_{m=n}^{N_{wind}} u_{n,m}\check{s}_m \right|^2 - \ln(Pr(\check{v}_n)) \right) + |u_{N_{wind},N_{wind}}|^2 \right\}, \tag{14.44}$$

where $u_{n,m}$ represents the specific element of the upper-triangular matrix $\mathbf{U}$ in row n and column m, while the arrow beneath the 'min' sign denotes the generation of $\check{\mathbf{s}}$ from a trial vector $\check{\mathbf{v}}$ using Equation (14.40). In order to solve efficiently the minimization problem of Equation (14.44), the MAP-MSDSD algorithm introduces a search radius R in order to reduce the search space, yielding

$$\hat{\mathbf{v}}_{MAP} = \min_{\check{\mathbf{v}} \to \check{\mathbf{s}} \in \chi} \left\{ \sum_{n=1}^{N_{wind}-1} \left(\left| \sum_{m=n}^{N_{wind}} u_{n,m}\check{s}_m \right|^2 - \ln(Pr(\check{v}_n)) \right) + |u_{N_{wind},N_{wind}}|^2 \leq R^2 \right\} \tag{14.45}$$

$$= \min_{\check{\mathbf{v}} \to \check{\mathbf{s}} \in \chi} \left\{ \sum_{n=1}^{N_{wind}-1} \left(\left| \sum_{m=n}^{N_{wind}} u_{n,m}\check{s}_m \right|^2 - \ln(Pr(\check{v}_n)) \right) \leq R^2 - |u_{N_{wind},N_{wind}}|^2 \right\} \tag{14.46}$$

$$= \min_{\check{\mathbf{v}} \to \check{\mathbf{s}} \in \chi} \left\{ \sum_{n=1}^{N_{wind}-1} \left(\left| \sum_{m=n}^{N_{wind}} u_{n,m}\check{s}_m \right|^2 - \ln(Pr(\check{v}_n)) \right) \leq \tilde{R}^2 \right\}, \tag{14.47}$$

where

$$\tilde{R}^2 \triangleq R^2 - |u_{N_{wind},N_{wind}}|^2. \tag{14.48}$$

Equivalently, the search space for each component symbol v_n, $n = 1, 2, \ldots, (N_{wind} - 1)$, is also confined in a manner described as

$$\mathcal{D}_n \triangleq \underbrace{\left| u_{n,n}\check{s}_{n+1}\check{\mathbf{v}}_n^* + \sum_{m=n+1}^{N_{wind}} \mathbf{U}_{nm}\check{s}_m \right|^2 - \ln(Pr(\check{\mathbf{v}}_n))}_{\triangleq \delta_n^2}$$
$$+ \underbrace{\sum_{l=n+1}^{N_{wind}-1} \left(\left| \sum_{m=l}^{N_{wind}} \mathbf{U}_{lm}\check{s}_m \right|^2 - \ln(Pr(\check{\mathbf{v}}_l)) \right) \leq \tilde{R}^2}_{\mathcal{D}_{n+1}}, \tag{14.49}$$

where $\mathcal{D}_n$ denotes the accumulated PED between the subvector candidate $[\check{v}_n, \check{v}_{n+1}, \ldots, \check{v}_{N_{wind}-1}]^T$ and the origin. Hence, the MAP-MSDSD scheme starts the search from the $(N_{wind} - 1)$th element of the symbol vector $\mathbf{v}$ by choosing a legitimate symbol candidate $\check{v}_{N_{wind}-1}$ from the constellation set of $\mathcal{M}_c$, which satisfies Equation (14.49), and then proceeds to search for the $(N_{wind} - 2)$th element, and so forth, until it reaches the $n = 1$ element. In other words, a trial vector $\check{\mathbf{v}} = [\check{v}_1, \check{v}_2, \ldots, \check{v}_{N_{wind}-1}]^T$ and the vector $\check{\mathbf{s}} = [\check{s}_1, \check{s}_2, \ldots, \check{s}_{N_{wind}}]$ generated using Equation (14.40) are found. Then, the search radius is updated to

$$R^2 = \mathcal{D}_1 = \| \mathbf{U}\check{\mathbf{s}} \|^2 - \ln[Pr(\check{\mathbf{v}})], \tag{14.50}$$

based on which the search is repeated by starting with the $n = 2$ component of the symbol vector $\mathbf{v}$. Therefore, the tree search employed by the MSDSD is carried out in a depth-first manner, which was the subject of a comprehensive discussion in Section 9.2.2. Finally, if the MSDSD of Figure 14.4 cannot find any legitimate symbol vector within the increasingly shrinking hypersphere-based search space, the previously obtained vector $\check{\mathbf{v}}$ is deemed to be the MAP solution of Equation (14.44).

14.3.2 Soft-Output Generation

Besides incorporating the a priori soft-bit information $L_{A1}(x_k)$ of Equation (14.33) delivered by the outer channel decoder of Figure 14.4, the iterative detection scheme requires the MAP-MSDSD to provide the a posteriori soft-bit information $L_{D1}(x_k)$ that will be used as a priori information by the decoder component, which can be calculated as follows:

$$L_{D1}(x_k) = \ln \frac{Pr(x_k = +1|\mathbf{y})}{Pr(x_k = -1|\mathbf{y})} \tag{14.51}$$

$$= \ln \frac{p(\mathbf{y}|x_k = +1)Pr(x_k = +1)/p(\mathbf{y})}{p(\mathbf{y}|x_k = -1)Pr(x_k = -1)/p(\mathbf{y})} \tag{14.52}$$

$$= \ln \frac{\sum_{\mathbf{x} \in \mathbb{X}_{k,+1}} p(\mathbf{y}|\mathbf{x})Pr(\mathbf{x})}{\sum_{\mathbf{x} \in \mathbb{X}_{k,-1}} p(\mathbf{y}|\mathbf{x})Pr(\mathbf{x})} \tag{14.53}$$

$$= \ln \frac{\sum_{\mathbf{x} \in \mathbb{X}_{k,+1}} \exp[-\mathbf{y}^H \Psi \mathbf{y} + \ln(Pr(\mathbf{x}))]}{\sum_{\mathbf{x} \in \mathbb{X}_{k,-1}} \exp[-\mathbf{y}^H \Psi \mathbf{y} + \ln(Pr(\mathbf{x}))]} \tag{14.54}$$

$$= \ln \frac{\sum_{\mathbf{x} \in \mathbb{X}_{k,+1}} \exp(-\|\mathbf{U}\mathbf{s}\|^2 + \ln[Pr(\mathbf{x})])}{\sum_{\mathbf{x} \in \mathbb{X}_{k,-1}} \exp(-\|\mathbf{U}\mathbf{s}\|^2 + \ln[Pr(\mathbf{x})])}, \tag{14.55}$$

where $\mathbb{X}_{k,+1}$ represents the set of $M_c^{N_{wind}}/2$ legitimate transmitted bit vectors $\mathbf{x}$ associated with $x_k = +1$, and, similarly, $\mathbb{X}_{k,-1}$ is defined as the set corresponding to $x_k = -1$.

To reduce the computational complexity imposed by the calculation of Equation (14.55), the *Jacobian logarithm* [553] may be employed to approximate the a posteriori LLRs, which can be expressed as

$$Jac[\ln(a_1, a_2)] = \ln(e^{a_1} + e^{a_2}) \tag{14.56}$$

$$= \max(a_1, a_2) + \ln(1 + e^{-|a_1 - a_2|}), \tag{14.57}$$

where the second term may be omitted in order further to approximate the original logarithmic function, since $\ln(1 + e^{-|a_1 - a_2|})$ can be regarded as a refinement or correction term of the coarse 'sum-max' approximation provided by the maximum, i.e. the first term of Equation (14.57). Explicitly, we have

$$max\_sum[\ln(a_1, a_2)] = \max(a_1, a_2). \tag{14.58}$$

Therefore, the so-called 'sum-max' approximation of the exact a posteriori LLR of Equation (14.55) can be reformulated with the aid of Equation (14.39) as

$$L_{D1}(x_k) = \ln \frac{\max_{\mathbf{x} \in \mathbb{X}_{k,+1}} \{\exp(-\|\mathbf{U}\mathbf{s}\|^2 + \ln[Pr(\mathbf{x})])\}}{\max_{\mathbf{x} \in \mathbb{X}_{k,-1}} \{\exp(-\|\mathbf{U}\mathbf{s}\|^2 + \ln[Pr(\mathbf{x})])\}} \tag{14.59}$$

$$= -\|\mathbf{U}\hat{\mathbf{s}}_{MAP}^{x_k=+1}\|^2 + \ln[Pr(\hat{\mathbf{x}}_{MAP}^{x_k=+1})]$$
$$+ \|\mathbf{U}\hat{\mathbf{s}}_{MAP}^{x_k=-1}\|^2 - \ln[Pr(\hat{\mathbf{x}}_{MAP}^{x_k=-1})], \tag{14.60}$$

where $\hat{\mathbf{s}}_{MAP}^{x_k=b}$ and $\hat{\mathbf{x}}_{MAP}^{x_k=b}$ represent the MAP-algorithm-based symbol vector estimation and MAP bit vector estimation, respectively, obtained by the MSDSD by fixing the kth bit value to b ($b = -1$ or $+1$). In the sequel, the extrinsic LLR, $L_{E1}(x_k)$, seen in Figure 14.4 can be obtained by excluding the corresponding a priori LLR, $L_{A1}(x_k)$, from the a posteriori LLR, $L_{D1}(x_k)$, as

$$L_{E1}(x_k) = L_{D1}(x_k) - L_{A1}(x_k), \tag{14.61}$$

which is exploited by the outer decoder after passing it through the interleaver.

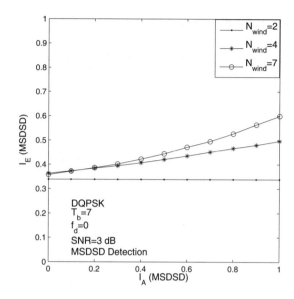

Figure 14.5: EXIT curves of the MSDSD for various observation window sizes N_{wind}. The Rayleigh fading envelope was constant for $T_b = 7$ symbols and was then randomly and independently faded for the next block.

14.3.3   Maximum Achievable Rate Versus the Capacity: An EXIT-Chart Perspective

In order to visualize the extrinsic information transfer characteristics of the iterative MSDSD scheme, we plot the EXIT curves associated with different observation window sizes of N_{wind} in Figure 14.5 by measuring the *extrinsic* mutual information, I_E, at the output of the MSDSD for a given input stream of bit LLRs, along with the a priori mutual information I_A at an SNR of 3 dB. A Rayleigh block-fading channel associated with a block length of $T_b = 7$ was assumed. Thus, the maximum value of N_{wind} that may be employed by the MSDSD is seven. As observed in Figure 14.5, the slope of the EXIT curve becomes increasingly steeper as the value of N_{wind} increases. More specifically, the EXIT curve associated with conventional differential detection (CDD) or $N_{wind} = 2$ is horizontal when Gray mapping is employed, indicating that no performance gains can be produced by the iterative detection mechanism. However, apart from having a higher starting point in the EXIT curve, a steeper slope is expected, when jointly and differentially detecting $(N_{wind} - 1) > 1$ data symbols using the MSDSD, leading to significantly increased iterative gains. In addition, according to the area properties of EXIT charts [579, 619], the area $\mathcal{A}$ under the bit-based EXIT curve of a soft detector/soft demapper is equal to the maximum possible code rate $R_{outer,max}$ of the outer channel code that can be employed to achieve near-error-free transmissions. Hence, the maximum achievable near-error-free transmission rate $R_{overall,max}$ of a differentially encoded system is computed as

$$R_{overall,max} = \left(\frac{T_b - 1}{T_b} \log_2 M_c \right) \cdot R_{outer,max}, \tag{14.62}$$

$$= \left(\frac{T_b - 1}{T_b} \log_2 M_c \right) \cdot \mathcal{A} \quad \text{bps/Hz}, \tag{14.63}$$

which may be improved with the aid of the MSDSD.

In the sequel, the maximum achievable rate of a differentially encoded system employing the MSDSD may be plotted versus the SNR, as shown in Figure 14.6, by evaluating the area under the

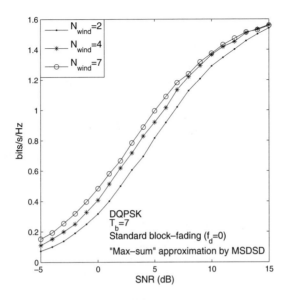

Figure 14.6: Maximum achievable rate of the SISO differentially encoded QPSK-modulated system using the MSDSD for various observation window sizes N_{wind}. The Rayleigh fading envelope was constant for $T_b = 7$ symbols and was then randomly and independently faded for the next block.

corresponding EXIT curve of the MSDSD. Observe in Figure 14.6 that a performance gain of about 2 dB may be attained by using the MSDSD associated with $N_{wind} = 7$ over the system employing the CDD of Section 12.1.1.1 for a wide range of SNRs, although naturally this is achieved at an increased complexity owing to the higher observation window size N_{wind} as well as the potential increased number of iterations between the MSDSD and the outer channel decoder.

On the other hand, according to the area properties of the EXIT chart [579], the area under the EXIT curve of an MAP detector/demapper is equal to the maximum possible code rate of the outer channel code, which can be employed in order to approach the DCMC capacity. In other words, the MAP-based MSDSD employing the highest possible observation window size N_{wind} can be regarded as the optimum differential detector in the interests of approaching the theoretically maximum transmission rate for a given differentially encoded modulation scheme. Figure 14.7 depicts the maximum achievable transmission rate of the system employing the MSDSD in comparison with the non-coherent DCMC capacity of Section 14.2 for various fading block lengths T_b, when aiming for a vanishingly low BER. Indeed, the former almost coincides with the latter for the Rayleigh block-fading channels associated with the three different T_b values considered. The slight gap between them is not unexpected, since the 'max-sum' approximation algorithm of Equation (14.60) is employed by the MSDSD in order to reduce the complexity imposed by computation of the a posteriori LLRs. Consequently, based on Equation (14.63), we have

$$C_{DCMC} = R_{overall,\max} \tag{14.64}$$

$$= \left(\frac{T_b - 1}{T_b} \log_2 M_c \right) \cdot \mathcal{A} \tag{14.65}$$

$$= \left(\frac{T_b - 1}{T_b} \log_2 M_c \right) \cdot R_{outer,\max} \tag{14.66}$$

when the MSDSD is employed with an observation window size of $N_{wind} = T_b$.

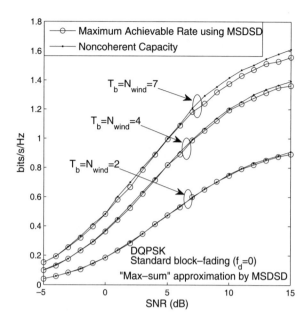

Figure 14.7: Maximum achievable rate of the SISO differentially encoded QPSK-modulated system using the MSDSD and the non-coherent DCMC capacity for various fading block lengths T_b. The Rayleigh fading envelope was constant for $T_b = 7$ symbols and was then randomly and independently faded for the next block.

14.4 Approaching the Capacity of the Differentially Modulated Cooperative Cellular Uplink

14.4.1 Relay-Aided Cooperative Network Capacity

14.4.1.1 Perfect-SR-Link DCMC Capacity

In Sections 14.2 and 14.3 we discussed the non-coherent DCMC capacity of the SISO direct-transmission-based system and the corresponding near-capacity MAP-based MSDSD detection scheme, respectively. Let us now continue by investigating the proposed near-capacity system design for the differentially modulated single-relay-aided cooperative system of Figure 14.1 by studying the corresponding network's DCMC capacity. We first define the two-hop, single-relay-aided network's capacity as the maximum achievable rate attained during the transmission of the source MS in the broadcast phase, namely phase I, which consists of L_s symbol periods, and an independent transmission by the RS during the relaying phase, namely phase II, when L_r symbols are transmitted. Initially a perfect SR link is assumed in order to guarantee 'error-free' relaying. Thus the above-mentioned network capacity is termed the cooperative system's DCMC capacity, which is not affected or constrained by the quality of the SR link. Hence we refer to it in parlance as the 'perfect-SR-link'-based capacity. By contrast, in Section 14.4.1.2 its 'imperfect-SR-link'-based counterpart will be investigated by considering the specific performance limitations imposed by the potentially error-prone SR link. According to the above definition, the corresponding network's 'perfect-SR-link'-based

DCMC capacity may be formulated as

$$C_{DCMC}^{coop}(\alpha, \text{SNR}_t^{overall}) = \alpha C_{DCMC}^{sd}(\text{SNR}_e^s) + (1-\alpha)C_{DCMC}^{rd}(\text{SNR}_e^r), \tag{14.67}$$

$$= \alpha C_{DCMC}(\text{SNR}_e^s)$$

$$+ (1-\alpha)C_{DCMC}[\text{SNR}_e^r + 10\log_{10}(\sigma_{rd}^2)], \tag{14.68}$$

where

$$\alpha \triangleq \frac{L_s}{L_s + L_r}. \tag{14.69}$$

In Equation (14.68) σ_{rd}^2 characterizes the reduced path-loss-related power gain, which was given by Equation (14.2), and $C_{DCMC}(\cdot)$ represents the SISO non-coherent DCMC capacity formula of Equation (14.16). Furthermore, SNR_e^s and SNR_e^r in Equation (14.68) represent the equivalent SNRs at the source and relay transmitters, respectively, which have the following relationship with the network's overall equivalent SNR, $\text{SNR}_e^{overall}$:

$$\text{SNR}_e^{overall} = \text{SNR}_e^s + \text{SNR}_e^r. \tag{14.70}$$

According to the simple cooperative resource allocation scheme mentioned in Section 14.1.1.1, namely the equal power allocation and the midpoint relay location, Equation (14.68) can be written as

$$C_{DCMC}^{coop}(\alpha, \text{SNR}_e^{overall}) = \alpha C_{DCMC}\left(\frac{\text{SNR}_e^{overall}}{2}\right)$$

$$+ (1-\alpha)C_{DCMC}\left(\frac{\text{SNR}_e^{overall}}{2} + 10\log_{10}(0.5^{-v})\right), \tag{14.71}$$

where v is the path-loss exponent. Furthermore, since the ratio of the differential-encoding frame lengths used by the source and relay is inversely proportional to the ratio of the channel code rate employed by the two, we have

$$\frac{L_s}{L_s + L_r} = \frac{R_r}{R_s + R_r} = \alpha. \tag{14.72}$$

Hence, Equation (14.71) may be reformulated as

$$C_{DCMC}^{coop}(\alpha, \text{SNR}_e^{overall}) = \frac{R_r}{R_s + R_r}C_{DCMC}\left(\frac{\text{SNR}_e^{overall}}{2}\right)$$

$$+ \frac{R_s}{R_s + R_r}C_{DCMC}\left(\frac{\text{SNR}_e^{overall}}{2} + 10\log_{10}(0.5^{-v})\right) \tag{14.73}$$

$$= \alpha C_{DCMC}\left(\frac{\text{SNR}_e^{overall}}{2}\right)$$

$$+ (1-\alpha)C_{DCMC}\left(\frac{\text{SNR}_e^{overall}}{2} + 10\log_{10}(0.5^{-v})\right). \tag{14.74}$$

Therefore, in contrast to the independence of the DCMC capacity of the channel code rate employed in the scenario of the conventional direct transmission system, the DCMC capacity of the relay-aided cooperative system is dependent on the ratio R_s/R_r of the channel code rates employed by the source and relay or, equivalently, dependent on α. In Figure 14.8 the cooperative system's DCMC capacity curves associated with different values of α are depicted based on Equation (14.74) for the 'perfect-SR-link' scenario at a *constant code-rate ratio* in comparison with the DCMC capacity curve of the direct-transmission-based system. As observed in Figure 14.8, the cooperative system's DCMC capacity is gradually decreased as α is increased. This is not unexpected, since the weight of the second term in Equation (14.74) decreases as that of the first term increases, while the second term is typically larger than the first term owing to the reduced path-loss-related power gain. Furthermore, since the source has

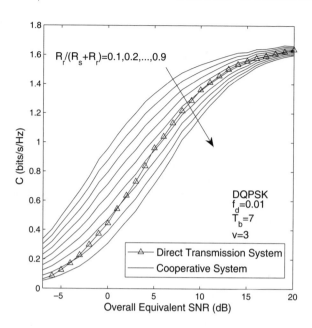

Figure 14.8: The single-relay-assisted cooperative system's constant code-rate-ratio-based DCMC capacity curves for the 'perfect-SR-link' scenario.

to remain silent, when the relay is transmitting during phase II, the system's constant code-rate-ratio-based 'perfect-SR-link' DCMC capacity may become even lower than that of the direct-transmission-based system, as seen in Figure 14.8, if both the overall equivalent SNR and α are sufficiently high. Naturally, the half-duplex constraint imposes a potentially substantial multiplexing loss. In other words, despite the reduced path-loss-related power gain, the single-relay-assisted cooperative system considered does not necessarily outperform its direct-transmission-based counterpart in terms of the maximum achievable transmission rate. In order to prevent this potential performance loss, a careful system design is required.

On the other hand, given the channel code rates (R_s, R_r) employed by the source and relay, the resultant bandwidth efficiency, η, may be expressed as

$$\eta = \frac{L_s}{L_s + L_r} R_s \frac{T_b - 1}{T_b} \log_2 M_c \tag{14.75}$$

$$= \frac{R_s R_r}{R_s + R_r} \frac{T_b - 1}{T_b} \log_2 M_c \tag{14.76}$$

$$= \alpha R_s \frac{T_b - 1}{T_b} \log_2 M_c. \tag{14.77}$$

Hence, by fixing the value of R_s and varying that of α, the resultant bandwidth efficiency η can be calculated using Equation (14.77). Based on Equation (14.77), the corresponding minimum overall equivalent SNR required by near-error-free transmissions may be found with the aid of the constant code-rate-ratio-based 'perfect-SR-link' DCMC capacity curves seen in Figure 14.8.

Consequently, the cooperative system's constant-R_s 'perfect-SR-link' DCMC capacity curves were plotted from Equation (14.77) based on Figure 14.8 for various values of R_s in Figure 14.9, where we observe that the capacity increases as R_s increases. However, all the constant-R_s-related capacity curves depicted in Figure 14.9 will intersect the capacity curve of the direct-transmission-based system plotted in Figure 14.8, if the overall equivalent SNR becomes sufficiently high. This results

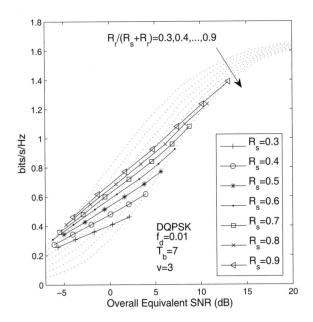

Figure 14.9: The single-relay-assisted cooperative system's constant-R_s DCMC capacity curves.

in a reduced maximum achievable transmission rate compared with the direct-transmission-based counterpart. Therefore, based on the observation of Figures 14.8 and 14.9, we may state that although the cooperative system's capacity increases steadily as the overall equivalent SNR increases, it might remain lower than that of its direct-transmission-based counterpart, even under the assumption of an idealized error-free SR link, if both R_s and α are of relatively high values. In other words, under the assumption of a perfect SR link, the single-relay-assisted DDF cooperative system is capable of exhibiting a higher capacity than its point-to-point transmission system, provided that the target throughput is low.

14.4.1.2 Imperfect-SR-Link DCMC Capacity

Until now the single-relay-assisted DDF cooperative system's capacity has been investigated under the assumption of an idealized error-free SR link. However, in practice the wireless channel connecting the source and relay MSs is typically far from perfect and its quality plays an important role in determining the overall cooperative network's achievable performance, as discussed in Chapter 13 in the context of uncoded scenarios. Furthermore, in order to create a near-capacity design for the overall cooperative system, near-capacity transmission over the potentially error-infested SR link during the broadcast phase I is a natural prerequisite, which in turn leads to an investigation of the performance limitations imposed by the SR link on the overall cooperative system.

Under the assumption of equal power allocation and a midpoint relay location, the non-coherent DCMC capacity of the SR link may be expressed as

$$C^{sr}_{DCMC}(\text{SNR}^{overall}_e) = C_{DCMC}\left(\frac{\text{SNR}^{overall}_e}{2} + 10\log_{10}(0.5^{-v})\right), \tag{14.78}$$

where $C_{DCMC}(\cdot)$ was formulated in Equation (14.16). Hence, the non-coherent DCMC capacity of the SR link may be plotted versus the overall equivalent SNR, as shown in Figure 14.10. Then, according to

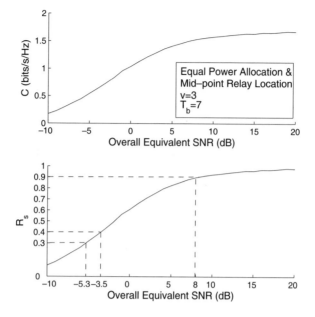

Figure 14.10: Non-coherent DCMC capacity of the SR link and its corresponding capacity-achieving channel code rate R_s employed by the source MS.

Equation (14.66), we can calculate the capacity-achieving channel code rate employed by the source as

$$R_{s,capacity}(\text{SNR}_e^{overall}) = \frac{T_b \cdot C_{DCMC}^{sr}(\text{SNR}_e^{overall})}{(T_b - 1) \cdot \log_2 M_c}, \quad (14.79)$$

which is also depicted versus the overall equivalent SNR in Figure 14.10. Therefore, the minimum overall equivalent SNR corresponding to a certain value of R_s, which facilitates near-error-free information delivery from the source to the relay, may be observed in Figure 14.10. These minimum overall equivalent SNRs characterize the performance limits imposed by the practical imperfect SR link on the entire cooperative system, when the corresponding rate R_s is employed by the source. Given these minimum overall equivalent SNRs associated with different values of R_s, we can now draw the cooperative system's 'imperfect-SR-link' DCMC capacity based on the constant-R_s 'perfect-SR-link' DCMC capacity curves of Figure 14.9. More specifically, observe in Figure 14.11 that in order to find, for example, the cooperative system's 'imperfect-SR-link' DCMC capacity for $R_s = 0.3$, we locate the particular point on the constant-R_s 'perfect-SR-link' DCMC capacity curve associated with $R_s = 0.3$, whose horizontal coordinate is equal to the corresponding minimum overall equivalent SNR of -5.3 dB found previously in the context of Figure 14.10. Then, the vertical coordinate of the point is the 'imperfect-SR-link' DCMC capacity of the cooperative system for $R_s = 0.3$ or when we have $\text{SNR}_e^{overall} = -5.3$ dB.

In order to gain an insight into the benefits of the single-relay-assisted DDF cooperative system over its conventional point-to-point direct-transmission-based counterpart from a pure capacity perspective, the 'imperfect-SR-link' DCMC capacity of the cooperative system associated with both $v = 2$ and $v = 3$ is depicted in comparison with that of the direct-transmission-based one in Figure 14.12. It can be seen in the figure that when the overall equivalent SNR is relatively low, the single-relay-assisted cooperative system exhibits a significantly higher capacity than its direct-transmission-based counterpart in typical urban cellular radio scenarios, e.g. when having a path-loss exponent of $v = 3$. However, the achievable capacity gain may be substantially reduced if we encounter a free-space propagation scenario, i.e. $v = 2$, since the reduced path-loss-related power gain achieved is

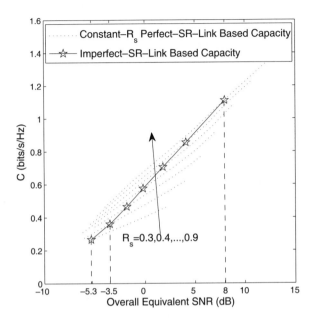

Figure 14.11: 'Imperfect-SR-link' DCMC capacity.

insufficiently high to compensate for the significant multiplexing loss inherent in the single-relay-aided half-duplex cooperative system. Moreover, as the overall equivalent SNR increases to a relatively high value, there is no benefit in invoking a single-relay-aided cooperative system, since its capacity becomes lower than that of the conventional point-to-point system.

14.4.2 Ir-DHCD Encoding/Decoding for the Cooperative Cellular Uplink

In conventional relay-aided DAF cooperative systems, the relay decodes the signal received from the source and re-encodes it using an identical channel encoder. Then the destination receives two versions of the same codeword, namely those from the source and relay, respectively, which may be viewed as a repetition code. Finally, the two replicas of the signal may be combined using maximal ratio combining (MRC) prior to the decoding. In order to enhance the coding gain achieved by the repetition code constituted by the relay-aided system, while maintaining the cooperative diversity gain, the classic turbo coding mechanism was introduced into the DF-aided cooperative system of [620], resulting in the so-called distributed turbo coding scheme. Specifically, according to the principle of parallel concatenated convolutional-code-based turbo coding, the data and their interleaved version are encoded in parallel, using two distinct recursive systematic convolutional (RSC) codes, respectively. Therefore, a distributed turbo code may be readily constructed at the relay by interleaving its received estimated source data prior to re-encoding. Owing to this interleaving at the RS, its encoded stream may be expected to be different from that of the source. Consequently, the direct encoding of the original bit stream takes place at the source, while the encoding of the interleaved sequence ensues at the RS in a distributed manner. A standard turbo decoder may be implemented at the destination. It was revealed in [620] that a significantly enhanced coding gain can be achieved by a distributed turbo code in comparison with that attained by a single convolutional code for transmission over two independently fading channels.

In order to improve the iterative decoder's achievable performance and hence achieve near-error-free transmissions between the source and relay, a unity-rate-coded (URC) three-stage serially

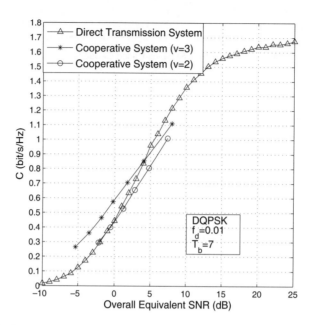

Figure 14.12: Capacity comparison of the single-relay-aided cooperative system and its direct-transmission-based counterpart.

concatenated transceiver employing the irregular convolutional codes (IrCCs) of Section 10.4.3 may be employed in the single-relay-aided DDF cooperative system. More specifically, since the URC has an infinite impulse response due to its recursive encoder architecture, the resultant EXIT curve of the URC-aided inner decoder is capable of reaching the point $(1, 1)$ of the EXIT chart, provided that the interleaver length is sufficiently high [619]. Furthermore, since the URC decoder employs the MAP decoding scheme, the extrinsic probability generated at the output of the URC decoder contains the same amount of information as the sequence at the input of the URC decoder [577, 621]. In other words, the area under the inner EXIT curve remains the same, regardless of the URC's employment. Hence, a higher end point of the EXIT curve leads to a lower starting point, implying a steeper slope for the EXIT curve, which in turn yields a reduced error floor and a higher SNR threshold, above which decoding convergence to a vanishingly low BER becomes possible, as we will demonstrate in the forthcoming sections.

Based on the above-mentioned arguments, the transmitter's architecture proposed for the source is depicted in Figure 14.13, where we use a conventional differential modulation scheme, such as DQPSK, amalgamated with the URC encoder in order to create a two-stage inner code, whereas an IrCC associated with an average code rate of R_s, namely IrCC$_s$, is employed as the outer code to achieve a performance that is close to the capacity of the SR link. Specifically, at the transmitter of the source in Figure 14.13, a block of L information bits u_1 is first encoded by the IrCC$_s$ encoder, in order to generate the coded bits c_1, which are interleaved by the interleaver π_{s1}. Then the resultant permuted bits u_2 are successively fed through the URC$_s$ encoder and the interleaver π_{s2}, yielding the interleaved double-encoded bits u_3, which are delivered to the bit-to-symbol differential modulator of Figure 14.13. Note that the labels u and c represent the uncoded and coded bits, respectively, corresponding to a specific module as indicated by the subscript. The corresponding URC-decoder-assisted three-stage receiver proposed for the relay is also portrayed in Figure 14.13 together with its two-stage transmitter schematic. Specifically, at the receiver of the relay, which comprises three modules, namely the MSDSD$_s$, the URC$_s$ decoder and the IrCC$_s$ decoder, the extrinsic information

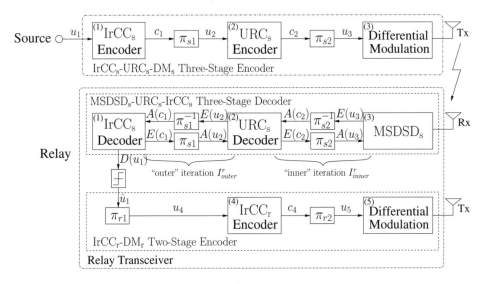

Figure 14.13: Schematic of the Ir-DHCD coding encoder.

is exchanged among the modules in a number of consecutive iterations. As shown in Figure 14.13, $A(\cdot)$ represents the a priori information expressed in terms of the LLRs, while $E(\cdot)$ denotes the corresponding *extrinsic* information. At the transmitter of the relay, the estimated data bit stream is fed through the interleaver π_{r1} prior to the IrCC_r encoder having an average code rate of R_r, as observed in Figure 14.13, in order to construct a distributed turbo code together with the source. Consequently, the proposed relay-aided cooperative system may be referred to here as an Ir-DHCD (Irregular Distributed Hybrid Concatenated Differential) coding scheme, under the assumption of an error-free decoding at the relay.

At the destination BS, according to the principles of the distributed turbo decoding mechanism proposed in [620], the novel iterative receiver of Figure 14.14 is used for decoding the Ir-DHCD coding scheme of Figure 14.13. To be specific, the first part of the iterative receiver is an amalgamated 'MSDSD_s–URC_s–IrCC_s' iterative decoder, which is used iteratively to decode the signal received directly from the source during phase I, while the second part consists of the MSDSD_r differential detector and the IrCC_r decoder, which is employed iteratively to decode the signal forwarded by the relay during phase II. Since the MSDSD_s–URC_s–IrCC_s decoder and the MSDSD_r–IrCC_r decoder may be regarded as two-component decoders of a turbo receiver, the extrinsic information exchange between them, which is referred to as the 'outer iteration', is expected to enhance significantly the achievable coding gain. In comparison with the conventional relay-aided cooperative system, where a simple repetition code is constructed, the extra coding gain achieved by the proposed Ir-DHCD coding scheme may be interpreted as the interleaving gain of the turbo code and the turbo processing gain of the outer iterations.

14.4.3 Approaching the Cooperative System's Capacity

In this section, we propose a practical framework capable of approaching the cooperative system's capacity. We propose a reduced-complexity near-capacity system design for the DQPSK-modulated single-relay-assisted DDF cooperative cellular uplink. We will consider two different propagation scenarios associated with a path-loss exponent of $v = 2$ and $v = 3$, respectively. Based on the proposed cooperative system design, we will verify in Section 14.4.4 that a single-relay-aided DDF cooperative system is not always superior to its conventional direct-transmission-based counterpart in

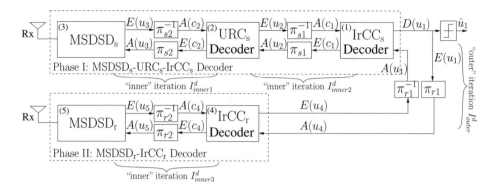

Figure 14.14: Iterative receiver at the destination BS.

Table 14.1: Summary of system parameters.

Single-relay-aided cooperative cellular uplink		
	Scenario I	Scenario II
Path-loss exponent	$v = 2$	$v = 3$
Doppler frequency		$f_d = 0.01$
Fading block size		$T = 7$
Tx at source or relay MS		1
Rx at relay MS or BS		1
Modulation		DQPSK
Detector/MAP		MSDSD
Channel code		IrCC (17 subcodes)
Code rate at source MS		$R_s = 0.5$
Power allocation		$P_s = P_r = \frac{1}{2}P = 0.5$
Relay location		$D_{sr} = D_{rd} = \frac{1}{2}D_{sd} = 0.5$

terms of the maximum achievable bandwidth efficiency. The time-selective block-fading channel model of Section 14.1.1.2 is employed in conjunction with a normalized Doppler frequency of $f_d = 0.01$. For the sake of simplicity, the equal power allocation and the midpoint relay location scenarios are assumed for the single-relay-aided cooperative system, where each terminal is equipped with a single Tx/Rx antenna. All other system parameters are summarized in Table 14.1.

14.4.3.1 Reduced-Complexity Near-Capacity Design at Relay MS

Without loss of generality, the average code rate R_s of the IrCC_s at the source is chosen to be 0.5. Based on Equation (14.79) of Section 14.4.1.2, the maximum possible code rate R_s that may be employed by the source to facilitate near-error-free transmissions between the source and relay is plotted in Figure 14.15 versus the corresponding overall equivalent SNR required for both Scenarios I and II of Table 14.1. According to Figure 14.15, the minimum overall equivalent SNRs required to support an SR-link transmission at an infinitesimally low BER are 1.1 and -1.9 dB for Scenarios I and II, respectively, when $R_s = 0.5$. The 3 dB SNR gain attained in Scenario II in comparison with Scenario I is due to the difference in the path-loss-related power gains/losses achieved in the two scenarios of

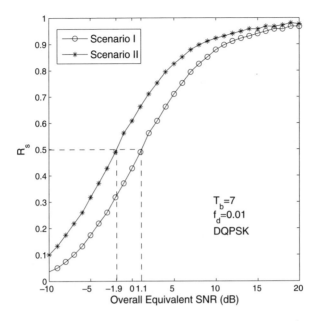

Figure 14.15: SR-link-capacity-achieving channel code rate R_s employed by the source versus the corresponding overall equivalent SNR required.

Table 14.1, namely:

$$10\log_{10}(0.5^{-3}) - 10\log_{10}(0.5^{-2}) = 3\,\text{dB}. \tag{14.80}$$

Theoretically, the SR link's non-coherent DCMC capacity can be achieved with the aid of an infinite number of iterations between the inner combined MSDSD_s–URC_s decoder and an ideally designed outer IrCC_s decoder at the relay's receiver of Figure 14.13, although naturally this would impose an excessive computational complexity. However, in practice, to avoid potentially excessive complexity at the relay, while approaching the capacity, a 'higher-than-necessary' EXIT curve may be ensured for the inner MSDSD_s–URC_s decoder by having a slightly higher overall equivalent SNR of 2.3 dB for Scenario I or -0.7 dB for Scenario II, as depicted in Figure 14.16 for a different number of iterations between the MSDSD_s and URC_s blocks. The observation window size of the MSDSD was set to its maximum value of $N_{wind} = 7$ in the time-selective block-fading channel having a fading block size of $T_b = 7$. It may be observed from Figure 14.16 that the iterative information exchange between the MSDSD_s and URC_s blocks approaches convergence as early as their second iteration. Hence, the number of iterations between the MSDSD_s and URC_s blocks was set to $I^r_{inner} = 2$ in our future simulations, in order to avoid any unnecessarily imposed complexity, while maintaining a near-capacity performance. Then, using the EXIT curve of the inner amalgamated MSDSD_s–URC_s decoder, the optimized weighting coefficients of the half-code-rate IrCC_s can be obtained with the aid of the EXIT-curve-matching algorithm of [581], resulting in a narrow but marginally open tunnel between the EXIT curves of the inner amalgamated MSDSD_s–URC_s decoder and the outer IrCC_s decoder, as seen in Figure 14.16.

On the other hand, in order to reduce further the complexity imposed by the MSDSD_s–URC_s decoder during the iterative decoding process at the relay, an adaptive-window-based scheme is proposed for the MSDSD_s, where the observation window size was initially set to the smallest value of $N_{wind} = 2$, which will be slightly increased as soon as the iterative decoding between the MSDSD_s–URC_s decoder and the IrCC_s decoder converges. The proposed adaptive-window-based scheme is characterized by Figure 14.17(a), where it is observed that the resultant bit-by-bit

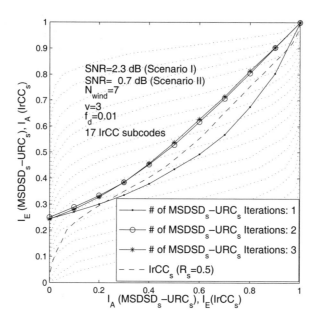

Figure 14.16: EXIT chart at the relay (design of the $IrCC_s$).

Monte Carlo simulation-based iterative decoding trajectory fails to reach the $(1, 1)$ point associated with $N_{wind} = 2$ and $N_{wind} = 4$. By contrast, it does indeed reach the $(1, 1)$ point of the EXIT chart for $N_{wind} = 7$. Thus, the original maximum achievable iteration gain corresponding to $N_{wind} = 7$ can be maintained by the adaptive-window-based scheme, despite having a reduced overall complexity imposed by the $MSDSD_s$. This is not unexpected, since although an increased number of iterations may be needed between the $MSDSD_s$–URC_s decoder and the $IrCC_s$ decoder to achieve the same amount of iteration gain when the adaptive-window-based scheme is employed, the complexity per iteration imposed by the $MSDSD_s$ using a reduced N_{wind} value is expected to be exponentially reduced, yielding a potentially reduced overall complexity. Indeed, the complexity imposed is significantly reduced by the adaptive-window-based scheme, as observed in Figure 14.17(b), where the complexity imposed by the $MSDSD_s$ in terms of the number of PED evaluations per bit is plotted versus the overall equivalent SNR for both systems operating with and without the adaptive-window-based scheme. To be specific, regardless of the employment of the adaptive-window-aided scheme, the complexity imposed by the $MSDSD_s$ arrangement during the iterative decoding process gradually decreases in Figure 14.17(b) as the SNR increases from about 2 dB for Scenario I, where a narrow but open tunnel is created by using $N_{wind} = 7$. This is because a reduced number of iterations are required in Figure 14.17(a) in the presence of an increasingly wider open EXIT tunnel. Remarkably, the complexity imposed by the $MSDSD_s$ in Scenario I is substantially reduced in Figure 14.17(b) with the aid of the adaptive-window-assisted scheme by as much as 75% at $SNR_t^{overall} = 2$ dB, when the open EXIT tunnel created by having $N_{wind} = 7$ becomes rather narrow. The attainable complexity reduction increases even further to about 83% when the open EXIT tunnel becomes wider at $SNR_t^{overall} = 4$ dB, as seen in Figure 14.17(b).

14.4.3.2 Reduced-Complexity Near-Capacity Design at Destination BS

In Section 14.4.3.1 the $IrCC_s$ decoder of Figure 14.13 was specifically designed to allow a near-capacity operation over the SR link with the aid of the EXIT-curve-matching algorithm of [581] carried out at the relay. Let us now consider the destination BS and optimize the weighting coefficients of the

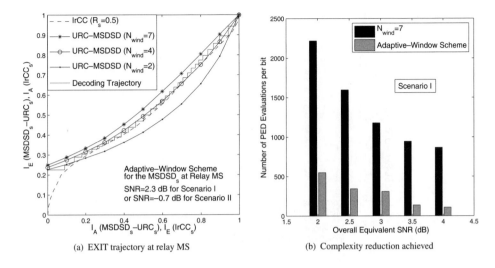

(a) EXIT trajectory at relay MS

(b) Complexity reduction achieved

Figure 14.17: Characterization of the adaptive-window-aided scheme for the MSDSD_s at the relay.

other irregular convolutional code employed by the RS's transmitter in Figure 14.13. In other words, we consider the IrCC_r design now, in order to approach the overall cooperative system's capacity quantified in Section 14.4.1.2, while maintaining a relatively low complexity. First of all, the EXIT curves of the amalgamated MSDSD_s–URC_s–IrCC_s decoder employed by the BS are depicted in Figure 14.18 for various values of N_{wind} for both Scenarios I and II associated with an overall equivalent SNR of 2.3 dB and -0.7 dB, respectively. The number of iterations between the MSDSD_s and the URC_s of Figure 14.14 as well as that between the combined MSDSD_s–URC_s decoder and the IrCC_s scheme are set to $I^d_{inner1} = 1$ and $I^d_{inner2} = 5$, respectively. It may be observed from Figure 14.18 that the desirable choice of the observation window size employed by the MSDSD_s at the BS for Scenario I is $N_{wind} = 4$, while $N_{wind} = 2$ for Scenario II, under our low-complexity near-capacity design criterion.

Then, based on the above-mentioned desirable choices of N_{wind}, we continue by determining the desirable number of iterations between the MSDSD_s and the URC_s arrangements as well as that required between the combined MSDSD_s–URC_s decoder and the IrCC_s, by plotting the EXIT curves of the amalgamated MSDSD_s–URC_s–IrCC_s decoder associated with various numbers of iterations in Figure 14.19. Specifically, as observed in Figure 14.19(a) for Scenario I, only a modest iteration gain may be attained by having more than a single iteration (i.e. $I^d_{inner1} = 1$) between the MSDSD_s and the URC_s blocks of Figure 14.14. By contrast, we need about $I^d_{inner2} = 5$ iterations between the combined MSDSD_s–URC_s decoder and the IrCC_s, while beyond $I^d_{inner2} = 5$ the increase of the area under the EXIT curve of the MSDSD_s–URC_s–IrCC_s decoder becomes marginal. Similarly, observe in Figure 14.19(b) that a sharply rising EXIT curve can be created for Scenario II when using our low-complexity near-capacity design criterion, since only a single iteration ($I^d_{inner1} = 1$) is required between the MSDSD_s and the URC_s, while $I^d_{inner2} = 6$ iterations may be necessitated between the combined MSDSD_s–URC_s decoder and the IrCC_s.

To determine the 17 optimized weighting coefficients of the IrCC_r employed by the transmitter of the relay seen in Figure 14.13, we commence by investigating the EXIT function of the amalgamated MSDSD_r–IrCC_r decoder associated with the 17 IrCC subcodes, which constitutes the second component decoder of the iterative receiver of Figure 14.14 employed at the BS. Since the previously considered equal power allocation and midpoint relay location scenarios are assumed, an identical EXIT chart will be created for the combined MSDSD_r–IrCC_r decoder at the BS for Scenarios I

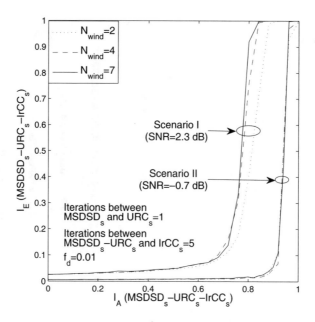

Figure 14.18: EXIT curves of the amalgamated $MSDSD_s$–URC_s–$IrCC_s$ decoder employed by the destination BS for various observation window sizes N_{wind}.

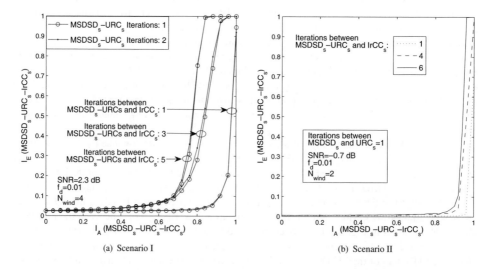

(a) Scenario I   (b) Scenario II

Figure 14.19: EXIT curves of the amalgamated $MSDSD_s$–URC_s–$IrCC_s$ decoder of Figure 14.14 employed by the destination BS for different numbers of subiterations between the $MSDSD_s$ and the URC_s and between the combined $MSDSD_s$–URC_s decoder and the $IrCC_s$.

and II associated with an overall equivalent SNR of 2.3 dB and -0.7 dB, respectively, as shown in Figure 14.20. The EXIT curves of the combined $MSDSD_r$–$IrCC_r$ decoder associated with each of the 17 IrCC subcodes are plotted in Figure 14.20(a) for the various values of N_{wind} employed by the $MSDSD_r$. It can be seen that the EXIT curves are shifted to the left on increasing the value of N_{wind}, which results in an increased *extrinsic* mutual information I_E evaluated at the output of the

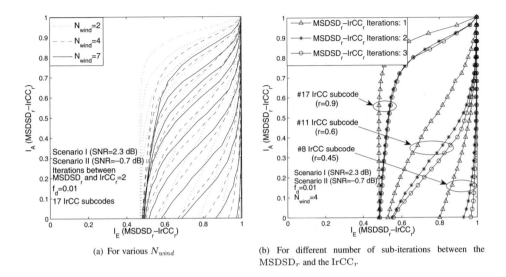

(a) For various N_{wind}

(b) For different number of sub-iterations between the MSDSD$_r$ and the IrCC$_r$

Figure 14.20: EXIT curves of the amalgamated MSDSD$_r$–IrCC$_r$ decoder employed by the destination BS.

combined MSDSD$_r$–IrCC$_r$ decoder of Figure 14.14 for a given input of a priori mutual information I_A. However, to achieve a near-capacity performance at a relatively low complexity, $N_{wind} = 4$ is considered to be a reasonable choice which strikes a compromise between the achievable performance and the complexity imposed. On the other hand, the EXIT curves of the combined MSDSD$_r$–IrCC$_r$ decoder associated with three selected IrCC subcodes are also depicted in Figure 14.20(b) for different numbers of iterations between the MSDSD$_r$ and the IrCC$_r$. As observed in Figure 14.20(b), although the EXIT curve can be shifted to the left by increasing the number of iterations between the MSDSD$_r$ and the IrCC$_r$, any further shifting of the EXIT curve starts to become rather difficult when the number of iterations exceeds $I_{inner3}^d = 2$. Hence, based on the low-complexity near-capacity design criterion, the number of iterations between the MSDSD$_r$ and the IrCC$_r$ blocks of Figure 14.14 is chosen to be $I_{inner3}^d = 2$. In the sequel, a group of EXIT curves corresponding to the MSDSD$_r$–IrCC$_r$ subcodes associated with $N_{wind} = 4$ and $I_{inner3}^d = 2$ iterations can be seen in Figure 14.21.

Finally, we use the EXIT-curve-matching algorithm of [581] in order to match the SNR-dependent EXIT curve of the combined MSDSD$_r$–IrCC$_r$ decoder of Figure 14.14 to the target EXIT curves of the amalgamated MSDSD$_s$–URC$_s$–IrCC$_s$ decoder of the BS portrayed in Figures 14.19(a) and 14.19(b) for Scenarios I and II of Table 14.1, respectively, as shown in Figure 14.21. As a result, to achieve a near-capacity performance while maintaining a moderate computational complexity, a 'wider-than-necessary' EXIT tunnel is created between the EXIT curve of the amalgamated MSDSD$_s$–URC$_s$–IrCC$_s$ decoder and that of the combined MSDSD$_r$–IrCC$_r$ decoder at the BS. Thus, the resultant average coding rates, R_r, of the IrCC$_r$ scheme designed for Scenarios I and II of Table 14.1 are equal to 0.6 and 0.5, respectively, which cannot be achieved by simply using one of 17 IrCC$_r$ subcodes having the same code rate, as observed in Figure 14.21, owing to the absence of an open EXIT tunnel. The corresponding Monte Carlo simulation-based decoding trajectory is also plotted in Figure 14.21 for both Scenarios I and II of Table 14.1, which reaches the $(1.0, 1.0)$ point of the EXIT chart, indicating the achievement of decoding convergence to an infinitesimally low BER at near-capacity SNRs for the Ir-DHCD coding scheme proposed in Section 14.4.2.

We have now completed the low-complexity near-capacity system design for the single-relay-aided cooperative system contrived for both Scenarios I and II of Table 14.1. Since the average code rate is fixed to $R_s = 0.5$ for the $IrCC_s$ at the transmitter of the source, and the resultant capacity-achieving

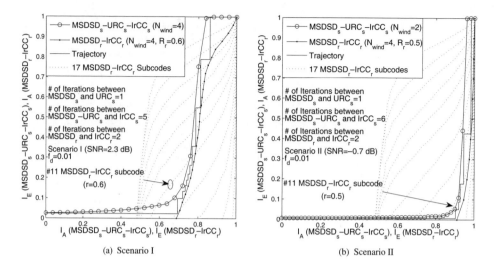

(a) Scenario I (b) Scenario II

Figure 14.21: Iterative decoding trajectory at the BS.

code rates R_r of the IrCC_r employed at the relay are 0.6 and 0.5 for Scenarios I and II, respectively, we can calculate the corresponding bandwidth efficiency using Equation (14.77) as follows:

$$\eta^{coop}_{\text{Scenario I}} = \frac{R_s R_r}{R_s + R_r} \frac{T_b - 1}{T_b} \log_2 M_c \tag{14.81}$$

$$= 0.4664 \quad \text{bps/Hz,} \tag{14.82}$$

and

$$\eta^{coop}_{\text{Scenario II}} = 0.4286 \quad \text{bps/Hz.} \tag{14.83}$$

14.4.4   Simulation Results and Discussion

In order to carry out a fair comparison between the cooperative system and its conventional direct-transmission-based counterpart, we also carry out a near-capacity system design for the latter in this section, which has exactly the same bandwidth efficiency as the former. To be more specific, according to Equation (14.62), we can obtain the required code rate of the outer IrCC decoder employed by the direct-transmission-based system in Scenario I as follows:

$$R_{outer,\text{Scenario I}} = \frac{T_b \cdot R_{overall}}{(T_b - 1) \cdot \log_2 M_c} \tag{14.84}$$

$$= \frac{T_b \cdot \eta^{coop}_{\text{Scenario I}}}{(T_b - 1) \cdot \log_2 M_c} \tag{14.85}$$

$$= 0.27; \tag{14.86}$$

and similarly for Scenario II

$$R_{outer,\text{Scenario II}} = \frac{T_b \cdot \eta^{coop}_{\text{Scenario II}}}{(T_b - 1) \cdot \log_2 M_c} \tag{14.87}$$

$$= 0.25. \tag{14.88}$$

Hence, the near-capacity design for the URC-aided three-stage direct-transmission-based system can be carried out for the target bandwidth efficiency, and the corresponding EXIT charts and Monte Carlo

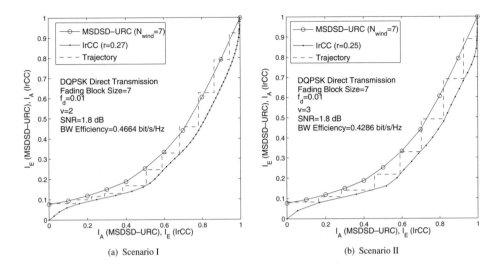

(a) Scenario I   (b) Scenario II

Figure 14.22: Near-capacity design of the direct transmission system for both Scenarios I and II.

simulation-based iterative decoding trajectories are plotted in Figure 14.22 for both Scenarios I and II of Table 14.1.

Let us now depict the BER versus the overall equivalent SNR curves for both the point-to-point transmission-based system and the single-relay-assisted cooperative system in Figure 14.23, which were previously designed to approach their corresponding capacity at a relatively low complexity. It is clearly shown in Figure 14.23 that, on using the near-capacity system design of Section 14.4.3, the proposed Ir-DHCD coding scheme becomes capable of performing within about 2 dB and 1.8 dB from the corresponding single-relay-aided DDF cooperative system's DCMC capacity in Scenarios I and II, respectively. Similarly, an infinitesimally low BER can be achieved by the point-to-point transmission-based system at an SNR of 1.6 dB and 1.9 dB from the corresponding SISO non-coherent DCMC capacity in Scenarios I and II of Table 14.1, respectively. Furthermore, in line with our predictions in Section 14.4.1.2, it is observed from Figure 14.23 that for a given target bandwidth efficiency, the single-relay-aided cooperative system does not necessarily guarantee a performance superior to that of the conventional direct-transmission-based system. More specifically, in Scenario I, where the path-loss exponent was set to $v = 2$ to simulate a free-space propagation environment, an SNR gain of 0.65 dB can be achieved by the direct-transmission-based system over its single-relay-aided cooperative counterpart, given a bandwidth efficiency of 0.4664 bps/Hz, as shown in Figure 14.23(a). However, when we have $v = 3$, in order to model the typical urban area cellular radio environment of Scenario II, the single-relay-aided cooperative system becomes capable of outperforming significantly the direct-transmission-based system, requiring an overall transmit power which is about 2.5 dB lower than that needed by the latter to achieve an infinitesimally low BER, while maintaining a bandwidth efficiency of 0.4286 bps/Hz.

14.5   Chapter Conclusions

In Section 14.1, we commenced by reviewing the SISO non-coherent DCMC capacity. More specifically, in Section 14.4.1 the single-relay-assisted DDF cooperative system's DCMC capacity was investigated and compared with that of the conventional direct-transmission-based system. For the sake of convenience, we repeat the non-coherent DCMC capacity curves of both the direct-transmission-based and of the cooperative systems in Figure 14.24. In order to create a near-capacity design for the

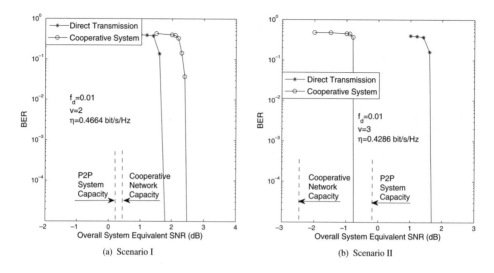

Figure 14.23: Achievable BER performance of the near-capacity designed single-relay-assisted cooperative system.

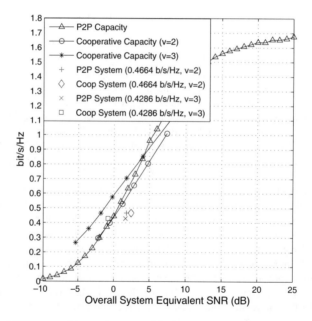

Figure 14.24: EXIT curves of the amalgamated MSDSD_s–URC_s–IrCC_s decoder employed by the destination BS for various observation window sizes N_{wind}.

cooperative system, in Sections 14.4.2 and 14.4.3 the so-called Ir-DHCD encoding/decoding schemes were proposed together with the adaptive-window-aided MSDSD scheme, respectively, which are capable of approaching the capacity at a relatively low complexity. In contrast to the conventional point-to-point system, the proposed near-capacity design of the single-relay-aided DDF cooperative system may be regarded as a joint source–relay mode design procedure, which was simplified in Section 14.4.3

to two EXIT-curve-matching problems in order to optimize the weighting coefficients of the IrCC decoders employed by both the source and the relay. A near-capacity performance can indeed be achieved by the proposed Ir-DHCD encoding/decoding schemes obeying our joint source–relay mode design, as demonstrated in Figure 14.24. More importantly, based on the capacity and on the practically achievable performance of classic direct transmission and single-relay-aided cooperative systems, we found in Figure 14.23(a) that the latter might be outperformed by the former, owing to the significant multiplexing loss inherent in the half-duplex single-relay-aided cooperative system. To be specific, in Figure 14.24 the single-relay-aided cooperative system was shown to be superior in comparison with its direct-transmission-based counterpart only in the specific scenario when the reduced path-loss-related power gain was sufficiently high in order to compensate for the multiplexing loss. In our future research we will consider successive relaying-aided arrangements [622] in order to mitigate the above-mentioned multiplexing loss.

Part III

Coherent SDM-OFDM Systems

List of Symbols in Part III

$\boldsymbol{A}^{\mathrm{T}}$	Matrix/vector transpose
$\boldsymbol{A}^{*}$	Matrix/vector/scalar complex conjugate
$\boldsymbol{A}^{\mathrm{H}}$	Matrix/vector Hermitian adjoint, i.e. complex conjugate transpose
$\boldsymbol{A}^{-1}$	Matrix inverse
$\boldsymbol{A}^{+}$	Moore–Penrose pseudo-inverse
$\mathbf{tr}\,(\boldsymbol{A})$	Trace of matrix, i.e. the sum of its diagonal elements
K	Number of OFDM subcarriers
n_{r}	Number of transmit antennas
m_{t}	Number of receive antennas
f_{D}	Maximum Doppler frequency
ν_{τ}	OFDM-symbol-normalized PDP tap drift rate
f_{c}	Carrier frequency
B	Signal bandwidth
T_s	OFDM FFT frame duration
τ_{rms}	RMS delay spread
T	OFDM symbol duration
L_f	Number of data frames per transmission burst
N_p	Number of pilot SDM-OFDM symbols in burst preamble
N_d	Number of data SDM-OFDM symbols per data frame
ζ	MIMO-CTF RLS tracking filter forgetting factor
η	PASTD-aided CIR tap tracking filter forgetting factor
β	RLS CIR tap prediction filter forgetting factor
ε	Pilot overhead
ρ	OHRSA search radius factor parameter
γ	OHRSA search resolution parameter
σ_w^2	Gaussian noise variance
E_b	Energy per transmitted bit
E_s	Energy per transmitted M-QAM symbol
r	Size of the transmitted bit-wise signal vector $\mathbf{t}$
$\mathcal{I}$	Mutual information
$\mathcal{C}$	Unconstrained capacity
$(i_{\mathrm{ce}}, i_{\mathrm{det}}, i_{\mathrm{dec}})$	Number of (channel estimation, detection, decoding) iterations
$\mathcal{H}$	Hadamard matrix
$\boldsymbol{I}$	Identity matrix
$\mathcal{M}$	Set of M-PSK/M-QAM constellation phasors
$\mathcal{L}$	Log-likelihood ratio value
$\mathbf{y}$	Received subcarrier-related SDM signal
$\mathbf{H}$	Subcarrier-related MIMO-CTF matrix
$\mathbf{d}$	Transmitted bit-wise signal

$\mathbf{s}$	Transmitted subcarrier-related SDM signal
$\mathbf{t}$	Transmitted subcarrier-related bit-wise SDM signal
$\check{\mathbf{s}}$	A priori signal vector estimate
$\hat{\mathbf{s}}$	A posteriori signal vector estimate
$\tilde{\mathbf{s}}$	Soft-information-aided signal vector estimate
$\hat{\mathbf{x}}$	Unconstrained a posteriori signal vector estimate
$\mathbf{w}$	Gaussian noise sample vector
$\mathrm{E}\{\cdot\}$	Expectation of a random variable
$\mathrm{Var}\{\cdot\}$	Variance of a random variable
$\mathrm{Cov}\{\cdot,\cdot\}$	Covariance of two random variables
$\mathrm{Ei}\{\cdot\}$	Exponential integral
$\mathrm{JacLog}(\cdot)$	Jacobian logarithm
$\mathrm{P}\{\cdot\}$	Probability density function
$\mathrm{rms}\{\cdot\}$	Root mean square value
$\|\cdot\|_2$	Second-order norm
κ	Channel estimation efficiency criteria
$\mathcal{R}\{\cdot\}$	Real part of a complex value
$\mathcal{I}\{\cdot\}$	Imaginary part of a complex value

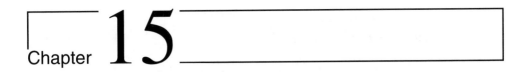

Chapter **15**

Multi-stream Detection for SDM-OFDM Systems

The principle of sphere decoding as a reduced-complexity near-ML SDMA MUD was first introduced in this volume in Chapter 9, where diverse SDs were studied comparatively. One of the important techniques considered was the OHRSA SDMA MUD. In this chapter no channel coding was employed – indeed, the sphere-decoding philosophy was originally conceived for hard-decision-aided MUDs, rather than soft-decision-assisted channel-coded systems. As a further enhanced SD, the centre-shifting philosophy was developed in Chapter 10 in the context of iterative detection-aided channel-coded SDMA systems. The SD of Chapter 10 was then further exploited throughout Part II of the book in diverse contexts, when a potentially complex detection task arose. For example, the SD was used for reducing the detection complexity of sphere-packing modulation-aided STBCs in Chapter 11 and for the complexity reduction of multiple-symbol differential detection in the cooperative wireless systems of Chapter 12.

By contrast, throughout Part III of the book, the application of SDs such as the OHRSA will be investigated in the context of SDM, rather than SDMA systems. The multi-stream detection task of SDM systems may be deemed to be more challenging than the SDMA MUD task, since in the case of SDM systems the antenna elements are likely to have a lower spatial separation than the geographically dispersed users of an SDMA system. Unless unique user/antenna-specific spreading codes are employed for differentiating the different data streams, the reduced-separation SDM antenna elements are likely to have similar impulse responses and hence the separation of the individual antennas' signals becomes a more challenging task than that in SDMA systems.

15.1 SDM/V-BLAST OFDM Architecture

In a simple SDM/V-BLAST OFDM architecture [326] the incoming data stream is demultiplexed into m_t parallel data streams. Each of the resultant data streams is independently channel encoded and OFDM modulated. The resultant m_t OFDM-modulated signals are processed by a bank of m_t synchronized transmitters, which operate within the same frequency band. Each of the m_t transmitters comprises a conventional OFDM transmitter having K subcarriers and an OFDM symbol period of T. In contrast to the D-BLAST scheme [312], the V-BLAST system configuration [326] imposes no special requirements on the particular structure of each of the multiple transmitters employed. Thus each of the transmitters can be thought of as a single-user transmitter employing a single transmit

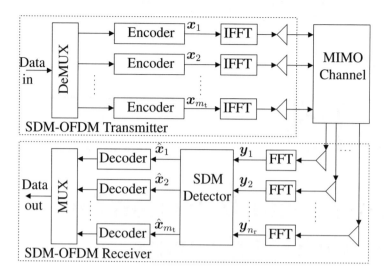

Figure 15.1: Schematic of an SDM-OFDM BLAST-type transceiver. In contrast to Figure 1.18, here the demultiplexed data substreams associated with different transmit antennas are channel encoded independently, which makes this system model equivalent to a multi-user SDMA system.

antenna. The SDM-OFDM architecture is illustrated in Figure 15.1. Observe that the structure of the SDM scheme depicted in Figure 15.1 is equally applicable to point-to-point SDM systems, as well as to systems supporting multiple users, each employing one or more transmit antennas. Consequently, the system configuration considered in this section is equivalent to the uplink multi-user SDMA-OFDM system discussed in [265].

15.2 Linear Detection Methods

The simple philosophy of the linear SDM detector is to recover the signal vector $\mathbf{x}[n, k] \in \mathbb{C}^{m_t}$ transmitted from the m_t elements of the transmit antenna array at time instance n and OFDM subcarrier k from the corresponding signal vector $\mathbf{y}[n, k] \in \mathbb{C}^{n_r}$, which is described by the received signal vector of Equation (1.25) recorded at the n_r elements of the receiver antenna array at time instance n and OFDM subcarrier k. More explicitly, we have

$$\hat{\mathbf{x}}[n, k] = \mathbf{W}^{\text{H}}[n, k]\mathbf{y}[n, k], \qquad (15.1)$$

where $\mathbf{W}[n, k] \in \mathbb{C}^{n_r \times m_t}$ is the corresponding linear SDM detector weight matrix, which is designed to yield the optimal linear estimate of the transmitted signal vector $\mathbf{x}[n, k]$, as detailed henceforth.

By substituting Equation (1.25) into (15.1) we have

$$\hat{x}_i = \mathbf{w}_i^{\text{H}}\mathbf{y}$$
$$= \mathbf{w}_i^{\text{H}}(\mathbf{Hx} + \mathbf{v})$$
$$= \underbrace{\mathbf{w}_i^{\text{H}}(\mathbf{H})_i x_i}_{\hat{x}_{i;\text{S}}} + \underbrace{\mathbf{w}_i^{\text{H}} \sum_{j=1; j \neq i}^{m_t} (\mathbf{H})_j x_j}_{\hat{x}_{i;\text{I}}} + \underbrace{\mathbf{w}_i^{\text{H}}\mathbf{v}}_{\hat{x}_{i;\text{N}}} \qquad (15.2)$$

$$= H_{ii;\text{eff}} x_i + v_{i;\text{eff}}, \qquad (15.3)$$

where $(\mathbf{H})_i$ is the ith column of the channel matrix $\mathbf{H}$, while $\mathbf{w}_i$ denotes the ith column of the weight matrix $\mathbf{W}$. We also define the corresponding additive components $\hat{x}_{i;S}$, $\hat{x}_{i;I}$ and $\hat{x}_{i;N}$ of the estimated signal $\hat{x}_i$ as suggested by Equation (15.2), where the subscripts S, I and N denote the Signal, Interference and the AWGN-related Noise signal components, respectively. Furthermore, we define the corresponding quantities seen in Equation (15.3) as in [265], namely

$$H_{ii;\text{eff}} = \mathbf{w}_i^H(\mathbf{H})_i \quad \text{and} \quad v_{i;\text{eff}} = \hat{x}_{i;I} + \hat{x}_{i;N}, \tag{15.4}$$

which are the effective channel coefficient and the effective interference-plus-noise component, respectively.

The choice of the particular linear SDM detector weight matrix $\mathbf{W}$ is dependent on the optimization criterion used. A number of examples of the relevant optimization criteria are discussed in [265] and include maximizing the Signal-to-Interference Ratio (SIR) as in the Least Squares (LS) method, maximizing the Signal-to-Interference-plus-Noise Ratio (SINR) as in the Minimum Mean Square Error (MMSE) technique, as well as maximizing the SIR, while ensuring a partial suppression of the AWGN as in the Minimum Variance (MV) method. When maximizing the SINR, which can be expressed as

$$SINR_i = \frac{\sigma_{i;S}^2}{\sigma_{i;I}^2 + \sigma_{i;N}^2}, \tag{15.5}$$

the associated MMSE method [265, Section 17.2.6.1] constitutes an optimal linear approach to the problem of the SDM detection. Thus, we will limit our discussion of the linear SDM detection methods here to the characterization of the MMSE SDM detector.

15.2.1 MMSE Detection

As advocated in [265], the problem of maximizing the $SINR$ of Equation (15.5) is equivalent to minimizing the mean square error at the output of the linear SDM detector of Equation (15.1). The MSE of the linear SDM detector of Equation (15.1) may be expressed as

$$\begin{aligned} MSE &= \mathrm{E}\{\Delta\mathbf{x}^H\Delta\mathbf{x}\} \\ &= \mathrm{E}\{(\mathbf{x} - \mathbf{W}^H\mathbf{y})^H(\mathbf{x} - \mathbf{W}^H\mathbf{y})\}. \end{aligned} \tag{15.6}$$

Differentiating the MSE of Equation (15.6) with respect to the elements of the liner SDM detector weight matrix $\mathbf{W}$ yields

$$\begin{aligned} \frac{\partial\, MSE}{\partial\mathbf{W}} &= \frac{\partial}{\partial\mathbf{W}}\sum_i \Delta x_i^* \Delta x_i \\ &= \mathrm{E}\left\{\sum_i\left(\frac{\partial}{\partial\mathbf{W}}\Delta x_i^*\,\Delta x_i + \Delta x_i^*\,\frac{\partial}{\partial\mathbf{W}}\Delta x_i\right)\right\} \\ &= -2\mathrm{E}\{\mathbf{y}(\mathbf{x} - \mathbf{W}^H\mathbf{y})^H\} \\ &= -2\mathrm{E}\{\mathbf{y}\Delta\mathbf{x}^H\} = \mathbf{0} \tag{15.7} \\ &= -2\mathrm{E}\{\mathbf{y}\mathbf{x}^H - \mathbf{y}\mathbf{y}^H\mathbf{W}\} \\ &= -2(\mathbf{R}_{yx} - \mathbf{R}_y\mathbf{W}) = \mathbf{0}, \tag{15.8} \end{aligned}$$

where $\mathbf{0} \in \mathbb{C}^{n_r \times m_t}$ is a zero matrix, while the cross-correlation and autocorrelation matrices R_{yx} and R_y of the transmitted and received signals, respectively, are given by

$$\begin{aligned} \mathbf{R}_{yx} &= \mathrm{E}\{(\mathbf{Hx} + \mathbf{v})\mathbf{x}^H\} \\ &= \mathbf{H}\mathrm{E}\{\mathbf{xx}^H\} = \mathbf{H}\mathbf{R}_x \tag{15.9} \end{aligned}$$

and

$$\begin{aligned}
\mathbf{R}_y &= \mathrm{E}\{(\mathbf{Hx} + \mathbf{v})(\mathbf{Hx} + \mathbf{v})^{\mathrm{H}}\} \\
&= \mathbf{H}\mathrm{E}\{\mathbf{xx}^{\mathrm{H}}\}\mathbf{H}^{\mathrm{H}} + \mathrm{E}\{\mathbf{vv}^{\mathrm{H}}\} \\
&= \mathbf{HR}_x\mathbf{H}^{\mathrm{H}} + \mathbf{R}_v \\
&= \mathbf{HR}_x\mathbf{H}^{\mathrm{H}} + \sigma_v^2\mathbf{I}.
\end{aligned} \tag{15.10}$$

Observe that Equation (15.7) represents the so-called *orthogonality principle* [330]. More specifically, the extremum of the cost function defined by the MSE of Equation (15.6) occurs when the estimation error signal $\Delta\mathbf{x}$ is orthogonal to the received signal $\mathbf{y}$. From Equation (15.8) we can deduce that

$$\mathbf{W}_{\mathrm{MMSE}} = (\mathbf{R}_y)^{-1}\mathbf{R}_{yx}. \tag{15.11}$$

Furthermore, substituting Equations (15.10) and (15.9) into (15.11) yields

$$\mathbf{W}_{\mathrm{MMSE}} = (\mathbf{HR}_x\mathbf{H}^{\mathrm{H}} + \sigma_v^2\mathbf{I})^{-1}\mathbf{HR}_x. \tag{15.12}$$

Equation (15.12) may be further expanded as follows:

$$\begin{aligned}
\mathbf{W} &= (\mathbf{R}_x^{-1}\mathbf{H}^{-1}(\mathbf{HR}_x\mathbf{H}^{\mathrm{H}} + \sigma_v^2\mathbf{I}))^{-1} \\
&= \left(\left(\mathbf{H}^{\mathrm{H}}\mathbf{H}\frac{1}{\sigma_v^2}\mathbf{R}_x + \mathbf{I}\right)\sigma_v^2\mathbf{R}_x^{-1}\mathbf{H}^{-1}\right)^{-1} \\
&= \frac{1}{\sigma_v^2}\mathbf{HR}_x\left(\mathbf{H}^{\mathrm{H}}\mathbf{H}\frac{1}{\sigma_v^2}\mathbf{R}_x + \mathbf{I}\right)^{-1}.
\end{aligned} \tag{15.13}$$

Finally, substituting the Hermitian transpose of the weight matrix $\mathbf{W}$ of Equation (15.13) into Equation (15.1) yields the MMSE SDM detector, which can be expressed as

$$\hat{\mathbf{x}} = (\mathbf{R}_{x;SNR}^{\mathrm{H}}\mathbf{H}^{\mathrm{H}}\mathbf{H} + \mathbf{I})^{-1}\mathbf{R}_{x;SNR}^{\mathrm{H}}\mathbf{H}^{\mathrm{H}}\mathbf{y}, \tag{15.14}$$

where we define the SNR-dependent autocorrelation matrix of the transmitted space-division signal vector $\mathbf{x}$ as $\mathbf{R}_{x;SNR} = (1/\sigma_v^2)\mathbf{R}_x$. In the typical case of mutually independent transmitted signal substreams $\mathbf{R}_{x;SNR}$ may be expressed as $\mathbf{R}_{x;SNR} = \mathrm{diag}\left(\sigma_i^2/\sigma_v^2\right)$, where σ_i^2 is the transmission power corresponding to the ith transmit antenna element. Furthermore, in the scenario where all the transmit antenna elements transmit the same power $\sigma_i^2 = \sigma_x^2/m_{\mathrm{t}}$, $i = 1, \ldots, m_{\mathrm{t}}$, we have

$$\mathbf{R}_{x;SNR} = \frac{\sigma_x^2}{m_{\mathrm{t}}\sigma_v^2}\mathbf{I} = \frac{\gamma}{m_{\mathrm{t}}}\mathbf{I}, \tag{15.15}$$

where, as before, γ is the average SNR value recorded at the receive antenna elements. Hence, the expression in Equation (15.14) can be further simplified by substituting Equation (15.15) into (15.14), yielding

$$\hat{\mathbf{x}} = \left(\frac{\gamma}{m_{\mathrm{t}}}\mathbf{H}^{\mathrm{H}}\mathbf{H} + \mathbf{I}\right)^{-1}\frac{\gamma}{m_{\mathrm{t}}}\mathbf{H}^{\mathrm{H}}\mathbf{y}. \tag{15.16}$$

15.2.1.1 Generation of Soft-Bit Information for Turbo Decoding

The BER associated with the process of communicating over a fading noisy MIMO channel can be dramatically reduced by means of employing channel coding. A particularly effective channel coding scheme is constituted by the *soft-input, soft-output* turbo coding method. Turbo coding, however, requires *soft* information concerning the bit decisions at the output of the SDM detector, in other words the a posteriori information regarding the confidence of the bit decision is required.

The derivation of an expression for the low-complexity evaluation of the soft-bit information associated with the bit estimates of the linear SDM detector's output characterized by Equation (15.16) is given in [265]. Here, we present a brief summary of the results deduced in [265].

The soft-bit value associated with the mth bit of the QAM symbol transmitted from the ith transmit antenna element is determined by the log-likelihood function defined in [623]

$$L_{im} = \ln \frac{P\{b_{im} = 1 | \hat{x}_i, H_{ii;\text{eff}}\}}{P\{b_{im} = 0 | \hat{x}_i, H_{ii;\text{eff}}\}}, \tag{15.17}$$

which is the logarithm of the a posteriori probabilities' ratio associated with the logical values of 1 and 0 of the mth bit corresponding to the QAM symbol transmitted from the ith transmit antenna. The term $\hat{x}_i$ in Equation (15.17) denotes the estimate of the transmitted signal $\mathbf{x}$ obtained by applying the linear SDM detection method considered, while $H_{ii;\text{eff}}$ is the effective channel coefficient defined by Equation (15.2), which can be evaluated as the ith element on the main diagonal of the effective channel matrix given by $\mathbf{H}_{\text{eff}} = \mathbf{W}^H \mathbf{H}$, where $\mathbf{W}$ is the linear SDM detector's weight matrix associated with the particular linear SDM detection method employed. More explicitly, in the case of the MMSE SDM detector of Equation (15.16) we have

$$\mathbf{H}_{\text{eff}} = \left(\frac{\gamma}{m_t} \mathbf{H}^H \mathbf{H} + \mathbf{I} \right)^{-1} \frac{\gamma}{m_t} \mathbf{H}^H \mathbf{H}. \tag{15.18}$$

The PDF of Equation (15.17) can be expressed as [265, Section 17.2.5]

$$P\{b_{im} = b \mid \hat{x}_i, H_{ii;\text{eff}}\} = \sum_{\check{x} \in \mathcal{M}_m^b} P\{\check{x}_i | \hat{x}_i, H_{ii;\text{eff}}\}, \tag{15.19}$$

where $\mathcal{M}_m^b$ denotes the specific subset of the total set $\mathcal{M}$ of constellation points associated with the modulation scheme employed, which have a logical value b at their mth bit position, i.e.

$$\mathcal{M}_m^b = \{\check{x} \mid \check{x} \in \mathcal{M}, b_m = b\}, \quad b \in \{0, 1\} \tag{15.20}$$

and b_m denotes the mth bit associated with the constellation point $\check{x}$. Furthermore, it is demonstrated in [265, Section 17.2.5] that

$$P\{\check{x}_i | \hat{x}_i, H_{ii;\text{eff}}\} = \frac{1}{(\pi \sigma_{v_i;\text{eff}}^2)} \exp\left(-\frac{1}{\sigma_{v_i;\text{eff}}^2} |\hat{x}_i - H_{ii;\text{eff}} \check{x}_i|^2 \right). \tag{15.21}$$

Consequently, substituting Equations (15.19) and (15.21) into (15.17) yields

$$L_{im} = \ln \frac{\sum_{\check{x} \in \mathcal{M}^1} \exp[-(1/\sigma_{v_i;\text{eff}}^2) |\hat{x}_i - H_{ii;\text{eff}} \check{x}_i^0|^2]}{\sum_{\check{x} \in \mathcal{M}^0} \exp[-(1/\sigma_{v_i;\text{eff}}^2) |\hat{x}_i - H_{ii;\text{eff}} \check{x}_i^1|^2]}. \tag{15.22}$$

15.2.1.2 Performance Analysis of the Linear SDM Detector

In this section, we present our simulation results for the SDM-OFDM system employing the MMSE SDM detection schemes described in Section 15.2.1.

Our simulations were performed in the baseband frequency domain and the system configuration characterized in Table 7.1 is to a large extent similar to that used in [260]. We assume a total bandwidth of 800 kHz. The OFDM system utilizes 128 QPSK-modulated orthogonal subcarriers. For Forward Error Correction (FEC) we use $\frac{1}{2}$-rate turbo coding [314] employing two constraint-length $K = 3$ Recursive Systematic Convolutional (RSC) component codes and the standard 124-bit WCDMA UMTS turbo code interleaver of [512]. The octally represented RCS generator polynomials of $(7, 5)$ were used. Finally, throughout this report we stipulate the assumption of perfect channel knowledge, where knowledge of the frequency-domain subcarrier-related coefficients $H[n, k]$ is deemed to be available in the receiver.

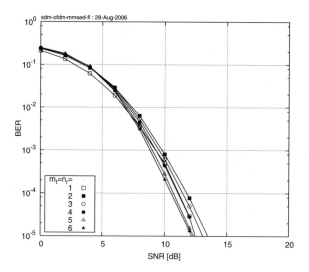

Figure 15.2: BER exhibited by the QPSK-modulated **SDM-OFDM** system employing an **MMSE** SDM detector of Equation (15.14) and $m_t = n_r = 1, \ldots, 6$ transmit and receive antennas. The abscissa represents the average SNR recorded at the receive antenna elements. The system parameters are summarized in Table 7.1.

Figure 15.2 demonstrates the ability of the SDM-OFDM system employing the MMSE SDM detector of Equation (15.16) to exploit the available MIMO channel capacity gain in the fully loaded system configuration, i.e. when the number of the transmit antenna elements m_t is equal to that of the receiver antenna elements n_r. Figure 15.2 depicts the achievable BER performance of the SDM-OFDM system considered as a function of the average SNR recorded at each of the receiver antenna elements. More explicitly, the results depicted in Figure 15.2 illustrate that the SDM-OFDM system employing $m_t = n_r = 6$ transmit and receive antennas, as well as the MMSE SDM detector of Equation (15.16), is capable of achieving an SNR gain of about 1 dB at the target BER of 10^{-3}, when compared with the same system employing a single antenna element at both the transmitter and receiver.

Figure 15.3 demonstrates the SDM-OFDM system's capability to detect the spatially multiplexed signals arriving from various numbers of transmit antennas, when employing the MMSE SDM detection method of Equation (15.16) and having a constant number of $n_r = 4$ receive antenna elements. Specifically, we aim at exploring the performance of the MMSE SDM detector in the overloaded system scenario, where the number of transmit antenna elements exceeds that of the receiver elements. Figure 15.3 demonstrates the achievable BER performance of the MMSE SDM detector considered as a function of the average SNR recorded at each of the receiver antenna elements. We can see that the MMSE SDM detector exhibits a relatively good performance whenever the number of transmit antenna elements is lower than or equal to the number of receiver antenna elements. As seen in Figure 15.3, the system exhibits a diversity gain of about 2 dB recorded in terms of the SNR at the target BER of 10^{-3}, when comparing the scenarios of $m_t = 3$ and $m_t = 4$ receiver antenna elements. On the other hand, however, the MMSE SDM detector of Equation (15.16) exhibits a severe performance degradation in the overloaded scenario, i.e. when we have $m_t > n_r$, which is confirmed by the curves corresponding to the scenarios of $m_t = 5$ and 6 characterized in Figure 15.3.

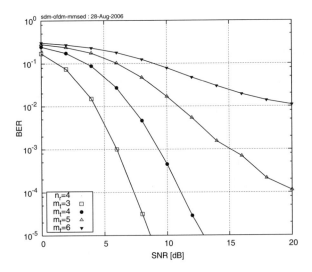

Figure 15.3: BER performance exhibited by the **SDM-QPSK-OFDM** system employing an **MMSE** SDM detector of Equation (15.14) and $m_t = 3, 4, 5$ and 6 transmit antennas, as well as $n_r = 4$ receive antennas. The abscissa represents the average SNR recorded at the receive antenna elements. The system parameters are summarized in Table 7.1.

15.3   Nonlinear SDM Detection Methods

In Section 15.2 we discussed the linear approach to the problem of SDM detection. The major advantage of the linear detection strategy is its conceptual simplicity and corresponding low computational complexity. Unfortunately, however, as is evident from our discussion in Section 15.2, the output of the linear SDM detector contains a substantial amount of residual interference.

In this section we explore a family of nonlinear SDM detection methods. We commence our discourse with the derivation of the Maximum Likelihood (ML) SDM detection method, which constitutes an optimal solution of the SDM detection problem from the ML sequence detection point of view. Unfortunately, the brute-force ML detection method does not provide a feasible solution to the generic SDM detection problem as a result of its excessive computational complexity. Nevertheless, it provides an important benchmark for the overall achievable performance of a generic SDM detector.

We then continue our discussions by considering two additional nonlinear SDM detection methods, which achieve a suboptimal performance at a realistic computational complexity. More explicitly, in Section 15.3.2 we consider the SIC-aided SDM detection method. Furthermore, in Section 15.3.3 we invoke the Genetic Algorithm (GA) aided MMSE SDM detector.

15.3.1   ML Detection

The ML method [193, 265, 299] constitutes an optimal SDM detection method in the sense of an a posteriori probability. The simple philosophy of ML detection is based on an exhaustive search through all possible values of the transmitted signal vector $\mathbf{x}$ with the aim of determining the value which is most likely to have been transmitted. Clearly, the major drawback of this strategy is its excessive computational complexity. Specifically, the number of potential candidate values of the signal vector $\mathbf{x}[n, k]$ of the m_t transmit antennas associated with the kth OFDM subcarrier of the nth OFDM symbol is given by $M^{m_t} = 2^{r m_t}$, where M is the number of phasor constellation points comprising the M-QAM/M-PSK constellation employed, while r is the corresponding number of bits per M-QAM/M-PSK modulated symbol. More explicitly, this relationship suggests that the number of potential signal vector candidates to be examined by the ML detector increases exponentially with the number of transmitter antennas, as well as with the number of bits per modulated symbol. The resultant computational complexity may become excessive for systems employing a high number of transmit antennas and/or high-level modulation schemes, which renders it unsuitable for practical applications. As noted above, however, the performance of the ML SDM detector constitutes an important benchmark for the performance evaluation of other, more practical SDM detection schemes.

Let us recall that our channel model described by Equation (1.25) was given by

$$\mathbf{y} = \mathbf{Hx} + \mathbf{w}, \tag{15.23}$$

where, as before, we omit the OFDM subcarrier and symbol indices k and n, respectively. As outlined above, the output of the ML SDM detector considered comprises a signal vector candidate $\hat{\mathbf{x}}$, which maximizes the a posteriori probability function

$$\hat{\mathbf{x}} = \arg\max_{\check{\mathbf{x}} \in \mathcal{M}^{m_t}} \mathrm{P}\{\check{\mathbf{x}}|\mathbf{y}, \mathbf{H}\}, \tag{15.24}$$

where $\mathcal{M}^{m_t}$ is the set of all possible candidate symbol values of the transmitted signal vector $\mathbf{x}$, i.e.

$$\mathcal{M}^{m_t} = \{\check{\mathbf{x}} = (\check{x}_1, \ldots, \check{x}_{m_t})^{\mathrm{T}};\ \check{x}_i \in \mathcal{M}\} \tag{15.25}$$

and $\mathcal{M}$ denotes the entire set of M complex constellation points associated with the particular M-QAM/M-PSK modulation scheme employed.

It follows from Bayes' theorem [623] that the conditional probability of Equation (15.24) can be expressed as

$$\mathrm{P}\{\check{\mathbf{x}}|\mathbf{y}, \mathbf{H}\} = \mathrm{P}\{\mathbf{y}|\check{\mathbf{x}}, \mathbf{H}\}\frac{\mathrm{P}\{\check{\mathbf{x}}\}}{\mathrm{P}\{\mathbf{y}\}}, \tag{15.26}$$

where all possible values of the transmitted signal vector $\check{\mathbf{x}}$ are assumed to be equally probable and therefore we have $\mathrm{P}\{\check{\mathbf{x}}\} = 1/M^{m_t} = \mathrm{const}$. Moreover, we have

$$\mathrm{P}\{\mathbf{y}\} = \sum_{\check{\mathbf{x}} \in \mathcal{M}^{m_t}} \mathrm{P}\{\mathbf{y}|\check{\mathbf{x}}, \mathbf{H}\}\mathrm{P}\{\check{\mathbf{x}}\} = \mathrm{const}, \tag{15.27}$$

which follows from the probability function normalization property of

$$\sum_{\check{\mathbf{x}} \in \mathcal{M}^{m_t}} \mathrm{P}\{\check{\mathbf{x}}|\mathbf{y}, \mathbf{H}\} \equiv 1. \tag{15.28}$$

We can therefore infer that

$$\hat{\mathbf{x}} = \arg\max_{\check{\mathbf{x}} \in \mathcal{M}^{m_t}} \mathrm{P}\{\check{\mathbf{x}}|\mathbf{y}, \mathbf{H}\} \quad \Leftrightarrow \quad \hat{\mathbf{x}} = \arg\max_{\check{\mathbf{x}} \in \mathcal{M}^{m_t}} \mathrm{P}\{\mathbf{y}|\check{\mathbf{x}}, \mathbf{H}\}. \tag{15.29}$$

As was pointed out in [265], the signal vector $\mathbf{y}$ recorded at the n_r receive antenna elements can be represented as a sample of multi-variate complex-Gaussian-distributed random variables with mean $\mathbf{Hx}$ and the covariance matrix given by Equation (15.10), which may be summarized as $\mathbf{y} \sim \mathcal{CN}(\mathbf{Hx}, \mathbf{R}_y)$, where we denote the complex-valued normal distribution having a mean given by the vector $\boldsymbol{\mu}$ and the covariance matrix $\mathbf{C}$ as $\mathcal{CN}(\boldsymbol{\mu}, \mathbf{C})$. The corresponding PDF can thus be expressed as in [330]

$$\mathrm{P}\{\mathbf{y}|\mathbf{x}, \mathbf{H}\} = \frac{1}{(\pi\sigma_w^2)^{n_r}} \exp\left(-\frac{1}{\sigma_w^2}\|\mathbf{y} - \mathbf{Hx}\|_2^2\right). \tag{15.30}$$

The PDF in Equation (15.30) has a form of $\mathrm{P}\{J\} = \alpha e^{-\beta J}$, where α and β are constants and we define $J(\check{\mathbf{x}}) = \|\mathbf{y} - \mathbf{H}\check{\mathbf{x}}\|_2^2$. Clearly, $\mathrm{P}\{J\}$ is a monotonically decreasing function of its argument J. Consequently, the maximum of the a posteriori probability function of Equation (15.24) can be substituted by the minimum of the corresponding argument $J(\breve{\mathbf{x}})$, such that

$$\hat{\mathbf{x}} = \arg\min_{\breve{\mathbf{x}} \in \mathcal{M}^{m_t}} J(\breve{\mathbf{x}}), \tag{15.31}$$

where, again, $J(\breve{\mathbf{x}})$ is defined as a Euclidean distance-based cost function, which may be expressed as

$$J(\breve{\mathbf{x}}) = \|\mathbf{y} - \mathbf{H}\breve{\mathbf{x}}\|_2^2 = \sum_{i=1}^{m_t} \left| y_i - \sum_{j=1}^{n_r} H_{ij}\breve{x}_j \right|^2. \tag{15.32}$$

15.3.1.1 Generation of Soft-Bit Information

Based on our arguments in Section 15.2.1.1, the soft-bit value associated with the mth bit of the QAM symbol transmitted from the ith transmit antenna element is determined by the log-likelihood function defined in [623]:

$$L_{mi} = \ln \frac{\sum_{\breve{\mathbf{x}} \in \mathcal{M}_{mi}^{1;m_t}} \mathrm{P}\{\mathbf{y}|\breve{\mathbf{x}}, \mathbf{H}\}}{\sum_{\breve{\mathbf{x}} \in \mathcal{M}_{mi}^{0;m_t}} \mathrm{P}\{\mathbf{y}|\breve{\mathbf{x}}, \mathbf{H}\}}; \tag{15.33}$$

where we define

$$\mathcal{M}_{mi}^{b;m_t} = \{\breve{\mathbf{x}} = (\breve{x}_1, \dots, \breve{x}_{m_t})^{\mathrm{T}}; \ \breve{x}_j \in \mathcal{M} \text{ for } j \neq i, \ \breve{x}_i \in \mathcal{M}_m^b\} \tag{15.34}$$

and $\mathcal{M}_m^b$ denotes the specific subset of the entire set $\mathcal{M}$ of constellation points of the modulation scheme employed, which comprises the bit value $b = \{0, 1\}$ at the mth bit position.

Substituting Equation (15.30) into (15.33) yields

$$L_{mi} = \ln \frac{\sum_{\breve{\mathbf{x}} \in \mathcal{M}_{mi}^{1;m_t}} \exp[-(1/\sigma_w^2)\|\mathbf{y} - \mathbf{H}\breve{\mathbf{x}}\|^2)]}{\sum_{\breve{\mathbf{x}} \in \mathcal{M}_{mi}^{0;m_t}} \exp[-(1/\sigma_w^2)\|\mathbf{y} - \mathbf{H}\breve{\mathbf{x}}\|^2)]}. \tag{15.35}$$

Note that Equation (15.35) involves summation over $2^{rm_t - 1}$ exponential functions. This operation may potentially impose an excessive computational complexity for large values of m_t and/or r. As demonstrated in [265], however, the expression in (15.35) may be closely approximated by a substantially simpler expression, namely by

$$L_{mi} \approx \frac{1}{\sigma_w^2}[\|\mathbf{y} - \mathbf{H}\breve{\mathbf{x}}_m^0\|^2 - \|\mathbf{y} - \mathbf{H}\breve{\mathbf{x}}_m^1\|^2], \tag{15.36}$$

where

$$\breve{\mathbf{x}}_m^b = \arg\min_{\breve{\mathbf{x}} \in \mathcal{M}_{mi}^{b;m_t}} \|\mathbf{y} - \mathbf{H}\breve{\mathbf{x}}\|^2, \quad b = 0, 1. \tag{15.37}$$

15.3.1.2 Performance Analysis of the ML SDM Detector

In this section, we present our simulation results characterizing the SDM-OFDM system employing the ML SDM detection schemes described in Section 15.3.1. Our simulation setup is identical to that described in Section 15.2.1.2 and the corresponding simulation parameters are summarized in Table 7.1.

Figure 15.4 demonstrates that the SDM-OFDM system employing the ML SDM detector of Equation (15.31) is capable of exploiting the available MIMO channel's multiplexing gain in the fully loaded system scenario, when the number of transmit antenna elements m_t is equal to that of receiver antenna elements n_r. More specifically, Figure 15.4 depicts the achievable BER performance of the SDM-OFDM ML detector considered as a function of the average SNR recorded at the receiver antenna elements.

The results depicted in Figure 15.4 illustrate that the SDM-OFDM ML detector having $m_t = n_r = 6$ transmit and receive antennas exhibits an SNR gain of 3 dB at the target BER of 10^{-3}, when compared

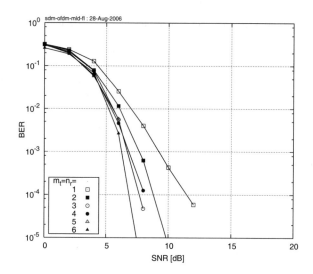

Figure 15.4: **BER** exhibited by the QPSK-modulated **SDM-OFDM** system employing an **ML** SDM detector of Equation (15.24) and $m_t = n_r = 1, \ldots, 6$ transmit and receive antennas. The abscissa represents the average SNR recorded at the receive antenna elements. The system parameters are summarized in Table 7.1.

with the same system employing a single antenna element at both the transmitter and receiver, as well as a higher throughput by a factor of six.

Additionally, Figure 15.5 characterizes the capability of the SDM-OFDM system employing the ML SDM detector of Equation (15.31) and having a constant number of $n_r = 4$ receive antenna elements, to detect the multiplexed signals arriving from various numbers of transmit antenna elements. Specifically, we aim at exploring the performance of the ML SDM detector in the overloaded system scenario, where the number of transmit antenna elements exceeds that of receiver elements and thus we have $m_t > n_r$. Figure 15.5 demonstrates the achievable BER performance of the SDM-OFDM system employing the ML SDM detector as a function of the average SNR recorded at the receive antenna elements. We can see that, as opposed to the MMSE SDM detector discussed in Section 15.3.1, the ML SDM detector exhibits a good performance both when $m_t \leq n_r$ and in the overloaded system scenario when the number of transmit antenna elements exceeds the number of receive antenna elements, i.e. when $m_t > n_r$.

15.3.2 SIC Detection

The SIC-assisted SDM detector was proposed by Foschini *et al.* in [312] and was discussed in further detail in [326, 327, 441, 624, 625].

In order to commence our discourse, let us recall the philosophy of the linear SDM detector discussed in Section 15.2, where the detection of the transmitted signal vector $\mathbf{x}[n, k]$ was performed using a linear transformation described by Equation (15.1), i.e. by

$$\hat{\mathbf{x}}[n, k] = \mathbf{W}^{\text{H}}[n, k]\mathbf{y}[n, k], \tag{15.38}$$

where $\mathbf{W}[n, k] \in \mathbb{C}^{n_r \times m_t}$ is the corresponding linear SDM detector weight matrix.

As was further inferred in Section 15.2, the corresponding SINR at the output of the linear SDM detector may vary considerably across different elements of the transmitted signal vector $\mathbf{x}[n, k]$, as substantiated by Equation (15.5). Consequently, as suggested in [265], the overall MSE at the output of the linear SDM detector employed is dominated by the SINR associated with the transmitted signal

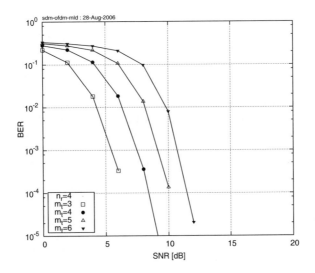

Figure 15.5: BER performance exhibited by the **SDM-QPSK-OFDM** system employing an **ML** SDM detector of Equation (15.24) and $m_t = 3, 4, 5$ and 6 transmit antennas, as well as $n_r = 4$ receive antennas. The abscissa represents the average SNR recorded at the receive antenna elements. The system parameters are summarized in Table 7.1.

component having the lowest signal power [265] determined by $\sum_j |H_{ij}|^2$. This observation suggests that a considerably higher performance can be achieved by employing SIC.

Following the SIC paradigm, the detection of the transmitted signal vector $\mathbf{x}[n, k]$ associated with the kth OFDM subcarrier of the nth OFDM symbol is performed in a successive manner, where at each detection iteration i we detect a single vector component $x_i[n, k]$ using the linear MMSE SDM detection method discussed in Section 15.2.1. We then modify the received signal vector $\mathbf{y}[n, k]$ by removing the remodulated interfering signal components and repeat the aforementioned linear detection process in order to estimate the next transmitted signal component $x_{j_{i+1}}$. The iterative process described above is then repeated until the transmitted signal components associated with all transmitter antenna elements are detected. In this section we will demonstrate that the successive structure of the detection process results in a substantially improved SIR for the weaker signal components. Note that in our forthcoming derivation we, once again, omit the OFDM symbol and subcarrier indices n and k, which does not restrict the generality of the results obtained, since the space-division detection process described is performed independently for each pair of time- and frequency-domain indices $[n, k]$.

More specifically, we commence our SIC detection process with a linear detection of the transmitted signal component x_{j_1}, as suggested by Equation (15.1), where we have

$$\hat{x}_{j_1} = \mathbf{w}_1^H \mathbf{y}_1, \tag{15.39}$$

and $\mathbf{w}_1 = (\mathbf{W})_{j_1}$ is the j_1th column of the SDM MMSE detector's weight matrix described by Equation (15.13), while $\mathbf{y}_1$ is assumed to be identical to the original received signal vector $\mathbf{y}$.

In the next step, the interference imposed by the just detected and remodulated signal component x_{j_1} is subtracted from the received signal $\mathbf{y}_1$, yielding

$$\mathbf{y}_2 = \mathbf{y}_1 - (\mathbf{H})_{j_1} Q(\hat{x}_{j_1}), \tag{15.40}$$

where $(\mathbf{H})_{j_1}$ is the j_1th column of the channel matrix $\mathbf{H}$, while $Q(x)$ represents the slicing or hard-decision operation performed in the receiver in order to estimate the transmitted information-carrying QAM/PSK symbol. The resultant partially decontaminated signal $\mathbf{y}_2$ comprises the contributions of a reduced number of interferers. In order to detect our next desired transmitted signal component

x_{j_2}, we have to calculate the updated linear SDM detector weight matrix $\mathbf{W}_2$, which may be readily achieved by substituting the effective channel matrix $\mathbf{H}_{\overline{j_1}}$, obtained by zeroing column j_1 of the original channel matrix $\mathbf{H}$, into Equation (15.13), yielding

$$\mathbf{W}_2 = \mathbf{H}_{\overline{j_1}}(\mathbf{H}_{\overline{j_1}}^{\mathrm{H}}\mathbf{H}_{\overline{j_1}} + m_t\sigma_w^2\mathbf{I})^{-1}, \tag{15.41}$$

where we follow the notation employed in [624] and correspondingly $\mathbf{H}_{\overline{j_i}}$ denotes the matrix obtained by zeroing columns $j_1, \ldots, j_i$ of the original matrix $\mathbf{H}$. By substituting the terms $\hat{x}_{j_1}$, $\mathbf{w}_1$ and $\mathbf{y}_1$ of Equation (15.39) by the corresponding terms $\hat{x}_{j_2}$, $\mathbf{y}_2$ of Equation (15.40) and $\mathbf{w}_2 = (\mathbf{W})_2$ of Equation (15.41), we arrive at the desired estimate of the next transmitted signal component. Finally, the iterative detection process described above is repeated until all desired transmitted signal components are successfully detected.

As argued in [624], the order in which the detection of the transmitted signal components $x_j[n]$, $j = 1, \ldots, m_t$, is performed is important for the overall performance of the detection process. Moreover, as demonstrated in [624], the optimal ordering arises if the 'best-first' successive detection strategy is applied, where the best possible performance is achieved when at each iteration i of the SIC detection process the desired signal component is selected according to the selection criterion of

$$j_{i+1} = \arg\max_j \|(\mathbf{H}_{\overline{j_i}})_j\|^2, \tag{15.42}$$

implying that the least-attenuated, i.e. the highest-power, antenna's signal is detected first.

The SDM SIC detection process employing the MMSE detection method of Section 15.2.1 is summarized in Algorithm 15.1.

Algorithm 15.1 MMSE-aided V-BLAST SIC SDM detector

$$\mathbf{y}_1 = \mathbf{y}[n]$$
$$\mathbf{W}_1 = \mathbf{H}(\mathbf{H}^{\mathrm{H}}\mathbf{H} + m_t\sigma_w^2\mathbf{I})^{-1}$$
$$j_1 = \arg\max_j \|(\mathbf{H})_j\|^2 \tag{15.43a}$$

$\quad$ for $\quad i = 1, 2, \ldots, m_t \quad$ do

$$\mathbf{w}_{j_i} = (\mathbf{W}_i)_{j_i} \tag{15.43b}$$
$$\hat{x}_{j_i}[n] = \mathbf{w}_{j_i}^{\mathrm{H}}\mathbf{y}_i \tag{15.43c}$$
$$\mathbf{y}_{i+1} = \mathbf{y}_i - (\mathbf{H}[n])_{j_i}Q(\hat{x}_{j_i}) \tag{15.43d}$$
$$\mathbf{W}_{i+1} = \mathbf{H}_{\overline{j_i}}(\mathbf{H}_{\overline{j_i}}^{\mathrm{H}}\mathbf{H}_{\overline{j_i}} + m_t\sigma_w^2\mathbf{I})^{-1} \tag{15.43e}$$
$$j_{i+1} = \arg\max_j \|(\mathbf{H}_{\overline{j_i}})_j\|^2 \tag{15.43f}$$

$\quad$ end $\quad$ for

15.3.2.1 Performance Analysis of the SIC SDM Detector

In this section we present our performance results for the SDM-OFDM system employing the SIC SDM detection scheme described in Section 15.3.2. The simulation setup is identical to that described in Section 15.2.1.2 and the corresponding simulation parameters are summarized in Table 7.1.

Figure 15.6 characterizes the ability of the SDM-OFDM system employing the SIC SDM detector of Algorithm 15.1 to exploit the available MIMO multiplexing gain in the fully loaded system configuration, when the number of transmit antenna elements m_t is equal to that of receiver antenna elements n_r. More explicitly, Figure 15.6 depicts the achievable BER performance of the SDM-OFDM

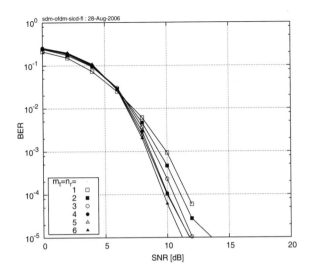

Figure 15.6: BER exhibited by the QPSK-modulated **SDM-OFDM** system employing an **SIC** SDM detector of Equation (15.38) and $m_t = n_r = 1, \ldots, 6$ transmit and receive antennas. The abscissa represents the average SNR recorded at the receive antenna elements. The system parameters are summarized in Table 7.1.

SIC system considered as a function of (1) the average SNR recorded at the receiver antenna elements, as well as (2) versus the corresponding E_b/N_0 value for various numbers of $m_t = n_r = 1, \ldots, 6$ transmit and receive antenna elements.

More specifically, the results portrayed in Figure 15.6 illustrate on the SNR scale that the SDM-OFDM SIC system having $m_t = n_r = 6$ transmit and receive antennas exhibits an SNR gain of about 2 dB at the target BER of 10^{-3}, when compared with the same system employing a single antenna element at both the transmitter and receiver.

Furthermore, Figure 15.7 illustrates the capability of the SDM-OFDM system employing the SIC SDM detector of Algorithm 15.1 and having a constant number of $n_r = 4$ receive antenna elements to detect the multiplexed signal arriving from various numbers of transmit antenna elements. Specifically, we aim at exploring the attainable performance of the SIC SDM detector in the overloaded system scenario, where the number of transmit antenna elements exceeds that of receiver antenna elements and thus we have $m_t > n_r$. Figure 15.7 demonstrates the achievable BER performance of the SDM-OFDM system employing the SIC SDM detector as a function of the average SNR recorded at the receive antenna elements. In can be seen by comparing Figures 15.7 and 15.3 that in an overloaded scenario the SIC SDM detector considered performs better than the MMSE SDM detector of Section 15.2.1. Nevertheless, note from the comparison of Figures 15.7 and 15.5 that a substantial performance degradation may still be observed in comparison with the ML SDM detector of Section 15.3.1. A more detailed comparison of the achievable performance corresponding to the various SDM detection methods considered will be carried out in Section 15.5.

15.3.3 GA-Aided MMSE Detection

Genetic algorithms [328, 454] constitute a family of optimization algorithms often utilized to find approximate solutions to optimization problems having irregular error surfaces associated with local minima, such as in interference, rather than noise-limited propagation environments [626]. GAs use biologically inspired search and optimization methods, such as inheritance, mutation, natural selection and recombination (or cross-over) of genes, each representing for example a bit string describing

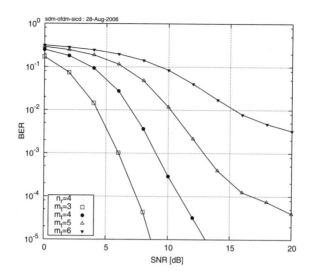

Figure 15.7: **BER** performance exhibited by the **SDM-QPSK-OFDM** system employing an **SIC** SDM detector of Equation (15.38) and $m_t = 3, 4, 5$ and 6 transmit antennas, as well as $n_r = 4$ receive antennas. The abscissa represents the average SNR recorded at the receive antenna elements. The system parameters are summarized in Table 7.1.

a potential candidate of the transmitted multiplexed signal vector. Again, the GA's individuals are represented as strings of discrete symbols, such as for instance zeros and ones, but using different encoding schemes is also possible. In each generation, pairs of parent individuals are selected from the current population based on their *fitness* properties. They are modified (mutated or recombined) to form a new population, which becomes the current population in the next iteration of the algorithm.

GAs were found to be highly efficient in numerous global search and optimization problems, especially when their solution using conventional methods is not feasible or otherwise would impose an excessive computational complexity. GAs were first applied to the problem of multi-user detection by Juntti *et al.* in [329] and Wang *et al.* in [627]. They were then documented in great detail in [626].

In our case, we explore the achievable performance of the GA-aided SDM detection method in the context of the SDM-OFDM system of Section 1.8.3. We employ an SDM-MMSE detector described in Section 15.2.1 as our solution in the initial population at the input of the GA-aided SDM detector. A detailed description of GA-aided detection and the particular configuration of the GA employed is beyond the scope of this report. The configuration of the GA employed here is identical to that described in much detail in [628]. The interested reader may also refer to [626] for further insight.

In the next section, we explore the achievable performance of the GA-aided SDM detector in the context of the SDM-OFDM system of Figure 15.1. The simulation setup of the SDM-OFDM system is identical to that described in Section 15.2.1.2, as summarized in Table 7.1, while details concerning the configuration of the GA-MMSE SDM detector employed can be found in Table 15.1.

15.3.3.1 Performance Analysis of the GA-MMSE SDM Detector

The achievable BER performance of the SDM-OFDM system of Figure 15.1 employing the GA-MMSE SDM detection method described in [628] is depicted in Figure 15.8. More explicitly, Figure 15.8 demonstrates the ability of the SDM-OFDM system employing the GA-MMSE SDM detector [628] to exploit the available MIMO capacity gain in the fully loaded system configuration, when the number of transmit antenna elements m_t is equal to that of the receiver antenna elements n_r. To elaborate a little further, Figure 15.8 depicts the achievable BER performance of the SDM-OFDM system considered

Table 15.1: The configuration parameters of the GA-aided SDM detector [628].

Population initialization method	Output of the MMSE MUD
Mating pool creation strategy	Pareto-optimality
Selection method	Fitness-proportionate
Cross-over operation	Uniform cross-over
Mutation operation	M-ary mutation
Elitism	Enabled
Incest prevention	Enabled
Population size X	Varied
Number of generations Y	Varied
Mutation probability p_m	0.1

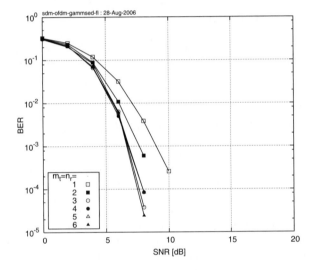

Figure 15.8: **BER** exhibited by the QPSK-modulated **SDM-OFDM** system employing an **SIC** SDM detector described in [628] and $m_t = n_r = 1, \ldots, 6$ transmit and receive antennas. The abscissa represents the average SNR recorded at the receive antenna elements. The system parameters are summarized in Table 7.1.

as a function of the average SNR recorded at each of the receiver antenna elements. As can be seen in Figure 15.8, the SDM-OFDM system employing the GA-MMSE SDM detector and $m_t = n_r = 6$ transmit and receive antennas exhibits an SNR gain of above 2 dB at the target BER of 10^{-3}, when compared with the same system of Table 7.1 employing a single antenna element at both the transmitter and receiver.

15.4 Performance Enhancement Using Space–Frequency Interleaving

Employing frequency-domain interleaving is common practice in OFDM transceivers [265], since it enables exploitation of the available Frequency-Domain (FD) diversity provided by a frequency-selective wireless fading channel. In this section we explore the further benefits of employing space–frequency interleaving in the context of the SDM-OFDM system architecture investigated.

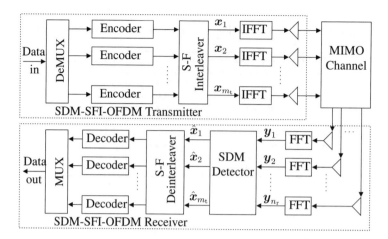

Figure 15.9: Schematic of an SDM-SFI-OFDM transceiver. In contrast to the SDM-OFDM scheme characterized in Figure 15.1, here the OFDM-subcarrier-related data substreams associated with different transmit antenna elements are SFI at the output of the channel encoder.

15.4.1 Space–Frequency-Interleaved OFDM

The structure of the Space–Frequency-Interleaved (SFI) SDM-OFDM system considered is illustrated in Figure 15.9. Observe that, in contrast to the system architecture portrayed in Figure 15.1, the set of OFDM-subcarrier-related data substreams at the outputs of the bank of channel encoders seen in Figure 15.9 are jointly interleaved, resulting in the SFI signal vectors $\mathbf{x}_i$, where $i = 1, \ldots, m_t$ is the index corresponding to the different transmit antenna elements. Correspondingly, at the SDM-SFI-OFDM receiver of Figure 15.9 the set of detected OFDM-subcarrier-related signal vectors $\hat{\mathbf{x}}_i$ is space–frequency deinterleaved, before they are processed by the bank of channel decoders portrayed in Figure 15.9. As a result, the impact of the channel impairments, such as fading and interference, is uniformly spread across the data substreams associated with the different transmit antenna elements. In other words, the SDM-SFI-OFDM system considered is capable of more efficiently exploiting both the space and frequency diversity benefits of the wireless MIMO channel. Consequently, we may expect that the SDM-SFI-OFDM system advocated will outperform the SDM-OFDM system of Section 15.1 in terms of the achievable BER performance.

15.4.1.1 Performance Analysis of the SFI-SDM-OFDM

As a test case for exploring the achievable performance of the SDM-SFI-OFDM scheme advocated, we employ the GA-MMSE SDM detector characterized in Section 15.3.3. Figure 15.10 demonstrates the ability of the SDM-SFI-OFDM system employing the GA-MMSE SDM detector of Section 15.3.3 to exploit the available MIMO channel capacity gain in the fully loaded system configuration, i.e. when the number of transmit antenna elements m_t is equal to that of receiver antenna elements n_r. Specifically, Figure 15.10 depicts the achievable BER performance of the SDM-OFDM system considered as a function of the average SNR recorded at each of the receiver antenna elements. Furthermore, the results depicted in Figure 15.10 illustrate that the SDM-OFDM system employing $m_t = n_r = 6$ transmit and receive antennas, as well as the GA-MMSE SDM detector, is capable of achieving an SNR gain of 3 dB at the target BER of 10^{-3}, when compared with the same system employing a single antenna element at both the transmitter and receiver.

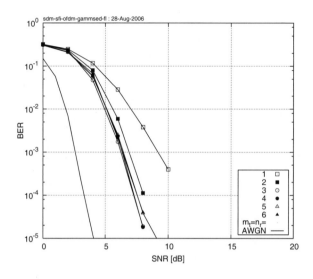

Figure 15.10: BER exhibited by the rate $\frac{1}{2}$ turbo-coded QPSK-modulated **SDM-SFI-OFDM** system employing the **GA-MMSE** SDM detector described in [628] and $m_t = n_r = 1, \ldots, 6$ transmit and receive antennas. The abscissa represents the average SNR recorded at the receive antenna elements. The achievable performance of the SDM-OFDM system employing the GA-MMSE detector was characterized in Figure 15.8. The OFDM system parameters are summarized in Table 7.1 and the corresponding GA configuration parameters are outlined in Table 15.1.

15.5 Performance Comparison and Discussion

In this section we compare the achievable performance of the SDM detection methods considered in Sections 15.2 and 15.3 in the context of both the SDM-OFDM and SDM-SFI-OFDM systems of Sections 15.1 and 15.4.1, respectively. More specifically, Figure 15.11 portrays the achievable BER performance of the SDM-MMSE detector of Section 15.2.1, as well as that of the ML, SIC and GA-MMSE SDM detectors described in Sections 15.3.1, 15.3.2 and 15.3.3, respectively. Figures 15.11(a) and 15.11(b) correspond to the scenarios of $m_t = n_r = 2$ and 6 transmit and receive antenna elements, respectively. Furthermore, the open markers in Figures 15.11(a) and 15.11(b) correspond to the SDM-OFDM scheme characterized in Figure 15.1, while the solid markers correspond to the SDM-SFI-OFDM arrangement portrayed in Figure 15.9.

It can be seen in Figures 15.11(a) and 15.11(b) that the SNR performance of the nonlinear SDM detection methods, namely that of the ML, SIC and GA-MMSE detectors of Sections 15.3.1, 15.3.2 and 15.3.3, respectively, is significantly higher than the corresponding performance of the linear MMSE SDM detector characterized in Section 15.2.1. This conclusion holds for the scenarios of both the SDM-OFDM and SDM-SFI-OFDM systems. Furthermore, the SNR performance of the GA-MMSE detector is within a 1 dB margin of the SNR performance exhibited by the ML SDM detector in both the SDM-OFDM and SDM-SFI-OFDM scenarios.

By comparing Figures 15.11(a) and 15.11(b) we can conclude that the SNR performance of all the SDM detection methods considered improves upon increasing the number of transmit and receive antenna elements. Additionally, Figure 15.11(b) suggests that for a high number of transmit and receive antennas, the achievable performance of the turbo-coded SDM-SFI-OFDM system employing the ML SDM detector of Section 15.3.1 and communicating over the dispersive fading channel categorized by Bug's channel model [346] approaches the performance attained over an AWGN channel. Specifically, in the scenario of $m_t = n_r = 6$ characterized in Figure 15.11(b), the SNR performance of the turbo-coded SDM-SFI-OFDM system communicating over the dispersive fading channel categorized

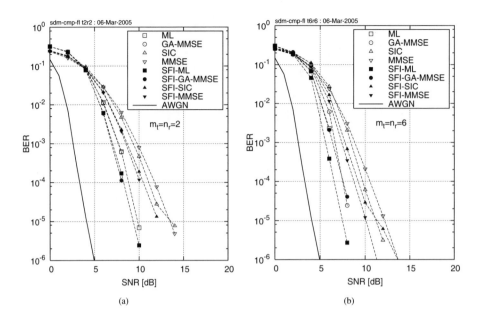

Figure 15.11: BER exhibited by the rate $\frac{1}{2}$ turbo-coded QPSK-modulated **SDM-OFDM** system of Section 15.1, as well as by the **SDM-SFI-OFDM** system employing the SDM detection methods of Sections 15.2 and 15.3. The abscissa represents the average SNR recorded at the receive antenna elements and the results correspond to the cases of (a) $n_\mathrm{r} = m_\mathrm{t} = 2$ and (b) $n_\mathrm{r} = m_\mathrm{t} = 6$. The OFDM system parameters are summarized in Table 7.1, while the corresponding GA configuration parameters are outlined in Table 15.1.

by Bug's channel model [346] is within a 2 dB margin of the corresponding performance in the AWGN channel.

Finally, it can be seen in Figures 15.11(a) and 15.11(b) that the SDM-SFI-OFDM system employing the SDM detectors considered outperforms its SDM-OFDM counterpart. Quantitatively, in the scenario of $m_\mathrm{t} = n_\mathrm{r} = 6$, the SDM-SFI-OFDM system employing the ML, SIC or GA-MMSE SDM detector exhibits an SNR gain of about 1 dB, when compared with its SDM-OFDM counterpart. In the case of employing the linear MMSE detector, the corresponding SNR difference between the SDM-SFI-OFDM and SDM-OFDM systems is about 2 dB at the target BER of 10^{-3}. It should be noted that the performance gains portrayed here are dependent on the particular channel model considered. The diversity gain associated with employing the SFI method becomes higher if the channel considered is less dispersive, i.e. the corresponding power delay profile characterizing the channel considered comprises fewer non-zero taps.

15.6   Conclusions

In this chapter we investigated the attainable performance benefits of employing multiple-antenna-aided SDM-OFDM architectures invoked in wireless communication systems in the context of a *point-to-point* system scenario, where two *peer* terminals employing multiple antennas communicate over a time-varying frequency-selective fading channel. We demonstrated that the linear capacity increase, predicted by the relevant information theoretical analysis [310], can indeed be achieved by employing a relatively low-complexity linear detection technique, such as the MMSE detector. We also showed that the ML detector is capable of attaining significant transmit diversity gains in *fully loaded* systems, where the number of transmit and receive antennas is identical. Furthermore, the ML detector is capable of adequately performing in an *overloaded* system configuration, where the number of transmit antennas exceeds that of receive antennas. Subsequently, we explored the potential of a range of additional advanced nonlinear SDM detection methods, which may potentially constitute an attractive compromise between the low complexity of the MMSE linear detector and the high performance of the ML detector. More explicitly, we demonstrated that the family of detection methods based on SIC as well as GA-aided MMSE detection are capable of satisfying these challenging requirements. Finally, we proposed a novel technique termed here SFI, which may be employed in the SDM system architecture advocated and may be beneficially combined with all the aforementioned detection techniques, resulting in a further SNR performance improvement of up to 2 dB.

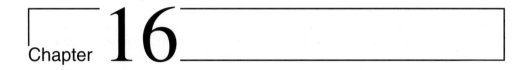

Chapter **16**

Approximate Log-MAP SDM-OFDM Multi-stream Detection

16.1 OHRSA-Aided SDM Detection[1]

As was pointed out in [265], the 'brute-force' ML detection method does not provide a feasible solution to the generic SDM detection problem as a result of its excessive computational complexity. Nevertheless, since typical wireless communication systems operate at moderate-to-high SNRs, a Reduced Search Algorithm (RSA) may be employed, which is capable of approaching the ML solution at a complexity considerably lower than that imposed by the ML detector of [265]. The most potent among the RSA methods found in the literature is constituted by the Sphere Decoder (SD) [333]. The SD was first proposed for employment in the context of space–time processing in [334], where it is utilized for computing the ML estimates of the modulated symbols transmitted simultaneously from multiple transmit antennas. The *complex* version of the SD, which is capable of approaching the channel capacity, was proposed by Hochwald and ten Brink in [335]. The subject was further investigated by Damen *et al.* in [336]. Subsequently, an improved version of the Complex Sphere Decoder (CSD) was advocated by Pham *et al.* in [337]. The issue of achieving near-capacity performance while reducing the associated complexity was revisited by Wang and Giannakis in [343, 629]. Further results on reduced-complexity CSD were published by Zhao and Giannakis in [342]. Finally, CSD-aided detection was considered by Tellambura *et al.* in a joint channel estimation and data detection scheme explored in [285], while a revised version of the CSD method, namely the so-called Multi-stage Sphere Decoding (MSD), was introduced in [338, 340].

In this chapter we introduce a novel Optimized Hierarchy Reduced Search Algorithm (OHRSA) aided SDM detection method, which may be regarded as an advanced extension of the CSD method portrayed in [337]. The algorithm proposed extends the potential range of applications of the CSD methods of [335] and [337], as well as reduces the associated computational complexity, rendering the algorithm attractive for employment in practical systems.

The method proposed, which we refer to as the Soft-output OPtimized HIErarchy (SOPHIE) algorithm, exhibits the following attractive properties:

---

[1]This chapter is partially based on ©IEEE Akhtman, Wolfgang, Chen & Hanzo 2007 [562].

1. The method can be employed in the so-called overloaded scenario, where the number of transmit antenna elements exceeds that of receive antenna elements. A particularly interesting potential application is found in a multiple-input, single-output scenario, where the system employs multiple transmit antennas and a single receive antenna. Moreover, the associated computational complexity is only moderately increased even in heavily overloaded scenarios and it is almost independent of the number of receive antennas.

2. As opposed to the conventional CSD schemes, the calculation of the sphere radius is not required and therefore the method proposed is robust to the particular choice of the initial parameters in terms of both the achievable performance and the associated computational complexity.

3. The method proposed allows for a selected subset of the transmitted information-carrying symbols to be detected, while the interference imposed by the undetected signals is suppressed.

4. The overall computational complexity required is only slightly higher than that imposed by the linear MMSE multi-user detector designed for detecting a similar number of users.

5. Finally, the associated computational complexity is fairly independent of the channel conditions quantified in terms of the Signal-to-Noise Ratio (SNR) encountered.

16.1.1   OHRSA-Aided ML SDM Detection

We commence our discourse by deriving an OHRSA-aided ML SDM detection method for a constant modulus modulation scheme, such as M-PSK, where the transmitted symbols s satisfy the condition of $|s|^2 = 1$, $\forall s \in \mathcal{M}$, and $\mathcal{M}$ denotes the set of M complex-valued constellation points. In the next section, we will demonstrate that the method derived is equally applicable for high-throughput multi-level modulation schemes, such as M-QAM.

Let us recall that our system model described in detail in Section 1.8 is given by

$$\mathbf{y} = \mathbf{Hs} + \mathbf{w}, \tag{16.1}$$

where we omit the OFDM subcarrier and symbol indices k and n, respectively. As outlined in [265], the ML SDM detector provides an m_t-antenna-based estimated signal vector candidate $\hat{\mathbf{s}}$, which maximizes the objective function defined as the conditional a posteriori probability function $P\{\check{\mathbf{s}}|\mathbf{y}, \mathbf{H}\}$ over the set $\mathcal{M}^{m_t}$ of legitimate solutions. More explicitly, we have

$$\hat{\mathbf{s}} = \arg\max_{\check{\mathbf{s}} \in \mathcal{M}^{m_t}} P\{\check{\mathbf{s}}|\mathbf{y}, \mathbf{H}\}, \tag{16.2}$$

where $\mathcal{M}^{m_t}$ is the set of all possible m_t-dimensional candidate symbol vectors of the m_t-antenna-based transmitted signal vector $\mathbf{s}$. More specifically, we have

$$\mathcal{M}^{m_t} = \{\check{\mathbf{s}} = (\check{s}_1, \ldots, \check{s}_{m_t})^{\mathrm{T}}; \check{s}_i \in \mathcal{M}\}. \tag{16.3}$$

Furthermore, it was shown in [265]that

$$P\{\check{\mathbf{s}}|\mathbf{y}, \mathbf{H}\} = A \exp\left[-\frac{1}{\sigma_w^2}\|\mathbf{y} - \mathbf{H}\check{\mathbf{s}}\|^2\right], \tag{16.4}$$

where A is a constant which is independent of any of the values $\{\check{s}_i\}_{i=1,\ldots,m_t}$. Thus, it may be shown [265] that the probability maximization problem of Equation (16.2) is equivalent to the corresponding Euclidean distance minimization problem. Specifically, we have

$$\hat{\mathbf{s}} = \arg\min_{\check{\mathbf{s}} \in \mathcal{M}^{m_t}} \|\mathbf{y} - \mathbf{H}\check{\mathbf{s}}\|^2, \tag{16.5}$$

where the probability-based objective function of Equation (16.2) is substituted by the objective function determined by the Euclidean distance between the received signal vector $\mathbf{y}$ and the corresponding product of the channel matrix $\mathbf{H}$ with the a priori candidate of the transmitted signal vector $\check{\mathbf{s}} \in \mathcal{M}^{m_t}$.

Consequently, our detection method relies on a specific observation, which may be summarized in the form of the following lemma.

Lemma 16.1. The ML solution of Equation (16.2) of a noisy linear problem described by Equation (16.1) is given by

$$\hat{\mathbf{s}} = \arg \min_{\check{\mathbf{s}} \in \mathcal{M}^{m_t}} \{\|\mathbf{U}(\check{\mathbf{s}} - \hat{\mathbf{x}})\|^2\}, \tag{16.6}$$

where $\mathbf{U}$ is an upper-triangular matrix having positive real-valued elements on the main diagonal and satisfying

$$\mathbf{U}^H\mathbf{U} = (\mathbf{H}^H\mathbf{H} + \sigma_w^2\mathbf{I}), \tag{16.7}$$

while

$$\hat{\mathbf{x}} = (\mathbf{H}^H\mathbf{H} + \sigma_w^2\mathbf{I})^{-1}\mathbf{H}^H\mathbf{y} \tag{16.8}$$

is the unconstrained MMSE estimate of the transmitted signal vector $\mathbf{s}$, which was derived in [265].

Note 1. Observe that Lemma 16.1 imposes no constraints on the dimensions or rank of the matrix $\mathbf{H}$ of the linear system described by Equation (16.1). This property is particularly important, since it enables us to apply our proposed detection technique to the scenario of *overloaded* systems, where the number of transmit antenna elements exceeds that of receive antenna elements.

Note 2. As substantiated by Equation (16.5), it is sufficient to prove that the following minimization problems are equivalent:

$$\hat{\mathbf{s}} = \arg \min_{\check{\mathbf{s}} \in \mathcal{M}^{m_t}} \|\mathbf{y} - \mathbf{H}\check{\mathbf{s}}\|^2 \tag{16.9}$$

$$\Leftrightarrow \quad \hat{\mathbf{s}} = \arg \min_{\check{\mathbf{s}} \in \mathcal{M}^{m_t}} \|\mathbf{U}(\check{\mathbf{s}} - \hat{\mathbf{x}})\|^2. \tag{16.10}$$

Proof of Lemma 16.1. It is evident that in contrast to the matrix $\mathbf{H}^H\mathbf{H}$, the matrix $(\mathbf{H}^H\mathbf{H} + \sigma_w^2\mathbf{I})$ of Equation (16.6) is always Hermitian and positive definite, regardless of the rank of the channel matrix $\mathbf{H}$ associated with the particular MIMO channel realization encountered. Consequently, it may be represented as the product of an upper-triangular matrix $\mathbf{U}$ and its Hermitian adjoint matrix $\mathbf{U}^H$ using for example the Cholesky factorization method [630].

Let $\mathbf{U}$ be the matrix generated by the Cholesky decomposition of the Hermitian positive definite matrix $(\mathbf{H}^H\mathbf{H} + \sigma_w^2\mathbf{I})$ of Equation (16.7). More specifically, we have

$$\mathbf{U}^H\mathbf{U} = (\mathbf{H}^H\mathbf{H} + \sigma_w^2\mathbf{I}), \tag{16.11}$$

where $\mathbf{U}$ is an upper-triangular matrix having positive real-valued elements on its main diagonal.

On expanding the objective function of Equation (16.6) and subsequently invoking Equation (16.7), we obtain

$$\begin{aligned}
J(\check{\mathbf{s}}) &= \|\mathbf{U}(\check{\mathbf{s}} - \hat{\mathbf{x}})\|^2 \\
&= (\check{\mathbf{s}} - \hat{\mathbf{x}})^H\mathbf{U}^H\mathbf{U}(\check{\mathbf{s}} - \hat{\mathbf{x}}) \\
&= (\check{\mathbf{s}} - \hat{\mathbf{x}})^H(\mathbf{H}^H\mathbf{H} + \sigma_w^2\mathbf{I})(\check{\mathbf{s}} - \hat{\mathbf{x}}) \\
&= \check{\mathbf{s}}^H(\mathbf{H}^H\mathbf{H} + \sigma_w^2\mathbf{I})\check{\mathbf{s}} - \hat{\mathbf{x}}^H(\mathbf{H}^H\mathbf{H} + \sigma_w^2\mathbf{I})\check{\mathbf{s}} \\
&\quad - \check{\mathbf{s}}^H(\mathbf{H}^H\mathbf{H} + \sigma_w^2\mathbf{I})\hat{\mathbf{x}} + \hat{\mathbf{x}}^H(\mathbf{H}^H\mathbf{H} + \sigma_w^2\mathbf{I})\hat{\mathbf{x}}.
\end{aligned} \tag{16.12}$$

Furthermore, substituting Equation (16.8) into (16.12) yields

$$\begin{aligned}
J(\check{\mathbf{s}}) &= \check{\mathbf{s}}^H\mathbf{H}^H\mathbf{H}\check{\mathbf{s}} - \mathbf{y}^H\mathbf{H}\check{\mathbf{s}} - \check{\mathbf{s}}^H\mathbf{H}^H\mathbf{y} \\
&\quad + \sigma_w^2\check{\mathbf{s}}^H\check{\mathbf{s}} + \mathbf{y}^H\mathbf{H}(\mathbf{H}^H\mathbf{H} + \sigma_w^2\mathbf{I})^{-1}\mathbf{H}^H\mathbf{y} \\
&= \|\mathbf{y} - \mathbf{H}\check{\mathbf{s}}\|^2 + \underbrace{\sigma_w^2\check{\mathbf{s}}^H\check{\mathbf{s}} + \mathbf{y}^H(\mathbf{H}(\mathbf{H}^H\mathbf{H} + \sigma_w^2\mathbf{I})^{-1}\mathbf{H}^H - \mathbf{I})\mathbf{y}}_{\psi}.
\end{aligned} \tag{16.13}$$

Observe that in the case of a system employing a constant modulus modulation scheme, such as M-PSK, where we have $\check{s}^H\check{s} = 1$, ψ of Equation (16.13) constitutes a real-valued scalar and its value does not depend on the argument $\check{s}$ of the minimization problem formulated in Equation (16.6). Consequently, the minimization of the objective function $J(\check{s})$ of Equation (16.13) can be reduced to the minimization of the term $\|y - H\check{s}\|^2$, which renders it equivalent to the minimization problem of Equation (16.9). This completes the proof. $\qquad\qquad\qquad\qquad\qquad\qquad\qquad\qquad\qquad\qquad\quad\Box$

Using Lemma 16.1, in particular the fact that the matrix U is an upper-triangular matrix, the objective function $J(\check{s})$ of Equation (16.13) may be reformulated as follows:

$$J(\check{s}) = \|U(\check{s} - \hat{x})\|^2$$
$$= (\check{s} - \hat{x})^H U^H U(\check{s} - \hat{x})$$
$$= \sum_{i=1}^{m_t}\left|\sum_{j=i}^{m_t} u_{ij}(\check{s}_j - \hat{x}_j)\right|^2 = \sum_{i=1}^{m_t}\phi_i(\check{s}_i), \qquad (16.14)$$

where $J(\check{s})$ and $\phi_i(\check{s}_i)$ are positive real-valued cost and subcost functions, respectively. Elaborating a little further we have

$$\phi_i(\check{s}_i) = \left|\sum_{j=i}^{m_t} u_{ij}(\check{s}_j - \hat{x}_j)\right|^2$$

$$= \left|u_{ii}(\check{s}_i - \hat{x}_i) + \underbrace{\sum_{j=i+1}^{m_t} u_{ij}(\check{s}_j - \hat{x}_j)}_{a_i}\right|^2. \qquad (16.15)$$

Note that the term a_i is a complex-valued scalar which is independent of the specific symbol value $\check{s}_i$ of the ith element of the a priori candidate signal vector $\check{s}$.

Furthermore, let $J_i(\check{s}_i)$ be a Cumulative Sub-Cost (CSC) function recursively defined as

$$J_{m_t}(\check{s}_{m_t}) = \phi_{m_t}(\check{s}_{m_t}) = |u_{m_t m_t}(\check{s}_{m_t} - \hat{x}_{m_t})|^2 \qquad (16.16a)$$
$$J_i(\check{s}_i) = J_{i+1}(\check{s}_{i+1}) + \phi_i(\check{s}_i), \quad i = m_t-1, \ldots, 1, \qquad (16.16b)$$

where we define the candidate subvector as $\check{s}_i = [\check{s}_i, \ldots, \check{s}_{m_t}]$. Clearly, $J_i(\check{s}_i)$ exhibits the following properties:

$$J(\check{s}) = J_1(\check{s}_1) > J_2(\check{s}_2) > \cdots > J_{m_t}(\check{s}_{m_t}) > 0 \qquad (16.17a)$$
$$J_i(\check{s}_i) = J_i(\{\check{s}_j\}, j = i, \ldots, m_t) \qquad (16.17b)$$

for all possible realizations of $\hat{x} \in \mathbb{C}^{m_t}$ and $\check{s} \in \mathcal{M}^{m_t}$, where the space $\mathbb{C}^{m_t}$ contains all possible unconstrained MMSE estimates $\hat{x}$ of the transmitted signal vector s.

Equations (16.17a) and (16.17b) enable us to employ a highly efficient RSA, which decreases the number of objective function evaluations of the minimization problem outlined in Equation (16.6) to a small fraction of the set $\mathcal{M}^{m_t}$. This reduced-complexity search algorithm is outlined in the next section.

16.1.1.1 Search Strategy

Example 16.1. OHRSA-ML 3 × 3 BPSK
Consider a BPSK system having $n_r = m_t = 3$ transmit and receive antennas, which is described by Equation (16.1). The transmitted signal s, the received signal y as well as the channel matrix H of Equation (16.1) are exemplified by the following values:

$$s = \begin{bmatrix} 1 \\ -1 \\ 1 \end{bmatrix}, \quad y = \begin{bmatrix} 0.2 \\ 0.8 \\ -1.2 \end{bmatrix}, \quad H = \begin{bmatrix} 0.5 & 0.4 & -0.2 \\ 0.4 & -0.3 & 0.2 \\ 0.9 & 1.8 & -0.1 \end{bmatrix}. \qquad (16.18)$$

Our task is to obtain the ML estimate of the transmitted signal vector s. Firstly, we evaluate the triangular matrix $\mathbf{U}$ of Equation (16.7) as well as the unconstrained MMSE estimate $\hat{\mathbf{x}}$ of Equation (16.8). The resultant quantities are given by

$$\mathbf{U} = \begin{bmatrix} 1.15 & 1.48 & -0.10 \\ 0 & 1.18 & -0.15 \\ 0 & 0 & 0.40 \end{bmatrix}, \quad \hat{\mathbf{x}} = \begin{bmatrix} 0.85 \\ -1.05 \\ -0.01 \end{bmatrix}. \tag{16.19}$$

Observe that the direct slicing of the MMSE estimate $\hat{\mathbf{x}}$ will result in an erroneously decided signal $\hat{\mathbf{s}} = \begin{bmatrix} 1 & -1 & -1 \end{bmatrix}^{\mathrm{T}}$. Subsequently, following the philosophy outlined in Section 16.1.1, for each legitimate candidate $\check{\mathbf{s}} \in \mathcal{M}^{m_t}$ of the m_t-antenna-based composite transmitted signal vector s we calculate the corresponding value of the cost function $J(\check{\mathbf{s}})$ of Equation (16.14) using the recursive method described by Equation (16.16). The search process performed is illustrated in Figure 16.1(a). Each evaluation step, i.e. each evaluation of the CSC function $J_i(\check{\mathbf{s}}_i)$ of Equation (16.16b), is indicated by an elliptic node in Figure 16.1(a). The label inside each node indicates the order of evaluation as well as the corresponding value $J_i(\check{\mathbf{s}}_i)$ of the CSC function inside the brackets. Furthermore, the branches corresponding to the two legitimate values of $\check{s}_i = -1$ and 1 are indicated using the dashed and solid edges and nodes, respectively.

More specifically, commencing from the top of Figure 16.1(a), at recursive step $i = 3$ we calculate the CSC function of Equation (16.16a) associated with all legitimate values of the last element of the signal vector s, where

$$J_3(\check{s}_3 = -1) = |u_{33}(\check{s}_3 - \hat{x}_3)|^2 = (0.40(-1 - (-0.01)))^2 = 0.15 \tag{16.20}$$

and

$$J_3(\check{s}_3 = 1) = (0.40(1 - (-0.01)))^2 = 0.16. \tag{16.21}$$

The corresponding values of $J_3(\check{s}_3 = -1) = 0.15$ and $J_3(\check{s}_3 = 1) = 0.16$ are indicated by the nodes 1 and 8 in Figure 16.1(a). Observe that the *recursive* nature of the search process considered suggests that the latter value of $J_3(\check{s}_3 = 1)$ is not considered until the entire search branch originating from the more promising node 1 associated with the lower CSC value of 0.15 is completed. Consequently, the value $J_3(\check{s}_3 = 1)$ is the eighth value of the CSC function to be evaluated, which is indicated by the corresponding node's index 8.

Furthermore, at recursive step $i = 2$ for each hypothesized value $\check{s}_3$ we calculate both the quantity a_2 of Equation (16.15) as well as the subcost function of Equation (16.15) and the corresponding CSC function of Equation (16.16b) associated with all legitimate values of the last-but-one element of the signal vector s. Explicitly, for $\check{s}_3 = -1$ we have

$$a_2 = u_{23}(\check{s}_3 - \hat{x}_3) = -0.15(-1 - (-0.01)) = 0.15 \tag{16.22}$$

and

$$\begin{aligned} J_2(\check{s}_2 = -1, \check{s}_3 = -1) &= J_3(\check{s}_3 = -1) + \phi_2(\check{s}_2 = -1, \check{s}_3 = -1) \\ &= J_3(\check{s}_3 = -1) + |u_{22}(\check{s}_2 - \hat{x}_2) + a_2|^2 \\ &= 0.15 + (1.18(-1 - (-1.05)) + 0.15) = 0.20 \\ J_2(\check{s}_2 = 1, \check{s}_3 = -1) &= J_3(\check{s}_3 = -1) + \phi_2(\check{s}_2 = 1, \check{s}_3 = -1) \\ &= 0.15 + (1.18(1 - (-1.05)) + 0.15) = 6.79. \end{aligned} \tag{16.23}$$

The corresponding values of $J_2(\check{s}_2 = [-1, -1]) = 0.20$ and $J_2(\check{s}_2 = [1, -1]) = 6.79$ are indicated by nodes 2 and 5 in Figure 16.1(a).

Finally, at recursive index $i = 1$ for each hypothesized subvector $\check{s}_2$ we calculate the quantity $a_1(\check{s}_2)$ and the subcost function $\phi_1(\check{s}_1)$ of Equation (16.15) as well as the corresponding *total* cost function $J(\check{s}_1 = -1, \check{s}_2)$ and $J(\check{s}_1 = 1, \check{s}_2)$ of Equation (16.14) associated with all legitimate values of the first element of the signal vector **s**. Specifically, for the leftmost search branch of Figure 16.1(a) corresponding to the a priori candidate $\check{s}_2 = [-1, -1]$ we have

$$a_1 = u_{12}(\check{s}_2 - \hat{x}_2) + u_{13}(\check{s}_3 - \hat{x}_3)$$
$$= 1.48(-1 - (-1.05)) + (-0.10)(-1 - (-0.01)) = 0.17 \tag{16.24}$$

and

$$J_1(\check{s}_1 = -1, \check{s}_2 = -1, \check{s}_3 = -1)$$
$$= J_2(\check{s}_2 = -1, \check{s}_3 = -1) + \phi_1(\check{s}_1 = -1, \check{s}_2 = -1, \check{s}_3 = -1)$$
$$= J_2(\check{s}_2 = -1, \check{s}_3 = -1) + |u_{11}(\check{s}_1 - \hat{x}_1) + a_1|^2$$
$$= 0.20 + (1.15(-1 - 0.85) + 0.17) = 4.03,$$
$$J_1(\check{s}_1 = 1, \check{s}_2 = -1, \check{s}_3 = 1)$$
$$= J_2(\check{s}_2 = -1, \check{s}_3 = -1) + \phi_2(\check{s}_1 = 1, \check{s}_2 = -1, \check{s}_3 = -1)$$
$$= 0.20 + (1.15(1 - 0.85) + 0.17) = 0.31. \tag{16.25}$$

On completing the entire search process outlined above we arrive at eight values of the total cost function $J(\check{s})$ associated with eight legitimate 3-bit solutions of the detection problem considered. The eight different candidate solutions are indicated by the eight bottommost elliptic nodes in Figure 16.1(a). Clearly, the ML solution is constituted by the search branch terminating at node 11 of Figure 16.1(a) and having the minimum value $J(\check{s}) = 0.19$ of the total cost function.

Observe that the difference between the values of $J_3(\check{s}_3 = -1)$ and $J_3(\check{s}_3 = 1)$ associated with nodes 1 and 8 in Figure 16.1(a) is quite small and thus the potential of finding the ML solution along either of the search branches commencing at nodes 1 and 8 in Figure 16.1(a) may not be recognized with a high degree of confidence. On the other hand, the difference between the values of the CSC function along two complementary search branches commencing at nodes 1 and 8 becomes substantially more evident if we apply the *best-first* detection strategy suggested in [631]. More specifically, we sort the columns of the channel matrix **H** in increasing order of their Euclidean or square norm. The resultant reordered channel matrix $\mathbf{H}'$ as well as the corresponding triangular matrix $\mathbf{U}$ and the unconstrained MMSE estimate $\hat{\mathbf{x}}'$ may be expressed as

$$\mathbf{H}' = \begin{bmatrix} -0.2 & 0.5 & 0.4 \\ 0.2 & 0.4 & -0.3 \\ -0.1 & 0.9 & 1.8 \end{bmatrix}, \quad \mathbf{U}' = \begin{bmatrix} 0.44 & -0.25 & -0.73 \\ 0 & 1.12 & 1.35 \\ 0 & 0 & 1.11 \end{bmatrix}, \quad \hat{\mathbf{x}}' = \begin{bmatrix} -0.01 \\ 0.85 \\ -1.05 \end{bmatrix}. \tag{16.26}$$

The search tree generated by applying the aforementioned search process and using the modified quantities $\mathbf{H}', \mathbf{U}'$ and $\hat{\mathbf{x}}'$ is depicted in Figure 16.1(b). Note the substantial difference between the values of the CSC function $J_3(\check{s}_3 = -1)$ and $J_3(\check{s}_3 = 1)$ associated with nodes 1 and 8. Moreover, by comparing the value of the CSC function $J_3(\check{s}_3)$ of node 8 with that of the total cost function $J(\check{s})$ of node 7 we can conclude that the search along the branch commencing at node 8 is in fact redundant.

In order to optimize our search process further, at recursive steps of $i = 3$ and 2 we first calculate the subcost functions $\phi_3(\check{s}_3 = \{-1, 1\})$ and $\phi_2(\check{s}_3, \check{s}_2 = \{-1, 1\})$ of Equation (16.15). We then compare the values obtained and continue with the processing of the specific search branch corresponding to the smaller value of the subcost function $\phi_i(\check{s}_i)$ first. The resultant search tree is depicted in Figure 16.1(c). Observe that in Figure 16.1(c) the minimum value of the total cost function $J(\check{s}) = 0.19$ is obtained faster, i.e. in three evaluation steps in comparison with seven steps required by the search tree of Figure 16.1(b).

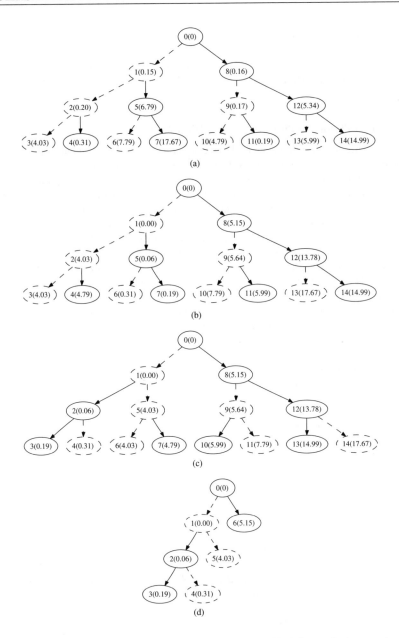

Figure 16.1: Examples of a search tree formed by the OHRSA-ML SDM detector in the scenario of a system employing BPSK modulation, $m_t = n_r = 3$ transmit and receive antennas and encountering average SNRs of 10 dB. The labels indicate the order of evaluation, as well as the corresponding value $J_i(\check{\mathbf{s}}_i)$ of the CSC function of Equation (16.16), as seen in the brackets. The dashed and solid arrows indicate the values of $\check{s}_i = -1$ and 1, respectively.

Finally, we discard all the search branches commencing at nodes having an associated value of the CSC function which is in excess of the minimum total cost function value obtained. Specifically, we discontinue the search branches commencing at nodes 5 and 8 having CSC function values in excess of

0.19, i.e. 4.03 and 5.15, respectively. The resultant reduced search tree is depicted in Figure 16.1(d). Note that the ML solution is obtained in six evaluation steps in comparison with the 14 steps required in the case of the exhaustive search of Figure 16.1(a). In conclusion, on performing the appropriate reordering of the obtained ML estimate, we arrive at the correct value of the transmitted signal vector $\hat{\mathbf{s}} = \begin{bmatrix} 1 & -1 & 1 \end{bmatrix}^{\mathrm{T}}$.

16.1.1.2   Generalization of the OHRSA-ML SDM Detector

Let us now generalize and substantiate further the detection paradigm derived in Example 16.1. Firstly, we commence the recursive search process with the evaluation of the CSC function value $J_{m_t}(\check{s}_{m_t})$ of Equation (16.16a). Secondly, at each recursive step i of the search algorithm proposed we stipulate a series of hypotheses concerning the value of the M-ary transmitted symbol s_i associated with the ith transmit antenna element and subsequently calculate the conditioned subcost function $J_i(\check{\mathbf{s}}_i)$ of Equation (16.16b), where $\check{\mathbf{s}}_i = (\check{s}_i, \ldots, \check{s}_{m_t})^{\mathrm{T}}$ denotes the subvector of the m_t-antenna-based candidate vector $\check{\mathbf{s}}$ comprising only indices higher than or equal to i. Furthermore, for each tentatively assumed value of $\check{s}_i$ we execute a successive recursive search step $i - 1$, which is conditioned on the hypotheses made in all preceding recursive steps $j = i, \ldots, m_t$. As substantiated by Equations (16.15) and (16.16b), the value of the CSC function $J_i(\check{\mathbf{s}}_i)$ is dependent only on the values of the elements $\{\check{s}_j\}_{j=i,\ldots,m_t}$ of the a priori candidate signal vector $\check{\mathbf{s}}$, which are hypothesized from step $j = m_t$ up to the present step i of our recursive process. At each arrival at step $i = 1$ of the recursive process, a complete candidate vector $\check{\mathbf{s}}$ is hypothesized and the corresponding value of the cost function $J(\check{\mathbf{s}})$ formulated in Equation (16.14) is evaluated.

   Observe that the recursive hierarchical search procedure described above may be employed to perform an exhaustive search through all possible values of the transmitted signal vector $\check{\mathbf{s}}$ and the resultant search process is guaranteed to arrive at the ML solution $\check{\mathbf{s}}_{\mathrm{ML}}$, which minimizes the value of the cost function $J(\check{\mathbf{s}})$ of Equation (16.14). Fortunately, however, as opposed to other ML search schemes, the search process described above can be readily optimized, resulting in a dramatic reduction of the associated computational complexity. Specifically, the potential optimization complexity gain originates from the fact that most of the hierarchical search branches can be discarded at an early stage of the recursive search process. The corresponding optimization rules proposed may be outlined as follows.

Rule 1.   We reorder the system model of Equation (16.1) as suggested in [631]. Specifically, we apply the *best-first* detection strategy outlined in [265, pp. 754–756], which implies that the transmitted signal vector components are detected in decreasing order of the associated channel quality. As advocated in [265, pp. 754–756], the quality of the channel associated with the ith element of the transmitted signal vector $\mathbf{s}$ is determined by the norm of the ith column of the channel matrix $\mathbf{H}$. Consequently, for applying the *best-first* detection strategy, the columns of the channel matrix $\mathbf{H}$ are sorted in increasing order of their norm. Thus, the resultant, column-reordered channel matrix $\mathbf{H}$ complies with the following criterion:

$$\|(\mathbf{H})_1\|^2 \leq \|(\mathbf{H})_2\|^2 \leq \cdots \leq \|(\mathbf{H})_{m_t}\|^2, \tag{16.27}$$

where $(\mathbf{H})_i$ denotes the ith column of the channel matrix $\mathbf{H}$. Note that the elements of the transmitted signal vector $\mathbf{s}$ are reordered correspondingly, but their original order has to be reinstated in the final stage of the detection process.

Rule 2.   At each recursive detection index $i = m_t, \ldots, 1$, the potential candidate values, namely $\{c_m\}_{m=1,\ldots,M} \in \mathcal{M}$ of the transmitted signal component s_i, are considered in increasing order of the corresponding value of the subcost function $\phi_i(\check{\mathbf{s}}_i) = \phi_i(c_m, \check{\mathbf{s}}_{i+1})$ of Equation (16.15), where

$$\phi_i(c_1, \check{\mathbf{s}}_{i+1}) < \cdots < \phi_i(c_m, \check{\mathbf{s}}_{i+1}) < \cdots < \phi_i(c_M, \check{\mathbf{s}}_{i+1}),$$

and according to Equation (16.15),

$$\phi_i(c_m, \check{s}_{i+1}) = |u_{ii}(c_m - \hat{x}_i) + a_i|^2$$

$$= u_{ii}\left|c_m - \hat{x}_i + \frac{a_i}{u_{ii}^2}\right|^2. \qquad (16.28)$$

Consequently, the more likely candidates c_m of the ith element of the transmitted signal vector s are examined first. Observe that the sorting criterion of Equation (16.28) may also be interpreted as a biased Euclidean distance of the candidate constellation point c_m from the unconstrained MMSE estimate $\hat{x}_i$ of the transmitted signal component s_i.

Rule 3. We define a *cut-off* value of the cost function $J_{\min} = \min\{J(\check{s})\}$ as the minimum value of the total cost function obtained up to the present point of the search process. Consequently, at each arrival at step $i = 1$ of the recursive search process, the *cut-off* value of the cost function is updated as follows:

$$J_{\min} = \min\{J_{\min}, J(\check{s})\}. \qquad (16.29)$$

Rule 4. Finally, at each recursive detection step i, only the high probability search branches corresponding to the highly likely symbol candidates c_m resulting in low values of the CSC function obeying $J_i(c_m) < J_{\min}$ are pursued. Furthermore, as follows from the sorting criterion of the optimization Rule 2, as soon as the inequality $J_i(c_m) > J_{\min}$ is encountered, the search loop at the ith detection step is discontinued.

An example of the search tree generated by the algorithm invoking Rules 1–4 described above is depicted in Figure 16.2. The search trees shown correspond to the scenario of using QPSK modulation and employing $m_t = n_r = 8$ antenna elements at both the transmitter and the receiver. The cases of encountering the average SNRs of (a) 10 and (b) 20 dB were considered. Each step of the search procedure is depicted as an ellipsoidal-shaped node. The label associated with each node indicates the order of visitation, as well as the corresponding value of the CSC function $J_i(\check{s}_i)$ formulated in Equation (16.16), as seen in the brackets. As suggested by the fact that QPSK modulation is considered, at each recursive step i, four legitimate search branches are possible. However, as can be seen in Figure 16.2(a), only a small fraction of the potential search branches are actually pursued. Observe that the rate of convergence of the algorithm proposed is particularly rapid at high values of SNR. In the case of encountering low SNR values, the convergence rate decreases. Nevertheless, the associated computational complexity is dramatically lower than that associated with an exhaustive ML search.

The pseudo-code summarizing the recursive implementation of the OHRSA-ML SDM detector proposed is depicted in Algorithm 16.1.

Given the cost functions of Equation (16.14) and the appropriately ordered matrix **H** of Equation (16.1), the proposed algorithm may be viewed as a specific manifestation of a tree search algorithm [632]. Another example of a tree search algorithm commonly employed in the design of communication systems constitutes the well-known Viterbi algorithm [314,633]. More specifically, the subcost function values of Equation (16.15) may be regarded as being analogous to the branch metrics, and the CSC values of Equation (16.16) to accumulated path metrics. It should be noted, however, that the OHRSA-ML algorithm described here and the classic tree-search-based Viterbi algorithm exhibit substantial differences. Specifically, the Viterbi algorithm assumes that the branch metric is a function of the system states constituting one particular state transition, which is equivalent to assuming a diagonal matrix **U** in Equation (16.7). Evidently, this requirement cannot be satisfied by our generic MIMO system. Consequently, the tree-search-based Viterbi algorithm cannot be applied to the search problem described above.

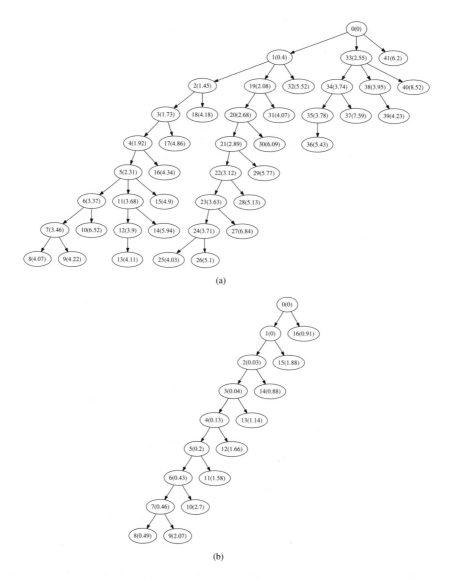

Figure 16.2: Examples of a search tree formed by the OHRSA-ML SDM detector in the scenario of a system employing QPSK modulation, $m_t = n_r = 8$ transmit and receive antennas and encountering average SNRs of (a) 10 dB and (b) 20 dB. The labels indicate the order of visitation, as well as the corresponding value $J_i(\check{s}_i)$ of the CSC function of Equation (16.16), as seen in the brackets. The ML solution is attained in (a) 41 and (b) 16 evaluation steps in comparison with the $4^8 = 65\,536$ evaluation steps required in the case of the exhaustive ML search.

16.1.2 Bit-wise OHRSA-ML SDM Detection

Example 16.2. OHRSA-ML QPSK 2 × 2

Let us now consider a QPSK system having $n_r = m_t = 3$ transmit and receive antennas, which is described by Equation (16.1). The transmitted signal **s**, the received signal **y** as well as the *best-first*

Algorithm 16.1 OHRSA-ML SDM detector

$$\text{Sort}\{\mathbf{H}\}, \text{ such that } \|(\mathbf{H})_1\|^2 \leq \cdots \leq \|(\mathbf{H})_{m_t}\|^2 \qquad (16.30\text{a})$$

$$\mathbf{G} = (\mathbf{H}^{\mathsf{H}}\mathbf{H} + \sigma_w^2 \mathbf{I}) \qquad (16.30\text{b})$$

$$\mathbf{U} = \text{CholeskyDecomposition}(\mathbf{G}) \qquad (16.30\text{c})$$

$$\hat{\mathbf{x}} = \mathbf{G}^{-1}\mathbf{H}^{\mathsf{H}}\mathbf{y} \qquad (16.30\text{d})$$

$$\text{Calculate } J_{m_t} \qquad (16.30\text{e})$$

$$\text{Unsort}\{\hat{\mathbf{s}}\} \qquad (16.30\text{f})$$

$$\text{function } \text{Calculate } J_i(\check{s}_i) \qquad (16.30\text{g})$$

$$a_i = \sum_{j=i+1}^{m_t} u_{ij}(\check{s}_j - \hat{x}_j) \qquad (16.30\text{h})$$

$$\text{Sort}\{c_m\}, \text{ such that } \phi_i(c_1) < \cdots < \phi_i(c_M), \qquad (16.30\text{i})$$

$$\text{where } \phi_i(c_m) = |u_{ii}(c_m - \hat{x}_i) + a_i|^2 \qquad (16.30\text{j})$$

$$\text{for } m = 1, 2, \ldots, M \text{ do}$$

$$\qquad \check{s}_i = c_m \qquad (16.30\text{k})$$

$$\qquad J_i(\check{s}_i) = J_{i+1}(\check{s}_{i+1}) + \phi_i(\check{s}_i) \qquad (16.30\text{l})$$

$$\qquad \text{if } J_i(\check{s}_i) < J_{\min} \text{ then} \qquad (16.30\text{m})$$

$$\qquad\qquad \text{if } i > 0 \text{ then } \text{Calculate } J_{i-1} \qquad (16.30\text{n})$$

$$\qquad\qquad \text{else}$$

$$\qquad\qquad\qquad J_{\min} = J(\check{s}) \qquad (16.30\text{o})$$

$$\qquad\qquad\qquad \hat{\mathbf{s}} = \check{s} \qquad (16.30\text{p})$$

$$\qquad\qquad \text{end if}$$

$$\qquad \text{end if}$$

$$\text{end for}$$

$$\text{end function}$$

---

reordered channel matrix $\mathbf{H}$ of Equation (16.1) are exemplified by the following values:

$$\mathbf{s} = \begin{bmatrix} 1 - 1\jmath \\ -1 - 1\jmath \end{bmatrix}, \quad \mathbf{y} = \begin{bmatrix} 0.2 + 1.1\jmath \\ 1.4 + 1.7\jmath \end{bmatrix},$$

$$\mathbf{H} = \begin{bmatrix} 0.1 - 0.2\jmath & -0.7 - 0.6\jmath \\ 0.3 + 0.4\jmath & -1.3 - 0.5\jmath \end{bmatrix}. \qquad (16.31)$$

As before, our task is to obtain the ML estimate of the transmitted signal vector $\mathbf{s}$. Firstly, we apply the OHRSA-ML method of Algorithm 16.1.

As suggested by Algorithm 16.1, we commence the detection process by evaluating the quantities $\mathbf{U}$ and $\hat{\mathbf{x}}$ of Equations (16.30c) and (16.30d) respectively, which yields

$$\mathbf{U} = \begin{bmatrix} 0.63 & -0.85 + 0.27\jmath \\ 0 & 1.45 \end{bmatrix}, \quad \hat{\mathbf{x}} = \begin{bmatrix} 0.43 - 0.34\jmath \\ -1.10 - 0.79\jmath \end{bmatrix}. \qquad (16.32)$$

Furthermore, we proceed by calculating *four* values of the CSC function $J_2(\check{s}_2 = c_m)$, $m = 1, \ldots, 4$, of Equation (16.30l) associated with the *four* different points c_m of the QPSK constellation. For instance, we have

$$J_2(\check{s}_2 = -1 - 1\jmath) = \phi_2(\check{s}_2 = -1 - 1\jmath) = |u_{22}(\check{s}_2 - \hat{x}_2)|^2$$
$$= |1.45(-1 - 1\jmath - (-1.10 - 0.79\jmath))|^2 = 0.12. \tag{16.33}$$

Subsequently, four QPSK symbol candidates c_m are sorted in the order of increasing subcost function $\phi_2(c_m)$, as described by Equation (16.30i) of Algorithm 16.1. For each hypothesized symbol value $\check{s}_2 = c_m$ we can now obtain *four* values of the total cost function $J(\check{s}) = J_1(\check{s}_1, \check{s}_2)$ of Equation (16.30l) associated with *four* legitimate values of $\check{s}_1 = c_m$. For instance, we have

$$J(\check{s}_1 = 1 - 1\jmath, \check{s}_2 = -1 - 1\jmath)$$
$$= J_2(\check{s}_2 = -1 - 1\jmath) + \phi_1(\check{s}_1 = 1 - 1\jmath, \check{s}_2 = -1 - 1\jmath)$$
$$= J_2(\check{s}_2 = -1 - 1\jmath) + |u_{11}(\check{s}_1 - \hat{x}_1)) + a_1|^2$$
$$= 0.12 + |0.63[1 - 1\jmath - (0.43 - 0.34\jmath)] + (-0.03 + 0.21\jmath)|^2 = 0.27, \tag{16.34}$$

where the quantity a_1 is given by Equation (16.30h) of Algorithm 16.1 as follows:

$$a_1(\check{s}_2 = -1 - 1\jmath) = u_{12}(\check{s}_2 - \hat{x}_2)$$
$$= (-0.85 + 0.27\jmath)[-1 - 1\jmath - (-1.10 - 0.79\jmath)] = -0.03 + 0.21\jmath. \tag{16.35}$$

As further detailed in Algorithm 16.1, we calculate the values of the total cost function $J(\check{s}_1, \check{s}_2)$ only for the specific hypothesis $\check{s}_2$, for which the value of the CSC function $J_2(\check{s}_2)$ is lower than the minimum value $J_{\min}$ obtained.

The resultant search tree is depicted in Figure 16.3(a), where, as before, each evaluation step, i.e. each evaluation of the CSC function $J_i(\check{s}_i)$ of Equation (16.30l), is indicated by an elliptic node. Moreover, the label inside each node indicates the order of evaluation as well as the corresponding value $J_i(\check{s}_i)$ of the CSC function inside the brackets. The branches corresponding to *four* legitimate values of the QPSK symbol are indicated by the specific type of the edges and nodes. Specifically, the *grey* and *black* lines indicate the value of the real part of the QPSK symbol $\mathcal{R}\{\check{s}_i\} = -1$ and 1, while the *dashed* and *solid* lines indicate the value of the imaginary part $\mathcal{I}\{\check{s}_i\} = -1$ and 1.

Example 16.3. Bit-wise OHRSA-ML QPSK 2 × 2

Let us consider a QPSK system identical to that described in Example 16.2 and attempt to derive an alternative way of finding the ML estimate of the transmitted signal vector s using the bit-based representation of the QPSK symbols. In order to describe this bit-based multi-user phasor constellation, let us develop a matrix- and vector-based mathematical model. Firstly, observe that each point of the QPSK constellation $c_m \in \mathcal{M}$ may be represented as the inner product $c_m = q^T d_m$ of a unique bit-based vector $d_m = [d_{m1}, d_{m2}]^T$, $d_{ml} = \{-1, 1\}$ and the vector $q = [1, 1\jmath]^T$. For instance, we have

$$c_1 = -1 - 1\jmath = q^T d_1 = \begin{bmatrix} 1 & 1\jmath \end{bmatrix} \cdot \begin{bmatrix} -1 \\ -1 \end{bmatrix}. \tag{16.36}$$

Furthermore, let us define a (4×2)-dimensional matrix

$$Q = I \otimes q = \begin{bmatrix} 1 & 1\jmath & 0 & 0 \\ 0 & 0 & 1 & 1\jmath \end{bmatrix}, \tag{16.37}$$

where I is (2×2)-dimensional identity matrix, while $\otimes$ denotes the *matrix direct product* [634]. Consequently, the QPSK-modulated signal vector s may be represented as

$$s = \begin{bmatrix} 1 - 1\jmath \\ -1 - 1\jmath \end{bmatrix} = Qt = \begin{bmatrix} 1 & 1\jmath & 0 & 0 \\ 0 & 0 & 1 & 1\jmath \end{bmatrix} \begin{bmatrix} 1 \\ -1 \\ -1 \\ -1 \end{bmatrix}, \tag{16.38}$$

where $\mathbf{t} = [\mathbf{t}_1^T, \mathbf{t}_2^T]^T$ is a column supervector comprising the two bit-based vectors $\mathbf{t}_1$ and $\mathbf{t}_2$ associated with the QPSK-modulated symbols s_1 and s_2, respectively.

Substituting Equation (16.38) into the system model of Equation (16.1) yields

$$\mathbf{y} = \mathbf{HQt} + \mathbf{w}. \tag{16.39}$$

Moreover, since $\mathbf{t}$ is a real-valued vector, we can elaborate a bit further and deduce a real-valued system model as follows:

$$\begin{bmatrix} \mathcal{R}\{\mathbf{y}\} \\ \mathcal{I}\{\mathbf{y}\} \end{bmatrix} = \begin{bmatrix} \mathcal{R}\{\mathbf{HQ}\} \\ \mathcal{I}\{\mathbf{HQ}\} \end{bmatrix} \mathbf{t} + \begin{bmatrix} \mathcal{R}\{\mathbf{w}\} \\ \mathcal{I}\{\mathbf{w}\} \end{bmatrix} = \tilde{\mathbf{H}}\mathbf{t} + \tilde{\mathbf{w}}, \tag{16.40}$$

where $\tilde{\mathbf{H}}$ is a real-valued (4×4)-dimensional bit-wise channel matrix, which may be expressed as

$$\tilde{\mathbf{H}} = \begin{bmatrix} \mathcal{R}\{\mathbf{HQ}\} \\ \mathcal{I}\{\mathbf{HQ}\} \end{bmatrix} = \begin{bmatrix} 0.1 & 0.2 & -0.7 & 0.6 \\ 0.3 & -0.4 & -1.3 & 0.5 \\ -0.2 & 0.1 & -0.6 & -0.7 \\ 0.4 & 0.3 & -0.5 & -1.3 \end{bmatrix}. \tag{16.41}$$

Thus, we arrive at the new system model of Equation (16.40), which may be interpreted as a (4×4)-dimensional BPSK-modulated SDM system. By applying the OHRSA-ML method of Algorithm 16.1 we have

$$\mathbf{U} = \begin{bmatrix} 0.63 & 0 & -0.85 & -0.27 \\ 0 & 0.63 & 0.27 & -0.85 \\ 0 & 0 & 1.45 & 0 \\ 0 & 0 & 0 & 1.45 \end{bmatrix}, \quad \hat{\mathbf{x}} = \begin{bmatrix} 0.43 \\ -0.34 \\ -1.10 \\ -0.79 \end{bmatrix}. \tag{16.42}$$

Furthermore, the first two steps of the recursive search process of Algorithm 16.1 are given by

$$J_4(\check{t}_4 = -1) = |u_{44}(\check{t}_4 - \hat{x}_4)|^2$$
$$= |1.45(-1 - (-0.79))|^2 = 0.10 \tag{16.43}$$

and

$$a_3(\check{t}_4 = -1) = u_{34}(\check{t}_4 - \hat{x}_4)$$
$$= 0(-1 - (-0.79)) = 0,$$
$$J_3(\check{t}_3 = -1, \check{t}_4 = -1) = |u_{33}(\check{t}_3 - \hat{x}_3) + a_3|^2$$
$$= |1.45(-1 - (-1.10)) + (0)|^2 = 0.12. \tag{16.44}$$

On completing the recursive search process of Algorithm 16.1 we arrive at the search tree depicted in Figure 16.3(b). As before, each evaluation step, i.e. each evaluation of the CSC function $J_i(\check{\mathbf{t}}_i)$ of Equation (16.30l), is indicated by an elliptic node. Moreover, the label inside each node indicates the order of evaluation as well as the corresponding value $J_i(\check{\mathbf{t}}_i)$ of the CSC function inside the brackets. The branches corresponding to two legitimate values $\check{t}_i = -1$ and 1 are indicated using the *dashed* and *solid* edges and nodes, respectively.

Observe that the ML estimates $\hat{\mathbf{s}}$ and $\hat{\mathbf{t}}$ of Figures 16.3(a) and 16.3(b) are obtained within the same number of evaluation steps. Nevertheless, the latter search procedure is constituted by lower-complexity real-valued operations. Furthermore, in contrast to the detection method considered in Example 16.2, the search method outlined in this QPSK-based example can be readily generalized for the scenario of M-QAM SDM systems, as demonstrated in the forthcoming section.

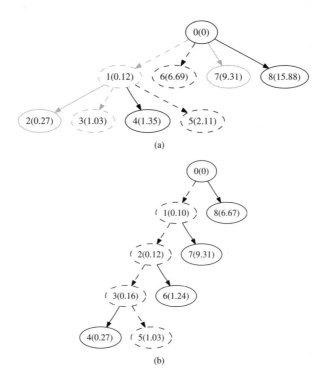

(a)

(b)

Figure 16.3: Examples of a search tree formed by the (a) OHRSA-ML and (b) BW-OHRSA-ML SDM detectors in the scenario of a system employing QPSK modulation, $m_t = n_r = 3$ transmit and receive antennas and encountering average SNRs of 10 dB. The labels indicate the order of execution, as well as the corresponding value $J_i(\tilde{s}_i)$ of the CSC function of Equation (16.16), as seen in the brackets.

16.1.2.1 Generalization of the BW-OHRSA-ML SDM Detector

In this section we generalize the result obtained in Section 16.1.1 to the case of systems employing hyper-rectangular modulation schemes, namely M-QAM, where each modulated symbol belongs to a discrete phasor constellation $\mathcal{M} = \{c_m\}_{m=1,\dots,M}$. It is evident that each phasor point c_m of an M-QAM constellation map may be represented as the inner product of a unique bit-based vector $d_m = \{d_{ml} = -1, 1\}_{l=1,\dots,b}$ and the corresponding *quantization vector* q. Specifically, we have

$$c_m = q^T d_m. \tag{16.45}$$

Some examples of the quantization vectors corresponding to the modulation schemes of BPSK, QPSK, 16-QAM as well as 64-QAM are portrayed in Table 16.1.

Furthermore, we define a $(bm_t \times m_t)$-dimensional *quantization matrix* $\mathbf{Q} = \mathbf{I} \otimes q$, where $\mathbf{I}$ is an $(m_t \times m_t)$-dimensional identity matrix and q is the aforementioned *quantization vector*, while $\otimes$ denotes the *matrix direct product* [634]. Consequently the M-QAM-modulated signal vector s may be represented as

$$s = \mathbf{Q}t, \tag{16.46}$$

where $t = [t_1^T, \dots, t_{m_t}^T]^T$ is a column supervector comprising the bit-based vectors t_i associated with each transmitted signal vector component s_i. Substituting Equation (16.46) into the system model of Equation (16.1) yields

$$y = \mathbf{H}\mathbf{Q}t + w. \tag{16.47}$$

Table 16.1: Examples of quantization vectors.

Modulation scheme	$\boldsymbol{q}^{\mathrm{T}}$
BPSK	$[1]$
QPSK	$\frac{1}{\sqrt{2}}[1,\, \jmath]$
16-QAM	$\frac{1}{\sqrt{10}}[1,\, 1\jmath,\, 2,\, 2\jmath]$
64QAM	$\frac{1}{\sqrt{42}}[1,\, 1\jmath,\, 2,\, 2\jmath,\, 4,\, 4\jmath]$

Moreover, since $\mathbf{t}$ is a real-valued vector, we can elaborate a bit further and deduce a real-valued system model as follows:

$$\tilde{\mathbf{y}} = \begin{bmatrix} \mathcal{R}\{\mathbf{y}\} \\ \mathcal{I}\{\mathbf{y}\} \end{bmatrix} = \begin{bmatrix} \mathcal{R}\{\mathbf{HQ}\} \\ \mathcal{I}\{\mathbf{HQ}\} \end{bmatrix} \mathbf{t} + \begin{bmatrix} \mathcal{R}\{\mathbf{w}\} \\ \mathcal{I}\{\mathbf{w}\} \end{bmatrix} = \tilde{\mathbf{H}}\mathbf{t} + \tilde{\mathbf{w}}, \tag{16.48}$$

where $\tilde{\mathbf{H}}$ is a real-valued $(2n_{\mathrm{r}} \times bm_{\mathrm{t}})$-dimensional bit-wise channel matrix. Observe in Equation (16.47) that the requirement of having constant modulus symbols is satisfied by the modified system model of Equation (16.47), since we have $|t_i|^2 = 1$ and thus the method described in Section 16.1.1 and summarized in Algorithm 16.1 is applicable for the evaluation of the bit-wise ML estimate $\hat{\mathbf{t}}$ of Equation (16.47). Consequently, we apply the following changes to Algorithm 16.1:

1. Include the evaluation of the bit-wise channel matrix $\tilde{\mathbf{H}}$ in (16.49a).

2. Adjust the number of candidate bit values of t_i to $d_m = \{-1, 1\}$ in (16.49l).

Hence we arrive at a new detection technique, namely the bit-wise OHRSA-ML SDM detector, which is summarized in Algorithm 16.2.

The operation of Algorithm 16.2 is illustrated by the search tree diagram depicted in Figure 16.4. Each circular node in the diagram represents a subvector candidate $\check{\mathbf{t}}_i = \{\check{t}_j\}_{j=i,\dots,r}$ of the transmitted bit-based signal vector $\mathbf{t}$. The solid and open nodes denote the binary values of the bit $\check{t}_i = \{-1, 1\}$ assumed in the current step of the recursive search process. The corresponding values of the CSC function $J_i(\check{\mathbf{t}}_i)$ are indicated by the tint and thickness of the transitions connecting each *child* or *descendant* node $\check{\mathbf{t}}_i$ with the corresponding *parent* node $\check{\mathbf{t}}_{i+1}$. The search tree diagram depicted in Figure 16.4 corresponds to the system scenario employing QPSK modulation, $m_{\mathrm{t}} = n_{\mathrm{r}} = 8$ operating at the average SNR of 6 dB. Observe that the ML solution is attained in 139 evaluation steps in comparison with the $2^{16} = 65\,536$ evaluation steps required by the exhaustive ML search.

16.1.3   OHRSA-Aided Log-MAP SDM Detection

It is evident [265] that the BER associated with the process of communicating over a noisy fading MIMO channel can be dramatically reduced by means of channel coding. A particularly effective channel coding scheme is constituted by the *soft-input, soft-output* turbo coding method [314]. Turbo coding, however, requires *soft* information concerning the bit decisions at the output of the SDM detector, in other words the a posteriori soft information regarding the confidence of the bit decision is required.

The derivation of an expression for the low-complexity evaluation of the soft-bit information associated with the bit estimates of the SDM detector's output characterized by Equation (16.5) is given in [265]. Here, we present a brief summary of the results deduced in [265].

Algorithm 16.2 Bit-wise OHRSA-ML SDM detector
©IEEE Akhtman, Wolfgang, Chen & Hanzo [562]

$$\tilde{\mathbf{H}} = \begin{bmatrix} \mathcal{R}\{\mathbf{HQ}\} \\ \mathcal{I}\{\mathbf{HQ}\} \end{bmatrix} \tag{16.49a}$$

$$\texttt{Sort}\{\tilde{\mathbf{H}}\}, \text{ such that } \|(\tilde{\mathbf{H}})_1\|^2 \leq \cdots \leq \|(\tilde{\mathbf{H}})_r\|^2 \tag{16.49b}$$

$$\mathbf{G} = (\tilde{\mathbf{H}}^{\mathrm{H}}\tilde{\mathbf{H}} + \sigma_w^2\mathbf{I}) \tag{16.49c}$$

$$\mathbf{U} = \texttt{CholeskyDecomposition}(\mathbf{G}) \tag{16.49d}$$

$$\hat{\mathbf{x}} = \mathbf{G}^{-1}\tilde{\mathbf{H}}^{\mathrm{H}}\tilde{\mathbf{y}} \tag{16.49e}$$

$$\texttt{Calculate} \quad J_r \tag{16.49f}$$

$$\texttt{Unsort}\{\hat{\mathbf{t}}\} \tag{16.49g}$$

$$\texttt{function} \quad \texttt{Calculate} \quad J_i \tag{16.49h}$$

$$a_i = \sum_{j=i+1}^{m_t} u_{ij}(\check{t}_j - \hat{x}_j) \tag{16.49i}$$

$$\texttt{Sort}\{\boldsymbol{d}\}, \text{ such that } \phi_i(d_1) < \phi_i(d_2), \tag{16.49j}$$

$$\texttt{where} \quad \phi_i(d) = |u_{ii}(d - \hat{x}_i) + a_i|^2 \tag{16.49k}$$

$$\texttt{for} \quad m = 1,2 \quad \texttt{do} \tag{16.49l}$$

$$\check{t}_i = d_m \tag{16.49m}$$

$$J_i = J_{i+1} + \phi_i(\check{t}_i) \tag{16.49n}$$

$$\texttt{if} \quad J_i < J_{\min} \quad \texttt{then} \tag{16.49o}$$

$$\texttt{if} \quad i > 0 \quad \texttt{then} \quad \texttt{Calculate} \quad J_{i-1} \tag{16.49p}$$

$$\texttt{else}$$

$$J_{\min} = J_0 \tag{16.49q}$$

$$\hat{\mathbf{t}} = \check{\mathbf{t}} \tag{16.49r}$$

$$\texttt{end if}$$

$$\texttt{end if}$$

$$\texttt{end for}$$

$$\texttt{end function}$$

The probability of the mth bit of the QAM symbol transmitted from the ith transmit antenna element is determined by the *likelihood* function, which may be expressed as follows [623]:

$$\mathcal{P}(b_{im}) = \sum_{\check{\mathbf{s}} \in \mathcal{M}_{im}^{1;m_t}} P(\check{\mathbf{s}})p(\mathbf{y}|\check{\mathbf{s}}, \mathbf{H}), \tag{16.50}$$

where we define

$$\mathcal{M}_{im}^{b;m_t} = \{\check{\mathbf{s}} = (\check{s}_1, \ldots, \check{s}_{m_t})^{\mathrm{T}}; \check{s}_j \in \mathcal{M} \text{ for } j \neq i, \check{s}_i \in \mathcal{M}_m^b\} \tag{16.51}$$

and $\mathcal{M}_m^b$ denotes the specific subset of the entire set $\mathcal{M}$ of modulation constellation phasors, which comprises the bit value $b = \{0,1\}$ at the mth bit position.

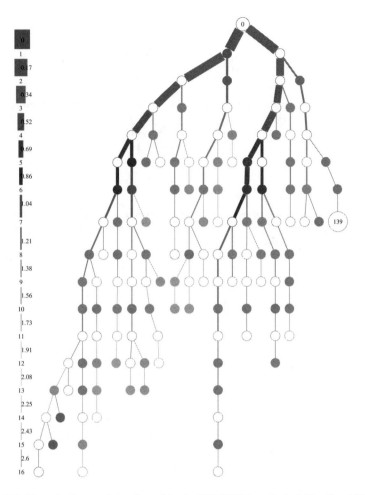

Figure 16.4: Example of a search tree formed by the BW-OHRSA method of Algorithm 16.2 in the scenario of QPSK, $m_{\mathrm{t}} = n_{\mathrm{r}} = 8$ and an average SNR of 6 dB. Each circular node in the diagram represents a subvector candidate $\breve{\mathbf{t}}_i = \{\breve{t}_j\}_{j=i,\dots,r}$ of the transmitted bit-based signal vector $\mathbf{t}$. The solid and open nodes denote the duo-binary values of the bit $\breve{t}_i = \{-1, 1\}$ assumed. The corresponding value of the CSC function $J_i(\breve{\mathbf{t}}_i)$ quantified in Equation (16.17b) is indicated by both the tint and the thickness of the transitions connecting each child node $\breve{\mathbf{t}}_i$ with the corresponding parent node $\breve{\mathbf{t}}_{i+1}$. The ML solution is attained in 139 evaluation steps in comparison with the $2^{16} = 65\,536$ evaluation steps required by the exhaustive ML search.

Correspondingly, the soft-bit value associated with the mth bit of the QAM symbol transmitted from the ith transmit antenna element is determined by the Log-Likelihood Ratio (LLR) values defined in [623] as

$$\mathcal{L}_{im} = \log \frac{\mathcal{P}(b_{im} = 1)}{\mathcal{P}(b_{im} = 0)} = \log \frac{\sum_{\breve{\mathbf{s}} \in \mathcal{M}_{im}^{1;m_{\mathrm{t}}}} P(\breve{\mathbf{s}}) p(\mathbf{y}|\breve{\mathbf{s}}, \mathbf{H})}{\sum_{\breve{\mathbf{s}} \in \mathcal{M}_{im}^{0;m_{\mathrm{t}}}} P(\breve{\mathbf{s}}) p(\mathbf{y}|\breve{\mathbf{s}}, \mathbf{H})}. \tag{16.52}$$

However, the direct calculation of the accumulated a posteriori conditional probabilities in the numerator and denominator of Equation (16.52) may have an excessive complexity in practice. Fortunately, as advocated in [265], the LLR values characterized in Equation (16.52) may be closely approximated as follows:

$$\mathcal{L}_{im} \approx \log \left(\frac{\max_{\check{s} \in \mathcal{M}_{im}^{1;m_t}} P(\check{s}) p(\check{s}|\mathbf{y}, \hat{\mathbf{H}})}{\max_{\check{s} \in \mathcal{M}_{im}^{0;m_t}} P(\check{s}) p(\check{s}|\mathbf{y}, \hat{\mathbf{H}})} \right), \tag{16.53}$$

where we assume equiprobable transmitted phasors $\check{s}$ and hence may elaborate a little further. That is, we have

$$\mathcal{L}_{im} \approx \log \frac{p(\mathbf{y}|\check{s}_{im}^1, \mathbf{H})}{p(\mathbf{y}|\check{s}_{im}^0, \mathbf{H})}, \tag{16.54}$$

where we define

$$\check{s}_{im}^b = \arg \max_{\check{s} \in \mathcal{M}_{im}^{b;m_t}} p(\mathbf{y}|\check{s}, \mathbf{H}), \quad b = 0, 1. \tag{16.55}$$

As suggested by the nature of Equation (16.54), the detection process employing the objective function determined by Equations (16.54) and (16.55) is usually referred to as the Logarithmic Maximum-A-Posteriori (Log-MAP) probability detector.

A practical version of the Log-MAP detector may be derived as follows. Substituting Equation (16.4) into Equations (16.52) and (16.50) yields

$$\mathcal{P}(b_{im}) = \sum_{\check{s} \in \mathcal{M}_{im}^{b;m_t}} \exp \left(-\frac{1}{n_r \sigma_w^2} \|\mathbf{y} - \mathbf{H}\check{s}\|^2 \right) \tag{16.56}$$

and

$$\mathcal{L}_{im} = \log \frac{\sum_{\check{s} \in \mathcal{M}_{im}^{1;m_t}} \exp[-(1/n_r \sigma_w^2)\|\mathbf{y} - \mathbf{H}\check{s}\|^2)]}{\sum_{\check{s} \in \mathcal{M}_{im}^{0;m_t}} \exp[-(1/n_r \sigma_w^2)\|\mathbf{y} - \mathbf{H}\check{s}\|^2)]}, \tag{16.57}$$

respectively. Note that Equation (16.57) involves two summations over 2^{rm_t-1} exponential functions. This operation may potentially impose an excessive computational complexity for large values of m_t and/or r. However, as demonstrated in [265], the expression in (16.57) may be closely approximated by a substantially simpler expression, namely

$$\mathcal{L}_{im} \approx \frac{1}{n_r \sigma_w^2} [\|\mathbf{y} - \mathbf{H}\check{s}_{im}^0\|^2 - \|\mathbf{y} - \mathbf{H}\check{s}_{im}^1\|^2], \tag{16.58}$$

where

$$\check{s}_{im}^b = \arg \min_{\check{s} \in \mathcal{M}_{im}^{b;m_t}} \|\mathbf{y} - \mathbf{H}\check{s}\|^2, \quad b = 0, 1, \tag{16.59}$$

and, again, $\mathcal{M}_{im}^{b;m_t}$ denotes the specific subset of the entire set $\mathcal{M}^{m_t}$ of signal vector candidates associated with the modulation scheme employed, which comprises the bit value $b = \{0, 1\}$ at the mth bit position of the ith signal vector component.

The Log-MAP detector defined by Equations (16.58) and (16.59) may be applied to obtain the soft-bit information associated with the bit-wise OHRSA-ML SDM detector derived in Section 16.1.2. Consequently, substituting the bit-wise system model of Equation (16.47) into (16.58) and (16.59) yields

$$\mathcal{L}_i \approx \frac{1}{n_r \sigma_w^2} [\|\mathbf{y} - \tilde{\mathbf{H}}\check{t}_{i;\min}^0\|^2 - \|\mathbf{y} - \tilde{\mathbf{H}}\check{t}_{i;\min}^1\|^2], \tag{16.60}$$

where

$$\check{t}_{i;\min}^b = \arg \min_{\check{t} \in \mathcal{D}_i^{m;r}} \|\mathbf{y} - \tilde{\mathbf{H}}\check{t}\|^2, \quad b = 0, 1 \tag{16.61}$$

and $\mathcal{D}_i^{b;r}$ denotes the subset of the entire set $\mathcal{D}^r$ of $(r = m_t \log_2 M)$-dimensional bit-wise vectors, which comprise the binary value $\check{t}_i = d_b = \{-1, 1\}$ at the ith bit position.

Furthermore, substituting the bit-wise objective function of Equation (16.58) into (16.60) yields

$$\mathcal{L}_i \approx \frac{1}{n_r \sigma_w^2}[J(\check{\mathbf{t}}_{i;\min}^0) + \phi - J(\check{\mathbf{t}}_{i;\min}^1) - \phi]$$

$$= \frac{1}{n_r \sigma_w^2}[J(\check{\mathbf{t}}_{i;\min}^0) - J(\check{\mathbf{t}}_{i;\min}^1)], \tag{16.62}$$

where $\check{\mathbf{t}}_{i;\min}^m$ and the corresponding cost function value $J(\check{\mathbf{t}}_{i;\min}^m)$ may be obtained by applying the constrained OHRSA-ML detection method derived in Section 16.1.2.

Consequently, the evaluation of the bit-wise Max-Log-MAP estimates of the transmitted bit-wise signal vector $\mathbf{t}$ involves repetitive evaluation of $2r$ constrained ML estimates $\check{\mathbf{t}}_{i;\min}^m$ along with the associated $2r$ values of the objective function $J(\check{\mathbf{t}}_{i;\min}^m)$.

Example 16.4. OHRSA-Log-MAP BPSK 3 × 3
Consider a BPSK system having $n_r = m_t = 3$ transmit and receive antennas, which is described by Equation (16.1). The transmitted signal $\mathbf{s}$, received signal $\mathbf{y}$ as well as the channel matrix $\mathbf{H}$ of Equation (16.1) are exemplified by the following values:

$$\mathbf{s} = \begin{bmatrix} -1 \\ 1 \\ 1 \end{bmatrix}, \quad \mathbf{y} = \begin{bmatrix} 0.2 \\ 0.3 \\ -0.5 \end{bmatrix}, \quad \mathbf{H} = \begin{bmatrix} 0.1 & -1 & 1.1 \\ -0.2 & 0.7 & -0.7 \\ 0.4 & 0.5 & -0.5 \end{bmatrix}. \tag{16.63}$$

Observe that the channel matrix $\mathbf{H}$ of Equation (16.63) happens to be *best-first* ordered and does not require any further reordering. Furthermore, in our scenario of BPSK modulation the channel matrix $\mathbf{H}$ of Equation (16.63) is equivalent to the bit-wise channel matrix $\tilde{\mathbf{H}}$ of Algorithm 16.3.

Subsequently, our task is to obtain the Log-MAP estimate of the transmitted signal vector $\mathbf{t} = \mathbf{s}$. We apply the OHRSA-Log-MAP method of Algorithm 16.3. Firstly, we evaluate the triangular matrix $\mathbf{U}$ of Equation (16.85d) as well as the unconstrained MMSE estimate $\hat{\mathbf{x}}$ of Equation (16.85e). The resultant quantities are given by

$$\mathbf{U} = \begin{bmatrix} 0.56 & -0.07 & 0.09 \\ 0 & 1.35 & -1.35 \\ 0 & 0 & 0.46 \end{bmatrix}, \quad \hat{\mathbf{x}} = \begin{bmatrix} -0.80 \\ -0.01 \\ 0.13 \end{bmatrix}. \tag{16.64}$$

Secondly, as further suggested by Algorithm 16.3, for each transmitted bit-wise symbol t_i we calculate the quantities $J(\check{\mathbf{t}}_{i;\min}^{-1})$ and $J(\check{\mathbf{t}}_{i;\min}^1)$ corresponding to the values of the cost function $J(\check{\mathbf{t}})$ of Equation (16.85o) associated with the constrained ML estimates of the transmitted bit-wise vector $\mathbf{t}$ with the ith bit component assuming values of -1 and 1, respectively.

For instance, the cost function value $J(\check{\mathbf{t}}_{1;\min}^{-1})$ associated with the ML estimate of the bit-wise signal vector $\mathbf{t}$ constrained by bit-component value $\check{t}_1 = -1$ may be calculated as follows:

$$J_3(\check{t}_3 = 1) = |u_{33}(\check{t}_3 - \hat{x}_3)|^2 = (0.46(1 - (0.13)))^2 = 0.16,$$

$$a_2(\check{t}_3 = 1) = u_{23}(\check{t}_3 - \hat{x}_3) = -1.35(1 - (0.13)) = -1.17,$$

$$J_2(\check{t}_2 = 1, \check{t}_3 = 1) = J_3(\check{t}_3 = 1) + |u_{22}(\check{t}_2 - \hat{x}_2) + a_2|^2$$

$$= 0.16 + |1.35(1 - (-0.01)) + (-1.17)|^2 = 0.20. \tag{16.65}$$

Furthermore, we have

$$a_1(\check{t}_2 = 1, \check{t}_3 = 1) = u_{12}(\check{t}_2 - \hat{x}_2) + u_{13}(\check{t}_3 - \hat{x}_3)$$

$$= -0.07(1 - (-0.01)) + 0.09(1 - (0.13)) = 0.00,$$

$$J(\check{\mathbf{t}}_{1;\min}^{-1}) = J_1(\check{t}_1 = -1, \check{t}_2 = 1, \check{t}_3 = 1)$$

$$= J_2(\check{t}_2 = 1, \check{t}_3 = 1) + |u_{11}(\check{t}_1 - \hat{x}_1) + a_1|^2$$

$$= 0.20 + |0.56(-1 - (-0.80)) + (0.00)|^2 = 0.21. \tag{16.66}$$

Observe that, for the sake of brevity, we omit the calculation of the CSC values outside the major search branch of Algorithm 16.3, i.e. outside the search branch leading to the constrained ML estimate. The corresponding search tree formed by the evaluation of the value of $J(\check{\mathbf{s}}_{1;\min}^{-1})$ using Algorithm 16.3 is depicted in Figure 16.5(a). Furthermore, Figures 16.5(b)–(f) illustrate the search trees formed by the search subprocesses of Algorithm 16.3 corresponding to the remaining *five* values $\{J(\check{\mathbf{s}}_{i;\min}^{b})\}_{i=1,\ldots,3}^{b=-1,1}$.

Finally, upon completing the calculation of all *six* values $\{J(\check{\mathbf{s}}_{i;\min}^{b})\}_{i=1,\ldots,3}^{b=-1,1}$ we arrive at the following matrix:

$$\hat{\mathbf{J}} = \{J(\check{\mathbf{s}}_{i;\min}^{b})\}_{i=1,\ldots,3}^{b=-1,1} = \begin{bmatrix} 0.21 & 1.21 \\ 0.33 & 0.21 \\ 0.33 & 0.21 \end{bmatrix}, \tag{16.67}$$

where the elements of the matrix $\hat{\mathbf{J}}$, which we refer to as the Minimum Cost Function (MCF) matrix, are defined as $\hat{J}_{ij} = J(\check{\mathbf{s}}_{i;\min}^{b_j})$. Consequently, the *soft-bit* vector representing the Log-MAP estimate of the transmitted bit-wise signal vector $\mathbf{t}$ may be expressed as

$$\mathbf{L} = \frac{1}{\sigma_w^2}[(\hat{\mathbf{J}})_1 - (\hat{\mathbf{J}})_2] = \begin{bmatrix} -9 \\ 1.2 \\ 1.2 \end{bmatrix}, \tag{16.68}$$

where $(\hat{\mathbf{J}})_j$ denotes the jth column of the MCF matrix $\hat{\mathbf{J}}$ defined in Equation (16.67).

Example 16.5. OHRSA Approximate Log-MAP BPSK 3 × 3

Again, consider a BPSK system identical to that described in Example 16.4. Specifically, we have a (3×3)-dimensional real-valued linear system described by Equation (16.1) with the corresponding transmitted signal $\mathbf{s}$, the received signal $\mathbf{y}$ and the channel matrix $\mathbf{H}$ described in Equation (16.63). In this example we want to demonstrate an alternative search paradigm, which avoids the repetitive process characterized by Algorithm 16.3 and exemplified in Figure 16.5 of Example 16.4, while obtaining a similar result.

Firstly, we apply the OHRSA-ML method of Algorithm 16.2. The triangular matrix $\mathbf{U}$ of Equation (16.49d) as well as the unconstrained MMSE estimate $\hat{\mathbf{x}}$ of Equation (16.49e) are similar to those evaluated in Example 16.4 and are characterized by Equation (16.64). The resultant search process is characterized by the search tree diagram portrayed in Figure 16.6(a).

Additionally, however, we define a (3×2)-dimensional MCF matrix $\hat{\mathbf{J}}$, which will be used for evaluation of the soft-bit information, and assign to it an initial value of $\hat{\mathbf{J}} = J_0 \mathbf{1}$, where $\mathbf{1}$ is a (3×2)-dimensional matrix of ones and $J_0 \gg \gamma$ is some large constant, which should be greater than the average SNR of $\gamma = 10$ encountered. For instance, let us assume $J_0 = 100$. Subsequently, the cost-function-related matrix $\hat{\mathbf{J}}$ is updated according to a procedure to be outlined below each time the search branch forming the search tree portrayed in Figure 16.6(a) is terminated, regardless of whether its termination occurred owing to reaching the final recursive index value of $i = 1$, or owing to exceeding the minimum value of the cost function $J_{\min}$. More specifically, we update the elements of the matrix $\hat{\mathbf{J}}$ corresponding to the bit-wise symbols $\check{t}_j$, $j = i, \ldots, 3$, constituting the bit-wise subvector candidate $\check{\mathbf{t}}_i$ associated with the particular search branch, as outlined below:

$$\hat{J}_{jb_j} = \min\{\hat{J}_{jb_j}, J_i(\check{\mathbf{t}}_i)\}, \quad j = i, \ldots, 3, \quad \check{t}_j = \{-1,1\}_{b_j}. \tag{16.69}$$

For instance, on completing the first, leftmost search branch depicted in Figure 16.6(a) and associated with the transmitted signal candidate $\check{\mathbf{t}} = \begin{bmatrix} -1 & 1 & 1 \end{bmatrix}^T$, i.e. on reaching node 3 of the search tree, the following update of the MCF matrix $\hat{\mathbf{J}}$ is performed:

$$\hat{J}_{11} = \min\{\hat{J}_{11}, J(\check{\mathbf{t}})\} = \min\{100, 0.21\} = 0.21$$
$$\hat{J}_{22} = \hat{J}_{32} = \min\{100, 0.21\} = 0.21. \tag{16.70}$$

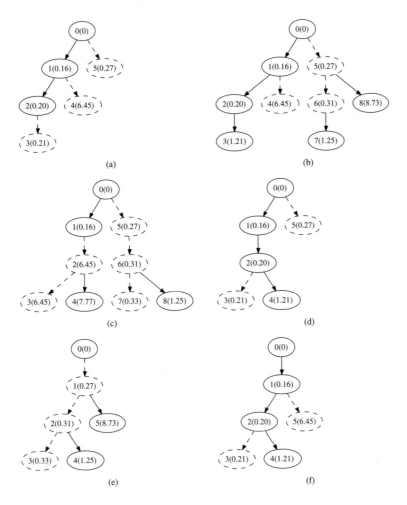

Figure 16.5: Example of search trees formed by the OHRSA-Log-MAP SDM detector of Algorithm 16.3 in the scenario of a system employing BPSK modulation, $m_t = n_r = 3$ transmit and receive antennas and encountering average SNRs of $10\,dB$. The labels indicate the order of visitation, as well as the corresponding value $J_i(\check{t}_i)$ of the CSC function of Equation (16.85o), as seen in the brackets.

Consequently, the matrix $\hat{\mathbf{J}}$ becomes

$$\hat{\mathbf{J}}(3) = \begin{bmatrix} 0.21 & 100 \\ 100 & 0.21 \\ 100 & 0.21 \end{bmatrix} \tag{16.71}$$

Furthermore, the states of the MCF matrix corresponding to the search steps 4, 5 and 6 of Figure 16.6(a) are

$$\hat{\mathbf{J}}(4) = \begin{bmatrix} 0.21 & 1.21 \\ 100 & 0.21 \\ 100 & 0.21 \end{bmatrix}, \quad \hat{\mathbf{J}}(5) = \begin{bmatrix} 0.21 & 1.21 \\ 6.45 & 0.21 \\ 100 & 0.21 \end{bmatrix}, \quad \hat{\mathbf{J}}(6) = \begin{bmatrix} 0.21 & 1.21 \\ 6.45 & 0.21 \\ 0.27 & 0.21 \end{bmatrix}. \tag{16.72}$$

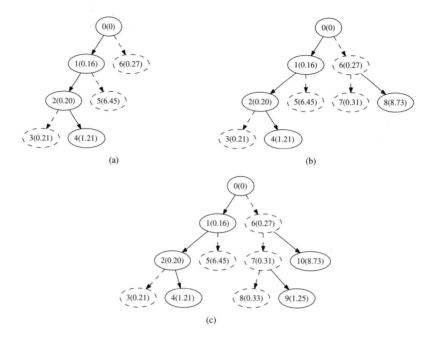

Figure 16.6: Example of the search trees formed by the modified OHRSA-ML SDM detector of Algorithm 16.2 using different values of the parameter ρ, namely (a) 1.0, (b) 1.3 and (c) 2.0. We consider a system employing BPSK modulation, $m_\mathrm{t} = n_\mathrm{r} = 3$ transmit and receive antennas and encountering an average SNR of 10 dB. The labels indicate the order of evaluation, as well as the corresponding value $J_i(\check{\mathbf{s}}_i)$ of the CSC function of Equation (16.16), as seen in the brackets.

Finally, by substituting the resultant value of the MCF matrix $\hat{\mathbf{J}}(6)$ into Equation (16.68) we obtain the following soft-bit estimate of the transmitted bit-wise signal vector $\mathbf{t}$:

$$\mathcal{L}_a = \begin{bmatrix} -9 \\ 62.39 \\ 0.60 \end{bmatrix}. \tag{16.73}$$

Observe that the soft-bit estimate $\mathcal{L}_a$ of Equation (16.73) appears to be considerably more reliable than the MMSE estimate $\hat{\mathbf{x}}$ of Equation (16.64). Specifically, as opposed to the MMSE estimate $\hat{\mathbf{x}}$ in Equation (16.19), the direct slicing of the soft-bit estimate $\mathcal{L}_a$ results in the correct signal vector $\mathbf{s}$ of Equation (16.63). Moreover, the soft-bit estimate $\mathcal{L}_a$ provides further information concerning the reliability of each estimated bit, albeit the resultant soft-bit information of Equation (16.73) substantially deviates from the more reliable exact Log-MAP estimate $\mathbf{L}$ given by Equation (16.68).

Fortunately, however, the precision of the soft-bit estimate $\mathcal{L}_a$ may be readily improved. Specifically, we introduce an additional parameter ρ, which will allow us to control the rate of convergence in the search process of Algorithm 16.2 by increasing the threshold value of the CSC function, which controls the passage of the recursive search process through *low-likelihood* search branches having CSC function values $J_i(\check{\mathbf{t}}_i)$ in excess of $\rho J_{\min}$, as opposed to $J_{\min}$ of Equation (16.49o) in Algorithm 16.2. Let us now execute the modified OHRSA-ML method of Algorithm 16.2, where the condition $J_i < J_{\min}$ of Equation (16.49o) is replaced by the corresponding condition of $J_i < \rho J_{\min}$.

The search trees formed by the execution of the modified Algorithm 16.2 in the scenarios of setting (b) $\rho = 1.3$ and (c) $\rho = 2.0$ are depicted in Figures 16.5(b) and 16.5(c), respectively. Furthermore, the convergence of the MCF matrix $\hat{\mathbf{J}}$ as well as the resultant soft-bit estimate $\mathbf{L}$ in both scenarios may be characterized as follows:

$$(b) \quad \hat{\mathbf{J}}(7) = \begin{bmatrix} 0.21 & 1.21 \\ 0.31 & 0.21 \\ 0.31 & 0.21 \end{bmatrix}, \quad \hat{\mathbf{J}}(8) = \begin{bmatrix} 0.21 & 1.21 \\ 0.31 & 0.21 \\ 0.31 & 0.21 \end{bmatrix}, \quad \mathcal{L}_b = \begin{bmatrix} -9 \\ 0.99 \\ 0.99 \end{bmatrix} \quad (16.74)$$

and

$$(c) \quad \hat{\mathbf{J}}(8) = \begin{bmatrix} 0.21 & 1.21 \\ 0.33 & 0.21 \\ 0.33 & 0.21 \end{bmatrix}, \quad \hat{\mathbf{J}}(10) = \begin{bmatrix} 0.21 & 1.21 \\ 0.33 & 0.21 \\ 0.33 & 0.21 \end{bmatrix}, \quad \mathcal{L}_c = \begin{bmatrix} -9 \\ 1.2 \\ 1.2 \end{bmatrix}, \quad (16.75)$$

where, as before, $\hat{\mathbf{J}}(n)$ denotes the state of the MCF matrix at search step n corresponding to the nth node of the search tree in Figures 16.5(b) and 16.5(c). Note that the search processes characterized by Figures 16.5(b) and 16.5(c) merely expand the search process portrayed in Figure 16.5(a). Consequently, for the sake of brevity, the corresponding Equations (16.74) and (16.75) depict only the extra states of the MCF matrix introduced by the expanded search procedure. For instance, the states $\hat{\mathbf{J}}(10)$ and $\hat{\mathbf{J}}(8)$ of Equation (16.75) complement the state $\hat{\mathbf{J}}(7)$ of Equation (16.74), as well as the states $\hat{\mathbf{J}}(6), \hat{\mathbf{J}}(5), \hat{\mathbf{J}}(4)$ and $\hat{\mathbf{J}}(3)$ of Equations (16.71) and (16.72), respectively.

Finally, by comparing the resultant soft-bit estimates $\mathcal{L}_a, \mathcal{L}_b$ and $\mathcal{L}_c$ of Equations (16.73), (16.74) and (16.75), corresponding to the scaling values of $\rho = 1.0, 1.3$ and 2.0, to the corresponding Log-MAP estimate $\mathbf{L}$ of Equation (16.68), we may hypothesize that the value of the soft-bit estimate obtained by the modified OHRSA-ML method of Algorithm 16.2 rapidly converges to the Log-MAP estimate of the OHRSA-Log-MAP method of Algorithm 16.3 on increasing the value of the parameter ρ. As expected, there is a trade-off between the accuracy of the soft-bit information obtained and the corresponding computational complexity associated with the particular choice of ρ. In the next section we will generalize the results obtained in this example and substantiate the aforementioned convergence-related hypothesis, as well as deduce the optimal value of the associated scaling parameter ρ.

16.1.4   Soft-Input, Soft-Output Max-Log-MAP SDM Detection

The OHRSA-aided Max-Log-MAP SDM detection method outlined in Section 16.1.3 may be easily adopted to exploit any a priori bit-related soft information available. More specifically, in Section 16.1.3 we assumed having equiprobable transmitted phasors $\check{\mathbf{s}}$. Correspondingly, in order to accommodate any available a priori probability information $P(\check{t}_i)$ associated with the estimated bit values $t_i, i = 1, \ldots, r$, Equations (16.60) and (16.61) may be modified as follows:

$$\mathcal{L}_i \approx \frac{1}{\sigma_w^2}[\|\mathbf{y} - \tilde{\mathbf{H}}\check{\mathbf{t}}_{i;\min}^0\|^2 - \|\mathbf{y} - \tilde{\mathbf{H}}\check{\mathbf{t}}_{i;\min}^1\|^2], \quad (16.76)$$

and

$$\check{\mathbf{t}}_{i;\min}^b = \arg\min_{\check{\mathbf{t}} \in \mathcal{D}_i^{m;r}} \left\{ -\log(P(\check{\mathbf{t}})) + \frac{\|\mathbf{y} - \tilde{\mathbf{H}}\check{\mathbf{t}}\|^2}{\sigma_w^2} \right\}, \quad b = 0, 1, \quad (16.77)$$

where, again, $\mathcal{D}_i^{b;r}$ denotes the subset of the entire set $\mathcal{D}^r$ of $(r = m_t \log_2 M)$-dimensional bit-wise vectors, which comprise the binary value $\check{t}_i = d_b = \{-1, 1\}$ at the ith bit position.

In practice, the probability-related soft information associated with the estimated bit values t_i is conveyed using the LLR values $\mathcal{L}_i$. Correspondingly, the logarithm of the probability value $\log(P(\check{\mathbf{t}}))$ of Equation (16.61) may be calculated as follows [314]:

$$\log(P(\check{\mathbf{t}})) = \sum_i P(\check{t}_i), \quad (16.78)$$

where

$$P(\check{t}_i = -1) = \mathtt{JacLog}(0, \mathcal{L}_i) \quad (16.79)$$

and

$$P(\check{t}_i = 1) = 1 - \texttt{JacLog}(0, \mathcal{L}_i), \tag{16.80}$$

where $\texttt{JacLog}(\cdot)$ denotes the Jacobian logarithm [635] defined as $\texttt{JacLog}(a, b) = \log(e^a + e^b)$.

The resultant a priori probability values $P(\check{t}_i)$ may be incorporated into the OHRSA SDM detector of Algorithm 16.2. That is, cost function constituent ϕ_i of Equation (16.49k) is redefined to accommodate the a priori log-probabilistic information $P(\check{t}_i)$ as follows:

$$\phi_i(d) = |u_{ii}(d - \hat{x}_i) + a_i|^2 - \sigma_w^2 P(\check{t}_i). \tag{16.81}$$

The pseudo-code describing the implementation of the bit-wise soft-input, soft-output OHRSA Max-Log-MAP SDM detector is summarized in Algorithm 16.3.

Clearly, the repetitive nature of the search process entailing Equations (16.85f, i–r) in Algorithm 16.3 and exemplified by Example 16.4 imposes a substantial increase on the associated computational complexity. Hence, in the next section we derive an OHRSA-aided approximate Log-MAP method which is capable of approaching the optimum Log-MAP performance, while avoiding the repetitive evaluation of Equation (16.85f) in Algorithm 16.3 and therefore imposes considerably reduced complexity requirements.

16.1.5   SOPHIE-Aided Approximate Log-MAP SDM Detection

Let us define the $(r \times 2)$-dimensional Bit-wise Minimum Cost (BMC) function matrix $\hat{\mathbf{J}}$ having elements as follows:

$$\hat{J}_{ib} = J(\hat{\mathbf{t}}_i^b), \ i = 1, \ldots, r, \ b = -1, 1, \tag{16.82}$$

where $\hat{\mathbf{t}}_i^b$ is defined by Equation (16.59). Using the BMC matrix of Equation (16.82), Equation (16.62) may also be expressed in vectorial form as

$$\mathbf{L} = \frac{1}{\sigma_w^2}[(\hat{\mathbf{J}})_1 - (\hat{\mathbf{J}})_2], \tag{16.83}$$

where, as before, $(\hat{\mathbf{J}})_b$ denotes the bth column of the matrix $\hat{\mathbf{J}}$ having elements defined by Equation (16.82).

Consequently, in order to evaluate the bit-related soft information we have to populate the BMC matrix $\hat{\mathbf{J}}$ of Equation (16.82) with the corresponding values of the cost function of Equation (16.82). Observe that the evaluation of the ML estimate $\hat{\mathbf{t}}$ will situate half the elements of the cost matrix $\hat{\mathbf{J}}$ with the corresponding minimum value of the cost function associated with the ML estimate, such that we have

$$J_{ib} = J(\hat{\mathbf{t}}), \ i = 1, \ldots, r, \ b = \hat{t}_i. \tag{16.84}$$

Subsequently, let us introduce the following adjustments to Algorithm 16.2. Firstly, we introduce an additional parameter ρ, which we refer to as the *search radius factor*. More specifically, the parameter ρ allows us to control the rate of convergence for the tree search process of Algorithm 16.2 and affects the cut-off value of a CSC function, which limits the passage of the recursive search process through *low-likelihood* search branches having a CSC function value $J_i(\check{\mathbf{t}}_i)$ in excess of $\rho J_{\min}$, as opposed to $J_{\min}$. Thus, the following rule replaces Rule 4 of Section 16.1.1.1.

Rule 4a. At each recursive detection level i, only the high-probability search branches corresponding to the highly likely symbol candidates c_m resulting in low values of the CSC function obeying $J_i(c_m) < \rho J_{\min}$ are pursued. Furthermore, as follows from the sorting criterion of the optimization in Rule 2, as soon as the inequality $J_i(c_m) > \rho J_{\min}$ is satisfied, the search loop at the ith recursive detection level is discontinued.

Secondly, we introduce an additional rule which facilitates the evaluation of the elements of the BMC matrix $\hat{\mathbf{J}}$ of Equation (16.82). Explicitly, we postulate Rule 5.

Algorithm 16.3 Bit-wise SISO-OHRSA-LogMAP SDM detector

$$\tilde{\mathbf{H}} = \begin{bmatrix} \mathcal{R}\{\mathbf{HQ}\} \\ \mathcal{I}\{\mathbf{HQ}\} \end{bmatrix} \tag{16.85a}$$

$$\texttt{Sort}\{\tilde{\mathbf{H}}\}, \text{ such that } \|(\tilde{\mathbf{H}})_1\|^2 \leq \cdots \leq \|(\tilde{\mathbf{H}})_{m_t}\|^2 \tag{16.85b}$$

$$\mathbf{G} = (\tilde{\mathbf{H}}^{\text{H}}\tilde{\mathbf{H}} + \sigma_w^2\mathbf{I}) \tag{16.85c}$$

$$\mathbf{U} = \texttt{CholeskyDecomposition}(\mathbf{G}) \tag{16.85d}$$

$$\hat{\mathbf{x}} = \mathbf{G}^{-1}\tilde{\mathbf{H}}^{\text{H}}\tilde{\mathbf{y}} \tag{16.85e}$$

$\texttt{for} \quad i = 1, \ldots, r$

$$\mathcal{L}_{im} = \frac{1}{\sigma_w^2}[J_{i;\min}^0 - J_{i;\min}^1] \tag{16.85f}$$

$\texttt{end} \quad \texttt{for}$

$$\texttt{Unsort}\{\mathcal{L}_i\}_{i=1,\ldots,r} \tag{16.85g}$$

$$\texttt{function} \quad \texttt{Calculate} \quad J_{k;i}^b \tag{16.85h}$$

$$a_i = \sum_{j=i+1}^{m_t} u_{ij}(\check{t}_j - \hat{x}_j) \tag{16.85i}$$

$\texttt{if} \quad i = k \quad \texttt{then}$

$$d_0 = \{-1, 1\}_b \tag{16.85j}$$

$\texttt{else}$

$$\texttt{Sort}\{d_m = -1, 1\}, \tag{16.85k}$$

$$\text{such that } \phi_i(d_0) < \phi_i(d_1), \tag{16.85l}$$

$$\text{where} \quad \phi_i(d_m) = |u_{ii}(d_m - \hat{x}_i) + a_i|^2 - \sigma_w^2 P(\check{t}_i) \tag{16.85m}$$

$\texttt{end} \quad \texttt{if}$

$\texttt{for} \quad m = 0, 1 \quad \texttt{do}$

$$\check{t}_i = d_m \tag{16.85n}$$

$$J_{k;i} = J_{k;i+1} + \phi_i(d_m) \tag{16.85o}$$

$\quad \texttt{if} \quad J_i < J_{\min} \quad \texttt{then}$

$$\quad\quad \texttt{if} \quad i > 0 \quad \texttt{then} \quad \texttt{Calculate} \quad J_{k;i-1}^b \tag{16.85q}$$

$\quad\quad \texttt{else}$

$$\quad\quad\quad J_{\min} = J_{k;\min}^b = J_{k;0}^b \tag{16.85r}$$

$\quad\quad \texttt{end} \quad \texttt{if}$

$\quad \texttt{end} \quad \texttt{if}$

$\quad \texttt{if} \quad i = k \quad \texttt{then break for loop}$

$\texttt{end} \quad \texttt{for}$

$\texttt{end} \quad \texttt{function}$

Rule 5. At each arrival at the bottom of the search tree, which corresponds to search level 1, the resultant value of the branch cost function $J(\check{t})$ is utilized to populate the elements of the BMC matrix

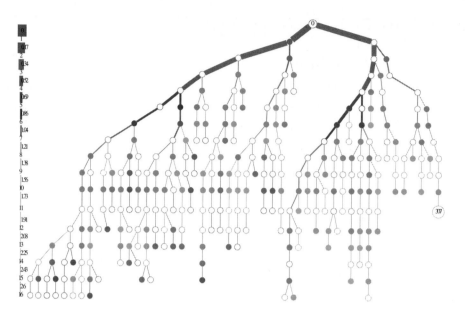

Figure 16.7: Example of a search tree formed by the SOPHIE SDM detector of Algorithm 16.2 in the scenario of QPSK, $m_t = n_r = 8$ and an average SNR of 6 dB. The approximate Log-MAP solution is attained in 307 evaluation steps in comparison with $32 \cdot 2^{15} = 1\,048\,576$ evaluation steps required by the exhaustive Log-MAP search. For more details on the notations employed in the diagram see the caption of Figure 16.4.

$\hat{\mathbf{J}}$, which correspond to the bit-wise signal components $\check{t}_i$ comprising the obtained signal candidate $\check{\mathbf{t}}$. That is, we have

$$\hat{J}_{ib} = \min\{\hat{J}_{ib}, J(\check{\mathbf{t}})\}, \quad i = 1, \ldots, r, \ b = \check{t}_i. \tag{16.86}$$

Subsequently, we suggest that the evaluation of the BMC matrix $\hat{\mathbf{J}}$, which is performed in the process of the ML search of Algorithm 16.2 extended by Rule 4a and using Rule 5, will allow us to provide reliable soft-bit information, while imposing a relatively low computational complexity. The main rationale of this assumption will be outlined in our quantitative complexity and performance analysis portrayed in Section 16.1.5.1.

As we will further demonstrate in Section 16.1.5.1, the resultant approximate Log-MAP SDM detector exhibits a particularly low complexity at high SNR values. On the other hand, at low SNR values the associated complexity substantially increases. Consequently, in order to control the computational complexity at low SNR values, we introduce the additional complexity control parameter γ. Our aim is to avoid the computationally demanding and yet inefficient detection of the specific signal components which have their signal energy well below the noise floor. More specifically, we modify Equation (16.49p) of Algorithm 16.2 according to Rule 6.

Rule 6. The branching of the tree search described by Algorithm 16.2 is truncated if the SNR associated with the corresponding signal component is lower than the value of the complexity control parameter γ. In other words, the search along a given branch is truncated if we have $\|\mathbf{H}_i\|^2/\sigma_w^2 < \gamma$.

On applying Rules 4, 5 and 6 in the context of the OHRSA-ML method of Algorithm 16.2, we arrive at an *approximate* OHRSA-Log-MAP SDM detector, which avoids the repetitive search required by the OHRSA-Log-MAP SDM detector of Section 16.1.3. The resultant OHRSA-aided approximate Log-MAP SDM detector, which we refer to as the SOPHIE (Soft-output OPtimized HIErarchy) SDM detector, is summarized in Algorithm 16.4 and exemplified in Figure 16.7.

Algorithm 16.4 SOPHIE approximate Log-MAP SDM detector
©IEEE Akhtman, Wolfgang, Chen & Hanzo [562]

$$\tilde{\mathbf{H}} = \begin{bmatrix} \mathcal{R}\{\mathbf{HQ}\} \\ \mathcal{I}\{\mathbf{HQ}\} \end{bmatrix} \tag{16.87a}$$

$\texttt{Sort}\{\tilde{\mathbf{H}}\}, \texttt{ such that } \|(\tilde{\mathbf{H}})_1\|^2 \leq \cdots \leq \|(\tilde{\mathbf{H}})_r\|^2$ (16.87b)

$\mathbf{G} = (\tilde{\mathbf{H}}^H \tilde{\mathbf{H}} + \sigma_w^2 \mathbf{I})$ (16.87c)

$\mathbf{U} = \texttt{CholeskyDecomposition}(\mathbf{G})$ (16.87d)

$\hat{\mathbf{x}} = \mathbf{G}^{-1} \tilde{\mathbf{H}}^H \tilde{\mathbf{y}}$ (16.87e)

$\texttt{Calculate } J_r$ (16.87f)

$\mathbf{L} = \dfrac{1}{\sigma_w^2}[(\hat{\mathbf{J}})_0 - (\hat{\mathbf{J}})_1]$ (16.87g)

$\texttt{Unsort}\{\mathcal{L}_i\}_{i=1,\ldots,r}$ (16.87h)

$\texttt{function Calculate } J_i$ (16.87i)

$a_i = \displaystyle\sum_{j=i+1}^{m_t} u_{ij}(\check{t}_j - \hat{x}_j)$ (16.87j)

$\texttt{Sort}\{\boldsymbol{d}\}, \texttt{ such that } \phi_i(d_1) < \phi_i(d_2),$ (16.87k)

$\texttt{where } \phi_i(d) = |u_{ii}(d - \hat{x}_i) + a_i|^2 - \sigma_w^2 P(\check{t}_i)$ (16.87l)

$\texttt{for } m = 1, 2 \texttt{ do}$ (16.87m)

$\quad \check{t}_i = d_m$ (16.87n)

$\quad J_i = J_{i+1} + \phi_i(\check{t}_i)$ (16.87o)

$\quad \texttt{if } J_i < \rho J_{\min} \texttt{ then}$ (16.87p)

$\qquad \texttt{if } i > 0 \texttt{ and } \dfrac{\|(\tilde{\mathbf{H}})_i\|^2}{\sigma_w^2} > \gamma \texttt{ then}$ (16.87q)

$\qquad\quad \texttt{Calculate } J_{i-1}$ (16.87r)

$\qquad \texttt{else}$

$\qquad\quad J_{\min} = \min(J_i, J_{\min})$ (16.87s)

$\qquad\quad \texttt{for } j = 1, \ldots, r$ (16.87t)

$\qquad\qquad \hat{J}_{j\check{t}_j} = \min\{\hat{J}_{j\check{t}_j}, J(\check{\mathbf{t}})\}, \; j = 1, \ldots, r$ (16.87u)

$\qquad\quad \texttt{end for}$ (16.87v)

$\qquad \texttt{end if}$

$\quad \texttt{end if}$

$\texttt{end for}$

$\texttt{end function}$

16.1.5.1 SOPHIE Algorithm Complexity Analysis

As pointed out in [265], 'the brute-force' ML SDM detection method does not provide a feasible solution to the generic SDM detection problem, as a result of the excessive associated computational complexity. More explicitly, the ML SDM detector advocated in [265]has a computational complexity of the order of

$$\mathcal{C}_{\mathrm{ML}} = O\{M^{m_t} \cdot (3n_r + 2n_r m_t)\}, \tag{16.88}$$

where $3n_r + 2n_r m_t$ is the complexity associated with a single search step, namely with the evaluation of the objective function value $\|\mathbf{H}\check{\mathbf{s}} - \mathbf{y}\|^2$, while M^{m_t} is the number of legitimate candidates of the transmitted signal vector $\mathbf{s}$. Clearly, the order of complexity imposed by Equation (16.88) becomes excessive for a large number of transmit antennas, e.g. in the case of employing 16-QAM and $m_t = n_r = 8$ transmit and receive antennas, where the computational complexity associated with ML detection is of the order of 10^7 complex operations per channel use, or 10^9 complex operations per OFDM symbol formed by $K = 128$ subcarriers. Furthermore, the evaluation of the soft-bit information required by an efficient turbo-decoder implementation imposes a further substantial increase on the associated computational complexity. Specifically, the soft-output Log-MAP SDM detector advocated in [265] has a computational complexity of the order of

$$\mathcal{C}_{\mathrm{LM}} = O\{m_t \log_2 M \cdot 2^{m_t \log_2 M - 1} \cdot (3n_r + 2n_r m_t)\}. \tag{16.89}$$

On the other hand, the MMSE SDM detector derived in [265]constitutes the low-complexity SDM detector. The complexity imposed by the MMSE SDM detector of [265]may be shown to be of the order of

$$\mathcal{C}_{\mathrm{MMSE}} = O\{m_t^3 + m_t n_r^2 + m_t^2 n_r + m_t n_r\}. \tag{16.90}$$

Clearly, the MMSE SDM detector's complexity is substantially lower than that associated with the ML or Log-MAP SDM detectors. Specifically, for example, only 1600 complex operations are required for detecting 16-QAM signals transmitted and received by $m_t = n_r = 8$ transmit and receive antennas. Unfortunately, however, as demonstrated in [265], the achievable performance exhibited by the linear MMSE SDM detector is considerably lower than that attained by the optimal Log-MAP SDM detector advocated in [265]. Moreover, linear SDM detectors such as the MMSE detector do not allow for the high-integrity detection of signals in the overloaded scenario, where the number of transmit antennas exceeds that of receive antennas.

Consequently, in Sections 16.1.2, 16.1.3 and 16.1.5 we have derived a family of methods which combine the advantageous properties of the ML and Log-MAP detection, while imposing a substantially lower complexity. In this section we demonstrate that the computational complexity associated with the SOPHIE-aided Log-MAP SDM detector of Algorithm 16.4 is in fact only slightly higher than that imposed by the low-complexity MMSE SDM detector advocated in [265], while its performance is virtually identical to the performance of the Log-MAP SDM detector [265].

The direct calculation of the complexity associated with the OHRSA methods of Algorithms 16.2, 16.3 and 16.4 is infeasible, since the complexity is a random variable which is a function of several parameters, such as the number m_t and n_r of transmit and receive antennas, the average SNR encountered as well as the value of the parameter ρ in Algorithm 16.4. Therefore, we perform the corresponding complexity analysis using computer simulations.

Figure 16.8(a) illustrates our comparison between the computational complexity required by different SDM detection methods, namely the linear MMSE detector advocated in [265], the SIC detector of [265, pp. 754–756], the exhaustive search-based ML and Log-MAP detectors of [265]as well as the OHRSA-aided ML, Log-MAP and SOPHIE SDM detectors of Algorithms 16.2, 16.3 and 16.4, respectively. The results depicted in Figure 16.8(a) correspond to the *fully loaded* scenario, where we have $m_t = n_r$ transmit and receive antennas. Observe that the complexity associated with both the OHRSA-ML and SOPHIE SDM detectors is only slightly higher than that imposed by the MMSE SDM detector and is in fact lower than the complexity imposed by the SIC SDM detector.

Furthermore, the achievable performance of the SDM-OFDM system employing the different SDM detection methods considered is depicted in Figure 16.8(b). Observe that both the OHRSA-Log-MAP and SOPHIE SDM detectors considerably outperform the linear MMSE detector. Moreover, the associated BER decreases upon increasing the number of transmit and receive antennas $m_t = n_r$, which suggests that, as opposed to both the MMSE and the SIC SDM detectors, the OHRSA-Log-MAP

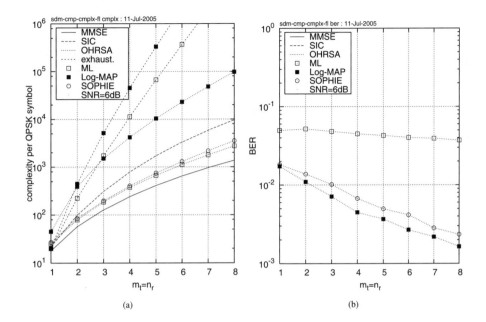

Figure 16.8: (a) **Computational complexity** quantified in terms of the total number of real multiplications and additions per detected QPSK symbol and (b) the corresponding **BER** exhibited by the rate $\frac{1}{2}$ turbo-coded **SDM-QPSK-OFDM** system employing the different SDM detection methods considered at SNR = 6 dB. The abscissa represents the number $m_t = n_r = 1, \ldots, 8$ of transmit and receive antenna elements. The corresponding system parameters are summarized in Table 7.1.

SDM detector is capable of achieving spatial diversity even in the *fully loaded* system. In other words, it is capable of simultaneously achieving both multiplexing and diversity gains, while maintaining a low computational complexity.

The relatively low performance of the OHRSA-ML SDM detector may be attributed to the fact that it produces no soft-bit information and therefore the efficiency of the turbo code employed is substantially degraded. Moreover, observe that while the SIC SDM detector outperforms its MMSE counterpart at high SNR values [265], the achievable performance of the two methods is fairly similar at low SNR values, such as 6 dB.

Additionally, Figure 16.9 illustrates the complexity imposed by the OHRSA methods of Algorithms 16.2, 16.3 and 16.4 as a function of the average SNR encountered. Figures 16.9(a) and 16.9(b) portray the average complexity encountered in the scenarios of $m_t = n_r = 8$ and $m_t = 8, n_r = 4$ transmit and receive antennas, respectively. Observe that the complexity associated with both the OHRSA-ML and SOPHIE methods of Algorithms 16.3 and 16.4 is mainly determined by the number m_t of transmit antennas employed. Furthermore, the complexity associated with the SOPHIE method closely matches that exhibited by the OHRSA-ML method at high SNR values and the complexity exhibited by both methods is only slightly higher than the complexity exhibited by the low-complexity MMSE SDM detector.

16.1.5.2 SOPHIE Algorithm Performance Analysis

In this section we present our simulation results characterizing the SDM-OFDM system employing the OHRSA-aided SDM detection schemes described in Section 16.1. Our simulations were performed in the baseband frequency domain and the system configuration characterized in Table 7.1 is to a large extent similar to that used in [260]. We assume a total bandwidth of 800 kHz. The OFDM system

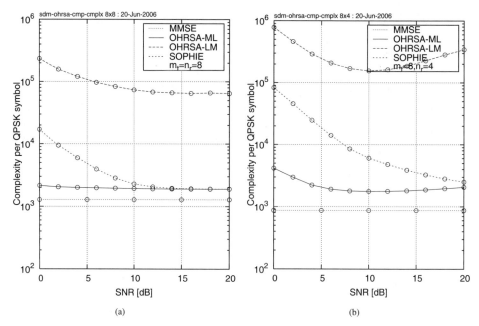

Figure 16.9: Computational complexity quantified in terms of the total number of real multiplications and additions per detected QPSK symbol. We consider the **OHRSA-ML, OHRSA-Log-MAP** and **SOPHIE** SDM detection methods of Algorithms 16.2, 16.3 and 16.4, respectively. Additionally, we show the corresponding computational complexity required by the low-complexity linear MMSE SDM detector as well as the optimum exhaustive Log-MAP detector. The abscissa represents the average SNR encountered.

utilizes 128 QPSK-modulated orthogonal subcarriers. For Forward Error Correction (FEC) we use rate $\frac{1}{2}$ turbo coding [314] employing two constraint-length $K = 3$ Recursive Systematic Convolutional (RSC) component codes and the standard 124-bit WCDMA UMTS turbo code interleaver of [512]. The octally represented RCS generator polynomials of $(7, 5)$ were used. Furthermore, we employ the eight-path urban non-line-of-sight Bug Rayleigh fading channel model characterized in [346]. Finally, throughout this report we stipulate the assumption of perfect channel knowledge, where knowledge of the frequency-domain subcarrier-related coefficients $H[n, k]$ is deemed to be available in the receiver.

Figure 16.10 characterizes the achievable performance as well as the associated computational complexity exhibited by the 4×4 16-QAM-SDM-OFDM system employing the SOPHIE SDM detector of Algorithm 16.4. More specifically, we analyse the associated performance versus complexity trade-offs of using various values of the complexity control parameters ρ and γ. In Figure 16.10(a) we can observe how the achievable BER performance (top) and the corresponding computational complexity depend on the value of the parameter γ. Using the results depicted in Figure 16.10(a) we may conclude that the optimum choice of the complexity control parameter γ lies in the range 0.5–0.8, where we have a minor BER performance degradation of less than 0.5 dB, while achieving a complexity reduction up to two orders of magnitude at low SNR values, when compared with the full-complexity SOPHIE algorithm assuming $\gamma = 0$.

On the other hand, Figure 16.10(b) portrays both the achievable BER performance and the associated complexity of the 4×4 16-QAM-SDM-OFDM system for different values of the complexity control parameter ρ. We may conclude that the optimum trade-off between the attainable BER performance and the associated complexity is achieved when the value of the complexity control parameter ρ lies in the range of 1.3–1.5, where the BER performance degradation imposed does not

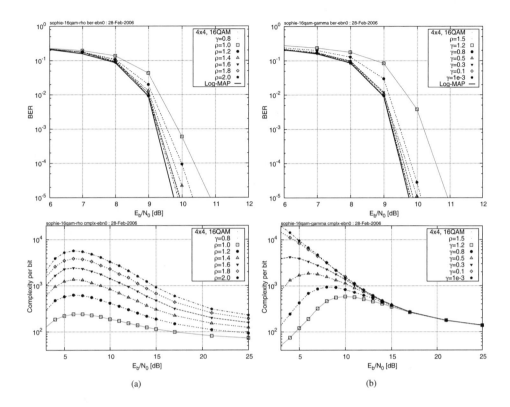

Figure 16.10: BER (top) and the associated **computational complexity per detected bit** (bottom) exhibited by the 4×4 **16-QAM-SDM-OFDM** system employing the **SOPHIE** SDM detector of Algorithm 16.4 and assuming different values of search radius and search resolution parameters (a) ρ and (b) γ. The abscissa represents the average E_b/N_0 recorded at the receive antenna elements. ©IEEE Akhtman, Wolfgang, Chen & Hanzo [562]

exceed $0.5\,\text{dB}$, while the associated computational complexity is reduced by more than an order of magnitude, when compared with large values of ρ, such as for instance $\rho = 2.0$.

Furthermore, Figure 16.11(a) demonstrates both the BER performance (top) and the associated computational complexity exhibited by the (8×8) 4-, 16- and 64-QAM-SDM-OFDM systems employing the SOPHIE SDM detector of Algorithm 16.4. Figure 16.11(b) characterizes the 16-QAM-SDM-OFDM system employing the SOPHIE SDM detector of Algorithm 16.4 and having a constant number of $n_r = 4$ receive antenna elements in terms of its ability to detect the multiplexed signals arriving from various numbers of transmit antenna elements. Specifically, we aim at exploring the performance of the SOPHIE SDM detector in the overloaded system scenario, where the number of transmit antenna elements exceeds that of receiver elements and thus we have $m_t > n_r$. Indeed, the BER curves portrayed in Figure 16.11 (top) confirm the near-Log-MAP performance of the SOPHIE SDM detector of Algorithm 16.4 in both systems employing high-throughput modulation schemes as well as in the overloaded system scenario.

Figure 16.12 characterizes the computational complexity imposed by the SOPHIE SDM detector of Algorithm 16.4 as a function of the number $m_t = n_r$ of transmit and receive antennas. More specifically, we consider three ranges of SNR values: low SNRs, the critical SNR, which corresponds to the 'waterfall' region of the BER versus SNR curve, as well as high SNRs, which correspond to the error-free detection region. In Figure 16.12 we can see that the computational complexity imposed by

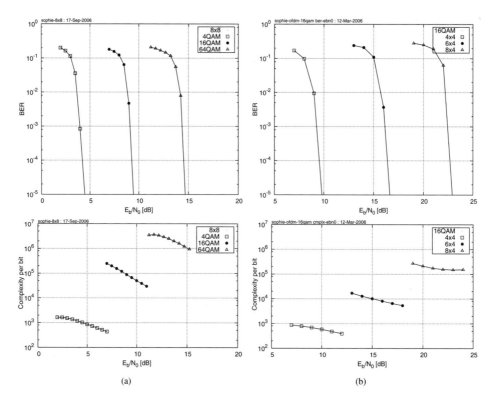

Figure 16.11: **BER** (top) and the associated **computational complexity per detected bit** (bottom) exhibited by the **SDM-OFDM** system employing the **SOPHIE** SDM detector of Algorithm 16.4 and assuming $\rho = 1.3$, $\gamma = 0.8$: (a) 8×8 system employing 4-, 16- and 64-QAM; (b) 16-QAM system employing a fixed number of four receive antennas, as well as four, six and eight transmit antennas. The abscissa represents the average E_b/N_0 recorded at the receive antenna elements. ©IEEE Akhtman, Wolfgang, Chen & Hanzo [562]

the SOPHIE detector increases according to a polynomial law as a function of the number of transmit antennas for both high and low SNRs.

Figure 16.13(a) demonstrates that the SDM-OFDM system employing the SOPHIE SDM detector of Algorithm 16.4 is capable of exploiting the available MIMO channel's multiplexing gain in the fully loaded system scenario, when the number of transmit antenna elements m_t is equal to that of receiver antenna elements n_r. More specifically, the results depicted in Figure 16.13(a) suggest that the SDM-OFDM SOPHIE SDM detector having $m_t = n_r = 8$ transmit and receive antennas exhibits an SNR-related diversity gain of 2 dB at the target BER of 10^{-4}, as well as a higher throughput by a factor of four, when compared with the same system employing two antennas at both the transmitter and receiver.

Additionally, Figure 16.13(b) characterizes the SDM-OFDM system employing the SOPHIE SDM detector of Algorithm 16.4 and having a constant number of $n_r = 4$ receive antenna elements in terms of its ability to detect the multiplexed signals arriving from various numbers of transmit antenna elements. Specifically, we aim at exploring the performance of the SOPHIE SDM detector in the overloaded system scenario, where the number of transmit antenna elements exceeds that of receiver elements and thus we have $m_t > n_r$. We can see that, as opposed to the MMSE SDM detector [265], the SOPHIE SDM detector exhibits a good performance both when we have $m_t \leq n_r$ and in the overloaded system scenario, when the number of transmit antenna elements exceeds the number of receive antenna elements, i.e. when we have $m_t > n_r$.

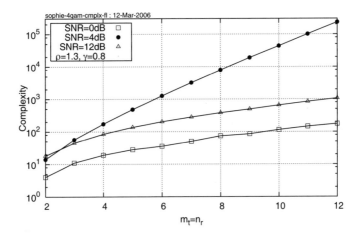

Figure 16.12: Computational complexity imposed by the MIMO-OFDM system employing the SOPHIE SDM detector of Algorithm 16.4 and assuming $\rho = 1.3$, $\gamma = 0.8$. The complexity is quantified in terms of the total number of real additions and multiplications as a function of the number $m_t = n_r$ of transmit and receive antennas. ©IEEE Akhtman, Wolfgang, Chen & Hanzo [562]

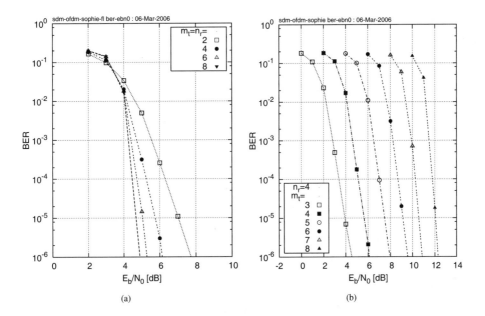

Figure 16.13: BER exhibited by the **SDM-QPSK-OFDM** system employing the **SOPHIE** SDM detector of Algorithm 16.4 in (a) fully loaded scenario with $m_t = n_r = 2, 4, 6$ and 8 transmit and receive antennas, as well as (b) overloaded scenario with a fixed number of $n_r = 4$ receive antennas and $m_t = 3, 4, \ldots, 8$ transmit antennas. The abscissa represents the average value of E_b/N_0 recorded at the receive antenna elements.

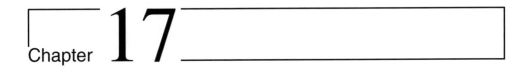

Chapter 17

Iterative Channel Estimation and Multi-stream Detection for SDM-OFDM

17.1 Iterative Signal Processing

Despite the immense interest of both the academic and the industrial research communities, the conception of a practical multiple-input, multiple-output (MIMO) transceiver architecture which is capable of approaching the MIMO channel's capacity in realistic channel conditions remains largely an open problem. An important overview encompassing most major aspects of broadband MIMO-OFDM wireless communications including both channel estimation and signal detection, as well as time- and frequency-domain synchronization, was contributed by Stüber *et al.* [295]. Other important publications considering MIMO systems operating in realistic channel conditions include those by Münster and Hanzo [297], Li *et al.* [294], Mai *et al.* as well as Qiao *et al.* [308]. Nevertheless, substantial contributions addressing all the major issues pertaining to the design of MIMO transceivers, namely error correction, space–time detection and channel estimation, in realistic channel conditions remain scarce.

Against this background, in this chapter we introduce an iterative, so-called *turbo* Multi-Antenna, Multi-Carrier (MAMC) receiver architecture. Our turbo receiver is illustrated in Figure 17.1. Following the philosophy of turbo processing [314], our turbo SDM-OFDM receiver comprises a succession of detection modules which iteratively exchange soft-bit-related information and thus facilitate a substantial improvement in the overall system performance.

More specifically, our turbo SDM-OFDM receiver comprises three major components, namely the soft-feedback decision-directed channel estimator detailed in Section 7.8, followed by the soft-input, soft-output OHRSA Log-MAP SDM detector derived in Section 16.1.3, and a classic parallel-concatenated soft-input, soft-output turbo code [345]. Consequently, in this chapter we analyse the achievable performance of each individual constituent of our turbo receiver, as well as the achievable performance of the entire iterative system. Our aim is to document the various design trade-offs, such as the achievable error-rate performance, the attainable data rate and the associated computational complexity.

MIMO-OFDM for LTE, Wi-Fi and WiMAX   Lajos Hanzo, Yosef Akhtman, Li Wang and Ming Jiang
© 2011 John Wiley & Sons, Ltd

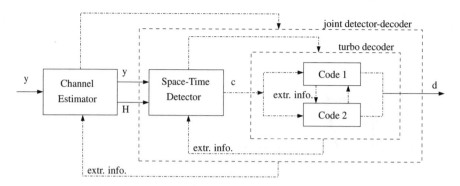

Figure 17.1: Schematic of an iterative turbo receiver employing an iterative decision-directed channel estimator as well as an iterative detection and decoding module.

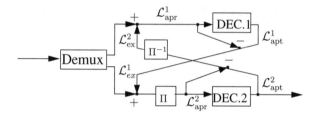

Figure 17.2: Schematic of an iterative turbo decoder employing two parallel-concatenated RSC codes.

17.2 Turbo Forward Error-Correction Coding

The family of the so-called *turbo* codes was first introduced by Berrou *et al.* [345, 636, 637]. The properties of turbo codes have been extensively studied in the context of various system architectures by a multiplicity of authors, for instance Benedetto [638], Battail [639], Ömer *et al.* [640] and Hanzo *et al.* [314]. The plausible conclusion of these studies was that turbo codes are capable of approaching capacity, while imposing a realistic computational complexity.

Consequently, at the first stage of our iterative turbo receiver architecture illustrated in Figure 17.1 we employ a turbo decoder. The detailed structure of the turbo decoder considered is depicted in Figure 17.2. More specifically, our turbo decoder consists of a pair of parallel-concatenated Soft-Input, Soft-Output (SISO) RSC decoders which iteratively exchange information-bit-related extrinsic information in the form of LLR values $\mathcal{L}_{\text{ex}}$ to attain the highest possible reliability of the decoded information-carrying bits. In this treatise we employed two rate $\frac{1}{2}$ punctured RSC codes [640]. Observe that the parallel-concatenated codes share the same information bits, while the corresponding parity bits at the output of the encoder are punctured, which results in the overall concatenated code rate of $\frac{1}{2}$. The octally represented RCS generator polynomials of $(7, 5)$ having a constraint length of 3 were used for both RCS codes. Note also that in the introduction of this treatise, namely in Figure 1.7, we depicted a serial-concatenated turbo decoder. In contrast, as seen in Figures 17.1 and 17.2, in this chapter we employed a parallel-concatenated code, reminiscent of that derived in [345]. Both the parallel and the serial versions of turbo codes are applicable in our system. Both methods were found to exhibit fairly similar performance, but in the rest of this chapter we will focus our attention on the former.

In this section we quantify the achievable performance of the turbo code considered in the context of increasingly more sophisticated systems communicating under increasingly more realistic channel conditions. We commence our discourse by characterizing the achievable BER performance of the turbo code in the uncorrelated Rayleigh fading in Figure 17.3. Subsequently, in Figure 17.4 we consider the

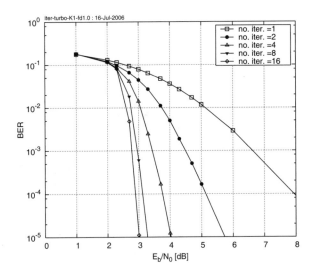

Figure 17.3: BER versus E_b/N_0 performance exhibited by the rate-$\frac{1}{2}$ parallel-concatenated turbo code in uncorrelated non-dispersive Rayleigh fading using single-antenna, single-carrier QPSK transmissions. All our additional system parameters are summarized in Table 1.11.

BER performance of the turbo code in the context of a 128-subcarrier OFDM system encountering both uncorrelated Rayleigh fading in the time domain as well as correlated fading having a time-domain correlation determined by the OFDM-symbol-normalized Doppler frequency spanning the range of $f_D = 0.1$ to 0.003.

More specifically, to characterize the achievable BER performance, in Figure 17.3 we portray the BER performance of the parallel-concatenated turbo decoder when encountering uncorrelated Rayleigh fading. More specifically, we considered a narrowband single-carrier QPSK-modulated system which employs a time-domain random block interleaver encompassing 1000 consecutive bits. Observe that the BER performance exhibited by the turbo decoder improves rapidly on increasing the number of decoding iterations performed. Furthermore, the decoder approaches its best possible performance after *eight* iterations. Consequently, in our further studies we consider performing $i_{\text{dec}} = 8$ iterations by the turbo decoder.

On a similar note, Figure 17.4 characterizes the achievable BER performance of the turbo decoder considered in the context of a QPSK-modulated OFDM system, while encountering a Rayleigh fading channel exhibiting various correlation properties. For benchmarking purposes, we contrast the performance of a narrowband system encountering uncorrelated Rayleigh fading as well as that of a $K = 128$-subcarrier OFDM system encountering a dispersive channel having uncorrelated time-domain Rayleigh fading taps specified by the COST207 Bad Urban (BU) seven-tap CIR [347]. In the frequency domain this CIR results in a corresponding correlated frequency-selective CTF. Furthermore, we also consider the more realistic scenario of a $K = 128$-subcarrier OFDM system encountering correlated time-domain Rayleigh fading having the Doppler frequencies of $f_D = 0.1$, 0.03 and 0.003 as well as a dispersive CIR characterized by the COST207 BU model [347].

From Figure 17.4 we conclude that, as expected, while our turbo decoder exhibits a good BER performance [266] in uncorrelated Rayleigh fading, the corresponding BER performance recorded in correlated fading is substantially degraded owing to the relatively low-memory 1000-bit turbo interleaver, which is unable to break up and randomize the long fading-induced error bursts.

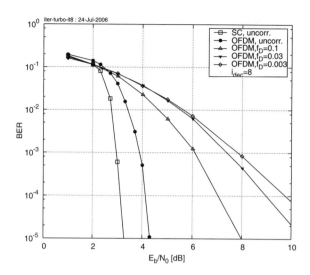

Figure 17.4: BER versus E_b/N_0 performance exhibited by the $K = 128$-subcarrier single-antenna QPSK-OFDM system employing a rate $\frac{1}{2}$ parallel-concatenated turbo code in correlated Rayleigh fading having the OFDM-symbol-normalized Doppler frequencies of $f_D = 0.1, 0.03$ and 0.003. The CIR was the seven-path COST207 BU model [347]. All additional system parameters are summarized in Table 1.11.

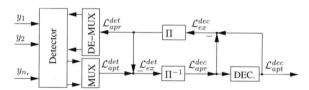

Figure 17.5: Schematic of a MIMO receiver employing iterative joint detection and decoding.

17.3 Iterative Detection – Decoding

Figure 17.5 portrays a schematic of the iterative space–time detector and decoder considered. Following the philosophy of iterative turbo detection, the incoming subcarrier-related signal vector $\mathbf{y}[n, k]$ is processed by SISO OHRSA Log-MAP detector of Algorithm 16.3, which delivers the bit-related a posteriori LLR values $\mathcal{L}_{apt}^{det}$. The resultant LLR values $\mathcal{L}_{apt}^{det}$ are then normalized and deinterleaved to generate the a priori bit-related LLR values $\mathcal{L}_{apr}^{dec}$, which may be utilized by the turbo decoder of Figure 17.5. Subsequently, the a posteriori LLR values $\mathcal{L}_{apt}^{dec}$ generated at the output of the decoder are normalized, interleaved and fed back to the SDM detector in the form of the LLR values $\mathcal{L}_{apr}^{det}$. This iterative detection process is continued for i_{det} number of detection iterations.

As a next step, we characterize the achievable performance of the iterative SDM detection and decoding scheme illustrated in Figure 17.5. Throughout this section we stipulate the idealistic assumption of having perfect knowledge of the OFDM-subcarrier-related CTF.

Firstly, for the sake of benchmarking, in Figure 17.6 we quantify the BER versus E_b/N_0 performance of the iterative SDM detection and decoding scheme of Figure 17.5 in the context of a rate $\frac{1}{2}$ turbo-coded 4×4 SDM-QPSK-OFDM system communicating over the uncorrelated time-domain Rayleigh fading channel characterized by the COST207 BU model [347]. We consider carrying out $i_{det} = 1, 2, 3$ and 4 iterations for the SDM detector, while performing $i_{dec} = 8$ iterations for the inner turbo decoder per iteration of the SDM detector. From Figure 17.6 we can see that an E_b/N_0 gain of

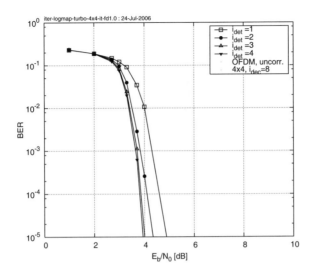

Figure 17.6: BER versus E_b/N_0 performance exhibited by the $K = 128$-subcarrier rate $\frac{1}{2}$ turbo-coded 4×4 SDM-QPSK-OFDM system employing the iterative SDM detection and decoding scheme of Figure 17.5 in uncorrelated time-domain Rayleigh fading channel characterized by the COST207 BU model [347]. The effective throughput of the system was $4 \cdot 2 \cdot \frac{1}{2} = 4$ bps/Hz. All additional system parameters are summarized in Table 1.11.

about 1 dB is achieved by invoking $i_{\text{det}} = 3$ iterations of the SDM detector and decoder in comparison with invoking a single detection iteration. By contrast, only a further minor improvement in E_b/N_0 may be achieved by invoking $i_{\text{det}} > 3$ number of iterations for the SDM detector and decoder complex of Figure 17.5.

Let us now consider the effects of realistic time-domain correlations encountered by our SDM-OFDM system employing the iterative SDM detection and decoding scheme of Figure 17.5. Figure 17.7 characterizes the achievable BER versus E_b/N_0 performance of the iterative SDM detection and decoding scheme, which assumes the iteration pattern of $(i_{\text{det}}, i_{\text{dec}}) = (3, 8)$, in the context of a rate $\frac{1}{2}$ turbo-coded 4×4 SDM-QPSK-OFDM system encountering the OFDM-normalized Doppler frequencies of $f_D = 0.1, 0.03$ and 0.003. The corresponding BER performance recorded in the uncorrelated time-domain Rayleigh fading conditions when using the seven-path COST207 BU CIR, is also shown for the sake of benchmarking. In contrast to the single-antenna scenario characterized in Figure 17.4, we may observe from Figure 17.7 that the BER performance exhibited by the system encountering a realistic OFDM-symbol-normalized Doppler frequency of $f_D = 0.003$ lies within an E_b/N_0 range of 0.8 dB from the corresponding BER curve exhibited by the system encountering idealistic uncorrelated time-domain Rayleigh fading conditions, when using the seven-path COST207 BU CIR. We may therefore conclude that, as expected, our 4×4-SDM-QPSK-OFDM system efficiently exploits the spatial diversity potential inherent in the MIMO channel.

This conclusion is further supported by the results depicted in Figure 17.8, where we plot the BER versus E_B/N_0 performance exhibited by the rate $\frac{1}{2}$ turbo-coded SDM-QPSK-OFDM system employing the iterative SDM detection and decoding scheme of Figure 17.5 and having $m_t = n_r = 1, 2, 4$ and 8 transmit and receive antennas. We assumed encountering an OFDM-symbol-normalized Doppler frequency of $f_D = 0.003$, while employing bit interleaving across $N_d = 10$ OFDM symbols. Observe that having an interleaved block of bits spanning the duration of $N_d T_s = 10 T_s$, which is substantially shorter than channel's coherence time of $1/f_D \approx 300 T_s$, corresponds to having virtually no time-domain diversity gain. In other words, a relatively short interleaver is unable to break up and randomize the long fading-induced error bursts. Consequently, we may conclude from Figure 17.8 that

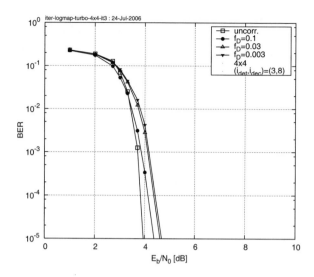

Figure 17.7: BER versus E_b/N_0 performance exhibited by the rate $\frac{1}{2}$ turbo-coded 4×4 SDM-QPSK-OFDM system employing the iterative SDM detection and decoding scheme of Figure 17.5 in realistic correlated Rayleigh fading conditions when using the seven-path COST207 BU CIR [347] and encountering the Doppler frequencies of $f_D = 0.1, 0.03$ and 0.003. The BER performance recorded in the case of uncorrelated time-domain Rayleigh fading is also shown for the sake of benchmarking. We invoked an iteration pattern of $(i_{\text{det}}, i_{\text{dec}}) = (3, 8)$. The overall throughput was $4 \cdot 2 \cdot \frac{1}{2} = 4$ bps/Hz. Additional system parameters are summarized in Table 1.11.

the BER performance exhibited by the single-antenna OFDM system is limited by the probability of occurrence of a precipitated burst of errors in some of the OFDM symbols, which we may refer to as an *outage* [266] inherent in single-antenna Rayleigh fading channels. On the other hand, SDM-OFDM systems operating in MIMO scenarios exhibit a BER performance which improves on increasing the number $m_t = n_r$ of transmit and receive antennas.

17.4 Iterative Channel Estimation – Detection and Decoding

In this section we consider the transmission of a sequence of consecutive SDM-OFDM transmission *bursts*, which are processed independently. In other words, each of the self-contained SDM-OFDM transmission bursts includes all the necessary data, such as for instance pilot signals, required for successful detection and decoding of the information accommodated by the OFDM transmission burst. Correspondingly, each SDM-OFDM transmission burst may be processed independently of the neighbouring bursts. This philosophy is reminiscent of the packet-based transmission scheme adopted, for example, in the IEEE 802.11 a/g WLAN standard [641]. The structure of a single SDM-OFDM transmission burst considered is depicted in Figure 17.9. More specifically, our OFDM transmission burst portrayed in Figure 17.9 commences with a channel-sounding preamble formed by N_p pure pilot SDM-OFDM symbols. Subsequently, our SDM-OFDM transmission burst accommodates a sequence of L_f so-called OFDM symbol frames. More explicitly, as seen in Figure 17.9, each OFDM symbol frame constitutes a single bit-interleaved turbo-encoded codeword and comprises a single full-pilot SDM-OFDM symbol followed by N_d information-carrying SDM-OFDM symbols.

For each SDM-OFDM transmission burst the detection process commences with initialization of the channel estimator by utilizing the pilot SDM-OFDM symbols constituting the burst's preamble, as seen

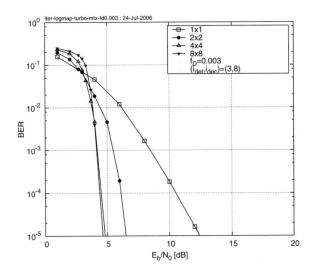

Figure 17.8: BER versus E_b/N_0 performance exhibited by the rate $\frac{1}{2}$ turbo-coded SDM-QPSK-OFDM system employing the iterative SDM detection and decoding scheme of Figure 17.5 and having $m_t = n_r = 1, 2, 4$ and 8 transmit and receive antennas. We invoked an iteration pattern of $(i_{\text{det}}, i_{\text{dec}}) = (3, 8)$. The seven-path COST207 BU channel model [347] was used and we assumed encountering the OFDM-symbol-normalized Doppler frequency of $f_D = 0.003$. The overall throughput was 1, 2, 4 and $8 \cdot 2 \cdot \frac{1}{2} = 8$ bps/Hz, respectively. Additional system parameters are summarized in Table 1.11.

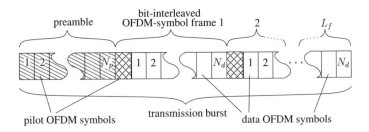

Figure 17.9: OFDM transmission burst structure comprising a preamble of N_p full-pilot OFDM symbols followed by a sequence of L_f data OFDM symbol frames. Each data OFDM symbol frame is preceded by a single full-pilot OFDM symbol followed by N_d information-carrying OFDM symbols. Consequently, our OFDM transmission burst accommodates a total number of $N_p + L_f$ full-pilot OFDM symbols as well as a total number of $L_f N_d$ information-carrying OFDM symbols.

in Figure 17.9. Specifically, both the received signals $\mathbf{y}[n]$ and the corresponding transmitted signals $\mathbf{s}[n]$ associated with the N_p pilot SDM-OFDM symbols constituting the burst preamble of Figure 17.9 are sequentially fed into the channel estimator of Figure 7.1 to attain an initial convergence for the three adaptive filters constituting the decision-directed channel estimator of Figure 7.1.

During the first iteration of the detection process, which is carried out for each subsequent N_d OFDM symbol data frame of Figure 17.9 that commences with a full-pilot SDM-OFDM symbol associated with the SDM-OFDM symbol index n, we perform a long-term prediction of the CIR-related taps using the CIR tap predictor of Figure 7.1. More specifically, we aim at predicting the CIR associated with the *last* OFDM symbol of the current OFDM symbol frame of Figure 17.9, i.e. the one associated with the SDM-OFDM symbol index of $(n + N_d)$. The CIRs associated with the remaining $(N_d - 1)$ SDM-OFDM symbols hosted by the current OFDM symbol frame are then obtained using linear interpolation between those associated with the nth pilot SDM-OFDM symbol preceding the

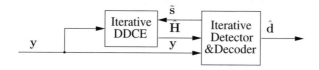

Figure 17.10: Schematic of an iterative turbo receiver employing the iterative decision-directed channel estimator of Figure 7.1 as well as the iterative detection and decoding module of Figure 17.5.

current OFDM symbol frame and the predicted CIR associated with the last $(n + N_d)$ data OFDM symbol.

The predicted and interpolated MIMO-CTF coefficients $\check{\mathbf{H}}[m]$, $m = n+1, \ldots, n+N_d$, are utilized to perform initial detection of the information-carrying data SDM-OFDM symbols $s[n]$. Observe that with the CIRs and the corresponding CTFs associated with the entire SDM-OFDM symbol frame, we are able to employ the iterative SDM detection and decoding scheme outlined in Section 17.3.

The resultant tentative estimates of the data bits $\mathbf{d}$, as well as the associated soft-bit information, corresponding to the entire data SDM-OFDM-symbol frame of Figure 17.9, are remodulated in order to generate the soft reference signal $\tilde{\mathbf{s}}[m]$, $m = n + 1, \ldots, n + N_d$, of Equation (7.98). The reference signal $\tilde{\mathbf{s}}[m]$ is fed back to the soft-input channel estimator of Algorithm 7.8 to refine the estimates of the CTF coefficients $\mathbf{H}[m]$, $m = n + 1, \ldots, n + N_d$. The interaction between the iterative channel estimator of Algorithm 7.8 and the iterative SDM detection and decoding module of Section 17.3 is illustrated in Figure 17.10. The iterative channel estimation–detection–decoding process portrayed in Figure 17.10 is repeated, until a sufficiently reliable detected SDM-OFDM symbol $\hat{\mathbf{s}}$ is generated.

17.4.1 Mitigation of Error Propagation

As we noted in Section 7.2, the main difficulty associated with the decision-directed approach to channel estimation is constituted by the potential error propagation, where the erroneous data decisions result in erroneous channel estimation, which inflicts further precipitated data decision errors etc. In other words, the reliability of the estimated CTF coefficients degrades rapidly in the presence of decision errors routinely occurring in the low-SNR region. The resultant degradation of the channel state information accuracy results in further decision errors and ultimately in divergence of the iterative channel estimation–data detection process and in a subsequent avalanche of decision errors. As we pointed out in Section 7.8.1.4, the *soft-feedback*-assisted RLS CTF estimator of Algorithm 7.7 is capable of substantially mitigating the effects of error propagation. Nevertheless, ensuring the stability of an iterative channel estimation–data detection system in the presence of data decision errors remains a challenging issue. Consequently, to mitigate the system's vulnerability to error-propagation-related instability effects we propose the following method.

Firstly, after each channel estimation and SDM detection iteration, which is performed on the N_d SDM-OFDM symbol data frame of Figure 17.9, we record the resultant MSE. The joint channel estimation and SDM detection MSE may be expressed as follows:

$$e^i[n] = \sum_{m=n+1}^{n+N_d} \sum_{k=1}^{K} \|\mathbf{y}[m, k] - \hat{\mathbf{H}}^i[m, k]\hat{\mathbf{s}}^i[m, k]\|^2, \tag{17.1}$$

where, as before, $\mathbf{y}[m, k]$ denotes the SDM signal associated with the kth subcarrier of the mth SDM-OFDM symbol and recorded at the n_r receive antennas, while $\hat{\mathbf{H}}^i[m, k]$ and $\hat{\mathbf{s}}^i[m, k]$ are the corresponding estimates of the CTF coefficient matrix and the transmitted signal vector, which were obtained after the ith iteration of the channel estimation and detection process.

Subsequently, after carrying out i_{ce} channel estimation iterations, we select the particular pair of CTF estimates $\hat{\mathbf{H}}^i[m, k]$ and data estimates $\hat{\mathbf{s}}^i[m, k]$ which correspond to the specific iteration resulting in the minimum MSE. More explicitly, the decision rule employed may be expressed as

$$\{\hat{\mathbf{H}}[m, k], \hat{\mathbf{s}}[m, k]\} = \arg \min_i e^i[n], \tag{17.2}$$

where we have $m = n + 1, \ldots, n + N_d$; $k = 1, \ldots, K$ and $i = 1, \ldots, i_{ce}$.

Let us now consider the scenario of encountering a large number of decision errors. Naturally, the decision errors in any of the iterations would result in a degraded channel estimation accuracy in the subsequent iteration and hence even more decision errors as well as an inevitable increase of the corresponding MSE $e^i[n]$. Consequently, invoking the final-decision rule of Equation 17.2 substantially mitigates the system's *avalanche*-like error propagation and hence improves the system's stability and robustness.

17.4.2 MIMO-PASTD-DDCE Aided SDM-OFDM Performance Analysis

17.4.2.1 Number of Channel Estimation–Detection Iterations

Firstly, we characterize the BER performance gain attained by the iterative MIMO-PASTD-DDCE in comparison with single-iteration channel estimation. More specifically, Figure 17.11 portrays the BER versus E_b/N_0 performance of the rate $1/2$ turbo-coded 4×4 SDM-QPSK-OFDM system employing the MIMO-PASTD-DDCE of Algorithm 7.8 and invoking $i_{ce} = 1, 2, 3$ and 4 channel estimation iterations as well as $i_{det} = 2$ SDM detector iterations and $i_{dec} = 4$ iterations of the parallel-concatenated turbo decoder per iteration of the channel estimator. We assumed employing the transmission burst structure depicted in Figure 17.9, where the corresponding parameters were given by $(L_f, N_p, N_d) = (8, 8, 10)$, which yields an overall pilot overhead of $\varepsilon = (N_p + L_f)/(L_f N_d) = 0.1$, or, in other words, 10%. The seven-path COST207 BU channel model was used and we assumed encountering the Doppler frequency of $f_D = 0.003$. As may be concluded from Figure 17.11, the SDM-OFDM system employing the iterative channel estimation scheme of Algorithm 7.8 exhibits an E_b/N_0 gain of about 2 dB when comparing three iterations and a single iteration of the channel estimator. Moreover, only a modest further E_b/N_0 gain may be achieved upon invoking a higher number of channel estimation iterations.

17.4.2.2 Pilot Overhead

In order to provide further insights, Figure 17.12 characterizes the achievable BER versus E_b/N_0 performance of the MIMO-PASTD-DDCE of Algorithm 7.8 in the context of employing different numbers m_t and n_r of transmit and receive antennas. Specifically, we consider the SDM-QPSK-OFDM turbo receiver of Figure 17.10, which invokes $(i_{ce}, i_{det}, i_{dec}) = (3, 2, 4)$ channel estimation, detection and decoding iterations, respectively, while employing $m_t = n_r = 1, 2, 4, 6$ and 8 transmit and receive antennas. Observe that the BER performance improves rapidly on increasing the number $m_t = n_r$ of transmit and receive antennas, as long as it does not exceed $m_t = n_r = 4$. Furthermore, the BER performance degrades slowly on further increasing the number of antennas according to $m_t = n_r > 4$. A simple explanation of this phenomenon is that, as expected, the SDM-OFDM system benefits from the increased spatial diversity associated with a higher number of antennas. On the other hand, as noted in Section 7.8, the channel estimation problem becomes increasingly more rank deficient and hence the estimation accuracy of the CIR taps as well as the corresponding subcarrier-related CTF coefficients degrades on increasing the number of independent spatial links constituting the MIMO channel. The overall system performance is determined by the associated trade-off between the beneficial diversity gain increase and the inevitable degradation of the estimated CTF accuracy. Ultimately, however, deterioration of the estimated CTF accuracy does not appear to constitute a major impediment. Quantitatively, as evidenced by the results of Figure 17.12, the BER performance exhibited by the high-complexity system having $m_t = n_r = 8$ antennas lies within a 1 dB margin in comparison with the corresponding BER performance curve associated with the system having $m_t = n_r = 4$ transmit and receive antennas. Observe that the 4×4 system exhibits the best recorded performance

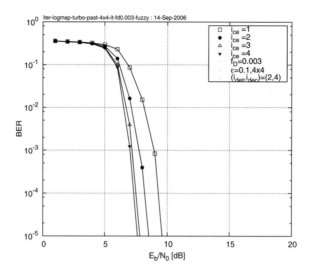

Figure 17.11: BER versus E_b/N_0 performance exhibited by the rate $\frac{1}{2}$ turbo-coded 4×4 SDM-4-QAM-OFDM iterative turbo receiver of Figure 17.10 employing the MIMO-PASTD-DDCE of Algorithm 7.8 and invoking $i_{ce} = 1, 2, 3$ and 4 channel estimation iterations as well as $(i_{det}, i_{dec}) = (2, 4)$ SDM detection and turbo decoding iterations, respectively. The seven-path COST207 BU channel model [347] was used and we assumed encountering the OFDM-symbol-normalized Doppler frequency of $f_D = 0.003$. The overall throughput was $4 \cdot 2 \cdot \frac{1}{2} = 4$ bps/Hz. All additional system parameters are summarized in Table 1.11.

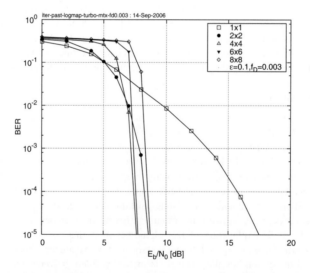

Figure 17.12: The BER versus E_b/N_0 performance exhibited by the rate $\frac{1}{2}$ turbo-coded SDM-QPSK-OFDM turbo receiver of Figure 17.10 employing the iterative MIMO-PASTD-DDCE of Algorithm 7.8 and using $m_t = n_r = 1, 2, 4, 6$ and 8 transmit and receive antennas. The corresponding effective throughputs were 1, 2, 4, 6 and $8 \cdot 2 \cdot \frac{1}{2} = 8$ bps/Hz, respectively. The seven-path COST207 BU channel model was used [347] and we assumed encountering the Doppler frequency of $f_D = 0.003$. The pilot overhead of 10% and the iteration pattern of $(i_{ce}, i_{det}, i_{dec}) = (3, 2, 4)$ were used. All additional system parameters are summarized in Table 1.11.

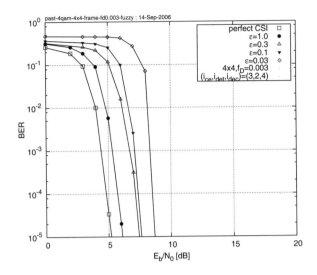

Figure 17.13: BER versus E_b/N_0 performance exhibited by the rate $\frac{1}{2}$ turbo-coded 4×4 SDM-QPSK-OFDM turbo receiver of Figure 17.10 employing the MIMO-PASTD-DDCE scheme of Algorithm 7.8. The pilot overhead was 3, 10, 30 or 100%, which corresponds to $\varepsilon = 0.03, 0.1, 0.3$ and 1.0, respectively, where we consider the idealistic scenario of having 100% pilots as well as the scenario of perfect channel state information for benchmarking purposes. The seven-path COST207 BU channel model [347] was used and we assumed encountering the Doppler frequency of $f_D = 0.003$. The iteration pattern of $(i_{\text{ce}}, i_{\text{det}}, i_{\text{dec}}) = (3, 2, 4)$ was used. The effective throughput was $4 \cdot 2 \cdot \frac{1}{2} = 4$ bps/Hz. All additional system parameters are summarized in Table 1.11.

and hence appears to represent an optimum trade-off between the beneficial spatial diversity gain and the degradation in accuracy of the system-size-related channel estimation.

17.4.2.3 Performance of a Symmetric MIMO System

Subsequently, we characterize the achievable BER performance exhibited by the SDM-QPSK-OFDM turbo receiver of Figure 17.10 employing the MIMO-PASTD-DDCE scheme of Algorithm 7.8 and using various densities of the dedicated pilot SDM-OFDM symbols. More specifically, in Figure 17.13 we have plotted the rate 1/2 turbo-coded QPSK-related BER exhibited by our SDM-OFDM system employing $m_t = n_r$ transmit and receive antennas. For benchmarking purposes we have included the BER versus E_b/N_0 performance of the SDM-OFDM system assuming perfect CIR knowledge, as well as assuming channel estimation based on the idealistic scenario of having 100% pilots. Furthermore, we present our results for the SDM-OFDM system using pilot overheads of 30, 10 and 3%, which correspond to the pilot overhead ratios of $\varepsilon = 0.3, 0.1$ and 0.003, respectively. We observe from Figure 17.13 that the 100% pilot-based channel estimation results in an approximately 1 dB E_b/N_0 degradation in comparison with the perfect CIR estimation scenario. Furthermore, the more realistic assumption of employing up to 10% dedicated SDM-OFDM pilot symbols results in a further E_b/N_0 degradation of about 1.5 dB in comparison with the 100% pilot-based scenario. Additionally, a further reduction of the pilot overhead to as low as 3% of pilots results in an E_b/N_0 degradation of 2.5 dB in comparison with the 100% pilot-based scenario.

17.4.2.4 Performance of a Rank-Deficient MIMO System

Similar phenomena may be observed in Figure 17.14, which characterizes the achievable BER performance exhibited by a rank-deficient 4×2 SDM-QPSK-OFDM system. The 4×2 MIMO

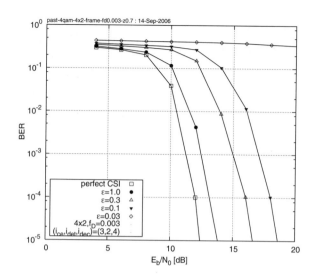

Figure 17.14: BER versus E_b/N_0 performance exhibited by the rank-deficient rate $\frac{1}{2}$ turbo-coded 4×2 SDM-QPSK-OFDM turbo receiver of Figure 17.10 employing the MIMO-PASTD-DDCE scheme of Algorithm 7.8. The pilot overhead was 3, 10, 30 or 100%, which corresponds to $\varepsilon = 0.03, 0.1, 0.3$ and 1.0, respectively, where we consider the idealistic scenario of having 100% pilots as well as the scenario of perfect channel state information for benchmarking purposes. The seven-path COST207 BU channel model was used [347] and we assumed encountering the Doppler frequency of $f_D = 0.003$. The iteration pattern of $(i_{ce}, i_{det}, i_{dec}) = (3, 2, 4)$ was used. The effective throughput was $4 \cdot 2 \cdot \frac{1}{2} = 4$ bps/Hz. All additional system parameters are summarized in Table 1.11.

scenario constitutes a particularly interesting detection problem. More specifically, let us consider the kth subcarrier of the nth SDM-OFDM symbol. The computational challenge lies in the fact that we have to estimate as many as *four* transmitted M-QAM symbols $s_j[n, k]$, $j = 1, \ldots, 4$, as well as the corresponding *eight* CTF coefficients $H_{ij}[n, k]$, $i = 1, 2$; $j = 1, \ldots, 4$, while utilizing merely the *two* recorded signal samples of $y_i[n, k]$, $i = 1, 2$. Consequently, similar to Figure 17.13, we have plotted the BER versus E_b/N_0 performance of the 4×2 SDM-QPSK-OFDM system assuming perfect CSI as well as assuming channel estimation based on the idealistic scenario of having 100% pilots. Furthermore, we have plotted the BER corresponding to scenarios using pilot overheads of 30, 10 and 3%. Similar to the 4×4 scenario, assuming 100% pilot-based channel estimation results in an approximately 1 dB E_b/N_0 degradation in comparison with the perfect CIR knowledge scenario. On the other hand, in contrast to the 4×4 scenario characterized in Figure 17.13, in Figure 17.14 we can see that the system employing 10% of dedicated SDM-OFDM pilot symbols results in nearly 6 dB E_b/N_0 degradation in comparison with the 100% pilot-based scenario. Furthermore, an additional reduction of the pilot overhead to 3% of pilots results in system instability and hence no satisfactory BER performance may be achieved, regardless of the SNR encountered.

17.5   Chapter Summary

In conclusion, in this chapter we documented the performance trends exhibited by the proposed turbo SDM-OFDM receiver of Figure 17.1, which comprises three main components, namely the soft-feedback decision-directed MIMO channel estimator derived in Section 7.8, the SISO OHRSA Log-MAP SDM detector of Section 16.1.3 and a SISO parallel-concatenated turbo code [345]. We analysed the achievable performance of each individual constituent component of our turbo receiver, as well as the achievable performance of the entire iterative system.

We found that our turbo SDM-OFDM system employing the MIMO-DDCE scheme of Section 7.8 as well as the OHRSA Log-MAP SDM detector of Section 16.1.3 remains effective in channel conditions associated with high terminal speeds up to 130 km/h, which corresponds to the OFDM-symbol-normalized Doppler frequency of 0.006. Additionally, in Figure 17.12 we reported a virtually error-free performance for a rate $1/2$ turbo-coded 8×8 QPSK-OFDM system, exhibiting an effective throughput of $8\,\text{MHz} \cdot 8\,\text{bps/Hz} = 64\,\text{Mbps}$ and having a pilot overhead of only 10% at an SNR of 7.5 dB and a normalized Doppler frequency of 0.003, which corresponds to a mobile terminal speed of about 65 km/h.[1]

---

[1] Additional system parameters are characterized in Table 1.11.

Chapter **18**

Summary, Conclusions and Future Research

18.1 Summary of Results

In this monograph we characterized a suite of iterative turbo receivers suitable for employment in a wide range of multi-antenna-aided multi-carrier systems operating in realistic, rapidly fluctuating channel conditions. More specifically, we reported the following major findings.

18.1.1 OFDM History, Standards and System Components

Chapter 2 provided a broad state-of-the-art review with reference to the open literature and detailed the general background of our studies throughout this book. The key milestones, events and contributions to the development of OFDM systems spanning several decades were summarized in Tables 1.1 and 1.2. Furthermore, in Section 1.1.1.1 an overview of the MIMO techniques was provided, followed by the introduction to MIMO-OFDM systems in Section 1.1.1.2, where the associated contributions found in the literature were outlined in Tables 1.3–1.5. Moreover, our review of the SDMA and SDMA-OFDM literature outlined in Section 1.1.1.3 was summarized in Tables 1.6 and 1.7. **Chapter 1** outlined the basic functional blocks of MIMO-OFDM systems.

18.1.2 Channel-Coded STBC-OFDM Systems

Our research in **Chapter 3** was conducted in the context of both single-user, single-carrier and single-user OFDM systems, which constitutes the introductory and background work for the following chapters. Specifically, in Section 3.2, a general outline of Alamouti's simple scheme [400] using two transmitters in conjunction with a number of receivers for communications in non-dispersive Rayleigh fading channels was provided, followed by a brief review of various Space–Time Block Codes (STBCs) proposed by Tarokh *et al.* [413, 414], namely the G_2, G_3, G_4, H_3 and H_4 codes, as well as their achievable performances over uncorrelated and correlated Rayleigh fading channels. The performances of all the space–time block codes considered are summarized in Table 3.2.

Following the above foundational work, in Section 3.3.1 the STBCs were combined with various Low-Density Parity Check (LDPC) codes in order to improve the system's performance. It was found that, provided the same modulation scheme was used, a lower-rate LDPC code benefited the system more than a space–time code of a higher-diversity order did for transmission over the uncorrelated

MIMO-OFDM for LTE, Wi-Fi and WiMAX   Lajos Hanzo, Yosef Akhtman, Li Wang and Ming Jiang
© 2011 John Wiley & Sons, Ltd

Rayleigh fading channels. On the other hand, when the same LDPC code was used, a lower-order modulation scheme tended to offer a higher performance improvement than an STBC of a higher diversity order did. The performance of different LDPC-aided STBC schemes was summarized in Table 3.5, where the half-rate LDPC-coded STBC G_2 was found to be the best option among all the LDPC-STBC-concatenated schemes investigated. Furthermore, in Section 3.3.2, we compared the performance of an LDPC-aided G_2-coded scheme with a half-rate Turbo Convolutional (TC) code, i.e. to that of a TC(2,1,4)-aided G_2-coded arrangement, which was found to be the best scheme in a range of TC-STBC-concatenated systems [314] designed for transmission over uncorrelated Rayleigh fading channels. From our coding gain versus complexity performance comparisons, it was concluded that the half-rate TC(2,1,4)-assisted STBC G_2 slightly outperforms the LDPC-assisted STBC schemes. However, the LDPC-STBC-concatenated schemes may be considered as better design options for complexity-sensitive systems, where the achievable performance does not necessarily have to be the highest possible, since the LDPC-aided schemes are capable of maintaining a lower complexity than the TC(2,1,4)-aided scheme is. In conclusion, the performance of the different schemes studied was summarized in Table 3.8.

Moreover, channel coding assisted STBC single-user OFDM systems were studied in Section 3.4. More specifically, in Section 3.4.1 Trellis-Coded Modulation (TCM), Turbo Trellis-Coded Modulation (TTCM), Bit-Interleaved Coded Modulation (BICM) and iterative decoded BICM (BICM-ID) of the Coded Modulation (CM) family were amalgamated with an STBC G_2-aided OFDM system for communicating over wideband channels. Our corresponding simulation results were summarized in Table 3.11, which was followed by Section 3.4.2, where the performance of the TTCM and LDPC-assisted G_2-coded single-user OFDM systems was documented. It was found that the coding gain performance of the TTCM-aided G_2 schemes surpasses that of the LDPC-aided G_2 schemes when the affordable complexity is higher than approximate 500, as observed in Figure 3.38. At a low complexity, i.e. below a value of about 500, however, the LDPC-aided schemes tend to achieve a higher coding gain than the TTCM-aided schemes. The associated results were summarized in Table 3.13.

18.1.3   Coded-Modulation-Assisted Multi-user SDMA-OFDM Using Frequency-Domain Spreading

In **Chapter 4**, our attention was focused on Space-Division Multiple Access (SDMA) based uplink multi-user OFDM systems, where the various CM-assisted SDMA-OFDM schemes using the frequency-domain subcarrier-based Walsh–Hadamard Transform Spreading (WHTS) were investigated. In this chapter, the Minimum Mean Square Error (MMSE) based Multi-User Detection (MUD) technique was considered. From the simulation results provided in Section 4.3, it was found that WHTS was capable of improving the system's performance, especially in uncoded scenarios, since the bursty subcarrier errors can be pseudo-randomly spread across the subcarriers of the entire WHT block. Furthermore, when we had two receiver antenna elements, the TTCM-aided WHTS-MMSE-SDMA-OFDM system achieved the best CodeWord Error Ratio (CWER) performance among all the CM- and WHTS-assisted schemes considered for transmission over the three-path SWATM channel of Figure 3.32. Comparing Figure 4.8 with Figure 4.10, we found that at the same user load level, e.g. at $\alpha_4 = \alpha_2 = 0.5$ or $\alpha_4 = \alpha_2 = 1.0$, the E_b/N_0 performance achieved by the four-receiver system was better than that of the two-receiver system, provided that the same CM-assisted scheme was used, since employing a higher number of SDMA receiver antennas offers a higher spatial diversity. Additionally, the performance of the various CM- and WHTS-assisted MMSE-SDMA-OFDM systems was evaluated in the context of the more dispersive 12-path COST207 HT channel of Figure 4.12 in Section 4.3.2.2, including the two- and four-receiver scenarios in Section 4.3.2.2.1 and Section 4.3.2.2.2, respectively. It was found that with the aid of WHTS, the TTCM-assisted scheme constituted the best design option in terms of both the Bit Error Ratio (BER) and CWER in comparison with the other three CM-aided schemes. Furthermore, the spreading-induced E_b/N_0 gain achieved by all the schemes over the COST207 HT channel was higher than that achieved over the SWATM channel. This may suggest that in highly dispersive environments the channel-coded SDMA-OFDM system's performance may be further improved by employing WHTS.

In Table 4.4 we summarized the E_b/N_0 values required by the various CM- and WHTS-assisted MMSE-SDMA-OFDM schemes for achieving a BER of 10^{-5}, also showing the corresponding gains attained by the WHTS-assisted schemes. We observed that, on one hand, when we had a specific user load, the spreading-induced E_b/N_0 gain achieved by a system having a lower diversity order was higher, regardless of the employment of CM. In other words, the subcarrier-based WHTS technique may be expected to attain a higher system performance improvement in relatively lower diversity-order scenarios. Moreover, if we supported a fixed number of users but varied the number of receiver antenna elements, similar conclusions might be drawn. A plausible explanation for this fact may be that in the SDMA-MIMO system, a potentially higher space diversity gain may be achieved, when a higher number of receiver antenna elements are employed, and thus the benefits of WHTS may be less substantial. This may be particularly true in the context of the CM-aided systems, since most of the attainable gain has already been achieved by using the channel codes. On the other hand, when a given number of receiver antenna elements was used, the spreading-induced E_b/N_0 gains achieved in the context of the fully loaded systems were higher than in the half-loaded systems, regardless of the employment of channel codes. This suggests that more benefits may arise from WHTS, especially in the fully loaded scenarios, where the MUD suffers from a relatively low efficiency in differentiating the different users' signals.

Additionally, we provided the E_b/N_0 crossing points and the corresponding total gain achieved by the various CM-WHTS-MMSE-SDMA-OFDM schemes at the BER of 10^{-5}, in Figures 4.19 and 4.20, respectively. It was demonstrated that the performance gap between the different CM-aided schemes increased as the user load increased. It was also found that when the user load was higher, i.e. when the Multi-User Interference (MUI) was increased, the performance degradation of the TTCM-aided schemes was lower than those of the TCM-, BICM- and BICM-ID-aided arrangements. By contrast, at a relatively low user load the various schemes provided a similar performance, because most of the attainable gain of the four-receiver SDMA-OFDM system had already been achieved.

The effects of using different WHT block sizes were studied in Section 4.3.2.3, in both the SWATM and COST207 HT channel scenarios, as seen in Figures 4.21 and 4.22, respectively. As expected, it was found that the system's performance was improved, while the WHT block size used was increased. In Section 4.3.2.4 the effects of the Doppler frequency were demonstrated. It was concluded that the maximum Doppler frequency does not significantly affect the performance of the WHTS-assisted MMSE-SDMA-OFDM system, regardless of the employment of CM, as for example portrayed in Figure 4.23.

From the investigations conducted in **Chapter 4**, it was concluded that the various CM schemes, namely TCM, TTCM, BICM and BICM-ID, are capable of substantially improving the achievable performance of SDMA-OFDM systems. The employment of WHTS has the potential of further enhancing the system's performance in highly dispersive propagation environments. As a result, the TTCM- and WHTS-assisted schemes were found to have the best CWER performance in all the scenarios investigated. Furthermore, this was also the best design option in terms of the achievable E_b/N_0 gain expressed in dB, when communicating in highly dispersive environments, e.g. over the COST207 HT channel of Figure 4.12, while carrying a high user load of $\alpha_P \geq 0.5$.

18.1.4 Hybrid Multi-user Detection for SDMA-OFDM Systems

Since the performance of the linear MMSE MUD employed in **Chapter 4** is limited, our work was carried forward in **Chapter 5** towards the design of more sophisticated MUDs for the multi-user SDMA-OFDM systems. More specifically, we proposed an MMSE-aided Genetic Algorithm (GA) based MUD for employment in a TTCM-assisted SDMA-OFDM scheme.

In Section 5.2.1 we provided a system overview of the proposed GA-assisted TTCM-MMSE-SDMA-OFDM system. The antenna-specific optimization metric of Equation (5.3) designed for the proposed GA MUD was derived in Section 5.2.2.1 from that of the optimum Maximum Likelihood

(ML) MUD. Moreover, in order to solve the decision conflict resulting from the P antenna-specific metrics, a joint metric of Equation (5.6) was employed in the GA MUD. Section 5.2.2.2 outlined the operation process of the concatenated MMSE-GA MUD, while its performance was evaluated in Section 5.2.3, where the GA-based schemes were shown to be capable of achieving a near-optimum performance. Additionally, a complexity comparison between the proposed GA MUD and the optimum ML MUD was provided in Section 5.2.4, where we showed that the complexity of the GA MUD was significantly lower than that of the ML MUD.

In order to enhance further the achievable performance of the TTCM-assisted MMSE-GA-SDMA-OFDM system, an improved GA MUD was proposed in Section 5.3. This was discussed in two steps. Firstly, we proposed the novel Biased Q-function-Based Mutation (BQM) scheme in Section 5.3.1. This consisted of a review of the conventional Uniform Mutation (UM) scheme in Section 5.3.1.1 and the detailed explanation of the BQM mechanism in Section 5.3.1.2. The proposed BQM-aided GA MUD exploits an effective mutation strategy, which significantly improves the mutation efficiency so that the GA's search space can be substantially compressed. Hence, the BQM-aided GA MUD is capable of achieving a better performance in comparison with its UM-aided counterpart, especially at high SNRs or high user loads, as evidenced by the simulation results given in Section 5.3.1.3. Furthermore, in Section 5.3.1.2.2 we pointed out that the BQM scheme can be readily simplified to the Closest-Neighbour Uniform Mutation (CNUM) scheme, which only considers a subset of all the theoretically possible mutation target symbols during the mutation process. The CNUM-related transition probability values associated with different modems were summarized in Table 5.4. As a result, the BQM-induced search space can be further reduced by the CNUM scheme without significantly degrading the achievable performance.

Secondly, an Iterative GA (IGA) aided MUD was introduced in Section 5.3.2. The theoretical foundations of the IGA MUD were presented in Section 5.3.2.1, where we characterized the IGA framework as well as its optimization potential. Specifically, the IGA MUD is capable of exploiting the benefits accruing both from improved TTCM-decoded initial symbol estimates and from the embedded two-dimensional joint optimization invoked in the codeword domain as well as the in the frequency domain, as illustrated in Figure 5.15. From the numerical results provided in Section 5.3.2.2, the combined BQM-IGA-aided system was found to give the best performance in all scenarios considered, while maintaining a modest computational complexity. More specifically, the underloaded and fully loaded scenarios were first investigated in Section 5.3.2.2.1, where the number of users was assumed to be less than or equal to the number of receiver antenna elements. In low-throughput fully loaded scenarios, e.g. in a six-user system employing a 4-QAM modem, a two-iteration BQM-IGA MUD associated with $X = 20$ and $Y = 5$ was capable of achieving the same performance as the optimum ML-aided system at a complexity of 200, which is only about 50% and 5% of the MUD-related complexity imposed by the conventional UM-aided single-iteration IGA MUD and by the optimum ML MUD, respectively. On the other hand, in high-throughput six-user systems where for example a 16-QAM modem was employed, the UM-aided GA or IGA MUDs suffered from a high residual error floor even when exploiting the iterative framework. By contrast, a two-iteration BQM-IGA MUD associated with $X = 40$ and $Y = 5$ achieved an E_b/N_0 gain of about 7 dB over the MMSE MUD benchmark at the BER of 10^{-5}.

Moreover, in Section 5.3.2.2.2 we demonstrated that the proposed BQM-IGA MUD is capable of providing a near-optimum performance even in the so-called overloaded scenarios, where the number of accommodated users is higher than the number of receiver antenna elements, while many conventional detection techniques suffer from an excessively high error floor. For instance, when we had $L = 8$ users and $P = 6$ receivers in Figure 5.20, the two-iteration-based BQM-IGA MUD reduced the BER recorded at an E_b/N_0 value of 3 dB by four orders of magnitude in comparison with the classic MMSE-MUD-aided benchmark system. This high robustness of the BQM-IGA MUD was achieved since the high MUI experienced in overloaded scenarios was effectively suppressed. Moreover, the associated E_b/N_0 gain was attained at a modest complexity of 400 objective function evaluations, which is only 0.002 38% of the excessive complexity imposed by the ML MUD that is prohibited from simulations in this case.

As a further investigation, we demonstrated in Section 5.3.2.2.3 that the proposed system was capable of achieving a satisfactory performance even in conjunction with imperfect channel information.

In Section 5.3.3, a complexity analysis of the proposed BQM-IGA scheme was carried out, where we concluded that the complexity of the proposed detection scheme is only moderately higher than that imposed by the linear MMSE MUD, and is substantially lower than that imposed by the optimum ML MUD, as shown in Figures 5.25 and 5.26. We also showed that in both the fully loaded scenario of Section 5.3.1.3.2 and the overloaded scenario of Section 5.3.2.2.2.2, the complexity of the BQM approach can be further reduced by employing its simplified version, namely the CNUM scheme, at the cost of a slightly degraded system performance. Additionally, in Table 5.6 we further demonstrated the attractive performance versus complexity balance of the BQM-IGA scheme, which reduced the BER by up to five orders of magnitude in comparison with the MMSE MUD at the cost of only a moderate complexity.

Note that the system parameters of the IGA framework, such as the number of TTCM iterations, the number of IGA MUD iterations and the GA-related parameter settings, are all readily configurable, enabling us to strike an attractive trade-off between the achievable performance and the complexity imposed. For specific scenarios, the TTCM scheme used in the system can also be conveniently substituted by other FEC schemes, e.g. the TC codes. Therefore, the facility provided by the proposed IGA MUD may make it possible for applications in multi-mode terminals, where good performance, low complexity and easy flexibility are all important criteria. It is also worth pointing out that the proposed BQM-aided IGA MUD can be readily incorporated into the multi-user CDMA systems, e.g. those of [38]. In this case, the initial detected signal supplied to the GA MUD for creating the first GA population is provided by the bank of matched filters installed at the CDMA BS, rather than by the MMSE MUD. However, the BQM scheme may remain unchanged.

18.1.5 DSS and SSCH-Aided Multi-user SDMA-OFDM Systems

As we saw in Chapter 5, in the so-called overloaded scenarios, the classic MMSE MUD results in an excessive residual BER, since the degree of freedom becomes insufficiently high. However, the GA-based MUDs are capable of supporting a higher number of users than the MMSE MUD. Furthermore, with the aid of a new Slow Frequency-Hopping (SFH) type of technique, more users can be accommodated by the SDMA-OFDM system. This constituted our main objective in **Chapter 6**, where we proposed a TTCM-assisted SDMA-OFDM system employing the MMSE-IGA MUD of Chapter 5 with the aid of both Direct-Sequence Spreading (DSS) and the so-called Slow SubCarrier-Hopping (SSCH) techniques.

Firstly, the conventional SDMA-OFDM systems were briefly reviewed in Section 6.1, where their two disadvantages were identified. On the one hand, conventional SDMA-OFDM is incapable of efficiently exploiting frequency diversity, because all the users simultaneously share the entire bandwidth for their own communications. On the other hand, in conventional SDMA-OFDM systems, supporting a high number of bandwidth-sharing users will result in a high MUI across the entire bandwidth, inevitably degrading all users' performance. Against this background, in Section 6.2 we introduced the concept of hybrid SFH/SDMA-OFDM as well as the proposed SSCH/SDMA-OFDM schemes and discussed their potential benefits, aiming at finding an effective solution to the above-mentioned problems. In Section 6.3 a comparison of conventional SDMA-OFDM, SFH/SDMA-OFDM and SSCH/SDMA-OFDM systems in terms of their frequency resource allocation strategies was given, where the SSCH scheme was considered to have more advantages than the others, such as for example its higher efficiency in exploiting frequency diversity.

Our design of the proposed hybrid DSS/SSCH SDMA-OFDM system was detailed in Section 6.4. More specifically, first an overview of the proposed hybrid system was provided in Section 6.4.1, followed by the design of the transmitter and receiver presented in Sections 6.4.1.1 and 6.4.1.2, respectively. Figures 6.3 and 6.4 provided an example of the architecture of the DSS/SSCH transmitter and receiver, respectively. In Section 6.4.2, we analysed two SSCH pattern assignment strategies, namely the Random SSCH (RSSCH) scheme of Section 6.4.2.1 and the Uniform SSCH (USSCH) scheme of Section 6.4.2.2, where the USSCH strategy was considered to be better. On the one hand, in order to mitigate the MUI, the USSCH strategy disperses rather than concentrates each user's high-MUI subcarriers across the FD. On the other hand, to combat FD fading, the USSCH scheme uniformly distributed the subcarriers activated by the same user across the entire bandwidth, rather than consecutively mapping them to a small fraction of the entire bandwidth, and thus making it possible to exploit frequency diversity efficiently. As a result of employing the USSCH technique of Section 6.4.2.2.1, the number of users activating each subcarrier becomes similar, which in turn minimizes the average MUI across the whole system bandwidth. Moreover, we pointed out in Section 6.4.2.4 that the USSCH patterns can be generated by offline pre-computation, since their choice is not based on any channel knowledge. This requires a substantially lower computational complexity than that imposed by other adaptive algorithms exploiting real-time channel knowledge. The acquired patterns can also be reused according to predefined reuse time intervals, which should be sufficiently long for ensuring efficient exploitation of frequency diversity. Additionally, to provide insight into the hybrid DSS/SSCH SDMA-OFDM system, in Section 6.4.3 the DSS despreading and SSCH demapping operations invoked at the SDMA receiver were presented, followed by a discussion of the MUD process in the context of the hybrid system in Section 6.4.4.

Section 6.5 compared the simulation results generated by the proposed DSS/SSCH SDMA-OFDM system with those of various other SDMA-OFDM systems. In Section 6.5.1, the performances of the DSS- and/or SSCH-aided SDMA-OFDM systems were compared, when employing either the classic MMSE MUD discussed in Chapter 4 or the MMSE-IGA MUD proposed in Chapter 5. More specifically, it was shown in Figure 6.8 that in the moderately overloaded scenario associated with a total system throughput of $B_T = 5120$ bits per OFDM symbol, most of the MUI encountered can be effectively suppressed by the MMSE-IGA MUD, while the overloaded MMSE MUD results in a high error floor. This result achieved in the context of the DSS/SSCH SDMA-OFDM system was consistent with our findings in Chapter 5, where the performance advantages of the MMSE-IGA MUD have been recognized.

In Section 6.5.2, we further compared the performances of various hybrid SDMA-OFDM as well as conventional SDMA-OFDM systems. More specifically, a moderately overloaded scenario was first considered in Section 6.5.2.1. It was observed that the Random SFH (RSFH), the Uniform SFH (USFH) and the RSSCH schemes suffered from the employment of random hopping patterns that resulted in an increased average BER due to the high MUI encountered at the subcarriers or subbands activated by an excessive number of users. By contrast, the schemes exploiting the uniform patterns generated by the algorithm of Section 6.4.2.2 were capable of providing an improved performance, as shown in Figure 6.9.

Furthermore, we pointed out that the proposed uniform pattern assignment algorithm is capable of providing more benefits in the scenarios associated with a higher throughput. This argument was validated in Section 6.5.2.2, where a highly overloaded scenario corresponding to a total system throughput of $B_T = 6144$ bits per OFDM symbol was considered. It was shown in Figure 6.10 that the USFH, the USSCH and the hybrid DSS/USSCH systems were capable of achieving a significantly better performance in comparison with the conventional SDMA-OFDM, the DSS/SDMA-OFDM as well as the RSFH/RSSCH-aided SDMA-OFDM systems, while the USSCH/SDMA-OFDM system was found to be the best option. The superior performance of the USSCH scheme was attributed to its specific characteristics, which successfully suppressed the MUI and thus eliminated the associated error floor. For example, we observed in Figure 6.10 that at an E_b/N_0 value of 10 dB, the USSCH/SDMA-OFDM system reduced the BER by about two, three and four orders of magnitude in comparison with the conventional SDMA-OFDM, the RSSCH/SDMA-OFDM and the RSFH/SDMA-OFDM systems, respectively. We also concluded that the frequency diversity benefits achieved by USSCH may be more significant than those of the time diversity attained by DSS, since the hybrid DSS/USSCH SDMA-OFDM system attained a performance between those of the USSCH/SDMA-OFDM and the DSS/SDMA-OFDM systems, while the DSS/SDMA-OFDM scheme achieved a similar performance to that of the conventional SDMA-OFDM system. This suggests that the fixed-bandwidth hybrid system should avoid using long DSS codes that result in a wider subcarrier bandwidth, so that a sufficiently high number of $Q > K$ subcarriers becomes available for maintaining a sufficiently high frequency diversity.

The proposed USSCH scheme's high robustness against MUI was further confirmed by Figures 6.11 and 6.12, where the USSCH/SDMA-OFDM system exhibited the best BER versus total system throughput performance as well as the best maximum total system throughput versus E_b/N_0 performance among the various SDMA-OFDM arrangements, respectively. For instance, as shown in Figure 6.12, at the E_b/N_0 value of 12 dB, a capacity increase of about 4%, 13% and 44% was achieved by the USSCH scheme compared with the USFH, the RSSCH and the RSFH arrangements at the target BER of 10^{-3}, and 4%, 14% and 78% at the BER of 10^{-5}, respectively. In Section 6.5.4, we demonstrated that the proposed hybrid system is capable of achieving an acceptable performance even without accurate channel knowledge. Finally, we pointed out that the superior performance of the USSCH-aided SDMA-OFDM system is achieved at a similar complexity to that of the conventional SDMA-OFDM arrangement, since the additional computational complexity imposed by the USSCH algorithm manifests itself in terms of offline pre-computation.

The capability of effectively exploiting frequency diversity and suppressing high MUI renders the channel-coded USSCH technique an attractive design option, lending a beneficial flexibility to practical applications. Furthermore, the USSCH/SDMA-OFDM scheme can be readily extended to a variable-rate system offering a high grade of flexibility required by future wireless multimedia services, supporting variable bit rates and different QoS requirements. Each user may activate a different number of subcarriers, depending on the type of service to be delivered or the bit rate to be supported. In such scenarios, the USSCH algorithm of Section 6.4.2.2 can be modified by taking into account the different rates. Moreover, in SSCH-based systems the different subcarriers do not have to be contiguously allocated, which is a further attractive property, especially in those scenarios where several systems operated by different service providers have to coexist and/or fractional bandwidths have to be exploited. In addition, the different types of hybrid SFH/SSCH-assisted SDMA-OFDM systems can be readily implemented by exploiting different numbers of subbands having different bandwidths, depending on the specific system requirements.

18.1.6 Channel Estimation for OFDM and MC-CDMA

The DDCE scheme proposed in **Chapter 7** is suitable for employment in both OFDM and MC-CDMA systems. We analysed the achievable performance of the estimation scheme considered in conjunction with a realistic dispersive Rayleigh fading channel model having a realistic Fractionally Spaced (FS) rather than an idealized Symbol-Spaced (SS) Power Delay Profile (PDP).

Specifically, in Section 7.4.1 we proposed the MMSE FD-CTF estimator, which is suitable for employment in both OFDM and MC-CDMA systems. In Section 7.4 we continued our discourse with the derivation of both sample-spaced as well as fractionally spaced CIR estimators. In Section 7.4.5 we performed a comparison between the two methods considered and demonstrated the advantages of the fractionally spaced scheme. Subsequently, in Section 7.5 we developed a parametric fractionally spaced CIR tap tracking technique, which facilitates low-complexity channel estimation in realistic channel conditions characterized by time-variant fractionally spaced power delay profiles. More specifically, we employ the deflation PAST method of Algorithm 7.2 for recursive tracking of the CTF's covariance matrix and for subsequent tracking of the corresponding CIR taps. We demonstrated that the PAST-aided DDCE scheme proposed exhibits a good performance over the entire range of practical propagation conditions.

In Section 7.6 we discussed two major CIR tap prediction strategies, namely the robust predictor, which was capable of guaranteeing a certain level of performance under specified worst-case PDP conditions, and the adaptive RLS predictor. In Figures 7.19 and 7.20 of Section 7.6.5 we characterized and compared the achievable performance of both methods considered and drew conclusions concerning their relative merits. Specifically, we demonstrated that the RLS prediction technique outperforms its robust counterpart over the entire range of the relevant channel conditions.

Subsequently, in Section 7.8 we addressed the problem of channel estimation in multi-antenna, multi-carrier systems. Specifically, we proposed a DDCE scheme which is suitable for employment in a wide range of multi-antenna, multi-carrier systems capable of operating over the entire range of practical channel conditions. In particular, we considered a generic MIMO-OFDM system employing K orthogonal frequency-domain subcarriers as well as having m_t and n_r transmit and receive antennas, respectively. The MIMO channel estimation scheme derived in Section 7.8 comprises an array of K per-subcarrier MIMO-CTF estimators, followed by a $(n_r \times m_t)$-dimensional array of parametric CIR estimators and a corresponding array of $(n_r \times m_t \times L)$ CIR tap predictors, where L is the number of CIR taps tracked per link of the MIMO channel.

In Section 7.8.1 we explored a family of recursive MIMO-CTF tracking methods, which were combined with the aforementioned PAST-aided CIR-tracking method of Section 7.5 as well as with the RLS CIR tap prediction method of Section 7.6.4 in order to create an efficient channel estimation scheme for MIMO-OFDM systems. More specifically, in Section 7.8.1 we considered both hard- and soft-feedback-assisted LMS and RLS CIR tap tracking algorithms as well as the modified RLS algorithm, which is capable of improved exploitation of the soft information associated with the decision-based estimates.

Finally, in Figures 7.25–7.28 of Section 7.8.1.5 we documented the achievable performance of the resultant MIMO-DDCE scheme employing the recursive CTF tracking of Section 7.8.1 followed by the parametric CIR tap tracking and CIR tap prediction techniques of Sections 7.5 and 7.6, respectively. We demonstrated that the MIMO-DDCE scheme proposed exhibits a good performance over the entire range of practical conditions. More specifically, both the BER as well as the corresponding MSE performance of the channel estimation scheme considered were characterized in the context of a turbo-coded MIMO-OFDM system in Figures 7.25–7.28. We demonstrated that the MIMO-DDCE scheme proposed remains effective in channel conditions associated with high terminal speeds of up to 130 km/h, which corresponds to the OFDM-symbol- normalized Doppler frequency of 0.006. Additionally, we reported a virtually error-free performance for a rate $\frac{1}{2}$ turbo-coded 8×8 QPSK-OFDM system, exhibiting a total bit rate of 8 bps/Hz and having a pilot overhead of only 10%, at an SNR of 10 dB and normalized Doppler frequency of 0.003, which corresponds to a mobile terminal speed of about 65 km/h.

In conclusion, the performance of the PAST-aided MIMO-DDCE scheme derived in Chapter 7 may be characterized based on the MSE performance results depicted in Figure 18.1. More specifically, the MSE σ_e^2 exhibited by the channel estimation scheme considered may be expressed as

$$\sigma_e^2 = \frac{1}{\kappa\gamma} \frac{Lm_t n_r}{K}, \tag{18.1}$$

where L is the number of estimated CIR taps, while m_t and n_r are the numbers of transmit and receive antennas, respectively. Correspondingly, $Lm_t n_r$ denotes the total number of independent channel-related parameters estimated, while γ is the average SNR encountered at the receiver. Furthermore, we employ the estimation efficiency factor κ of Equation 7.110. The value of the parameter κ was determined empirically using Equation 7.110, yielding $\kappa = 4$ dB.

18.1.7   Joint Channel Estimation and MUD for SDMA-OFDM

The assumption of perfect channel estimation has been used in most of the previous chapters to simplify the constraints. However, in practical MIMO-OFDM systems accurate channel estimation is required at the receiver for invoking both coherent demodulation and interference cancellation. This task is more challenging than that in the SISO scenario, owing to an increased number of independent transmitter–receiver channel links as well as an increased interference level, as a result of employing multiple transmitter antennas. Aiming to overcome the channel estimation challenge in MIMO-OFDM systems,

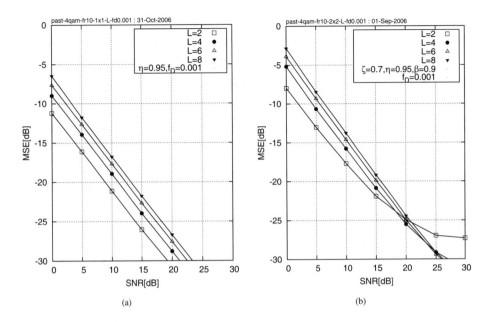

Figure 18.1: The **MSE** exhibited by (a) the 1×1 and (b) the 2×2 **4-QAM-OFDM** system employing the **PASTD** CIR estimator of Algorithm 7.2 and tracking $L = 2, 4, 6$ and 8 CIR taps. The value of the PASTD forgetting factor was $\eta = 0.95$. The OFDM-symbol-normalized Doppler frequency was $f_D = 0.001$. The abscissa represents the average SNR recorded at the receive antenna elements.

in **Chapter 8** we proposed a GA-assisted iterative Joint Channel Estimation and Multi-User Detection (GA-JCEMUD) scheme which constitutes an attractive and effective solution to this problem.

In Section 8.1, various channel estimation approaches designed for MIMO-OFDM were outlined, where we pointed out that none of the methods proposed for the BLAST- or SDMA-type multi-user MIMO-OFDM systems in the open literature allowed the number of users to exceed that of the receiver antennas, which we refer to as an overloaded scenario, because the corresponding singular channel matrix results in an insufficient degree of freedom at the detector. Explicitly, in this overloaded scenario channel estimation becomes more challenging, especially in the context of decision-directed channel estimators which are more vulnerable to error propagation. Against this background, we proposed an iterative joint channel estimation and symbol detection scheme benefitting from the GA's optimization power, which was illustrated in Figure 8.1 of Section 8.2. A full discussion of the proposed iterative GA-JCEMUD approach was provided in Section 8.3, where the structure of the GA-JCEMUD was detailed in Figure 8.2. The joint optimization method first uses pilot OFDM symbols for generating initial Frequency-Domain CHannel Transfer Function (FD-CHTF) estimates, which are then improved by the time-domain filters, as elaborated on in Section 8.3.1. The improved channel estimates are used to assist the Optimized Hierarchy Reduced Search Algorithm (OHRSA) of Section 8.3.2 in invoking a first-stage MUD. The initially detected user signals and the FD-CHTF estimates associated with the previous OFDM symbol duration are then forwarded to the proposed GA-JCEMUD printed in the grey block of Figure 8.2 for jointly optimizing the estimates of both the FD-CHTFs and multi-user data symbols.

The GA-aided joint optimization process was detailed in Section 8.3.3. More specifically, we outlined the mathematical representation of the GA individuals according to the MIMO channel's structure in Section 8.3.3.1 and described the initialization of the joint optimization in Section 8.3.3.2. Furthermore, since the data symbols belong to the discrete legitimate constellation symbol set $\mathcal{M}_c$, while the FD-CHTFs are continuous valued and can be potentially infinite across the complex plane $\mathbb{C}$, separate cross-over/mutation methods have to be invoked for the symbol chromosome and the channel chromosome, as discussed in Sections 8.3.3.3.1 and 8.3.3.3.2. Additionally, in order to overcome the limitations imposed by the conventional GA-aided MUDs, which can only provide single-individual-based hard-decoded symbol estimates, a new algorithm was derived in Section 8.3.3.4 to assist the GA in generating soft outputs based on the entire population, where we also argued that the proposed GA generating the above-mentioned population-based soft outputs imposes only a modest complexity increase in comparison with its conventional hard-decision-aided individual-based counterparts.

In Section 8.4 the numerical results generated by computer simulations were provided. As a preliminary investigation, in Section 8.4.1 we first identified the effect of the maximum mutation step size λ_{max} used in the GA-JCEMUD. It was shown in Figures 8.3 and 8.4 that the choice of λ_{max} has a significant effect on the system's performance, especially in high-SNR scenarios. In order to achieve the best possible performance, typically a higher λ_{max} value is required, when a higher OFDM-symbol-normalized Doppler frequency F_D is encountered, so that the rapid changes of the channel fades can be successfully captured by the GA-JCEMUD. Based on Figure 8.4, the recommended values of λ_{max} in conjunction with different values of F_D in terms of achieving the best attainable BER performance can be identified. In Figure 8.5 of Section 8.4.2, the effect of different Doppler frequencies on the proposed scheme was evaluated, where it was noticed that the performances of both the uncoded and coded GA-JCEMUD-aided systems degraded as F_D was increased. Nonetheless, the proposed GA-JCEMUD-aided SDMA-OFDM system using a modest pilot overhead of $\epsilon = 2.5\%$ was capable of achieving a performance close to the perfect-CSI-aided optimum ML MUD at $F_D = 0.001$, as seen in Figure 8.5, while the system employing the MMSE MUD completely failed even with the aid of perfect CSI. Furthermore, it was shown in Section 8.4.3 that the system's performance can be improved by invoking a higher number of GA-JCEMUD iterations, both with and without channel coding. Observe in Figure 8.7 that at an SNR value of about 13 dB, the GA-JCEMUD using $\epsilon = 2.5\%$ pilot overhead approached the best-case FD-CHTF estimation MSE performance associated with $\epsilon = 100\%$. In an effort to investigate the effect of the pilot overhead ϵ, we argued that a satisfactory performance can be achieved with the aid of a pilot overhead as low as $\epsilon = 1.5\%$ to 2.5%, as evidenced by Figure 8.8 of Section 8.4.4.

To evaluate further the benefits of the proposed joint optimization approach, in Section 8.4.5 we compared the performances of the GA-JCEMUD and its counterpart, namely the GACE-OHRSA-MUD, which serially concatenates the OHRSA MUD with the stand-alone GA-aided channel estimator. It was shown in Figure 8.9 that the GA-JCEMUD outperformed the GACE-OHRSA-MUD, especially in high-Doppler scenarios. This demonstrates the superiority of the joint optimization mechanism over conventional 'serially concatenated' detection architectures. In Section 8.4.6, the merits of the proposed soft GA-JCEMUD scheme were identified in terms of its ability of providing soft outputs. Observe in Figure 8.10 that, with the advent of FEC codes, the proposed population-based soft-decoded GA was capable of significantly outperforming the conventional individual-based hard-decoded GA, especially in high-Doppler scenarios, since it exhibited a higher robustness against fast-fading channels than the conventional GAs. Last but not least, in Section 8.4.7 we examined the robustness of the GA-JCEMUD in MIMO scenarios. As portrayed in Figure 8.11, the proposed iterative GA-JCEMUD was capable of simultaneously capturing the fading envelope changes of each individual user–receiver link, regardless of its instantaneous fading envelope value, and thus achieving an equally good performance over all the user–receiver links. This result potently demonstrates the robustness of the GA-JCEMUD in MIMO scenarios.

18.1.8   Sphere Detection for Uncoded SDMA-OFDM

18.1.8.1   Exploitation of the LLRs Delivered by the Channel Decoder

We demonstrated in **Chapter 9** that although the conventional LSD is capable of achieving a near-MAP performance while imposing a significantly reduced complexity in comparison with the exact-MAP

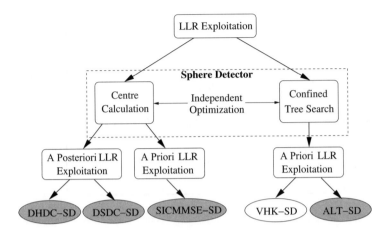

Figure 18.2: Independent search centre calculation and search phase operations that may be used by the SD. Note that the shaded ellipses represent the novel schemes contrived in this treatise.

detector, the complexity of the LSD may still become excessive in the channel-coded iterative-detection-aided MIMO system supporting a high number of users/transmit antennas and/or employing high-order modulation schemes, such as 16-QAM. This is because the size of the transmitted MIMO symbol candidate list generated by the LSD has to remain sufficiently large in the above-mentioned scenarios, in order to deliver sufficiently accurate soft-bit information, i.e. LLRs, to the channel decoder during the iterative detection process, thus achieving a high iteration gain.

Based on the above-mentioned observations of **Chapter 9** we demonstrated in **Chapter 10** that the LSD's EXIT curve may even decay upon increasing the a priori information fed back by the channel decoder, as observed in Figure 10.2 of Section 10.2.1.3 as a consequence of having an insufficiently large candidate list size. This results in a significantly reduced iteration gain. This is because the inner SD and outer channel decoder of Figure 10.1 exchange flawed information owing to a shortage of candidate solutions – more particularly, owing to the absence of the ML solution in the candidate list, as revealed in Section 10.2.1.3. Thus, in order to reduce further the computational complexity imposed by the conventional LSD, we optimized the LSD algorithm based on the exploitation of the LLRs gleaned from the channel decoder during the iterative detection process by devising two enhanced SD schemes, namely the centre-shifting-based SD of Section 10.2 and the ALT-based SD of Section 10.3.

More specifically, the SD procedure may be divided into two successive phases, namely the search-centre calculation phase followed by the confined tree search phase, as shown in Figure 18.2. The operations in both of these phases may be optimized independently by exploiting the LLRs provided by the channel decoder during the iterative detection process. For the centre calculation phase, conventional SDs employ either the LS or the MMSE algorithm for computation of the search centre $\hat{c}$, as discussed in Section 10.2.2, which is carried out only once during the first iteration. However, during our investigations in Section 10.2.2, we realized that it would be desirable to set the SD's search centre to a multi-user signal constellation point, which may be obtained using more sophisticated algorithms rather than the LS or MMSE scheme and thus is expected to be closer to the real MAP solution than both the LS and MMSE solutions, because this would allow us to reduce the SD's search space and hence its complexity.

Then, based on the centre-shifting theory, we proposed a generic centre-shifting-aided SD scheme in Section 10.2.2, as portrayed in Figure 18.3, which may be expected to become significantly more powerful if it is employed in an iterative-detection-aided channel-coded system, since the process of generating a more accurate search centre is further aided by the channel decoder, which substantially

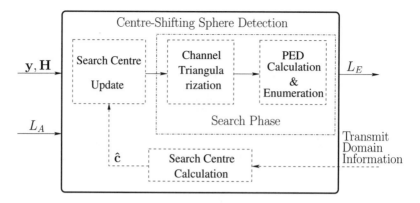

Figure 18.3: The structure of the generic iterative centre-shifting based SD scheme.

contributes towards the total error-correction capability of the iterative receiver. Hence, in this treatise, three particular centre-shifting-based SD schemes were devised, as shown in Figure 18.2, which are the DHDC-aided centre-shifting SD of Section 10.2.3.1, the DSDC-aided centre-shifting SD of Section 10.2.3.2 and the SIC-MMSE-aided centre-shifting SD of Section 10.2.3.3. The former two simply update the search centre of the SD by the hard and soft decisions of the corresponding transmitted MIMO symbols, respectively, which are obtained based on the a posteriori LLRs at the output of the channel decoder, while the latter one exploits the slightly more sophisticated SIC-MMSE algorithm in order to update the search centre based on the a priori LLRs provided by the channel decoder. The operations of the proposed centre-shifting schemes are summarized in Table 18.1 in contrast to those of the conventional SD using no centre shifting. Note that in contrast to the LSD dispensing with the centre-shifting scheme, which may generate the candidate list only once at the very beginning of the entire iterative detection process, our proposed iterative centre-shifting-based SDs have to regenerate the candidate list for the following soft-bit-information calculation, as long as the search centre is updated during the iterative detection process. However, the computational complexity imposed by each candidate list generation of the centre-shifting-based SD may be exponentially reduced, since the increasingly accurate search centres generated during the iterative detection process allow us to rely on a significantly reduced candidate list size, while maintaining a near-MAP performance. Hence, the overall complexity imposed by the iterative centre-shifting-based SD is expected to be significantly reduced in comparison with the conventional SD dispensing with the centre-shifting scheme, despite having an increased number of list generations during the entire iterative detection process.

According to the iterative centre-shifting-based SD receiver design of Sections 10.2.3.1–10.2.3.3, the centre-shifting SD-aided receiver design principles can be summarized as follows:

1.The search-centre calculation is based on the soft-bit information provided by the channel decoder.

2. The search-centre update can be carried out in a more flexible manner by activating the proposed centre-shifting scheme, whenever the system requires its employment during the iterative detection process in order to maximize the achievable iterative gain.

3. The search-centre update is flexible, since it may be carried out by any of the well-known linear or nonlinear detection techniques.

The key simulation results obtained in Section 10.2 for the BER performance and complexity of the three proposed iterative centre-shifting-based SDs are summarized and quantified in Table 10.7, in comparison with the conventional SD-aided iterative receiver using no centre-shifting scheme. Specifically, in the challenging (8×4)-element 4-QAM SDMA-OFDM system, the DHDC-aided,

Table 18.1: Operations of various iterative centre-shifting-based SD schemes in comparison with that of the conventional SD.

	Conventional SD	DHDC-SD	DSDC-SD	SI-CMMSE-SD
Center-update algorithm	No centre update	Direct hard decision	Direct soft decision	SIC-MMSE
Type of exploited LLR in centre-update phase	No LLR exploitation	A posteriori LLR fed by outer decoder	A posteriori LLR fed by outer decoder	A priori LLR fed by outer decoder
When to update centre	Compute centre only once at the first iteration	Centre update is activated when iteration converges	Identical to DHDC	Centre update activated at the beginning for each iteration
When to regenerate candidate list	No need to regenerate during iterations	Regenerate every time the centre is updated	Identical to DHDC	Identical to DHDC
When to recalculate the output LLRs	Recalculate for each iteration	Identical to conventional SD	Identical to conventional SD	identical to conventional SD

Table 18.2: Comparison of the conventional K-best SD and our proposed iterative centre-shifting-based K-best SDs in the scenario of an (8×4)-element rank-deficient SDMA-OFDM system. Note that the computational complexity of the SD, i.e. the list generation complexity of the SD, is calculated in terms of the total number of PED evaluations, while that of the soft information generation by the SD/MAP detector is quantified on the basis of Equation (10.16) in terms of the total number of OF evaluations corresponding to the two terms in Equation (10.14).

BER	Centre shifting	$N_{cand}(K)$	Iterations	SNR	Memory	SD compl.	MAP compl.
10^{-5}	None	1024	3	10.5	8196	13652	49152
		128	3	11.2	1024	2388	6144
		64	2	12	512	1364	2048
		32	2	12.8	256	724	1024
		16	2	15	128	404	512
	DHDC	64	2 + 2	11.3	512	4092	4096
		32	2 + 2	12.7	256	2172	2048
	DSDC	64	2 + 2	10.5	512	4092	4096
		32	2 + 2	12.2	256	2172	2048
	SIC-MMSE	64	3	10.2	512	4092	3072
		32	3	10.2	256	2172	1536
		16	3	11	128	1212	768

the DSDC-assisted and the SIC-MMSE-aided centre-shifting SDs of Sections 10.2.3.1–10.2.3.3 using $K = N_{cand} = 32$ are capable of achieving a BER of 10^{-5} by requiring 0.1, 0.6 and 2.6 dB lower transmit power or SNR than that necessitated by the conventional SD dispensing with the centre-shifting scheme and using the same values of K as well as N_{cand}. Remarkably, the SIC-MMSE centre-shifting SD scheme of Section 10.2.3.3 using $K = N_{cand} = 32$ may enable the iterative receiver to exhibit a near-MAP performance which is achieved by the conventional SD using no centre shifting in conjunction with a significantly larger candidate list size of $K = N_{cand} = 1024$. This near-MAP performance is achieved despite imposing a reduced complexity relating to detection candidate list generation, which is about an order of magnitude lower than that exhibited by the list SD dispensing with the proposed centre-shifting scheme. As a further benefit, the computational complexity associated with the *extrinsic* LLR calculation was reduced by a factor of about 64. The associated memory requirements were also reduced by a factor of 64.

As shown in Figure 18.2, the soft-bit information delivered by the channel decoder may be exploited for both the tree search phase and the centre calculation phase of the SD. The LSD proposed by Vikalo, Hassibi and Kailath (VHK-SD) [642] was the first one to exploit the a priori LLRs provided by the

Table 18.3: Comparison of the Centre Shifting (CS) and the ALT-aided SDs of Sections 10.3 and 10.2.

	ALT	CS
LLR exploitation based	Yes	Yes
Type of exploited LLR	A priori LLR	A posteriori or a priori LLR, depending on the employed centre calculation algorithm
Optimization target	Tree search phase	Centre calculation phase
Centre recalculation	No	Yes
Candidate list regeneration	Yes, for each iteration	Yes, when the centre is updated
Achievable benefits	(1) Significant overall reduced detection complexity. (2) Significant reduced memory requirements	More pronounced achievements in detection complexity and memory requirements than ALT
Overheads (side effects)	(1) Performance sensitive to threshold choice. (2) Non-Gaussian output LLRs	Increased complexity in centre calculation phase for each iteration
CS–ALT combination	Significantly more detection complexity reductions can be achieved	
Applications	Applicable to both coherent SDs and non-coherent MSDSDs of Chapter 12	Only works for coherent SDs, no centre-update is needed for MSDSDs (centred at the origin)

channel decoder in the confined tree search process, which was arranged by including the effect of the a priori LLRs in the OF of the SD in a similar manner to that seen in Equation (14.44) of Section 14.3.1, where the exploitation of the soft-bit information by the MSDSD was highlighted. In Section 10.3, another reduced-complexity technique termed the ALT-aided SD scheme was devised by exploiting the reliability of the bit decision conveyed by the a priori LLRs. More specifically, the philosophy of the ALT-aided SD is to assume perfect knowledge of a particular bit, i.e. 0 or 1, and then test whether the absolute value of the corresponding a priori LLR is higher than the preset threshold (ALT), followed by pruning the branch associated with the opposite bit value, before the tree search continues. Thus, a better pruning search tree may be formed as seen for example in Figure 10.23 of Section 10.3.1, resulting in an improved performance and a reduced complexity, as observed in Figure 10.24 of Section 10.3.2.1 and Figure 10.25 of Section 10.3.2.2, respectively. As demonstrated in Section 10.3.2.3, the ALT threshold has to be carefully adjusted to achieve the target performance as a function of the SNR encountered. Furthermore, Section 10.3.2.4 demonstrated that the non-Gaussian distribution of the LLRs at the output of the ALT-aided SD during the iterative detection process limits its capacity and imposes difficulties in predicting the EXIT-chart-assisted performance. On the other hand, although the proposed ALT scheme is capable of providing useful performance improvements, which are slightly less significant than those achieved by the SIC-MMSE-assisted centre-shifting-based SD scheme of Section 10.2.3.3, an attractive performance–complexity trade-off may be achieved by the combination of the two, as seen in Section 10.3.3. More particularly, the detection complexity imposed by the SIC-MMSE-assisted centre-shifting SD can be halved with the aid of the ALT technique as observed in Figure 10.32, despite suffering a modest performance loss of about 0.5 dB, as demonstrated in Figure 10.31 in the (8×4)-element 4-QAM SDMA-OFDM scenario. The characteristics of a range of centre-shifting-aided SDs are summarized in Table 18.2, while the features of the proposed ALT technique are summarized in comparison with those of the centre-shifting scheme in Table 18.3. Note that the ALT scheme is applicable to both the coherent SD and the non-coherent SDs (MSDSD), whereas the centre-shifting scheme works only for the coherent SDs. This is because the equivalent search centre of the MSDSD of Chapter 12 is the origin, which is independent of the transmitted signal, thus requiring no updates.

Table 18.4: Generic adaptive SD mechanism.

Adaptive parameter choice	As the parameter increases: (1) the SD's complexity exhibits an exponential growth; (2) the corresponding EXIT curve is shifted increasingly higher.
Operations	(1) Set the parameter to the smallest value; (2) slightly increase it as soon as the iterative decoding converges; (3) candidate list regeneration is required if the value of the parameter is changed.
Achievable benefits	Significant complexity reduction, as seen in Figure 14.17(b)
Overheads	Increased number of total detection iterations
Applications	Applicable to both the coherent SD and the non-coherent SD

18.1.8.2 EXIT-Chart-Aided Adaptive SD Mechanism

As a further evolution of the LLR-exploitation-based complexity-reduction schemes contrived for coherent SD-aided iterative receivers in Sections 10.2 and 10.3, an EXIT-chart-aided adaptive mechanism was proposed to reduce the complexity imposed by an SD-aided near-capacity system in the context of non-coherently detected SD-aided cooperative systems. Specifically, an adaptive-window-duration-based MSDSD scheme was devised in Section 14.4.3.1. The philosophy of the proposed adaptive-window-based scheme, which is characterized by the EXIT chart of Figure 14.17(a) in Section 14.4.3.1, is based on the observation that the intersection point of the EXIT curves of the inner and outer decoders may be gradually pushed towards the $(1, 1)$ point by increasing the observation window size N_{wind} at the cost of imposing an exponentially increased computational complexity per iteration. Thus, in order to reduce the overall detection complexity while maintaining a near-capacity performance, the observation window size of the adaptive-window-duration-based MSDSD was initially set to the smallest value of $N_{wind} = 2$, which would be slightly increased as soon as the iterative decoding between the inner and outer decoders converged. The significantly reduced overall complexity observed in Figure 14.17(b) of Section 14.4.3.1 is due to the exponentially reduced complexity imposed by the early-stage iterations using small values of N_{wind}, despite having an increased total number of iterations required by the decoding trajectory to reach the $(1, 1)$ point. Remarkably, the complexity imposed by the MSDSD_s is substantially reduced, as seen in Figure 14.17(b) of Section 14.4.3.1, which is a benefit of the adaptive-window-assisted scheme. Quantitatively, observe in Figure 14.17(b) that this complexity reduction may be as high as 75% at $SNR_t^{overall} = 2$ dB, when the open EXIT tunnel created by having $N_{wind} = 7$ becomes rather narrow. The attainable complexity reduction increases even further to about 83% when the open EXIT tunnel becomes wider at $SNR_t^{overall} = 4$ dB, as also seen in Figure 14.17(b).

Importantly, we should point out here that the simple yet powerful adaptive SD mechanism proposed for the MSDSD in Section 14.4.3.1 may be also applied to the coherent SD-aided iterative receiver of Chapter 10 for reducing the overall iterative detection complexity. More specifically, the candidate list size N_{cand} may be adaptively increased based on the same philosophy used by the adaptive-window-duration-aided scheme during the iterative detection process, since the EXIT curve is shifted upwards as the value of N_{cand} increases, as observed for example in Figure 10.2 of Section 10.2.1.3. The principles of the generic adaptive SD mechanism invoked in iterative receivers are summarized in Table 18.4 along with its major characteristics.

18.1.9   Transmit Diversity Schemes Employing SDs

A multi-layer tree search mechanism was proposed for SDs in **Chapter 11** in order to facilitate its application in STBC-SP-assisted MU-MIMO systems in the interest of achieving a near-MAP performance at a low complexity. Based on the philosophy of the multi-layer tree search, we also adopted the SD algorithm at the BS for the differentially modulated user-cooperation-based cellular

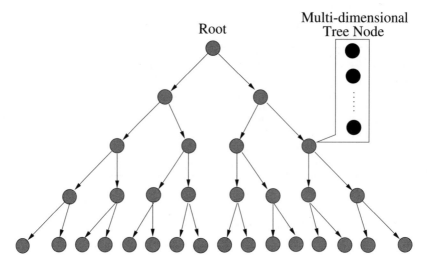

Figure 18.4: Generic multi-layer search tree.

uplink scenario of Chapter 12 for jointly and non-coherently detecting the multi-path signals transmitted by the source and relay nodes. In the ensuing Section 18.1.9.1 the idea of the multi-layer tree-search-aided SD will be briefly echoed, followed by a summary of its applications in the above-mentioned transmit-diversity-oriented systems in Section 18.1.9.2.

18.1.9.1   Generalized Multi-layer Tree Search Mechanism

In comparison with the tree search of Section 9.2 conducted by the conventional SD algorithm, where each tree node is a single dimension associated with the trial symbol point transmitted from a specific transmit antenna, the tree node of the generic multi-layer search tree depicted in Figure 18.4 may consist of a number of symbol candidates transmitted from multiple spatially separated antennas either of a particular user or of different users, and thus may be of multi-dimensional nature. Actually, the generic multi-layer tree search mechanism of Section 11.3.2.3 includes the conventional single-layer tree search as a special case, when the signals transmitted from different antennas are uncorrelated with each other. In other words, when the signals transmitted from different antennas are jointly designed, as, for example, experienced in the STBC-SP-aided MU-MIMO system of Section 11.2.2.1, or exhibit correlations, as, for example, encountered in the user-cooperation-based systems of Section 12.3.1, the multi-layer tree search of Section 11.3.2.3 has to be invoked.

18.1.9.2   Spatial Diversity Schemes Using SDs

It was argued in Section 11.2.2.1 that combining orthogonal transmit diversity designs with the principle of SP is capable of maximizing the achievable coding advantage. The resultant STBC-SP scheme was capable of outperforming the conventional orthogonal-design-based STBC schemes in the SU-MIMO scenario. Specifically, the STBC-SP scheme of Section 11.2.2.1 combines the signals transmitted from multiple antennas into a joint ST design using the SP scheme, as demonstrated in Figure 11.4. However, existing MU-MIMO designs make no attempt to do so, owing to the associated complex detection. We solved this complexity problem by further developing the SD for detection of SP modulation using the proposed multi-layer tree search mechanism in Section 11.3, because SP may offer a substantial SNR reduction, although at a potentially excessive complexity, which can be reduced by the multi-dimensional tree-search-based SD of Section 11.3.2. Explicitly, the K-best SD algorithm designed for a four-dimensional SP modulation scheme was summarized in Section 11.3.2.3. The enhanced coding advantage achieved by the STBC-SP is indicated by the increased area under the corresponding EXIT curve in contrast to that associated with the conventional STBC scheme, as observed in Figure 11.9 of Section 11.3.3. More particularly, as a benefit of employing the SP modulation, performance gains of 1.5 dB and 3.5 dB can be achieved by 16-SP and 256-SP modulated systems in *Scenario I* and *Scenario II* of Table 11.2, respectively, in comparison with their identical-throughput QAM-based counterparts, given a target BER of 10^{-4} and $N_{cand} = 128$, as observed from Figures 11.10(a) and 11.10(b) of Section 11.3.3.

Although co-located multiple-transmit-antenna-aided diversity techniques are capable of mitigating the deleterious effects of fading, as discussed in Chapter 11, it is often impractical for a pocket-sized terminal to employ a number of antennas owing to its limited size and cost constraint. Fortunately, another type of transmit diversity, namely the so-called cooperative diversity relying on cooperation among multiple single-antenna-assisted terminals, may be achieved in multi-user wireless systems.

Hence, more realistic user-cooperation mechanisms requiring no CSI were advocated in **Chapter 12**, which was based on differentially encoded transmissions and on non-coherent detection techniques. Hence they circumvent the potentially excessive-complexity channel estimation as well as the high pilot overhead encountered by conventional coherent-detection-aided cooperative systems, especially in mobile environments associated with relatively rapidly fluctuating channel conditions. The ML-MSDD technique of Section 12.2.1 was introduced in support of user cooperation in the context of the multi-layer tree-search-based SD algorithm of Section 12.3.3.4, for rendering the system robust to time-selective propagation environments at an affordable complexity, leading to an MSDSD-aided differential user-cooperation-based system. The characteristics of the uncoded MSDSD-aided cooperative systems using both the DAF and DDF schemes of Sections 12.3.2.1 and 12.3.2.2 are compared and summarized in Table 18.5. Note that during the MSDSD design procedure adopted for differentially encoded cooperative systems in Section 12.3.3, we relied on the assumptions that both the signal and noise received at the BS in the DAF-aided system obey complex-Gaussian distributions and that the source's signal can always be error-freely decoded prior to forwarding it to the BS in the DDF-aided system. This allowed us to simplify significantly the associated MSDSD design problems, while still being able to construct a powerful non-coherent detector which is substantially more robust to the effects of mobile environments than the CDD. Finally, since the SD devised for both the STBC-SP-based non-cooperative MU-MIMO system of Chapter 11 and the differentially encoded user-cooperation-based system of Chapter 12 employs the multi-layer tree search mechanism, the salient features concerning the multi-dimensional search tree are summarized in Table 18.6.

18.1.10 SD-Aided MIMO System Designs

18.1.10.1 Resource-Optimized Hybrid Cooperative System Design

Although it was shown in **Chapter 12** that the maximum attainable spatial diversity gain can usually be achieved by the differentially modulated user cooperative uplink system, the achievable end-to-end BER performance may significantly depend on the specific choice of the cooperative protocol employed and/or on the quality of the relay channel. Therefore, the resource allocation arrangements employed by the cooperative cellular uplink, namely the transmit power allocation and the RS's geometric location, play a vital role in achieving the best possible performance. In order to achieve the best possible BER performance, a flexible resource-optimized adaptive hybrid cooperation-aided system was designed in **Chapter 13**.

The corresponding system design procedure based on the major findings of each section of Chapter 13 is summarized in Table 18.7. More specifically, the associated theoretical performance analysis was carried out for both the DAF- and DDF-aided cooperative systems in Sections 13.2.1 and 13.2.2, respectively. The derived exact end-to-end BER expression of Equation (13.40) for the DAF-aided system was significantly simplified by assuming a high SNR and by using the same technique as in [604], resulting in the tight BER lower bound of Equation (13.48), as characterized in Figure 13.5, which was valid for high SNRs. As for the DDF-aided system, the BER upper bound of

Table 18.5: Characteristics of the DAF- and DDF-aided cooperative systems of Sections 12.3.2.1 and 12.3.2.2.

	DAF	DDF	Remarks
Sensitivity to SR link	Modest	Strong	See Figure 12.14
Potential drawbacks	Noise amplification	Error propagation	See Equation (12.62) for DAF & Figure 12.13 for DDF
MSDSD design assumptions	Gaussian-distributed received signal and noise	Error-free de-coding at RS	Reasonable assumptions significantly simplify the MSDSD design problem
Complexity at relay	Low (amplify)	High (decode and re-encode)	E2E performance improves as more computational efforts put into RS, see Figure 12.15
SR link vs. RD link	Equally important	SR link is more important	Exchangeability of SR and RD links for DAF system is seen in Equation (12.60) and Figure 12.19; SR link's importance for DDF is seen in Figure 12.22
Performance gain achieved by MSDSD	Error floor completely eliminated	Error floor completely eliminated	See Figure 12.18 for DAF, see Figure 12.21 for DDF

Table 18.6: Generic multi-layer tree-search-based SD in two particular applications.

	STBC-SP-based MU-MIMO system (Chapter 11)	Differential user-cooperation-based system (Chapter 12)
Tree node representation	The SP-aided joint ST designed signal transmitted by each user, see Equation (11.52)	The equivalent user-cooperation-based ST signal transmitted by the source and relays, see Equation (12.46)
Tree node dimension	Number of antennas per user (M_u)	Number of cooperating users (U)
Tree node structure	Column vector	Diagonal signal matrix

Equation (13.56) was derived for the single-relay-assisted system as an example based on the worst-case PLR of Equation (13.52) at the RS, which was shown to be capable of closely capturing the dependency of the BER on the SNR, as seen in Figure 13.6.

The above-mentioned power allocation and RS location selection schemes were investigated for both the DAF- and DDF-aided systems in Section 13.3 with the aid of our theoretical BER results of Section 13.2. More particularly, based on the minimum BER criterion, the APC schemes of the DAF- and DDF-aided system were devised in Sections 13.3.1.1 and 13.3.2 respectively, which were capable of achieving significant performance gains by finding the optimum power allocation for a given RS location arrangement, as depicted in Figures 13.9(a) and 13.14(a). On the other hand, based on the observation of Figures 13.9(b) and 13.14(b), we found that the achievable BER is proportional to the distance between the cooperating MS and the optimum RS location. Hence the CUS schemes, which were contrived for both the DAF- and DDF-aided systems in Sections 13.3.1.2 and 13.3.2, respectively, simply activate the RS closest to the optimum location from the available RS candidate pool, resulting in substantial performance improvements, as observed in Figures 13.11 and 13.17. To achieve the globally optimum resource allocation, iterative power versus RS location optimization was also carried out in Sections 13.3.1.2 and 13.3.2, as illustrated in Figures 13.10 and 13.15. Remarkably, apart from having an enhanced BER performance, the complexity imposed by the MSDSD of Chapter 12 can also be significantly reduced by employing the CUS and APC schemes in the context of rapidly fading channels, as observed in Figures 13.12(b) and 13.13(b).

Table 18.7: Resource-optimized differentially modulated hybrid cooperative system design procedure.

Theoretical BER performance analysis (Section 13.2)	1. For DAF systems, the simplified high-SNR-based BER lower bound of Equation (13.48) was found to be tight in Figure 13.5 2. For DDF systems, based on the worst-case PLR of Equation (13.52) at the RS, the BER upper bound of Equation (13.56) closely captured the BER's dependency on the SNR in Figure 13.6
↓ Resource allocation optimization (Section 13.3)	1. Criterion: minimum BER 2. Optimized resources: power allocation and RS location 3. Benefits: performance gain (see Figures 13.9 and 13.14) and detection-complexity reduction (see Figure 13.12(b)), which may be enhanced by iterative power versus RS location optimization schemes (see Figures 13.11 and 13.17)
↓ Comparative studies of resource-optimized DAF and DDF systems (Section 13.4.1)	1. Sensitivity to the SR link quality: DDF system's performance degrades more rapidly for a poor SR link quality (see Figure 13.18) 2. Effect of the packet length: the DAF system's performance is independent of the employed packet length L_f, while its DDF counterpart's performance is sensitive to L_f (see Figure 13.16) 3. Resource allocation: DAF and DDF systems exhibit complementarity (see Figure 13.19 and Tables 13.5 and 13.7)
↓ Resource-optimized hybrid system design (Section 13.4.2)	1. Goal: exploit the complementarity of the DAF and DDF systems in order to design a more flexible resource-optimized hybrid cooperative system 2. Mechanism: cooperative protocol employed by the activated RS is adaptively chosen in the interest of achieving the best BER performance 3. Benefits: improved performance (see Figure 13.21)

Owing to the different levels of sensitivity to the quality of the SR link as seen in Figure 13.18, the optimum resource allocation arrangements corresponding to the two above-mentioned systems may be quite different, as revealed by our comparative studies in Section 13.4.1. Specifically, as indicated by Table 13.5, it is desirable that the activated cooperating MSs are roaming in the vicinity of the source MS for the DDF-aided system, while the cooperating MSs roaming in the neighbourhood of the BS are preferred for its DAF-aided counterpart. Additionally, in comparison with the former system, a larger portion of the total transmit power should be allocated to the source MS in the context of a DAF-aided system. Furthermore, in order to exploit the complementarity of the above-mentioned cooperative systems, a more flexible resource-optimized adaptive hybrid cooperation-aided system was proposed in Section 13.4.2, where the protocol employed by a specific cooperating MS may also be adaptively selected in the interest of achieving the best possible BER performance. Thus, the DAF and DDF cooperative protocols may coexist in the same cooperative network. As an example, the operations of the hybrid cooperative cellular uplink system are summarized as follows:

1. Determine the DAF and DDF areas between the source MS and the BS by calculating the globally optimum RS locations via the proposed iterative power versus RS location optimization scheme.

2. In order to exploit the complementarity of the DAF and DDF schemes, activate an RS in each of the above-defined areas which is situated closest to the globally optimum location.

3. Adaptively calculate the power allocation solution based on the actual locations of the activated RSs.

18.1.10.2   Near-Capacity Cooperative and Non-cooperative System Designs

To achieve a near-capacity performance, in **Chapter 14** we devised a low-complexity near-capacity system design with the aid of near-optimum SDs for both the non-cooperative and cooperative MIMO systems in Section 10.4 and Section 14.4, respectively. The near-capacity design of the former system, which is reduced to an EXIT-curve-matching problem, serves as the fundamental method of achieving the cooperative network's capacity for the latter system, since the joint source–relay mode design

Table 18.8: Near-capacity design for non-cooperative coherent-detection-aided MIMO systems.

	MIMO detector	Unity-rate code	Channel decoder
Three-stage receiver structure	Low-complexity near-capacity centre-shifting LSD of Section 10.2.3.3	Has an IIR, hence efficiently spreads the extrinsic information and improves the iteration gain	IrCCs [580], fixed average code rate, shaped EXIT curve
Key method		EXIT-curve-matching algorithm of [581]	
Design steps	1. Set the effective target throughput		
	2. Find the theoretical minimum SNR supporting the target transmission rate		
	3. In order to achieve a near-capacity performance at low complexity, generate the combined inner URC-SD decoder's EXIT curve at higher-than-necessary SNR		
	4. Design the IrCCs for the average target rate using EXIT-curve matching		

procedure of the single-relay-aided cooperative system can be decoupled into two separate EXIT-curve-matching problems.

For the non-cooperative MIMO system, the near-capacity system design procedure is summarized in Table 18.8 along with its corresponding URC-aided three-stage receiver structure. To be specific, in order to approach the channel capacity of the non-cooperative MIMO system, we demonstrated in Section 10.4.1 that the iterative decoding convergence of this two-stage system may be improved by incorporating a URC having an infinite impulse response which improves the efficiency of extrinsic information exchange, as observed in Figure 10.35(a). More particularly, the URC-aided inner decoder's EXIT curve of Figure 10.35(a) is capable of reaching the $(1, 1)$ point by having a lower starting point, which in turn yields a reduced error floor and a higher SNR threshold, above which decoding convergence to a vanishingly low BER becomes possible. Furthermore, this slightly more complex three-stage system architecture allows us to use a low-complexity SD having a significantly reduced candidate list size N_{cand}. Alternatively, a reduced SNR is required. For example, as depicted in Figure 18.5(a), given a target BER of 10^{-5} and $N_{cand} = 32$ for the SD, the three-stage receiver is capable of achieving a performance gain of 2.5 dB over its two-stage counterpart in a rank-deficient SDMA-OFDM 4-QAM system supporting $U = 8$ co-channel users and employing $N = 4$ receive antennas at the BS, i.e. in an (8×4)-element system. For further enhancing the three-stage concatenated receiver, the proposed iterative centre-shifting SD scheme of Section 10.2.3.3 and IrCCs of Section 10.4.3 are intrinsically amalgamated, leading to an additional performance gain of 2 dB, as also observed in Figure 18.5(a).

Figure 18.5(b) depicts the computational complexity – which is quantified in terms of the number of PED evaluations corresponding to the term ϕ of Equation (9.25) – imposed by the SD versus the E_b/N_0 value for the above-mentioned receivers. The number of PED evaluations carried out per channel use by the system dispensing with the centre-shifting scheme remains as high as 13 652, regardless of the SNR and the number of iterations, since we assume a sufficiently large buffer size to store the resultant candidate list in order to eliminate the need for list regeneration. On the other hand, in the presence of the centre-shifting scheme, the candidate list has to be regenerated at each iteration; nonetheless, the total complexity imposed by the centre-shifting-based SD of the two-stage receiver is substantially reduced, as seen in Figure 18.5(b). We can also observe from Figure 18.5(b) that the centre-shifting K-best SD employed by the IrCC-aided three-stage system using the near-capacity design of Section 10.4.1 imposes a computational complexity, which is even below that of its centre-shifting-aided two-stage counterpart, while achieving a performance gain of 2 dB at the target BER of 10^{-5}, as seen in Figure 18.5(a). Hence, the significant complexity reduction facilitated by the proposed SD scheme in the context of the three-stage receiver outweighs the relatively small additional complexity cost imposed by the URC, which only employs a two-state trellis, leading to an overall reduced complexity. Furthermore, in addition to the complexity reduction achieved by the proposed scheme, another benefit is the attainable memory reduction, since there is no need to store the resultant

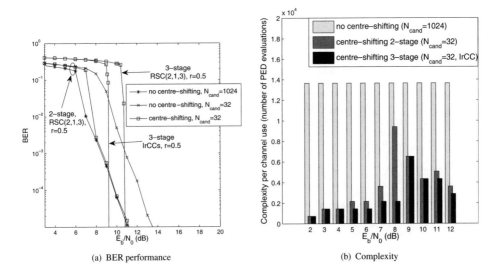

(a) BER performance (b) Complexity

Figure 18.5: Near-capacity design for the coherent-detection-aided (8×4)-element SDMA system. All the system parameters were given in Table 10.1.

candidate list for the forthcoming iterations. As a result, the memory size required can be substantially reduced by having significantly reduced values of N_{cand} and K.

Prior to outlining the near-capacity design principles for the DDF-aided user-cooperation-based system of Section 14.4.3, the corresponding DCMC capacity was quantified in Section 14.4.1 in comparison with that of its classic direct-transmission-based counterpart in order to answer the grave fundamental question of whether it is worth introducing cooperative mechanisms into the development of wireless networks, such as the cellular voice and data networks. This is because, when a cooperative wireless communication system is designed to approach the maximum achievable spectral efficiency by taking the cooperation-induced multiplexing loss into account, it is not obvious whether or not the repetition-based relay-aided system becomes superior to its direct-transmission-based counterpart, especially when advanced channel coding techniques are employed. It was observed in Figure 14.12 of Section 14.4.1 that when the overall equivalent SNR is relatively low, the single-relay-assisted cooperative system exhibits a significantly higher capacity than its direct-transmission-based counterpart in typical urban cellular radio scenarios, e.g. when having a path-loss exponent of $v = 3$. However, the achievable capacity gain may be substantially reduced if we encounter a free-space propagation scenario [607], i.e. $v = 2$, since the reduced-path-loss-related power gain achieved is insufficiently high to compensate for the significant multiplexing loss inherent in the single-relay-aided half-duplex cooperative system. Moreover, as the overall equivalent SNR increases to a relatively high value, there is no benefit in invoking a single-relay-aided cooperative system, since its capacity becomes lower than that of the conventional point-to-point system.

Then, based on the investigation of the single-relay-assisted DDF-based cooperative system's DCMC capacity detailed in Section 14.4.1, we proposed a practical framework for designing a cooperative system which is capable of performing close to the network's corresponding non-coherent DCMC capacity. Specifically, based on our low-complexity near-capacity design criterion, a novel Ir-DHDC coding scheme was contrived, which was depicted in Figure 14.13 in Section 14.4.2 for the DDF-aided cooperative system employing the low-complexity SISO iterative MSDSD scheme of Section 14.3. The SISO MSDSD was shown to achieve capacity for direct transmission over time-selective block-fading channels, as shown in Figure 14.7, provided that the observation window size

Table 18.9: Near-capacity design for non-coherent-detection-aided cooperative MIMO systems.

Transceiver	Ir-DHCD coding (source+relay, Figure 14.13)	Source's transmitter: URC-aided three-stage transmitter employing IrCC$_s$ with code rate R_s, in order to achieve near-capacity SR transmission Relay's receiver: URC-aided three-stage receiver employing the MSDSD Relay's transmitter: two-stage receiver employing IrCC$_r$ having a code rate of R_r; an interleaver is added before the IrCC$_r$ to facilitate distributed turbo decoding at the BS
Structure	Ir-DHCD decoding (destination, Figure 14.14)	Destination's receiver: 1. Constituted of two parts: first part is a three-stage receiver identical to relay's receiver iteratively decoding the signal received in broadcast phase second part is a two-stage receiver corresponding to relay's transmitter, iteratively decoding the signal received in relay phase 2. Extrinsic information exchanges between the first and second parts
Design steps		1. Choose a specific network's effective throughput, based on which R_s and R_r are calculated 2. Find the theoretical minimum overall equivalent SNR supporting the target overall transmission rate according to the imperfect-SR-link- based network's DCMC capacity of Figure 14.12 3. Carry out the near-capacity design for the SR link's transmission following the design steps for the P2P system of Table 18.8 4. Determine $I^d_{inner\,1}$ and $I^d_{inner\,2}$ of Figure 14.14 based on the low-complexity near-capacity criterion, as shown in Figure 14.19 5. Similarly, determine $I^d_{inner\,3}$ of Figure 14.14 as shown in Figure 14.20 6. Design IrCC$_r$ for the average rate R_r, so that a narrow but open EXIT tunnel emerges between the EXIT curves associated with the first and second parts of the BS's receiver

N_{wind} employed was equal to the fading block length T_b. On the other hand, in order to enhance the coding gain achieved by the repetition code constituted by the relay-aided system, while maintaining a high cooperative diversity gain, the classic turbo coding mechanism was introduced into the DF-aided cooperative system by interleaving the RS's estimated source data prior to re-encoding them, as seen in Figure 14.13 of Section 14.4.2, resulting in the so-called distributed turbo coding philosophy. Furthermore, in order to achieve near-capacity transmissions between the source and relay, a URC-aided three-stage serially concatenated transceiver employing the IrCCs of Section 10.4.3 may be employed in the single-relay-aided DDF cooperative system in Figure 14.13 together with the corresponding three-stage receiver employed by the RS. According to the principles of the distributed turbo decoding mechanism proposed in [620], at the destination BS the novel iterative receiver of Figure 14.14 is used for decoding the Ir-DHCD coded signal received from the source and relay nodes. More explicitly, we have to decode iteratively the signal received during the broadcast phase and the relay phase, respectively, followed by the classic extrinsic information exchange between the two. The near-capacity single-relay-assisted cooperative system design procedure of Section 14.4.3 is summarized in Table 18.9 along with the proposed Ir-DHCD coding scheme of Section 14.4.2.

It was clearly shown in Figure 14.23 that, on using the near-capacity system design of Section 14.4.3, the proposed Ir-DHCD coding scheme becomes capable of performing within about 2 dB and 1.8 dB from the corresponding single-relay-aided DDF cooperative system's DCMC capacity for the free-space environment associated with $v = 2$ and for the typical urban scenario associated with $v = 3$, respectively. Furthermore, an SNR gain of 0.65 dB can be achieved in a free-space scenario by the direct-transmission-based system over its single-relay-aided cooperative counterpart, given a bandwidth efficiency of 0.47 bps/Hz, as shown in Figure 14.23(a). However, in a typical urban area cellular radio environment, the single-relay-aided cooperative system becomes capable of significantly outperforming the direct-transmission-based system, requiring an overall transmit power which is about 2.5 dB lower than that necessitated by the latter in order to achieve an infinitesimally low BER, while maintaining a bandwidth efficiency of 0.43 bps/Hz, as depicted in Figure 14.23(b). Therefore, in line with our predictions made in Section 14.4.1.2, it was observed from Figures 14.23 and 14.24 that for a given target bandwidth efficiency, the single-relay-aided cooperative system does not necessarily guarantee a performance superior to that of the conventional direct-transmission-based system.

18.1.11 Multi-stream Detection in SDM-OFDM Systems

In **Chapter 15**, we presented an overview of several popular SDM detection methods available in the literature. Specifically, in Section 15.2.1 we demonstrated that the linear increase in capacity, which was predicted by the information theoretical analysis of [266], may indeed be achieved by employing a relatively low-complexity linear SDM detection method, such as the MMSE SDM technique [330]. Secondly, in Section 15.3.1 we showed that a substantially better performance can be achieved by employing the higher-complexity nonlinear Maximum Likelihood (ML) SDM detector [311,331,332], which constitutes the optimal detection method from a probabilistic sequence-estimation point of view. To elaborate a little further, the ML SDM detector is capable of attaining transmit diversity in *fully loaded* systems, where the number of transmit and receive antennas is equal. Moreover, as opposed to the linear detection schemes considered, the ML SDM detector is capable of operating in the *rank-deficient* system configuration, when the number of transmit antennas exceeds that of receive antennas. Unfortunately, however, the excessive computational complexity associated with the exhaustive search employed by the ML detection method renders it inapplicable to practical implementation in systems having a high number of transmit antennas. Subsequently, in Sections 15.3.2 and 15.3.3 we explored a range of advanced nonlinear SDM detection methods, namely both an SIC and a GA-aided MMSE detector, respectively, where the latter may constitute an attractive compromise between the low complexity of the linear SDM detector and the high performance of the ML SDM detection schemes. Indeed, we demonstrated in Section 15.3.3 that both the SDM detection method based on the SIC and the GA-MMSE detector [329] are capable of satisfying these requirements.

18.1.12 Iterative Channel Estimation and Multi-stream Detection in SDM-OFDM Systems

In **Chapter 16** we focused our attention on a family of potent Reduced Search Algorithm (RSA) aided space–time processing methods, the members of which exhibit a particularly advantageous trade-off between the achievable performance and the associated computational complexity, namely the family of SD-aided SDM detection methods. Consequently, a set of novel OHRSA-aided SDM detection methods was outlined in Section 16.1. Specifically, in Section 16.1.1 we derived the OHRSA-aided ML SDM detector, which benefits from the optimal performance of the ML SDM detector [265], while exhibiting a relatively low computational complexity which is only slightly higher than that required by the low-complexity MMSE SDM detector [265]. To elaborate a little further, in Section 16.1.2 we derived a bit-based OHRSA-aided ML SDM detector which allows us to apply the OHRSA method of Section 16.1 in high-throughput systems which employ multi-level modulation schemes, such as M-QAM [265].

In Section 16.1.3 we deduced the OHRSA-aided Max-Log-MAP SDM detector which allows for efficient evaluation of the soft-bit information and therefore results in highly efficient turbo decoding. Unfortunately, however, in comparison with the OHRSA-aided ML SDM detector of Section 16.1.2 the OHRSA-aided Max-Log-MAP SDM detector of Section 16.1.3 still exhibits a substantially higher complexity. Consequently, in Section 16.1.5 we derive an approximate Max-Log-MAP method, namely the SOPHIE SDM detector. This detector combines the advantages of both the OHRSA-aided ML and OHRSA-aided Log-MAP SDM detectors of Sections 16.1.2 and 16.1.3, respectively. Specifically, it exhibits a similar performance to that of the optimal Max-Log-MAP detector, while imposing a modest complexity which is only slightly higher than that required by the low-complexity MMSE SDM detector [265]. The computational complexity as well as the achievable performance of the SOPHIE SDM detector of Section 16.1.5 were analysed and quantified in Sections 16.1.5.1 and 16.1.5.2, respectively.

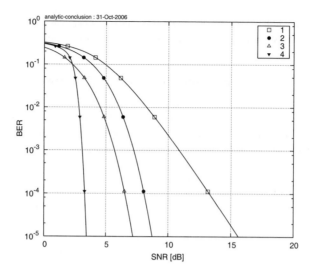

Figure 18.6: BER versus SNR performance of an iterative multi-antenna, multi-carrier system in dispersive Rayleigh fading channel. We consider the scenarios of: 1, low diversity rank; 2, high diversity rank and suboptimum SDM detector; 3, high diversity rank and optimum SDM detector; and 4, high diversity rank and iterative optimum SDM detector and decoder.

To elaborate a little further, based on Figure 16.11 we reported achieving a BER of 10^{-4} at SNRs of $\gamma = 4.2$, 9.2 and 14.5 in high-throughput 8×8 rate $\frac{1}{2}$ turbo-coded $M = 4$-, 16- and 64-QAM systems communicating over a dispersive Rayleigh fading channel. Additionally, recall from Figure 16.10 that we reported achieving a BER of 10^{-4} at SNRs of $\gamma = 9.5$, 16.3 and 22.8 in high-throughput rank-deficient 4×4, 6×4 and 8×4 rate $\frac{1}{2}$ turbo-coded 16-QAM systems, respectively.

18.1.13   Approximate Log-MAP SDM-OFDM Multi-stream Detection

In **Chapter 17** we derived an iterative, so-called *turbo* Multi-Antenna, Multi-Carrier (MAMC) receiver architecture. Following the philosophy of turbo processing [314], our turbo SDM-OFDM receiver of Figure 17.1 comprises a succession of detection modules which iteratively exchange soft-bit-related information and thus facilitate a substantial improvement in the overall system performance.

More specifically, our turbo SDM-OFDM receiver comprises three major components, namely the soft-feedback decision-directed channel estimator discussed in detail in Section 7.8, followed by the soft-input, soft-output OHRSA Log-MAP SDM detector derived in Section 16.1.3 and a soft-input, soft-output serially concatenated turbo code [345]. Consequently, in Figures 17.3–17.14 of Chapter 17 we analysed the achievable performance of each individual constituent component of our turbo receiver, as well as the achievable performance of the entire amalgamated iterative system. We aimed at identifying the optimum system configuration while considering various design trade-offs, such as the achievable BER performance, the attainable data rate and the associated computational complexity.

In Section 17.4.2.4 we demonstrated that our turbo SDM-OFDM system employing the MIMO-DDCE scheme of Section 7.8 as well as the OHRSA Max-Log-MAP SDM detector of Section 16.1.3 remains effective in channel conditions associated with high mobile speeds of up to 130 km/h, which corresponds to the OFDM-symbol-normalized Doppler frequency of 0.006. Additionally, in Figure 17.13 we reported a virtually error-free performance for a rate $1/2$ turbo-coded 8×8 QPSK-OFDM system, exhibiting an effective throughput of 8 MHz $\cdot$ 8 bps/Hz $= 64$ Mbps and having a pilot overhead of only 10% at an SNR of 7.5 dB and a normalized Doppler frequency of 0.003, which corresponds to a mobile terminal speed of about 65 km/h.

In conclusion, we offer the following important observations. The potential performance gain achievable by an iterative multi-antenna, multi-carrier system may be dissected into several major regions, where we may identify the *diversity gain* region, the *detection gain* region and the *iterative gain* region. Consider the BER versus SNR performance curves depicted in Figure 18.6:

- Firstly, the diversity gain region may be associated with the interval spanning the SNR values of Figure 18.6, which lie between performance curves 1 and 2 corresponding to the scenarios of low and high diversity ranks,[1] respectively. Correspondingly, the achievable diversity gain may be realized by attaining a sufficient diversity rank contributed by the combination of the channel and waveform parameters. This phenomenon is exemplified, for instance, by Figure 17.8 of Section 17.3.

- The detection gain region may be identified as the region of the SNR values located between performance curves 2 and 3 of Figure 18.6, which correspond to the systems employing for example a linear MMSE detector and a near-optimum Max-Log-MAP detector, respectively. The achievable detection gain may be realized by means of employing an efficient MIMO detection method reminiscent of the OHRSA method derived in Chapter 16. This phenomenon is exemplified, for instance, by Figure 15.11 of Section 15.5.

- Finally, the iterative gain region corresponds to the interval of the SNR values located between performance curves 3 and 4 of Figure 18.6, which correspond to the systems employing a single as well as eight detection and decoding iterations. Correspondingly, the attainable iterative gain may be realized by employing iterative detection and decoding, which invokes the iterative exchange of soft-bit-related information and thus facilitates the efficient exploitation of the diversity rank available. This phenomenon is exemplified, for instance, by Figures 17.3 and 17.6 of Sections 17.2 and 17.3, respectively.

18.2 Suggestions for Future Research

In this section, based on the work reported in this monograph, a few suggestions for future research are summarized.

18.2.1 Optimization of the GA MUD Configuration

The GA-based MUDs have found applications in diverse SDMA-OFDM systems investigated in this book. It is worth pointing out that the proposed GAs may be further improved in various ways. On the one hand, the operators and parameters are all tunable, implying that the configurations we have invoked may not necessarily be optimum. For example, the value of the mutation probability p_m can be adapted according to:

- The number of users. A higher value of p_m can be used when more users' signals have to be detected, which may increase the chances of finding the optimum solution.

- The GA's generation index. During the early generations, p_m can be set to a high value to assist the GA in identifying the more promising search branches. As the search process progresses, p_m can be reduced to achieve a better local fine resolution of the search region.

The maximum mutation step size λ_{max} of the GA-JCEMUD of Chapter 8 can also be adjusted adaptively in accordance with the Doppler frequency, as discussed in Section 8.3.3.3.2. Other GA parameter control methods, e.g. those of [466], may also be employed to optimize the GA's configurations.

On the other hand, in the context of generating the GA's soft output, in this book we only exploited the individuals of the final generation for calculating the soft information. However, a

---

[1]Quantitatively speaking, the low diversity rank channel is a channel where the distribution of the total channel energy is reminiscent of a χ^2_{2D} distribution, with $D \approx 1$. Correspondingly, the high diversity channels are channels where we have $D \gg 1$.

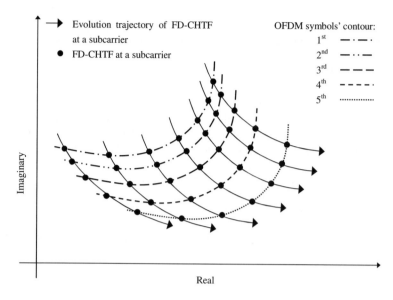

Figure 18.7: Illustration of the time-domain correlation of FD-CHTFs associated with $n = 1, \ldots, 5$ consecutive OFDM symbols at a specific user–receiver channel link. Similar to Figure 8.11, each black dot represents the FD-CHTF at a specific subcarrier, while each of the five different dashed or dotted curves denotes the virtual contour of the corresponding OFDM symbol, respectively. The solid curves with arrows show the track and direction of the FD-CHTF variation at each subcarrier in the context of the time domain.

performance improvement may be achieved by introducing an 'individual log', which stores the OSs of all meritorious individuals throughout all generations and thus assists in improving the reliability of the GA's soft output.

Additionally, the GA individual's symbol chromosome, which consists of the multiple users' hard-decoded symbol estimates, may also be represented by the soft-bit estimates, enabling the GA to benefit from the soft information provided by the channel decoders during the external iterative processing. This is expected to improve the performance of the iterative GAs, such as the IGA MUD of Section 5.3.2 and the GA-JCEMUD of Chapter 8.

18.2.2 Enhanced FD-CHTF Estimation

While the proposed GA-JCEMUD scheme of Chapter 8 uses the previous FD-CHTF estimates as a reference for the current FD-CHTF estimates, its mutation operator, namely the step mutation of Section 8.3.3.3.2, invokes the mutation process based on a uniform probability for all directions around the original FD-CHTF to be mutated, as implied by Equation (8.26). In other words, it does not exploit the 'directional' information of the time-domain correlation between the FD-CHTFs associated with the subcarriers of consecutive OFDM symbols. Here we recall Figure 8.11, which compared the real/imaginary values' evolution of the perfect as well as the GA-estimated FD-CHTFs. In the same spirit, an example of the above-mentioned directional correlation is illustrated in Figure 18.7, where the virtual contours of $n = 1, \ldots, 5$ OFDM symbols observed in five consecutive OFDM symbol durations at a specific user–receiver channel link are portrayed, which are constituted by the different dashed curves. Similar to Figure 8.11, each black dot seen in Figure 18.7 represents the FD-CHTF at a specific subcarrier. The solid curves with arrows show the direction of evolution for the FD-CHTFs at each subcarrier in the context of the time domain.

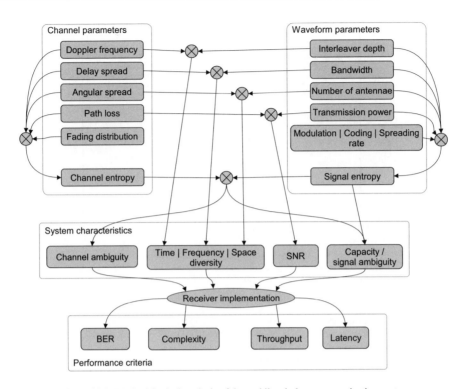

Figure 18.8: Methodological analysis of the mobile wireless communication system.

Therefore, by borrowing the idea of the BQM approach proposed in Section 5.3.1.2, it would be beneficial to design a mutation operator which is capable of tracking the direction of the FD-CHTF variation at each subcarrier in consecutive OFDM symbol durations. Similarly, the directional frequency-domain correlation between consecutive subcarriers in the same OFDM symbol can also be exploited. The mutation operator combining both the time-domain and the frequency-domain direction-tracking strategies may be expected to improve the performance of the GA-JCEMUD scheme.

Moreover, the joint channel estimation and symbol detection approach can be further enhanced by introducing a Soft-Input, Soft-Output (SInSOu) mechanism. More specifically, not only the GA-optimized FD-CHTF estimates but also the GA-optimized symbol estimates can be forwarded to the OHRSA MUD of Section 8.3.2 to assist in the initial symbol detection invoked in the next iteration. In this way the OHRSA MUD would be capable of benefitting from the GA in terms of both perspectives, thus improving the reliability of the initial symbol estimates, which in turn may be expected to result in a better overall system performance.

18.2.3   Radial-Basis-Function-Assisted OFDM

Generally speaking, GAs belong to the family of Evolutionary Algorithms (EAs) [643, 644], which invokes the principles of natural evolution. With the aid of the advances in Artificial Intelligence (AI), a range of other problem-solving methods have also emerged. One of these techniques is constituted by the family of Neural Networks (NNs) [645, 646], which is based on the models that mimic the operation of how biological neurons are connected in the human brain. Furthermore, EAs/GAs have been applied in constructing and training NNs in numerous research fields [538, 647, 648]. In the context of wireless communications, NNs have also found employment in various fields, e.g. in channel equalization [569, 649, 650]. Moreover, NNs are also applicable to the field of multi-user detection.

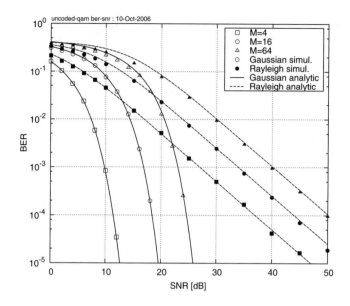

Figure 18.9: BER versus SNR performance of uncoded M-QAM in Gaussian and Rayleigh channels. The markers characterize the simulated results corresponding to $M = 4$, 16 and 64. The solid and dashed lines show the corresponding calculated BER versus SNR for Gaussian and Rayleigh channels, respectively, obtained using the semi-analytical model.

For example, the so-called Radial Basis Function (RBF) [569, 651, 652] based NN has been proposed for multi-user detection in CDMA-type systems [653–656]. However, hardly any research has been conducted in the context of RBF-assisted multi-user OFDM systems [657–660]. Explicitly, exploiting RBF with or without the aid of EAs/GAs for employment in OFDM, MIMO-OFDM and SDMA-OFDM systems, especially for highly complicated nonlinear applications such as for example joint detection schemes, constitutes a wide research area. Alternatively, Minimum Bit Error Rate (MBER) types of receivers [202, 661] may also be employed, which are capable of supporting up to twice the number of users in comparison with the classic MMSE MUD.

18.2.4 Non-coherent Multiple-Symbol Detection in Cooperative OFDM Systems

The novel aspects of the cooperative OFDM systems investigated may be summarized as follows:

- **A Priori LLR-Threshold-Assisted MSDSD in Channel-Coded Cooperative Systems:** In Chapter 10 the ALT scheme was proposed for the coherent SD-aided non-cooperative MIMO-OFDM system in order to achieve the required complexity reduction. Similar ideas can be employed in the non-coherent SD-aided cooperative system for further reducing the complexity imposed.

- **DSTBC/DSFBC-SP-Aided Cooperative Systems:** DSTBC/DSFBC schemes can be employed in the MSDSD-aided cooperative OFDM system in order to improve the attainable transmission efficiency further. Moreover, the sphere-packing scheme of Chapter 11 can also be employed jointly to design the ST signals transmitted from distributed multiple antennas for further improving the achievable performance.

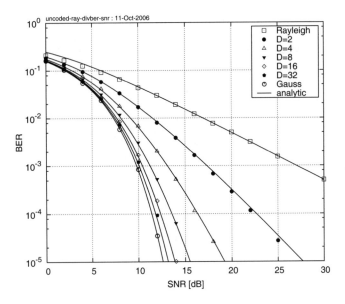

Figure 18.10: BER versus SNR performance of an uncoded QPSK system communicating over a χ_D^2-distributed flat-fading channel. The markers portray the simulated results associated with the diversity ranks $D = 1, 2, \ldots, 32$. The solid lines show the corresponding calculated BER versus SNR curves obtained using the semi-analytical model.

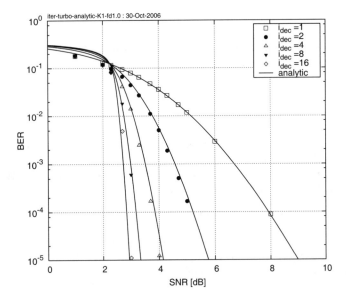

Figure 18.11: BER versus SNR performance of a turbo code in uncorrelated flat Rayleigh channel. The markers characterize the simulated results while invoking 1 to 16 iterations of the turbo decoder. The lines show the corresponding calculated SNR versus BER obtained using the semi-analytical model.

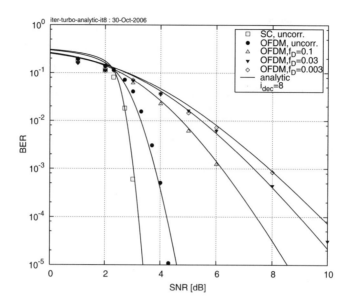

Figure 18.12: BER versus SNR performance exhibited by the $K = 128$-subcarrier single-antenna QPSK-OFDM system employing a rate $\frac{1}{2}$ parallel-concatenated turbo code in a correlated Rayleigh fading having the OFDM-symbol-normalized Doppler frequencies of $f_D = 0.1$, 0.03 and 0.003. The CIR was the seven-path COST207 BU model [347]. All additional system parameters are summarized in Table 1.11. The markers characterize the simulated results while invoking $i_{\text{dec}} = 8$ turbo decoder iterations. The solid lines show the corresponding calculated SNR versus BER obtained using the semi-analytical model.

- **Resource Allocation for Channel-Coded Near-Capacity Differential Cooperative Systems:** Given the importance of resource allocation in the uncoded differential cooperative system of Chapter 13, it is worthwhile investigating cooperative resource allocation schemes designed for the channel-coded differentially encoded cooperative systems.

- **Successive Relaying-Aided Differential Cooperative Systems:** As revealed in Chapter 14, the single-relay-assisted DDF cooperative system does not guarantee a superior performance in comparison with that achieved by the conventional direct-transmission-based system, owing to the significant multiplexing loss inherent in the half-duplex relaying mechanism. In order to recover this multiplexing loss, a successive relaying cooperative protocol may be introduced in the differentially modulated cooperative system.

- **Interference-Limited Multi-user Differential Cooperative Systems:** In this treatise we considered a single-source differentially encoded cooperative system in order to investigate the achievable diversity gains. However, it is worth investigating how to improve the fundamental trade-offs between the achievable multi-path diversity gain and multiplexing gain in the context of interference-limited multi-user scenarios relying on half-duplex relay networks.

18.2.5   Semi-Analytical Wireless System Model

The family of state-of-the-art communication systems invokes a conglomerate of complex mathematical algorithms. The analytical expressions describing the behaviour of these algorithms are often hard to derive. Correspondingly, the performance of complex systems is typically evaluated using extensive software simulations. Unfortunately, however, the multiplicity of effects imposed by the different

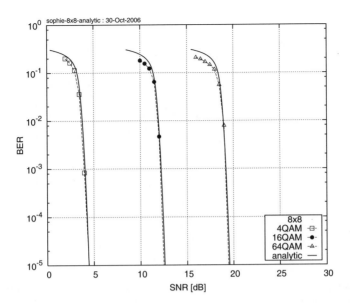

Figure 18.13: BER versus SNR performance exhibited by a rate $\frac{1}{2}$ turbo-coded 8×8 OFDM system employing 4-, 16- and 64-QAM and communicating over a dispersive Rayleigh fading channel. The markers portray the simulated results, while the solid lines show the corresponding results obtained using the semi-analytical model.

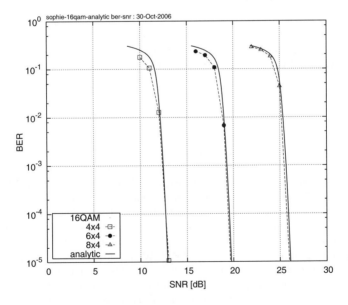

Figure 18.14: BER versus SNR performance exhibited by a turbo-coded 16-QAM-SDM-OFDM system using $m_t = 4, 6$ and 8 transmit antennas as well as 4 receive antennas. The markers characterize the simulated results. The solid lines show the corresponding calculated SNR versus BER obtained using the semi-analytical model.

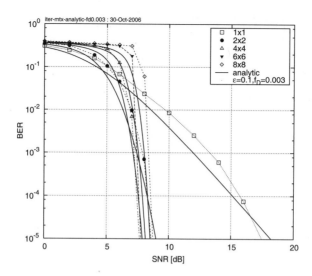

Figure 18.15: BER versus SNR performance exhibited by a rate $\frac{1}{2}$ turbo-coded QPSK-MIMO-OFDM system employing numbers $n_r = m_t = 1, 2, 4, 6$ and 8 of transmit and receive antennas and using the PAST-MIMO-DDCE scheme of Section 7.8, while communicating over a dispersive Rayleigh fading channel. The markers portray the simulated results, while the solid lines show the corresponding results obtained using the semi-analytical model.

phenomena in the complex systems considered tends to obscure the important trends and trade-offs which have to be considered in the process of system design and optimization.

Consequently, we propose a semi-analytic methodology which facilitates the prediction of the performance achievable by a system characterized by a specific ensemble of system and channel parameters.

The proposed semi-analytical technique attempts to dissect the complex problem of system performance analysis into a set of factors originating from different aspects of both the channel and the waveform characteristics, thus exposing the various trends and trade-offs inherent in the design of an efficient wireless mobile smart-antenna-aided multi-carrier communication system.

Let us consider the system analysis methodology characterized in the stylized illustration of Figure 18.8, where we identify two sets of parameters which characterize our system. Firstly, on the left of the figure we have a set of *channel parameters*, which comprise the Doppler frequency f_D, the RMS delay spread τ_{rms}, the angular spread σ_a^2 as well as the AWGN variance σ_w^2. Additionally, we have to consider the statistical distribution of the CIR taps-related fading coefficients. Secondly, for each channel-related parameter, we have the corresponding *waveform parameter*, as seen on the right of Figure 18.8. That is, we have the bit interleaver depth T, the signal bandwidth B, the numbers m_t and n_r of transmit and receive antennas as well as the SNR γ. Additionally, we have the statistical distribution of the energy associated with the transmitted symbols, which is determined by the particular coding, spreading and modulation scheme. Some examples of the possible symbol–power distributions include the constant power in the case of a PSK modulation, the quantized multi-level uniform distribution in the case of M-QAM as well as the near-Rayleigh power distribution in the cases of CDMA and OFDM.

Consequently, our aim is to derive a set of semi-analytical expressions which would describe the interdependencies between the aforementioned system parameters and a set of criteria characterizing the performance of the mobile wireless communication system considered. Specifically, we choose four major performance criteria, which form the performance metric depicted in Figure 18.8, namely the *BER*, *complexity*, *throughput* and *latency*.

We have completed a feasibility study and our preliminary results suggest that a semi-analytical model may be devised for characterizing the various phenomena, which is capable of accounting for the majority of the effects featuring in Figures 18.9–18.14, which determine the performance of a complex mobile wireless communication system. Some examples of these aspects, which may be taken into account in a corresponding model, include:

- Modulation scheme, e.g. 4-, 16-, 64-QAM (Figures 18.9 and 18.12).

- Coding scheme, e.g. block, convolutional, turbo code with a given number of decoding iterations (Figure 18.11).

- MIMO system dimensions, i.e. number of transmit and receive antennas (Figure 18.12).

- Multi-user environment, i.e. number of coherent and non-coherent users (Figures 18.12 and 18.13).

- Channel correlation properties, i.e. Doppler frequency, delay spread (Figure 18.10).

- MIMO detection complexity (Figures 18.13 and 18.14).

- Imperfect channel estimation (Figure 18.15).

Appendix A

Appendix to Chapter 5

A.1 A Brief Introduction to Genetic Algorithms

The GAs [328,453–456] were first introduced by Holland [453] during the 1960s. Since then, a growing interest in GAs has resulted in a rapid development in this area [328, 662, 663], since GAs have been shown to perform well in numerous robust global search and optimization problems, which may not be conveniently be solved by using traditional search methods.

In this section, we will briefly introduce the GAs in the context of multi-user SDMA-OFDM systems, in order to offer a better understanding of our proposed GA-based MUD of Chapter 5. Figure A.1 shows a flowchart of the GA MUD employed in our multi-user SDMA-OFDM system, while the actions of the GA-based search procedure during one generation are detailed in Figure A.2. Both of these figures will be referred to during our further discourse in this section.

Population Initialization. At the beginning of the GA-based search, an initial *population* consisting of X so-called *individuals* is created, each representing a possible solution of the optimization problem considered, either on a random basis or with the aid of a priori knowledge concerning the optimum solution. In the context of the SDMA-OFDM MUD, an individual is represented by a symbol vector containing L complex symbols, each of which belongs to one of the L users at the specific subcarrier considered. More specifically, the xth individual of the yth *generation* is expressed as

$$\tilde{\mathbf{s}}_{(y,x)} = [\tilde{s}_{(y,x)}^{(1)}, \tilde{s}_{(y,x)}^{(2)}, \dots, \tilde{s}_{(y,x)}^{(L)}], \quad x = 1, \dots, X, \ y = 1, \dots, Y, \tag{A.1}$$

where $\tilde{s}_{(y,x)}^{(l)} \in \mathcal{M}_c$ denotes the lth ($l = 1, \dots, L$) *gene* of the xth individual, while $\mathcal{M}_c$ is the set containing the 2^m legitimate constellation symbols. For example, if the 4-QAM modem constellation given in Figure A.3 is used, each gene $\tilde{s}_{(y,x)}^{(l)}$ of an individual illustrated in Figure A.4 can be represented by the indices of the 4-QAM constellation symbols, namely $\tilde{s}_{(y,x)}^{(l)} \doteq \{1, \dots, 4\}$. For example, given $L = 6$ users, each employing 4-QAM, the composite multi-user signal of each OFDM subcarrier may assume $4^6 = 4096$ possible symbol combinations. In this case the ML-aided MUD would have to find the most likely combination by invoking a full search of all the 4096 metric evaluations. By contrast, the GA-aided MUD typically finds this optimum solution by searching only a fraction of this search space. The population consisting of X individuals then forms the starting point of the optimization process, which is referred to as the $y = 1$st generation, as depicted in Figure A.2.

Fitness Value Evaluation. The GA's task is to find an individual which is considered optimum or near optimum in terms of minimizing the joint metric of Equation (5.6), which is based on the

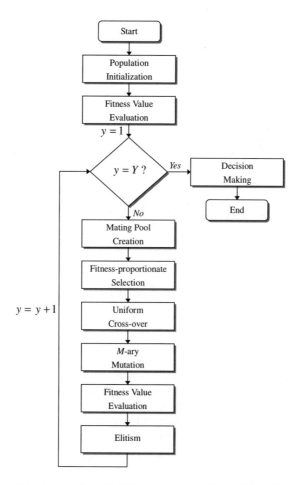

Figure A.1: The structure of the GA MUD used in the multi-user SDMA-OFDM system.

Objective Function (OF) given in Equation (5.3) in the context of a P-antenna SDMA-OFDM system. More explicitly, for each individual, an associated *Objective Score (OS)* can be derived by evaluating Equation (5.6). The OS is then converted to a corresponding *fitness* value, which indicates the fitness of the specific individual. This fitness calculation is carried out in the 'Fitness Value Evaluation' block of Figure A.1.

Mating Pool Creation. Based on the fitness values, T individuals that have the highest fitness values may be selected for creating the so-called *mating pool*, as shown in Figure A.2. However, for the SDMA-OFDM system, a better strategy for creating the mating pool is to follow the principle of *Pareto-Optimality* [328]. This strategy favours the so-called *non-dominated* individuals and ignores the so-called *dominated* individuals [38]. More specifically, the uth L-symbol individual is considered to be dominated by the vth individual, if we have [465]

$$\forall i \in \{1, \ldots, P\} : \Omega_i(\tilde{\mathbf{s}}_{(y,v)}) \leq \Omega_i(\tilde{\mathbf{s}}_{(y,u)})$$
$$\wedge \; \exists j \in \{1, \ldots, P\} : \Omega_j(\tilde{\mathbf{s}}_{(y,v)}) < \Omega_j(\tilde{\mathbf{s}}_{(y,u)}), \tag{A.2}$$

where $\Omega_p(\cdot)$ is the OF associated with the pth receiver antenna element, which is defined by Equation (5.3). If an individual is not dominated in the sense of Equation (A.2) by any other individuals

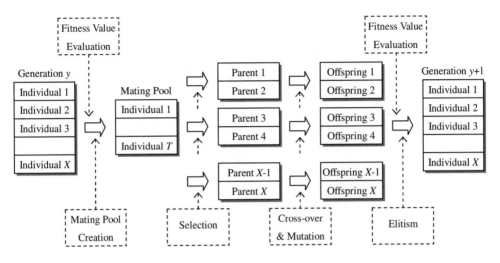

Figure A.2: The GA-based search procedure during one generation.

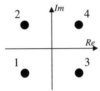

Figure A.3: The complex constellation symbols of a 4-QAM modem, each of which corresponds to a gene of the L-gene GA individual.

$\tilde{S}^{(1)}_{(y,x)}$	$\tilde{S}^{(2)}_{(y,x)}$		$\tilde{S}^{(L)}_{(y,x)}$

Figure A.4: An individual that contains L genes, each of which is one of the constellation symbols shown in Figure A.3.

in the population, then it is considered to be non-dominated. All the non-dominated individuals are then selected and placed in the mating pool, which will have a size of $2 < T \leq X$ [38].

Selection. In order to evolve the population throughout the consecutive generations, the individuals in the mating pool are then selected as *parents* for producing *offspring*. The selection process is based on the so-called *fitness-proportionate* algorithm employed by the 'Fitness-proportionate Selection' block of Figure A.1, which is widely used in the literature [664]. According to fitness-proportionate selection, each of the T individuals in the mating pool is first assigned a selection probability proportionate to its fitness value. More precisely, the individuals having higher fitness values will be assigned higher selection probabilities, based on which $X/2$ pairs of parents are selected, as illustrated in Figure A.2. Moreover, during the selection process the so-called *incest prevention* [665] technique can be invoked, which requires that the two individuals selected to form a pair of parents are different. This can effectively prevent the GA from premature convergence.

Cross-over. For each pair of the $X/2$ selected parents, a genetic operation referred to as *cross-over* [454] is invoked, as shown in Figure A.2. The cross-over operation is a process during which

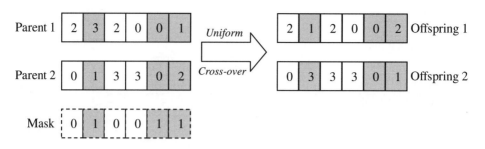

Figure A.5: An example of the uniform cross-over operation.

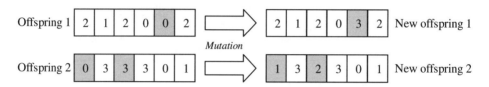

Figure A.6: An example of the mutation operation.

some of the genes of a parent are exchanged with those of the other parent, thus creating two offspring. In our proposed GA MUD, the well-known *uniform cross-over* [666] scheme corresponding to the 'Uniform Cross-over' block of Figure A.1 is employed, as illustrated in Figure A.5. Suppose we have $L = 6$ users; hence each individual will have six genes accordingly, as portrayed in Figure A.5. Two individuals are then selected as parents from the mating pool, followed by the creation of a binary cross-over mask, which contains $L = 6$ randomly generated ones and zeros. The genes, represented for example by the indices of the 4-QAM constellation symbols given in Figure A.3, are then exchanged between the selected pair of parents at positions associated with a one in the cross-over mask, giving birth to two offspring.

Mutation. After the cross-over operation is applied to each pair of parents, X offspring are produced, which are then subjected to the so-called *mutation* [454] operation and some of the offspring's genes may be changed. More specifically, any gene of an offspring may be mutated to another legitimate gene under the control of the specific mutation strategy employed. Furthermore, activation of the mutation process is governed by the so-called mutation probability p_m. An example of the mutation procedure following the cross-over operation of Figure A.5 is plotted in Figure A.6. Therefore, following the cross-over and mutation blocks illustrated in Figure A.1, the new population consists of X mutated offspring. It is worth pointing out that the mutation operation is critical to the success of the genetic evolution, since it ensures that sufficient diversity is maintained in the population, thus preventing the GA's search from being trapped at local optima.

Elitism. While the cross-over and mutation operations offer the opportunity to improve the average fitness of the population, they do not guarantee that each of the offspring is better than their parents in terms of their fitness values. In other words, the better individuals associated with the higher fitness values found in the yth generation may not be retained in the $(y + 1)$th generation. To ensure that the high-fitness individuals are not lost from generation to generation, the best or a few of the best individuals of the parent population are copied into the new population, replacing the worst offspring. This technique is known as *elitism* [454], as illustrated in Figures A.1 and A.2.

    The genetic operation cycle mentioned above forms the basis of the GA-aided optimization, yielding an offspring population having an improved average fitness. This evolution continues until the generation index reaches its maximum. Then the operation of the GA is terminated and the highest-fitness individual of the last population will be considered as the final solution, which is the specific length-L symbol vector that contains the detected transmitted symbols of the L users at the specific OFDM subcarrier considered. If the population size X and/or the number of generations Y is sufficiently high, the GA's final solution approaches the optimum [38].

A.2 Normalization of the Mutation-Induced Transition Probability

Without loss of generality, let us first provide the following definitions:

$\mathcal{M}_c$: The set containing the 2^m legitimate constellation symbols $\hat{s}_i^{(l)}$ ($i = 1, \ldots, 2^m$).

$\hat{s}_1^{(l)}$: The original gene (constellation symbol) of the lth user, which is subjected to mutation.

$\hat{s}_2^{(l)}$: The target gene (constellation symbol) of the lth user, which is the destination of mutation.

A: The event that the *current* gene is $\hat{s}_1^{(l)}$.

B: The event that the *next* gene is *not* $\hat{s}_1^{(l)}$.

C: The event that the *next* gene is $\hat{s}_2^{(l)}$.

According to conditional probability theory [468], we have

$$\begin{cases} P(ABC) = P(A) \cdot P(B|A) \cdot P(C|AB) \\ P(ABC) = P(A) \cdot P(BC|A). \end{cases} \tag{A.3}$$

Based on Equation (A.3), the *normalized transition probability* $\tilde{p}_{mt}^{(12)}$ can be derived with the aid of the 2D transition probabilities:

$$\tilde{p}_{mt}^{(12)} = P(C|AB) \tag{A.4}$$

$$= \frac{P(BC|A)}{P(B|A)} \tag{A.5}$$

$$= \frac{P(C|A)}{P(B|A)} \tag{A.6}$$

$$= \frac{p_{mt}^{(12)}}{\sum_{j=2}^{2^m} p_{mt}^{(1j)}} \tag{A.7}$$

$$= \frac{p_{mt}^{(12)}}{1 - p_{mt}^{(11)}}, \tag{A.8}$$

where $p_{mt}^{(12)}$ and $p_{mt}^{(11)}$ are the corresponding 2D transition probabilities, which are given for example by Equations (5.16) and (5.17) in the context of 4-QAM, respectively. Similarly, we can derive the normalized transition probabilities $\tilde{p}_{mt}^{(ij)}$ for all constellation symbols $\hat{s}_i^{(l)}$ ($i = 1, \ldots, 2^m$) of $\mathcal{M}_c$, although these calculations are not included here for reasons of space economy.

Glossary

E_b/N_0	Ratio of bit energy to noise power spectral density
μGA	Micro Genetic Algorithm
16QAM	16-level Quadrature Amplitude Modulation
1D	One-Dimensional
2D	Two-Dimensional
3G	Third Generation
4G	The Fourth Generation
64QAM	64-level Quadrature Amplitude Modulation
8PSK	8-level Phase-Shift Keying
AAA	Authentication, Authorization, and Accounting
AAS	Adaptive Antenna System
ADC	Analogue-to-Digital Converter
ADSL	Asymmetric Digital Subscriber Line
AF	Amplify-and-Forward
AGC	Automatic Gain Control
AI	Artificial Intelligence
ALT	A priori-LLR-Threshold
AMC	Adaptive Modulation and Coding
AMPS	Advanced Mobile Phone Service
ANSI	American National Standards Institute
APA	Adaptive-Power-Allocation
APC	Adaptive Power Control
APP	A Posteriori Probability
ARQ	Automatic Re-transmission Request
ASN	Access Service Network
AWGN	Additive White Gaussian Noise
BCH	Bose–Chaudhuri–Hocquenghem, a class of FEC codes
BCH	Broadcast Channel
BE	Best Effort
BER	Bit Error Rate
BER	Bit Error Ratio
BICM	Bit-Interleaved Coded Modulation
BICM-ID	Iteratively Decoded Bit-Interleaved Coded Modulation
BLAST	Bell Labs Layered Space–Time
BLAST	Bell Labs Layered Space–Time architecture
BPS	Bits Per Symbol
BPSK	Binary Phase-Shift Keying
BQM	Biased Q-function Based Mutation

BRAN	Broadband Radio Access Network
BS	Base Station
BSS	Basic Service Set
BTC	Block Turbo Coding
BWA	Broadband Wireless Access
BWMA	Broadband Wireless Multiple Access
CATV	Cable Television
CC	Convolutional Codes
CCI	Co-Channel Interference
CDD	Conventional Differential Detection
CDD	Cyclic Delay Diversity
CDF	Cumulative Distribution Function
CDMA	Code Division Multiple Access
CFO	Carrier Frequency Offset
CIR	Channel Impulse Response
CL	Closed-Loop
ClOFDM	Clustered Orthogonal Frequency Division Multiplexing
CLS	Constrained Least-Squares
CM	Coded Modulation
CNUM	Closest-Neighbour Uniform Mutation
COFDM	Coded Orthogonal Frequency Division Multiplexing
COST207 HT	COST207 Hilly Terrain channel
CPS	Cooperative-Protocol-Selection
CRC	Cyclic Redundancy Check
CSC	Cumulative Sub-Cost
CSCF	Cumulative Sub-Cost Function
CSD	Complex Sphere Decoder
CSI	Channel State Information
CSN	Connectivity Service Network
CTC	Convolutional Turbo Coding
CTF	Channel Transfer Function
CUS	Cooperating-User-Selection
CWER	CodeWord Error Ratio
D-BLAST	Diagonal BLAST
DAB	Digital Audio Broadcasting
DAC	Digital-to-Analogue Converter
DAF	Differential Amplify-and-Forward
DAPSK	Differential Amplitude and Phase-Shift Keying
DCMC	Discrete-input Continuous-output Memoryless Channel
DDCE	Decision Directed Channel Estimation
DDF	Differential Decode-and-Forward
DF	Decode-and-Forward
DFDD	Decision-Feedback Differential Detection
DFT	Discrete Fourier Transform
DFTS-OFDMA	Discrete Fourier Transform Spread OFDMA
DHDC	Direct-Hard-Decision-Center-Shifting
DL	Downlink
DMT	Discrete MultiTone

DS-CDMA	Direct-Sequence Code Division Multiple Access
DSDC	Direct-Soft-Decision-Center-Shifting
DSL	Digital Subscriber Line
DSS	Direct-Sequence Spreading
DSSS	Direct-Sequence Spread Spectrum
DVB	Digital Video Broadcasting
DVB-H	Digital Video Broadcasting for Handheld terminals
DVB-T	Digital Video Broadcasting for Terrestrial television
E-MBS	Enhanced Multicast and Broadcast Services
E-UTRAN	Evolved Universal Terrestrial Radio Access Network
EA	Evolutionary Algorithm
EM	Expectation Maximization
ertPS	extended real-time Polling Service
ETSI	European Telecommunications Standards Institute
FBSS	Fast BS Switching
FCH	Feedback Channel
FD	Frequency Domain
FD-CHTF	Frequency-Domain CHannel Transfer Functions
FD-CTF	Frequency-Domain Channel Transfer Function
FDD	Frequency Division Duplex
FDMA	Frequency Division Multiple Access
FEC	Forward Error Correction
FFH	Fast Frequency-Hopping
FFT	Fast Fourier Transform
FH	Frequency-Hopping
FH/SSMA	Frequency-Hopped SSMA
FHSS	Frequency-Hopping Spread Spectrum
FIR	Finite Impulse Response
FS	Fractionally-Spaced
FS-CIR	Fractionally-Spaced CIR
FUSC	Full Usage of Subchannels
GA	Genetic Algorithm
GA-JCEMUD	Joint Channel Estimation and Multi-User Detection
GA-MMSE	Genetic Algorithm-aided MMSE
GACE-OHRSA-MUD	GA-based Channel Estimation assisted OHRSA Multi-User Detection
GMD	Geometric Mean Decomposition
GSD	Generalized Sphere Detection
GSM	Global System for Mobile communications
H-FDD	Half-duplex FDD
H-NSP	Home Network Service Provider
HARQ	Hybrid Automatic Repeat Request
HDSL	High-bit-rate Digital Subscriber Line
HIHO	Hard-Input Hard-Output
HO	HandOver
HSDPA	High Speed Downlink Packet Access
HT	Hilly Terrain

ICI	Inter-Carrier Interference
IE	Information Element
IEEE	The Institute of Electrical and Electronics Engineers
IFFT	Inverse Fast Fourier Transform
IGA	Iterative Genetic Algorithm
IIR	Infinite-Impulse-Response
IR	Incremental Redundancy
Ir-DHCD	Irregular Distributed Hybrid Concatenated Differential
IrCCs	Irregular Convolutional Codes
ISI	Inter-Symbol Interference
ISR	Initial Search Radius
ITU-R	International Telecommunications Union - Radio Communications Sector
IWHT	Inverse Walsh-Hadamard Transform
L1	Layer-1
L2	Layer-2
L3	Layer-3
LAN	Local Area Network
LBS	Location Based Service
LDPC	Low Density Parity Check
LDPC	Low-Density Parity Check codes
LLR	Log-Likelihood Ratio
Log-MAP	Logarithmic Maximum A Posteriori
LOS	Line-Of-Sight
LPF	Low-Pass Filter
LS	Least Squares
LTE	Long-Term Evolution
MAC	Media Access Control
MAC	Medium Access Control
MAMC	multi-antenna-multi-carrier
MAN	Metropolitan Area Network
MAP	Maximum-A-Posteriori
MBER	Minimum Bit Error Rate
MBMS	Multimedia Broadcast Multicast Service
MBS	Multicast and Broadcast Service
MC-CDMA	Multi-Carrier Code Division Multiple Access
MCF	Minimum Cost Function
MD	M-Dimensional
MDHO	Macro Diversity HandOver
MED	Minimum Euclidean Distance
MGF	Joint Moment Generating Function
MI	Mutual Information
MIB	Management Information Base
MIH	Media Independent Handover
MIMO	Multiple-Input Multiple-Output
ML	Maximum Likelihood
ML	Maximum-Likelihood
MLD	Maximum Likelihood Detection

MMSE	Minimum Mean-Square Error
MRC	Maximum Ratio Combining
MS	Mobile Station
MSDD	Multiple-Symbol Differential Detection
MSDSD	Multiple-Symbol Differential Sphere Detection
MSE	Mean-Square Error
MSI	Multi-Stream-Interference
MU-MIMO	Multi-User MIMO
MUD	Multi-User Detection/Detector
MUD	Multi-User Detection
MUI	Multi-User Interference
MUI	Multiple-Access-Interference
MV	Minimum Variance
NLOS	Non-Line-Of-Sight
NMSE	Normalized Mean Square Error
NN	Neural Network
NRM	Network Reference Model
nrtPS	non-real-time Polling Service
O-QAM	Orthogonal Quadrature Amplitude Modulation
OF	Objective Function
OFDM	Orthogonal Frequency Division Multiplexing
OFDMA	Orthogonal Frequency Division Multiple Access
OHRSA	Optimized Hierarchy Reduced Search Algorithm
OL	Open-Loop
OS	Objective Score
OSTBC	Orthogonal STBC
PAPR	Peak-to-Average Power Ratio
PAST	Projection Approximation Subspace Tracking
PDF	Probability Density Function
PDP	Power Delay Profile
PED	Partial Euclidean Distance
PHY	Physical layer
PIC	Parallel Interference Cancellation
PLR	Packet Loss Ratio
PSD	Power Spectral Density
PSK	Phase-Shift Keying
PSP	Per-Survivor Processing
PUSC	Partial Usage of Subchannels
QAM	Quadrature Amplitude Modulation
QoS	Quality-of-Service
QPSK	Quadrature Phase-Shift Keying
QRD-M	QR Decomposition combined with the M-algorithm
RA	Rural Area
RAT	Radio Access Technology
RBF	Radial Basis Function

RF	Radio Frequency
RG	802.16m Rapporteur Group
RLS	Recursive Least-Squares
RMS	Root Mean Square
RRM	Radio Resource Management
RS	Relay Station
RSA	Reduced Search Algorithm
RSC	Recursive Systematic Convolutional
RSFH	Random Slow Frequency-Hopping
RSSCH	Random Slow SubCarrier-Hopping
RTG	Receive Transition Gap
rtPS	real-time Polling Service
S-OFDMA	Scalable OFDMA
SAE	System Architecture Evolution
SC	Single-Carrier
SCF	Sub-Cost Function
SCH	SubCarrier-Hopping
SD	Sphere Decoder
SD	Sphere Detection
SDD	802.16m System Description Document
SDM	Space-Division Multiplexing
SDMA	Space-Division Multiple Access
SDMD	SDM Detection
SFH	Slow Frequency-Hopping
SIC	Successive Interference Cancellation
SIC-MMSE	Soft Interference Cancellation aided MMSE
SINR	Signal to Noise and Interference Ratio
SINR	Signal-to-Interference-plus-Noise Ratio
SInSOu	Soft-Input Soft-Output
SIR	Signal-to-Interference Ratio
SISO	Single-Input Single-Output
SM	Spatial Multiplexing
SM	Step Mutation
SNR	Signal-to-Noise Ratio
SOPHIE	Soft-output OPtimized HIErarchy
SOS	Second-Order Statistics
SP	Sphere Packing
SRD	802.16m System Requirement Document
SS	Spread-Spectrum
SS	Subscriber Station
SS	Symbol-Spaced
SSCH	Slow SubCarrier-Hopping
SSD	Sorted Sphere Detection
SSD-TT	Termination-Threshold-Aided Sorted SD
SSD-UB	Updated-Bound-Aided Sorted SD
SSMA	Spread-Spectrum Multiple Access
STBC	Space–Time Block Code
STBC	Space–Time Block Coding
STC	Space–Time Code

STTC	Space–Time Trellis Code
SU-MIMO	Single-User MIMO
SWATM	Short Wireless Asynchronous Transfer Mode channel
TBCH	Turbo Bose–Chaudhuri–Hocquenghem codes
TC	Turbo Convolutional codes
TCM	Trellis Coded Modulation
TD	Time-Domain
TD-CDM-OFDM	Time-Division Code-Division Multiplexing Orthogonal Frequency Division Multiplexing
TD-CDMA	Time Division Code Division Multiple Access
TD-SCDMA	Time-Division Synchronous Code Division Multiple Access
TDD	Time Division Duplex
TDM	Time Division Multiplexing
TDMA	Time Division Multiple Access
TGe	Task Group e
TGm	Task Group m
TH	Typical Urban
TOA	Time-of-Arrival
TT	Termination Threshold
TTA	Telecommunications Technology Association
TTCM	Turbo Trellis Coded Modulation
TTG	Transmit Transition Gap
TUSC	Tile Usage of Subchannels
UGS	Unsolicited Grant Service
UL	Uplink
UM	Uniform Mutation
UMTS	Universal Mobile Telecommunications System
UNII	Unlicensed National Information Infrastructure
URC	Unity-Rate Code
USFH	Uniform Slow Frequency-Hopping
USSCH	Uniform Slow SubCarrier-Hopping
UTRAN	Universal Terrestrial Radio Access Network
UWB	Ultra WideBand
V-BLAST	Vertical Bell Labs Layered Space–Time architecture
VA	Viterbi Algorithm
VAA	Virtual Antenna Array
VDSL	Very-high-speed Digital Subscriber Line
VoIP	Voice over Internet Protocol
WAN	Wide Area Network
WATM	Wireless Asynchronous Transfer Mode
WCDMA	Wideband Code Division Multiple Access
WG	Working Group
WHT	Walsh-Hadamard Transform
WHTS	Walsh-Hadamard Transform Spreading
WiBro	Wireless Broadband
WiMAX	Worldwide Interoperability for Microwave Access

WLAN	Wireless Local Area Network
WMAN	Wireless Metropolitan Area Network
WSS	Wide Sense Stationary
ZF	Zero-Forcing

Bibliography

[1] J. A. C. Bingham, 'Multicarrier modulation for data transmission: an idea whose time has come', *IEEE Communications Magazine*, vol. 28, pp. 5–14, May 1990.

[2] L. Hanzo, W. Webb, and T. Keller, *Single- and Multi-carrier Quadrature Amplitude Modulation*. Chichester: IEEE Press and John Wiley & Sons, Ltd, 2nd edn, 2000.

[3] L. Hanzo, M. Münster, B. J. Choi, and T. Keller, *OFDM and MC-CDMA for Broadband Multi-user Communications, WLANs and Broadcasting*. Chichester: IEEE Press and John Wiley & Sons, Ltd, 2003.

[4] R. V. Nee and R. Prasad, *OFDM for Wireless Multimedia Communications*. London: Artech House, 2000.

[5] J. L. Holsinger, 'Digital communication over fixed time-continuous channels with memory – with special application to telephone channels', Technical Report No. 366, MIT – Lincoln Laboratory, Cambridge, MA, October 1964.

[6] J. M. Cioffi, *A Multicarrier Primer*. ANSI T1E1.4/91-157, November 1991.

[7] ANSI Committee T1-Telecommunications, *A Technical Report on High-Bit-Rate Digital Subscriber Lines (HDSL)*, Technical Report No. 28, February 1994.

[8] *Very-high-speed Digital Subscriber Lines: System Requirements*, ANSI T1E1.4 VDSL SR: 98-043R5, September 1998.

[9] European Telecommunications Standards Institute, *Transmission and Multiplexing (TM); Access transmission systems on metallic access cables; Very high speed Digital Subscriber Line (VDSL); Part 1: Functional requirements*, ETSI TS 101 270-1 V1.1.2, June 1998.

[10] European Telecommunications Standards Institute, *Digital Audio Broadcasting (DAB); DAB to mobile, portable and fixed Receivers*, ETSI ETS 300 401 ed.1, February 1995.

[11] European Telecommunications Standards Institute, *Digital Video Broadcasting (DVB); Framing structure, channel coding and modulation for digital terrestrial television (DVB-T)*, ETSI ETS 300 744 ed.1, March 1997.

[12] European Telecommunications Standards Institute, *Digital Video Broadcasting (DVB); Transmission system for handheld terminals (DVB-H)*, ETSI EN 302 304 V1.1.1, November 2004.

[13] European Telecommunications Standards Institute, *Radio Equipment and Systems (RES); HIgh PErformance Radio Local Area Network (HIPERLAN) Type 1; Functional specification*, ETSI ETS 300 652 ed.1, October 1996.

[14] European Telecommunications Standards Institute, *Broadband Radio Access Networks (BRAN); Inventory of broadband radio technologies and techniques*, ETSI TR 101 173 V1.1.1, May 1998.

[15] Institute of Electrical and Electronics Engineers, *IEEE Standard 802.11: Wireless LAN Medium Access Control (MAC) and Physical Layer (PHY) Specifications*, 18 November 1997.

[16] Institute of Electrical and Electronics Engineers, *IEEE Standard 802.16: Air Interface for Fixed Broadband Wireless Access Systems*, 2004.

[17] M. Doelz, E. Heald, and D. Martin, 'Binary data transmission techniques for linear systems', *Proceedings of the IRE*, vol. 45, pp. 656–661, May 1957.

[18] R. W. Chang, 'Synthesis of band-limited orthogonal signals for multichannel data transmission', *Bell System Technical Journal*, vol. 45, pp. 1775–1796, December 1966.

[19] B. R. Saltzberg, 'Performance of an efficient parallel data transmission system', *IEEE Transactions on Communications*, vol. 15, pp. 805–811, December 1967.

[20] R. W. Chang and R. A. Gibby, 'A theoretical study of performance of an orthogonal multiplexing data transmission scheme', *IEEE Transactions on Communications*, vol. 16, pp. 529–540, August 1968.

[21] R. W. Chang, *US Patent No. 3,488,445: Orthogonal Frequency Division Multiplexing*, filed November 14, 1966, issued January 6, 1970.

[22] S. B. Weinstein and P. M. Ebert, 'Data transmission by frequency-division multiplexing using the discrete Fourier transform', *IEEE Transactions on Communications*, vol. 19, pp. 628–634, October 1971.

[23] A. Peled and A. Ruiz, 'Frequency domain data transmission using reduced computational complexity algorithms', in *IEEE International Conference on Acoustics, Speech, and Signal Processing (ICASSP'80)*, (Denver, USA), vol. 5, pp. 964–967, 9–11 April 1980.

[24] L. Hanzo, S. X. Ng, T. Keller, and W. T. Webb, *Quadrature Amplitude Modulation: From Basics to Adaptive Trellis-Coded, Turbo-Equalised and Space-Time Coded OFDM, CDMA and MC-CDMA Systems*. Chichester: IEEE Press and John Wiley & Sons, Ltd, 3rd edn, 2004.

[25] W. E. Keasler, D. L. Bitzer, and P. T. Tucker, *US Patent No. 4,206,320: High-speed Modem Suitable for Operating with a Switched Network*, filed August 21, 1978, issued June 3, 1980.

[26] B. Hirosaki, 'An analysis of automatic equalizers for orthogonally multiplexed QAM systems', *IEEE Transactions on Communications*, vol. 28, pp. 73–83, January 1980.

[27] B. Hirosaki, 'An orthogonally multiplexed QAM system using the discrete Fourier transform', *IEEE Transactions on Communications*, vol. 29, pp. 982–989, July 1981.

[28] B. Hirosaki, S. Hasegawa, and A. Sabato, 'Advanced groupband data modem using orthogonally multiplexed QAM technique', *IEEE Transactions on Communications*, vol. 34, pp. 587–592, June 1986.

[29] L. J. Cimini Jr, 'Analysis and simulation of a digital mobile channel using orthogonal frequency division multiplexing', *IEEE Transactions on Communications*, vol. 33, pp. 665–675, July 1985.

[30] I. Kalet, 'The multitone channel', *IEEE Transactions on Communications*, vol. 37, pp. 119–124, February 1989.

[31] M. Alard and R. Lassalle, 'Principles of modulation and channel coding for digital broadcasting for mobile receivers', *EBU Technical Review*, pp. 168–190, August 1987.

[32] K. Fazel and G. Fettweis, *Multi-carrier Spread Spectrum and Related Topics*. Boston, MA: Kluwer, 2000.

[33] T. Keller and L. Hanzo, 'Adaptive multicarrier modulation: a convenient framework for time-frequency processing in wireless communications', *Proceedings of the IEEE*, vol. 88, pp. 611–640, May 2000.

[34] L. Hanzo, B. J. Choi, and M. Münster, 'A stroll along Multi-Carrier Boulevard towards Next-Generation Plaza – OFDM background and history', *IEEE VTS News*, vol. 51, pp. 4–10, November 2004.

[35] L. Hanzo, B. J. Choi, and M. Münster, 'A stroll along Multi-Carrier Boulevard towards Next-Generation Plaza – space-time coded adaptive OFDM and MC-CDMA comparison', *IEEE VTS News*, vol. 51, pp. 10–19, November 2004.

[36] R. Steele and L. Hanzo, *Mobile Radio Communications: Second and Third Generation Cellular and WATM Systems*. New York: IEEE Press and John Wiley & Sons, Inc., 2nd edn, 1999.

[37] A. J. Viterbi, *CDMA: Principles of Spread Spectrum Communication*. Reading, MA: Addison-Wesley, 1995.

[38] L. Hanzo, L.-L. Yang, E.-L. Kuan, and K. Yen, *Single- and Multi-Carrier DS-CDMA: Multi-User Detection, Space-Time Spreading, Synchronisation and Standards*. Chichester: IEEE Press and John Wiley & Sons, Ltd, 2003.

[39] L. Wang and L. Hanzo, 'The amplify-and-forward cooperative uplink using multiple-symbol differential sphere-detection', *IEEE Signal Processing Letters*, vol. 16, pp. 913–916, October 2009.

[40] K. Zigangirov, *Theory of Code Division Multiple Access Communication*. New York: IEEE Press and John Wiley & Sons, Inc., 2004.

[41] L. E. Miller and J. S. Lee, *CDMA Systems Engineering Handbook*. London: Artech House, 1998.

[42] J. S. Lee, 'Overview of the technical basis of QUALCOMM's CDMA cellular telephone system design: a view of North American TIA/EIA IS-95', in *International Conference on Communications Systems (ICCS)*, (Singapore), pp. 353–358, 1994.

[43] I. Koffman and V. Roman, 'Broadband wireless access solutions based on OFDM access in IEEE 802.16', *IEEE Communications Magazine*, vol. 40, pp. 96–103, April 2002.

[44] R. Laroia, S. Uppala, and J. Li, 'Designing a mobile broadband wireless access network', *IEEE Signal Processing Magazine*, vol. 21, pp. 20–28, September 2004.

[45] P. Xia, S. Zhou, and G. B. Giannakis, 'Bandwidth- and power-efficient multicarrier multiple access', *IEEE Transactions on Communications*, vol. 51, pp. 1828–1837, November 2003.

[46] Z. Cao, U. Tureli, and Y. Yao, 'Deterministic multiuser carrier-frequency offset estimation for interleaved OFDMA uplink', *IEEE Transactions on Communications*, vol. 52, pp. 1585–1594, September 2004.

[47] R. Bercovich, 'OFDM enhances the 3G high-speed data access', in *GSPx 2004 Conference*, (Santa Clara, USA), 27–30 September 2004. http://www.techonline.com/pdf/pavillions/gspx/2004/1084.pdf.

[48] T. May, H. Rohling, and V. Engels, 'Performance analysis of Viterbi decoding for 64-DAPSK and 64-QAM modulated OFDM signals', *IEEE Transactions on Communications*, vol. 46, pp. 182–190, February 1998.

[49] L. Lin, L. J. Cimini Jr., and C.-I. Chuang, 'Comparison of convolutional and turbo codes for OFDM with antenna diversity in high-bit-rate wireless applications', *IEEE Communications Letters*, vol. 9, pp. 277–279, September 2000.

[50] P. H. Moose, 'A technique for orthogonal frequency division multiplexing frequency offset correction', *IEEE Transactions on Communications*, vol. 42, pp. 2908–2914, October 1994.

[51] Institute of Electrical and Electronics Engineers, *IEEE Standard 802.11a: Wireless LAN Medium Access Control (MAC) and Physical Layer (PHY) Specifications: high-speed physical layer in the 5 GHz band*, 1999.

[52] Institute of Electrical and Electronics Engineers, *IEEE Standard 802.11g: Wireless LAN Medium Access Control (MAC) and Physical Layer (PHY) Specifications*, 2003.

[53] 'OFDM for Mobile Data Communications', White Paper, Flarion Technologies, Inc., Bedminster, NY, USA, March 2003.

[54] 'FLASH-OFDM for 450 MHz – Advanced Mobile Broadband Solution for 450 MHz Operators', White Paper, Flarion Technologies, Inc., Bedminster, NY, USA, November 2004.

[55] Institute of Electrical and Electronics Engineers, 'IEEE Standard for Local and Metropolitan Area Networks – Part 16: Air Interface for Fixed Broadband Wireless Access Systems (Revision of 802.16–2001)', IEEE 802.16d-2004, 2004.

[56] Institute of Electrical and Electronics Engineers, 'Draft IEEE Standard for Information Technology – Telecommunications and Information Exchange Between Systems – Local and Metropolitan Area Networks – Specific Requirements – Part 11: Wireless LAN Medium Access Control (MAC) and Physical Layer (PHY) Specifications – Amendment 4: Enhancements for Higher Throughput', IEEE P802.11n/D4, 2008.

[57] 3GPP. http://www.3gpp.org/.

[58] M. Jiang and L. Hanzo, 'Multi-user MIMO-OFDM for next-generation wireless', *Proceedings of the IEEE*, vol. 95, pp. 1430–1469, July 2007.

[59] L. Hanzo and B. Choi, 'Near-instantaneously adaptive HSDPA-style OFDM and MC-CDMA transceivers for WiFi, WiMAX and next-generation systems', *Proceedings of the IEEE*, vol. 95, pp. 2368–2392, December 2007.

[60] W. D. Warner and C. Leung, 'OFDM/FM frame synchronization for mobile radio data communication', *IEEE Transactions on Vehicular Technology*, vol. 42, pp. 302–313, August 1993.

[61] T. Pollet, M. V. Bladel, and M. Moeneclaey, 'BER sensitivity of OFDM systems to carrier frequency offset and Wiener phase noise', *IEEE Transactions on Communications*, vol. 43, pp. 191–193, February/March/April 1995.

[62] A. E. Jones, T. A. Wilkinson, and S. K. Barton, 'Block coding scheme for reduction of peak to mean envelope power ratio of multicarrier transmission schemes', *Electronics Letters*, vol. 30, pp. 2098–2099, December 1994.

[63] S. J. Shepherd, P. W. J. V. Eetvelt, C. W. Wyatt-Millington, and S. K. Barton, 'Simple coding scheme to reduce peak factor in QPSK multicarrier modulation', *Electronics Letters*, vol. 31, pp. 1131–1132, July 1995.

[64] D. Wulich, 'Reduction of peak to mean ratio of multicarrier modulation using cyclic coding', *Electronics Letters*, vol. 32, pp. 432–433, February 1996.

[65] D. Wulich, 'Peak factor in orthogonal multicarrier modulation with variable levels', *Electronics Letters*, vol. 32, pp. 1859–1861, September 1996.

[66] X. Li and L. J. Cimini Jr, 'Effects of clipping and filtering on the performance of OFDM', in *Proceedings of the IEEE 47th Vehicular Technology Conference (VTC 1997 Spring)*, (Tokyo, Japan), vol. 3, pp. 1634–1638, 4–7 May 1997.

[67] X. Li and L. J. Cimini Jr, 'Effects of clipping and filtering on the performance of OFDM', *IEEE Communications Letters*, vol. 2, pp. 131–133, May 1998.

[68] S. Hara and R. Prasad, 'Overview of multicarrier CDMA', *IEEE Communications Magazine*, vol. 35, pp. 126–133, December 1997.

[69] Y. Li, L. J. Cimini Jr, and N. R. Sollenberger, 'Robust channel estimation for OFDM systems with rapid dispersive fading channels', *IEEE Transactions on Communications*, vol. 46, pp. 902–915, July 1998.

[70] Y. Li and N. R. Sollenberger, 'Adaptive antenna arrays for OFDM systems with cochannel interference', *IEEE Transactions on Communications*, vol. 47, pp. 217–229, February 1999.

[71] S. Armour, A. Nix, and D. Bull, 'Pre-FFT equaliser design for OFDM', *Electronics Letters*, vol. 35, pp. 539–540, April 1999.

[72] S. Armour, A. Nix, and D. Bull, 'Performance analysis of a pre-FFT equalizer design for DVB-T', *IEEE Transactions on Consumer Electronics*, vol. 45, pp. 544–552, August 1999.

[73] S. Armour, A. Nix, and D. Bull, 'Complexity evaluation for the implementation of a pre-FFT equalizer in an OFDM receiver', *IEEE Transactions on Consumer Electronics*, vol. 46, pp. 428–437, August 2000.

[74] B. Y. Prasetyo and A. H. Aghvami, 'Simplified frame structure for MMSE-based fast burst synchronisation in OFDM systems', *Electronics Letters*, vol. 35, pp. 617–618, April 1999.

[75] B. Y. Prasetyo, F. Said, and A. H. Aghvami, 'Fast burst synchronisation technique for OFDM-WLAN systems', *IEE Proceedings – Communications*, vol. 147, pp. 292–298, October 2000.

[76] C. Y. Wong, R. S. Cheng, K. B. Lataief, and R. D. Murch, 'Multiuser OFDM with adaptive subcarrier, bit, and power allocation', *IEEE Journal on Selected Areas in Communications*, vol. 17, pp. 1747–1758, October 1999.

[77] B. Lu, X. Wang, and K. R. Narayanan, 'LDPC-based space-time coded OFDM systems over correlated fading channels: performance analysis and receiver design', in *Proceedings of the 2001 IEEE International Symposium on Information Theory*, (Washington, DC, USA), vol. 1, p. 313, 24–29 June 2001.

[78] B. Lu, X. Wang, and K. R. Narayanan, 'LDPC-based space-time coded OFDM systems over correlated fading channels: performance analysis and receiver design', *IEEE Transactions on Communications*, vol. 50, pp. 74–88, January 2002.

[79] B. Lu, X. Wang, and Y. Li, 'Iterative receivers for space-time block coded OFDM systems in dispersive fading channels', in *IEEE Global Telecommunications Conference (GLOBECOM'01)*, (San Antonio, USA), vol. 1, pp. 514–518, 25–29 November 2001.

[80] B. Lu, X. Wang, and Y. Li, 'Iterative receivers for space-time block-coded OFDM systems in dispersive fading channels', *IEEE Transactions on Wireless Communications*, vol. 1, pp. 213–225, April 2002.

[81] O. Simeone, Y. Bar-Ness, and U. Spagnolini, 'Pilot-based channel estimation for OFDM systems by tracking the delay-subspace', *IEEE Transactions on Wireless Communications*, vol. 3, pp. 315–325, January 2004.

[82] J. Zhang, H. Rohling, and P. Zhang, 'Analysis of ICI cancellation scheme in OFDM systems with phase noise', *IEEE Transactions on Broadcasting*, vol. 50, pp. 97–106, June 2004.

[83] M. C. Necker and G. L. Stüber, 'Totally blind channel estimation for OFDM on fast varying mobile radio channels', *IEEE Transactions on Wireless Communications*, vol. 3, pp. 1514–1525, September 2004.

[84] A. Doufexi, S. Armour, A. Nix, P. Karlsson, and D. Bull, 'Range and throughput enhancement of wireless local area networks using smart sectorised antennas', *IEEE Transactions on Wireless Communications*, vol. 3, pp. 1437–1443, September 2004.

[85] E. Alsusa, Y. Lee, and S. McLaughlin, 'Channel-adaptive sectored multicarrier packet based systems', *Electronics Letters*, vol. 40, pp. 1194–1196, September 2004.

[86] C. Williams, M. A. Beach, and S. McLaughlin, 'Robust OFDM timing synchronisation', *Electronics Letters*, vol. 41, pp. 751–752, June 2005.

[87] R. Fischer and C. Siegl, 'Performance of peak-to-average power ratio reduction in single- and multi-antenna OFDM via directed selected mapping', *IEEE Transactions on Communications*, vol. 57, pp. 3205–3208, November 2009.

[88] G. Mileounis, N. Kalouptsidis, and P. Koukoulas, 'Blind identification of Hammerstein channels using QAM, PSK, and OFDM inputs', *IEEE Transactions on Communications*, vol. 57, pp. 3653–3661, December 2009.

[89] S. Huang and C. Hwang, 'Improvement of active interference cancellation: avoidance technique for OFDM cognitive radio', *IEEE Transactions on Wireless Communications*, vol. 8, pp. 5928–5937, December 2009.

[90] H. Chen, W. Gao, and D. Daut, 'Spectrum sensing for OFDM systems employing pilot tones', *IEEE Transactions on Wireless Communications*, vol. 8, December 2009.

[91] S. Talbot and B. Farhang-Boroujeny, 'Time-varying carrier offsets in mobile OFDM', *IEEE Transactions on Communications*, vol. 57, pp. 2790–2798, September 2009.

[92] J. H. Winters, 'Optimum combining in digital mobile radio with cochannel interference', *IEEE Journal on Selected Areas in Communications*, vol. 2, pp. 528–539, July 1984.

[93] J. H. Winters, *US Patent No. 4,639,914: Wireless PBX/LAN System with Optimum Combining*, filed December 6, 1984, issued January 27, 1987.

[94] J. Salz, 'Digital transmission over cross-coupled linear channels', *AT&T Technical Journal*, vol. 64, pp. 1147–1159, July–August 1985.

[95] J. H. Winters, 'On the capacity of radio communication systems with diversity in a Rayleigh fading environment', *IEEE Journal on Selected Areas in Communications*, vol. 5, pp. 871–878, June 1987.

[96] J. H. Winters, 'Optimum combining for indoor radio systems with multiple users', *IEEE Transactions on Communications*, vol. 35, pp. 1222–1230, November 1987.

[97] S. Cheng and S. Verdu, 'Gaussian multiaccess channels with ISI: capacity region and multiuser water-filling', *IEEE Transactions on Information Theory*, vol. 39, pp. 773–785, May 1993.

[98] A. Duel-Hallen, 'Equalizers for multiple input/multiple output channels and PAM systems with cyclostationary input sequences', *IEEE Journal on Selected Areas in Communications*, vol. 10, pp. 630–639, April 1992.

[99] J. H. Winters, J. Salz, and R. D. Gitlin, 'The impact of antenna diversity on the capacity of wireless communication systems', *IEEE Transactions on Communications*, vol. 5, pp. 1740–1751, February/March/April 1994.

[100] J. Yang and S. Roy, 'On joint transmitter and receiver optimization for multiple-input-multiple-output (MIMO) transmission systems', *IEEE Transactions on Communications*, vol. 42, pp. 3221–3231, December 1994.

[101] J. Yang and S. Roy, 'Joint transmitter-receiver optimization for multi-input multi-output systems with decision feedback', *IEEE Transactions on Information Theory*, vol. 40, pp. 1334–1347, September 1994.

[102] J. H. Winters, 'The diversity gain of transmit diversity in wireless systems with Rayleigh fading', *IEEE Transactions on Vehicular Technology*, vol. 47, pp. 119–123, February 1998.

[103] J. H. Winters and J. Salz, 'Upper bounds on the bit-error rate of optimum combining in wireless systems', *IEEE Transactions on Communications*, vol. 46, pp. 1619–1624, December 1998.

[104] G. G. Raleigh and J. M. Cioffi, 'Spatio-temporal coding for wireless communications', in *IEEE Global Telecommunications Conference, 1996 (GLOBECOM'96)*, (London, UK), vol. 3, pp. 1809–1814, 18–22 November 1996.

[105] G. J. Foschini, 'Layered space-time architecture for wireless communication in a fading environment when using multi-element antennas', *Bell Labs Technical Journal*, Autumn, pp. 41–59, 1996.

[106] G. G. Raleigh and J. M. Cioffi, 'Spatio-temporal coding for wireless communication', *IEEE Transactions on Communications*, vol. 46, pp. 357–366, March 1998.

[107] G. J. Foschini and M. J. Gans, 'On limits of wireless communications in a fading environment when using multiple antennas', *Wireless Personal Communications*, vol. 6, pp. 311–335, March 1998.

[108] G. J. Foschini, G. D. Golden, R. A. Valenzuela, and P. W. Wolniansky, 'Simplified processing for high spectral efficiency wireless communication employing multi-element arrays', *IEEE Journal on Selected Areas in Communications*, vol. 17, pp. 1841–1852, November 1999.

[109] B. Lu and X. Wang, 'Iterative receivers for multiuser space-time coding systems', *IEEE Journal on Selected Areas in Communications*, vol. 18, pp. 2322–2335, November 2000.

[110] S. Y. Kung, Y. Wu, and X. Zhang, 'Bezout space-time precoders and equalizers for MIMO channels', *IEEE Transactions on Signal Processing*, vol. 50, pp. 2499–2514, October 2002.

[111] F. Petré, G. Leus, L. Deneire, M. Engels, M. Moonen, and H. D. Man, 'Space-time block coding for single-carrier block transmission DS-CDMA downlink', *IEEE Journal on Selected Areas in Communications*, vol. 21, pp. 350–361, April 2003.

[112] L. Zhang, L. Gui, Y. Qiao, and W. Zhang, 'Obtaining diversity gain for DTV by using MIMO structure in SFN', *IEEE Transactions on Broadcasting*, vol. 50, pp. 83–90, March 2004.

[113] X. Zhu and R. D. Murch, 'Layered space-frequency equalization in a single-carrier MIMO system for frequency-selective channels', *IEEE Transactions on Wireless Communications*, vol. 3, pp. 701–708, May 2004.

[114] M. R. McKay and I. B. Collings, 'Capacity and performance of MIMO-BICM with zero-forcing receivers', *IEEE Transactions on Communications*, vol. 53, pp. 74–83, January 2005.

[115] J. Hoadley, 'Building Future Networks with MIMO and OFDM'.
http://telephonyonline.com/wireless/technology/mimo_ofdm_091905/, 19 September 2005.

[116] A. J. Paulraj, D. A. Gore, R. U. Nabar, and H. Bölcskei, 'An overview of MIMO communications – a key to gigabit wireless', *Proceedings of the IEEE*, vol. 92, pp. 198–218, February 2004.

[117] 'Using MIMO-OFDM Technology To Boost Wireless LAN Performance Today', White Paper, Datacomm Research Company, St Louis, USA, June 2005.

[118] H. Sampath, S. Talwar, J. Tellado, V. Erceg, and A. J. Paulraj, 'A fourth-generation MIMO-OFDM broadband wireless system: design, performance, and field trial results', *IEEE Communications Magazine*, vol. 40, pp. 143–149, September 2002.

[119] Airgo Networks. http://www.airgonetworks.com/.

[120] Institute of Electrical and Electronics Engineers, *IEEE Candidate Standard 802.11n: Wireless LAN Medium Access Control (MAC) and Physical Layer (PHY) Specifications*, 2004.
http://grouper.ieee.org/groups/802/11/Reports/tgn_update.htm.

[121] H. Bölcskei, D. Gesbert, and A. J. Paulraj, 'On the capacity of OFDM-based spatial multiplexing systems', *IEEE Transactions on Communications*, vol. 50, pp. 225–234, February 2002.

[122] A. Ganesan and A. M. Sayeed, 'A virtual input-output framework for transceiver analysis and design for multipath fading channels', *IEEE Transactions on Communications*, vol. 51, pp. 1149–1161, July 2003.

[123] R. S. Blum, Y. Li, J. H. Winters, and Q. Yan, 'Improved space-time coding for MIMO-OFDM wireless communications', *IEEE Transactions on Communications*, vol. 49, pp. 1873–1878, November 2001.

[124] H. E. Gamal, A. R. Hammons Jr, Y. Liu, M. P. Fitz, and O. Y. Takeshita, 'On the design of space-time and space-frequency codes for MIMO frequency-selective fading channels', *IEEE Transactions on Information Theory*, vol. 49, pp. 2277–2292, September 2003.

[125] P. Dayal, M. Brehler, and M. K. Varanasi, 'Leveraging coherent space-time codes for noncoherent communication via training', *IEEE Transactions on Information Theory*, vol. 50, pp. 2058–2080, September 2004.

[126] W. Su, Z. Safar, M. Olfat, and K. J. R. Liu, 'Obtaining full-diversity space-frequency codes from space-time codes via mapping', *IEEE Transactions on Signal Processing*, vol. 51, pp. 2905–2916, November 2003.

[127] W. Su, Z. Safar, and K. J. R. Liu, 'Full-rate full-diversity space-frequency codes with optimum coding advantage', *IEEE Transactions on Information Theory*, vol. 51, pp. 229–249, January 2005.

[128] J. H. Moon, Y. H. You, W. G. Jeon, K. W. Kwon, and H. K. Song, 'Peak-to-average power control for multiple-antenna HIPERLAN/2 and IEEE802.11a systems', *IEEE Transactions on Consumer Electronics*, vol. 49, pp. 1078–1083, November 2003.

[129] Y. L. Lee, Y. H. You, W. G. Jeon, J. H. Paik, and H. K. Song, 'Peak-to-average power ratio in MIMO-OFDM systems using selective mapping', *IEEE Communications Letters*, vol. 7, pp. 575–577, December 2003.

[130] S. H. Han and J. H. Lee, 'An overview of peak-to-average power ratio reduction techniques for multicarrier transmission', *IEEE Wireless Communications*, vol. 12, pp. 56–65, April 2005.

[131] Y. Li, 'Simplified channel estimation for OFDM systems with multiple transmit antennas', *IEEE Transactions on Wireless Communications*, vol. 1, pp. 67–75, January 2002.

[132] I. Barhumi, G. Leus, and M. Moonen, 'Optimal training design for MIMO OFDM systems in mobile wireless channels', *IEEE Transactions on Signal Processing*, vol. 51, pp. 1615–1624, June 2003.

[133] M. Shin, H. Lee, and C. Lee, 'Enhanced channel-estimation technique for MIMO-OFDM systems', *IEEE Transactions on Vehicular Technology*, vol. 53, pp. 262–265, January 2004.

[134] Y. Li, J. H. Winters, and N. R. Sollenberger, 'MIMO-OFDM for wireless communications: signal detection with enhanced channel estimation', *IEEE Transactions on Communications*, vol. 50, pp. 1471–1477, September 2002.

[135] L. Giangaspero, L. Agarossi, G. Paltenghi, S. Okamura, M. Okada, and S. Komaki, 'Co-channel interference cancellation based on MIMO OFDM systems', *IEEE Wireless Communications*, vol. 9, pp. 8–17, December 2002.

[136] J. Li, K. B. Letaief, and Z. Cao, 'Co-channel interference cancellation for space-time coded OFDM systems', *IEEE Transactions on Wireless Communications*, vol. 2, pp. 41–49, January 2003.

[137] S. Y. Park and C. G. Kang, 'Complexity-reduced iterative MAP receiver for interference suppression in OFDM-based spatial multiplexing systems', *IEEE Transactions on Vehicular Technology*, vol. 53, pp. 1316–1326, September 2004.

[138] G. L. Stüber, J. R. Barry, S. W. McLaughlin, Y. Li, M. A. Ingram, and T. G. Pratt, 'Broadband MIMO-OFDM wireless communications', *Proceedings of the IEEE*, vol. 92, pp. 271–294, February 2004.

[139] C. Dubuc, D. Starks, T. Creasy, and Y. Hou, 'A MIMO-OFDM prototype for next-generation wireless WANs', *IEEE Communications Magazine*, vol. 42, pp. 82–87, December 2004.

[140] R. J. Piechocki, P. N. Fletcher, A. Nix, N. Canagarajah, and J. P. McGeehan, 'Performance evaluation of BLAST-OFDM enhanced Hiperlan/2 using simulated and measured channel data', *Electronics Letters*, vol. 37, pp. 1137–1139, August 2001.

[141] S. Catreux, V. Erceg, D. Gesbert, and R. W. Heath Jr, 'Adaptive modulation and MIMO coding for broadband wireless data networks', *IEEE Communications Magazine*, vol. 40, pp. 108–115, June 2002.

[142] R. Piechocki, P. Fletcher, A. Nix, N. Canagarajah, and J. McGeehan, 'A measurement based feasibility study of space-frequency MIMO detection and decoding techniques for next generation wireless LANs', *IEEE Transactions on Consumer Electronics*, vol. 48, pp. 732–737, August 2002.

[143] A. F. Molisch, M. Z. Win, and J. H. Winters, 'Space-time-frequency (STF) coding for MIMO-OFDM systems', *IEEE Communications Letters*, vol. 6, pp. 370–372, September 2002.

[144] A. Stamoulis, S. N. Diggavi, and N. Al-Dhahir, 'Intercarrier interference in MIMO OFDM', *IEEE Transactions on Signal Processing*, vol. 50, pp. 2451–2464, October 2002.

[145] A. Doufexi, M. Hunukumbure, A. Nix, M. A. Beach, and S. Armour, 'COFDM performance evaluation in outdoor MIMO channels using space/polarisation-time processing techniques', *Electronics Letters*, vol. 38, pp. 1720–1721, December 2002.

[146] H. Bölcskei, M. Borgmann, and A. J. Paulraj, 'Impact of the propagation environment on the performance of space-frequency coded MIMO-OFDM', *IEEE Journal on Selected Areas in Communications*, vol. 21, pp. 427–439, April 2003.

[147] J. Cai, W. Song, and Z. Li, 'Doppler spread estimation for mobile OFDM systems in Rayleigh fading channels', *IEEE Transactions on Consumer Electronics*, vol. 49, pp. 973–977, November 2003.

[148] G. Leus and M. Moonen, 'Per-tone equalization for MIMO OFDM systems', *IEEE Transactions on Signal Processing*, vol. 51, pp. 2965–2975, November 2003.

[149] R. J. Piechocki, A. Nix, J. P. McGeehan, and S. M. D. Armour, 'Joint blind and semi-blind detection and channel estimation', *IEE Proceedings – Communications*, vol. 150, pp. 419–426, December 2003.

[150] P. Xia, S. Zhou, and G. B. Giannakis, 'Adaptive MIMO-OFDM based on partial channel state information', *IEEE Transactions on Signal Processing*, vol. 52, pp. 202–213, January 2004.

[151] D. Huang and K. B. Letaief, 'Symbol-based space diversity for coded OFDM systems', *IEEE Transactions on Wireless Communications*, vol. 3, pp. 117–127, January 2004.

[152] M. R. G. Butler and I. B. Collings, 'A zero-forcing approximate log-likelihood receiver for MIMO bit-interleaved coded modulation', *IEEE Communications Letters*, vol. 8, pp. 105–107, February 2004.

[153] B. Lu, G. Yue, and X. Wang, 'Performance analysis and design optimization of LDPC-coded MIMO OFDM systems', *IEEE Transactions on Signal Processing*, vol. 52, pp. 348–361, February 2004.

[154] A. V. Zelst and T. C. W. Schenk, 'Implementation of a MIMO OFDM-based wireless LAN system', *IEEE Transactions on Signal Processing*, vol. 52, pp. 483–494, February 2004.

[155] A. Pascual-Iserte, A. I. Pérez-Neira, and M. A. Lagunas, 'On power allocation strategies for maximum signal to noise and interference ratio in an OFDM-MIMO system', *IEEE Transactions on Wireless Communications*, vol. 3, pp. 808–820, May 2004.

[156] Y. Zeng and T. S. Ng, 'A semi-blind channel estimation method for multiuser multiantenna OFDM systems', *IEEE Transactions on Signal Processing*, vol. 52, pp. 1419–1429, May 2004.

[157] B. Alien, R. Brito, M. Dohler, and A. H. Aghvami, 'Performance comparison of spatial diversity array topologies in an OFDM based wireless LAN', *IEEE Transactions on Consumer Electronics*, vol. 50, pp. 420–428, May 2004.

[158] J. Tan and G. L. Stüber, 'Multicarrier delay diversity modulation for MIMO systems', *IEEE Transactions on Wireless Communications*, vol. 3, pp. 1756–1763, September 2004.

[159] X. Wang, Y. R. Shayan, and M. Zeng, 'On the code and interleaver design of broadband OFDM systems', *IEEE Communications Letters*, vol. 8, pp. 653–655, November 2004.

[160] Y. Pan, K. B. Letaief, and Z. Cao, 'Dynamic spatial subchannel allocation with adaptive beamforming for MIMO/OFDM systems', *IEEE Transactions on Wireless Communications*, vol. 3, pp. 2097–2107, November 2004.

[161] C. Tepedelenlioĝlu and R. Challagulla, 'Low-complexity multipath diversity through fractional sampling in OFDM', *IEEE Transactions on Signal Processing*, vol. 52, pp. 3104–3116, November 2004.

[162] M. S. Baek, M. J. Kim, Y. H. You, and H. K. Song, 'Semi-blind channel estimation and PAR reduction for MIMO-OFDM system with multiple antennas', *IEEE Transactions on Broadcasting*, vol. 50, pp. 414–424, December 2004.

[163] G. Barriac and U. Madhow, 'Space-time communication for OFDM with implicit channel feedback', *IEEE Transactions on Information Theory*, vol. 50, pp. 3111–3129, December 2004.

[164] J. Zhang, A. Kavcic, and K. M. Wong, 'Equal-diagonal QR decomposition and its application to precoder design for successive-cancellation detection', *IEEE Transactions on Information Theory*, vol. 51, pp. 154–172, January 2005.

[165] Y. Yao and G. B. Giannakis, 'Blind carrier frequency offset estimation in SISO, MIMO, and multiuser OFDM systems', *IEEE Transactions on Communications*, vol. 53, pp. 173–183, January 2005.

[166] K. Zheng, L. Huang, W. Wang, and G. Yang, 'TD-CDM-OFDM: evolution of TD-SCDMA toward 4G', *IEEE Communications Magazine*, vol. 43, pp. 45–52, January 2005.

[167] H. Yang, 'A road to future broadband wireless access: MIMO-OFDM-based air interface', *IEEE Communications Magazine*, vol. 43, pp. 53–60, January 2005.

[168] Y. Zhang and K. B. Letaief, 'An efficient resource-allocation scheme for spatial multiuser access in MIMO/OFDM systems', *IEEE Transactions on Communications*, vol. 53, pp. 107–116, January 2005.

[169] X. Ma, M. K. Oh, G. B. Giannakis, and D. J. Park, 'Hopping pilots for estimation of frequency-offset and multiantenna channels in MIMO-OFDM', *IEEE Transactions on Communications*, vol. 53, pp. 162–172, January 2005.

[170] M. Fozunbal, S. W. McLaughlin, and R. W. Schafer, 'On space-time-frequency coding over MIMO-OFDM systems', *IEEE Transactions on Wireless Communications*, vol. 4, pp. 320–331, January 2005.

[171] S. Nanda, R. Walton, J. Ketchum, M. Wallace, and S. Howard, 'A high-performance MIMO OFDM wireless LAN', *IEEE Communications Magazine*, vol. 43, pp. 101–109, February 2005.

[172] K. J. Kim, J. Yue, R. A. Iltis, and J. D. Gibson, 'A QRD-M/Kalman filter-based detection and channel estimation algorithm for MIMO-OFDM systems', *IEEE Transactions on Wireless Communications*, vol. 4, pp. 710–721, March 2005.

[173] Y. Qiao, S. Yu, P. Su, and L. Zhang, 'Research on an iterative algorithm of LS channel estimation in MIMO OFDM systems', *IEEE Transactions on Broadcasting*, vol. 51, pp. 149–153, March 2005.

[174] H. Sampath, V. Erceg, and A. Paulraj, 'Performance analysis of linear precoding based on field trials results of MIMO-OFDM system', *IEEE Transactions on Wireless Communications*, vol. 4, pp. 404–409, March 2005.

[175] F. Rey, M. Lamarca, and G. Vazquez, 'Robust power allocation algorithms for MIMO OFDM systems with imperfect CSI', *IEEE Transactions on Signal Processing*, vol. 53, pp. 1070–1085, March 2005.

[176] Y. Sun, Z. Xiong, and X. Wang, 'EM-based iterative receiver design with carrier-frequency offset estimation for MIMO OFDM systems', *IEEE Transactions on Communications*, vol. 53, pp. 581–586, April 2005.

[177] A. Lodhi, F. Said, M. Dohler, and A. H. Aghvami, 'Performance comparison of space-time block coded and cyclic delay diversity MC-CDMA systems', *IEEE Wireless Communications*, vol. 12, pp. 38–45, April 2005.

[178] Z. Wang, Z. Han, and K. J. R. Liu, 'A MIMO-OFDM channel estimation approach using time of arrivals', *IEEE Transactions on Wireless Communications*, vol. 4, pp. 1207–1213, May 2005.

[179] C. K. Wen, Y. Y. Wang, and J. T. Chen, 'A low-complexity space-time OFDM multiuser system', *IEEE Transactions on Wireless Communications*, vol. 4, pp. 998–1007, May 2005.

[180] W. Su, Z. Safar, and K. J. R. Liu, 'Towards maximum achievable diversity in space, time, and frequency: performance analysis and code design', *IEEE Transactions on Wireless Communications*, vol. 4, pp. 1847–1857, July 2005.

[181] M. Tan, Z. Latinović, and Y. Bar-Ness, 'STBC MIMO-OFDM peak-to-average power ratio reduction by cross-antenna rotation and inversion', *IEEE Communications Letters*, vol. 9, pp. 592–594, July 2005.

[182] K. W. Park and Y. S. Cho, 'An MIMO-OFDM technique for high-speed mobile channels', *IEEE Communications Letters*, vol. 9, pp. 604–606, July 2005.

[183] L. Shao and S. Roy, 'Rate-one space-frequency block codes with maximum diversity for MIMO-OFDM', *IEEE Transactions on Wireless Communications*, vol. 4, pp. 1674–1687, July 2005.

[184] T. C. W. Schenk, X. Tao, P. F. M. Smulders, and E. R. Fledderus, 'On the influence of phase noise induced ICI in MIMO OFDM systems', *IEEE Communications Letters*, vol. 9, pp. 682–684, August 2005.

[185] M. Borgmann and H. Bölcskei, 'Noncoherent space-frequency coded MIMO-OFDM', *IEEE Journal on Selected Areas in Communications*, vol. 23, pp. 1799–1810, September 2005.

[186] A. Tarighat and A. H. Sayed, 'MIMO OFDM receivers for systems with IQ imbalances', *IEEE Transactions on Signal Processing*, vol. 53, pp. 3583–3596, September 2005.

[187] Y. Jiang, J. Li, and W. W. Hager, 'Joint transceiver design for MIMO communications using geometric mean decomposition', *IEEE Transactions on Signal Processing*, vol. 53, pp. 3791–3803, October 2005.

[188] J. Choi and R. W. Heath Jr, 'Interpolation based transmit beamforming for MIMO-OFDM with limited feedback', *IEEE Transactions on Signal Processing*, vol. 53, pp. 4125–4135, November 2005.

[189] M. S. Baek, H. J. Kook, M. J. Kim, Y. H. You, and H. K. Song, 'Multi-antenna scheme for high capacity transmission in the digital audio broadcasting', *IEEE Transactions on Broadcasting*, vol. 51, pp. 551–559, December 2005.

[190] M. Fakhereddin, M. Sharif, and B. Hassibi, 'Reduced feedback and random beamforming for OFDM MIMO broadcast channels', *IEEE Transactions on Communications*, vol. 57, pp. 3827–3835, December 2009.

[191] P. De, T. Chang, and C. Chi, 'Linear prediction based semiblind channel estimation for multiuser OFDM with insufficient guard interval', *IEEE Transactions on Wireless Communications*, vol. 8, pp. 5728–5736, December 2009.

[192] L. Haring, S. Bieder, and A. Czylwik, 'Fine frequency synchronization in the uplink of multiuser OFDM systems', *IEEE Transactions on Communications*, vol. 57, pp. 3743–3752, December 2009.

[193] P. Vandenameele, L. V. D. Perre, and M. Engels, *Space Division Multiple Access for Wireless Local Area Networks*. London: Kluwer, 2001.

[194] I. P. Kovalyov, *SDMA for Multipath Wireless Channels: Limiting Characteristics and Stochastic Models*. Berlin: Springer, 1st edn, 2004.

[195] D. Tse and P. Viswanath, *Fundamentals of Wireless Communication*. Cambridge: Cambridge University Press, 2005. ISBN-13 978-0-521-84527-4.

[196] M. Cooper and M. Goldburg, 'Intelligent antennas: spatial division multiple access', *ArrayComm: Annual Review of Communications*, pp. 999–1002, 1996.

[197] P. Vandenameele, L. V. D. Perre, M. Engels, B. Gyselinckx, and H. D. Man, 'A novel class of uplink OFDM/SDMA algorithms: a statistical performance analysis', in *Proceedings of the IEEE Vehicular Technology Conference (VTC 1999 Fall)*, (Amsterdam, The Netherlands), vol. 1, pp. 324–328, 19–22 September 1999.

[198] P. Vandenameele, L. V. D. Perre, M. Engels, B. Gyselinckx, and H. D. Man, 'A combined OFDM/SDMA approach', *IEEE Journal on Selected Areas in Communications*, vol. 18, pp. 2312–2321, November 2000.

[199] S. Thoen, L. V. D. Perre, M. Engels, and H. D. Man, 'Adaptive loading for OFDM/SDMA-based wireless networks', *IEEE Transactions on Communications*, vol. 50, pp. 1798–1810, November 2002.

[200] S. Thoen, L. Deneire, L. V. D. Perre, M. Engels, and H. D. Man, 'Constrained least squares detector for OFDM/SDMA-based wireless networks', *IEEE Transactions on Wireless Communications*, vol. 2, pp. 129–140, January 2003.

[201] A. T. Alastalo and M. Kahola, 'Smart-antenna operation for indoor wireless local-area networks using OFDM', *IEEE Transactions on Wireless Communications*, vol. 2, pp. 392–399, March 2003.

[202] M. Y. Alias, A. K. Samingan, S. Chen, and L. Hanzo, 'Multiple antenna aided OFDM employing minimum bit error rate multiuser detection', *Electronics Letters*, vol. 39, pp. 1769–1770, November 2003.

[203] X. Dai, 'Carrier frequency offset estimation for OFDM/SDMA systems using consecutive pilots', *IEE Proceedings – Communications*, vol. 152, pp. 624–632, October 2005.

[204] Y. S. Yeh and D. Reudink, 'Efficient spectrum utilization for mobile radio systems using space diversity', *IEEE Transactions on Communications*, vol. 30, pp. 447–455, March 1982.

[205] K. T. Ko and B. Davis, 'A space-division multiple-access protocol for spot-beam antenna and satellite-switched communication network', *IEEE Journal on Selected Areas in Communications*, vol. 1, pp. 126–132, January 1983.

[206] S. C. Swales, M. A. Beach, and D. J. Edwards, 'Multi-beam adaptive base-station antennas for cellular land mobile radio systems', in *Proceedings of the IEEE 39th Vehicular Technology Conference (VTC 1989 Spring)*, (San Francisco, USA), vol. 1, pp. 341–348, 1–3 May 1989.

[207] S. C. Swales, M. A. Beach, D. J. Edwards, and J. P. McGeehan, 'The performance enhancement of multibeam adaptive base-station antennas for cellular land mobile radio systems', *IEEE Transactions on Vehicular Technology*, vol. 39, pp. 56–67, February 1990.

[208] B. G. Agee, S. V. Schell, and W. A. Gardner, 'Spectral self-coherence restoral: a new approach to blind adaptive signal extraction using antenna arrays', *Proceedings of the IEEE*, vol. 78, pp. 753–767, April 1990.

[209] S. Anderson, M. Millnert, M. Viberg, and B. Wahlberg, 'An adaptive array for mobile communication systems', *IEEE Transactions on Vehicular Technology*, vol. 40, pp. 230–236, February 1991.

[210] P. Balaban and J. Salz, 'Optimum diversity combining and equalization in digital data transmission with applications to cellular mobile radio. Part I: Theoretical considerations', *IEEE Transactions on Communications*, vol. 40, pp. 885–894, May 1992.

[211] P. Balaban and J. Salz, 'Optimum diversity combining and equalization in digital data transmission with applications to cellular mobile radio. Part II: Numerical results', *IEEE Transactions on Communications*, vol. 40, pp. 895–907, May 1992.

[212] G. Xu, H. Liu, W. J. Vogel, H. P. Lin, S. S. Jeng, and G. W. Torrence, 'Experimental studies of space-division-multiple-access schemes for spectral efficient wireless communications', in *IEEE International Conference on Communications (ICC 1994)*, (New Orleans, USA), vol. 2, pp. 800–804, 1–5 May 1994.

[213] S. Talwar, M. Viberg, and A. Paulraj, 'Blind estimation of multiple co-channel digital signals using an antenna array', *IEEE Signal Processing Letters*, vol. 1, pp. 29–31, February 1994.

[214] A. J. V. D. Veen, S. Talwar, and A. Paulraj, 'Blind estimation of multiple digital signals transmitted over FIR channels', *IEEE Signal Processing Letters*, vol. 2, pp. 99–102, May 1995.

[215] B. H. Khalaj, A. Paulraj, and T. Kailath, 'Spatio-temporal channel estimation techniques for multiple access spread spectrum systems with antenna arrays', in *IEEE International Conference on Communications (ICC 1995)*, (Seattle, USA), vol. 3, pp. 1520–1524, 18–22 June 1995.

[216] K. Anand, G. Mathew, and V. U. Reddy, 'Blind separation of multiple co-channel BPSK signals arriving at an antenna array', *IEEE Signal Processing Letters*, vol. 2, pp. 176–178, September 1995.

[217] H. Liu and G. Xu, 'Smart antennas in wireless systems: uplink multiuser blind channel and sequence detection', *IEEE Transactions on Communications*, vol. 45, pp. 187–199, February 1997.

[218] G. Tsoulos, M. A. Beach, and J. McGeehan, 'Wireless personal communications for the 21st century: European technological advances in adaptive antennas', *IEEE Communications Magazine*, vol. 35, pp. 102–109, September 1997.

[219] L. Deneire and D. T. M. Slock, 'Blind channel identification based on cyclic statistics', *IEE Proceedings – Radar, Sonar and Navigation*, vol. 145, pp. 58–62, February 1998.

[220] G. Tsoulos, J. McGeehan, and M. A. Beach, 'Space division multiple access (SDMA) field trials. Part I: Tracking and BER performance', *IEE Proceedings – Radar, Sonar and Navigation*, vol. 145, pp. 73–78, February 1998.

[221] G. Tsoulos, J. McGeehan, and M. A. Beach, 'Space division multiple access (SDMA) field trials. Part II: Calibration and linearity issues', *IEE Proceedings – Radar, Sonar and Navigation*, vol. 145, pp. 79–84, February 1998.

[222] V. A. N. Barroso, J. M. F. Moura, and J. Xavier, 'Blind array channel division multiple access (AChDMA) for mobile communications', *IEEE Transactions on Signal Processing*, vol. 46, pp. 737–752, March 1998.

[223] F. Demmerle and W. Wiesbeck, 'A biconical multibeam antenna for space-division multiple access', *IEEE Transactions on Antennas and Propagation*, vol. 46, pp. 782–787, June 1998.

[224] B. Lindmark, S. Lundgren, J. R. Sanford, and C. Beckman, 'Dual-polarized array for signal-processing applications in wireless communications', *IEEE Transactions on Antennas and Propagation*, vol. 46, pp. 758–763, June 1998.

[225] B. Suard, G. Xu, H. Liu, and T. Kailath, 'Uplink channel capacity of space-division-multiple-access schemes', *IEEE Transactions on Information Theory*, vol. 44, pp. 1468–1476, July 1998.

[226] S. S. Jeng, G. Xu, H. P. Lin, and W. J. Vogel, 'Experimental studies of spatial signature variation at 900 MHz for smart antenna systems', *IEEE Transactions on Antennas and Propagation*, vol. 46, pp. 953–962, July 1998.

[227] P. Petrus, R. B. Ertel, and J. H. Reed, 'Capacity enhancement using adaptive arrays in an AMPS system', *IEEE Transactions on Vehicular Technology*, vol. 47, pp. 717–727, August 1998.

[228] J. M. F. Xavier, V. A. N. Barroso, and J. M. F. Moura, 'Closed-form blind channel identification and source separation in SDMA systems through correlative coding', *IEEE Journal on Selected Areas in Communications*, vol. 16, pp. 1506–1517, October 1998.

[229] C. Farsakh and J. A. Nossek, 'Spatial covariance based downlink beamforming in an SDMA mobile radio system', *IEEE Transactions on Communications*, vol. 46, pp. 1497–1506, November 1998.

[230] G. V. Tsoulos, 'Smart antennas for mobile communication systems: benefits and challenges', *Electronics and Communication Engineering Journal*, vol. 11, pp. 84–94, April 1999.

[231] F. Piolini and A. Rolando, 'Smart channel-assignment algorithm for SDMA systems', *IEEE Transactions on Microwave Theory and Techniques*, vol. 47, pp. 693–699, June 1999.

[232] G. M. Galvan-Tejada and J. G. Gardiner, 'Theoretical blocking probability for SDMA', *IEE Proceedings – Communications*, vol. 146, pp. 303–306, October 1999.

[233] G. M. Galvan-Tejada and J. G. Gardiner, 'Theoretical model to determine the blocking probability for SDMA systems', *IEEE Transactions on Vehicular Technology*, vol. 50, pp. 1279–1288, September 2001.

[234] G. V. Tsoulos, 'Experimental and theoretical capacity analysis of space-division multiple access (SDMA) with adaptive antennas', *IEE Proceedings – Communications*, vol. 146, pp. 307–311, October 1999.

[235] U. Vornefeld, C. Walke, and B. Walke, 'SDMA techniques for wireless ATM', *IEEE Communications Magazine*, vol. 37, pp. 52–57, November 1999.

[236] P. Djahani and J. M. Kahn, 'Analysis of infrared wireless links employing multibeam transmitters and imaging diversity receivers', *IEEE Transactions on Communications*, vol. 48, pp. 2077–2088, December 2000.

[237] F. Shad, T. D. Todd, V. Kezys, and J. Litva, 'Dynamic slot allocation (DSA) in indoor SDMA/TDMA using a smart antenna base station', *IEEE/ACM Transactions on Networking*, vol. 9, pp. 69–81, February 2001.

[238] R. Kuehner, T. D. Todd, F. Shad, and V. Kezys, 'Forward-link capacity in smart antenna base stations with dynamic slot allocation', *IEEE Transactions on Vehicular Technology*, vol. 50, pp. 1024–1038, July 2001.

[239] S. S. Jeon, Y. Wang, Y. Qian, and T. Itoh, 'A novel smart antenna system implementation for broad-band wireless communications', *IEEE Transactions on Antennas and Propagation*, vol. 50, pp. 600–606, May 2002.

[240] S. Bellofiore, C. A. Balanis, J. Foutz, and A. S. Spanias, 'Smart-antenna systems for mobile communication networks. Part 1: Overview and antenna design', *IEEE Antennas and Propagation Magazine*, vol. 44, pp. 145–154, June 2002.

[241] S. Bellofiore, C. A. Balanis, J. Foutz, and A. S. Spanias, 'Smart-antenna systems for mobile communication networks. Part 2: Beamforming and network throughput', *IEEE Antennas and Propagation Magazine*, vol. 44, pp. 106–114, August 2002.

[242] X. Fang, 'More realistic analysis for blocking probability in SDMA systems', *IEE Proceedings – Communications*, vol. 149, pp. 152–156, June 2002.

[243] A. Arredondo, K. R. Dandekar, and G. Xu, 'Vector channel modeling and prediction for the improvement of downlink received power', *IEEE Transactions on Communications*, vol. 50, pp. 1121–1129, July 2002.

[244] C. M. Walke and T. J. Oechtering, 'Analytical expression for uplink C/I-distribution in interference-limited cellular radio systems', *Electronics Letters*, vol. 38, pp. 743–744, July 2002.

[245] T. Zwick, C. Fischer, and W. Wiesbeck, 'A stochastic multipath channel model including path directions for indoor environments', *IEEE Journal on Selected Areas in Communications*, vol. 20, pp. 1178–1192, August 2002.

[246] S. A. Zekavat, C. R. Nassar, and S. Shattil, 'Oscillating-beam smart antenna arrays and multicarrier systems: achieving transmit diversity, frequency diversity, and directionality', *IEEE Transactions on Vehicular Technology*, vol. 51, pp. 1030–1039, September 2002.

[247] J. L. Pan and P. M. Djurić, 'Multibeam cellular mobile communications with dynamic channel assignment', *IEEE Transactions on Vehicular Technology*, vol. 51, pp. 1252–1258, September 2002.

[248] C. C. Cavalcante, F. R. P. Cavalcanti, and J. C. M. Mota, 'Adaptive blind multiuser separation criterion based on log-likelihood maximisation', *Electronics Letters*, vol. 38, pp. 1231–1233, September 2002.

[249] H. Yin and H. Liu, 'Performance of space-division multiple-access (SDMA) with scheduling', *IEEE Transactions on Wireless Communications*, vol. 1, pp. 611–618, October 2002.

[250] M. Rim, 'Multi-user downlink beamforming with multiple transmit and receive antennas', *Electronics Letters*, vol. 38, pp. 1725–1726, December 2002.

[251] I. Bradaric, A. P. Pertropulu, and K. I. Diamantaras, 'Blind MIMO FIR channel identification based on second-order spectra correlations', *IEEE Transactions on Signal Processing*, vol. 51, pp. 1668–1674, June 2003.

[252] Q. H. Spencer, A. L. Swindlehurst, and M. Haardt, 'Zero-forcing methods for downlink spatial multiplexing in multiuser MIMO channels', *IEEE Transactions on Signal Processing*, vol. 52, pp. 461–471, February 2004.

[253] J. Li, K. B. Letaief, and Z. Cao, 'A reduced-complexity maximum-likelihood method for multiuser detection', *IEEE Transactions on Communications*, vol. 52, pp. 289–295, February 2004.

[254] L. U. Choi and R. D. Murch, 'A pre-BLAST-DFE technique for the downlink of frequency-selective fading MIMO channels', *IEEE Transactions on Communications*, vol. 52, pp. 737–743, May 2004.

[255] W. Ajib and D. Haccoun, 'An overview of scheduling algorithms in MIMO-based fourth-generation wireless systems', *IEEE Network*, vol. 19, pp. 43–48, September/October 2005.

[256] K. M. Nasr, F. Costen, and S. K. Barton, 'A wall imperfection channel model for signal level prediction and its impact on smart antenna systems for indoor infrastructure WLAN', *IEEE Transactions on Antennas and Propagation*, vol. 53, pp. 3767–3775, November 2005.

[257] P. Höher, S. Kaiser, and P. Robertson, 'Pilot-symbol-aided channel estimation in time and frequency', in *IEEE Global Telecommunications Conference: The Mini-Conference*, (Phoenix, USA), pp. 90–96, November 1997.

[258] P. Höher, S. Kaiser, and P. Robertson, 'Two-dimensional pilot-symbol-aided channel estimation by Wiener filtering', in *IEEE International Conference on Acoustics, Speech and Signal Processing*, (Munich, Germany), pp. 1845–1848, April 1997.

[259] O. Edfords, M. Sandell, J.-J. van de Beek, S. Wilson, and P. Börjesson, 'OFDM channel estimation by singular value decomposition', *IEEE Transactions on Communications*, vol. 46, pp. 931–939, July 1998.

[260] Y. Li, L. Cimini, and N. Sollenberger, 'Robust channel estimation for OFDM systems with rapid dispersive fading channels', *IEEE Transactions on Communications*, vol. 46, pp. 902–915, April 1998.

[261] Y. Li, 'Pilot-symbol-aided channel estimation for OFDM in wireless systems', *IEEE Transactions on Vehicular Technology*, vol. 49, pp. 1207–1215, July 2000.

[262] B. Yang, Z. Cao, and K. Letaief, 'Analysis of low-complexity windowed DFT-based MMSE channel estimation for OFDM systems', *IEEE Transactions on Communications*, vol. 49, pp. 1977–1987, November 2001.

[263] M. Münster and L. Hanzo, 'RLS-adaptive parallel interference cancellation assisted decision-directed channel estimation for OFDM', in *IEEE Wireless Communications and Networking (WCNC 2003)*, (New Orleans, USA), vol. 1, pp. 50–54, 16–20 March 2003.

[264] R. Otnes and M. Tüchler, 'Iterative channel estimation for turbo equalization of time-varying frequency-selective channels', *IEEE Transactions on Wireless Communications*, vol. 3, no. 6, pp. 1918–1923, 2004.

[265] L. Hanzo, M. Münster, B. J. Choi, and T. Keller, *OFDM and MC-CDMA for Broadband Multi-User Communications, WLANs and Broadcasting*. Chichester: IEEE Press and John Wiley & Sons, Ltd, 2003.

[266] A. Goldsmith, S. A. Jafar, N. Jindal, and S. Vishwanath, 'Capacity limits of MIMO channels', *IEEE Journal on Selected Areas in Communications*, vol. 21, pp. 684–702, June 2003.

[267] M. Münster, 'Antenna diversity-assisted adaptive wireless multiuser OFDM systems', PhD thesis, University of Southampton, 2002.

[268] M. Morelli and U. Mengali, 'A comparison of pilot-aided channel estimation methods for OFDM systems', *IEEE Transactions on Signal Processing*, vol. 49, pp. 3065–3073, December 2001.

[269] M.-X. Chang and Y. Su, 'Model-based channel estimation for OFDM signals in Rayleigh fading', *IEEE Transactions on Communications*, vol. 50, pp. 540–544, April 2002.

[270] J.-J. van de Beek, O. Edfors, M. Sandell, S. Wilson, and P. Börjesson, 'On channel estimation in OFDM systems', in *Proceedings of IEEE Vehicular Technology Conference*, (Chicago, USA), vol. 2, pp. 815–819, July 1995.

[271] V. Mignone and A. Morello, 'A novel demodulation scheme for fixed and mobile receivers', *IEEE Transactions on Communications*, vol. 44, pp. 1144–1151, September 1996.

[272] Y. Li and N. Sollenberger, 'Clustered OFDM with channel estimation for high rate wireless data', *IEEE Transactions on Communications*, vol. 49, pp. 2071–2076, December 2001.

[273] M. Münster and L. Hanzo, 'Second-order channel parameter estimation assisted cancellation of channel variation-induced inter-subcarrier interference in OFDM systems', in *EUROCON'2001, International Conference on Trends in Communications*, (Bratislava, Slovakia), vol. 1, pp. 1–5, 4–7 July 2001.

[274] M. Münster and L. Hanzo, 'MMSE channel prediction assisted symbol-by-symbol adaptive OFDM', in *Proceedings of IEEE International Conference on Communications*, (New York, USA), vol. 1, pp. 416–420, 28 April–2 May 2002.

[275] M. Sandell, C. Luschi, P. Strauch, and R. Yan, 'Iterative channel estimation using soft decision feedback', in *IEEE Global Telecommunications Conference, GLOBECOM'98: The Bridge to Global Integration*, (Sydney, Australia), vol. 6, pp. 3728–3733, 1998.

[276] M. Valenti, 'Iterative channel estimation for turbo codes over fading channels', in *IEEE Wireless Communications and Networking Conference (WCNC 2000)*, (Chicago, USA), vol. 3, pp. 1019–1024, 23–28 September 2000.

[277] B.-L. Yeap, C. Wong, and L. Hanzo, 'Reduced complexity in-phase/quadrature-phase m-QAM turbo equalization using iterative channel estimation', *IEEE Transactions on Wireless Communications*, vol. 2, no. 1, pp. 2–10, 2003.

[278] S. Song, A. Singer, and K.-M. Sung, 'Turbo equalization with an unknown channel', in *Proceedings of International Conference on Acoustics, Speech, and Signal Processing, 2002 (ICASSP'02)*, (Hong Kong), vol. 3, 2002.

[279] S. Song, A. Singer, and K.-M. Sung, 'Soft input channel estimation for turbo equalization', *IEEE Transactions on Signal Processing*, vol. 52, pp. 2885–2894, October 2004.

[280] R. Otnes and M. Tüchler, 'Soft iterative channel estimation for turbo equalization: comparison of channel estimation algorithms', in *The 8th International Conference on Communication Systems (ICCS 2002)*, (Amsterdam, The Netherlands), vol. 1, pp. 72–76, 2002.

[281] N. Seshadri, 'Joint data and channel estimation using blind trellis search techniques', *IEEE Transactions on Communications*, vol. 42, pp. 1000–1011, February/March/April 1994.

[282] A. Knickenberg, B.-L. Yeap, J. Hàmorský, M. Breiling, and L. Hanzo, 'Non-iterative joint channel equalisation and channel decoding', *IEE Electronics Letters*, vol. 35, pp. 1628–1630, 16 September 1999.

[283] C. Cozzo and B. Hughes, 'Joint channel estimation and data detection in space-time communications', *IEEE Transactions on Communications*, vol. 51, pp. 1266–1270, August 2003.

[284] T. Cui and C. Tellambura, 'Joint channel estimation and data detection for OFDM systems via sphere decoding', in *Proceedings of IEEE Global Telecommunications Conference*, (Dallas, USA), vol. 6, pp. 3656–3660, 29 November–3 December 2004.

[285] T. Cui and C. Tellambura, 'Joint data detection and channel estimation for OFDM systems', *IEEE Transactions on Communications*, vol. 54, no. 4, pp. 670–679, 2006.

[286] C. Antón-Haro, J. Fonolossa, and J. Fonolossa, 'Blind channel estimation and data detection using hidden Markov models', *IEEE Transactions on Signal Processing*, vol. 45, pp. 241–247, January 1997.

[287] D. Boss, K. Kammeyer, and T. Petermann, 'Is blind channel estimation feasible in mobile communication systems? A study based on GSM', *IEEE Journal on Selected Areas of Communications*, vol. 16, pp. 1479–1492, October 1998.

[288] T. Endres, S. Halford, C. Johnson, and G. Giannakis, 'Blind adaptive channel equalization using fractionally-spaced receivers: a comparison study', in *Proceedings of the Conference on Information Sciences and Systems*, (Princeton, USA), 20–22 March 1996.

[289] G. Giannakis and S. Halford, 'Asymptotically optimal blind fractionally spaced channel estimation and performance results', *IEEE Transactions on Signal Processing*, vol. 45, pp. 1815–1830, July 1997.

[290] S. Zhou and G. Giannakis, 'Finite-alphabet based channel estimation for OFDM and related multicarrier systems', *IEEE Transactions on Communications*, vol. 49, pp. 1402–1414, August 2001.

[291] M. Necker and G. Stüber, 'Totally blind channel estimation for OFDM over fast varying mobile channels', in *Proceedings of IEEE International Conference on Communications*, (New York, USA), April 28–May 2 2002.

[292] S. Haykin, *Adaptive Filter Theory*. Englewood Cliffs, NJ: Prentice Hall, 1996.

[293] D. Schafhuber and G. Matz, 'MMSE and adaptive prediction of time-varying channels for OFDM systems', *IEEE Transactions on Wireless Communications*, vol. 4, pp. 593–602, March 2005.

[294] Y. Li, J. Winters, and N. Sollenberger, 'MIMO-OFDM for wireless communications: signal detection with enhanced channel estimation', *IEEE Transactions on Communications*, vol. 50, no. 9, pp. 1471–1477, 2002.

[295] G. Stüber, J. Barry, S. McLaughlin, Y. Li, M. Ingram, and T. Pratt, 'Broadband MIMO-OFDM wireless communications', *Proceedings of the IEEE*, vol. 92, pp. 271–294, February 2004.

[296] X. Deng, A. Haimovich, and J. Garcia-Frias, 'Decision directed iterative channel estimation for MIMO systems', in *Proceedings of IEEE International Conference on Communications*, (Anchorage, USA), vol. 4, pp. 2326–2329, 11–15 May 2003.

[297] M. Münster and L. Hanzo, 'Parallel-interference-cancellation-assisted decision-directed channel estimation for OFDM systems using multiple transmit antennas', *IEEE Transactions on Wireless Communications*, vol. 4, pp. 2148–2162, September 2005.

[298] S. Yatawatta and A. Petropulu, 'Blind channel estimation in MIMO OFDM systems with multiuser interference', *IEEE Transactions on Signal Processing*, vol. 54, pp. 1054–1068, March 2006.

[299] P. Vandenameele, L. van der Perre, M. Engels, B. Gyselinckx, and H. D. Man, 'A combined OFDM/SDMA approach', *IEEE Journal on Selected Areas in Communications*, vol. 18, pp. 2312–2321, November 2000.

[300] V. Jungnickel, T. Haustein, E. Jorswieck, V. Pohl, and C. von Helmolt, 'Performance of a MIMO system with overlay pilots', in *Global Telecommunications Conference (GLOBECOM'01)*, (San Antonio, USA), vol. 1, pp. 594–598, 25–29 November 2001.

[301] H. Bolcskei, R. Heath, and A. Paulraj, 'Blind channel identification and equalization in OFDM-based multiantenna systems', *IEEE Transactions on Signal Processing*, vol. 50, pp. 96–109, January 2002.

[302] H. Zhu, B. Farhang-Boroujeny, and C. Schlegel, 'Pilot embedding for joint channel estimation and data detection in MIMO communication systems', *IEEE Communications Letters*, vol. 7, pp. 30–32, January 2003.

[303] Y. Li, N. Seshardi, and S. Ariyavisitakul, 'Channel estimation for OFDM systems with transmit diversity in mobile wireless channels', *IEEE Journal on Selected Areas in Communications*, vol. 17, pp. 461–471, March 1999.

[304] Y. Li, 'Simplified channel estimation for OFDM systems with multiple transmit antennas', *IEEE Transactions on Wireless Communications*, vol. 1, pp. 67–75, January 2002.

[305] M. Münster and L. Hanzo, 'Multi-user OFDM employing parallel interference cancellation assisted decision-directed channel estimation', in *Proceedings of IEEE Vehicular Technology Conference (VTC 2002 Fall)*, (Vancouver, Canada), vol. 3, pp. 1413–1417, 24–28 September 2002.

[306] A. Grant, 'Joint decoding and channel estimation for linear MIMO channels', in *IEEE Wireless Communications and Networking Conference, WCNC 2000*, (Chicago, USA), vol. 3, pp. 1009–1012, 23–28 September 2000.

[307] H. Mai, Y. Zakharov, and A. Burr, 'Iterative B-spline channel estimation for space-time block coded systems in fast flat fading channels', in *IEEE Vehicular Technology Conference (VTC 2005 Spring)*, (Dallas, USA), vol. 1, pp. 476–480, 30 May–1 June 2005.

[308] X. Qiao, Y. Cai, and Y. Xu, 'Joint iterative decision feedback channel estimation for turbo coded v-BLAST MIMO-OFDM systems', in *IEEE International Symposium on Communications and Information Technology*, (Beijing, China), vol. 2, pp. 1384–1388, 2005.

[309] H. Mai, A. Burr, and S. Hirst, 'Iterative channel estimation for turbo equalization', in *IEEE International Symposium on Personal, Indoor and Mobile Radio Communications (PIMRC 2004)*, (Barcelona, Spain), vol. 2, pp. 1327–1331, 2004.

[310] G. Foschini Jr. and M. Gans, 'On limits of wireless communication in a fading environment when using multiple antennas', *Wireless Personal Communications*, vol. 6, pp. 311–335, March 1998.

[311] A. van Zelst, R. van Nee, and G. Awater, 'Space Division Multiplexing (SDM) for OFDM systems', in *Proceedings of IEEE Vehicular Technology Conference*, (Tokyo, Japan), vol. 2, pp. 1070–1074, 15–18 May 2000.

[312] G. Foschini, 'Layered space-time architecture for wireless communication in a fading environment when using multi-element antennas'', *Bell Labs Technical Journal*, Autumn, pp. 41–59, 1996.

[313] J. Blogh and L. Hanzo, *Third-Generation Systems and Intelligent Networking*. Chichester: IEEE Press and John Wiley & Sons, Ltd, 2002. http://www-mobile.ecs.soton.ac.uk.

[314] L. Hanzo, T. H. Liew, and B. L. Yeap, *Turbo Coding, Turbo Equalisation and Space-Time Coding*. Chichester: IEEE Press and John Wiley & Sons, Ltd, 2002. http://www-mobile.ecs.soton.ac.uk.

[315] S. X. Ng, B. L. Yeap, and L. Hanzo, 'Full-rate, full-diversity adaptive space time block coding for transmission over Rayleigh fading channels', in *Proceedings of IEEE Vehicular Technology Conference (VTC'05 Fall)*, (Dallas, USA), 25–28 September 2005.

[316] S. Alamouti, 'A simple transmit diversity technique for wireless communications', *IEEE Journal on Selected Areas in Communications*, vol. 16, pp. 1451–1458, October 1998.

[317] V. Tarokh, N. Seshadri, and A. Calderbank, 'Space-time codes for high data rate wireless communication: performance criterion and code construction', *IEEE Transactions on Information Theory*, vol. 44, pp. 744–765, March 1998.

[318] V. Tarokh, A. Naguib, N. Seshadri, and A. Calderbank, 'A space-time coding modem for high-data-rate wireless communications', *IEEE Journal on Selected Areas in Communications*, vol. 16, pp. 1459–1477, October 1998.

[319] V. Tarokh, A. Naguib, N. Seshadri, and A. Calderbank, 'Space-time codes for high data rate wireless communication: performance criteria in the presence of channel estimation errors, mobility and multiple paths', *IEEE Transactions on Communications*, vol. 47, pp. 199–207, February 1999.

[320] V. Tarokh, H. Jafarkhani, and A. Calderbank, 'Space-time block coding for wireless communications: performance results', *IEEE Transactions on Communications*, vol. 17, pp. 451–460, March 1999.

[321] V. Tarokh, A. Naguib, N. Seshadri, and A. R. Calderbank, 'Combined array processing and space-time coding', *IEEE Transactions on Information Theory*, vol. 45, pp. 1121–1128, May 1999.

[322] V. Tarokh, H. Jafarkhani, and A. Calderbank, 'Space-time block codes from orthogonal designs', *IEEE Transactions on Information Theory*, vol. 45, pp. 1456–1467, July 1999.

[323] V. Tarokh and H. Jafarkhani, 'A differential detection scheme for transmit diversity', *IEEE Journal on Selected Areas in Communications*, vol. 18, pp. 1169–1174, July 2000.

[324] S. Ariyavisitakul, J. Winters, and I. Lee, 'Optimum space-time processors with dispersive interference: unified analysis and required filter span', *IEEE Transactions on Communications*, vol. 47, pp. 1073–1083, July 1999.

[325] S. Ariyavisitakul, J. Winters, and N. Sollenberger, 'Joint equalization and interference suppression for high data rate wireless systems', *IEEE Journal on Selected Areas in Communications*, vol. 18, pp. 1214–1220, July 2000.

[326] G. Golden, G. Foschini, R. Valenzuela, and P. Wolniansky, 'Detection algorithms and initial laboratory results using V-BLAST space-time communication architecture', *IEE Electronics Letters*, vol. 35, pp. 14–16, January 1999.

[327] C. Hassell-Sweatman, J. Thompson, B. Mulgrew, and P. Grant, 'A comparison of detection algorithms including BLAST for wireless communication using multiple antennas', in *Proceedings of IEEE International Symposium on Personal, Indoor and Mobile Radio Communications*, (London, UK), vol. 1, pp. 698–703, 18–21 September 2000.

[328] D. E. Goldberg, *Genetic Algorithms in Search, Optimization, and Machine Learning*. Reading, MA: Addison-Wesley, 1989.

[329] M. J. Juntti, T. Schlösser, and J. O. Lilleberg, 'Genetic algorithms for multiuser detection in synchronous CDMA', in *IEEE International Symposium on Information Theory – ISIT'97*, (Ulm, Germany), p. 492, 1997.

[330] S. M. Kay, *Fundamentals of Statistical Signal Processing*. Englewood Cliffs, NJ: Prentice Hall, 1998.

[331] R. van Nee, A. van Zelst, and G. Awater, 'Maximum likelihood decoding in a space-division multiplexing system', in *Proceedings of IEEE Vehicular Technology Conference*, (Tokyo, Japan), vol. 1, pp. 6–10, 15–18 May 2000.

[332] G. Awater, A. van Zelst, and R. van Nee, 'Reduced complexity space division multiplexing receivers', in *Proceedings of IEEE Vehicular Technology Conference*, (Tokyo, Japan), vol. 1, pp. 11–15, 15–18 May 2000.

[333] U. Fincke and M. Pohst, 'Improved method for calculating vector of short length in a lattice, including a complexity analysis', *Mathematics of Computation*, vol. 44, pp. 463–471, April 1985.

[334] M. O. Damen, A. Chkeif, and J.-C. Belfiore, 'Lattice code decoder for space-time codes', *IEEE Communications Letters*, pp. 161–163, May 2000.

[335] B. Hochwald and S. ten Brink, 'Achieving near-capacity on a multiple-antenna channel', *IEEE Transactions on Communications*, vol. 51, no. 3, pp. 389–399, 2003.

[336] M. O. Damen, H. E. Gamal, and G. Caier, 'On maximum-likelihood detection and the search for closest lattice point', *IEEE Transactions on Information Theory*, vol. 49, pp. 2389–2402, October 2003.

[337] D. Pham, K. R. Pattipati, P. K. Willett, and J. Luo, 'An improved complex sphere decoder for V-BLAST systems', *IEEE Signal Processing Letters*, vol. 11, pp. 748–751, September 2004.

[338] T. Cui and C. Tellambura, 'Approximate ML detection for MIMO systems using multistage sphere decoding', *IEEE Signal Processing Letters*, vol. 12, pp. 222–225, March 2005.

[339] M. Damen, K. Abed-Meraim, and J.-C. Belfiore, 'Generalised sphere decoder for asymmetrical space-time communication architecture', *Electronics Letters*, vol. 36, no. 2, pp. 166–167, 2000.

[340] T. Cui and C. Tellambura, 'An efficient generalized sphere decoder for rank-deficient MIMO systems', *IEEE Communications Letters*, vol. 9, no. 5, pp. 423–425, 2005.

[341] Z. Yang, C. Liu, and J. He, 'A new approach for fast generalized sphere decoding in MIMO systems', *IEEE Signal Processing Letters*, vol. 12, pp. 41–44, January 2005.

[342] W. Zhao and G. Giannakis, 'Sphere decoding algorithms with improved radius search', *IEEE Transactions on Communications*, vol. 53, no. 7, pp. 1104–1109, 2005.

[343] R. Wang and G. Giannakis, 'Approaching MIMO channel capacity with soft detection based on hard sphere decoding', *IEEE Transactions on Communications*, vol. 54, pp. 587–590, April 2006.

[344] H. Vikalo, B. Hassibi, and T. Kailath, 'Iterative decoding for MIMO channels via modified sphere decoding', *IEEE Transactions on Wireless Communications*, vol. 3, pp. 2299–2311, November 2004.

[345] C. Berrou, A. Glavieux, and P. Thitimajshima, 'Near Shannon limit error-correcting coding and decoding: turbo codes', in *Proceedings of IEEE International Conference on Communications*, (Geneva, Switzerland), pp. 1064–1070, May 1993.

[346] S. Bug, C. Wengerter, I. Gaspard, and R. Jakoby, 'Channel model based on comprehensive measurements for DVB-T mobile applications', in *IEEE Instrumentation and Measurements Technology Conference*, (Budapest, Hungary), 21–23 May 2001.

[347] M. Failli, 'Digital land mobile radio communications COST 207', Technical Report, European Commission, 1989.

[348] R. Steele and L. Hanzo, eds, *Mobile Radio Communications*. New York: IEEE Press and John Wiley & Sons, Inc., 2nd edn, 1999.

[349] W. Jakes Jr, ed., *Microwave Mobile Communications*. New York: John Wiley & Sons, Inc., 1974.

[350] M. C. Jeruchim, P. Balaban, and K. S. Shanmugan, *Simulation of Communication Systems*. New York: Kluwer, 2nd edn, 2000.

[351] L. Hanzo, C. Wong, and M. Yee, *Adaptive Wireless Transceivers*. Chichester: IEEE Press and John Wiley & Sons, Ltd, 2002. http://www-mobile.ecs.soton.ac.uk.

[352] M.-S. Alouini and A. J. Goldsmith, 'Capacity of Rayleigh fading channels under different adaptive transmission and diversity-combining techniques', *IEEE Transactions on Vehicular Technology*, vol. 49, pp. 1165–1181, July 1999.

[353] L. Zheng and N. Tse, 'Diversity and multiplexing: a fundamental tradeoff in multiple-antenna channels', *IEEE Transactions on Information Theory*, vol. 49, pp. 1073–1096, May 2003.

[354] C. E. Shannon, 'A mathematical theory of communication', *Bell System Technical Journal*, vol. 27, pp. 379–423 and 623–656, June and October 1948.

[355] B. Yang, 'Projection approximation subspace tracking', *IEEE Transactions on Signal Processing*, vol. 43, pp. 95–107, January 1995.

[356] W. C. Chung, N. J. August, and D. S. Ha, 'Signaling and multiple access techniques for ultra wideband 4G wireless communication systems', *IEEE Wireless Communications*, vol. 12, pp. 46–55, April 2005.

[357] Wi-Fi Alliance, 'Wi-Fi brand awareness and consumer affinity continues to grow worldwide', September 2008. http://www.wi-fi.org/pressroom_overview.php?newsid=711/.

[358] Institute of Electrical and Electronics Engineers, 'IEEE Standard for Information Technology – Telecommunications and Information Exchange Between Systems – Local and Metropolitan Area Networks – Specific Requirements – Part 11: Wireless LAN Medium Access Control (MAC) and Physical Layer (PHY) Specifications', IEEE 802.11–1997, 1997.

[359] Institute of Electrical and Electronics Engineers, 'IEEE Standard for Information Technology – Telecommunications and Information Exchange Between Systems – Local and Metropolitan Area Networks – Specific Requirements – Part 11: Wireless LAN Medium Access Control (MAC) and Physical Layer (PHY) Specifications', IEEE 802.11–1999, 1999.

[360] Institute of Electrical and Electronics Engineers, 'Supplement to IEEE Standard for Information Technology – Telecommunications and Information Exchange Between Systems – Local and Metropolitan Area Networks – Specific Requirements – Part 11: Wireless LAN Medium Access Control (MAC) and Physical Layer (PHY) Specifications: High-speed Physical Layer in the 5 GHz Band', IEEE 802.11a-1999, 1999.

[361] Institute of Electrical and Electronics Engineers, 'Supplement to IEEE Standard for Information Technology – Telecommunications and Information Exchange Between Systems – Local and Metropolitan Area Networks – Specific Requirements – Part 11: Wireless LAN Medium Access Control (MAC) and Physical Layer (PHY) Specifications: Higher-speed Physical Layer Extension in the 2.4 GHz Band', IEEE 802.11b-1999, 1999.

[362] Institute of Electrical and Electronics Engineers, 'IEEE Standard for Information Technology – Telecommunications and Information Exchange Between Systems – Local and Metropolitan Area Networks – Specific Requirement – Part 11: Wireless LAN Medium Access Control (MAC) and Physical Layer (PHY) Specification – Amendment 3: Specifications for Operation in Additional Regulatory Domains', IEEE 802.11d-2001, 2001.

[363] Institute of Electrical and Electronics Engineers, 'IEEE Standard for Information Technology – Telecommunications and Information Exchange Between Systems – Local and Metropolitan Area Networks – Specific Requirements – Part 11: Wireless LAN Medium Access Control (MAC) and Physical Layer (PHY) Specifications – Amendment 4: Further Higher Data Rate Extension in the 2.4 GHz Band', IEEE 802.11g-2003, 2003.

[364] Institute of Electrical and Electronics Engineers, 'IEEE Standard for Information Technology – Telecommunications and Information Exchange Between Systems – Local and Metropolitan Area Networks – Specific Requirements – Part 11: Wireless LAN Medium Access Control (MAC) and Physical Layer (PHY) Specifications – Amendment 5: Spectrum and Transmit Power Management Extensions in the 5 GHz band in Europe', IEEE 802.11h-2003, 2003.

[365] Institute of Electrical and Electronics Engineers, 'IEEE Standard for Information Technology – Telecommunications and Information Exchange Between Systems – Local and Metropolitan Area Networks – Specific Requirements – Part 11: Wireless LAN Medium Access Control (MAC) and Physical Layer (PHY) Specifications – Amendment 6: Medium Access Control (MAC) Security Enhancements', IEEE 802.11i-2004, 2004.

[366] Institute of Electrical and Electronics Engineers, 'IEEE Standard for Information Technology – Telecommunications and Information Exchange Between Systems – Local and Metropolitan Area Networks – Specific Requirements – Part 11: Wireless LAN Medium Access Control (MAC) and Physical Layer (PHY) Specifications – Amendment 7: 4.9 GHz-5 GHz Operation in Japan', IEEE 802.11j-2004, 2004.

[367] Institute of Electrical and Electronics Engineers, 'IEEE Standard for Information Technology – Telecommunications and Information Exchange Between Systems – Local and Metropolitan Area Networks – Specific Requirements – Part 11: Wireless LAN Medium Access Control (MAC) and Physical Layer (PHY) Specifications – Amendment 8: Medium Access Control (MAC) Quality of Service Enhancements', IEEE 802.11e-2005, 2005.

[368] Institute of Electrical and Electronics Engineers, 'IEEE Standard for Information Technology – Telecommunications and Information Exchange Between Systems – Local and Metropolitan Area Networks – Specific Requirements – Part 11: Wireless LAN Medium Access Control (MAC) and Physical Layer (PHY) Specifications', IEEE 802.11–2007, 2007.

[369] Institute of Electrical and Electronics Engineers, 'IEEE Standard for Information Technology – Telecommunications and Information Exchange Between Systems – Local and Metropolitan Area Networks – Specific Requirements – Part 11: Wireless LAN Medium Access Control (MAC) and Physical Layer (PHY) Specifications – Amendment 1: Radio Resource Measurement of Wireless LANs (Amendment to IEEE 802.11–2007)', IEEE 802.11k-2008, 2008.

[370] Institute of Electrical and Electronics Engineers, 'IEEE Standard for Information Technology – Telecommunications and Information Exchange Between Systems – Local and Metropolitan Area Networks – Specific Requirements – Part 11: Wireless LAN Medium Access Control (MAC) and Physical Layer (PHY) Specifications – Amendment 2: Fast Basic Service Set (BSS) Transition (Amendment to IEEE 802.11–2007 and IEEE 802.11k-2008)', IEEE 802.11r-2008, 2008.

[371] Institute of Electrical and Electronics Engineers, 'Draft IEEE Standard for Information Technology – Telecommunications and Information Exchange Between Systems – Local and Metropolitan Area Networks – Specific Requirements – Part 11: Wireless LAN Medium Access Control (MAC) and Physical Layer (PHY) Specifications – Amendment 3: 3650–3700 MHz Operation in USA (Draft Amendment to IEEE 802.11–2007)', IEEE P802.11y/D10, 2008.

[372] 3GPP, 'UTRA-UTRAN Long Term Evolution (LTE) and 3GPP System Architecture Evolution (SAE)', May 2008. http://www.3gpp.org/Highlights/LTE/.

[373] 3GPP, 'Overview of 3GPP Release 6: Summary of all Release 6 features (Version TSG #33)', 3GPP Rel-6, 2006.

[374] E. Dahlman, S. Parkvall, J. Sköld, and P. Beming, *3G Evolution: HSPA and LTE for Mobile Broadband*. Oxford: Academic Press, 2007.

[375] 3GPP, '3rd Generation Partnership Project; Technical Specification Group Radio Access Network; Physical layer aspects for evolved Universal Terrestrial Radio Access (UTRA) (Release 7)', 3GPP TR 25.814 V7.1.0, September 2006.

[376] J. G. Andrews, A. Ghosh, and R. Muhamed, *Fundamentals of WiMAX: Understanding Broadband Wireless Networking*. Englewood Cliffs, NJ: Prentice Hall, 2007.

[377] S. A. M. Ilyas, *WiMAX Applications*. Boca Raton, FL: CRC Press, 2007.

[378] WiMAX Forum, 'Frequently Asked Questions'. http://www.wimaxforum.org/technology/faq/.

[379] Institute of Electrical and Electronics Engineers, 'IEEE Standard for Local and Metropolitan Area Networks – Part 16: Air Interface for Fixed Broadband Wireless Access Systems', IEEE 802.16–2001, 2001.

[380] Institute of Electrical and Electronics Engineers, 'IEEE Standard for Local and Metropolitan Area Networks – Part 16: Air Interface for Fixed Broadband Wireless Access Systems – Amendment 1: Detailed System Profiles for 10–66 GHz', IEEE 802.16c-2002, 2002.

[381] Institute of Electrical and Electronics Engineers, 'IEEE Standard for Local and Metropolitan Area Networks – Part 16: Air Interface for Fixed Broadband Wireless Access Systems – Amendment 2: Medium Access Control Modifications and Additional Physical Layer Specifications for 2–11 GHz', IEEE 802.16a-2003, 2003.

[382] Institute of Electrical and Electronics Engineers, 'IEEE Standard for Local and Metropolitan Area Networks – Part 16: Air Interface for Fixed and Mobile Broadband Wireless Access Systems – Amendment 2: Physical and Medium Access Control Layers for Combined Fixed and Mobile Operation in Licensed Bands and Corrigendum 1', IEEE 802.16e-2005, 2005.

[383] Institute of Electrical and Electronics Engineers, 'IEEE Standard for Local and Metropolitan Area Networks – Part 16: Air Interface for Fixed Broadband Wireless Access Systems – Amendment 1: Management Information Base', IEEE 802.16f-2005, 2005.

[384] Institute of Electrical and Electronics Engineers, 'IEEE Standard for Local and Metropolitan Area Networks Media Access Control (MAC) Bridges – Amendment 5: Bridging of IEEE 802.16', IEEE 802.16k-2007, 2007.

[385] Institute of Electrical and Electronics Engineers, 'IEEE Standard for Local and Metropolitan Area Networks – Part 16: Air Interface for Fixed and Mobile Broadband Wireless Access Systems – Amendment 3: Management Plane Procedure and Services', IEEE 802.16g-2007, 2007.

[386] Institute of Electrical and Electronics Engineers, 'Draft IEEE Standard for Local and Metropolitan Area Networks – Part 16: Air Interface for Fixed and Mobile Broadband Wireless Access Systems: Improved Coexistence Mechanisms for License-Exempt Operation (Amendment to IEEE 802.16d-2004)', IEEE P802.16h/D7, 2008.

[387] Institute of Electrical and Electronics Engineers, 'Draft IEEE Standard for Local and Metropolitan Area Networks – Part 16: Air Interface for Fixed and Mobile Broadband Wireless Access Systems Multihop Relay Specification', IEEE P802.16j/D5, 2008.

[388] G. S. V. Rhada K. Rao and G. Radhamani, *WiMAX – A Wireless Technology Revolution*. Boca Raton, FL: Auerbach, 2007.

[389] Institute of Electrical and Electronics Engineers, 'Draft IEEE Standard for Local and Metropolitan Area Networks – Part 16: Air Interface for Broadband Wireless Access Systems (Revision of IEEE 802.16d-2004 and consolidates material from IEEE 802.16e-2005, IEEE 802.16d-2004/Cor1-2005, IEEE 802.16f-2005 and IEEE 802.16g-2007)', IEEE P802.16Rev2/D5, 2008.

[390] Institute of Electrical and Electronics Engineers, 'WirelessHUMAN Tutorial', IEEE 802.16h-00/08, July 2000.

[391] L. Nuaymi, *WiMAX: Technology for Broadband Wireless Access*. Chichester: John Wiley & Sons, Ltd, 2007.

[392] Institute of Electrical and Electronics Engineers, 'IEEE Standard for Local and Metropolitan Area Networks – Media Access Control (MAC) Bridges', IEEE 802.1D-2004, 2004.

[393] M. Hancock, 'WiMAX Questions and Answers', July 2008. http://www.wimaxforum.org/news/wimax_faq_10–2007.pdf.

[394] European Telecommunications Standards Institute, 'ETSI HiperMAN'. http://www.etsi.org/WebSite/homepage.aspx.

[395] WiMAX Forum, 'The Relationship Between WiBro and Mobile WiMAX', 2006. http://www.wimaxforum.org/technology/downloads/.

[396] C. Koo, '2.3 GHz Portable Internet – WiBro', July 2004. http://www.knom.or.kr/tutorial/Tutorial2004/knom_T_1.pdf.

[397] S. Q. Lee, N. Park, C. Cho, H. Lee, and S. Ryu, 'The wireless broadband (wibro) system for broadband wireless internet services', *Communications Magazine, IEEE*, vol. 44, pp. 106–112, 2006.

[398] WiMAX Forum, 'Mobile WiMAX – Part I: A Technical Overview and Performance Evaluation', February 2006. http://www.wimaxforum.org/technology/downloads/.

[399] H. Yaghoobi, 'Scalable OFDMA physical layer in IEEE 802.16 WirelessMAN', *Intel Technology Journal*, vol. 8, pp. 201–212, August 2004.

[400] S. M. Alamouti, 'A simple transmit diversity technique for wireless communications', *IEEE Journal on Selected Areas in Communications*, vol. 16, pp. 1451–1458, October 1998.

[401] International Telecommunication Union, 'Draft report on requirements related to technical system performance for IMT-Advanced radio interface(s) [IMT.TECH]', ITU-R Document 5D/TEMP/89r1Cleanv2, June 2008.

[402] S. Ortiz, '4G wireless begins to take shape', *Computer*, vol. 40, pp. 18–21, November 2007.

[403] Institute of Electrical and Electronics Engineers, 'IEEE 802.16m System Requirements', IEEE 802.16m-07/002r4, October 2007.

[404] WiMAX Forum, 'WiMAX Forum Mobile System Profile, Release 1.0 Approved Specification (Revision 1.4.0: 19 2007-05-02)', May 2007. http://www.wimaxforum.org/technology/documents/.

[405] Institute of Electrical and Electronics Engineers, 'Draft IEEE Standard for Local and Metropolitan Area Networks: Media Independent Handover Services)', IEEE P802.21/D13, 2008.

[406] Institute of Electrical and Electronics Engineers, 'IEEE 802.16m System Description Document', IEEE 802.16m-08/003r4, July 2008.

[407] R. B. Marks, M. Lynch, and K. McCabe, 'IEEE 802.16 WirelessMAN Standard enters IMT-2000 Family for International Mobile Telecommunications', October 2007. http://standards.ieee.org/announcements/PR_802.16MT2000.html.

[408] WiMAX Forum, 'WiMAX Forum WiMAX Technology Forecast (2007–2012)', June 2008. http://www.wimaxforum.org/technology/downloads/.

[409] WiMAX Forum, 'WiMAX Forum announces first certified MIMO 2.5 GHz Mobile WiMAX products', June 2008. http://www.wimaxforum.org/news/.

[410] P. Chaudhury, W. Mohr, and S. Onoe, 'The 3GPP Proposal for IMT-2000', *IEEE Communications Magazine*, vol. 37, pp. 72–81, December 1999.

[411] B. Glance and L. Greestein, 'Frequency-selective fading effects in digital mobile radio with diversity combining', *IEEE Transactions on Communications*, vol. COM-31, pp. 1085–1094, September 1983.

[412] F. Adachi and K. Ohno, 'BER performance of QDPSK with postdetection diversity reception in mobile radio channels', *IEEE Transactions on Vehicular Technology*, vol. 40, pp. 237–249, February 1991.

[413] V. Tarokh, H. Jafarkhani, and A. R. Calderbank, 'Space-time block codes from orthogonal designs', *IEEE Transactions on Information Theory*, vol. 45, pp. 1456–1467, May 1999.

[414] V. Tarokh, H. Jafarkhani, and A. R. Calderbank, 'Space-time block coding for wireless communications: performance results', *IEEE Journal on Selected Areas in Communications*, vol. 17, pp. 451–460, March 1999.

[415] V. Tarokh, N. Seshadri, and A. R. Calderbank, 'Space-time codes for high data rate wireless communication: performance criterion and code construction', *IEEE Transactions on Information Theory*, vol. 44, pp. 744–765, March 1998.

[416] V. Tarokh, A. Naguib, N. Seshadri, and A. R. Calderbank, 'Space-time codes for high data rate wireless communication: performance criteria in the presence of channel estimation errors, mobility, and multiple paths', *IEEE Transactions on Communications*, vol. 47, pp. 199–207, February 1999.

[417] V. Tarokh, A. Naguib, N. Seshadri, and A. Calderbank, 'Space-time codes for high data rate wireless communications: mismatch analysis', in *Proceedings of IEEE International Conference on Communications (ICC'97)*, (Montreal, Canada), vol. 1, pp. 309–313, June 1997.

[418] N. Seshadri, V. Tarokh, and A. Calderbank, 'Space-Time codes for high data rate wireless communications: code construction', in *Proceedings of the IEEE Vehicular Technology Conference (VTC 1997 Spring)*, (Phoenix, USA), vol. 2, pp. 637–641, May 1997.

[419] R. Gallager, 'Low density parity check codes', *IEEE Transactions on Information Theory*, vol. 8, pp. 21–28, January 1962.

[420] C. Berrou, A. Glavieux, and P. Thitimajshima, 'Near Shannon limit error-correcting coding and decoding: turbo codes', in *Proceedings of the International Conference on Communications*, (Geneva, Switzerland), pp. 1064–1070, May 1993.

[421] C. Berrou and A. Glavieux, 'Near optimum error correcting coding and decoding: turbo codes', *IEEE Transactions on Communications*, vol. 44, pp. 1261–1271, October 1996.

[422] G. Bauch, 'Concatenation of space-time block codes and turbo-TCM', in *Proceedings of IEEE International Conference on Communications*, (Vancouver, Canada), vol. 2, pp. 1202–1206, June 1999.

[423] W. Koch and A. Baier, 'Optimum and sub-optimum detection of coded data distributed by time-varying inter-symbol interference', in *Proceedings of IEEE GLOBECOM'90*, (San Diego, USA), vol. 3, pp. 1679–1684, 2–5 December 1990.

[424] J. Erfanian, S. Pasupathy, and G. Gulak, 'Reduced complexity symbol detectors with parallel structures for ISI channels', *IEEE Transactions on Communications*, vol. 42, pp. 1661–1671, February/March/April 1994.

[425] P. Robertson, E. Villebrun, and P. Höher, 'A comparison of optimal and sub-optimal MAP decoding algorithms operating in the log domain', in *Proceedings of the International Conference on Communications*, (Seattle, USA), pp. 1009–1013, June 1995.

[426] D. J. C. Mackay and R. M. Neal, 'Near Shannon limit performance of low density parity check codes', *Electronics Letters*, vol. 33, pp. 457–458, March 1997.

[427] M. G. Luby, M. Mitzenmacher, M. A. Shokrollahi, and D. A. Spielman, 'Improved low density parity check codes using irregular graphs and belief propagations', in *Proceedings of the IEEE International Symposium on Information Theory*, (Cambridge, MA, USA), p. 117, 1998.

[428] M. C. Davey, 'Error-correction using low density parity check codes', PhD thesis, University of Cambridge, 1999.

[429] M. Chiani, A. Conti, and A. Ventura, 'Evaluation of low-density parity-check codes over block fading channels', *IEEE International Conference on Communications, ICC'2000*, (New Orleans, USA), vol. 3, pp. 1183–1187, 2000.

[430] J. Hou, P. H. Siegel, and L. B. Milstein, 'Performance analysis and code optimization of low density parity-check codes on Rayleigh fading channels', *IEEE Journal on Selected Areas in Communications*, vol. 19, pp. 924–934, May 2001.

[431] F. Guo, S. X. Ng, and L. Hanzo, 'LDPC assisted block coded modulation for transmission over Rayleigh fading channels', in *Proceedings of IEEE Vehicular Technology Conference (VTC'03 Spring)*, (Jeju, Korea), vol. 3, pp. 1867–1871, April 2003.

[432] J. Hagenauer, E. Offer, and L. Papke, 'Iterative decoding of binary block and convolutional codes', *IEEE Transactions on Information Theory*, vol. 42, pp. 429–445, March 1996.

[433] M. Y. Alias, F. Guo, S. X. Ng, T. H. Liew, and L. Hanzo, 'LDPC and turbo coding assisted space-time block coded OFDM', in *Proceedings of IEEE Vehicular Technology Conference (VTC'03 Spring)*, (Jeju, Korea), vol. 4, pp. 2309–2313, April 2003.

[434] G. Ungerböeck, 'Channel coding with multilevel/phase signals', *IEEE Transactions on Information Theory*, vol. IT-28, pp. 55–67, January 1982.

[435] P. Robertson and T. Wörz, 'Bandwidth efficient turbo trellis-coded modulation using punctured component codes', *IEEE Journal on Selected Areas in Communications*, vol. 16, pp. 206–218, February 1998.

[436] E. Zehavi, '8-PSK trellis codes for a Rayleigh fading channel', *IEEE Transactions on Communications*, vol. 40, pp. 873–883, May 1992.

[437] X. Li and J. A. Ritcey, 'Bit-interleaved coded modulation with iterative decoding using soft feedback', *Electronics Letters*, vol. 34, pp. 942–943, May 1998.

[438] Z. Guo and W. Zhu, 'Performance study of OFDMA vs. OFDM/SDMA', in *Proceedings of IEEE Vehicular Technology Conference (VTC 2002 Spring)*, (Birmingham, AL, USA), vol. 2, pp. 565–569, 6–9 May 2002.

[439] S. Verdu, *Multiuser Detection*. Cambridge: Cambridge University Press, 1998.

[440] C. Z. W. H. Sweatman, J. S. Thompson, B. Mulgrew, and P. M. Grant, 'A comparison of detection algorithms including BLAST for wireless communication using multiple antennas', in *Proceedings of IEEE International Symposium on Personal, Indoor and Mobile Radio Communications*, (London, UK), vol. 1, pp. 698–703, 18–21 September 2000.

[441] M. Münster and L. Hanzo, 'Co-channel interference cancellation techniques for antenna array assisted multiuser OFDM systems', in *Proceedings of IEE 3G 2000 Conference*, (London, UK), vol. 1, pp. 256–260, 27–29 March 2000.

[442] M. Sellathurai and S. Haykin, 'A simplified diagonal BLAST architecture with iterative parallel-interference cancellation receivers', in *Proceedings of IEEE International Conference on Communications*, (Helsinki, Finland), vol. 10, pp. 3067–3071, 11–14 June 2001.

[443] M. Münster and L. Hanzo, 'Performance of SDMA multiuser detection techniques for Walsh-Hadamard-spread OFDM schemes', in *Proceedings of IEEE Vehicular Technology Conference (VTC'01 Fall)*, (Atlantic City, USA), vol. 4, pp. 2319–2323, 7–11 October 2001.

[444] Commission of the European Communities, 'COST207, Digital Land Mobile Radio Communications', Final Report, Luxembourg, 1989.

[445] U. Fincke and M. Pohst, 'Improved method for calculating vector of short length in a lattice, including a complexity analysis', *Mathematics of Computation*, vol. 44, pp. 463–471, April 1985.

[446] E. Viterbo and J. Boutros, 'A universal lattice code decoder for fading channels', *IEEE Transactions on Information Theory*, vol. 45, pp. 1639–1642, July 1999.

[447] M. O. Damen, A. Chkeif, and J.-C. Belfiore, 'Lattice code decoder for space-time codes', *IEEE Communications Letters*, vol. 4, pp. 161–163, May 2000.

[448] B. M. Hochwald and S. ten Brink, 'Achieving near-capacity on a multiple-antenna channel', *IEEE Transactions on Communications*, vol. 51, pp. 389–399, March 2003.

[449] L. Brunel, 'Multiuser detection techniques using maximum likelihood sphere decoding in multicarrier CDMA systems', *IEEE Transactions on Wireless Communications*, vol. 3, pp. 949–957, May 2004.

[450] D. Pham, K. R. Pattipati, P. K. Willett, and J. Luo, 'An improved complex sphere decoder for V-BLAST systems', *IEEE Signal Processing Letters*, vol. 11, pp. 748–751, September 2004.

[451] T. Cui and C. Tellambura, 'Approximate ML detection for MIMO systems using multistage sphere decoding', *IEEE Signal Processing Letters*, vol. 12, pp. 222–225, March 2005.

[452] P. W. Wolniansky, G. J. Foschini, G. D. Golden, and R. A. Valenzuela, 'V-BLAST: an architecture for realizing very high data rates over the rich-scattering wireless channel', in *URSI International Symposium on Signals, Systems, and Electronics (ISSSE'98)*, (Pisa, Italy), pp. 295–300, 29 September–2 October 1998.

[453] J. Holland, *Adaptation in Natural and Artificial Systems*. Ann Arbor, MI: University of Michigan Press, 1975.

[454] M. Mitchell, *An Introduction to Genetic Algorithms*. Cambridge, MA: MIT Press, 1996.

[455] D. Whitley, 'A genetic algorithm tutorial', *Statistics and Computing*, vol. 4, pp. 65–85, June 1994.

[456] S. Forrest, 'Genetic algorithms: principles of natural selection applied to computation', *Science*, vol. 261, pp. 872–878, August 1993.

[457] M. J. Juntti, T. Schlösser, and J. O. Lilleberg, 'Genetic algorithms for multiuser detection in synchronous CDMA', in *IEEE International Symposium on Information Theory (ISIT'97)*, (Ulm, Germany), p. 492, 29 June–4 July 1997.

[458] X. F. Wang, W.-S. Lu, and A. Antoniou, 'A genetic-algorithm-based multiuser detector for multiple-access communications', in *Proceedings of the 1998 IEEE International Symposium on Circuits and Systems*, (Monterey, CA, USA), vol. 4, pp. 534–537, 31 May–3 June 1998.

[459] C. Ergün and K. Hacioglu, 'Multiuser detection using a genetic algorithm in CDMA communications systems', *IEEE Transactions on Communications*, vol. 48, pp. 1374–1383, August 2000.

[460] K. Yen and L. Hanzo, 'Antenna-diversity-assisted genetic-algorithm-based multiuser detection schemes for synchronous CDMA systems', *IEEE Transactions on Communications*, vol. 51, pp. 366–370, March 2003.

[461] K. Yen and L. Hanzo, 'Genetic algorithm assisted joint multiuser symbol detection and fading channel estimation for synchronous CDMA systems', *IEEE Journal on Selected Areas in Communications*, vol. 19, pp. 985–998, June 2001.

[462] S. Abedi and R. Tafazolli, 'Genetically modified multiuser detection for code division multiple access systems', *IEEE Journal on Selected Areas in Communications*, vol. 20, pp. 463–473, February 2002.

[463] U. Fawer and B. Aazhang, 'A multiuser receiver for code division multiple access communications over multipath channels', *IEEE Transactions on Communications*, vol. 43, pp. 1556–1565, February/April 1995.

[464] T. Blickle and L. Thiele, 'A comparison of selection schemes used in evolutionary algorithms', *Evolutionary Computation*, vol. 4, pp. 361–394, January 1996.

[465] E. Zitzler and L. Thiele, 'Multiobjective evolutionary algorithms: a comparative case study and the strength Pareto approach', *IEEE Transactions on Evolutionary Computation*, vol. 3, pp. 257–271, November 1999.

[466] A. E. Eiben, R. Hinterding, and Z. Michalewicz, 'Parameter control in evolutionary algorithms', *IEEE Transactions on Evolutionary Computation*, vol. 3, pp. 124–141, July 1999.

[467] J. G. Proakis, *Digital Communications*. New York: McGraw-Hill, 4th edn, 2001.

[468] W. Ledermann and E. Lloyd, *Handbook of Applicable Mathematics, Volume II: Probability*. Chichester: John Wiley & Sons, Ltd, 1980.

[469] M. Jiang and L. Hanzo, 'Multi-user MIMO-OFDM Systems Using Subcarrier Hopping', *IEE Proceedings on Communications*, vol. 153, pp. 802–809, December 2006.

[470] M. K. Simon, J. K. Omura, R. A. Scholtz, and B. K. Levitt, *Spread Spectrum Communications: Volume I.* Rockville, MD: Computer Science Press, 1985.

[471] M. K. Simon, J. K. Omura, R. A. Scholtz, and B. K. Levitt, *Spread Spectrum Communications: Volume II.* Rockville, MD: Computer Science Press, 1985.

[472] M. K. Simon, J. K. Omura, R. A. Scholtz, and B. K. Levitt, *Spread Spectrum Communications: Volume III.* Rockville, MD: Computer Science Press, 1985.

[473] R. E. Ziemer and R. L. Peterson, *Digital Communications and Spread Spectrum System.* New York: Macmillan, 1985.

[474] G. Einarsson, 'Address assignment for a time-frequency-coded, spread-spectrum system', *Bell System Technical Journal*, vol. 59, pp. 1241–1255, September 1980.

[475] D. J. Goodman, P. S. Henry, and V. K. Prabhu, 'Frequency-hopped multilevel FSK for mobile radio', *Bell System Technical Journal*, vol. 59, pp. 1257–1275, September 1980.

[476] A. W. Lam and D. P. Sarwate, 'Time-hopping and frequency-hopping multiple-access packet communications', *IEEE Transactions on Communications*, vol. 38, pp. 875–888, June 1990.

[477] U. Fiebig, 'Iterative interference cancellation for FFH/MFSK MA systems', *IEE Proceedings – Communications*, vol. 143, pp. 380–388, December 1996.

[478] L.-L. Yang and L. Hanzo, 'Slow frequency-hopping multicarrier DS-CDMA for transmission over Nakagami multipath fading channels', *IEEE Journal on Selected Areas in Communications*, vol. 19, pp. 1211–1221, July 2001.

[479] L.-L. Yang and L. Hanzo, 'Blind joint soft-detection assisted slow frequency-hopping multicarrier DS-CDMA', *IEEE Transactions on Communications*, vol. 48, pp. 1520–1529, September 2000.

[480] E. A. Geraniotis, 'Coherent hybrid DS-SFH spread-spectrum multiple-access communications', *IEEE Journal on Selected Areas in Communications*, vol. 3, pp. 695–705, September 1985.

[481] J. Wang and H. Huang, 'Multicarrier DS/SFH-CDMA systems', *IEEE Transactions on Vehicular Technology*, vol. 51, pp. 867–876, September 2002.

[482] M. Jankiraman and R. Prasad, 'A novel solution to wireless multimedia application: the hybrid OFDM/CDMA/SFH approach', in *11th IEEE International Symposium on Personal, Indoor and Mobile Radio Communications (PIMRC 2000)*, (London, UK), vol. 2, pp. 1368–1374, 18–21 September 2000.

[483] K. Hamaguchi and L. Hanzo, 'Time-frequency spread OFDM/FHMA', in *Proceedings of the IEEE Vehicular Technology Conference, 2003 (VTC 2003 Spring)*, (Jeju, Korea), vol. 2, pp. 1248–1252, 22–25 April 2003.

[484] E. A. Geraniotis, 'Noncoherent hybrid DS-SFH spread-spectrum multiple-access communications', *IEEE Transactions on Communications*, vol. 34, pp. 862–872, September 1986.

[485] Y. Li and N. R. Sollenberger, 'Clustered OFDM with channel estimation for high rate wireless data', *IEEE Transactions on Communications*, vol. 49, pp. 2071–2076, December 2001.

[486] B. Daneshrad, L. J. Cimini Jr, M. Carloni, and N. Sollenberger, 'Performance and implementation of clustered OFDM for wireless communications', *ACM Journal on Mobile Networks and Applications (MONET) special issue on PCS*, vol. 2, no. 4, pp. 305–314, 1997.

[487] H. Niu, M. Shen, J. A. Ritcey, and H. Liu, 'Performance of clustered OFDM with low density parity check codes over dispersive channels', in *Conference Record of the 36th Asilomar Conference on Signals, Systems and Computers*, (Pacific Grove, CA, USA), pp. 1852–1856, November 2002.

[488] G. Parsaee and A. Yarali, 'OFDMA for the 4th generation cellular networks', in *Canadian Conference on Electrical and Computer Engineering*, (Calgary, Canada), vol. 4, pp. 2325–2330, 2–5 May 2004.

[489] H. Sari and G. Karam, 'Orthogonal frequency-division multiple access and its application to CATV network', *European Transactions on Telecommunications*, vol. 9, p. 507–516, November/December 1998.

[490] J. Jang and K. B. Lee, 'Transmit power adaptation for multiuser OFDM systems', *IEEE Journal on Selected Areas in Communications*, vol. 21, pp. 1747–1758, February 2003.

[491] C. Y. Wong, C. Y. Tsui, R. S. Cheng, and K. B. Lataief, 'A real-time sub-carrier allocation scheme for multiple access downlink OFDM transmission', in *Proceedings of the IEEE 50th Vehicular Technology Conference (VTC 1999 Fall)*, (Amsterdam, The Netherlands), vol. 2, pp. 1124–1128, 19–22 September 1999.

[492] D. Kivanc, G. Li, and H. Liu, 'Computationally efficient bandwidth allocation and power control for OFDMA', *IEEE Transactions on Wireless Communications*, vol. 2, pp. 1150–1158, November 2003.

[493] D. Kivanc and H. Liu, 'Subcarrier allocation and power control for OFDMA', in *Proceedings of the 34th Asilomar Conference on Signals, Systems and Computers, 2000*, (Pacific Grove, CA, USA), vol. 1, pp. 147–151, 29 October–1 November 2000.

[494] I. Kim, H. L. Lee, B. Kim, and Y. H. Lee, 'On the use of linear programming for dynamic subchannel and bit allocation in multiuser OFDM', in *Proceedings of the IEEE Global Telecommunications Conference (GLOBECOM 2001)*, (San Antonio, USA), vol. 6, pp. 3648–3652, 25–29 November 2001.

[495] W. Rhee and J. M. Cioffi, 'Increase in capacity of multiuser OFDM system using dynamic subchannel allocation', in *Proceedings of the IEEE 51st Vehicular Technology Conference (VTC 2000 Spring)*, (Tokyo, Japan), vol. 2, pp. 1085–1089, 15–18 May 2000.

[496] S. Das and G. D. Mandyam, 'An efficient sub-carrier and rate allocation scheme for M-QAM modulated uplink OFDMA transmission', in *Proceedings of the 37th Asilomar Conference on Signals, Systems and Computers, 2003*, (Pacific Grove, CA, USA), vol. 1, pp. 136–140, 9–12 November 2003.

[497] S. Pietrzyk and G. J. M. Janssen, 'Multiuser subcarrier allocation for QoS provision in the OFDMA systems', in *Proceedings of the IEEE 56th Vehicular Technology Conference (VTC 2002 Fall)*, (Birmingham, AL, USA), vol. 2, pp. 1077–1081, 24–28 September 2002.

[498] Z. Hu, G. Zhu, Y. Xia, and G. Liu, 'Multiuser subcarrier and bit allocation for MIMO-OFDM systems with perfect and partial channel information', in *Proceedings of the IEEE Wireless Communications and Networking Conference (WCNC 2004)*, (Atlanta, USA), vol. 2, pp. 1188–1193, 21–25 March 2004.

[499] S. Zhou, G. B. Giannakis, and A. Scaglione, 'Long codes for generalized FH-OFDMA through unknown multipath channels', *IEEE Transactions on Communications*, vol. 49, pp. 721–733, April 2001.

[500] Z. Cao, U. Tureli, and P. Liu, 'Optimum subcarrier assignment for OFDMA uplink', in *The 37th Asilomar Conference on Signals, Systems and Computers*, (Pacific Grove, CA, USA), vol. 1, pp. 708–712, 9–12 November 2003.

[501] Y. H. Kim, K. S. Kim, and J. Y. Ahn, 'Iterative estimation and decoding for an LDPC-coded OFDMA system in uplink environments', in *IEEE International Conference on Communications*, (Paris, France), vol. 4, pp. 2478–2482, 20–24 June 2004.

[502] H. Sari, Y. Levy, and G. Karam, 'An analysis of orthogonal frequency-division multiple access', in *IEEE Global Telecommunications Conference*, (San Francisco, USA), vol. 3, pp. 1635–1639, 3–8 November 1997.

[503] T. Kurt and H. Delic, 'Collision avoidance in space-frequency coded FH-OFDMA', in *IEEE International Conference on Communications*, (Paris, France), vol. 1, pp. 269–273, 20–24 June 2004.

[504] H. H. Chen, Y. C. Yeh, C. H. Tsai, and W. H. Chang, 'Uplink synchronisation control technique and its environment-dependent performance analysis', *Electronics Letters*, vol. 39, pp. 1755–1757, November 2003.

[505] J.-J. V. D. Beek, P. O. Börjesson, M.-L. Boucheret, D. Landström, J. M. Arenas, P. Ödling, C. Östberg, M. Wahlqvist, and S. K. Wilson, 'A time and frequency synchronization scheme for multiuser OFDM', *IEEE Journal on Selected Areas in Communications*, vol. 17, pp. 1900–1914, November 1999.

[506] M. Pompili, S. Barbarossa, and G. B. Giannakis, 'Channel-independent non-data aided synchronization of generalized multiuser OFDM', in *2001 IEEE International Conference on Acoustics, Speech, and Signal Processing (ICASSP'01)*, (Salt Lake City, USA), vol. 4, pp. 2341–2344, 7–11 May 2001.

[507] S. T. Wu and K. C. Chen, 'Programmable multiuser synchronization for OFDM-CDMA', in *Proceedings of the IEEE 53rd Vehicular Technology Conference (VTC 2001 Spring)*, (Bratislava, Slovakia), vol. 2, pp. 830–834, 6–9 May 2001.

[508] S. Lipschutz and M. L. Lipson, *Theory and Problems of Probability*, Schaum's Outline Series. New York: McGraw-Hill, 2nd edn, 2000.

[509] M. K. Simon, J. K. Omura, R. A. Scholtz, and B. K. Levitt, *Spread Spectrum Communications: Volumes I, II, III*. Rockville, MD: Computer Science Press, 1985.

[510] A. Papoulis, *Probability, Random Variables, and Stochastic Processes*. New York: McGraw-Hill, 2nd edn, 1984.

[511] J. Lee and K. B. Bae, 'Numerically stable fast sequential calculation for projection approximation subspace tracking', in *Proceedings of the 1999 IEEE International Symposium on Circuits and Systems*, (Orlando, USA), vol. 3, pp. 484–487, 30 May–2 June 1999.

[512] H. Holma and A. Toskala, eds, *WCDMA for UMTS: Radio Access for Third Generation Mobile Communications*. Chichester: John Wiley & Sons, Ltd, 2000.

[513] B. Yang, K. Letaief, R. Cheng, and Z. Cao, 'Channel estimation for OFDM transmission in multipath fading channels based on parametric channel modeling', *IEEE Transactions on Communications*, vol. 49, pp. 467–479, March 2001.

[514] J. Yang and M. Kaveh, 'Adaptive eigensubspace algorithm for direction or frequency estimation and tracking', *IEEE Transactions on Acoustics, Speech and Signal Processing*, vol. 36, pp. 241–251, 1988.

[515] S. L. Marple, *Digital Spectral Analysis with Applications*. Englewood Cliffs, NJ: Prentice Hall, 1987.

[516] D. Schafhuber, G. Matz, and F. Hlawatsch, 'Adaptive prediction of time-varying channels for coded OFDM systems', in *Proceedings of the IEEE International Conference on Acoustics, Speech, and Signal Processing*, (Washington, DC, USA), vol. 3, pp. III-2549–III-2552, May 2002.

[517] M. Jiang, J. Akhtman, and L. Hanzo, 'Iterative Joint Channel Estimation and Multi-User Detection for Multiple-Antenna Aided OFDM Systems', *IEEE Transactions on Wireless Communications*, vol. 6, pp. 2904–2914, August 2007.

[518] J. Xavier, V. A. N. Barroso, and J. M. F. Moura, 'Closed-form correlative coding (CFC2) blind identification of MIMO channels: isometry fitting to second order statistics', *IEEE Transactions on Signal Processing*, vol. 49, pp. 1073–1086, May 2001.

[519] W. Nabhane and H. V. Poor, 'Blind joint equalization and multiuser detection in dispersive MC-CDMA/MC-DS-CDMA/MT-CDMA channels', in *Proceedings of MILCOM 2002*, (Anaheim, CA, USA), vol. 2, pp. 814–819, 7–10 October 2002.

[520] Y. Li, N. Seshadri, and S. Ariyavisitakul, 'Channel estimation for OFDM systems with transmitter diversity in mobile wireless channels', *IEEE Journal on Selected Areas in Communications*, vol. 17, pp. 461–471, March 1999.

[521] H. Minn, D. I. Kim, and V. K. Bhargava, 'A reduced complexity channel estimation for OFDM systems with transmit diversity in mobile wireless channels', *IEEE Transactions on Communications*, vol. 50, pp. 799–807, May 2002.

[522] F. W. Vook and T. A. Thomas, 'MMSE multi-user channel estimation for broadband wireless communications', in *IEEE Global Telecommunications Conference (GLOBECOM'01)*, vol. 1, (San Antonio, USA), pp. 470–474, 25–29 November 2001.

[523] K. J. Kim and R. A. Iltis, 'Joint detection and channel estimation algorithms for QS-CDMA signals over time-varying channels', *IEEE Transactions on Communications*, vol. 50, pp. 845–855, May 2002.

[524] F. Horlin and L. V. D. Perre, 'Optimal training sequences for low complexity ML multi-channel estimation in multi-user MIMO OFDM-based communications', in *2004 IEEE International Conference on Communications*, (Paris, France), vol. 4, pp. 2427–2431, 20–24 June 2004.

[525] T. Cui and C. Tellambura, 'Joint channel estimation and data detection for OFDM systems via sphere decoding', in *IEEE Global Telecommunications Conference (GLOBECOM'04)*, (Dallas, USA), vol. 6, pp. 3656–3660, 29 November–3 December 2004.

[526] H. Zhu, B. Farhang-Boroujeny, and C. Schlegel, 'Pilot embedding for joint channel estimation and data detection in MIMO communication systems', *IEEE Communications Letters*, vol. 7, pp. 30–32, January 2003.

[527] J. Wang and K. Araki, 'Pilot-symbol aided channel estimation in spatially correlated multiuser MIMO-OFDM channels', in *Proceedings of the IEEE 60th Vehicular Technology Conference (VTC 2004 Fall)*, (Milan, Italy), vol. 1, pp. 33–37, 26–29 September 2004.

[528] J. Siew, J. Coon, R. J. Piechocki, A. Dowler, A. Nix, M. A. Beach, S. Armour, and J. McGeehan, 'A channel estimation algorithm for MIMO-SCFDE', *IEEE Communications Letters*, vol. 8, pp. 555–557, September 2004.

[529] S. Chen and Y. Wu, 'Maximum likelihood joint channel and data estimation using genetic algorithms', *IEEE Transactions on Signal Processing*, vol. 46, pp. 1469–1473, May 1998.

[530] Y. S. Zhang, Y. Du, W. Zhang, X. Z. Wang, and J. Li, 'A data-aided time domain channel estimation method', in *Proceedings of the 2004 Joint Conference of the 10th Asia-Pacific Conference on Communications and the 5th International Symposium on Multi-Dimensional Mobile Communications*, (Beijing, China), vol. 1, pp. 469–473, 29 August–1 September 2004.

[531] C. E. Tan and I. J. Wassell, 'Near-optimum training sequences for OFDM systems', in *The 9th Asia-Pacific Conference on Communications (APCC 2003)*, (Malaysia), vol. 1, pp. 119–123, 21–24 September 2003.

[532] K. Yen and L. Hanzo, 'Genetic-algorithm-assisted multiuser detection in asynchronous CDMA communications', *IEEE Transactions on Vehicular Technology*, vol. 53, pp. 1413–1422, September 2004.

[533] X. Wu, T. C. Chuah, B. S. Sharif, and O. R. Hinton, 'Adaptive robust detection for CDMA using a genetic algorithm', *IEE Proceedings – Communications*, vol. 150, pp. 437–444, 10 December 2003.

[534] J. Akhtman and L. Hanzo, 'Reduced-complexity maximum-likelihood detection in multiple-antenna-aided multicarrier systems', in *Proceedings of the 5th International Workshop on Multi-Carrier Spread Spectrum Communications*, (Oberpfaffenhofen, Germany), 14–16 September 2005. http://www.ecs.soton.ac.uk/~mj02r/t/mcss-05.pdf.

[535] J. Akhtman and L. Hanzo, 'Novel optimized-hierarchy RSA-aided space-time processing method', Mobile VCE Core 3 Programme – Wireless Enablers 2.2: ICR-WE2.2.1, University of Southampton, May 2005.

[536] N. Seshadri, 'Joint data and channel estimation using blind trellis search techniques', *IEEE Transactions on Communications*, vol. 42, pp. 1000–1011, February–April 1994.

[537] R. Raheli, A. Polydoros, and C. K. Tzou, 'Per-survivor processing: a general approach to MLSE in uncertain environments', *IEEE Transactions on Communications*, vol. 43, pp. 354–364, February/April 1995.

[538] R. L. Haupt and S. E. Haupt, *Practical Genetic Algorithms*. Hoboken, NJ: John Wiley & Sons, Inc., 2nd edn, 2004.

[539] Z. Michalewicz, *Genetic Algorithms + Data Structures = Evolution Programs*. New York: Springer, 2nd edn, 1994.

[540] T. K. Moon and W. C. Stirling, *Mathematical Methods and Algorithms for Signal Processing*. Englewood Cliffs, NJ: Prentice Hall, 2002.

[541] S. Kay, *Fundamentals of Statistical Signal Processing, Estimation Theory*. Englewood Cliffs, NJ: Prentice Hall, 1993.

[542] L. Hanzo, M. Münster, B. J. Choi, and T. Keller, *OFDM and MC-CDMA for Broadband Multi-User Communications, WLANs and Broadcasting*. Chichester: John Wiley & Sons, Ltd, 2003.

[543] U. Fincke and M. Pohst, 'Improved method for calculating vector of short length in a lattice, including a complexity analysis', *Mathematics of Computation*, vol. 44, pp. 463–471, April 1985.

[544] E. Viterbo and J. Boutros, 'A universal lattice code decoder for fading channels', *IEEE Transactions on Information Theory*, vol. 45, pp. 1639–1642, July 1999.

[545] A. M. Chan and I. Lee, 'A new reduced-complexity sphere decoder for multiple antenna systems', *IEEE International Conference on Communications (ICC 2002)*, (New York, USA), vol. 1, pp. 460–464, April 2002.

[546] K. Su and I. J. Wassell, 'A new ordering for efficient sphere decoding', in *IEEE International Conference on Communications (ICC 2005)*, (Seoul, Korea), vol. 3, pp. 1906–1910, May 2005.

[547] D. Pham, K. R. Pattipati, P. K. Willett, and J. Luo, 'An improved complex sphere decoder for v-BLAST systems', *IEEE Signal Processing Letters*, vol. 11, pp. 748–751, September 2004.

[548] Q. Liu and L. Yang, 'A novel method for initial radius selection of sphere decoding', *IEEE 60th Vehicular Technology Conference (VTC 2004 Fall)*, (Milan, Italy), vol. 2, pp. 1280–1283, September 2004.

[549] Z. Guo and P. Nilsson, 'Algorithm and implementation of the K-best sphere decoding for MIMO detection', *IEEE Journal on Selected Areas in Communications*, vol. 24, pp. 491–503, March 2006.

[550] K. Wong, C. Tsui, R. Cheng, and W. Mow, 'A VLSI architecture of a K-best lattice decoding algorithm for MIMO channels', *IEEE International Symposium on Circuits and Systems (ISCAS 2002)*, (Scottsdale, AZ, USA), vol. 3, May 2002.

[551] J. Pons and P. Duvaut, 'New approaches for lowering path expansion complexity of k-best MIMO detection algorithms', in *IEEE International Conference on Communications (ICC'09)*, (Dresden), pp. 1–6, June 2009.

[552] B. Kim, H. Kim, and K. Choi, 'An adaptive k-best algorithm without SNR estimation for MIMO systems', in *IEEE Vehicular Technology Conference (VTC Spring 2008)*, (Singapore), pp. 817–821, May 2008.

[553] B. M. Hochwald and S. ten Brink, 'Achieving near-capacity on a multiple-antenna channel', *IEEE Transactions on Communications*, vol. 51, pp. 389–399, March 2003.

[554] M. Pohst, 'On the Computation of lattice vectors of minimal length, successive minima and reduced bases with applications', *Proceedings of ACM SIGSAM*, pp. 37–44, 1981.

[555] U. Fincke and M. Pohst, 'Improved methods for calculating vectors of short length in a lattice, including a complexity analysis', *Mathematics of Computation*, vol. 44, pp. 463–471, April 1985.

[556] M. O. Damen, K. Abed-Meraim, and J. C. Belfiore, 'Generalised sphere decoder for asymmetrical space-time communication architecture', *Electronics Letters*, vol. 36, pp. 166–167, January 2000.

[557] T. Cui and C. Tellambura, 'An efficient generalized sphere decoder for rank-deficient MIMO systems', *IEEE 60th Vehicular Technology Conference, VTC2004-Fall*, (Milan, Italy), vol. 5, pp. 3689–3693, September 2004.

[558] M. O. Damen, K. Abed-Meraim, and M. S. Lemdani, 'Further results on the sphere decoder', *Proceedings of IEEE International Symposium on Information Theory, 2001*, (San Diego, USA), June 2001.

[559] J. Li and Y. Yang, 'Ordered sphere detector for high rate spatial multiplexing architecture', *Proceedings of the IEEE 6th Circuits and Systems Symposium on Emerging Technologies: Frontiers of Mobile and Wireless Communication, 2004*, (Shanghai, China), vol. 2, pp. 417–420, May 2004.

[560] O. Damen, A. Chkeif, and J. C. Belfiore, 'Lattice code decoder for space-time codes', *IEEE Communications Letters*, vol. 4, pp. 161–163, May 2000.

[561] G. D. Golden, C. J. Foschini, R. A. Valenzuela, and P. W. Wolniansky, 'Detection algorithm and initial laboratory results using v-BLAST space-time communication architecture', *Electronics Letters*, vol. 35, pp. 14–16, January 1999.

[562] J. Akhtman, A. Wolfgang, S. Chen, and L. Hanzo, 'An optimized-hierarchy-aided approximate log-MAP detector for MIMO systems', *IEEE Transactions on Wireless Communications*, vol. 6, pp. 1900–1909, May 2007.

[563] J. Akhtman and L. Hanzo, 'An optimized-hierarchy-aided maximum likelihood detector for MIMO-OFDM', *IEEE 63rd Vehicular Technology Conference, VTC 2006-Spring*, (Melbourne, Australia), vol. 3, pp. 1526–1530, 2006.

[564] L. Wang, O. Alamri, and L. Hanzo, 'Sphere packing modulation in the SDMA uplink using k-best sphere detection', *IEEE Signal Processing Letters*, vol. 16, pp. 291–294, April 2009.

[565] C. Berrou, A. Glavieux, and P. Thitimajshima, 'Near Shannon limit error-correcting coding and decoding: turbo-codes. 1', in *IEEE International Conference on Communications (ICC 1993)*, (Geneva, Switzerland), vol. 2, pp. 1064–1070, May 1993.

[566] C. Berrou and A. Glavieux, 'Near optimum error correcting coding and decoding: turbo-codes', *IEEE Transactions on Communications*, vol. 44, pp. 1261–1271, October 1996.

[567] S. Benedetto, D. Divsalar, G. Montorsi, and F. Pollara, 'Analysis, design, and iterative decoding of double serially concatenated codes with interleavers', *IEEE Journal on Selected Areas in Communications*, vol. 16, pp. 231–244, February 1998.

[568] J. Li, K. B. Letaief, and Z. Cao, 'Space-time turbo multiuser detection for coded MC-CDMA', *IEEE Transactions on Wireless Communications*, vol. 4, pp. 538–549, March 2005.

[569] L. Hanzo, C. H. Wong, and M. S. Yee, *Adaptive Wireless Transceivers*. Chichester: IEEE Press and John Wiley & Sons, Ltd, 2002.

[570] S. Baro, J. Hagenauer, and M. Witzke, 'Iterative detection of MIMO transmission using a list-sequential (LISS) detector', *IEEE International Conference on Communications, ICC'03*, (Anchorgae, USA), vol. 4, pp. 2653–2657, May 2003.

[571] M. S. Yee, 'Max-log-MAP sphere decoder', *Proceedings of IEEE International Conference on Acoustics, Speech, and Signal Processing (ICASSP'05)*, (Philadelphia, USA), vol. 3, March 2005.

[572] M. Tuchler, A. C. Singer, and R. Koetter, 'Minimum mean squared error equalization using a priori information', *IEEE Transactions on Acoustics, Speech, and Signal Processing*, vol. 50, pp. 673–683, March 2002.

[573] L. Xu, S. Tan, S. Chen, and L. Hanzo, 'Iterative minimum bit error rate multiuser detection in multiple antenna aided OFDM', in *IEEE Wireless Communications and Networking Conference (WCNC 2006)*, (Las Vegas, USA), vol. 3, pp. 1603–1607, April 2006.

[574] L. Xu, S. Tan, S. Chen, and L. Hanzo, 'Iterative minimum bit error rate multiuser detection in multiple antenna aided OFDM', *IEEE Wireless Communications and Networking Conference (WCNC 2006)*, (Las Vegas, USA), vol. 3, pp. 1603–1607, April 2006.

[575] J. Wang, S. X. Ng, L. L. Yang, and L. Hanzo, 'Combined serially concatenated codes and MMSE equalization: an EXIT chart aided perspective', in *IEEE Vehicular Technology Conference (VTC 2006 Fall)*, (Montreal, Canada), pp. 1–5, September 2006.

[576] J. Wang, S. X. Ng, A. Wolfgang, L. L. Yang, S. Chen, and L. Hanzo, 'Near-capacity three-stage MMSE turbo equalization using irregular convolutional codes', in *4th International Symposium on Turbo Codes in connection with 6th International ITG Conference on Source and Channel Coding (ISTC'06)*, (Munich, Germany), April 2006.

[577] J. Kliewer, S. X. Ng, and L. Hanzo, 'Efficient computation of exit functions for non-binary iterative decoding', *IEEE Transactions on Communications*, vol. 54, pp. 2133–2136, December 2006.

[578] J. Kliewer, A. Huebner, and D. J. Costello, 'On the achievable extrinsic information of inner decoders in serial concatenation', in *Proceedings of the IEEE International Symposium on Information Theory*, (Seattle, USA), pp. 2680–2684, July 2006.

[579] S. X. Ng, J. Wang, and L. Hanzo, 'Unveiling near-capacity code design: the realization of Shannon's communication theory for MIMO channels', *IEEE International Conference on Communications, 2008*, pp. 1415–1419, May 2008.

[580] M. Tüchler and J. Hagenauer, 'Exit charts of irregular codes', in *Proceedings of Conference on the Information Science and Systems [CD-ROM]*, (Princeton University, USA), pp. 20–22, March 2002.

[581] M. Tüchler, 'Design of serially concatenated systems depending on the block length', *IEEE Transactions on Communications*, vol. 52, pp. 209–218, February 2004.

[582] R. W. Heath and A. J. Paulraj, 'Switching between diversity and multiplexing in MIMO systems', *IEEE Transactions on Communications*, vol. 53, pp. 962–968, June 2005.

[583] J. H. Conway and N. J. A. Sloane, *Sphere Packings, Lattices and Groups*. Berlin: Springer, 3rd edn, 1999.

[584] W. F. Su, Z. Safar, and K. J. R. Liu, 'Space-time signal design for time-correlated Rayleigh fading channels', *IEEE International Conference on Communications 2003*, (Anchorage, USA), vol. 5, pp. 3175–3179, May 2003.

[585] S. M. Alamouti, 'A simple transmit diversity technique for wireless communications', *IEEE Journal on Selected Areas in Communications*, vol. 16, pp. 1451–1458, October 1998.

[586] V. Tarokh, H. Jafarkhani, and A. R. Calderbank, 'Space-time block codes from orthogonal designs', *IEEE Transactions on Information Theory*, vol. 45, pp. 1456–1467, July 1999.

[587] W. Su and X.-G. Xia, 'On space-time block codes from complex orthogonal designs', *Wireless Personal Communications*, vol. 25, pp. 1–26, April 2003.

[588] A. F. Naguib, N. Seshadri, and A. R. Calderbank, 'Applications of space-time block codes and interference suppression for high capacity and high data rate wireless systems', *The 32nd Asilomar Conference on Signals, Systems & Computers*, (Pacific Grove, CA, USA), vol. 2, pp. 1803–1810, November 1998.

[589] X.-B. Liang and X.-G. Xia, 'On the nonexistence of rate-one generalized complex orthogonal designs', *IEEE Transactions on Information Theory*, vol. 49, pp. 2984–2988, November 2003.

[590] O. R. Alamri, B. L. Yeap, and L. Hanzo, 'A turbo detection and sphere-packing-modulation-aided space-time coding scheme', *IEEE Transactions on Vehicular Technology*, vol. 56, pp. 575–582, March 2007.

[591] M. El-Hajjar, 'Multi-functional MIMO schemes employing iterative detected multi-dimensional sphere packing modulation', Doctor of Philosophy Mini Thesis, University of Southampton, 2007.

[592] S. ten Brink, 'Convergence behavior of iteratively decoded parallel concatenated codes', *IEEE Transactions on Communications*, vol. 49, pp. 1727–1737, October 2001.

[593] L. Hanzo, L.-L. Yang, E.-L. Kuan, and K. Yen, *Single and Multi-Carrier DS-CDMA: Multi-User Detection, Space-Time Spreading, Synchronisation, Networking and Standards*. Chichester: IEEE Press and John Wiley & Sons, Ltd, 2003.

[594] G. J. Foschini and M. J. Gans, 'On limits of wireless communications in a fading environment when using multiple antennas', *Wireless Personal Communications*, vol. 6, pp. 311–335, March 1998.

[595] J. N. Laneman, D. N. C. Tse, and G. W. Wornell, 'Cooperative diversity in wireless networks: efficient protocols and outage behavior', *IEEE Transactions on Information Theory*, vol. 50, pp. 3062–3080, December 2004.

[596] L. Hanzo, O. Alamri, M. El-Hajjar, and N. Wu, *Near-Capacity Multi-Functional MIMO Systems: Sphere-packing, iterative detection, and cooperation*. Chichester: John Wiley & Sons, Ltd, 2009.

[597] L. Lampe, R. Schober, V. Pauli, and C. Windpassinger, 'Multiple-symbol differential sphere decoding', *IEEE Transactions on Communications*, vol. 12, pp. 1981–1985, December 2005.

[598] V. Pauli, L. Lampe, and R. Schober, '"Turbo DPSK" using soft multiple-symbol differential sphere decoding', *IEEE Transactions on Information Theory*, vol. 52, pp. 1385–1398, April 2006.

[599] V. Pauli, L. Lampe, and J. Huber, 'Differential space-frequency modulation and fast 2-d multiple-symbol differential detection for MIMO-OFDM', *IEEE Transactions on Vehicular Technology*, vol. 57, pp. 297–310, January 2008.

[600] V. Pauli and L. Lampe, 'Multiple-symbol differential sphere decoding for unitary space-time modulation', *IEEE Global Telecommunications Conference*, (St Louis, USA), vol. 3, p. 6, November 2005.

[601] D. Divsalar and M. K. Simon, 'Multiple-symbol differential detection of MPSK', *IEEE Transactions on Communications*, vol. 38, pp. 300–308, March 1900.

[602] D. Divsalar and M. K. Simon, 'Maximum-likelihood differential detection of uncoded and trellis-coded amplitude phase modulation over AWGN and fading channels–metrics and performance', *IEEE Transactions on Communications*, vol. 42, pp. 76–89, January 1994.

[603] M. O. Damen, H. E. Gamal, and G. Caire, 'On maximum-likelihood detection and the search for the closest lattice point', *IEEE Transactions on Information Theory*, vol. 49, pp. 2389–2402, October 2003.

[604] T. Himsoon, W. Su, and K. J. R. Liu, 'Differential transmission for amplify-and-forward cooperative communications', *IEEE Signal Processing Letters*, vol. 12, pp. 597–600, September 2005.

[605] T. Himsoon, W. P. Siriwongpairat, W. Su, and K. J. R. Liu, 'Differential modulation with threshold-based decision combining for cooperative communications', *IEEE Transactions on Signal Processing*, vol. 55, pp. 3905–3923, July 2007.

[606] L. Wang and L. Hanzo, 'The resource-optimized differentially modulated hybrid AF/DF cooperative cellular uplink using multiple-symbol differential sphere detection', *IEEE Signal Processing Letters*, vol. 16, pp. 965–968, November 2009.

[607] T. S. Rappaport, *Wireless Communications Principles and Practice*. Hong Kong: Pearson Education Asia and Publishing House of Electronics Industry, 2nd edn, 2002.

[608] M. K. Simon and M. S. Alouini, 'A unified approach to the probability of error for noncoherent and differentially coherent modulations over generalized fading channels', *IEEE Transactions on Communications*, vol. 46, pp. 1625–1638, December 1998.

[609] T. Himsoon, W. P. Siriwongpairat, W. Su, and K. J. R. Liu, 'Differential modulations for multinode cooperative communications', *IEEE Transactions on Signal Processing*, vol. 56, pp. 2941–2956, July 2008.

[610] I. S. Gradshteyn and I. M. Ryzhik, *Table of Integrals, Series, and Products*. Berlin: Springer, 7th edn, 2007.

[611] Q. T. Zhang and X. W. Cui, 'A closed-form expression for the symbol-error rate of M-ary DPSK in fast Rayleigh fading', *IEEE Transactions on Communications*, vol. 53, pp. 1085–1087, July 2005.

[612] S. X. Ng, Y. Li, and L. Hanzo, 'Distributed turbo trellis coded modulation for cooperative communications', in *IEEE International Conference on Communications (ICC'09)*, (Dresden, Germany), pp. 1–5, June 2009.

[613] K. Lee and L. Hanzo, 'Iterative detection and decoding for hard-decision forwarding aided cooperative spatial multiplexing', in *IEEE International Conference on Communications (ICC'09)*, (Dresden, Germany), pp. 1–5, June 2009.

[614] T. L. Marzetta and B. M. Hochwald, 'Capacity of a mobile multiple-antenna communication link in Rayleigh flat fading', *IEEE Transactions on Information Theory*, vol. 45, pp. 139–157, January 1999.

[615] L. Z. Zheng and D. N. C. Tse, 'Communication on the Grassmann manifold: a geometric approach to the noncoherent multiple-antenna channel', *IEEE Transactions on Information Theory*, vol. 48, pp. 359–383, February 2002.

[616] Y. Liang and V. V. Veeravalli, 'Capacity of noncoherent time-selective Rayleigh-fading channels', *IEEE Transactions on Information Theory*, vol. 50, pp. 3095–3110, December 2004.

[617] T. M. Cover and J. A. Thomas, *Elements of Information Theory*. Hoboken, NJ: John Wiley & Sons, Inc., 2nd edn, 2006.

[618] R. R. Chen, R. Koetter, U. Madhow, and D. Agrawal, 'Joint noncoherent demodulation and decoding for the block fading channel: a practical framework for approaching Shannon capacity', *IEEE Transactions on Communications*, vol. 51, pp. 1676–1689, October 2003.

[619] A. Ashikhmin, G. Kramer, and S. ten Brink, 'Extrinsic information transfer functions: model and erasure channel properties', *IEEE Transactions on Information Theory*, vol. 50, pp. 2657–2673, November 2004.

[620] B. Zhao and M. C. Valenti, 'Distributed turbo coded diversity for relay channel', *Electronics Letters*, vol. 39, pp. 786–787, May 2003.

[621] J. Kliewer, A. Huebner, and D. J. Costello, 'On the achievable extrinsic information of inner decoders in serial concatenation', in *IEEE International Symposium on Information Theory, 2006*, (Seattle, USA), pp. 2680–2684, July 2006.

[622] Y. Fan, C. Wang, J. Thompson, and H. V. Poor, 'Recovering multiplexing loss through successive relaying using repetition coding', *IEEE Transactions on Wireless Communications*, vol. 6, pp. 4484–4493, December 2007.

[623] T. Moon and W. Stirling, *Mathematical Methods and Algorithms for Signal Processing*. Englewood Cliffs, NJ: Prentice Hall, 2000.

[624] P. W. Wolniansky, G. J. Foschini, G. D. Golden, and R. A. Valenzuela, 'V-BLAST: an architecture for realizing very high data rates over the rich scattering wireless channels', in *Proceedings of IEEE ISSSE-98*, (Pisa, Italy), Invited Paper, September 1998.

[625] A. Bhargave, R. Figueiredo, and T. Eltoft, 'A detection algorithm for the V-BLAST system', in *Proceedings of IEEE Global Telecommunications Conference*, (San Antonio, USA), vol. 1, pp. 494–498, 25–29 November 2001.

[626] L. Hanzo, L.-L. Yang, E.-L. Kuan, and K. Yen, *Single- and Multi-Carrier DS-CDMA*. Chichester: IEEE Press and John Wiley & Sons, Ltd, 2003.

[627] X. F. Wang, W. S. Lu, and A. Antoniou, 'A genetic algorithm-based multiuser detector for multiple-access communications', in *IEEE International Symposium on Circuits and System – ISCAS'98*, (Monterey, CA, USA), pp. 534–537, 1998.

[628] M. Jiang, S. Ng, and L. Hanzo, 'Hybrid iterative multiuser detection for channel coded space division multiple access OFDM systems', *IEEE Transactions on Vehicular Technology*, vol. 55, pp. 115–127, January 2006.

[629] R. Wang and G. Giannakis, 'Approaching MIMO channel capacity with reduced-complexity soft sphere decoding', in *Proceedings of IEEE Wireless Communications and Networking Conference*, (Atlanta, USA), vol. 3, pp. 1620–1625, 21–25 March 2004.

[630] J. E. Gentle, *Numerical Linear Algebra for Applications in Statistics*. Berlin: Springer, 1998.

[631] M. K. Varanasi, 'Decision feedback multiuser detection: a systematic approach', *IEEE Transactions on Information Theory*, vol. 45, pp. 219–240, January 1999.

[632] G. Valiente, *Algorithms on Trees and Graphs*. New York: Springer, 2002.

[633] A. Viterbi, *CDMA: Principles of Spread Spectrum Communication*. Reading MA: Addison-Wesley, June 1995.

[634] R. D. Schafer, *An Introduction to Nonassociative Algebras*. New York: Dover, 1996.

[635] J. Hagenauer, E. Offer, and L. Papke, 'Iterative decoding of binary block and convolutional codes', *IEEE Transactions on Information Theory*, vol. 42, pp. 429–445, March 1996.

[636] C. Berrou and A. Glavieux, 'Near optimum error-correcting coding and decoding: turbo codes', *IEEE Transactions on Communications*, vol. 44, pp. 1261–1271, October 1996.

[637] C. Berrou, 'Some clinical aspects of turbo codes', in *Proceedings of the International Symposium on Turbo Codes & Related Topics*, (Brest, France), pp. 26–31, 3–5 September 1997.

[638] S. Benedetto, R. Garello, and G. Montorsi, 'A search for good convolutional codes to be used in the construction of turbo codes', *IEEE Transactions on Communications*, vol. 46, pp. 1101–1105, September 1998.

[639] G. Battail, 'A conceptual framework for understanding turbo codes', *IEEE Journal on Selected Areas in Communications*, vol. 16, no. 2, pp. 245–254, 1998.

[640] O. Açikel and W. Ryan, 'Punctured turbo-codes for BPSK/QPSK channels', *IEEE Transactions on Communications*, vol. 47, pp. 1315–1323, September 1999.

[641] IEEE LAN/MAN Standards Committee, *Wireless LAN Medium Access Control (MAC) and Physical Layer (PHY) Specifications*, IEEE 802.11g ed., 2003.

[642] H. Vikalo, B. Hassibi, and T. Kailath, 'Iterative decoding for MIMO channels via modified sphere decoding', *IEEE Transactions on Wireless Communications*, vol. 3, pp. 2299–2311, November 2004.

[643] T. Bäck, *Evolutionary Algorithms in Theory and Practice: Evolution Strategies, Evolutionary Programming, Genetic Algorithms*. New York: Oxford University Press, 1996.

[644] D. B. Fogel, 'An introduction to simulated evolutionary optimization', *IEEE Transactions on Neural Networks*, vol. 5, pp. 3–14, January 1994.

[645] S. Haykin, *Neural Networks*. Englewood Cliffs, NJ: Prentice Hall, 2nd edn, 1999.

[646] L. Fausett, *Fundamentals of Neural Networks: Architectures, Algorithms and Applications*. Englewood Cliffs, NJ: Prentice Hall, 1994.

[647] V. W. Porto and D. B. Fogel, 'Alternative neural network training methods', *IEEE Expert*, vol. 10, pp. 16–22, June 1995.

[648] K. Chellapilla and D. Fogel, 'Evolving an expert checkers playing program without using human expertise', *IEEE Transactions on Evolutionary Computation*, vol. 5, pp. 422–428, August 2001.

[649] G. Kechriotis, E. Zervas, and E. S. Manolakos, 'Using recurrent neural networks for adaptive communication channel equalization', *IEEE Transactions on Neural Networks*, vol. 5, pp. 267–278, March 1994.

[650] R. Parisi, E. D. D. Claudio, G. Orlandi, and B. D. Rao, 'Fast adaptive digital equalization by recurrent neural networks', *IEEE Transactions on Signal Processing*, vol. 45, pp. 2731–2739, November 1997.

[651] M. D. Buhmann, *Radial Basis Functions: Theory and Implementations*. Cambridge: Cambridge University Press, 2003.

[652] P. V. Yee and S. Haykin, *Regularized Radial Basis Function Networks: Theory and Applications*. New York: John Wiley & Sons, Inc., 2001.

[653] U. Mitra and H. V. Poor, 'Neural network techniques for adaptive multiuser demodulation', *IEEE Journal on Selected Areas in Communications*, vol. 12, pp. 1460–1470, December 1994.

[654] K. Ko, S. Choi, C. Kang, and D. Hong, 'RBF multiuser detector with channel estimation capability in a synchronous MC-CDMA system', *IEEE Transactions on Neural Networks*, vol. 12, pp. 1536–1539, November 2001.

[655] C. Ahn and I. Sasase, 'Adaptive array antenna based on radial basis function network as multiuser detection for WCDMA', *Electronics Letters*, vol. 38, pp. 1208–1210, September 2002.

[656] H. Wei and L. Hanzo, 'Reduced-complexity genetic algorithm aided and radial basis function assisted multiuser detection for synchronous CDMA', in *Proceedings of European Signal Processing Conference (EUSIPCO 2004)*, (Vienna, Austria), pp. 157–160, September 2004.

[657] X. Zhou and X. Wang, 'Channel estimation for OFDM systems using adaptive radial basis function networks', *IEEE Transactions on Vehicular Technology*, vol. 52, pp. 48–59, January 2003.

[658] G. Charalabopoulos, P. Stavroulakis, and A. H. Aghvami, 'A frequency-domain neural network equalizer for OFDM', in *IEEE Global Telecommunications Conference (GLOBECOM'03)*, (San Francisco, USA), vol. 2, pp. 571–575, 1–5 December 2003.

[659] S. Lerkvaranyu, K. Dejhan, and Y. Miyanaga, 'M-QAM demodulation in an OFDM system with RBF neural network', in *The 2004 47th Midwest Symposium on Circuits and Systems (MWSCAS'04)*, (Hiroshima, Japan), vol. 2, pp. II-581–II-584, 25–28 July 2004.

[660] T. Cui and C. Tellambura, 'Channel estimation for OFDM systems based on adaptive radial basis function networks', in *Proceedings of the IEEE 60th Vehicular Technology Conference (VTC 2004 Fall)*, (New Orleans, USA), vol. 1, pp. 608–611, 26–29 September 2004.

[661] S. Chen, A. K. Samingan, B. Mulgrew, and L. Hanzo, 'Adaptive minimum-BER linear multiuser detection for DS-CDMA signals in multipath channels', *IEEE Transactions on Signal Processing*, vol. 49, pp. 1240–1247, June 2001.

[662] H. Mühlenbein, *Foundations of Genetic Algorithms*, ed. G. Rawlins. San Francisco: Morgan Kaufmann, 1991.

[663] J. J. Grefenstette and J. E. Baker, 'How genetic algorithms work: a critical look at implicit parallelism', in *Proceedings of the Third International Conference on Genetic Algorithms*, ed. J. D. Schaffer, (Fairfax, VA, USA), pp. 20–27, 1989.

[664] B. L. Miller and D. E. Goldberg, 'Genetic algorithms, selection schemes and the varying effects of noise', *Evolutionary Computation*, vol. 4, pp. 113–131, 1996.

[665] L. J. Eshelman and J. D. Schaffer, 'Preventing premature convergence in genetic algorithms by preventing incest', in *Proceedings of the Fourth International Conference on Genetic Algorithms*, ed. R. K. Belew and L. B. Booker, (San Diego, USA), pp. 115–122, 1991.

[666] G. Syswerda, 'Uniform crossover in genetic algorithms', in *Proceedings of the Third International Conference on Genetic Algorithms*, ed. J. D. Schaffer, (Fairfax, VA, USA), pp. 2–9, 1989.

Subject Index

Author Index